D1626270

THE NEUROBIOLOGY OF THE AMYGDALA

ADVANCES IN BEHAVIORAL BIOLOGY

THE NEUROBIOLOGY
OF THE AMYGDALA

The Proceedings of a Symposium on the Neurobiology of the
Amygdala, Bar Harbor, Maine, June 6-17, 1971

Basil E. Eleftheriou

Staff Scientist
The Jackson Laboratory
Bar Harbor, Maine

℗ **PLENUM PRESS • NEW YORK-LONDON • 1972**

149133

Library of Congress Catalog Card Number 77-188921
ISBN 0-306-37902-3

© 1972 Plenum Press, New York
A Division of Plenum Publishing Corporation
227 West 17th Street, New York, N.Y. 10011

United Kingdom edition published by Plenum Press, London
A Division of Plenum Publishing Company, Ltd.
Davis House (4th Floor), 8 Scrubs Lane, Harlesden, London,
NW10, 6SE, England

Printed in the United States of America

In Memory
Geoffrey W. Harris and
Tryphena Humphrey

ACKNOWLEDGMENTS

The Editor wishes to express his sincere appreciation and gratitude to Dr. Earl L. Green, Director, The Jackson Laboratory, for his support; Mr. Thomas Hyde and Mrs. Susan Myers for their untiring efforts in various phases of the Conference.

Sincere thanks are extended also to Dr. R. L. Norman, Department of Anatomy, School of Medicine, University of California, Los Angeles, and Dr. Mark Kristal, The Jackson Laboratory, for assistance in compiling this volume.

Finally, appreciation and gratitude are expressed to the speakers and all other participants who came to Bar Harbor from all corners of our globe, often with very little or no direct financial support from the Conference. Truly, all were amygdalophiles above and beyond the call of duty.

PREFACE

In recent years, great interest has been focused on the field of neurobiology. In the last decade, various international and regional meetings, symposia, seminars and workshops have been organized to discuss brain regions such as the hypothalamus, cerebellum, medulla, cortex and hippocampus. A number of books have been published as a consequence of these gatherings. Uniquely and singularly absent from these conclaves has been a truly interdisciplinary discussion of the amygdala.

The various chapters of this book represent the formal talks presented at The Advanced Study Institute held at the Jackson Laboratory, Bar Harbor, Maine, from June 6 to 17, 1971, with funds made available from the Scientific Affairs Committee of the North Atlantic Treaty Organization and the National Science Foundation.

The speakers and participants are grateful to these two institutions for being given the opportunity to gather and discuss their respective works that represent years of experimental and clinical research centering on the amygdala.

It is hoped that the experiments discussed in this book will act as a major stimulus to other scientists to initiate complementary and supplementary experiments for the better understanding of the specific role of the amygdala.

Bar Harbor, Maine, 1972   Basil E. Eleftheriou

ATTENDANTS

E. S. Barratt, Ph. D., Department of Neurology and Psychiatry, University of Texas Medical Branch, Galveston, Texas

N. Bedworth, Hampshire College, Amherst, Massachusetts

M. A. B. Brazier, D. Sc., Department of Anatomy, School of Medicine, University of California, Los Angeles, California

K. A. Brown, Ph. D., Department of Physiology, School of Medicine, University of California, Los Angeles, California

N. P. Chapanis, Ph. D., Department of Psychiatry and Behavioral Science, Johns Hopkins University School of Medicine, Baltimore, Maryland

C. M. Christenson, Ph. D., Department of Natural Science, Indiana University, Jeffersonville, Indiana

J. S. de Olmos, Ph. D., College of Medicine, Department of Anatomy, The University of Iowa, Iowa City, Iowa

J. S. De Sisto, Jr., Ph. D., Department of Psychology, Colby College, Waterville, Maine

J. J. Dreifuss, M. D., Ph. D., Departement de Physiologie de l'Université, Geneva, Switzerland

D. Egger, Ph. D., Department of Anatomy, Yale University School of Medicine, New Haven, Connecticut

E. Eidelberg, M. D., Division of Neurobiology, Barrow Neurological Institute, Saint Joseph's Hospital, Phoenix, Arizona

B. E. Eleftheriou, Ph. D., The Jackson Laboratory, Bar Harbor, Maine

M. F. Elias, Ph. D., Center for the Study of Aging and Human Development, Duke University Medical Center, Duke University, Durham, North Carolina

F. W. Ellendorff, Ph. D., Department of Anatomy, School of Medicine, University of California, Los Angeles, California

H. Erskine, Ph. D., Department of Justice, Washington, D. C.

F. R. Ervin, M. D., Department of Psychiatry, Harvard University Medical School, Boston, Massachusetts

J. L. Gerlach, Ph. D., Rockefeller University, New York, New York

M. Girgis, M. D., Ph. D., Department of Psychiatry and Neurology, Missouri Institute of Psychiatry, St. Louis, Missouri

P. Gloor, M. D., Department of Neurology and Neurosurgery, McGill University, Montreal, Canada

G. Goddard, Ph. D., Department of Psychology, Dalhousie University, Halifax, Nova Scotia, Canada

L. D. Grant, Ph. D., Department of Psychiatry, University of North Carolina, Chapel Hill, North Carolina

A. Grasso, Ph. D., Department of Neurobiology, Laboratory of Cellular Biology, Rome, Italy

M. C. Green, Ph. D., The Jackson Laboratory, Bar Harbor, Maine

S. P. Grossman, Ph. D., Department of Psychology, University of Chicago, Chicago, Illinois

E. Hall, M. D., Ph. D., Department of Anatomy, Faculty of Medicine, University of Ottawa, Ottawa, Canada

J. P. Hanig, Ph. D., Bureau of Drugs, Food and Drug Administration, Washington, D. C.

G. W. Harris, C. B. E., Sc. D., M. D., F. R. S., Department of Human Anatomy Oxford University, Oxford, England

J. N. Hayward, M. D., University of California School of Medicine, Los Angeles, California

T. Humphrey, M. D., Ph. D., Department of Anatomy, School of Medicine, University of Alabama, Birmingham, Alabama

L. C. Iorio, Ph. D., Department of Neuropharmacology, Sandoz-
    Wander, Incorporated, Hanover, New Jersey

B. Kaada, M. D., Institute of Neurophysiology, Univeristy of
    Oslo, Oslo, Norway

K. J. Kant, Ph. D., Department of Physiology, State University
    of New York, Buffalo, New York

P. Karli, M. D., Centre de Neurochimie, Laboratoire de
    Neurophysiologie, Strasbourg, France

A. Kling, M. D., Department of Psychiatry, Rutgers Medical School,
    New Brunswick, New Jersey

M. B. Kristal, M. S., Department of Psychology, Kansas State
    University, Manhattan, Kansas

D. T. Krieger, M. D., Department of Medicine, Mount Sinai School
    of Medicine, New York, New York

H. Krieger, M. D., Department of Neurology, Mount Sinai School of
    Medicine, New York, New York

F. J. Kuhn, M. D., Department of Pharmacology, C. H. Boehringer
    Sohn, Rhein, West Germany

C. W. Laird, Ph. D., Biology Department, Research and Development,
    Hycel, Incorporated, Houston, Texas

H. J. Lammers, M. D., Medical Faculty, Catholic University,
    Laboratory of Anatomy and Embryology, Nijmegen, Holland

R. C. Leaf, Ph. D., Department of Psychology, Rutgers University,
    New Brunswick, New Jersey

H. J. Lescault, M. D., Department of Anatomy, Faculty of Medicine,
    University of Ottawa, Ottawa, Canada

J. F. Lubar, Ph. D., Department of Psychology, University of
    Tennessee, Knoxville, Tennessee

V. H. Mark, M. D., Director of Neurosurgery, The Boston City
    Hospital, Boston, Massachusetts

M. H. Miller, Ph. D., Department of Psychiatry, Rutgers Medical
    School, New Brunswick, New Jersey

J. T. Murphy, M. D., Ph. D., Department of Physiology, University
    of Toronto, Toronto, Ontario, Canada

R. E. Musty, Ph. D., Department of Psychology, University of Vermont, Burlington, Vermont

H. Narabayashi, M. D., Department of Neurology and Stereotaxy, Neurologica Clinic, Jutendo Medical School, Meguro, Tokyo, Japan

R. L. Norman, Ph. D., Department of Anatomy, School of Medicine, University of California, Los Angeles, California

L. J. Pellegrino, Ph. D., Department of Psychology, Middlebury College, Middlebury, Vermont

D. W. Pfaff, Ph. D., Rockefeller University, New York, New York

P. H. Platenius, Ph. D., Department of Psychology, Queen's University, Kingston, Ontario, Canada

S. Prelevic, M. D., Department of Neurophysiology, McGill University, Montreal, Canada

G. Raisman, Ph. D., B. M., B. Ch., Department of Human Anatomy, University of Oxford, Oxford, England

D. J. Reis, M. D., Department of Neurology, Cornell University Medical College, The New York Hospital, New York, New York

I. Samuels, Ph. D., Department of Psychology, Brandeis University, Brookline, Massachusetts

M. Sar, Ph. D., Department of Reproductive Biology, University of North Carolina, Chapel Hill, North Carolina

C. J. Sawyer, Ph. D., Department of Anatomy, School of Medicine, University of California, Los Angeles, California

W. E. Stumpf, M. D., Ph. D., The Laboratories for Reproductive Biology, The University of North Carolina, Chapel Hill, North Carolina

L. Van Atta, Ph. D., Department of Psychology, Oberlin College, Oberlin, Ohio

G. W. Van Hoesen, Ph. D., Department of Anatomy, Harvard Medical School, Boston, Massachusetts

M. Vergnes, Sc. D., Centre de Neurochimie, Laboratoire de Neurophysiologie, Strasbourg, France

C. L. Wakefield, Ph. D., Department of Anatomy, Faculty of Medicine, University of Ottawa, Ottawa, Canada

A. M. Welch, M. S., Department of Pharmacology, School of Medicine, Johns Hopkins University, Baltimore, Maryland

B. L. Welch, Ph. D., Department of Psychiatry, Johns Hopkins University, Baltimore, Maryland

J. G. Wepsic, M. D., Massachusetts General Hospital and Harvard Medical School, Boston, Massachusetts

A. S. Wilson, Ph. D., Department of Research, Veterans Administration Center, Wood, Wisconsin

D. Wnek, Department of Psychology, Rutgers University, New Brunswick, New Jersey

P. Wright, Ph. D., Department of Psychology, Edinburgh University, Edinburgh, Scotland

A. W. Zbrozyna, M. D., Department of Physiology, The University of Birmingham, The Medical School, Birmingham, England

H. J. Zeier, Ph. D., Department of Biology, Swiss Federal Institute of Technology, Zurich, Switzerland

R. E. Zigmond, Ph. D., Department of Neurobiology, Rockefeller University, New York, New York

A. Zolovick, Ph. D., Department of Psychiatry, Yale University School of Medicine, New Haven, Connecticut

SPEAKERS

Mary A. B. Brazier, D. Sc., Professor of Anatomy and Physiology, Brain Research Institute, University of California Medical Center, Los Angeles, California

José de Olmos, Ph. D., Research Associate, College of Medicine, Department of Anatomy, The University of Iowa, Iowa City, Iowa

J. J. Dreifuss, M. D., Ph. D., Professor of Physiology, Institut de Physiologie de l'Université, Ecole de Medicine, Geneva, Switzerland

David Egger, Ph. D., Associate Professor, Department of Anatomy, Yale University School of Medicine, New Haven, Connecticut

E. Eidelberg, M. D., Chairman, Division of Neurobiology, Barrow Neurological Institute, St. Joseph's Hospital, Phoenix, Arizona

Basil E. Eleftheriou, Ph. D., Staff Scientist, The Jackson Laboratory, Bar Harbor, Maine

Frank R. Ervin M. D., Associate Professor of Psychiatry, Harvard University Medical School, Director, S. Cobb Laboratories for Psychiatric Research, Massachusetts General Hospital, Boston, Massachusetts

Pierre Gloor, M. D., Professor, Electroencephalography and Clinical Neurophysiology, Montreal Neurological Institute, Department of Neurology and Neurosurgery, McGill University, Montreal, Quebec, Canada

Graham V. Goddard, Ph. D., Associate Professor, Department of Psychology, Dalhousie University, Halifax, Nova Scotia, Canada

Margaret C. Green, Ph. D., Senior Staff Scientist, The Jackson
    Laboratory, Bar Harbor, Maine

Sebastian P. Grossman, Ph. D., Professor of Biopsychology,
    Department of Psychology, University of Chicago, Chicago,
    Illinois

Elizabeth Hall, M. D., Ph. D., Professor, Department of Anatomy,
    Faculty of Medicine, University of Ottawa, Ottawa, Canada

Geoffrey W. Harris, C. B. E., M. A., Sc. D., M. D., D. M.,
    F. R. S., Professor, Department of Human Anatomy, Oxford
    University, Oxford, England

James N. Hayward, M. D., Professor of Anatomy and Neurology,
    University of California, School of Medicine, Los Angeles,
    California

Tryphena Humphrey, M. D., Ph. D., Professor of Anatomy, Department
    of Anatomy, School of Medicine, University of Alabama,
    Birmingham, Alabama

Birgir Kaada, M. D., Professor and Head, Institute of
    Neurophysiology, University of Oslo, Oslo, Norway

Pierre Karli, Professor, Centre National de la Recherche
    Scientifique, Centre de Neurochimie, Laboratoire de
    Neurophysiologie, Strasbourg, France

Arthur Kling, M. D., Professor of Psychiatry, Department of
    Psychiatry, Rutgers Medical School, New Brunswick, New Jersey

H. J. Lammers, M. D., Professor and Chairman, Medical Faculty,
    Catholic University, Laboratory of Anatomy and Embryology,
    Nijmegen, Holland

V. H. Marks, M. D., M. Sc., F. A. C. S., Associate Professor of
    Surgery, Harvard Medical School, Director of Neurosurgery,
    The Boston City Hospital, Clinical Associate, The
    Massachusetts General Hospital, Boston, Massachusetts

J. T. Murphy, Ph. D., M. D., Department of Physiology, University
    of Toronto, Toronto, Ontario, Canada

H. Narabayashi, M. D., Professor and Director, Department of
    Neurology and Stereotaxy, Neurological Clinic, Jutendo
    Medical School, Meguro, Tokyo, Japan

Geoffrey Raisman, Ph. D., B. M., B. Ch., Department of Human
    Anatomy, University of Oxford, Oxford, England

Donald J. Reis, M. D., Associate Professor of Neurology, Department
    of Neurology, Cornell University Medical College, The New
    York Hospital, New York, New York

Charles H. Sawyer, Ph. D., Professor of Anatomy, Department of
    Anatomy, School of Medicine, University of California, The
    Center for Health Sciences, Los Angeles, California

Walter E. Stumpf, M. D., Ph. D., Associate Professor of Anatomy
    and Pharmacology, The Laboratories for Reproductive Biology,
    The University of North Carolina, Chapel Hill, North Carolina

Loche Van Atta, Ph. D., Professor of Psychology, Department of
    Psychology, Oberlin College, Oberlin, Ohio

James G. Wepsic, M. D., Neurosurgeon, Massachusetts General
    Hospital, Instructor in Surgery (Neurosurgery), Harvard
    Medical School, Boston, Massachusetts

A. W. Zbrozyna, M. D., Department of Physiology, The University of
    Birmingham, The Medical School, Birmingham, England

Andrew J. Zolovick, Ph. D., Department of Psychiatry, Yale
    University School of Medicine, New Haven, Connecticut

CONTENTS

EMBRYOLOGY

ANATOMY

PSYCHOLOGY - BEHAVIOR - PSYCHIATRY

PHARMACOLOGY

NEUROENDOCRINOLOGY

MOLECULAR BIOLOGY

THE NEUROBIOLOGY OF THE AMYGDALA

INTRODUCTORY REMARKS

Basil E. Eleftheriou

The Jackson Laboratory

Bar Harbor, Maine

When I organized this conference, I had hoped to accomplish
three things: One was to gather together experimental and clinical researchers working on the amygdala who under normal circumstances do not meet each other because of the varied backgrounds
of their specialized fields and their varied society affiliations.
Second, I had hoped that the idyllic locale of the conference
would induce a relaxed environment for maximum interaction among
the participants ultimately leading to elucidation of some of
the knotty problems associated with the functional role of the
amygdala. Third, I had hoped that, in organizing an interdisciplinary meeting with biomedical scientists of varied interests,
we would apply a new stimulus to research on the amygdala. Truly,
I feel that all three of the original intentions were realized.
However, as can be seen in the papers that follow, the general
conclusion, as with many other biological phenomena, was that we
need extensive and further systematic work, at all levels, into
the mechanisms of mediation of amygdaloid functions. This
represents only an initial approach into the problems that con-
front all of us when dealing with experimental work on the
amygdala. It must be kept in mind, however, that we now at
least possess an interdisciplinary and comprehensive compilation
of the available experimental and clinical data derived from
years of research with the amygdala.

Generally, the data helped to establish the amygdala as a
major regional component of everyday neural life dealing directly
or indirectly with such varied biomedical phenomena as epilepsy,
hyperkinesis, heart regulation, emotionality, olfaction, hormone
regulation, defense reaction, learning, territoriality, rage,

sleep, awakening, neurotransmitter phenomena, sexual behavior,
aggression, ovulation, and a multitude of other components of
daily living. The fact that there is disagreement regarding the
exact role of the amygdala in regulating each and every one of
these phenomena is not disturbing. The new available techniques
and approaches for studying the central nervous system only
recently make it possible to study extensively and systematically
the role of the amygdala. Regardless of the inherent problems
in studying any neural function, it can be seen that we possess
extensive basis for continuing our research into this often-
ignored, but truly fascinating, brain region.

Beginning with the impressive and exhaustive work of the late
Tryphena Humphrey, we can follow the tortuous, intricate, and
rather involved development of the human amygdaloid complex which
represents the initial portion of the striatal complex to appear
during embryonic development. In her early work, the differentiation of the nuclei of the amygdaloid complex is traced from the
initial migration of the neuroblast outward from the medullary
epithelium, at the time that the telencephalic hemispheres first
begin to evaginate, to 8.5 weeks of menstrual age when all but
one nucleus of the complex could be identified. In order to
bring all the attendants of the conference up to date, the late
Dr. Humphrey presented new and hitherto unpublished data which
demonstrate the shift in position of the amygdala from the time
that the primordial cell mass appears to the oldest available age
(24.5 weeks). Additionally, her presentation, diagrams, and
photomicrographs, which due to length have been reduced in numbers
in this volume, helped us all to understand the developmental
intricacies of the amygdaloid nuclear complex. Indeed, her work
will forever remain as a true classic in the embryology of the
amygdala.

Raisman's presentation gives us a novel strategy of approach
into the study of the amygdaloid complex. His work with neuropil
analysis for characterization of regions in the central nervous
system presents us with lucid and novel methods of neuroanatomical analysis to demonstrate functional interrelationships
between the amygdala and other brain regions. An early dividend
of this approach is the finding that there are highly specific
differences between the area of termination of amygdaloid fibers
in the preoptic area and in the tuberal hypothalamus. Using the
shaft/spine ratio as a means of quantitating responses to various
treatments, he demonstrates sexual differences in the preoptic
area, but not in the ventromedial hypothalamic nucleus. The
latter analysis reveals that this sexual difference lies in the
mode of termination of synapses, although it is yet unknown
which fibers give rise to their synapses. One of the basic
differences between the two sexes is that the preoptic area of

the female possesses more synapses upon dendritic spines than
does the male. Although this observation is only suggestive, a
clear and definitive answer to some of the problems will be given
once current parallel Golgi studies are terminated.

By virtue of the sequence in presentation, Hall's work is
the first in emphasizing the utter complexity of the amygdaloid
nuclear groups and the difficulties encountered in cytoarchitectural studies due to species variability and the qualitative
approach of several investigators. Indeed, it may well be the
latter two distinctive problems which have contributed to the
vast disagreement among investigators dealing with electrophysiological studies of the amygdaloid complex in the same or
a number of varied species. Unless we can arrive at a common
agreement, and establish unequivocally the number of nuclei and
subnuclei within the amygdala, our problems and disagreements
will only be compounded in the future. Employing Nissl, Golgi,
and chemoarchitectural techniques, Hall suggests that the lateral,
basal, cortical, and central amygdaloid nuclei are more heterogeneous than usually reported by European and North American
investigators, but that they probably consist of fewer subdivisions than those reported by the Japanese. The suggestion is
further made that due to the shifting series of groupings and regroupings that result from the application of different techniques,
such terms as basolateral and corticomedial be eliminated altogether.

The comparative anatomy of the mammalian amygdaloid complex
is discussed most concisely and clearly by Lammers, who brings
together his own contributions to this field, as well as the work
of new investigators. Possibly of great interest and primary
concern to a number of scientists should be the discussion on
the connections of the amygdala with the secondary olfactory area.
Recently, great interest has been aroused in the field of mammalian
olfaction specifically dealing with pheromones. Thus, we have been
introduced to pheromones that facilitate ovulation, regulate estrous cycle and behavior, enhance or inhibit implantation, enhance
or inhibit implantation, enhance or inhibit learning, and regulate
a number of other biological phenomena. Because of the importance
of the amygdala as a subcortical center of coordination of olfactory impulses with other sensory input, the specific projection of
the lateral olfactory tract to the cortical amygdaloid nucleus,
and the interrelationships of the hypothalamus and the amygdala,
in my opinion, the amygdala probably regulates, coordinates, and
undoubtedly modulates all pheromonal phenomena. Those scientists
dealing with olfaction and biological phenomena may derive an
hitherto unrealized dividend by extending their work to the
amygdaloid complex. Conversely, however, many more problems may
arise than be solved by such an approach, since the nucleus of

the lateral olfactory tract is apparently present even in the
anosmatic porpoise.

The presentation by de Olmos has given us a clear and con-
cise view of the amygdaloid projection field in the rat brain.
Using a cupric-silver method of staining, de Olmos established,
beyond any reasonable doubt, that in the rat the stria terminalis
constitutes not only the major efferent pathway linking directly
the corticomedial nuclear amygdaloid complex with the ipsilateral
medial hypothalamus, but also with telencephalic formations of
both hemispheres. Of great significance is the strial efferent
connections with the ipsilateral accessory olfactory bulb, the
pars medialis of the anterior olfactory nucleus, the ventromedial
hypothalamus nucleus, and with the contralateral olfactory
tubercle and prepiriform cortex. Of great significance to neuroendocrinologists is the finding by de Olmos that the central
amygdaloid nucleus appears to emit fibers which become incorporated into the compact division of the ventral amygdalofugal
pathways and form a continuous field of terminals along the
nucleus itself which extends along the subventricular portion
of the substantia innominata as far as the ventral postcommissural portion of the bed nucleus of the stria terminalis. In
addition, lesions damaging the central amygdaloid nucleus in its
total extent, but which encroach upon the caudolateral end of the
sublenticular portion of the substantia innominate, were associated with abundant fiber degeneration in the medial forebrain
bundle and consequent heavy terminal degeneration in the lateral
hypothalamic area and nuclei gemini. Such degenerative changes
in the MFB and its terminal field were never so pronounced after
extensive lesions of the periamygdaloid cortex or of the anterior
amygdaloid area as here defined. These observations may help us
understand the peculiar interrelationships between hypothalamus
and amygdala for the regulation of hypothalamic releasing factors
and the regulation of hypophyseal tropic hormone secretion.

Kaada's singularly outstanding and comprehensive review
makes one wonder as to the reasons that the amygdaloid nuclear
complex has not merited a great deal more scientific interest and
attention, but seem to have been overshadowed by the hypothalamus
and other regions, and still remains, for a great majority of
scientists, an esoterically oriented region of misunderstanding.
Possibly, the term, amygdala, should be altered to a more colorful
and appealing one. Perhaps in no other section of this conference
are we more impressed than with Kaada's extensive defense of the
amygdala and its involvement in almost every single critical
phase of life of the mammalian organism.

Generally, the electrophysiological-neurophysiological data
presented by Dreifuss, Egger, van Atta, Murphy and Brazier tend to
support strongly and clarify some of the functions and interre-

lationships of the amygdala and other brain regions. Perhaps one of the most significant and outstanding factors, among many, is the clear evidence for species variability in a number of electro-physiological events dealing with various phases of stimulation, elaboration, and propagation of electrical activity within the amygdala, and between the latter and brain regions such as the septum, dorsal medial nuclear group of the thalamus, hippocampus, and hypothalamus. In short, the major established findings indicate that the amygdala and hypothalamus are connected reciprocally, and the ventromedial nucleus of the hypothalamus is a critical focus of amygdaloid input. Additionally, there exists a rudimentary organization of topographical projection of the amygdala onto the hypothalamus. Generally, it is agreed that the amygdala acts as a biaser or modulator, rather than a controller, in influencing hypothalamic neurons. Possibly, the amygdala acts as an intermediate gray region between cortical regions and the hypothalamus, and modulates as well as clocks important functions over which the hypothalamus ultimately exerts the controlling integration and elaboration of neural, behavioral, or neuroendocrine responses. As with other brain regions, the amygdala may stimulate or inhibit the type of changes depending on the rate of impulse transmission. The problem is somewhat complicated in that amygdaloid influences on the hypothalamus may be modified (minimized or maximized) by interactive effects exerted by a number of other limbic structures upon groups of cells in the hypothalamus. Thus, these data suggest that a functional compartmentalization of the amygdala is premature and needs further extensive research. We need additional electro-physiological information on the amygdala and its relationships to other brain regions, especially using chronically implanted electrodes during wakefulness, sleep, drug influences, as well as spontaneously occurring electrical seizure activity.

Further clarification of some of the electrophysiological-neurophysiological interrelationships of the amygdala and other brain regions is provided in the presentations of Gloor, Mark and Narabayashi. Of great significance and invaluable insight into the functional role of the amygdala in everyday living as well as clinical malfunction is the data presented in their respective chapters. Thus, based on human clinical as well as experimental animal studies, Gloor points to the possible role of the amygdala, along with other limbic structures, as one of a link between the master storehouse of information that is laid down in the neocortex and the fundamental motivational drive mechanisms centered upon the hypothalamus. Neural activity in these temporo-amygdaloid motivational systems apparently represents the substrate for subjectively experienced emotions. These are the subjective counterpart of neural activity being directed towards neuronal pools of the hypothalamus which are in command of the fundamental drive mechanisms of the organism. Clinical and experimental animal studies, representing the work of Mark and colleagues, based on behavioral

changes following stimulation without detectable alteration in
local EEG point to the possibility of influence of cellular components at a distance from the recording electrodes by prolonged
release of the exciting neurotransmitters or exhaustion of
antagonists. Furthermore, data from their epileptic patients
with behavior disorders indicate certain interictal mood and
behavioral aberrations may indeed be associated with overactivity
of neuronally separate but physically proximate monoaminergic
systems which undoubtedly interacts reciprocally with the
cholinergic system. To what extent neuronal overactivity and
prolonged neurotransmitter secretion play a role in the paroxysmal
patients of Narabayashi is not clear. However, it is of significance that the application of stereotaxic amygdalotomy produces a
calming effect and reduction of paroxysmal activity in his
patients. Metabolically as well as behaviorally it is of considerable significance that, after treatment, there is a significant reduction in dosages of barbiturates required to induce
sleep. The findings of Narabayashi regarding the calming effects
exhibited by hyperkinetic patients, after stereotaxic amygdalectomy,
may be of considerable value, but certainly acceptance of the operative procedure remains to be considered by the clinicians.

Based on the discussion presented by the neuroendocrinologists,
there is unequivocal evidence that the amygdala modulates and parti-
cipates actively in the regulation of hormonal and neurohumoral
secretions either directly by interacting with the hypothalamus, or
indirectly through reverberating circuits with other limbic structures such as the hippocampus and the limbic midbrain areas. Thus,
Zolovick, with the presentation of a summary of the existing data
as well as a discussion of his own data, gives us a vista of the
vast role of the amygdaloid complex in the regulation of such
hormones as luteinizing hormone, adrenocorticotropin, corticosterone, estrogen, thyrotropin, somatotropin and several others.
His theoretical proposal based on endocrine data that the amygdala may be divided functionally into two divisions, medial and
lateral, may simplify the neuroendocrine approach, but, certainly,
complicates the issues involved in cytochemistry and neurophysiology. However, bringing together these three different areas is
not incompatible or insurmountable, additional data are necessary.

There is no doubt now that the amygdala must be accepted in
the general scheme of endocrine function as an important integra-
tor of events dealing with the neural milieu, adaptation, homeostasis, reproduction as modified by behavioral and neurophysiological events. Indeed, further support to this view is given by
Sawyer with his presentation of the role of the amygdala in the
feedback actions of gonadal steroid hormones. There is strong
evidence that the amygdala exerts an inhibitory influence on the
hypothalamus for the secretion of gonadotropins. How and at what

stage of the feedback sequence the amygdala exerts its influence
certainly needs to be clarified. Zolovick's electrophysiological-
endocrine data point to the distinct possibility that the amygdala
does not exert its gonadotropic regulating influence nor does it
become sensitive to estrogen until after the initial secretion of
luteinizing hormone. Indeed, it appears that the amygdala may be
a modulator of on-going hypothalamic activity and/or modifier of
neuroendocrine events that are related externally or internally
to a number of varied exteroceptive behavioral or physiological
mechanisms.

The singularly outstanding autoradiographic technique of
Stumpf contributes greatly to the concept of amygdaloid regulation
of hormonal events. The finding that the amygdaloid nuclear
groups possess neurons which actively bind steroid hormones in
appreciable concentrations reinforces the concept of a specific
neuroendocrine modulating center within the amygdala for the
regulation of a number of steroid hormones. The concept of
hormone-neuron circuits contributes significantly to amygdaloid
neuroendocrine role and correlates highly with existing anatomical,
neurophysiological and behavioral data for the establishment of
possible neuronal routes of influence.

Finally, with the presentation by Hayward, the role of the
amygdala expands to possible control of the neurohypophysis.
Although a direct link between the amygdala and the preoptic-
thermoregulatory effector mechanisms for regulation of water
balance is unproven, there is ample evidence that the amygdala,
along with the olfactory bulb, olfactory tubercle and the preoptic
area, is part of the secondary forebrain osmoreceptor system of
Sawyer. Thus, a number of varied noxious stimuli, pain, emotional
stress, hypoxia, hypertonicity of the carotid blood, may activate
limbic interneurons with resultant vasopressin release.

Historically, the behaviorists have dealt with functions of
the amygdala and attempts to elucidate the behavioral role of the
amygdaloid complex for a much longer period of time than all other
scientists with the possible exception of the anatomists and
embryologists. As a result, the behaviorists have an overabun-
dance of data involving the amygdala in almost all phases of
behavior dealing with aggression, defense, predatory attack,
active and passive avoidance, hyperphagia, hyperdipsia, hypersexuality, hyposexuality and general mating behavior, social
behavior and a host of other behavior too numerous to mention.
Because of the interdisciplinary nature of the meeting and the
time allotted for each of the major areas discussed, all types of
behaviors modulated by the amygdala could not be covered in
breadth. Thus, only a small sample of the different behaviors
are presented and discussed. Beginning with the presentation by

Kling, we are reminded, once again, of the significant role of
the amygdala in maternal behavior, maternal-infant social interactions, grooming among juvenile and adult primates, and a host
of other social behavioral phenomena. Although it appears that
certain passive social interactions are reduced after amygdalectomy, in general, amygdalectomy heightens rough and tumble
play as well as increases sexual and oral behavior. The
reduction of social behavior in caged animals appears to be
accentuated under field conditions where amygdalectomized primates become complete isolates, fearful and withdrawn from any
group, whether they had previously interacted with such group
or not. There is considerable agreement between primate and
human data in this respect, and Kling's data on primates
parallels Narabayashi's observations of the "taming" effect
on human patients. Further comparative compilation of such
correlative data is very desirable if, ultimately, we are to
understand the functions of the amygdaloid complex.

The hypothalamus and the brain stem comprise a final integrating center which activates autonomic and postural changes in
an adequate defense pattern. Zbrozyna's presentation emphasizes
the role of the amygdala in defense as one which provides refinement in the control of intensity and timing of the display of
the defense reaction. In contrast to this type of behavior,
Karli outlined and discussed the rather dominant contribution
by the amygdala to the mechanisms that facilitate "mouse-killing"
behavior in the rat. Thus, bilateral amygdaloid lesions abolish
this behavior of the rat.

Considerable evidence also was presented to support the
view that cholinergic components of the amygdaloid nuclear complex
may be involved in the mediation of escape-avoidance behavior.
Stimulation of such pathways leads to interference with inhibition
of aggressive reaction to stimulation that ultimately leads to an
inability to develop normal escape-avoidance reactions. This
latter view was proposed by Grossman who provided an impressive
array of data to give support and credence to this theory.

Finally, with the presentation by Goddard, we are exposed to
a new role of the amygdala - that of a kindling effect. Although
this effect can be obtained from stimulation of areas outside of
the amygdala, responsive areas are largely restricted to the
limbic system and related structures and, within the former
system, the amygdala is particularly responsive. The rather
intriguing aspect of these studies is the suggestion that the
responsiveness to kindling effects by particular areas is related directly to the extent that those areas connect anatomically with the amygdala.

Wepsic presents data to support his proposal that marked reduction in activation of amygdaloid neurons by septal stimulation after amphetamine may be due to the fact that this drug acts in the amygdala to decrease electrical excitability. However, the possibility exists that this decreased excitability is secondary to effects in other areas which project to the amygdala. The demonstration that direct application of d-amphetamine to the amygdala also depressed electrical activation of amygdaloid neurons, however, tends to support the original hypothesis of a direct pharmacologic effect. Basically, Eidelberg's finding of amygdaloid neurons that increase their firing rate after parenteral administration of d-amphetamine tends to support the general concept of only excitability of amygdaloid neurons after the drug administration. Although there may be some disagreement among the various pharmacologic studies regarding the specific role played by the amygdala, once again, we are reminded of the myriad of roles that the amygdala exhibits in its functions.

We have only begun to scratch the surface. Perhaps we have confused some and stimulated others, but whatever the generalized effect, we have made an initial attempt to unify the data and to discuss them in their proper perspective. Controversies have always been the inducers that lead to ultimate understnading in science. We all have tried our best with such inducers.

PROLOGUE

Graham V. Goddard

Department of Psychology
Dalhousie University
Halifax, N. S., Canada

In a conference on the neurobiology of the amygdala it is to
be expected that, following a description of the morphology of
the structure, much of the discussion will be devoted to its
functions, and ultimately this will lead to a consideration of
its role in behavior. Often it is expected that a simple description of a selective psychological ability, like the ability
to recognize an appropriate sex partner, will describe what the
amygdala does, i.e., that this ability is its function, or at
least that one function of the amygdala is to participate in that
ability.

In discussions of the amygdala it is common to hear much talk
about emotions, motivation, reward, mood, awareness, hallucinations;
and some question of how activity in the amygdala relates to events
in time. It is not known how much activity is determined by the
traces of past experience, or to what extent the amygdala is involved in laying down memories for the future. Such questions
are very vague; sometimes on the brink of mentalism, often based
on imprecise technique, hardly material to nurture rigorous
scientific enquiry.

There is good reason for difficulty. Consider the anatomical
location of the amygdala and ask what or how any behavioral technique could be applied to study it. Although every sensory
system has a pathway into the amygdala, that pathway is long and
indirect. It first passes through various relays, including
sensory cortex, and the information which arrives at the amygdala
is abstract, almost abstruse. On the output side, the amygdala
is equally remote from the external world. It plays its part in

11

the organization of behaviour, but its role is far from the scene
of action. No particular response or act that an organism makes
can be said to be the direct responsibility of the amygdala. No
response is lost when the amygdala has been removed. The subject's
disposition may change but the component behaviour patterns remain
intact.

In this dream world of the amygdala, the internal environment
of the body receives equal representation with the outside world.
Autonomic and endocrine systems may be sensed by the amygdala and,
in some way, they are controlled by the amygdala - not directly,
of course, but indirectly, as with the external environment.

The amygdala is a complex group of subcortical forebrain
nuclei which may have evolved with the cyclostomes, is apparent
in the tailless amphybia, and has become relatively stable in
mammals. It perhaps reaches its most complex form in modern man
(Humphrey, this volume). Classically, the fiber tracts of the
amygdala were seen to be extremely widespread. The diversity of
these tracts led Johnston to comment that "while the amygdaloid
complex in mammals is a compact collection of cell masses occupy-
ing a restricted area in the temporal pole, it is a complex of
many diverse elements which have been brought together by
mechanical forces and have no primary functional unity" (Johnston,
1915, p. 419). Although Johnston's conclusion may still retain
some validity, several of the classic impressions of amygdaloid
tracts were in error. Of particular interest, in the context of
behaviour, was the error regarding the degree of direct connection
from the olfactory bulbs: the fact that olfactory input to the
amygdala is indirect, passing predominantly through the pyriform
cortex, makes it easier to understand why anosmic animals, such
as the dolphin, have a well developed amygdala, and why destruction of the amygdala has not been found to prevent olfactory discrimination. The surviving known connections of the amygdala are
extremely widespread and complex, being both cortical and subcortical (see de Olmos, Lammers, this volume).

It will be helpful, from a functional point of view, to divide
the amygdala into smaller groups of nuclei, with distinctive projection systems (see Hall, Kaada, this volume). Before much can
be made of these divisions either theoretically or experimentally,
it will be essential to know much more about connections that
exist between amygdaloid nuclei. Virtually nothing is known about
them at present and, until such information becomes available, we
must be content with electrophysiological data.

Sensory stimulation from all modalities, and electrical stimulation of many brain areas, have been shown to trigger evoked
potentials and alter unit activity in the amygdala. These re-

sponses are usually of long latency, often unreliable, and sometimes continue after the end of the stimulation. Convergence
from many sources is often encountered on the same cell, yet many
cells do not respond to any of the inputs that have been tested.
One aspect of the convergence is elegantly demonstrated by the
work of Caruthers and his colleagues (1964) who have shown that
evoked potentials from various sources can be blocked for several
hundred milliseconds following an initial input from either the
cortex or the hypothalamus.

One of the more tantalizing characteristics of amygdaloid
units is that some of them will respond selectively to complex
environmental stimuli: meows rather than clicks, mice rather
than flashes. Furthermore, using chronically implanted microelectrodes in freely moving cats, O'Keefe and Bouma (1969)
observed four amygdaloid units which were selectively responsive
to complex inputs and which maintained a high frequency response
for 30 seconds or longer after the stimulus had been removed.
But this work can be very misleading. Apart from the difficulties
involved in sampling bias, the dangers of generalizing from the
very small number of such cells encountered, and the extraordinary
difficulty of knowing when the vast array of possible stimulus
conditions has been appropriately tested, there is a danger that
the results may lead to thinking in terms of "tuna-fish detectors."

Most of the other recent neurophysiological work has concentrated on details of the amygdaloid projection to the hypothalamus. This symposium contains much about the manner in which
cells in different hypothalamic nuclei are excited and inhibited
by activity in the amygdala (Dreifuss, Egger, Murphy, Van Atta,
this volume). These details are of great importance and will
provide a basis for the understanding of output mechanisms of
amygdaloid function. Perhaps their greatest contribution will
be to the understanding of the extensive amygdaloid control of
hormonal and autonomic regulation (Koikegami, 1964, and articles
in this volume, Eleftheriou, Hayward, Raisman, Sawyer, Stumpf,
Zbrozyna, Zolovick). It is to be expected that the latter direction will yield the most rapid and unambiguous understanding of
amygdaloid function since the amygdalo-hypophyseal pathway is
the shortest and most direct exit from the central nervous
system. In contrast, the possible role of the amygdala in
controlling behavior, such as sexual behavior, is far more indirect. The exit from amygdala to observable overt behavior
must involve many additional neural structures.

It used to be thought that amygdaloid lesions resulted in
hypersexuality: males with amygdaloid lesions were thought to
copulate more frequently, and with a greater diversity of sex
partners than intact animals. However, many of the earlier con-

clusions were based on animals with lesions that were not restricted to the amygdala (e.g. Kling, this volume). Particularly
misleading were studies on male cats which <u>normally</u> can become
hypersexual if tested repeatedly in the same limited environment.
Many authors have failed to find hypersexuality following amygdaloid lesions, and in a recent study of the male rat, Giantonio
and co-workers (1970) found a small reduction in promptness to
ejaculate following lesions of stria terminalis or the medial
half of the amygdala. This would be more consistent with the
testicular degeneration reported to follow amygdalectomy.

Emotionality and responsiveness to noxious stimuli have been
the most controversial and frequently studied phenomena to be
linked with amygdaloid function. The lesion studies which bear
on this problem most frequently report placidity but sometimes
report savageness. The placidity involves a loss of aggression,
a reduced responsiveness to normally noxious stimuli and a fear-
less curiosity for dangerous or threatening objects including
members of other species (see also Karli, Kling, Narabayashi,
this volume). A number of hypotheses have been considered in an
attempt to explain the discrepancy between these studies and the
few which have reported rage. It seems that differences in
species, surgical technique, size or location of lesion, involvement of different extra-amygdaloid structures and pre- or postoperative experience cannot be considered as complete explanations
(for review see Goddard, 1964).

Kling and co-workers have produced rage in a few of their cats
using exactly the same technique as that which produced placidity
in other cats. Furthermore, they were unable to find any histological differences in these animals that correlated with the
behavioral differences. Green and co-workers (1957) also obtained
some cats that were placid and others that showed rage. The rage
was found to develop only if the lesions involved the hippocampus,
and then only if the animals developed periodic seizures. It has
never been demonstrated, but it may be reasonable to suggest,
that in all cases where rage follows amygdalectomy it is due to a
discharging epileptogenic focus on the periphery of the lesion.

Connection between emotional responses and amygdaloid seizure
activity has been reported in a number of situations. Psychomotor
epilepsy with the seizure focus in or near the amygdala is accompanied by fear or rage far more frequently than epileptic seizures
originating from other areas of the brain. Similarly, an experimental tungstic acid focus results in seizures accompanied by
aggressive behaviour in cats (Blum and Liban, 1960). Furthermore,
Grossman (1963) reported that a single injection of carbachol
into the cat amygdala resulted in seizures which recurred two or
three times daily for 5 months, the animals remained savage and

unapproachable throughout this period. Other chemical stimulation
studies that have elicited aggression have also used chemicals
that are able to cause seizure discharge.

For many years it has been known that rage can be produced by
electrical stimulation of the amygdala (Kaada, Reis, this volume).
Traditionally, this was thought to imply that in the normal brain
the amygdala controls rage. The theory was extended by tractotomy
experiments showing that stimulation-induced rage was mediated by
activity conducted into the amygdala from the stria terminalis
and out of the amygdala through the hypothalamus and tegmentum.
If true, of course, the theory would certainly account for the
rage reactions that are associated with seizure activity of the
amygdala. However, there is reason to think that amygdaloid
stimulation does not cause rage unless it also causes seizure
activity. Most of the animal experiments in which rage reactions
have been reported did not include EEG recordings, so we cannot
be sure of this, but the other behaviors that were reported (such
as eye closing and facial twitching) are known to be associated
with seizure activity. In fact, the threshold intensity for producing flight or rage reactions was usually reported to be above
that for some of the known seizure-associated responses.

The main problem is simply that seizures propagate to involve
other structures, and it is impossible to know the extent of
propagation at the moment the behavior appears. Thus it is
impossible to gauge whether the amygdala is involved in this
behavior or not. All that can be concluded with safety is that
the amygdala projects to structures that are themselves important
for the rage reactions.

It was mentioned earlier that one difficulty with such
studies comes from the anatomical location of the amygdala. In
order to observe a behavioral response to amygdaloid stimulation
it is necessary to stimulate strongly enough to force a pattern
of activity out of the amygdala. But consider what that pattern
must be forced through. Many other brain structures must mediate
that pattern before it is observed in behavior. Each mediator
will alter the pattern depending on its various other inputs at
the time. It is not surprising that seizure thresholds are ex-
ceeded before direct behavioral responses are evoked.

Perhaps for this reason other authors have attempted to super-
impose amygdaloid stimulation upon behavioral responses that are
controlled by environmental manipulations. That is, these studies
have used either electrical or chemical stimulation of the amygdala
to interfere with the acquisition or performance of conditioned
responses, usually avoidance responses. The intensity of stimulation in these experiments is usually kept below that which would
force out any particular behavioral response. The typical results

of these studies is that the stimulation interferes with the
learning or performance of some aspect of the avoidance task.
Unfortunately, even here, it is not clear that results have been
obtained without contamination by localized or propagated seizure
activity. Many authors have deliberately used stimulation which
caused seizure, others may have employed seizures unwittingly
(Lidsky et al., 1970). Studies have varied on the specificity of
the behavioral responses that were found to be disrupted by
stimulation. It might be expected that such variation depends
entirely on the extent of seizure propagation. One of the most
important studies of this type has been reported by Nakao (1962,
1967). Nakao was not dealing with learning, but with a well
established escape response. Hippocampal after-discharge did
not interfere with the response unless it propagated to involve
the amygdala. But the same type of after-discharge, when started
in the amygdala, did not interfere with the response until it
propagated to the hippocampus. Thus we cannot know whether the
effect was due to a combined seizure in both amygdala and hippocampus or a seizure in some third structure not being monitored.

The problem of seizure propagation is even further compounded
by the so-called "kindling" effect (Goddard, this volume). The
extent of seizure propagation from any given activation of the
amygdala does not remain constant but changes as a function of
repetition. Even intensities of stimulation which initially
cause no after-discharge at all can come to trigger extensive
seizures if repeated once each day. Thus, even from one trial
to another, it is very difficult to specify which behavioral
effects were associated with seizure unless electrographic recordings are made on every trial.

Experiments using lesions of the amygdala have tended to be
a little more informative when the lesion effect is evaluated by
investigations of the learning abilities of the animal. Although
these studies have provided no clue about mechanism, they have
identified aspects of behavior in which the amygdala has some
influence. My own impression, based on a partial survey of the
literature, is that in approach learning situations removal of
the amygdala has little effect except to decrease attention to
task and impair the ability to withhold responses. Reversal
learning is commonly disrupted. In the avoidance learning situation, all types of response may be impaired, but CER and passive
avoidance are the hardest hit. Unfortunately, individual animals
which show a marked deficit on a particular avoidance measure
often have lesions that cannot be distinguished in size or
location from individuals that show no such deficit (see also
Grossman, this volume).

It is often thought that much more will be learned about

functions of the amygdala from stimulation studies in humans. The patient is almost thought of as a talking preparation that can introspect and see what is being experienced during the amygdaloid stimulation. Besides the philosophical dualism which sometimes impedes interpretation of these data, there is the additional difficulty that all observations are obtained from patients with an abnormal (usually epileptic) temporal lobe - otherwise there would not have been justification for the surgery.

In practice the technique has yielded a bewildering array of illusions, hallucinations, mood changes and reports of subjective confusion. Clinically useful information is obtained when the electrical stimulation evokes auras and other components of the psychomotor epilepsy. Two points of major interest are that the patients rarely respond or report experiences unless the stimulation causes propagated abnormal activity, and that these psychic responses occur only in patients with an epileptic focus within that same temporal lobe. Rasmussen (1967) has reported that, at the Montreal Neurological Institute, complex hallucinatory or interpretive responses have never been evoked from the temporal lobe in patients whose clinical attacks were arising from above the Sylvian fissure. He also has reported that, when experiences are triggered by stimulation of the amygdala, they are always accompanied by extensive changes in activity of the temporal cortex, frequently involving an organized epileptiform after-discharge.

Thus, as with the animal studies of electrical stimulation, recording, and lesions of the amygdala, only very limited conclusions can be drawn from the human data. Certainly, some knowledge has been gained and certain inferences are apparent. Gloor, for example, has presented an excellent hypothetical framework of amygdaloid function to which most of the experimental data can be fitted (Gloor, this volume). Gloor's account is necessarily vague and, unfortunately, as it must, omits any statement of mechanism. It is the best we have at present.

REFERENCES

BLUM, B., & LIBAN, E. Experimental basotemporal epilepsy in the cat. Neurology, 1960, 10, 546.

CARUTHERS, R., MULLER, A. K., MULLER, H. F., & GLOOR, P. Interaction of evoked potentials of neocortical and hypothalamic origin in the amygdala. Science, 1964, 144, 422.

GIANTONIO, G. W., LUND, N. L., & GERALL, A. A. Effect of dien-
    cephalic and rhinencephalic lesions on the male rat's sexual
    behavior. Journal of Comparative and Physiological Psychol-
    ogy, 1970, 73, 38.

GODDARD, G. V. Functions of the amygdala. Psychological Bulletin,
    1964, 62, 89.

GREEN, J. D., CLEMENTE, C. D., & DeGROOT, J. Rhinencephalic
    lesions and behaviour in cats. Journal of Comparative
    Neurology, 1957, 108, 505.

GROSSMAN, S. P. Chemically induced epileptiform seizures in the
    cat. Science, 1963, 142, 409.

JOHNSTON, J. B. The cell masses in the forebrain of the turtle,
    Cistudo Carolina. Journal of Comparative Neurology, 1915,
    25, 393.

KOIKEGAMI, H. Amygdala and other related limbic structures - ex-
    perimental studies on the anatomy and function. II. Func-
    tional experiments. Acta Medica et Biologica, 1964, 12, 73.

LIDSKY, T. I., LEVINE, M. S., KREINICK, C. J., & SWARTZBAUM, J. S.
    Retrograde effects of amygdaloid stimulation on conditioned
    suppression (CER) in rats. Journal of Comparative and
    Physiological Psychology, 1970, 73, 135.

NAKAO, H. The spread of hippocampal after-discharges and the
    performance of switch-off behavior motivated by hypothalamic
    stimulation in cats. Folia Psychiatrica et Neurologica
    Japonica (Niigata), 1962, 16, 168.

NAKAO, H. Facilitation and inhibition in centrally induced
    switch-off behavior in cats. Progress Brain Research, 1967,
    27, 128.

O'KEEFE, J., & BOUMA, H. Complex sensory properties of certain
    amygdala units in the freely moving cat. Experimental
    Neurology, 1969, 23, 384.

RASMUSSEN, T. Comments in Discussion, L. F. Chapman et al.
    Memory changes induced by stimulation of hippocampus or
    amygdala in epilepsy patients with implanted electrodes.
    Transactions of the American Neurological Association,
    1967, 92, 50.

EMBRYOLOGY

THE DEVELOPMENT OF THE HUMAN AMYGDALOID COMPLEX

Tryphena Humphrey*

Department of Anatomy, Medical Center, University of
Alabama in Birmingham

Birmingham, Alabama

INTRODUCTION

In a recent paper (Humphrey, 1968), the human amygdala (or
better the amygdaloid complex) was identified as the first portion
of the striatal complex to appear during development, although
other parts of the corpus striatum are identifiable soon afterward.
In this paper, the differentiation of the nuclei of the amygdaloid
complex was traced from the initial migration of the neuroblasts
outward from the medullary epithelium, at the time that the telencephalic hemispheres first begin to evaginate, to 8.5 weeks of
menstrual age when all but one nucleus of the complex could be
identified. At the present time, when the amygdaloid complex, or
archistriatum, is being investigated from so many approaches, it
seems appropriate to review briefly this early period of development previously reported, and to include the additional stages in
its differentiation necessary to reach the age level at which
function has probably begun.

MATERIALS AND METHODS

The data for the preparation of the 15 specimens on which the
earlier paper (Humphrey, 1968) was based is given in Table 1 of
that publication. Of the older fetuses available in the Hooker-
Humphrey collection, photographs were made of the amygdaloid
complex of 12 and stages illustrating representative changes
chosen from them for the illustrations. Only these fetuses and

* Deceased August, 1971

TABLE I

Specimens Illustrated or Described

No. in collection	Crown rump (CR) length	Menstrual Age in weeks[1]	Technique	Plane & Thickness of sections[2]
Sen-2[3][4]	4.8	?	Hematoxylin & eosin	Transverse
02[3]	10.0	6	Stotler's silver intensifier	Transverse
Sen-3N[3][4]	13.5	6.5	Hematoxylin, eosin, orange G and analin blue	Sagittal
113	14.0	6.5	Protargol[5]	Transverse
P[3]	14.0	6.5	Erythrosin & Toluidin blue	Transverse
142	18.0	7+	Protargol	Transverse
93A	20.7	7.5	Protargol	Transverse
130	22.2	8-	Protargol	Transverse
152	24.5	8+	Thionin	Sagittal
A[3]	27.4	8.5	Thionin & erythrosin	Transverse
126	33.8	9.5	Sperry's silver method	Transverse
101	38.2	10	Stotler's silver intensifier	Transverse
103	40.7	10	Thionin & erythrosin	Transverse
N3	42.0	10.5	Thionin & eosin	Sagittal
119	48.6	11	Thionin & erythrosin	Transverse
148	56.0	11.5	Thionin & eosin	Transverse (15μ)
132	60.5	12	Thionin & erythrosin	Transverse (15μ)
118	79.0	13.5	Thionin & eosin	Coronal
157	89.0	14	Thionin & erythrosin	Coronal (15μ)
147	114.0	16	Thionin & erythrosin	Coronal (15μ)
98M	144.0	18.5	Thionin & erythrosin	Coronal (15μ)
117	216.0	24.5	Thionin & erythrosin Right hemisphere Left hemisphere	Coronal (20μ) Sagittal (25μ)

[1] The menstrual age was estimated by Davenport Hooker from the crown rump length (CR) by the tables of Streeter (1920).

[2] All of the specimens were serially sectioned at 10μ except for those for which the thickness is given after the plane of sectioning.

[3] These specimens were measured after fixation. The numbered specimens, except those from the collection of Dr. E. Carl Sensenig, were measured prior to fixation and are part of the series for which reflex activity was tested by Dr. Davenport Hooker (1952, 1958 and elsewhere) and motion picture recordings made if the fetuses were motile.

[4] From the collection of Dr. E. Carl Sensenig, Department of Anatomy, University of Alabama in Birmingham.

[5] All of the protargol series listed in this table were prepared by the method of Bodian (1937).

those used for the illustrations in the 1968 paper and/or
mentioned in the description of the amygdaloid complex are
included in the table of the material (Table 1). However, the
amygdaloid complex was examined for more than twice this number
of specimens.

The embryos and fetuses listed in this table were sectioned
transversely in toto until after 12 weeks of menstrual age.
Sectioning the fetus transversely provides a more or less horizontal plane through the forebrain at 7+ to 12 weeks. Such
sections through the amygdala are difficult to interpret without
orientation with reference to low power sections and a surface
view as in Figure 1. For the fetuses over 12 weeks, the brain
was removed and sectioned separately, either sagittally or in
the customary coronal plane as shown in Figure 2.

The drawings in Figure 1 illustrate the general plane of
the sections, the position of the amygdaloid complex with reference to the surface of the telencephalon, and the location of
both the corpus striatum and the amygdala from about 18.0 to
40 mm as seen in the sections 7+ to 10 wks). The increased
development of the forebrain, and the consequent shifts in its
various parts between 10 and 12 weeks, give rise to the additional
changes in position shown in Figures 9D, 11A and 12A. After 12
weeks the brain was always removed and sectioned either in the
coronal plane, as indicated in Figure 2A, or in the sagittal
plane. Inasmuch as the shifts in position of the amygdaloid
complex during development constitute a significant phase in its
development, an explanation of the planes of sectioning is essential to an understanding of its differentiation.

Throughout the present paper, the embryonic ages used are
menstrual age based on the tables of Streeter (1920). In
referring to the work of other investigators, the age mentioned
is also based on these tables as determined by the crown rump
length (CR) given by the author in question. This has been done
to avoid confusion in making comparisons or correlations with
the observations reported in the literature.

RELEVANT LITERATURE

The literature on the early development of the human striatal
complex was reviewed in an earlier paper in some detail (Humphrey,
1968), so only the references specifically related to the development of the amygdala will be repeated here. It is of interest,
however, that although the striatal complex was identified in
embryos as small as 7 mm by other investigators (7-15 mm, Kodama,
1926a, 1926b, 1927; 8 mm, Sharp, 1959; and 8.5 to 10 mm, Cooper,
1946), the subdivision represented earliest in development --

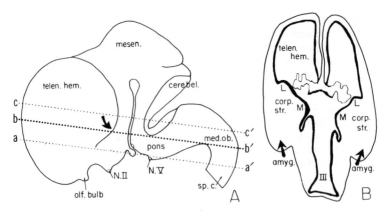

Fig. 1.* Drawings to illustrate the location of the amygdaloid complex (large arrows) and the general plane of the sections photographed for Figs 4 and 6 to 8. A. Outline of the lateral surface of the brain of a 27.0 mm fetus as illustrated by Hochstetter (1919). The dotted lines indicate the general plane of sectioning from the anteroinferior to the posterosuperior levels of the amygdaloid complex (line a - a´ to line c - c´). B. Outline of a section through the amygdaloid complex of the 22.2 mm embryo illustrated in Fig. 6E at about the level of the heavy dotted line (b - b´) on part A. (Fig. 5 from Humphrey, 1968, J. Comp. Neur., 132: 135-165, reproduced through the courtesy of The Wistar Press).

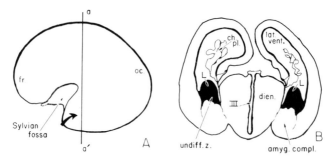

Fig. 2. Outline drawings to show the location of the amygdaloid complex after removal of the brain. A. The lateral surface of the left cerebral hemisphere of a fetus approximately 14 weeks of menstrual age as modified from Retzius, 1896. The plane of the serial sections passing from frontal (fr.) to occipital levels (oc.) is shown by the line a - a´. The large arrow indicates the location of the amygdala. B. A coronal section through the forebrain of a fetus of 14 weeks to show the location of the amygdaloid complex as seen in similar sections after removal of the brain (see Fig. 14 for photomicrographs of the amygdaloid complex of this fetus).

*   For abbreviations to all figures, consult list of Abbreviations at the end of this paper.

paleostriatum, archistriatum or neostriatum -- was not suggested.
Indeed, in the papers of Smith-Agreda (13 mm, 1963), Hewitt
(15 mm, 1961), Johnston (18 mm, 1923) and Hines (19.1 mm, 1922),
the caudate nucleus, a part of the neostriatum, was the region
of the striatal complex to be identified earliest in development.
The paleostriatum (globus pallidus) and the archistriatum (amygdala) were identified by Johnston (1923), by Macchi (1951), and
by Smith-Agreda (1961, 1962, 1963) at the same stage in development (23 mm, 25 mm and 26.0 mm respectively). In the 1968 paper
of Humphrey, however, it was concluded, on the basis of the
location with reference to the primordial hippocampal formation,
that the first neuroblasts to migrate out from the medullary
epithelium on the posterior aspect of the lateral ventricle when
it first begins to evaginate consititute a primordial amygdaloid
complex (Fig. 3). Still other investigators either did not
mention the amygdala at all (e.g., Cooper, 1946; Hewitt, 1961)
or referred to it as an undifferentiated cell mass (at 23 mm,
Macchi, 1948, 1951) identifiable only on the basis of its position.
Even as late during development as 45.0 mm, Hewitt (1958) referred
to this area as the "primordium of the amygdaloid complex" and
Hochstetter (1919) called it the anlage of the amygdala in a
46.5 mm fetus (fetuses 11 weeks of menstrual age).

It was not until 13.5 weeks (80 mm CR) that any of the
specific amygdaloid nuclei were identified by the investigators
just mentioned. At that age, Macchi (1948, 1951) recognized five
areas -- the medial, cortical, central, basal and lateral amygdaloid nuclei. At 16.5 weeks (120 mm CR), however, Escolar (1959)
mentioned only a deep part and a periamygdaloid portion of the
amygdala. Nevertheless, at 18.5 weeks (145, CR), Johnston (1923)
identified all of the nuclei of the amygdaloid complex except
the accessory basal nucleus and the nucleus of the lateral
olfactory tract. At 21 weeks (175 mm CR), the nucleus of the
lateral olfactory tract was recognized also by Johnston. The
amygdala was referred to by Hewit (1958) as being developed
completely in a fetus of 26 weeks (230 mm CR), but no description
was given. Three portions of the amygdala were mentioned by
Hilpert (1928) at about 33.5 weeks (300 mm CR) - ventral, central
and dorsal parts. Hilpert also recognized the three parts of the
periamygdaloid cortex of Rose (1926), but identified this layer
of neurons over the surface of the amygdala as the cortical amygdaloid nucleus. Neither the cortical nucleus of Hilpert (1928)
and of Johnston (1923), however, nor the cortical nucleus of
Crosby and Humphrey (1941) in adult human amygdala include all
of the periamygdaloid cortex of Rose. Indeed, the periamygdaloid
cortex of Rose extends frontally farther and includes also part
of the corticoamygdaloid transition area of Crosby and Humphrey
(1941, p. 338). Following Johnston's contribution (1923), the
cell layer over the surface of the amygdala commonly has been

included as a part of the amygdaloid complex in most papers on
the mammalian amygdala. The development of the amygdaloid complex
has been investigated in only a few mammals, aside from man. In
mouse embryos, Völsch (1910) identified the amygdala as a group of
medium-sized cells constituting a sinking-in of part of the piriform area. In later autoradiographic investigations, on the
development of the amygdaloid nuclei in the mouse, Sidman and
Angevine (1962) demonstrated that neurons "destined for" the
medial and central nuclei arise on day 10, earlier than those
for the other nuclei. Surprisingly enough, the authors found that
the ventral part of the lateral nucleus began on the same day,
although it did not reach its peak until two days later (day 12).
According to these investigators, both the basal and cortical
nuclei begin on day 11, later than the lateral nucleus, and the
peak of development was not reached until days 12 and 13. Concerning their study, Sidman and Angevine stated that "the method
gives unequivocal data on time of neuron origin."

More recently, the development of the amygdaloid complex was
studied by Brown (1967) in insectivorous bats, using Mallory's
quadriple stain and protargol silver material. In this investigation, primordial corticomedial and basolateral cell masses were
recognized before specific nuclei. The cortical nucleus was
identified as early in development as the medial nucleus, and the
accessory basal nucleus was recognized equally early (7 mm
embryos), whereas the central nucleus was not identified until
9 mm. Distinct lateral and basal nuclei appeared still later
(10 mm), after the accessory basal nucleus. Evidently, the basal
nucleus is identifiable before the lateral nucleus, but Brown
found that the nucleus of the lateral olfactory tract is the
latest of all of the amygdaloid nuclei to appear. The early
appearance of the lateral amygdaloid nucleus in mouse embryos,
reported by Sidman and Angevine (1962), and the late development
in isectivorous bat embryos, found by Brown (1967), might be due
to species differences, but also might be related to differences
in identification of the lateral nucleus in its early stages.

The mammalian amygdala was recognized by a number of investigators prior to the classical work of Johnston (1923). This early
work is reviewed in the paper of Johnston and additional data on
the mammalian amygdaloid complex is to be found in the several
published papers and will not be reviewed here (Obenchain, 1925;
van der Sprenkel, 1926; Loo, 1931; Humphrey, 1936; Young, 1936;
Brockhaus, 1938; Crosby and Humphrey, 1941, 1944; Fox, 1940;
Jeserich, 1945; Lauer, 1945, 1949; Jansen and Jansen, 1953; and
Hamel, 1966).

During human development, Macchi (1948, 1951) classified the
amygdaloid nuclei, identified by him, into a centromedial complex

consisting of a central, and a medial nucleus, and a basolateral complex that included basal and lateral nuclei. He also recognized an anterior amygdaloid area and the intercalated cell masses of Johnston (1923), as did later investigators of the human amygdala (Crosby and Humphrey, 1941), but did not identify the nucleus of the lateral olfactory tract at any age studied.

Johnston (1923, p. 456) subdivided the amygdaloid nuclei into two groups based on his interpretation of their phylogenetic and embryologic development in human fetuses. His more primitive cell group includes the medial, central and cortical nuclei and the nucleus of the lateral olfactory tract. The basal and lateral nuclei were classed as recently developing structures formed by an infolding of the surface cortex. Although included in his more primitive nuclei, Johnston also considered the cortical nucleus to be derived from the piriform lobe cortex.

In their report on the adult human amygdala, Crosby and Humphrey (1941) followed the classification of the amygdaloid nuclei introduced by Johnston (1923), as modified by Humphrey (1936), and used since by many other observers of the adult mammalian amygdala. It was adopted also by Brown (1967) and Humphrey (1968) in their investigations on the development of the bat and of the human amygdalae respectively. This grouping of the nuclei is as follows: (1) a corticomedial amygdaloid complex consisting of the cortical, medial and central nuclei and the nucleus of the lateral olfactory tract; (2) a basolateral complex made up of the lateral, the basal, and the accessory basal nuclei; (3) the anterior amygdaloid area consisting of the relatively less differentiated areas not allocated to any specific nucleus. In addition, intercalated cell masses have been recognized between and along the surface of the various nuclei of the basolateral complex.

In their study of the amygdaloid complex in the shrew (Crosby and Humphrey, 1944), a comparison was made between the position of its constituent nuclei in the shrew and in man. It was pointed out at that time that due to the development of the human temporal region the amygdala rotates medially through approximately 140°, as compared with its position in the shrew. As a result, the nuclei which are lateral and dorsal in the shrew are located ventrally (and medially) in the human amygdaloid complex. The fundamental pattern and nuclear relationships of the mammalian amygdaloid complex is retained in the human amygdala, however. Johnston (1923, p. 452) noted that in a fetus of the "eighth month" (i.e., 28 to 32 wks) the human amygdaloid complex already had rotated "downward and inward" enough to bring the nuclei situated laterally at 21 weeks (175 mm CR) to a ventral (or basal) position.

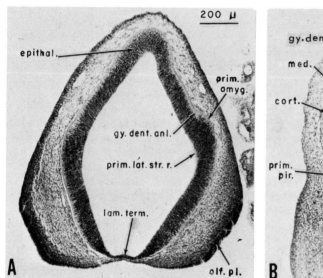
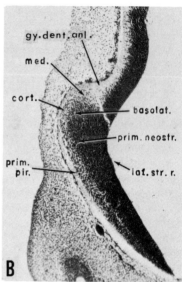

Fig. 3. Photomicrographs illustrating the early development of
the human amygdaloid complex. The magnification scale applies to
both <u>A</u> and <u>B</u>. <u>A</u>, section through the forebrain of a human embryo
in the older levels of Streeter's horizon XV (1948; 6.0 to 9.5 mm)
in which the evaginations for the cerebral hemispheres are beginning (Right side of figure). The small area of cell proliferation and migration from the ependymal layer, or medullary epithelium, constitutes the primordial amygdala. Embryo from the
collection of Dr. E. Carl Sensenig (Sen-2, section 3-2-4). Fig. 1
from Humphrey, 1968, J. Comp. Neur., 132: 135-165, reproduced
through the courtesy of The Wistar Press. <u>B</u>. Section through one
side of the forebrain of a 10.1 mm human embryo (No. 02, section
6-3-2) with a more distinct evagination of the telecephalic
hemispheres. Both the medial and cortical nuclei of the corticomedial complex are already identifiable, but no parts of the
basolateral amygdaloid complex have appeared. (Adapted from
part B of Fig. 2 of Humphrey, 1968, J. Comp. Neur., 132: 135-165
and used with the permission of The Wistar Press).

A major objective of the present paper, then, is to show the shift in position of the amygdala from the time that the primordial cell mass appears to the oldest age available (24.5 wks). Other aims include the determination of the age level at which the different nuclei of the complex become recognizable, the sequence in which the nuclei develop, and when new cells are no longer added to the complex.

OBSERVATIONS

The Primordial Amygdala and the Development of its Major Subdivisions

From an earlier study (Humphrey, 1968) it was concluded that the early migrating neuroblasts from the medullary epithelium (or ependymal layer) of the ventrocaudal wall of the interventricular foramen, identified by Sharp (1959) as the primitive striatum, actually constitute a primordial amygdala (archistriatum). This conclusion was based, in part, on the location of these migrating cells adjacent to the cortical area previously identified as the primordial hippocampal formation (Humphrey, 1966a) and, more specifically, the anlage of the gyrus dentatus (Humphrey, 1966b). This initial cell migration first appears at the time that the telencephalic vesicles begin to evaginate (the upper levels of Streeter's Horizon XV, 1948, 8 to 10 mm embryos). No subdivisions of the amygdala are identifiable at first (Fig. 3A), although it might be pointed out that the migrating cells near the anlage of the gyrus dentatus tend to be more separated from the underlying medullary epithelium than those more distant from the primordial hippocampal formation.

Almost at once, however, enough neuroblasts have separated from ependymal layer to enable identification of both the corticomedial complex and two of its nuclei, the medial nucleus adjacent to the anlage of the gyrus dentatus and the cortical nucleus lateral to it (Fig. 3B). The cell layer forming the cortical nucleus becomes less distinct as it becomes continuous with the primordial piriform cortex which overlies the cell masses beginning to form the neostriatum. The additional cells that are separating from the ependymal layer deep to the corticomedial complex, but are still in continuity with it, represent the basolateral complex of the amygdala. In this subdivision of the amygdala, however, no separate nuclei are as yet recognizable. Additional neuroblasts which are separating from the ependymal layer deep to the area identified as primordial piriform cortex have been identified as the primordial neostriatum.

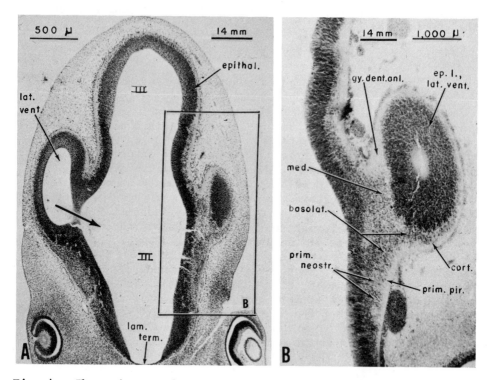

Fig. 4. Photomicrographs illustrating the amygdaloid complex at approximately the same level of two different 14.0 mm human embryos, but at different magnifications and at a slightly different plane of sectioning. A. The amygdala as it lies posterior to the interventricular foramen (large arrow on left side). Note the degree of development of the telencephalic hemispheres with reference to that of the diencephalon. This photograph is from the section in Fig. 4A of Humphrey, 1968, but at a higher magnification (No. 113, section 15-1-6). B. An area comparable to that enclosed in rectangle B on part A, but from another 14.0 mm (Embryo P, section 12-3-6) to demonstrate the uniformity in the degree of differentiation of the amygdaloid complex at this age.

THE DEVELOPMENT OF THE SPECIFIC AMYGDALOID NUCLEI

6.5 weeks (13.5 and 14.0 mm). In two embryos of 14.0 mm
(6.5 weeks, Fig. 4), the size of the cell mass constituting the
amygdala has more than doubled that present at 10 mm. The medial
nucleus is defined more clearly at both low (Fig. 4A) and higher
magnifications (Fig. 4B) in the usual plane of sectioning and in
the sagittal sections (13.5 mm, Fig. 5A). The cortical nucleus is
less distinct, however, probably due to the increased size of the
basolateral complex underlying it without a commensurate increase
in neuroblasts within the cortical nucleus. The other nuclei
(central and nucleus of the lateral olfactory tract) in the
corticomedial complex have not yet appeared. In the basolateral
amygdaloid complex, neither basal nor lateral nuclei are recognizable, but the characteristic continuity of this complex with
the primordial neostriatum is clear in both planes of sectioning
(Figs. 4 and 5A). Likewise, the continuity with the anlage of
the gyrus dentatus is equally distinct.

18.0 to 22.2 mm (7+ to 8- wks). In embryos under 13.0 mm in
length, only one striatal ridge (or elevation) has been recognized.
Johnston (1923, p. 359) identified two striatal ridges at 13.0 mm
and at 14.8 mm Hochstetter (1919) recognized both medial and
lateral elevations in the region of the interventricular foramen.
The two striatal elevations are separated by a sulcus that is
best designated the interstriatal sulcus both embryologically
(Brown, 1967; Humphrey, 1968) and phylogenetically (Schnitzlein
and Crosby, 1967) because this name implies no functional relationship as do the terms strio-caudate sulcus (Johnston, 1923) and
fissura paleo-neo-striatica (Ariëns Kappers, 1929). As in the
earlier paper on the embryogenesis of the human amygdala (1968),
interstriatal sulcus will be used in the present account. Although the medial striatal elevation has been recognized as early
as 13.0 mm, it has not developed far enough caudalward from its
area of origin near the interventricular foramen to come into
relation with the amygdala in the 18.0 and 20.7 mm embryos
(Fig. 6A-C). In the 22.2 mm embryo (Fig. 6D-E) and later (Fig. 7),
however, the medullary epithelium of both the medial and the
lateral striatal elevations contribute neuroblasts to the developing amygdaloid complex. Later in development, cell migration into
the amygdala is mainly from the lateral striatal ridge.

During the age period under consideration, there is a greater
amount of differentiation in the corticomedial division of the
amygdaloid complex than in the basolateral one, although the
increase in size of the former is less. In addition, a less well
developed area, identified as the anterior amygdaloid area at
18.0 mm, is present thereafter (Humphrey, 1968). The basolateral
complex is developing directly out of the medullary epithelium at

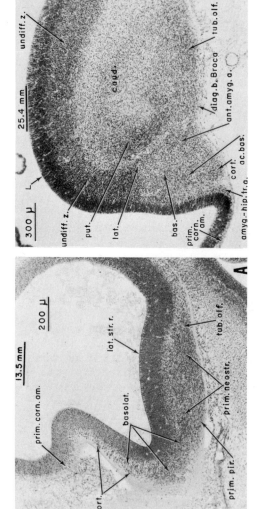

Fig. 5. Photomicrographs of sagittal sections through the amygdaloid complex of two human embryos illustrating the difference in the position of this complex with reference to the lateral ventricle at 13.5 mm (A) and 24.5 mm (B). Note that in the 13.5 mm embryo the amygdaloid complex lies directly posterior to the lateral ventricle, whereas at 24.5 mm the ventral extension from the lateral ventricle that develops into its inferior horn carries the amygdaloid complex forward so that it lies anterior to the ventrally directed inferior horn. The sections are oriented with the posterior aspect toward the left in each case. A. Sagittal section of a 13.5 mm embryo from the collection of Dr. E. Carl Sensenig(Sen-3N, section 14-2-2) shown in Fig. 3B of Humphrey, 1968, J. Comp. Neur. 132: 135-165 at a different magnification and orientation. Reproduced here with the permission of The Wistar Press. B. Sagittal section near the middle of the amygdaloid complex of 24.5 mm embryo at a level that shows the inferior horn of the lateral ventricle beginning to turn frontally (No. 152, section 23-1-2). (Fig. 9B from Humphrey, 1968, J. Comp. Neur., 132: 135-165, reproduced with the permission of The Wistar Press).

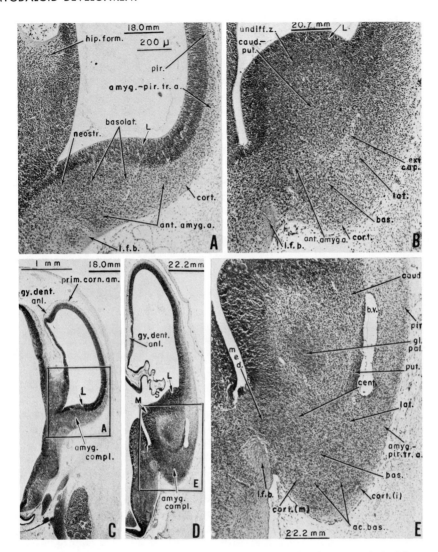

Fig. 6. Photomicrographs of sections through the amygdaloid complex
of three different embryos at the level where the size of the complex is greatest. These illustrations are reproduced from the Figs.
6A-B, 7B, 8C and the right half of Figs. 6A and 8A of Humphrey,
1968, J. Comp. Neur., 132: 135-165 with the permission of The Wistar
Press. The magnification scale of A, B, and E is given on A and
that for C and D is on C. The length of the embryo is given on each
photograph. The area A enclosed in the rectangle in C is shown in
part A and the area in rectangle E is reproduced in part D. Note
the rapid increase in size and differentiation of the amygdaloid
complex at 20.7 mm (part B) and 22.2 mm (part E) as compared with
that at 18.0 mm (part A). A and C, No. 142, sections 15-4-2 and
15-3-7 respectively; B, No. 93A, section 31-1-6; D and E, No. 130.
section 21-3-6.

18.0 mm (Fig. 6A) but in the 20.7 mm embryo an undifferentiated
zone (Brown, 1967; Humphrey, 1968) intervenes between the ependymal
layer and the amygdaloid nuclei. No differentiation of the basolateral complex into the different nuclei has been seen at 18.0 mm,
but the basal amygdaloid nucleus has appeared at 20.7 mm and there
is a small anlage of the lateral nucleus laterally in the region
where the external capsule develops later (Fig. 6B). By this age,
also, a dense cell mass posterior to the Fig. 6B that is characterized by small cell clusters within it has been identified as
the accessory basal nucleus (Fig. 7C of Humphrey, 1968). Thus,
the greater part of the basolateral complex in this embryo is
formed by the basal amygdaloid nucleus. Although better represented at 22.2 mm, the lateral nucleus is still small (Fig. 6E).
The accessory basal nucleus has grown proportionally more in size
than the lateral nucleus and, as before, is characterized by the
small cell clumps within it, but the basal nucleus still constitutes the major mass of the basolateral complex.

     In spite of remaining smaller than the basolateral complex,
the corticomedial subdivision of the amygdala differentiates to a
considerable degree during this period. The central nucleus was
identified at 22.2 mm and the medial nucleus (Fig. 6E) has increased somewhat in size. The contiguity of the medial nucleus
with the hippocampal formation is retained, as is also the compact
character of this cell mass. The cortical amygdaloid nucleus
exhibits less development in the thickness of the cell layer that
forms it than in its mediolateral dimension and anteroposterior
extent. At 18.0 mm it is small anteriorly, but near the posterior
pole of the basolateral complex spreads across the surface of this
cell mass from the primordial hippocampal area medially to the
amygdalopiriform transition area laterally (Fig. 6D of Humphrey,
1968). Although poor in cellular content, minute differences
from the medial to the lateral side of this nucleus lead to the
conclusion that medial, intermediate and lateral parts are already
developing at 18.0 mm and become progressively more clearly represented at 20.7 mm and at 22.2 mm (Figs., 7C-D and 8C-D of Humphrey,
1968). Because the cortical nucleus is better developed posteriorly and the sections shown in Figure 6 pass through the best developed part of the amygdala, subdivisions of the cortical nucleus
are present only in Figure 6E.

     25.4 and 27.4 mm (8+ and 8.5 weeks). A rather surprising
amount of additional development has taken place in the amygdaloid
complex at this age level. The changes include a considerable increase in size, a clearer delineation of the individual nuclei,
and a definite shift in position. Even in the 22.2 mm embryo, the
inferior horn of the lateral ventricle is beginning to turn
ventrally. At 25.4 mm, this ventral outpocketing from the body of
the lateral ventricle is quite prominent (Fig. 5B) and its tip
even points anteriorly a trifle. This forward extension that

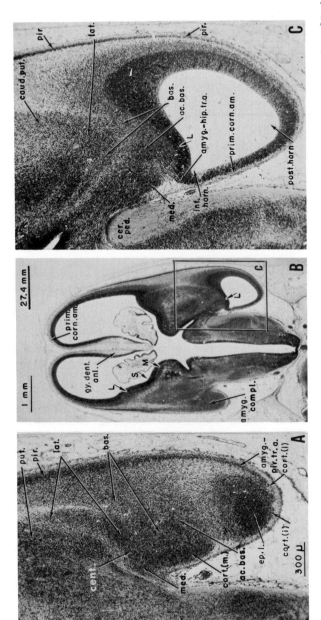

Fig. 7.   Photomicrographs of two sections (A and C) through the amygdaloid complex of a 27.4 mm human fetus (Fetus A, sections 30-3-7 and 28-1-4 respectively) and a low power photograph (B of 28-1-4) for orientation.   Reproduced from Figs. 10C, and 11A and 11C from Humphrey, 1968, J. Comp. Neur., 132: 135-165 with the permission of The Wistar Press.   At this age both the size and the degree of differentiation of the amygdaloid complex is much greater than heretofore.   The lateral position of the amygdaloid complex with reference to the outpocketing for the inferior horn of the lateral ventricle is shown here whereas in the sagittal section at 24.5 mm (Fig. 3B) the location anterior to the developing inferior horn is illustrated.   The magnification scale on A applies also to C, the area enclosed in the rectangle on B.   The photograph in A is from an area comparable to C, the area enclosed in the rectangle on B, but is from the opposite side.

develops into the inferior horn of the lateral ventricle is
quite distinct at 27.4 mm as it extends forward from the medial
border of the ventricle along the medial side of the lateral
striatal ridge (Fig. 7B-C). As a result of the development of
the inferior horn of the lateral ventricle and the concomitant
increase in growth of the hemisphere posteriorly, the position
of the amygdala is altered. Whereas originally it lies posterior
to the interventricular foramen (Fig. 4) and so at the posterior
end of the developing lateral ventricle (Fig. 5A), as the inferior
horn develops, it is carried forward. Consequently, it begins to
shift anteriorly with the developing inferior horn by 8 to 8.5
weeks (Figs. 5B and 7B-C). It then lies largely lateral to the
inferior horn of the ventricle as well as anterior to it (Figs. 8
to 12). Later, the extensive development of the temporal lobe
cortex changes the position still more.

All of the amygdaloid nuclei are present in the fetus of
8+ weeks (Fig. 5B) except the nucleus of the lateral olfactory
tract which is absent also at 8.5 weeks (Fig. 7A). As in the
preceding period of development, the basolateral nuclear complex
has increased the most in size and the corticomedial complex the
least. Both the basal and accessory basal nuclei are developing
their medial (more dense) and a lateral (less dense) mass of
neurons that become pars medialis and pars lateralis of these
nuclei later in fetal life and in the adult human amygdala (Crosby
and Humphrey, 1941). Where best developed (Fig. 7A-B), the
lateral amygdaloid nucleus extends along the lateral surface of
the putamen as a prominent tongue-like or wedge-shaped cell mass.
It is less well represented posteriorly, and becomes continuous
with the caudate nucleus (Fig. 11B of Humphrey, 1968), then even
more posteriorly with a common caudate-putamen cell mass (Fig. 7C).
Finally, near the posterior end of the lateral striatal ridge, a
single mass of cells representing a common putamen-caudate-amyg-
daloid complex is separating directly from the medullary epithelium (Fig. 11D of Humphrey, 1968).

In the corticomedial complex, there is no significant change
in the central nucleus (Fig. 7A). The medial nucleus is somewhat
larger, but its cells continue to be compactly arranged (Fig. 7A)
and it retains its close relationship with the anlage of the gyrus
dentatus posteriorly (Fig. 7C). The three parts of the cortical
nucleus differ more from each other structurally where they overlie the basolateral amygdaloid complex. The medial part is continuous with the medial nucleus and resembles the medial amygdaloid
nucleus in the density of its cells and in the thickness of the
cell layer. The intermediate part is least well developed and is
still poorly separated from the underlying cells, but the lateral
part is taking on the characteristics of the cortical plate in
the amygdalopiriform transition area (Fig. 7A).

When it first appears (Fig. 3A), the primordial amygdala is associated with the anlage of the gyrus dentatus medially and, as soon as a primordial piriform cortex is identifiable (Fig. 3B), it is bordered by this region laterally. As development and differentiation proceed, a transition area appears between the amygdala and each of these cortical regions. The transition area that links the amygdala with the piriform cortex has been designated the amygdalopiriform transition area both in bat (Brown, 1967) and in human embryos (Humphrey, 1968). It is the corticoamygdaloid transition area of Crosby and Humphrey (1941) that links the cortical nucleus with the piriform area laterally (Figs. 6 and 7). Another area, the amygdalohippocampal transition area (Humphrey, 1968), relates the hippocampal formation with the amygdala medially. This second transition area is either with the medial amygdaloid nucleus (Fig. 7C) or the medial part of the cortical nucleus (Fig. 5B). The tendency at this age level is for the cortical nucleus to be more closely associated with the primordial cornu ammonis and the medial nucleus with the anlage of the gyrus dentatus (Figs. 9 and 11 of Humphrey, 1968).

    33.8 to 42.0 mm (9.5 to 10.5) weeks. Although the internal capsule has been identified as early as 22.0 mm (His, 1904) and is distinct at 27.4 mm (Humphrey, 1968, Fig. 10D) and the cerebral peduncle is present at 37.0 mm (Cooper, 1950; Humphrey, 1960), the great increase in size of this fiber bundle constitutes the most striking change in the area of the striatal complex during this age period. Where the fibers form a fan-like mass as they pass through the striatal complex, they separate the amygdala from the putamen, the caudate nucleus, and the globus pallidus. Nevertheless, at 9.5 and 10 weeks (Fig. 8) the configuration of the amygdaloid complex is definitely similar to that at 8.5 weeks as shown in Figure 7A, although the borders of the individual nuclei in the best developed areas are uniformly more distinct (Fig. 8A-B).

    The inferior horn of the lateral ventricle has grown forward (or anteriorly) an appreciable amount at 9.5 weeks (Fig. 8) and at 10 weeks has widened out as well (Fig. 9). This forward growth of the inferior horn is medial to the amygdaloid complex where it occupies the lateral striatal ridge posterior to the internal capsule. Thus, the major contribution of new cells to the amygdaloid complex during this period is either from the germinal epithelium or from the undifferentiated zone of the lateral striatal ridge. Anterior to the internal capsule to a slight degree, however, and posterior to it to a little greater extent (Fig. 8C), the undifferentiated zone of the medial striatal ridge also contributes to the amygdaloid complex, mainly to the medial nucleus and to the accessory basal nucleus. Another outstanding characteristic of the amygdaloid complex, during this period, is

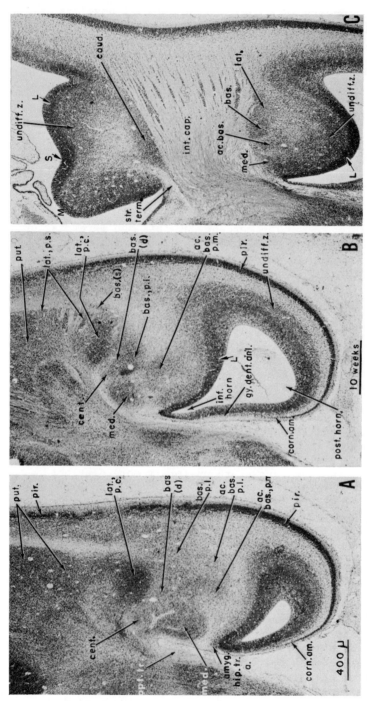

Fig. 8. Photomicrographs of three sections through the amygdaloid complex of a fetus of 10 weeks of menstrual age (No. 101, 38.2 mm CR length, sections 69-2-6, 66-2-4 and 62-1-3 respectively). <u>A</u>, Section through the anteroinferior part of the complex. <u>B</u>, Section through the well developed middle of the amygdaloid complex. <u>C</u>, Section through the less developed posterosuperior part of the amygdala. The magnification scale on <u>A</u> applies to all parts of the figure.

the remarkable distinctness of the borders of the individual
nuclei in the best differentiated region (Fig. 8).

The change in the ventricles coupled with a variation in the
degree of curvature of fetuses at 10 to 10.5 weeks and with
differences in orientation for sectioning transversely (through
the trunk) changes the plane of the sections through the forebrain enough in some instances to cut the developing tip of the
temporal region separately (Fig. 9A-B) and give an almost horizontal section through the forebrain (Fig. 9D), a plane distinctly unfavorable for identification of the amygdaloid nuclei.
Consequently, the nuclei in the amygdaloid complex of one fetus
may be clearly delineated (Fig. 8, 38.2 mm CR) and almost
impossible to identify in another fetus of almost the same size
(Fig. 9, 40.7 mm CR).

CORTICOMEDIAL COMPLEX

The medial nucleus has the same compact structure as at 8
weeks and occupies the dorsomedial angle of the amygdala adjacent
to the tip of the inferior horn of the ventricle and to the
developing hippocampal formation (Fig. 8A-B). The medial part
of the cortical nucleus lies inferior to the medial nucleus
(Fig. 10A) and is better shown in the plane of sectioning that
cuts the telencephalon more horizontally where all three of its
parts are identifiable (Fig. 9A-B). Unlike the medial nucleus,
the cortical nucleus has increased in size in every dimension.
The cell layer is thicker and greater in extent, in the antero-
posterior and in the mediolateral diameter. In addition, the
three parts differ more from each other than heretofore. The
intermediate part is the least distinct and less characteristic
than the lateral and medial parts. The lateral part usually is
a continuous layer that might be called cortex-like and is
similar to the piriform cortex (Figs. 9A-B and 10C-D). The
medial part, however, is thickest and most compact so that it
resembles the medial amygdaloid nucleus (Figs. 9B and 10A-B),
especially near that nucleus, but differs in more inferior areas
where it appears to have a layer of cells migrating toward its
deep surface (Fig. 9A).

The central amygdaloid nucleus is enclosed in a semicircle
consisting of the medial, the basal (pars medialis) and the
lateral nucleus from its medial to lateral aspects. It has
changed less in size than the other nuclei of the corticomedial
complex. Indeed, except that it is now sharply delineated there
is little change from the appearance at 8.5 weeks. A characteristic nucleus of the lateral olfactory tract could not be identified. However, in one fetus of 10 weeks (#101), where the
lateral olfactory tract approaches the prepiriform cortex, a

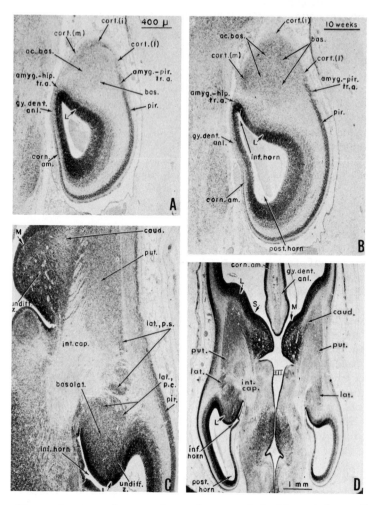

Fig. 9. Photomicrographs to show the amygdaloid complex of another fetus of 10 weeks of menstrual age (No. 103, 40.7 mm CR,
sections 57-1-4, 55-1-3, 42-2-6 and 47-1-2 respectively). Although the crown-rump length is close to that of the 10-week fetus
for which the amygdaloid complex is shown in Fig. 8, slight differences in the degree of curvature of the fetus, in the orientation
for sectioning, and in the forward growth of the temporal pole result in a somewhat different plane through the amygdaloid complex
so that the individual nuclei of the amygdaloid complex are identified much less satisfactorily. In addition, the tip of the temporal
pole is sectioned separately in these more horizontal sections
through the telencephalon as the inferior surface of the brain is
approached (see 11A). The magnification scale on A applies to B too.

small cluster of cells lies deep to this tract, between it and the
external capsule component of the anterior commissure, for a distance of 180μ on one side, but for only a short distance on the
other side. This region lies close to the anterior portion of the
anterior amygdaloid area, the region of the amygdala in which the
nucleus of the lateral olfactory tract was found in the cat (Fox,
1940), and the mink (Jeserich, 1945) and the major portion of it
in the panda (Lauer, 1949). With the forward growth of the human
amygdala that occurs later in development, this cell mass could
be incorporated easily in the anterior amygdaloid area or even
between the parts of the cortical nucleus. In the sagittal
sections at 10.5 weeks (Fig. 10D) this nucleus is found adjacent
to the anterior border of the anterior amygdaloid area.

BASOLATERAL COMPLEX

The lateral nucleus has increased more in size by 9.5 weeks
than either the basal or the accessory basal nuclei. Because of
the greater mass of fibers in the internal capsule, the lateral
nucleus is now partially separated from the putamen by these
crossing fibers as well as distinguished by a more compact cell
arrangement (Figs. 8A-B and 9C). Since some internal capsule
fibers cut across the lateral nucleus, the bands of cells in the
adult amygdala formed by the crossing fibers are already present
(Figs. 8B and 9C). Therefore, we now have a compact part of the
lateral nucleus and a striped portion which, for brevity in
description, will be referred to as pars compacta and pars
striatalis respectively. Both parts are in continuity with the
putamen, pars compacta anteroinferiorly (Fig. 8A) and pars
striatalis more posteriorly and superiorly (Figs. 8B and 9C).

The cells of the basal and accessory basal nuclei are less
densely arranged than those of the lateral nucleus, both at 9.5
and at 10 weeks. In both nuclei, however, cells located more
medially are more closely packed together than those in the
lateral region. As a result, the medial and lateral parts of
these nuclei are easily distinguishable when the plane of
sectioning is favorable (Fig. 8A-B). In addition, pars medialis
of the basal nucleus has developed a small superficial portion
lateral to the lateral amygdaloid nucleus, labeled bas.(s) in
Figure 8B. The greater portion of nucleus basalis pars medialis
is situated deeply between the compact part of nucleus lateralis
and pars medialis of the accessory basal nucleus. It is labeled
nuc. bas. (d) in Figure 8A-B).

Attention should be called to the conspicuous migration of
cells from the undifferentiated zone underlying the lateral striatal ridge into the nuclei of the basolateral amygdaloid complex
(Figs. 8 and 9) and even directly from the ependymal layer about

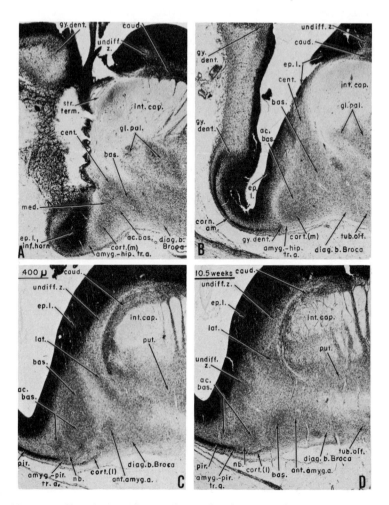

Fig. 10. Sagittal sections through the amygdaloid complex and
the adjacent areas of the telencephalon and dorsal thalamus of a
human fetus of 10.5 weeks of menstrual age (Fetus N3, sections
135-1-3, 136-2-4, 141-2-2 and 143-1-1 respectively). As followed
from A through D, the sections pass from near the medial side
toward the lateral side of the amygdaloid complex. Note that the
ependymal (or germinal) layer from which the cells migrate into
the amygdaloid nuclei is now narrow medially (see B) in the area
that gives rise largely to the corticomedial complex, but wide
laterally (C and D) where the cells are migrating in great
numbers into both the basal and lateral nuclei of the basolateral complex.

the ventricle in some areas (Figs. 8C and 9C). This relation is
well shown in the sagittal sections of a 10.5-week fetus (Fig. 10).
The contribution of cells from the ependymal layer to the lateral
nucleus is extensive here (Fig. 10C-D), but this layer is still
wide in their area of origin. Cells are also migrating into the
basal nucleus (Fig. 10C-D) and into the accessory basal nucleus
(Fig. 10B-D), although apparently in lesser numbers. Cell
migration into the medial nucleus is not shown in the figures,
but cells are obviously streaming into the central nucleus in
Figure 10B. In the area from which the cells migrating into the
central nucleus arise the ependymal layer is thin and the surface
broken (between forks of leader labeled ep. l. and in the area
crossed by the leader labeled cent.). Thus the area of the
ependymal layer giving rise to the central nucleus and the medial
nucleus (area between A and B of Figure 10), is depleted of cells
and thin, whereas that from which the late developing lateral
nucleus arises remains wide at this age (Figs. 10C-D). The area
of origin for the basal and accessory basal nuclei is less wide
(Fig. 10C) but more so than that contributing cells to the medial
and central nuclei. This relationship to the ependymal layer may
be seen also in Figures 8A-C and 9A-B, although less clearly.

RELATIONS WITH CAUDATE PUTAMEN COMPLEX

As is obvious from Figures 8A-B, 9C-D and 10C-D, the lateral
amygdaloid nucleus is directly continuous with the putamen. In
the less differentiated areas (Figs. 9C) separate basal and
lateral amygdaloid nucleui are not identifiable so the relationships are not clear. Also, where the tail of the caudate nucleus
swings forward it is in continuity with the lateral nucleus
(medial to Fig. 10C) and almost indistinguishable from it although
more laterally (Fig. 10C-D) the two nuclei are easily defined.

CORTICOAMYGDALOID TRANSITION AREAS

Transition areas lie between the amygdala and the hippocampal
formation medially (amygdalohippocampal transition area), and
between the amygdala and the piriform cortex laterally (amygdalopiriform transition area). The amygdalohippocampal transition
area appears to unite the gyrus dentatus anlage, rather than the
cornu ammonis, with either the medial amygdaloid nucleus or with
pars medialis of the cortical nucleus (Figs. 8B, 9A-B and 10A-B).
The transition from the amygdala to the piriform cortex is through
pars lateralis of the cortical nucleus (Figs. 9A-B and 10C-D). At
10.5 weeks, however, additional cells (nb in Figs. 10C-D) are
joining the deep surface of the amygdalopiriform transition area
to unite it with the accessory basal nucleus and in some areas as
well as with the cortical nucleus (Fig. 10C). In some places,
these migrating cells already give a layer-like appearance to

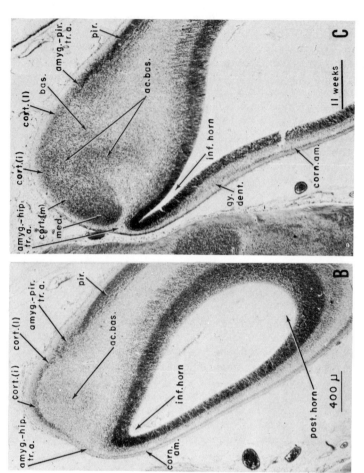

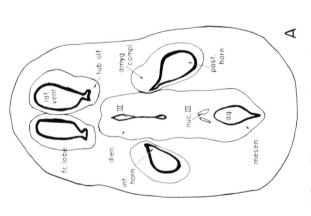

Fig. 11. Two photomicrographs (B and C) of sections through temporal pole of the cerebral hemisphere of a human fetus of 11 weeks of menstrual age (No. 119, sections 96-2-1 and 90-1-1 respectively) and a low power drawing (A) of a transverse section through the brain of the same fetus to show the topographic relations of this region to the remainder of the forebrain. Note that the inferior horn of the lateral ventricle is beginning to rotate medially as compared with its position at 9 weeks (see Fig. 9D).

this transition zone such as has been seen in the adult human amygdala (Cort.-amyg. tr.a., Fig. 19) and usually also in later fetal development.

11 to 12 weeks (48.6 to 60.5 mm). The available fetuses in this age group were sectioned in toto in the transverse plane and present a picture of the amygdala comparable to that illustrated for the 10-week fetus in Figure 9. The major changes in the amygdala consist of an increase in size of the nuclei of the basolateral amygdaloid complex, additional forward growth of the inferior horn of the lateral ventricle which carries the amygdala farther anteriorly in front of it, and a beginning medial rotation of the temporal pole. Because of these changes in position of the amygdala and the increase in size of the internal capsule, the amygdaloid complex is separated completely from the medial striatal ridge. Consequently, all new cells that join the complex are from the lateral striatal ridge.

CORTICOMEDIAL COMPLEX

There is little change in the medial nucleus during this period, and the central nucleus is not identified satisfactorily in the sections cut in this plane. The cortical nucleus (Figs. 11B-C and 12B-C) has increased greatly, both in size and in the degree of differentiation. The cells being added to it appear to be derived entirely from the lateral striatal ridge and therefore migrate beyond the basal and accessory basal amygdaloid nuclei to reach the surface. Both the medial and intermediate parts of the cortical nucleus overlie the accessory basal nucleus and the lateral part overlies the basal nucleus (Figs. 11B-C and 12C-D). Where best represented, the intermediate part also over-lies the basal nucleus (Figs. 12C-D). At the tip of the temporal pole, at 12 weeks, the surface layer of cells extends beyond the borders of the underlying basal and accessory basal amygdaloid nuclei (Fig. 12B). Possibly, the cell layer in this area is comparable to that portion of the periamygdaloid cortex of Rose (1926) which occupies a similar position in the adult and it has been so named in this figure.

In both the 11- and the 11.5-week fetuses included here, but less clearly at 12 weeks, there is some evidence for the nucleus of the lateral olfactory tract. A small cluster of cells situated between its lateral and intermediate parts but deep to the cortical nucleus is present bilaterally at 11 weeks. A less clear but similar group of cells is identifiable in one amygdaloid complex at 11.5 weeks and two such cell clusters in the other, the second one between the medial and intermediate parts of the cortical nucleus. The orientation of the small cells of these clusters indicate either cell migration, or that they are aligned along

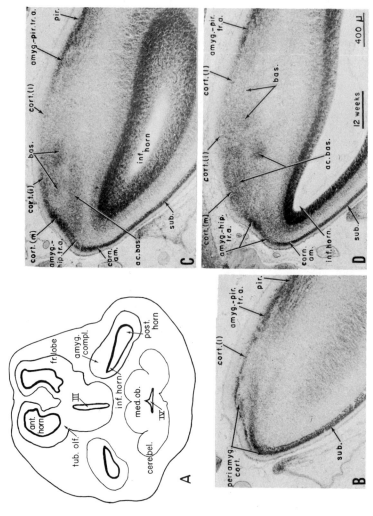

Fig. 12. Three photomicrographs (B to D) showing the temporal pole of the cerebral hemisphere of a human fetus of 12 weeks of menstrual age and drawing (A) of a comparable section through the brain of a fetus of 11.5 weeks (No. 148) to show the location of the amygdaloid complex with reference to the remainder of the forebrain (No. 132, photomicrographs from sections 150-1-2, 145-2-3 and 143-1-2 respectively). Note the more marked medial rotation of the inferior horn of the lateral ventricle, as compared with its position in Figure 11A, and the accompanying medial rotation of the amygdaloid complex.

fibers entering and/or leaving the cluster, an arrangement common
to the nucleus of the lateral olfactory tract. A similar location
deep to the cortical and medial amygdaloid nuclei has been reported for part of this nucleus in the shrew (Crosby and Humphrey,
1944) in the adult bat (Humphrey, 1936) and in embryonic bat
brains (Brown, 1967).

CHANGES IN POSITION

At 10 weeks, as shown in Figure 9, the inferior horn of the
lateral ventricle has grown forward and carried the amygdaloid
complex with it so that an appreciable part of the amygdala lies
anterior to the ventricle. However, the alignment of the amygdala
with reference to the midline has changed very little if at all
from that seen at 8.5 weeks (compare Figs. 7B and 9D). At 11
weeks, however, the tip of the temporal pole has begun to rotate
inward, or medially, so that the long axis of the inferior horn
forms an acute angle (about 35°) with the midline (Fig. 11A). At
11.5 weeks, this acute angle has almost doubled due to the added
medial rotation of the temporal pole. Probably, both this medial
rotation and the plane of sectioning contribute to the difficulty
in identifying the nuclei in the basolateral complex during this
period.

BASOLATERAL COMPLEX

Again, the plane of sectioning makes recognition of the individual nuclei difficult. Consequently, the subdivisions of the
nuclei already present clearly at 9.5 and 10 weeks (Fig. 8) cannot
be identified even where the relations are best. The most rapid
growth is taking place in the basal and accessory basal nuclei at
the tip of the temporal pole (Figs. 11B-C and 12C-D). There is
relatively less increase in the size of the lateral amygdaloid
nucleus, and its subdivisions are not recognizable. There is no
discernible change in the relations with the caudate-putamen
complex.

CORTICOAMYGDALOID TRANSITION AREAS

The amygdalohippocampal transition area between the cornu
ammonis and the medial part of the cortical nucleus is distinct in
some areas (Figs. 11B and 12C-D). The gyrus dentatus, however,
is associated more closely with the medial nucleus, although in
some areas also with the medial part of the cortical near the
medial nucleus (Fig. 11C). As at 10 weeks, the intermediate part
of the cortical nucleus is least distinct and the lateral part,
although the cell layer is thinner, the most like the cortical
plate of the piriform cortex.

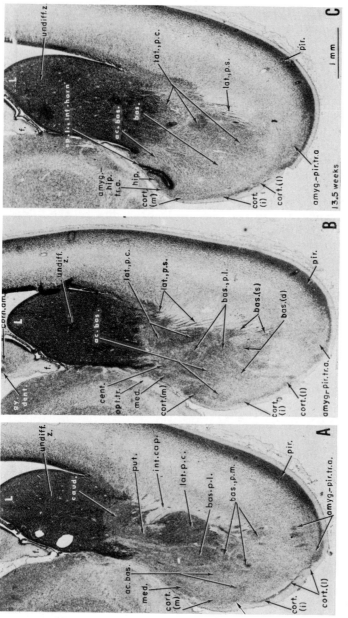

Fig. 13. Photomicrographs of two coronal sections (A and B) through the best differentiated portion of the amygdaloid complex of a human fetus of 13.5 weeks of menstrual age and of another section (C) near its posterior end (No. 118, sections 102-2-3, 105-1-1 and 107-3-3 respectively. Note the large undifferentiated zone constituting the lateral striatal ridge (L) and the massive migration of cells from it into the nuclei of the basolateral complex. The plane of sectioning and location of the amygdaloid complex in the hemisphere are shown in Figure 2.

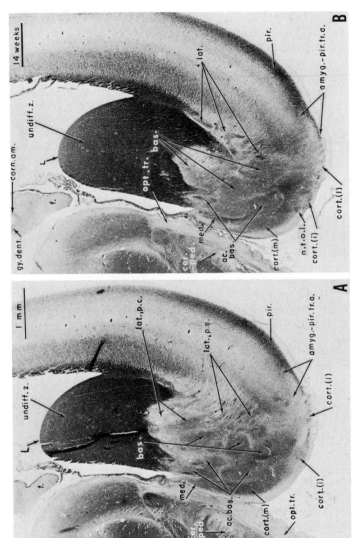

Fig. 14. Photomicrographs of two coronal sections through the amygdaloid complex of a human fetus of 14 weeks of menstrual age (No. 157, sections 80-1-2 and 81-1-3 respectively) to illustrate the individual areas of rapid cell proliferation and migration from the undifferentiated zone of the lateral striatal ridge (L) into the several nuclei of the basolateral complex. A, Section near the middle of the amygdala. B. Section toward the caudal part of the amygdaloid complex. The location of the amygdaloid complex with reference to the remainder of the hemisphere at this age is shown in Figure 2.

13.5 and 14 weeks (79.0 and 89.0 mm). After 12 weeks, the
coronal plane of the sections through the forebrain (see Fig. 2)
makes it possible to identify, once more, all of the nuclei of the
amygdaloid complex and also the parts of those that have subdivisions. Although the amygdala already lies adjacent to the optic
tract at 10 weeks (Fig. 8A), this relationship, where the optic
tract lies between the amygdala and the cerebral peduncle, is
typical at both 13.5 (Fig. 13B) and 14 weeks (Fig. 14A) of that
seen in the adult brain. As mentioned for the previous age period,
the amygdaloid complex is cut off completely from the medial
striatal ridge and cell migration into it is exclusively from the
lateral ridge posterior the internal capsule. Indeed, the most
characteristic feature of the amygdaloid complex is the prominent
lateral striatal ridge from which cells are migrating in great
numbers directly into the nuclei of the basolateral complex, into
the amygdalopiriform transition area and the cortical nucleus as
well (Figs. 13 and 14). The basolateral complex, therefore, is
markedly larger than heretofore.

CORTICOMEDIAL COMPLEX

This subdivision of the amygdala now constitutes a very minor
part of the complex. The central (Fig. 13B) and medial nuclei
(Figs. 13A-B and 14) have increased very little. However, the
greater medial rotation of the amygdala has brought them into a
more dorsal position. The cortical nucleus, however, is much
more extensive than at 10 weeks and its three parts even more
distinctive than at 10 weeks (Figs. 13 and 14). Nevertheless,
the medial part has the thickest cell layer and is much like the
medial nucleus, whereas the cell layer of the intermediate part
is thin although a more distinct layer than at 11 to 12 weeks.
The cell lamina of the lateral part of the cortical nucleus tends
to be broken into cell clusters (Fig. 13A-B) that are more distinct at 14 weeks (Fig. 14).

The nucleus of the lateral olfactory tract continues to be a
puzzle. Anteriorly at 13.5 weeks, where only the anterior amygdaloid area and the cortical nucleus are present, a tiny cell
cluster superficial to the lateral part of the cortical nucleus
and near the amygdalopiriform transition area appears characteristic of this nucleus as it lies deep to fibers of the lateral
olfactory tract. It extended for 20 sections on one side and 24
on the other in the same location but was not as large. A
comparable second part of this nucleus was not found on either
side, but in some areas of the intermediate part of the cortical
nucleus or between this part and the medial part a small cluster
of cells slightly deeper than the cortical nucleus itself might
be a second nucleus of the lateral olfactory tract. In the
14-week fetus, a cell cluster on each side, identified as the

nucleus of the lateral olfactory tract, was observed but not in the same position as in the 13.5-week fetus. On one side (Fig. 14B), a deeply situated cell cluster between the medial and intermediate parts of the cortical nucleus was so identified, at a level relatively far posterior in the amygdaloid complex. On the other side, no equally definite cell cluster was found.

BASOLATERAL COMPLEX

The characteristic feature of the large basolateral complex is the massive migration of cells from the large undifferentiated lateral striatal ridge into the three nuclei; medially the migrating cells join the accessory basal nucleus, and laterally they contribute to the lateral amygdaloid nucleus. Between these two nuclei, the cells leaving the lateral striatal ridge join the basal nucleus. The migration evidently is very rapid for the border with the germinal layer constituting the lateral ridge is streaked with lighter lines formed by the better developed cells leaving the ridge, especially posteriorly (Fig. 13B-C) although not anteriorly (Fig. 13A) at 13.5 weeks. Posterior to the sections illustrated in Figure 14, the same striated appearance is present but not for so great a distance. Anteriorly, at 13.5 weeks (Fig. 13A), and in both levels shown at 14 weeks (Fig. 14A-B), the area of junction of the basolateral nuclei with the germinal cell mass has a scalloped appearance where the rounded border of each nucleus is represented. These distinct areas, and the medial to lateral arrangement of the accessory basal, basal and lateral nuclei respectively, indicate that each nucleus takes from its own specific area of the lateral striatal ridge.

Although it was possible to recognize medial and lateral parts of the accessory basal nucleus at 10 weeks (Fig. 8A), they were not identified satisfactorily in either fetus in this age group (Figs. 13 and 14). However, at 13.5 weeks both medial and lateral parts of the basal nucleus were recognized (Fig. 13A-B) and where best represented (Fig. 13B) both superficial, bas (s), and deep, bas. (d), parts of nucleus basalis pars medialis. The superficial part, as in the adult amygdala, extends from the deep part around the border of the lateral nucleus to lie on its outer surface. The lateral nucleus is large in all dimensions but is not present as far anteriorly as the basal complex. Where crossed by fibers of the anterior commissure, a small pars striatalis as well as large compacta is present.

AMYGDALOCORTICAL TRANSITION AREAS

The migration of cells into the amygdalopiriform area is very conspicuous during this period. Sometimes clusters of cells join this transition area (Figs. 13A-B and 14A). In other areas, a broad band of cells migrate into it (Fig. 14B). Occasionally, also, the cells migrate into the area in waves (Fig. 13C). There is no evidence whatsoever of an infolding of the cortical plate in this region, however, as believed by Johnston (1923). The amygdalohippocampal transition areas is shown only in Figure 13C where the tip of the inferior horn of the lateral ventricle lies medial to the amygdala. The presence of a cell layer in the hippocampal formation adjacent to the transition zone indicates that the cornu ammonis is linked with pars medialis of the cortical nucleus in this region, a relationship which is even more definite farther posteriorly. The gyrus dentatus is represented too poorly to determine any relationship with the amygdala in this fetus. At 14 weeks, also, the amygdalohippo-campal transition area lies between the cornu ammonis and the cortical nucleus, but apparently pars intermedialis for pars medialis is no longer present.

RELATION WITH CAUDATE PUTAMEN COMPLEX

Anteriorly, the putamen comes into continuity with the anterior amygdaloid area at 13.5 weeks and with strands of cells that are migrating forward to extend the lateral and basal amyg-daloid nuclei farther anteriorly. More caudally, the internal capsule partially separates the tail of the caudate from the putamen (Fig. 13A) which is directly continuous with the lateral nucleus. The relations at 14 weeks are comparable, but are not illustrated in the figures. In certain respects, the relations are comparable to those seen at 10 weeks (Figs. 8B and 9C-D) where the continuity of the putamen with the lateral amygdaloid nucleus is more obvious, partly due to the difference in the plane of sectioning, but also probably to the lesser degree of development of the internal capsule.

16 and 18.5 weeks (114.0 and 144 mm). The most striking difference between the amygdaloid complex in this age level and that at 13.5 to 14 weeks is the rapid growth anteriorly of the lateral and basal nuclei (Figs. 15A-B and 16B). The next most conspicuous change is the added medial rotation. In this respect, it should be observed that the greater amount of medial rotation illustrated at 16 weeks (Fig. 15D) than at 18.5 weeks (Fig. 16A) is due to dorsoventral compression of the paraffin sections (Fig. 15D) by the microtome knife resulting from too soft a paraffin. The brain at 18.5 weeks (Fig. 16A) was sec-tioned from the lateral surface and compression appears to be minimal.

CORTICOMEDIAL COMPLEX

Of the nuclei constituting this subdivision of the amygdala,

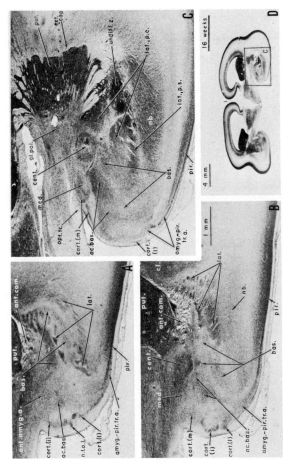

Fig. 15. Photomicrographs of three coronal sections (A to C) through the amygdaloid complex of a human fetus of 16 weeks of menstrual age (No. 147, sections 72-2-2, 75-2-2 and 84-3-1 respectively) and a low power photograph (D) of a section (85-3-1) through the entire forebrain for orientation. A. A section near the anterior end of the amygdala where the basal and lateral nuclei are intermingled in their rapid growth forward at this age. B. The amygdaloid complex more posteriorly where the various nuclei are better represented. C. A section near the enclosed area in D that is approaching the inferior horn of the lateral ventricle posteriorly where the undifferentiated zone of the lateral striatal ridge is contributing cells to the amygdaloid complex. This series was sectioned dorsoventrally rather than from side to side and the brain and the amygdaloid complex are compressed dorsoventrally (15D) as compared with the sections shown in Figures 16A and 18A.

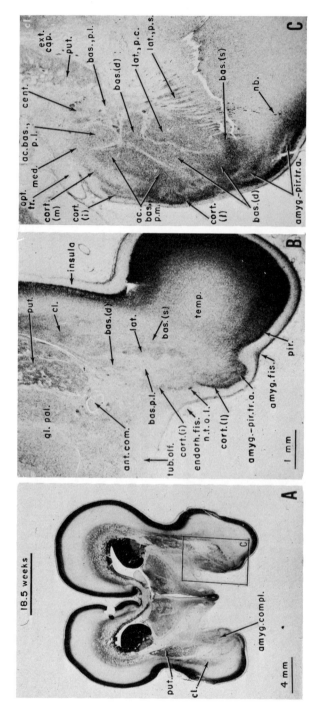

Fig. 16. Photomicrographs illustrating the amygdaloid complex of a human fetus at 18.5 weeks of menstrual age (No. 98M). A. Low power photograph (section 214-2) to show the topographic relations of the amygdaloid complex to the other parts of the telencephalon. The enclosed area labeled C illustrates the general area enlarged in part C (section 231-1), but at a more anterior level of the amygdala. The photograph in B shows the anterior portion of the amygdaloid complex (section 154-1) into which cells joining the lateral and basal amygdaloid still are migrating forward rapidly as also at 16 weeks (Fig. 15).

the central nucleus has increased the most in size (Figs. 15B-C
and 16C). The medial nucleus is poorly defined anteriorly (Fig.
15B) and posteriorly, and it may be overlaid superficially by
the adjacent medial part of the cortical nucleus (Fig. 15C). As
in the previous age level, the cell layer of the cortical nucleus
has increased in thickness and in area. The medial part is relatively less conspicuous (Figs. 15B-C and 16C) and the intermediate
part variable (Figs. 15A-C, 16B-C). Although the cell layer of
the lateral part is thicker than at 14 weeks, its mediolateral
extent is relatively less, apparently, because of the more extensive amygdalopiriform transition area (Figs. 15A-C and 16B-C).

     The nucleus of the lateral olfactory tract is illustrated at
16 weeks between the intermediate and lateral parts of the cortical nucleus (Fig. 15A). In this area, it is superficial to the
cell lamina forming the cortical nucleus, but farther posteriorly
it shifts to the deep aspect of this lamina and is definitely
larger. Here the alignment of its cells with fibers passing
dorsally is characteristic of the nucleus of the lateral olfactory
tract in mammals. The nucleus extends for a distance of approximately 300µ anteroposteriorly, but with varying size and degree
of distinctness. On the other side, the nucleus of the lateral
olfactory is slightly less extensive anteroposteriorly (280µ)
and is located somewhat more anteriorly within the pars intermedialis of the cortical nucleus. Instead of a rounded cell
mass, the nucleus is much elongated in the plane of the fibers
from it that join the anterior commissure. On this side, this
nucleus is situated anterior to the level in Figure 15A where
the accessory basal nucleus is still absent, the anterior amygdaloid area large, and the bands of cells constituting the
lateral and basal amygdaloid nuclei are less well represented.
On both sides, the nucleus of the lateral olfactory tract is
delineated sharply by surrounding fibers, presumably lateral
olfactory tract fibers terminating in it, and the stria terminalis
component of the anterior commissure originating there.

     At 18.5 weeks, at least two nuclei of the lateral olfactory
tract are identifiable. The more anterior one is small and sharply
circumscribed (Fig. 16B). On both sides, it is located between
pars lateralis and pars intermedialis of the cortical nucleus and
at its outer surface. Although a little more extensive on one
side than on the other, this anterior nucleus of the lateral
olfactory tract is not over 150µ in anteroposterior extent on
either side. This part of the nucleus is located even relatively
farther anteriorly than that at 16 weeks (Fig. 15A), at a level
where only the anterior amygdaloid area, the cortical nucleus
and the cell bands of the lateral and basal amygdaloid nucleus
are present (Fig. 16B). Posterior to this region, but still
anterior to the region in which all nuclei of the basolateral

complex are fully represented, there are one or two other small
circumscribed clusters of cells, usually between pars lateralis
and pars intermedialis but one at least medial to pars intermedialis of the cortical nucleus that have all of the characteristics of the nucleus of the lateral olfactory tract. They
extend through only a few sections, but probably constitute additional nuclei of the lateral olfactory tract.

BASOLATERAL COMPLEX

The most astonishing change in this age group, and the most
difficult to interpret, is in this division of the amygdala and
involves particularly the basal and lateral nuclei. The rapid
growth forward of the temporal pole and its medial rotation, together with the great increase in growth of the amygdala, bring
its anterior portion increasingly farther from the source of new
cells in the lateral striatal ridge posterior to the internal
capsule. The best differentiated portion of the amygdala (Figs.
15C and 16C), its midportion at all stages of development, is
bordered anteriorly and posteriorly by progressively less well
developed regions. Posteriorly, where the lateral striatal ridge
appears, the arrangement of the amygdaloid nuclei at 18.5 weeks
is similar to that at 14 weeks in Figure 14, with marked specific
areas of cell migration into the nuclei of the basolateral complex.
Due to differences in the plane of sectioning, or to the degree of
development, or both, the undifferentiated area of the lateral
striatal ridge extends forward to encroach on the well differentiated part of the amygdala (Fig. 15C) in such a manner that the
areas from which each nucleus of the basolateral complex takes
origin is much obscured. Of the basolateral complex, the
lateral nucleus extends the least far posteriorly, the basal
next and the accessory basal nucleus the farthest posteriorly at
16 weeks and probably also at 18.5 weeks although the material is
less satisfactory for determining this point.

Anteriorly, the anterior amygdaloid area is the only portion
of the amygdala identifiable. The accessory basal nucleus differentiates out of this area and the cortical nucleus appears on its
surface. At 16 weeks, irregular bands of cells that have been
identified as forward extensions of the basal and lateral amygdaloid nuclei extend even farther forward than the accessory basal
nucleus to invade the anterior amygdaloid area and are prominent
after the accessory basal nucleus develops (Fig. 15A). At 18.5
weeks, these bands of cells extend even farther anteriorly and
replace rapidly the anterior amygdaloid area. In such areas, it
is almost impossible to determine with certainty which cell bands
belong to the lateral and which to the basal nucleus for they
appear to intermingle, especially anteriorly (Fig. 15A). Farther
posteriorly, the more compact and more deeply staining cell masses

are continuous with the lateral nucleus and are part of it, where
the lighter staining and less compact areas join the basal nucleus
(Fig. 15B-C). At 18.5 weeks, a small strand of the lighter staining cells is found superficial to the lateral nucleus and continues around its ventral aspect into deeply situated cells of
the basal nucleus (Fig. 16B). Evidently, we have here the superficial and deep parts of pars medialis of the basal nucleus with
pars lateralis adjacent to the deep part and overlaid by pars
intermedialis of the cortical nucleus. In sections through the
well differentiated part of the amygdala at 18.5 weeks (Fig. 16C),
a nuclear arrangement more like that in the adult is present. The
major difference is due to the incomplete medial rotation of the
amygdaloid complex, so that the lateral parts of the basal and
accessory basal nuclei are dorsally situated rather than laterally.
It is evident that sufficient medial rotation of these nuclei will
bring them into the lateral position in which they are found in
the adult. The continuity of the superficial and deep parts of
pars medialis around the border of the lateral nucleus is
especially clear in some of these sections (Fig. 16C). As in
the adult amygdala, pars medialis has a more compact cell arrangement and is deeper staining. The division of lateral nucleus
into a pars compacta and a pars striatalis, where the fibers of
the anterior commissure cross through it, is especially clear.

TRANSITION AREAS

An amygdalohippocampal transition at 16 weeks is not identifiable with certainty. The hippocampal formation is poorly developed at this age where it swings forward toward the amygdala. If
present at all, this transition area is via an undifferentiated
area that probably develops into the accessory basal nucleus and
lies between a minute hippocampal formation medially and the
piriform cortex laterally as illustrated by Crosby and Humphrey
(Fig. 11, 1941) for the adult human amygdala. Although the hippocampal formation is well developed in this fetus at more posterior
levels (Humphrey, 1966, Fig. 14A) it is poorly represented where
it approaches the amygdala and does not appear to come into
relation with it at all. However, the material in this region
is not adequate for determining this point satisfactorily.

The amygdalopiriform transition area has become progressively larger and more distinct at 16 weeks and even more so at 18.5
weeks. In some areas, at 16 weeks, there is a tendency toward
layer formation (Figs. 15A-B). In others, the migration of cells
into this area is even more extensive so that the appearance of
infolding adjacent to the cortical nucleus (pars lateralis),
noted by Johnston (1923) is very marked. Here, the cell migration
into this area is almost massive especially in Figure 15B where a
broad band of cells (nb.) are streaming from the lateral part of

the undifferentiated lateral striatal ridge into this region.

At 18.5 weeks, the amygdalopiriform transition area is even
more extensive than at 16 weeks. It makes up most of the eminence
madial to the amygdaloid fissure (Fig. 16B) where it is distinguished from a small pars lateralis of the cortical nucleus by
the waves of cells migrating into it that provide a laminated
appearance. Through the well developed part of the amygdala,
however, the amygdalopiriform transition area also has the
appearance of layer formation (Fig. 16C). It is a much sider
zone of cells mediolaterally and the layers that form it are
thicker. Cell migration into this transition area is markedly
reduced, but find strands (nb., Fig. 16C) may be seen joining it,
laterally, at the junction with the piriform cortex in most
sections through the well developed part of the amygdala.
Posteriorly, the fine strands of cells (nb. in Fig. 16C) disappear
but there is a massive migration into the medial side of this
transition area (left fork of leader labeled amyg.-pir.tr.a. in
Fig. 16C) connecting it with pars striatalis of the lateral
nucleus. At its caudal end, only the accessory basal nucleus
remains. Mediodorsally, it passes over into the hippocampal
formation and ventrolaterally into the piriform cortex. In
each case, these transitions are to more differentiated parts
of these cortical areas, where layers are present in the piriform
region and possibly to the subiculum of the hippocampal formation.

RELATIONS WITH CAUDATE-PUTAMEN COMPLEX

Anteriorly, the amygdala is separated from the putamen by the
fibers of the anterior commissure (Fig. 15A), but comes into continuity with the basal and central nuclei more posteriorly (Fig.
15B) and with the lateral nucleus still farther posteriorly
toward the inferior horn of the ventricle (Fig. 15C). Still
nearer the posterior pole of the amygdala, the tail of the caudate
nucleus comes into relation with the bed nucleus of the stria
terminalis, at 16 weeks, posterior to the area where the putamen
is in continuity with the basal nucleus after the lateral nucleus
is no longer present. Where the caudate joins the central nucleus
and, more posteriorly, the bed nucleus of the stria terminalis,
however, it is in continuity with the putamen rather than separated
by fibers of the internal capsule. Anteriorly, at both ages, the
ventral claustrum comes into continuity with the bands of cells
constituting the lateral and basal amygdaloid nuclei.

24.5 weeks (216 mm CR). Additional development in the
temporal lobe has brought the amygdaloid complex sufficiently far
forward to almost overlap the lateral part of the tuberculum
olfactorium (Fig. 17C). It also has brought about an additional
rotation medialward of the amygdaloid complex. The portion
anterior to the inferior horn of the lateral ventricle is best

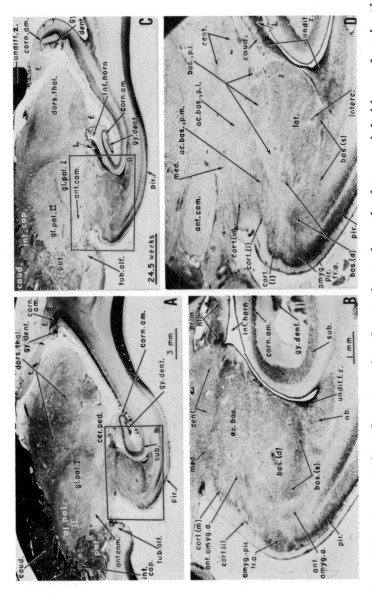

Fig. 17. Photomicrographs of two sagittal sections through the amygdaloid complex in the left hemisphere of a human fetus of 24.5 weeks of menstrual age (No. 117). A and C are low power photographs to illustrate the position of the amygdaloid complex with reference to other fore-brain structures and B and D are photographs at a higher magnification to show the individual amygdaloid nuclei. The photographs in A and B (section 251-1) are medial to those in C and D (section 320-1) by approximately 2 mm.

demonstrated in sagittal sections (Fig. 17) where the more anterosuperior position of the amygdala and the more posteroinferior
location of the hippocampal formation are well shown. The corticomedial complex has increased relatively little in size but the
basolateral complex is strikingly larger and resembles this
complex in the adult brain much more closely.

CORTICOMEDIAL COMPLEX

The medial nucleus is now the least extensive in size of all
the nuclei except the nucleus of the lateral olfactory tract. It
also is more diffuse in character (Figs. 17B & D and 18B) and consequently more difficult to identify. The central nucleus also
is diffuse (Fig. 18B) and its greater size anteroposteriorly is
shown well in sagittal sections (Fig. 17B). The cortical nucleus
appears relatively less extensive mediolaterally (Fig. 18B),
probably because of the marked medial (Fig. 18B) and the antero-
posterior (Fig. 17) increase in the piriform cortex. All three
of its parts are recognizable, however, both in sagittal (Fig. 17)
and in coronal (Fig. 18) sections. Moreover, pars medialis continues to resemble the medial nucleus and pars lateralis (Figs.
17B and 18B) is more like the piriform cortex. Pars intermedialis
is variable, since it is an ill-defined single layer (Figs. 17D
and 18B), a double layer (Figs. 17B) or even may have a somewhat
scalloped appearance.

The nucleus of the lateral olfactory tract is the most elusive
of all the nuclei at this age. Small clusters of well differentiated cells in pars intermedialis of the cortical nucleus are
set off by fibers as observed in sagittal sections (Fig. 17B)
and, probably, constitute two or even more nuclei of the lateral
olfactory tract in the left hemisphere. In the coronal sections,
through the right hemisphere, this nucleus is represented by a
single larger, more superficially placed, discrete cell mass
nearer the junction of pars lateralis and pars intermedialis of
the cortical nucleus and at approximately the same anteroposterior
level of the amygdala as this nucleus was identified at 16 and at
18.5 weeks. Thus, it lies at the level of the anterior amygdaloid
area into which bands of cells of the basal and lateral nuclei
are developing as at 18.5 weeks (Fig. 16C).

BASOLATERAL COMPLEX

Both the two basal nuclei and the lateral nucleus have increased greatly in size between 18.5 and 24.5 weeks, the basal
nuclei more than the lateral nucleus (compare Fig. 17A-B with
17C-D). All parts of the basal and accessory basal nuclei are
present (Fig. 17B and D) including the superficial and deep parts
of pars medialis [bas. (s) and bas. (d) of Figures 17B and 18B].

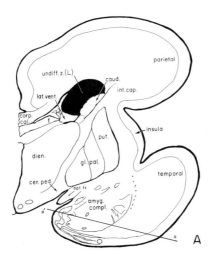

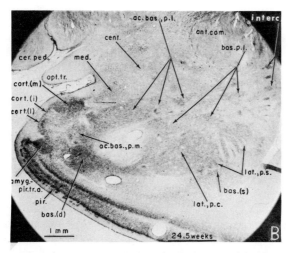

Fig. 18. **A.** Drawing of a coronal section through the right hemisphere at the level of the main mass of the amygdaloid complex of the human fetus of 24.5 weeks of menstrual age for which sagittal sections through the left hemisphere are given in Figure 17 (No. 117, section 200-1). The line a - a´ shows the orientation of the amygdaloid complex with reference to the other portions of the hemisphere and the degree of its rotation medially at this age as compared with that of the amygdaloid complex in the adult brain (see Fig. 20C). **B.** Photograph of the same section from which the drawing was made to show the individual amygdaloid nuclei at this level. The area of degeneration under the label (ac. bas., p. m.) is evidently secondary since the characteristic topographic relations of the individual nuclei are retained.

The lateral nucleus is crossed conspicuously in its lateral part
by fibers joining the anterior commissure so a conspicuous pars
striatalis is present as well as a large pars compacta (Fig. 18B).
The superficial part of nucleus basalis pars medialis lies ventral
to nucleus amygdalae lateralis rather than medially as in the
adult since the medial rotation of the amygdaloid complex is not
yet complete (see Figs. 19 and 20). At this age, tiny clusters
of small cells may be seen embedded in the lateral nucleus (Fig.
17D) or among the fibers at its surface. These are the intercalated cell masses that are found in the adult amygdala.

Anteriorly, there is a large anterior amygdaloid area out of
which first the accessory basal nucleus develops a little more
posteriorly. Next, the basal nucleus appears and still farther
posteriorly the lateral nucleus. In the right hemisphere
(coronally sectioned, Fig. 18), an area of secondary degeneration
is present in the midportion of the basal nucleus (both anteroposteriorly and mediolaterally (Fig. 18B), but it is limited to
the basal nucleus and does not appear to materially affect the
relations with the other amygdaloid nuclei. Posteriorly, the
lateral nucleus increases in size and remains proportionally
large as the basal and accessory basal nuclei become progressively
smaller. As the anterior tip of the inferior horn appears, the
amygdala is largely dorsal and posterior to it and the hippocampal
formation more ventral and inferior. There is only a little contribution of cells from the small undifferentiated area into the
accessory basal nucleus but farther posteriorly the undifferentiated zone is larger and more cells are joining the basal nucleus,
especially the superficial part of its pars medialis. A considerable number of neuroblasts are also becoming incorporated into
the lateral nucleus still farther posteriorly.

TRANSITION AREAS

The amygdalohippocampal transition area is more distinct at
this period than earlier, probably largely because of the greater
development of the hippocampal formation in the temporal region
(Figs. 17A and C). On the lateral aspect of this transition
area, a cell mass constituting a poorly developed gyrus dentatus
becomes continuous with the central amygdaloid nucleus (Fig. 17B)
and slightly more medially with the medial amygdaloid nucleus.
Still more medially, this poorly represented gyrus dentatus comes
into continuity with the accessory basal nucleus. Farther
medially, the cornu ammonis portion of the poorly developed
hippocampal formation becomes continuous with the accessory basal
nucleus and also the basal and cortical nuclei. In the hemisphere
that is coronally sectioned, the relations are similar but less
clear, possibly, because of the defect in the basal amygdaloid
nucleus although perhaps also due to the different plane of
sectioning.

The amygdalopiriform transition area is conspicuous at almost all levels of the amygdala and in both planes of sectioning. In some regions, this transitionarea is think and the migration of cells into it not clear (Fig. 17B). In others, cells appear to be streaming into the amygdalopiriform transition area over a wide area (Fig. 17D). In still other instances, a more deeply situated mass of cells lying deep to the piriform cortex and the cortical nucleus constitute the transition area (Fig. 18B). It is such areas that give the impression of cortical infolding to form the basolateral amygdaloid complex proposed by Johnston (1923). In some regions also the amygdalopiriform transition area has a layered appearance like that illustrated at 18.5 weeks (Fig. 16B-C).

RELATIONS WITH CAUDATE-PUTAMEN COMPLEX

In sagittal sections, the relations between the caudate nucleus and the putamen on the one hand and the amygdaloid complex on the other are demonstrable more clearly than in the coronal sections. Laterally, both the caudate and the putamen become continuous with the lateral nucleus, the putamen with pars compacta and the caudate with pars striatalis. A little farther medially there is no separation between the putamen and the tail of the caudate nucleus and the putamen part of the complex continues into the basal amygdaloid nucleus. Here the caudate nucleus is separated from the lateral amygdaloid nucleus by undifferentiated cells that are migrating away from the striatal ridge. Somewhat more medially, the common caudate-putamen cell mass that is largely putamen becomes continuous with nucleus amygdalae basalis, pars lateralis. Cells typical of the tail of the caudate more laterally disappear more medially but yet small cells that join it are separating from the much depleted undifferentiated lateral striatal ridge (Fig. 17C-D), and the inferior end of this less developed cell band lies adjacent to the lateral amygdaloid nucleus.

DISCUSSION AND CONCLUSIONS

The amygdala, or archistriatum, is the first part of the human striatal complex to appear embryologically, but neither the first part to finish differentiating, nor the simplest portion structurally. Indeed, morphologically, it is the most complex of all of the striatal subdivisions, although undoubtedly outranked in size by both parts of the neostriatum (the caudate nucleus and the putamen) as well as by the paleostriatum (globus pallidus). The very first neuroblasts to migrate outward from the ependymal layer when the telencephalic hemisphere begins to develop constitute the primordial amygdala. The location of the primordial

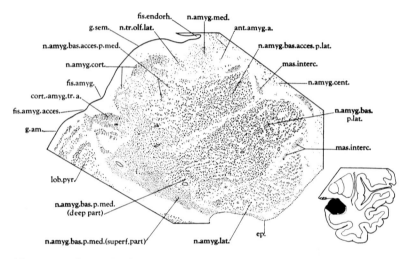

Fig. 19. Drawing of the adult human amygdaloid complex as seen
in a coronal section through the right hemisphere at a level
where the development is at about its highest level. The black
area of the small inset at the left of the main figure shows the
position and orientation of the amygdaloid complex in the adult
brain. Toluidin blue preparation About X 6. (Fig. 10 from Crosby
and Humphrey, 1941, reproduced with the permission of The Wistar
Press.)

amygdala, between the area in which the anlage of the gyrus
dentatus develops medially (Fig. 3A) and the primordial piriform
cortex which appears laterally a little later in development
(Fig. 3B) is maintained in the adult amygdala (Fig. 19) and
indeed is present throughout phylogeny as well (Crosby et al.,
1966).

The two major subdivisions of this complicated telencephalic
nucleus, more correctly designated as the amygdaloid complex,
appear almost at once -- the corticomedial complex and the basolateral complex (10 mm., Fig. 3B). In the material studied, the
two major nuclei of the corticomedial complex, which is considered to be the older part phylogenetically, are identifiable
soon after the primordial amygdala is recognizable. These two
nuclei, the cortical and the medial, although small in cell
numbers, are formed by a surface layer of cells and a cell mass
respectively, as is characteristic in the adult human amygdala
and throughout phylogeny. At the same age, however, the cells
of the basolateral division of the amygdala are associated more
closely with the germinal layer of the developing striatal ridge
and its constituent nuclei are not identifiable, although the
distinctive positional relations with the primordial piriform
cortex and the primordial neostriatum are present (Fig. 3B).
Indeed, none of the nuclei that form the basolateral complex
have been identified until the embryo has doubled its length
(20.7 mm, Fig. 6B) when all three of the nuclei are represented.
Thus, the nuclei of the phylogenetically younger basolateral
complex do not develop until appreciably later than do the two
major nuclei of the corticomedial complex.

In the basolateral complex, it seems probable that the basal
nucleus develops slightly ahead of the accessory basal nucleus
and that both of these nuclei appear before the lateral nucleus,
although in the material studied all three were identified in
the same embryo (Humphrey, 1968). This conclusion is based on
the greater size of the basal as compared with the accessory
basal nucleus when they are first identified, and the minute
representation of the lateral nucleus in this embryo. In this
respect, it may be of significance also that cells continue to
join the lateral nucleus and the superficial part of pars
medialis of nucleus basalis later in development than they join
the remainder of the basolateral complex for this part of the
basal nucleus is slow to appear also and is located outside the
lateral nucleus (Fig. 8). Evidently, then, not only do the
basolateral nuclei develop later, but the latest appearing
portions of the complex continue to increase in size for a
longer period in fetal life.

Although the central nucleus of the corticomedial complex
could not be identified as early in development as the medial and

cortical nuclei, probably it is represented by scattered cells
earlier than it is identifiable. Before a definitive cell cluster
can be recognized, some scattered cells are undoubtedly near the
origin of the stria terminalis fibers, the characteristic relationship of the central nucleus. Until these fibers collect into a
bundle, however, the cell mass does not become large enough to
form a recognizable nucleus (Humphrey, 1968) such as was first
seen at 22.2 mm (Fig. 6E). In insectavorous bat embryos, Brown
(1967) also found the central nucleus to develop relatively late,
but earlier than the lateral nucleus. This later embryonic time
of appearance of the lateral nucleus in the bat as compared with
its appearance in man is probably related to its small size and
poor development in the adult bats as compared with the large
size and greater development of the adult human amygdala. The
late development of the central nucleus embryonically is in
harmony with its appearance phylogenetically for the central
nucleus has not been recognized below reptiles (Crosby et al.,
1966).

The nucleus of the lateral olfactory tract is the latest of
all of the amygdaloid nuclei to develop in both man (Humphrey,
1968) and in insectivorous bats (Brown, 1967). Like the central
nucleus, this tiny nucleus associated with the lateral olfactory
tract has not been recognized below reptiles (Crosby et al.,
1966). Its late development was discussed in detail in the
author's 1968 paper on the early development of the amygdaloid
complex where it was pointed out that because of its dual
relationship with the anterior commissure (stria terminalis
component) and the lateral olfactory tract this nucleus is not
recognizable until after the anterior commissure has developed
its connections between the two amygdalae. Although presumably
dependent on lateral olfactory tract as well as anterior
commissure connections, this nucleus is present apparently even
in the anosmatic porpoise (Breathnach and Goldby, 1954) as well
as in all other mammals in which the amygdala has been studied
(Coenolasters, Obenchain, 1925; rat, Gurdjian, 1928; rabbit,
Young, 1936; bat, Humphrey, 1936 and Brown, 1967; cat, Fox, 1940;
adult man, Crosby and Humphrey, 1941; the short tailed shrew,
Crosby and Humphrey, 1944; the mink, Jeserich, 1945; the macaque,
Lauer, 1945; the giant panda, Lauer, 1949; the fin whale, Jansen
and Jansen, 1953; the opossum, Volker and Hamel, 1966; the
kangaroo, Hamel, 1966).

In the adult man, the nucleus of the lateral olfactory tract
was identified in its characteristic position between the medial
and the cortical amygdaloid nuclei (Crosby and Humphrey, 1941).
Its location is highly variable, however, ranging from a position
between the piriform cortex and the preoptic area in the rabbit
(Young, 1936), through a position in the anterior amygdaloid area

(rostral part in shrew, Crosby and Humphrey, 1944; panda, Lauer, 1949; mink, Jeserich, 1945), to a position within the medial nucleus or between the medial and cortical nuclei (bat, Humphrey, 1936). Moreover, in the macaque (Lauer, 1945) the posterior part was found deep to the cortical and between the cortical and medial nuclei and the anterior part between the cortical and medial nuclei.

The nucleus of the lateral olfactory tract may have medial and lateral parts (rabbit, Young, 1936) or rostral and caudal parts (bat, Humphrey, 1936; shrew, Crosby and Humphrey, 1944; macaque, Lauer, 1945). One part may be deep and the other more superficial (shrew, Crosby and Humphrey, 1944). As a suggested earlier (Humphrey, 1968), this variability in position and the frequent occurrence of two parts probably are due to variability in the size of its two major connections.

In spite of a careful check for the nucleus of the lateral olfactory tract, its identification is uncertain until relatively late in development. At 10 weeks, there is a mass of small cells deep to the lateral olfactory where that fiber bundle approaches the prepiriform cortex (to which it distributes fibers in mammals), but this small cell mass does not border the amygdala. However, on the deep surface of this cell cluster some fibers from the prepiriform cortex and many of the external capsule component of the anterior commissure pass close to it. In this fetus (No. 101, 38.2 mm CR), a few fibers of the anterior commissure are crossing in the midline. A similar cell cluster is present at 9.5 weeks (No. 126, 33.8 mm CR), but no commissural fibers have reached the midline. It has been concluded that this cell mass becomes the nucleus of the lateral olfactory tract in later fetal life through differential growth changes. The great forward (or anterior) development of the basolateral division of the amygdaloid complex and the overlying cortical nucleus, as well as the development of the connections of the basolateral complex and the piriform cortex through the anterior commissure, should bring about the incorporation of this cell cluster in the amygdaloid complex. As previously suggested, the relative size of the connection with the anterior commissure and the lateral olfactory tract probably determine the deep or superficial position of this nucleus. The variability, in its anteroposterior locations, probably is related to the degree of differentiation of the amygdaloid nuclei anteriorly. The variability in mediolateral position might be more specifically related to the degree of development of the cortical nucleus. At any rate, in the development of the human amygdala, the nucleus of the lateral olfactory has been found only within the cortical nucleus after it has become incorporated in the amygdaloid complex.

All nuclei of the amygdala take origin from the germinal,
or ependymal, layer of the lateral ventricle. The area of so-
called cortical infolding that Johnston (1923) considered to contribute cells to the amygdala from the cortex actually is formed
by cells that have not completed their migration into either
the piriform cortex or into the cortical nucleus, possibly
because they become too well differentiated to continue migration.
This area, designated the corticoamygdaloid transition area by
Crosby and Humphrey (1941, 1944) and more specifically the amygdalopiriform transition area by Brown (1967) and Humphrey (1968)
is a poorly delineated part of the superficial cell layer in
early development but by 11 weeks the area is easily distinguished
from either the piriform cortex or the lateral part of the cortical
nucleus by the extensive cell migration into it. Thereafter, the
amygdalopiriform transition area is conspicuous. Often the cell
migration is in waves or layers of cells but strands of cells
are frequent also. Not until the oldest age level studied does
the appearance of infolding become prominent (Fig. 18B). The
amygdalopiriform transition area is almost exclusively with pars
lateralis of the cortical nucleus. However, at 13.5 weeks and
thereafter there is conspicuous continuity also with the superficial part of nucleus basalis pars medialis, which in turn
unites with the deep part, and at other levels with the lateral
amygdaloid nucleus.

A second transition area between cortex and amygdala is
formed at the junction with the hippocampal formation. This
transition may be with either the medial nucleus or pars medialis
of the cortical nucleus of the amygdala. In early developmental
stages, the relationship is either with the anlage of the gyrus
dentatus or with the primordial cornu ammonis. Later, before
the hippocampal formation is better developed near the amygdala,
the nature of the transition is not clear. By 24.5 weeks, however, when the hippocampal formation is developing rapidly near
the amygdala, both a poorly developed gyrus dentatus area and
the cornu ammonis come into continuity with the amygdala. This
transition is never a clear-cut one but is always with nuclei of
the corticomedial complex, the medial nucleus, pars medialis of
the cortical nucleus and, later in development, the central
nucleus.

It already has been mentioned that the cortical and medial
amygdaloid nuclei of the corticomedial complex can be distinguished before any of the other amygdaloid nuclei. In later development, there are other evidences of their early development as
well as the time of origin. Indeed, the general sequence in
which the amygdaloid nuclei develop is shown by their relation
to the ependymal layer from which they are derived. Thus, in
sagittal sections of fetuses as young as 8+ weeks (Fig. 5B) and

even more clearly at 10.5 weeks (Fig. 10) the areas of the ependymal layer from which cells are migrating into the medial,
cortical and the accessory basal nuclei are narrow as compared
with the regions from which cells are migrating into the basal
nucleus, and especially into the lateral nucleus. At 10.5 weeks,
this relationship is even more clear. Thus, the source of new
cells for the corticomedial complex is depleted earlier than
that for the basolateral complex.

Accounts in the literature differ concerning which striatal
ridge gives rise to the amygdala, medial or lateral. Hewitt
(1958) stated that the lateral striatal ridge provided the major
source of cells with some arising from the medial ridge. From
the study of development through 8.5 weeks (Humphrey, 1968), it
was concluded that the major source of cells early in development
is from the lateral striatal ridge. After the medial striatal
ridge develops, cells are derived from it also, but the amount
decreased after the internal capsule crosses the caudate-putamen
complex. After the inferior horn of the lateral ventricle
develops and carries the amygdala anteriorly, the amygdala is
associated only with the lateral striatal ridge. Further contributions to the amygdaloid nuclei are then derived solely from
the lateral ridge. Although it cannot be concluded with certainty, the corticomedial complex is derived in part from both
striatal ridges, the lateral one early in development and the
medial ridge after it appears. The additions of cells, after the
internal capsule becomes prominent, is from the lateral ridge
alone.

From the lateral striatal ridge, at least, cell migration
into the amygdaloid nuclei has a definite pattern, shown most
clearly at 13.5 and 14 weeks in the available material. Here,
cells that are joining the medial, the central and the adjacent
part of the cortical nuclei are taking origin from the medial
striatal ridge, those joining the accessory basal nucleus a little
more laterally, those adding to the basal nucleus next laterally
and those migrating into the lateral nucleus most laterally of
all. The migration of cells into the amygdalopiriform transition
area take origin at the border of the lateral striatal ridge with
the wall of the hemisphere. The migration into the amygdala, at
this age level, is evidently a massive one, but almost exclusively
into the basolateral complex, although cells also join the medial
and intermediate parts of the cortical nucleus. Although it is
not surprising that there should be so definite an arrangement
for the derivation of individual nuclei, it is astonishing to
see them so sharply defined.

The amygdala begins its development before the inferior horn
of the lateral horn develops and lies in its anterior wall when

the outgrowth that will become the inferior horn grows ventral-
ward. As the inferior horn turns anteriorly, the amygdala is
carried forward along its medial side (Figs. 7 to 9). With
further development in the temporal area, the amygdala is
rotated medially as well. The final position of the amygdala
then is anterosuperior to the inferior horn of the ventricle
whereas the hippocampal formation bulges into the ventricle
inferolaterally. Because the amygdala is carried anteriorly by
these growth changes, during at least part of the period of its
most rapid increase in size, the source of new cells, the undifferentiated cell mass forming the lateral striatal ridge, is
far removed from its anterior pole. At this time, compact
strands or bands of cells (Fig. 15A and 16A) make up a conspicuously large part of the anterior portion of the amygdala
although still further anteriorly there is a uniformly arranged
looser cell mass typical of the anterior amygdaloid area. These
cell bands join the basal and lateral amygdaloid nuclei posteriorly and have been interpreted as extensions of these nuclei
that are growing rapidly into its anterior pole. Some of these
strands are dense and more deeply staining and appear to join
the lateral nucleus. Others, that are less dense and lighter
staining, have been allocated to the basal nucleus. In some
areas, a cell band outside the lateral nucleus is continuous
with another deep to it and so represent the superficial and
deep parts of nucleus amygdalae basalis pars medialis. By
24.5 weeks, the zone of undifferentiated cells adjacent to the
amygdala has almost disappeared (Fig. 17B and D), but a considerable mass remains along the tail of the caudate nucleus
(Fig. 18A). Evidently, then, the amygdala completes its differentiation while it is still possible for new cells to be added
to the tail of the caudate nucleus.

It already has been mentioned that during its embryonic
development the human amygdala rotates as was pointed out earlier
by Crosby and Humphrey (1941, 1944) and by Humphrey (1968) must
be the case. This medial rotation is the greatest in man of
all of the mammals for which the amygdala has been studied
adequately. The location of the medial nucleus deep to the
endorhinal fissure puts it in a pivotal position (Humphrey, 1968)
so that it changes position the least as the amygdala is rotated
medially. This medial rotation from lower mammals to man was
discussed by Crosby and Humphrey (1944) in comparing the amygdala
of the short tailed shrew with that of adult man. One of the
drawings of the adult human amygdala is reproduced here (Fig. 19)
for comparison with the relative position of the nuclei during
development. Although the amygdala is carried anteriorly as
soon as the inferior horn of the ventricle grows forward, its
shift medialward does not become apparent until 11 to 12 weeks
(Figs. 11A and 12A). By 13.5 ot 14 weeks, however, the position

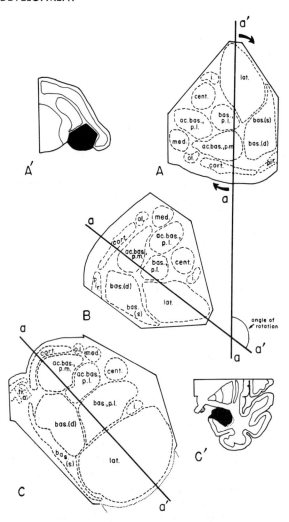

Fig. 20. Diagrammatic illustrations of the amygdaloid complex of
the short tailed shrew (Blarina brevicauda) and of the adult human
amygdaloid complex to illustrate the rotation of the amygdala
medialward phylogenetically from lower mammals to man without any
essential change in the relative position of the individual nuclei.
A and A′ show the amygdala of the shrew in its normal position. B
shows the amygdaloid complex of the shrew rotated medially through
an angle of approximately 130 to 140 degrees. C and C′ are drawings of the adult human amygdaloid complex to show the position
and orientation of the amygdala and the relative location of the
individual amygdaloid nuclei. The line a - a′ in A, B, and C has
been drawn through the amygdaloid complex in essentially the same
plane in order to compare the normal orientation of the adult human
amygdala of the shrew and show the rotation phylogenetically. Comparison with a similar line through the amygdaloid complex of the
24.5-week human fetus (Fig. 18A) demonstrates that the medial
rotation at 24.5 weeks, although well advanced, is not yet complete.

of the nuclei is fairly comparable to that in the shrew, the bat
and other mammals with a smooth hemisphere, in which the cortical
nucleus lies along the ventral surface. By 16 to 18.5 weeks, the
temporal pole has rotated medially enough to bring the cortical
nucleus to the ventromedial surface of the amygdala (Figs. 16 and
17). In the oldest fetus included in this study, the cortical
nucleus is entirely medial in position (Fig. 18B), but has not yet
completed the rotation found in the adult human brain where this
nucleus is dorsomedially situated. The difference is evident
on comparing Figures 18 and 19. The rotation during mammalian
phylogeny was emphasized by Crosby and Humphrey (1944) in their
diagram reproduced here in Figure 20. This figure shows that,
if the amygdala of the shrew were to be rotated medialward on its
axis for approximately 130° to 140°, the position of the nuclei
would then be comparable to those found in adult man. Comparison
of the axis of the amygdaloid complex at 24.5 weeks with that in
adult man (Figs. 18A and 20C) shows that rotation is not yet complete at that fetal age. In the full term infant, however, the
insula is covered by the temporal operculum (Conel, 1939), so
the medial rotation should be completed.

ACKNOWLEDGMENTS

This investigation was supported by a Public Health Service
research career program award, 5-K6-NS-16716, from the National
Institute of Neurological Diseases and Stroke and aided by grant
HD-00230, National Institute of Child Health and Human Development,
National Institutes of Health. This paper is publication No. 61
in a series of physiologic and morphologic studies on human pre-
natal development begun in 1932 under the direction of Dr.
Davenport Hooker. The data on which this paper is based were
collected during support in the past by grants from The Penrose
Fund of the American Philosophical Society, The Carnegie Corpora-
tion of New York, The University of Pittsburgh, The Sarah Mellon
Scaife Foundation of Pittsburgh, and Grant B-394 from the
National Institute of Neurological Diseases and Blindness to
Davenport Hooker and/or to the author.

REFERENCES

ARIËNS KAPPERS, C. U. The Evolution of the Nervous System in
        Invertebrates, Vertebrates and Man. Haarlem, Bohn, 1929.

BODIAN, D. The staining of paraffin sections of nervous tissues
        with activated protargol. The role of fixatives. Anatomical
        Record, 1937, 69, 153-162.

BREATHNACH, A. S., & GOLDBY, F. The amygdaloid nuclei, hippocampus and other parts of the rhinencephalon in the porpoise
(Phocaena phocaena). Journal of Anatomy (London), 1954, 88,
267-291.

BROCKHAUS, H. Zur normalen und pathologischen Anatomie des Mandel-
kerngebietes. Journal of Psychology and Neurology (Leipzig),
1938, 49, 1-136.

BROWN, J. W. The development of the amygdaloid complex in
insectivorous bat embryos. Alabama Journal of Medical
Science, 1967, 4, 399-415.

CONEL, J. L. The Postnatal Development of the Human Cerebral
Cortex. Vol. 1. The Cortex of the Newborn. Cambridge:
Harvard University Press, 1939.

COOPER, E. R. A. The development of the human red nucleus and
corpus striatum. Brain, 1946, 69, 34-44.

COOPER, E. R. A. The development of the thalamus. Acta Anatomica,
1950, 9, 201-226.

CROSBY, E. C., & HUMPHREY, T. Studies of the vertebrate telencephalon. II. The nuclear pattern of the anterior olfactory
nucleus, tuberculum olfactorium and the amygdaloid complex in
adult man. Journal of Comparative Neurology, 1941, 74,
309-352.

CROSBY, E. C., & HUMPHREY, T. Studies of the vertebrate telencephalon. III. The amygdaloid complex in the shrew (Blarina
brevicauda). Journal of Comparative Neurology, 1944, 81,
285-305.

CROSBY, E. C., HUMPHREY, T., & LAUER, E. W. Correlative Anatomy of
the Nervous System. New York, Macmillan Co., 1962.

CROSBY, E. C., DEJONGE, B. R., & SCHNEIDER, R. C. Evidence for
some of the trends in the phylogenetic development of the
vertebrate telencephalon. In R. Hassler and H. Stephan (Eds.),
Evolution of the Forebrain. Stuttgart: George Thieme Verlag,
1966. Pp. 117-135.

ESCOLAR, J. El complejo amigdalino en relación con el allocortex,
considerado ontogénica y filogenicamente. Anales de Anatomia,
1959, 8, 215-231.

FOX, C. A. Certain basal telencephalic centers in the cat.
Journal of Comparative Neurology, 1940, 72, 1-62.

GANSER, S. Vergleichend - anatomische Studien über das Gehirn
    des Maulwurfs. Morphologisches Jahrbuch, 1882, 7, 591.

GURDJIAN, E. S. The corpus striatum of the rat. Studies on the
    brain of the rat, No. 3. Journal of Comparative Neurology,
    1928, 45, 249-281.

HAMEL, E. G., JR. The amygdaloid complex in the kangaroo and the
    North and South American opossum. Anatomical Record, 1966,
    154, 353 (Abstract).

HEWITT, W. The development of the human caudate and amygdaloid
    nuclei. Journal of Anatomy (London), 1958, 92, 377-382.

HEWITT, W. The development of the human internal capsule and
    lentiform nucleus. Journal of Anatomy (London), 1961, 95,
    191-199.

HILPERT, P. Der Mandelkern des Menschen. I. Cytoarchitektonik
    und Faserverbindunger. Journal of Psychology and Neurology
    (Leipzig), 1928, 36, 44-74.

HINES, M. Studies in the growth and differentiation of the telen-
    cephalon in man. The fissura hippocampi. Journal of Compara-
    tive Neurology, 1922, 34, 73-171.

HOCHSTETTER, F. Beiträge zur Entwicklungsgeschichte des mensch-
    lichen Gehirns, Vol. 1. Deuticke, Leipzig und Wien, 1919.

HOOKER, D. The Prenatal Origin of Behavior. Lawrence: University
    of Kansas Press, 1952. (Reprinted by Hafner Publishing Co.,
    New York, 1969)

HOOKER, D. Evidence of prenatal function of the central nervous
    system in man. James Arthur Lecture on The Evolution of the
    Human Brain for 1957, American Museum of Natural History,
    New York, 1958.

HUMPHREY, T. The telencephalon of the bat. I. The non-cortical
    nuclear masses and certain pertinent fiber connections.
    Journal of Comparative Neurology, 1936, 65, 603-711.

HUMPHREY, T. The development of the pyramidal tracts in human
    fetuses, correlated with cortical differentiation. In D. V.
    Tower and J. P. Schadé (Eds.), Structure and Function of the
    Cerebral Cortex. Amsterdam: Elsevier, 1960. Pp. 93-103.

HUMPHREY, T. The development of the human hippocampal formation
    correlated with some aspects of its phylogenetic history.

In R. Hassler and H. Stephan (Eds.) Evolution of the Forebrain. Stuttgart: George Thieme Verlag, 1966a. Pp. 104-116.

HUMPHREY, T. Correlations between the development of the hippocampal formation and the differentiation of the olfactory bulbs. Alabama Journal of Medical Science, 1966b, 3, 235-269.

HUMPHREY, T. The development of the human amygdala during early embryonic life. Journal of Comparative Neurology, 1968, 132, 135-165.

JANSEN, J., JR., & JANSEN, J. A note on the amygdaloid complex in the fin whale (Balaenoptera physalus L.). Hvalrådets Skrifter No. 39, 1-13.

JESERICH, M. W. The nuclear pattern and the fiber connections of certain non-cortical areas of the telencephalon of the mink (Mustela vision). Journal of Comparative Neurology, 1945, 83, 173-211.

JOHNSTON, J. B. Further contributions to the study of the evolution of the forebrain. Journal of Comparative Neurology, 1923, 35, 337-481.

KODAMA, S. Über die sogenannte Basalganglien (Morphogenetische und pathologisch-anatomische Untersuchungen). Schweizer Archiv für Neurologie und Psychiatrie, 1926a, 19, 152-177.

KODAMA, S. Über die sogenannten Basalganglien (Morphogenetische und pathologisch-anatomische Untersuchungen). Schweizer Archiv für Neurologie und Psychiatrie, 1926b, 18, 179-246.

KODAMA, S. Über die Entwicklung des striären Systems beim Menschen, Schweizer Archiv für Neurologie und Psychiatrie, 1927, 20, 1-98.

LAUER, E. W. The nuclear pattern and fiber connections of certain basal telencephalic centers in the macaque. Journal of Comparative Neurology, 1945, 82, 215-254.

LAUER, E. W. Certain olfactory centers of the forebrain of the giant panda (Ailuropoda melanoleuca). Journal of Comparative Neurology, 1949, 90, 213-241.

LOO, Y. T. The forebrain of the opossum, Didelphis virginiana. Part II. Histology. Journal of Comparative Neurology, 1931, 52, 1-148.

MACCHI, G. Sviluppo ontogenetico del nucleo amigdaloideo dell' Uomo. Achivio Italiano di Anatomio e di Embriologia, 1948, 53, 207-248.

MACCHI, G. The ontogenetic development of the olfactory telen-
    cephalon in man. Journal of Comparative Neurology, 1951,
    95, 245-305.

OBENCHAIN, J. B. The brains of the South American marsupials,
    Caenolestes and Orolestes. Field Museum of Natural History,
    Zoology Series, XIV, 175-232.

RETZIUS, G. Das Menschenhirn Studien in der makroskopischen
    Morphologie, Vol. II. Stockholm: Tafeln, Norstedt und
    Söner, 1896.

ROSE, M. Der Allocortex bei Tier und Mensch, I. Teil. Journal
    of Psychology and Neurology (Leipzig), 1926, 34, 1-111.

SCHNITZLEIN, H. N., & CROSBY, E. C. The telencephalon of the
    lungfish, Protopterus. Zeitschrift für Hirnforschung, 1967,
    9, 105-149.

SHARP, J. A. The junctional region of cerebral hemisphere and
    third ventricle in mammalian embryos. Journal of Anatomy
    (London), 1959, 93, 159-168.

SIDMAN, R. L., & ANGEVINE, J. B. Autoradiographic analysis of
    time of origin of nuclear versus cortical components of
    mouse telencephalon. Anatomical Record, 1962, 142, 326-327
    (Abstract).

SMITH-AGREDA, J. Relación a lo largo del desarrollo entre
    fascicúlos epithalámicos y subtalámicos (estudio en el
    hombre). Anales de Anatomia, 1961, 10, 205-229.

SMITH-AGREDA, J. Matriz y emigraciones del encéfalo humano en
    un embrión de 25 mm. Anales de Anatomia, 1962, 417-428.

SMITH-AGREDA, J. Aportacion al estudio del epitálamo humanos (un
    estudio de la topographiá del substrato diencefálico desde
    el punto de vista ontogénico humano). Anales de Anatomia,
    1963, 12, 229-263.

VAN DER SPRENKEL, H. BERKELBACH. Stria terminalis and amygdala in
    the brain of the opossum (Didelphis virginiana). Journal of
    Comparative Neurology, 1926, 42, 211-254.

STREETER, G. L. Weight, sitting height, head size, foot length,
    and menstrual age of the human embryo. Carnegie Institution
    Washington Publication No. 274. Contributions to Embryology,
    1920, 11, 143-170.

STREETER, G. L. Developmental horizons in human embryos. Description of age groups XV, SVI, SVII, and SVIII, being the third issue of a survey of the Carnegie collection. Carnegie Institution Washington Publication No. 575. Contributions to Embryology, 1948, 32, 133-203.

VOLKER, V. S., & HAMEL, E. G., JR. Teh nuclear configuration and cytoarchitecture of the amygdaloid complex in Didelphis virginiana. Alabama Journal of Medical Science, 1966, 3, 54-69.

VÖLSCH, M. Zur vergleichenden Anatomie des Mandelkerns und seiner Nachbargebilde, II. Teil. Archiv für Mikroskopische Anatomie, 1910, 76, 373-523. (Quoted from Landau, 1919, p. 358.)

LIST OF ABBREVIATIONS

ac.bas., nucleus amygdalae basalis accessorius
ac.bas.,p.l., nucleus amygdalae basalis accessorius pars
            lateralis
ac.bas.,p.m., nucleus amygdalae basalis accessorius pars medialis
amyg., amygdala
amyg.compl., amygdaloid complex
amyg.fis., fissura circularis amygdalae
amyg.-hip.tr.a., amygdalohippocampal transition area
amyg.-pir.tr.a., amygdalopiriform transition area
ant.amyg.a., anterior amygdaloid area
ant.com., commissura anterior
ant.horn, anterior horn of lateral ventricle
aq., cerebral aqueduct
bas., nucleus amygdalae basalis
bas.(d), nucleus amygdalae basalis, pars medialis (deep portion)
bas.,p.l., nucleus amygdalae basalis pars lateralis
bas.,p.m., nucleus amygdalae basalis pars medialis
bas.(s), nucleus amygdalae basalis, pars medialis (superficial
            portion)
basolat., basolateral amygdaloid complex
b.v., blood vessel
caud., nucleus caudatus
caud.-put., caudate-putamen complex
cent., nucleus amygdalae centralis
cerebel., cerebellum
cer.ped., cerebral peduncle
ch.pl., choroid plexus
cl., claustrum
corn.am., cornu ammonis
corp.cal., corpus callosum
corp.str., corpus striatum
cort., nucleus amygdalae corticalis
cort.(i), cort.(l) and cort.(m), nucleus amygdalae corticalis,
        pars intermedialis, pars lateralis and pars medialis
        respectively
diag.b.Broca, diagonal band of Broca (and its nucleus)
dien., diencephalon
dors.thal., dorsal thalamus
endorh.fis., fissura endorhinalis

epithal.,   epithalamus
ep.l., or ep.l., lat.vent.,   ependymal layer of lateral ventricle
ext.cap.,   capsula externa
f., fimbria, or fornix
fr.,   frontal pole of hemisphere
fr.lobe,   frontal lobe
gl.pal.I and II, globus pallidus, deep and superficial divisions
          respectively
gy.dent.,   gyrus dentatus
gy.dent.anl.,   anlage of gyrus dentatus
hip. or hip.form.,   hippocampal formation
hypothal.,   hypothalamus
i.,   massa intercalata
inf.horn,   inferior horn of lateral ventricle
int.cap.,   capsula interna
interc.,   massa intercalata
L,   lateral striatal ridge
lam.term.,   lamina terminalis
l.f.b.,   lateral forebrain bundle, ventral peduncle
lat.,   nucleus amygdalae lateralis
lat.p.c.,   nucleus amygdalae lateralis, pars compacta
lat.,p.s.,   nucleus amygdalae lateralis pars striatalis
lat.str.r.,   lateral striatal ridge
lat.vent.,   ventriculus lateralis
M, medial striatal ridge
med.,   nucleus amygdalae medialis
med.ob.,   medulla oblongata
mesen.,   mesencephalon
n.t.o.l.,   nucleus tractus olfactorius lateralis
N.II,   nervus opticus
N.V.,   nervus trigeminus
nb.,   neuroblasts migrating into amygdalopiriform transition area
neostr.,   neostriatum
nuc.III, nucleus oculomotorius
oc.,   occipital pole of telencephalon
ol.,   nucleus tractus olfactorius lateralis
olf.bulb,   bulbus olfactorius
olf.pl.,   olfactory placode
opt.tr.,   tractus opticus
periamyg.cort.,   periamygdaloid cortex
pir.,   piriform cortex
post.horn,   posterior horn of lateral ventricle
prim.amyg.,   primordial amygdala
prim.corn.am.,   primordial cornu ammonis
prim.hip.,   primordial hippocampal formation
prim.lat.str.r.,   primordial lateral striatal ridge
prim.neostr.,   primordial neostriatum
prim.pir.,   primordial piriform cortex
put.,   putamen

S, sulcus interstriatalis
sp.c., spinal cord
str.term., stria terminalis
sub., subiculum
telen., telencephalon
telen.hem., telencephalic hemisphere
temp., temporal lobe
tr.a., amygdalopiriform transition area
tub.olf., tuberculum olfactorium
undiff.z., undifferentiated zone of striatal ridge
III or III vent., third ventricle
IV, fourth ventricle

ANATOMY

FUNCTIONAL IMPLICATIONS OF A QUANTITATIVE
ULTRASTRUCTURAL ANALYSIS OF SYNAPSES IN THE
PREOPTIC AREA AND VENTROMEDIAL NUCLEUS

G. Raisman and P. M. Field

Department of Human Anatomy

University of Oxford, Oxford, England

The present experimental anatomical investigations have been
stimulated by a desire to examine some of the central neural mechanisms underlying the control of reproductive functions, and in
particular will refer to the control of ovulation and of mating
behaviour in the rat. The general assumption has been that the
analysis of patterns of synaptic connections may afford information which can be correlated with observations from experiments
based on endocrine and electrophysiological techniques. A
comprehensive survey of the relevant functional evidence is not
within the scope of this presentation but an attempt has been made
to show which points have been important in guiding the course of
the anatomical investigations.

The central nervous system plays a crucial role in the
control of gonadotrophin secretion and in the manifestation of
mating behaviour. In the regulation of pituitary secretion,
neural structures in the hypothalamus act as essential final
links in the chain of connections. Among extrahypothalamic
structures which have been most often implicated in the control
of gonadotrophin secretion are the hippocampus and the amygdala
(for a review see Raisman and Field, 1971); to the neuroanatomist
this is not surprising, as these 'limbic' areas are the source of
the two major fibre pathways to the medial hypothalamus - the
medial cortico-hypothalamic tract which passes from the hippocampus
to the arcuate nucleus, and the stria terminalis which passes from
the amygdala through the preoptic area and anterior hypothalamus
to the ventromedial hypothalamic nucleus (Raisman, 1970). While
not excluding the importance of other connections of these tracts,
or other afferent connections to the hypothalamus, the present work

has concentrated upon two aspects of the projection of the stria
terminalis - the terminations in the peripheral parts of the
ventromedial hypothalamic nucleus and in the preoptic area.

The probable localisation of function within this system may
be illustrated by considering the effects of lesions in the brains
of female rats at various levels above the pituitary (Fig. 1).
If all nervous and vascular connections between the hypothalamus
and pituitary gland are severed (a), the animal does not ovulate
and shows ovarian atrophy (Harris and Campbell, 1966). If, however, a circumscribed lesion (b) is made such that the mediobasal
('tuberal') portion of the hypothalamus is left in contact with
the median eminence, and hence with the pituitary, gonadal atrophy
does not ensue, although the ovaries are polyfollicular and
ovulation does not occur (Halász, 1969). It is postulated that
the tuberal hypothalamus contains a neural apparatus capable of
maintaining a basal secretion of gonadotrophins, although on its
own this island of tissue cannot initiate the burst of gonadotrophin output required to produce ovulation (Barraclough, 1967).
Should the anterior border of the lesion be extended forward (c)- an
operation associated with a very high mortality and morbidity - so

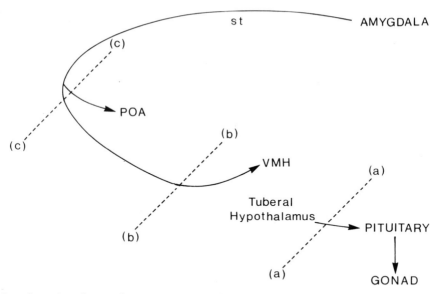

Fig. 1. A schematic representation of amygdaloid projections
through the stria terminalis (st) to the preoptic area (POA) and
ventromedial nucleus (VMH), and their postulated relationship to
the levels of the lesions (a-a, b-b, c-c) described in the discussion of the neural control of pituitary gonadotrophin
secretion.

as to leave the preoptic area in continuity with the hypothalamic
island, the surviving animals may exhibit 'spontaneous' ovulation,
although a return to regular cycles has not been unequivocally
demonstrated (Halász, 1969). This suggests that the neural
mechanism required for the 'triggering' of the preovulatory surge
of gonadotrophins lies in the preoptic area, and it corresponds
well with earlier evidence that destruction of the preoptic area
prevents spontaneous ovulation in the rat (see Harris and Campbell,
1966).

Lesions severing the dorsal connections of the preoptic area
(Taleisnik et al., 1970), or complete bilateral section of the
stria terminalis and the fimbria in the dorsal part of their
course (Raisman and Brown-Grant, unpublished observations) would
interrupt the major direct fibre tracts from the forebrain to the
medial hypothalamus. After both types of lesion regular cyclic
ovulation continued (following an initial dioestrous period).
This indicates that these limbic connections are not essential for
ovulation in the rat. It therefore leaves open the question of
what role they may play in modulating gonadotrophin secretion.
That such a role does exist is suggested by many published
observations that either lesions or stimulation of the amygdala
and hippocampus may affect reproductive functions. For example,
stimulation of the amygdala can induce ovulation in rats in which
spontaneous ovulation has been blocked by drugs or constant
light, and this induction of ovulation by amygdaloid stimulation
is prevented by lesions of the stria terminalis but not of the
ventral amygdalo-fugal pathway (Velasco and Taleisnik, 1969).
Lesions of the amygdala or of the stria terminalis also advance
the time of onset of puberty in the rat (Critchlow and Bar-Sela,
1967). More recently, Kalra and Sawyer (1970) have taken advantage of the fact that when spontaneous ovulation is blocked
by Nembutal, ovulation can be induced in the female rat by copulation, and they have used this situation to show that if a lesion
is made at the anterior border of the preoptic area, such coital
induction of ovulation is prevented. These lines of experiment
suggest that the amygdala and its projection pathway through the
stria terminalis are implicated in some way in gonadotrophin
control, but that the precise role will require some quite subtle
testing situation for its elucidation. In view of the evidence
that oestrogen is involved in the initiation of mating behaviour
and the timing of ovulation, it seems significant that in an
autoradiographic study Stumpf (1970) has shown that the neurons
whose nuclei retain tritiated oestradiol are located in the
amygdala and in those parts of the diencephalon which correspond
fairly closely with the distribution of the fibres of the stria
terminalis.

In the present series of experiments, which have been
carried out in rats, lesions were made in the stria terminalis in

the middle part of its course by means of a stereotaxically
guided knife entering the brain from its dorsal aspect. This
lesion has the advantage of being well away from the preoptic
area and the hypothalamus, but is complicated by the fact that
it also transects the fimbria. However, discrete control lesions
placed in the amygdala or hippocampus, as well as in the fimbria
alone, establish that the terminal synaptic fields investigated in
this study are derived solely from the strial fibres. Light
microscopy of orthograde degeneration has established that fibres
of the stria terminalis pass through the preoptic area and anterior hypothalamus and form a dense plexus in and around the
ventromedial hypothalamic nucleus. At the ultrastructural level,
Heimer and Nauta (1969) have shown that true terminal degeneration occurs in the peripheral shell of the ventromedial nucleus
and the observations described here show that the stria terminalis
also forms synapses in the preoptic area. In the present study,
representative samples of neuropil from the preoptic area and the
shell of the ventromedial nucleus (adjacent to the arcuate nucleus)
have been selected for electron microscopic analysis both in normal
rats and in animals in which the stria terminalis had been transected two days prior to sacrifice. Ultrathin sections from the
selected areas were mounted on uncoated grids whose mesh served
to divide the section up into convenient sampling areas of about
1800 square microns. In each animal all the synapses on at least
20 grid squares were counted, and each synapse was classified by
several different ultrastructural features (site of termination,
synaptic thickening, types of synaptic vesicles, etc). For the
purposes of the present study, the most useful diagnostic feature
of the synapsès has been their site of termination. A small
minority of synapses terminate directly upon cell bodies (axosomatic synapses) whereas the remainder terminate either upon
dendritic shafts or else upon the spines of dendrites. The ratio
of the number of synapses terminating upon dendritic shafts to
those on dendritic spines has been found to be a distinctive
feature of the neuropil; it will be referred to as the 'shaft/spine
ratio.' At a survival time of two days after a lesion of the stria
terminalis, a majority of axon terminals belonging to fibres in the
stria undergo a readily recognisable form of orthograde degeneration, involving increased electron density and collapse of the
terminal (Fig. 2). By the use of this reaction, the samples of
synapses drawn from the preoptic area and from the ventromedial
nucleus have been divided into degenerating synapses of amygdaloid
origin and non-degenerating synapses of non-amygdaloid origins.
The origins of this second group of axon terminals are at present
unknown.

One of the most important findings (based on counts of over
30,000 synapses) has been that both the shaft/spine ratio and the
proportion of synapses degenerating are remarkably uniform from

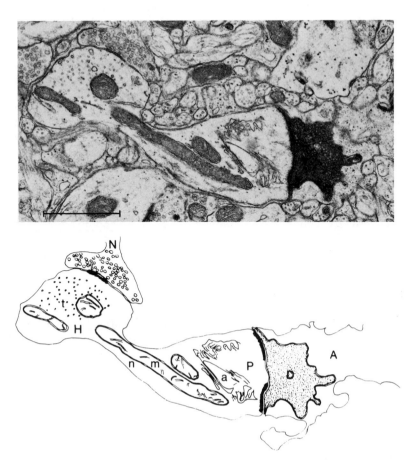

Fig. 2. An electron micrograph showing two synapses in the ventromedial nucleus two days after a lesion of the stria terminalis. The non-degenerating (i.e. non-amygdaloid) synapse involves an axon terminal (N) containing synaptic vesicles and making contact with a transversely sectioned dendritic shaft (H) which can be recognised on account of its microtubules (t). The degenerating synapse involves an axon terminal (D) of amygdaloid origin, which makes contact at two areas of synaptic thickening with a dendritic spine (P). As compared with the normal terminal, the degenerating terminal is more electron dense and is collapsed and indented by an adjacent phagocytic astrocytic process (A). The dendritic spine is characterised by a large 'spine apparatus' (a) and is connected to the shaft by a narrower neck region (n) containing an elongated mitochondrion (m). Calibration bar = 1 micron.

one grid square to another on the same section and also from one
animal to another for the same brain area. They may therefore be
employed to characterise the neuropil of the areas studied. This
is most clearly seen by contrasting the neuropil analyses from the
two different areas. In the <u>ventromedial nucleus</u> the majority of
the total number of synapses are borne upon dendritic shafts, although quite a large minority contact dendritic spines, giving a
shaft/spine ratio of 3:1. The degenerating synapses account for
up to 20 per cent of the total number of contacts, but differ from
the non-strial synapses in showing a marked preference for dendritic spines, so that the shaft/spine ratio for non-amygdaloid
synapses is 4:1 and that for amygdaloid synapses is 1:4. As a
consequence, although 20 per cent of the total population of
synapses are degenerating, this figure is partitioned unequally,
35 per cent of the spine synapses undergoing degeneration, and
only 3 per cent of the shaft synapses. In the <u>preoptic area</u>, the
overall shaft/spine ratio is 13:1 - i.e. there are relatively far
fewer synapses upon dendritic spines than in the ventromedial
nucleus. Furthermore, the degenerating synapses form a smaller
proportion of the whole (some 10 per cent). Although the amygdaloid fibres still account for a large proportion of the
dendritic spine synapses, a far larger proportion end on dendritic
shafts in the preoptic area than in the ventromedial nucleus, so
that the shaft/spine ratio of the amygdaloid synapses is 3:2. By
contrast the shaft/spine ratio of the non-amygdaloid synapses is
16:1. Thus the proportion of the total number of spine synapses
which undergo degeneration is as high as 24 per cent, and the
proportion of shaft synapses degenerating is less than 3 per cent.

These observations not only confirm the potential value of
neuropil analysis for the characterisation of regions in the central nervous system, but also indicate that there are quite specific
differences between the areas of termination of amygdaloid fibres
in the preoptic area and in the tuberal hypothalamus. In view of
the evidence (a) that the stria terminalis is involved in gonadotrophic function, and (b) that the preoptic area and the tuberal
hypothalamus can be correlated respectively with the cyclic and
basal control of gonadotrophins, it seems possible that the
quantitative differences in the neuropil of these two areas
reflect in some way these different functional roles. At this
point in the investigations it therefore seemed a possibility that,
by comparing the brains of male and female rats, it might be
possible to detect some anatomical differences, at this level of
analysis, between the sexes. Figure 3 shows a ranked series of
shaft/spine ratios taken from the ventromedial nucleus and the
preoptic area of 10 male and 8 female rats. This clearly reflects the characteristic difference already described between
the shaft/spine ratios of the two regions. In addition, it shows
that sexual differences occur in the preoptic area but not in

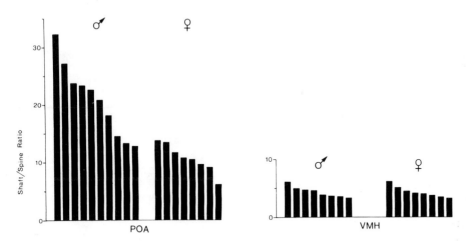

Fig. 3.   Shaft/spine ratios of non-amygdaloid synapses of 10
male and 8 female rats, ranked in order of magnitude.   POA =
preoptic area; VMH = ventromedial nucleus.

the ventromedial nucleus.   Considering the non-degenerating
synapses, it can be seen that the shaft/spine ratios range from
3.3 to 6.2 in the samples from the ventromedial nuclei, and 6.2
to 32.3 in the preoptic area.   In the preoptic area the samples
drawn from the males are generally higher than those of the
females (with a slight overlap); in the ventromedial nucleus no
such differences exist.   The observation that sexual differences
occur in the preoptic area correlates well with the functional
evidence that the neural trigger mechanism for the cyclic preovulatory surge of gonadotrophins lies in the preoptic area,
since this mechanism is present in the female but not in the
male.   The amygdaloid (strial) fibres form synapses in both the
preoptic area and the ventromedial nucleus, but in neither area
do these synapses show any significant difference between the
sexes.   Thus the anatomical evidence indicates that whereas the
stria terminalis may not itself have sexually differentiated
synapses, it does terminate in a sexually differentiated zone
in the preoptic area but not in the ventromedial nucleus.   These
results are shown schematically in Figure 4.

    While the neuropil analysis does not reveal the identify
(i.e. cells of origin etc.) of the postsynaptic elements in the

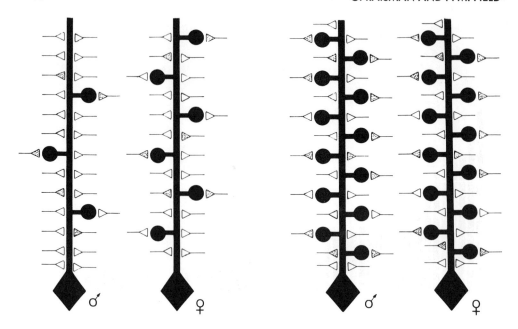

POA   VMH

Fig. 4. A schematic representation of segments of dendrites from
the preoptic area (POA) and ventromedial nucleus (VMH) of males
and females. Dendritic spines are shown by the black circles
borne on the ends of stalks. Synaptic contacts may be upon
dendritic spines or shafts, and are made by axon terminals
represented by white triangles (non-degenerating - i.e. non-
amygdaloid) or grey triangles (degenerating - i.e. of amygdaloid origin).

Note: i)   The number of dendritic spines is greater in the
           ventromedial nucleus (where the male and female are
           identical) than in the preoptic area of either sex.

      ii) In the preoptic area, the female has twice as many
           spines as the male.

      iii) In all areas a large proportion of the dendritic
           spine synapses are of amygdaloid origin.

      iv) In the ventromedial nucleus there are very few
           amygdaloid synapses upon dendritic shafts, but in
           the preoptic area of both male and female the
           amygdaloid synapses contact dendritic spines and
           shafts in roughly equal numbers.

preoptic area, it does show that the sexual difference lies in
the mode of termination of synapses. It is not known which
fibres give rise to the sexually differentiated synapses - they
may consist of the terminals of other extrinsic fibre systems
or of intrinsic short axon cells or axon collaterals of neurons
in the preoptic area itself. The suggestion that it is the
input rather than some intrinsic property of the cells of the
preoptic area which is sexually differentiated has already been
made on functional grounds (Everett, 1969). Thus, neither the
castrated adult male rat transplanted with ovaries nor the
adult female neonatally treated with testosterone has a spontaneous trigger for inducing ovulation, but in both cases
electrical stimulation of the preoptic area can induce ovulation.
Also supporting the view that the sexual difference lies at some
level between the amygdala and the preoptic area is the observation by Velasco and Taleisnik (1969) that stimulation of the
preoptic area causes a rise in plasma luteinising hormone
(assayed by the ovarian ascorbic acid depletion method in both
the male and the female, but that stimulation in the amygdala
causes a rise in the female only.

All the observations have so far dealt with adult rats.
The adult female pattern of gonadotrophin release (i.e. the
ability to produce a cyclic preovulatory surge of gonadotrophins)
is not solely determined by the genetic sex of the animal but
depends upon the presence or absence of androgens (or oestrogen)
during a critical period of development, which in the rat includes the first two weeks after birth (Harris, 1964). The adult
male pattern (i.e. the absence of a cyclic surge of gonadotrophins)
occurs in the normal intact male (which is exposed to the action
of androgens secreted by its own testes) or in the female treated
during the critical postnatal period with testosterone. Conversely, the adult female pattern - i.e. the ability to elicit a
periodic ovulatory surge of gonadotrophins - occurs either in the
female or in the genetic male which has been castrated at birth
and is therefore not exposed to androgens from its own testes. In
such a neonatally castrated male, the inhibitory effect of androgens on the development of the cyclic neural trigger for ovulation
can be demonstrated by showing that treatment with testosterone
(i.e. androgen replacement) during the postnatal period can also
prevent development of the adult female pattern. That the crucial
difference between the male and female patterns lies in the
central nervous system has been established by experiments in
which the pituitary gland and gonads have been transplanted (Harris
and Campbell, 1966), and the evidence of lesion experiments (such
as those quoted above) localises this difference to the preoptic
area. This implies that the part of the brain which is acted upon
by androgens during the critical postnatal period is in fact the
preoptic area. Direct support for such a contention is offered

by the findings of Clayton et al. (1970) who have shown that in
the neonatal female rat administration of testosterone causes a
general depression of the uptake of radioactive uridine in all
brain areas except the preoptic area and the medial amygdala
(areas which are linked by the stria terminalis). In order to
assess the effects of neonatal hormonal manipulations upon the
neuropil of the preoptic area, we have embarked on a series of
neuropil analyses in adult females which have been treated with
androgen during the postnatal period, and in adult males castrated
at birth. Observations are as yet only available for a preliminary
group of adult female rats which were treated with 1.25 mg of
testosterone on the fourth postnatal day, and which were anovulatory
at the time of sacrifice. These suggest that the shaft/spine ratios
of these rats are indeed somewhat higher than in the control
females, although rather lower than that found in most of the males.
This conclusion must remain tentative until further material becomes available.

The basic difference in the preoptic area appears to be that
the female possesses more synapses upon dendritic spines than does
the male. This conclusion is at present only suggestive, as the
shaft/spine ratio of course only measures the relative numbers of
the two types of synapses, so that the same result could be achieved
if the male had more shaft synapses. A definitive answer to this
problem may be forthcoming from the parallel Golgi studies which are
at present under way in this laboratory. The suggestion that
dendritic spines may be a modifiable feature of the neuron is in
agreement with observations on the visual cortex of neonatally
light deprived animals (Globus and Scheibel, 1967; Valverde, 1967)
which indicate that the adult pattern of dendritic spines may be
altered by manipulations of specific afferent (in this case
visual) input during a critical postnatal period.

Putting together the above observations, it may be helpful to
outline the sort of working hypothesis upon which we are currently
designing further experiments, although accepting that this is at
best only a tentative model of events. Firstly, it is assumed
that in the adult the neural mechanism for the cyclic preovulatory
surge of gonadotrophins either resides in or is intimately involved
with the neuropil of that part of the preoptic area which receives
connections through the stria terminalis. In the neonate, it is
proposed that these cells are sensitive to circulating gonadal
steroid hormones, and that this sensitivity is manifested in a
permanent modification of the ultimate pattern of spine synapse
development. In the adult, a comparable sensitivity to gonadal
steroids is reflected in the effects of gonadal steroids upon
the timing of ovulation, and upon the initiation of mating
behaviour.

ACKNOWLEDGMENTS

This work was supported by the Medical Research Council
(G 970/668/B) and the Foundations Fund for Research in Psychiatry
(70-472).

REFERENCES

BARRACLOUGH, C. A. Modifications in reproductive function after
    exposure to hormones during the prenatal and early postnatal
    period. In L. Martini and W. F. Ganong (Eds.), Neuroendo-
    crinology, Vol. 2. New York: Academic Press, 1967.
    Pp. 61-99.

CLAYTON, R. B., KOGURA, J., & KRAEMER, H. C. Sexual differentia-
    tion of the brain: effects of testosterone on brain RNA
    metabolism in newborn female rats. Nature, 1970, 226,
    810-811.

CRITCHLOW, B. V., & BAR-SELA, M. E. Control of the onset of
    puberty. In L. Martini and W. F. Ganong (Eds.), Neuro-
    endocrinology, Vol. 2. New York: Academic Press, 1967.
    Pp. 101-162.

EVERETT, J. W. Neuroendocrine aspects of mammalian reproduction.
    Annual Review of Physiology, 1969, 31, 383-416.

GLOBUS, A., & SCHEIBEL, A. B. The effect of visual deprivation
    on cortical neurons: A Golgi study. Experimental Neurology,
    1967, 19, 331-345.

HALÁSZ, B. The endocrine effects of isolation of the hypothalamus
    from the rest of the brain. In W. F. Ganong and L. Martini
    (Eds.), Frontiers in Neuroendocrinology. New York: Oxford
    University Press, 1969. Pp. 307-342.

HARRIS, G. W. Sex hormones, brain development and brain function.
    Endocrinology, 1964, 75, 627-648.

HARRIS, G. W., & CAMPBELL, H. G. The regulation of the secretion
    of luteinizing hormone and ovulation. In G. W. Harris and
    B. T. Donovan (Eds.), The Pituitary Gland, Vol. 2. London:
    Butterworths, 1966. Pp. 99-165.

HEIMER, L., & NAUTA, W. J. H. The hypothalamic distribution of
    the stria terminalis in the rat. Brain Research, 1969,
    13, 284-297.

KALRA, S. P., & SAWYER, C. H. Blockade of copulation-induced
    ovulation in the rat by anterior hypothalamic deafferenta-
    tion. Endocrinology, 1970, 87, 1124-1128.

RAISMAN, G. An evaluation of the basic pattern of connections
    between the limbic system and the hypothalamus. American
    Journal of Anatomy, 1970, 129, 197-202.

RAISMAN, G., & FIELD, P. M. Anatomical considerations relevant
    to the interpretation of neuroendocrine experiments. In
    L. Martini and W. F. Ganong (Eds.), Frontiers in Neuro-
    endocrinology, Vol. 2. New York: Oxford University Press,
    1971, in press.

STUMPF, W. E. Estrogen-neurons and estrogen-neuron systems in
    the periventricular brain. American Journal of Anatomy,
    1970, 129, 207-218.

TALEISNIK, S., VELASCO, M. E., & ASTRADA, J. J. Effect of hypo-
    thalamic deafferentation on the control of luteinizing
    hormone secretion. Journal of Endocrinology, 1970, 46, 1-7.

VALVERDE, F. Apical dendritic spines of the visual cortex and
    light deprivation in the mouse. Experimental Brain Research,
    1967, 3, 337-352.

VELASCO, M. E., & TALEISNIK, S. Release of gonadotropins induced
    by amygdaloid stimulation in the rat. Endocrinology, 1969,
    84, 132-139.

SOME ASPECTS OF THE STRUCTURAL ORGANIZATION

OF THE AMYGDALA

Elizabeth Hall

Department of Anatomy

University of Ottawa, Canada

INTRODUCTION

Historically it has been the cytoarchitectural studies of the central nervous system which have provided the foundation upon which concepts concerning the structural organization of specific nuclear regions have been built. In most areas such investigations have included the study not only of Nissl but also of Golgi preparations. In this regard the amygdala must be considered an exception, as investigators have relied almost solely on the Nissl method in determining its nuclear subdivisions.

Unfortunately, the amygdala does not lend itself easily to this approach, firstly because of the differences encountered from species to species (for details see Koikegami, 1963), and secondly because of the transitional zones occurring between adjacent nuclei. The latter feature especially allows a subjective quality to enter into the description of the amygdala, because one may describe many or few subdivisions according to the significance one attaches to minor variations in the size or intensity of the staining of a particular group of cells. A striking example of differences in interpretation may be seen in the descriptions of the amygdala in the guinea pig given by Uchida (1950b) and Johnson (1957). The seven subdivisions of the amygdala superficialis described by the former author correspond to two nuclei, the cortical and the medial, described by the latter.

The differences of opinion concerning the number of nuclei and subnuclei within the amygdala are reflected in the two terminologies most commonly employed. In general it can be said

95

that the Japanese investigators describe more subdivisions and base their terms on those of the early German scientists, while most European and North American workers describe fewer nuclei and follow the terminology of Johnston (1923)[1].

The inconsistencies of the anatomical descriptions based on the study of Nissl preparations have forced both physiologists and psychologists to take a rather simplified view of the organization of the amygdala which in turn has prevented a precise correlation of their results with specific nuclei or subnuclei. Obviously, it would be a great advantage if agreement could be reached regarding these smaller structural units of the amygdala. It would seem essential, therefore, to bring together the results of Nissl and Golgi studies, chemoarchitectural investigations, and experiments on the connections of the amygdala in the hope that this approach might lead to a greater uniformity of concepts regarding the structural units of the amygdala than the study of Nissl alone.

It will be seen that each of these techniques in fact yields different groupings of the nuclear subunits, so that several different patterns of organization emerge. It may be that within these different groupings lie important clues as to the overall structural and functional organization of the amygdala.

CYTOARCHITECTURE

a) Nissl Stain:

Many articles have appeared concerning the nuclei of the amygdala as determined by Nissl stains, including a comprehensive review of the amygdala in mammals, birds and reptiles by Koikegami (1963). Thus, it is not the intention of the present author to provide a detailed description of each nucleus, but rather to focus attention on the heterogeneity of certain regions, specifically the lateral, basal, cortical and central nuclei, and provide a background for the consideration of results obtained by other histological methods.

[1] Johnston (1923) divided the amygdala into basolateral and cortico-medial groups. The first consisted of the lateral nucleus and of the large and small-celled parts of the basal nucleus; the second was composed of the cortical, medial and central nuclei and of the nucleus of the lateral olfactory tract. Later investigators, beginning with Gurdjian (1928), divided the central nucleus of Johnston (1923) into two parts, one of which retained the name central nucleus while the other was called the anterior amygdaloid area. On the basis of their anatomical position, the anterior amygdaloid area and the nucleus of the lateral olfactory tract were brought together by some authors to form the anterior group of nuclei. (Footnote continues on page 98.)

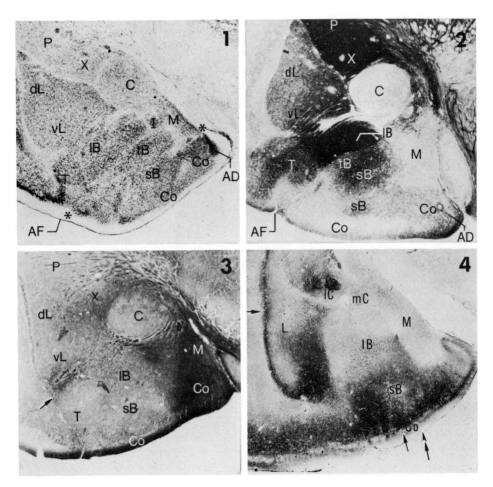

Figs. 1 - 3 illustrate approximately the same frontal level of the amygdala of the guinea pig stained with the Nissl, AChE and MAO methods respectively (from Hall and Geneser-Jensen, 1971).

Fig. 4. Frontal section of the amygdala of the cat stained by the Timm (1958) method (from Hall et al., 1969).

The lateral nucleus lies immediately ventral to the putamen
and medial to the external capsule (Fig. 1). Its internal
structure is not constant from species to species. In some
animals it has been considered a homogeneous mass (the opossum,
Johnston 1923; the bat, Humphrey 1936; the cat, Fox 1940; man,
Crosby and Humphrey 1941; the shrew, Crosby and Humphrey 1944;
the monkey, Lauer 1945). In others, differences in size and/or
density of the cell population have led to a subdivision of the
nucleus into two and, occasionally, three parts (the rat, Gurdjian
1928, Brodal 1947, and Uchida 1950a; the rabbit, Young 1936 and
Uchida 1950b; the mink, Jeserich 1945; the guinea pig, Uchida
1950b and Johnson 1957). Recently Hall and Geneser-Jensen (1971)
have confirmed that the lateral nucleus of the guinea pig consists
of a dorsal small-celled and a ventral larger-celled part (Fig. 1)
and, in agreement with Johnson (1957), they noted that the two
areas are not separated by a sharp border.

Koikegami (1963) and his co-workers also subdivided the
lateral nucleus not only in some but in all of the many animals
they investigated, and in the monkey they described as many as
seven parts. Further, Koikegami (1963) suggested that the dorsal
tip of the lateral nucleus is not simply a part of the dorsal subnucleus but is a separate entity in a number of species.

The basal nucleus lies between the lateral and medial nuclei
and inferior to the central nucleus (Fig. 1). Its large and small-
celled subdivisions are much more distinct than those of the

Footnote 1 Cont.

On comparing these three groups of nuclei with those of
Uchida (1950a, 1950b) one finds that the basolateral complex has
an exact equivalent in the amygdala propria, a group of cells
divided into lateral, intermediate and medial nuclei which are
comparable to the lateral nucleus and the large and small-celled
parts of the basal nucleus respectively. In addition, the amygdala
superficialis of Uchida (1950a, 1950b) corresponds to the corticomedial complex in that it consists of several subdivisions that
are equivalent to the cortical and medial nuclei. However, the
third main component of his amygdala superficialis is the nucleus
of the lateral olfactory tract rather than his equivalent of the
central nucleus. The latter, together with the anterior amygdaloid
area form Uchida's (1950a, 1950b) supraamygdala.

Koikegami (1963) refers to the nuclei of the basolateral
complex as the lateral, intermediate and medial principal nuclei
and subdivides each of them into several subunits. His term for
the medial nucleus is medial superficial nucleus and for the
central nucleus, dorsal central nucleus. However, he has adopted
the term cortical nucleus.

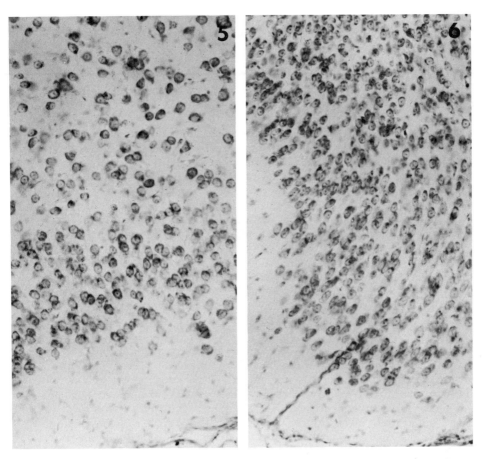

Fig. 5.   Lateral part of the cortical nucleus of the guinea pig.
Nissl Stain.   x 120.

Fig. 6.   Medial part of the cortical nucleus of the guinea pig.
Nissl Stain.   x 120.

lateral nucleus and in the terminology of Koikegami (1963) the
two parts are separate nuclei within the deep or principal cell
mass of the amygdala. In attempting to correlate his additional
subdivisions of these nuclei with those of other authors,
Koikegami (1963) made the important observation that there is
frequently a graded change in the size of cells in the junctional
region between the large and small-celled parts of the basal
nucleus and that in some cases this group of neurons is sufficiently circumscribed to be considered a special subnucleus.
Leaving aside considerations regarding terminology, Hall and
Geneser-Jensen (1971) have supported this concept in their study
of the amygdala in the guinea pig, where they observed a transitional zone between the large-celled and the proper small-celled
part of the basal nucleus (Fig. 1).

The cortical nucleus lies superficial to the basal nucleus
and extends from the amygdaloid fissure to the medial nucleus
(Fig. 1). It has been subdivided only occasionally by investigators outside the Japanese school. Young (1936), for example,
subdivided it into superficial and deep parts in the rabbit. More
recently Hall and Geneser-Jensen (1971) have noted that the
cortical nucleus of the guinea pig can be divided into lateral
and medial regions. The former is similar to the adjacent pyriform cortex, presenting a layer of relatively compact cells and a
deeper layer of more scattered cells (Fig. 5). The latter is not
organized so distinctly into layers coursing parallel to the pial
surface. Instead, the neurons of this region are often arranged
into irregular columns that appear to be aligned parallel to
fibers of the stria terminalis (Fig. 6)[2].

The central nucleus lies dorsally in the amygdala, bounded
superiorly by the globus pallidus and laterally by the putamen
(Fig. 1). In most species it consists of a homogeneous group of
relatively small cells which are similar in appearance to those
of the putamen. However, in a few species (the cat, Fox 1940;
the mink, Jeserich 1945; the rat, Brodal 1947), this same group
of small cells has been called the lateral part of the central
nucleus to distinguish it from a group of larger cells on its
medial aspect called the medial part of the central nucleus. Hall
and Geneser-Jensen (1971) noted a similar group of larger cells
inferomedial to the central nucleus of the guinea pig but considered it a posterior extension of the anterior amygdaloid area rather
than a medial subdivision of the central nucleus. The central
nucleus of the guinea pig like that of other rodents stands out

---

[2] These two parts of the cortical nucleus correspond, according to
Hall and Geneser-Jensen (1971), to the periamygdaloid cortical
regions PAM 2 and PAM 3 of Rose (1929).

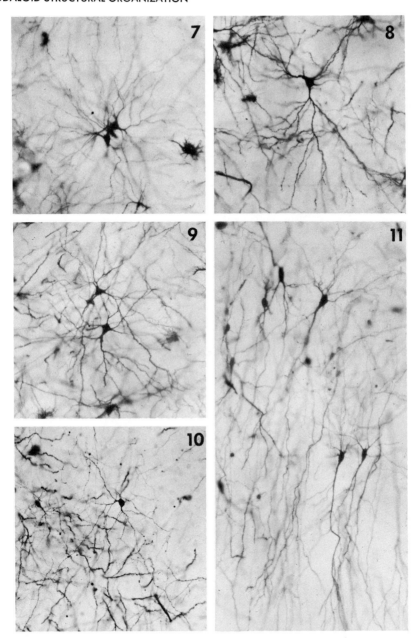

Figs. 7-14. Golgi preparations of the amygdala of the cat. Fig.
7 - Two type P cells in the lateral nucleus; horizontal section.
Fig. 8 - Type P cell in the magnocellular part of the basal nucleus; horizontal section. Fig. 9 - Type P cell in the parvocellular part of the basal nucleus; horizontal section. Fig. 10 -
Type S cell in the lateral nucleus; frontal section. Fig. 11 -
Pyramidal cells in the medial part of the cortical nucleus;
frontal section. All figures magnified 130 times.

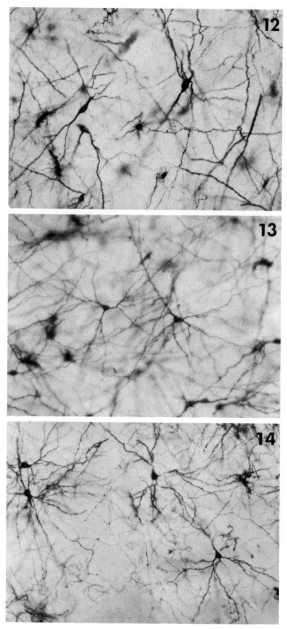

Fig. 12.   Cells in the medial part of the central nucleus.
Horizontal Section.   x 150.

Fig. 13.   Cells in the lateral part of the central nucleus.
Sagittal Section.   x 150.

Fig. 14.   Cells in the putamen.   Note the similarity of the
cells in Figs. 13 and 14.   Sagittal Section.   x 150.

clearly at most levels as a circular group of small cells
separated from the adjacent putamen by a cell-poor zone (Fig. 1).
It should be noted that the portion of the putamen immediately
lateral to the cell-poor zone appears different from the rest of
the putamen in that its cells are not broken up into clusters or
irregular columns as there are no large bundles of fibers traversing it. This region has been designated area X by Hall and
Geneser-Jensen (1971).

In other species, the absence of the cell-poor zone noted
above makes the separation of the central nucleus from the putamen
extremely difficult. Johnston (1923) noted that the putamen is
much richer in myelinated fibers than the central nucleus and used
this feature to determine the boundary between them. Although Fox
(1940) did not accept this definition unreservedly, it has been
generally accepted by investigators using the cat as their experimental animal. It is possible, therefore, that the small-celled
lateral part of the central nucleus of the cat may correspond
not only to the central nucleus of the guinea pig, but also to
area X.[3]

b) Golgi Stain:

Very few reports have been published concerning the cytoarchitecture of the amygdaloid nuclei as seen in Golgi preparations. Gurdjian (1928) in his investigation of the forebrain of
the rat presented a few drawings of cells from some of the amygdaloid nuclei and confirmed earlier observations regarding the
origin of the stria terminalis. Valverde (1962, 1963) presented
brief descriptions of the cell types in the amygdala of the mouse,
including a few comments on their dendritic arborizations, and
later (1965) published data on both the pyriform cortex and the
amygdala, bringing together observations on the mouse, rat and,
to a lesser extent, the cat. The main emphasis of Valverde's
work, however, was on axons and their collaterals. More recently,
the present author (1971) has investigated the amygdaloid nuclei
of the cat employing the Ramon-Moliner (1958) modification of the
Golgi-Cox method.

In the cat, the cells of the lateral and basal nuclei can be
divided into two main types on the basis of their dendritic
arborization and the size of their somata. The more common of

---

3 In his illustration of the amygdala of the cat, Koikegami
(Fig. 12, 1963) distinguishes medial and lateral parts of the
central nucleus which appear to correspond to the medial part of
the central nucleus and intercalated mass of other authors. An
area roughly comparable to the lateral part of the central nucleus
is labelled caudate-putamen nucleus.

the two bears some resemblance to a pyramidal cell in that it
presents three to five parent stem dendrites of medium caliber,
and one of larger caliber and of slightly greater length which
is reminiscent of an apical dendrite (Figs. 7, 8, 9). It will
be referred to as type P. Together, these primary dendritic
branches influence the shape of the soma to a greater or lesser
degree so that the cell body appears triangular or pyriform in
shape. The medium-sized primary dendrites usually divide into
two secondary branches of equal diameter close to the cell body
and these in turn further subdivide. The thick "apical" dendrite
often gives off one or more fine branches before dividing into
two of equal diameter. There is a moderate number of spines on
the dendritic arborization.

The second type of cell seen in the two nuclei (Fig. 10) is
smaller in size and is scattered sparsely amongst the type P
neurons. It will be referred to as type S because of its similarity to the stellate cell of the cortex. Although the primary
dendrites of these cells may be as numerous as those of the type P
neuron, they are of much finer caliber and do not alter the basically oval to round shape of the small soma. In addition, they
rarely undergo more than two subdivisions. Usually the dendrites
are beaded and have virtually no spines.

The cortical nucleus displays a somewhat greater variety of
neuronal configurations, but the majority can be classified as
pyramidal, modified pyramidal or stellate cells. Laterally they
are organized into relatively distinct superficial and deep
layers, but, unlike cells of the cortex, the pyramidal cells are
disposed at any angle to the pial surface. Medially, the layers
are less clear-cut but the apical dendrites of the pyramidal cells
are rigidly orientated in parallel with the stria terminalis
(Fig. 11).

One of the most striking features observed on scanning from
the lateral part of the cortical nucleus through the basal into
the lateral nucleus is the gradual modification of the typical
pyramidals of the former to the P cell of the latter. The modification of the pyramidal type of cell is so gradual and the dendritic trees are so intermingled that the individual nuclei cannot be
determined with certainty unless reference is made to surrounding
structures. Even the transitions in the size of the somata are
of relatively little assistance.

Neurons in the two parts of the central nucleus of the cat
are not only quite different from those of the other three nuclei,
but are also quite different from each other. The larger cells
of the medial division (Fig. 12) give off two to four primary
dendrites which undergo only one or two divisions. These extend

in such a manner as to give the somata either a triangular or fusiform shape. A moderate number of spines are present on the dendritic tree. Anteriorly these cells extend into the anterior amygdaloid area which is composed of similar, but on the whole slightly smaller, neurons. It is difficult to establish the limits of these two regions from each other.

Cells in the lateral part of the central nucleus are identical with those of the putamen and caudate nucleus (compare Fig. 13 with Fig. 14). The cell body is relatively small and is round to oval in shape. As many as six fine primary dendrites may arise from the cell body and these are disposed more or less evenly around and away from the cell body except where they border the lateral nucleus. In this region the dendrites often emerge from opposite poles of the soma and display themselves parallel to the line of nuclear apposition. All the primary dendrites undergo at least one and often three or four divisions. The outstanding characteristic of the neurons is their dendritic spine population which is the heaviest of all the cell types in the amygdala.

Thus on the basis of observations presented above, one could suggest that the lateral, basal and cortical nuclei be grouped together because of the similarity of their cell types and their apparent continuity with each other. On a similar basis one might group the medial part of the central nucleus with the anterior amygdaloid area. The lateral small-celled part of the central nucleus must be considered unique amongst the amygdaloid nuclei in its similarity to the striatum.

CHEMOARCHITECTURE

Enzyme stains provide additional methods by which the heterogeneity of the amygdaloid nuclei can be determined.

a) <u>Acetylcholinesterase</u> Stain.

A number of investigators who have carried out surveys on the location of acetylcholinesterase (AChE) in the central nervous system have commented briefly on the differential distribution of this enzyme within the amygdala (Koelle, 1954; Gerebtzoff, 1959; Shute and Lewis, 1963; Krnjevic and Silver, 1965; Ishii and Friede, 1967). More detailed information has been presented by De Giacomo (1960), Girgis (1967, 1968, 1969), Yu (1969), and Hall and Geneser-Jensen (1971).

The last authors observed that the lateral nucleus of the guinea pig does not stain in a uniform manner. Ventromedially, the reaction was of moderate intensity while dorsolaterally it was weaker (Fig. 2). It is rather surprising that these two areas

could not be correlated precisely with those identified in the
Nissl preparations. Girgis (1969) reported a similar distribution
of AChE within the lateral nucleus of the Galago. However, in
the Grivet monkey the lateral nucleus was unstained and in the
coypu rat it was the dorsolateral rather than the ventromedial
region which appeared darker (Girgis 1967, 1968). The last
finding was also reported by Yu (1969) in his study of the
amygdala in the rat.

There is almost universal agreement that the large-celled
part of the basal nucleus is stained very intensely by the AChE
method (Fig. 2). Further, Girgis (1969) and Hall and Geneser-
Jensen (1971) have noted that the dark staining is present in
both the somata and neuropil. The small-celled region has been
described as unstained in the Grivet monkey (Girgis, 1968) and
weakly stained in the rat (Yu, 1969). In the guinea pig, however,
this subnucleus reacts differently (Fig. 2), the proper small-
celled part staining moderately and the transitional zone appearing distinctly paler (Hall and Geneser-Jensen, 1971).

Girgis (1967, 1968, 1969) reported a slight staining reaction
in the superficial part of the molecular layer of the cortical
nucleus in the coypu rat and no staining at all in the Galago and
the monkey. Hall and Geneser-Jensen (1971) also observed that
the reaction of the cortical nucleus was weak but reported a
slight difference between the lateral and medial segments, the
former appearing slightly paler than the latter (Fig. 2). Thus,
their findings lend modest support to the subdivision they describe
in Nissl and Golgi preparations.

The small-celled central nucleus shows an extremely low level
of AChE in the rat (Yu, 1969), the Galago (Girgis, 1969), the
Grivet monkey (Girgis, 1968) and the guinea pig (Hall and Geneser-
Jensen, 1971). However, it has been described as moderately
stained in the coypu rat (Girgis, 1967) and the cat (Krnjevic
and Silver, 1965). In the guinea pig, the area X observed by
Hall and Geneser-Jensen (1971) stains as intensely as the putamen
(Fig. 2), a point in favour of considering it a part of that
nucleus rather than a lateral extension of the central nucleus
in this species.

On the basis of the intensity of the AChE staining in the
guinea pig, one could divide the subnuclei into a weakly stained
group, consisting of the dorsolateral part of the lateral nucleus,
the transitional zone of the basal nucleus and the cortical
nucleus, and a moderately stained group consisting of the ventromedial part of the lateral nucleus and the small-celled part of
the basal nucleus. The large-celled part of the basal nucleus
and area X must be considered unique amongst the regions under

consideration in that it stains as intensely as the striatum.

b) Monoamine Oxidase Stain:

Few reports have appeared on the distribution of monoamine oxidase (MAO) in the amygdaloid complex. These consist primarily of brief comments by Shimizu et al. (1959) and the more specific descriptions of Hashimoto et al. (1962), Manocha et al. (1967) and more recently Hall and Geneser-Jensen (1971). The last authors found no significant differences in the distribution of monoamine oxidase throughout the lateral and basal nuclei (Fig. 3) so that with this technique, as well as the Golgi method, the basolateral complex appears more homogeneous than in Nissl preparations.

In the cortical nucleus, however, these authors observed a marked variation in the intensity of the stain that corresponded to the two segments they identified in Nissl preparations. Laterally, the reaction was weak and of about the same intensity as that of the basolateral complex. Medially, it was much more intense (Fig. 3).

The central nucleus appeared pale while the adjacent area X was darker than either the central nucleus or the putamen (Fig. 3).

Thus, the results obtained with the monoamine oxidase stain suggest that both parts of the lateral and basal nuclei, the lateral part of the cortical nucleus and the central nucleus could be grouped together as weakly stained areas. The medial part of the cortical nucleus which reacts much more intensely would belong in a subdivision that included the medial nucleus. Area X would then fall into an intermediate group.

c) Dithizone and Timm Stains:

Fleischhauer and Horstmann (1957) and Koikegami (1963) observed that the amygdala gives a positive reaction with the dithizone method and Hirata (cited by Koikegami, 1963) commented briefly on the differential staining obtained with the silver sulphide method of Timm (1958). Both these techniques are considered to demonstrate the presence of zinc, and Haug (1967) has shown that the particles precipitated in the hippocampus by a modified Timm procedure are located in terminal boutons. It appears relevant, therefore, to review some aspects of the study carried out by Hall et al. (1969) on the amygdala of the cat employing these two techniques.

With both methods a differential staining was observed in the lateral and basal nuclei. In the lateral nucleus there was a

ventromedial region of intense staining that was continuous on its
lateral aspect with a dark narrow band bounding the whole lateral
convexity of the nucleus (Fig. 4). This band extended beyond
the most lateral cell somata and thus encroached upon the external
capsule. As the dendrites of the type P cells project into this
area it was considered that the stain was due to "boutons de
passage" or short collaterals from ascending fibers synapsing in
this region. The rest of the dorsolateral region of the lateral
nucleus gave a very weak staining reaction.

In the basal nucleus, the large-celled area was very pale,
the dorsal part of the small-celled area slightly darker, and the
ventral part of the small-celled area darkest. Although no
transitional zone was reported in the basal nucleus of the cat by
Fox (1940) the differential staining reaction suggests that such
an area may exist in this species. It should also be noted that
the observations described above do not correlate with the subdivisions of either the lateral or basal nuclei as illustrated
by Koikegami (Fig. 12, 1963) in the cat.

Although no difference in staining was reported within the
cortical nucleus, it is of interest to note that the intensity of
the stain in this region was similar to that in the adjacent
small-celled part of the basal and much more intense than that in
the medial nucleus (Fig. 4).

Some difficulty was encountered in interpreting the exact
location of a densely stained region dorsal to the basolateral
complex. It appeared to be located predominantly in the lateral
part of the central nucleus but did not occupy the whole of this
region and there was some question as to whether or not it extended into the adjacent putamen. One might speculate on the possibility of this darkly-stained region being homologous to area X
in the guinea pig.

Thus, on the basis of the dithizone and Timm stains the
lateral rim and ventromedial part of the lateral, the ventral part
of the small-celled region of the basal, the cortical and part of
the lateral subdivision of the central nucleus could be grouped
together as intensely stained regions, while the dorsolateral
part of the lateral, the large-celled part of the basal and the
medial subdivision of the central could be brought together as
weakly stained areas. The dorsal region of the small-celled
basal nucleus would occupy an intermediate position between the
two groups.

AFFERENT CONNECTIONS OF THE AMYGDALA

a) Neocortical Afferents:

A number of investigators have reported that the amygdala receives afferents from the temporal cortex. Whitlock and Nauta (1956) observed that the projection of the inferior temporal gyrus to the amygdala was distributed mainly to the basolateral complex with a few fibers going to the central nucleus. Conversely, Powell et al. (1965) found no neocortical projections coursing to the amygdala of the rat. Recently, Lescault (1969, 1971) and Druga (1969) have reported that the anterior and posterior sylvian gyri project primarily to the dorsolateral part of the lateral nucleus with few if any fibers terminating ventromedially. Lescault (1971) has also noted that the anterior and posterior ectosylvian gyri project to this area, while the orbital gyrus projects only to the more ventromedial segment. Thus, the neocortical projections uphold a division of the lateral nucleus in the cat, even though two parts have not been identified in Nissl preparations in this particular species. In addition, these two authors observed degenerating preterminals in the large-celled part of the basal nucleus, although there are discrepancies regarding the exact site of origin of these fibers. Druga (1969) found them only when he made lesions of the anterior sylvian gyrus, while Lescault (1969, 1971) observed such fibers only when he made a large lesion in the posterior ectosylvian and posterior sylvian gyri.

Neither of these investigators reported neocortical projections to either the small-celled part of the basal nucleus or the cortical nucleus.

In agreement with the observation of Whitlock and Nauta (1956), Druga (1969) and Lescault (1969, 1971) described preterminal degeneration in the small-celled lateral part of the central nucleus. With the exception of auditory cortex (Lescault, 1971) these fibers arose from the same neocortical areas they each described projecting to the lateral nucleus. In addition, the illustrations of Lescault (1971) indicate that the degeneration is heaviest in the lateral extreme of this subnucleus.

Preliminary electron microscopic observations have verified Lescault's findings regarding the temporal cortical projections (Hall and Prym, 1971). Following large lesions involving this region, degenerated axons and terminal boutons were identified in the dorsolateral part of the lateral nucleus, the lateral part of the central nucleus and the magnocellular part of the basal nucleus (Figs. 15-18), where they were abundant, less numerous and rare respectively. The degenerating profiles were frequently still in contact with the post-synaptic structures which were ex-

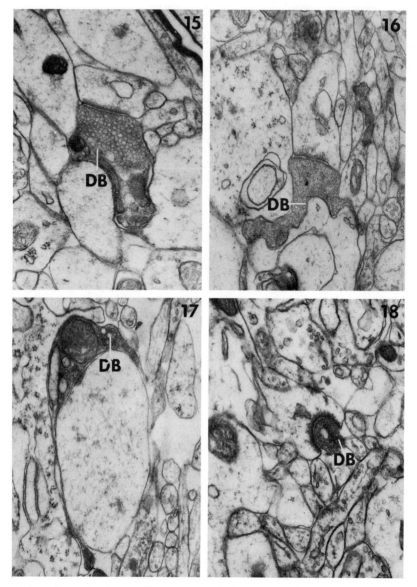

Fig. 15.   Bouton in early stage of degeneration.   Lateral
nucleus.   Survival time 3 days.   x 30,000.

Fig. 16.   Bouton in more advanced stage of degeneration.
Lateral nucleus.   Survival time 3 days.   x 30,000.

Fig. 17.   Degenerating bouton in the magnocellular part of the
basal nucleus.   Survival time 7 days.   x 30,000.

Fig. 18.   Degenerating bouton in the lateral part of the
central nucleus.   Survival time 7 days.   x 30,000.

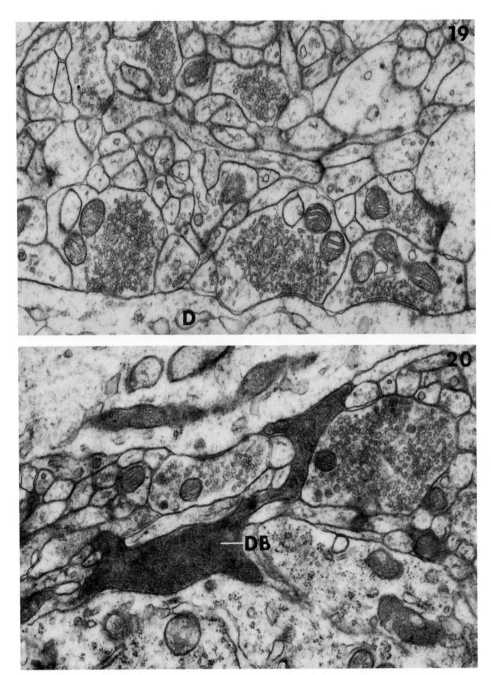

Fig. 19. Boutons containing flat vesicles making symmetrical synaptic contact with a dendritic shaft. Normal central nucleus.

Fig. 20. Degenerating terminal containing flat vesicles. Note absence of post-synaptic thickening. Medial part of the central nucleus. Survival time 5 days. Both figures x 30,000.

clusively dendritic spines or small to medium-sized dendrites
(Figs. 15-18). There was usually a distinct postsynaptic thickening (Figs. 15-18) and the synaptic vesicles, when still distinguishable, were usually round to ovoid in shape (Fig. 15). It
would appear, therefore, that the terminals from the temporal
cortex are most probably of the B_1 or B_2 type described by Hall
(1968) in the normal lateral nucleus.

b) Subcortical Afferents:

Some data are also available concerning the subcortical
afferents to the amygdala. Nauta (1958) noted that following a
lesion in the lateral preoptic region of the cat degeneration
could be identified in the anterior, medial, central and basal
amygdaloid nuclei, but none reached the lateral or the cortical
nucleus. Cowan et al. (1965) reported similar results following
lesions of the preoptic nucleus in the rat. However, they
observed little if any degeneration in the central nucleus, and
a small amount in the lateral nucleus.

In a continuing light and electron microscopic study of the
subcortical amygdaloid afferents in the cat, Wakefield (1971) has
placed a series of lesions in the lateral and medial preoptic
areas and throughout the hypothalamus. She has found that in the
cat as in the rat (Cowan et al., 1965) only lesions of the lateral
preoptic region give rise to degeneration in the amygdala. In
agreement with Nauta's (1958) report, she observed degeneration
in the basal, central and medial nuclei. Neither the Nauta (1957)
nor the Fink and Heimer (1967) techniques provided convincing
evidence of degeneration in the lateral nucleus. It is of special
interest therefore that with the electron microscope she not
only confirmed the presence of degenerating boutons in the above
three nuclei, but also observed them in the lateral nucleus as
well.

In her initial examination of the normal material, Wakefield
(1971) noted that in both the lateral and medial parts of the
central nucleus a high proportion of boutons contained flat
vesicles (Fig. 19) and formed symmetrical synaptic contacts
(Colonnier, 1968) primarily with dendritic shafts and sometimes
with somata. Following lesions of the lateral preoptic region
it was mainly these boutons which underwent degeneration (Figs. 20,
22-24) although occasionally a degenerating type B_1 or B_2 bouton
was also observed (Fig. 21).

Thus, in regard to the four nuclei under consideration here,
it is the dorsolateral part of the lateral, the magnocellular
part of the basal and the lateral part of the central nucleus
which receive neocortical afferents and both parts of the basal
and central nuclei and the lateral nucleus which receive sub-

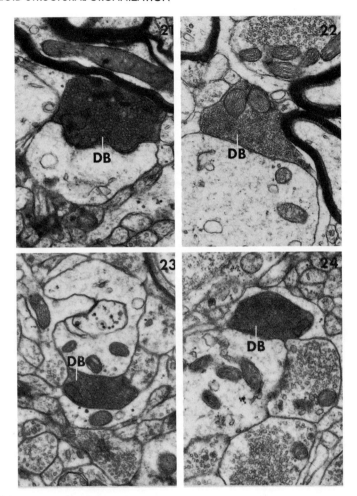

Fig. 21. Degenerating bouton containing round vesicles. Note post-synaptic thickening. Medial part of the central nucleus. Survival time 5 days. x 30,000.

Fig. 22. Degenerating bouton containing flat vesicles. Medial part of the central nucleus. Survival time 5 days. x 30,000.

Fig. 23. Degenerating bouton containing flat vesicles. Lateral part of the central nucleus. Survival time 7 days. x 30,000.

Fig. 24. Degenerating bouton containing flat vesicles. The pre- and post-synaptic membranes are cut obliquely except at the extreme left limit of the synaptic contact. Lateral part of the central nucleus. Survival time 7 days. x 30,000. (Figs. 19-24 courtesy of Wakefield, 1971)

cortical afferents. As yet there is no convincing evidence that the cortical nucleus receives fibers from either of these two sources.

CONCLUSIONS

Throughout the foregoing presentation, two main problems have been considered: firstly, whether each of the four nuclei is a homogeneous unit or whether it can be divided into subnuclei; secondly, whether certain of the nuclei or subnuclei can be grouped together on the basis of structural similarities. One might argue that if the amygdala does not lend itself easily to such an anlysis with the Nissl technique, neither can it be sub-divided readily in a uniform manner with other methods. Never-theless, it is the opinion of the author that certain conclusions can be drawn which may prove useful in future investigations of this complex region.

Observations made employing the Golgi technique suggest that the lateral, basal and cortical nuclei form a relative homogeneous mass in which the nuclear boundaries are indefinite. These find-ings emphasize Johnston's (1923) statement that the three nuclei share a common origin and may indicate that they process data in a similar manner.

The monoamine oxidase stain was relatively homogeneous within the same group of nuclei with the exception of the medial part of the cortical nucleus which stained much more darkly and in this respect appeared more closely related to the medial nucleus and the anterior amygdaloid area (Hall and Geneser-Jensen, 1971).

With all other methods there was a degree of heterogeneity not only within the cortical nucleus, but within the lateral and basal nuclei as well. In some instances, the more detailed sub-divisions described in Nissl preparations receive support, in others the less extensive subdivisions are upheld.

On the basis of the acetylcholinesterase and zinc stains and the distribution of neocortical afferents it appears relatively certain that the lateral nucleus contains two structurally differ-ent areas. However, no evidence was obtained confirming Koikegami's (1963) suggestion that the dorsalmost tip of the nucleus is a separate entity. On the contrary, the cells of this area appear to form an integral part of the dorsal subdivision of the lateral nucleus. In addition, it must be emphasized that although the nucleus has two parts these are not necessarily homologous from species to species, as indicated by the differ-ences in the concentration of acetylcholinesterase and the dis-tribution of neocortical fibers from species to species.

It is tempting, on the basis of the acetylcholinesterase stain and more especially on the concentration of this enzyme within the perikarya, to consider the large-celled part of the basal nucleus as a separate nucleus. Indeed, it has more features in common with the dorsal part of the lateral nucleus in regard to the distribution of neocortical afferent fibers and its appearance in Timm (1958) stained sections than it has with the rest of the basal nucleus. For this reason the terminology of Koikegami (1963), which recognizes the large-celled part of the basal nucleus as a separate intermediate nucleus within the principal mass of the amygdala, appears appropriate. However, his additional subdivisions may fit more meaningfully into other nuclei, or alternatively should be recognized as distinct transitional areas which must be considered separate entities in themselves.

The small-celled part of the basal nucleus shares a number of features in common with the immediately adjacent lateral part of the cortical nucleus. However, as it stains moderately with acetylcholinesterase and since it has been proven to receive afferent fibers from the preoptic region, the present author considers it sufficiently different to be regarded as a separate structural unit.

From the observations presented above, the subdivision of the cortical nucleus into two parts would appear justified. Of particular significance is the orientation of the medial part of the cortical nucleus in parallel with the stria terminalis. Further work concerning the afferent and efferent projections of this nucleus is required before its relationship to the periamygdaloid cortex, the small-celled part of the basal amygdaloid nucleus and the medial nucleus can be clearly defined.

Finally, the central nucleus presents an interesting problem with regard to its homologous counterpart in different species. The comments which follow must be limited to the two species with which the author has personal research experience. In the guinea pig, Hall and Geneser-Jensen (1971) observed a group of cells immediately lateral to the central nucleus which, according to previous investigations (Uchida, 1950b; Johnson, 1957), belongs to the striatum. A similar group of cells in the cat is considered part of the lateral subdivision of the central nucleus as there is no cell-poor area between them.

The supposition that in the guinea pig this cellular region, or area X, is part of the striatum is supported by the fact that it stains intensely with the acetylcholinesterase method. However, it stains differently from both the striatum and central nucleus with the monoamine oxidase technique.

In the cat, it is the most lateral extreme of the central nucleus that receives the majority of neocortical afferent fibers, and in addition stains intensely with the Timm (1958) method.

On the basis of these data it is suggested that the lateral part of the central nucleus in the cat may be homologous with the central nucleus plus area X in the guinea pig. Further, one might speculate that this special group of cells is a transitional zone between the central nucleus and putamen that provides a bridge between the amygdala proper and the extrapyramidal system.

In conclusion, it would appear that the lateral, basal, cortical and central nuclei are more heterogeneous than usually reported by European and North American investigators but that they probably consist of fewer subdivisions than reported by the Japanese. Further, within this heterogeneity there is a dynamically shifting series of groupings and regroupings that emerges with the application of different techniques. On this basis, the author suggests that such terms as basolateral and corticomedial complex have outlived their usefulness. It would appear more appropriate to recognize the different patterns of nuclear and subnuclear grouping and attempt to determine whether these patterns have special significance for particular aspects of function within the broad concept of emotional expression.

REFERENCES

BRODAL, A. The amygdaloid nucleus in the rat. Journal of Comparative Neurology, 1947, 87, 1.

COLONNIER, M. Synaptic patterns on different cell types in the different laminae of the cat visual cortex. An electron microscope study. Brain Research, 1968, 9, 268.

COWAN, W. M., RAISMAN, G., & POWELL, T. P. S. The connexions of the amygdala. Journal of Neurology, Neurosurgery and Psychiatry, 1965, 28, 137.

CROSBY, E. C., & HUMPHREY, T. Studies of the vertebrate telencephalon. II. The nuclear pattern of the anterior olfactory nucleus, tuberculum olfactorium, and the amygdaloid complex in adult man. Journal of Comparative Neurology, 1941, 74, 309.

CROSBY, E. C. & HUMPHREY, T. Studies of the vertebrate telencephalon. III. The amygdaloid complex in the shrew (Blarina brevicauda). Journal of Comparative Neurology, 1944, 81, 285.

DRUGA, R. Neocortical projections to the amygdala (An experimental study with the Nauta method). Journal für Hirnforschung (Berlin), 1969, 11, 467.

FINK, R. P., & HEIMER, L. Two methods for selective silver impregnation of degenerating axons and their synaptic endings in the central nervous system. Brain Research, 1967, 4, 369.

FLEISCHHAUER, K., & HORSTMANN, E. Intravitale Dithizonfärbung homologer Felder der Ammonsformation von Säugern. Zietschrift für Zellforschung und Mikroskopische Anatomie (Berlin), 1957, 46, 598.

FOX, C. A. Certain basal telenchephalic centers in the cat. Journal of Comparative Neurology, 1940, 72, 1.

GEREBTZOFF, M. A. Cholinesterases; a Histochemical Contribution to the Solution of Some Functional Problems. London: Pergamon Press, 1959.

GIACOMO, P. DE. Attivita colinesterasica nel complesso amigdaloideo della cavia. Lavord Neuropsichiatrice, 1960, 27, 3.

GIRGIS, M. Distribution of cholinesterase in the basal rhinencephalic structures of the coypu (Myocastor coypus). Journal of Comparative Neurology, 1967, 129, 85.

GIRGIS, M. Distribution of cholinesterase in the basal rhinencephalic structures of the Grivet monkey (Cercopithecus aethiops aethiops). Acta Anatomica, 1968, 70, 568.

GIRGIS, M. Distribution of cholinesterase in the basal rhinencephalic structures of the Senegal bush baby (Galago senegalensis senegalensis). Acta Anatomica, 1969, 72, 94.

GURDJIAN, E. S. The corpus striatum of the rat. Journal of Comparative Neurology, 1928, 45, 249.

HALL, E. Some observations on the ultrastructure of the amygdala. Zeitschrift für Zellforschung und Mikroskopische Anatomie (Berlin), 1968, 92, 169.

HALL, E. The amygdala of the cat: A Golgi study. Submitted for publication, 1971.

HALL, E., & GENESER-JENSEN, F. A. Distribution of acetylcholin-
    esterase and monoamine oxidase in the amygdala of the guinea
    pig. Zeitschrift für Zellforschung und Mikroskopische
    Anatomie (Berlin), 1971, in press.

HALL, E., & PRYM, U., 1971, in preparation.

HALL, E., HAUG, F.-M.S., & URSIN, H. Dithizone and sulphide
    silver staining of the amygdala in the cat. Zeitschrift für
    Zellforschung und Mikroskopische Anatomie (Berlin), 1969,
    102, 40.

HASHIMOTO, P. H., MAEDA, T., TORII, K., & SHIMIZU, N. Histo-
    chemical demonstration of autonomic regions in the central
    nervous system of the rabbit by means of a monoamine oxidase
    staining. Medical Journal of Osaka University, 1962, 12, 425.

HAUG, F.-M.S. Electron microscopical localization of the zinc in
    hippocampal mossy fibre synapses by a modified sulphide silver
    procedure. Histochemie, 1967, 8, 355.

HUMPHREY, T. The telencephalon of the bat. I. The non-cortical
    nuclear masses and certain pertinent fiber connections.
    Journal of Comparative Neurology, 1936, 65, 603.

ISHII, T., & FRIEDE, R. L. A comparative histochemical mapping of
    the distribution of acetylcholinesterase and nicotinamide
    adenine dinucleotide-diaphorase activities in the human brain.
    International Review of Neurobiology, 1967, 10, 231.

JESERICH, M. W. The nuclear pattern and the fiber connections of
    certain non-cortical areas of the telencephalon of the mink
    (Mustela vison). Journal of Comparative Neurology, 1945,
    83, 173.

JOHNSON, T. N. Studies on the brain of the guinea pig. I. The
    nuclear pattern of certain basal telencephalic centers.
    Journal of Comparative Neurology, 1957, 107, 353.

JOHNSTON, J. B. Further contributions to the study of the evolu-
    tion of the forebrain. Journal of Comparative Neurology,
    1923, 35, 337.

KOELLE, G. B. The histochemical localization of cholinesterases
    in the central nervous system of the rat. Journal of
    Comparative Neurology, 1954, 100, 211.

KOIKEGAMI, H. Amygdala and other related limbic structures; ex-
    perimental studies on the anatomy and function. I. Anatomi-
    cal researches with some neurophysiological observations.

Acta Medica et Biologica (Niigata), 1963, 10, 161.

KRNJEVIC, K., & SILVER, A. A histochemical study of cholinergic
    fibres in the cerebral cortex. Journal of Anatomy, 1965,
    99, 711.

LAUER, E. W. The nuclear pattern and fiber connections of certain
    basal telencephalic centers in the macaque. Journal of
    Comparative Neurology, 1945, 82, 215.

LESCAULT, H. Some neocortico-amygdaloid connections in the cat.
    Proceedings of the Canadian Federation of Biological
    Societies, 1969, 12, 24.

LESCAULT, H. Some neocortico-amygdaloid connections in the cat.
    Thesis, University of Ottawa, 1971.

MANOCHA, S. L., SHANTHA, T. R., & BOURNE, G. H. Histochemical
    mapping of the distribution of monoamine oxidase in the
    diencephalon and basal telencephalic centers of the brain
    of squirrel monkey (Saimiri sciureus). Brain Research,
    1967, 6, 570.

NAUTA, W. J. H. Silver impregnation of degenerating axons. In
    W. F. Windle (Ed.) New Research Techniques of Neuroanatomy.
    Springfield, Illinois: Charles C. Thomas. 1957. Pp. 17-26.

NAUTA, W. J. H. Hippocampal projections and related neural
    pathways to the mid-brain in the cat. Brain, 1958, 81, 319.

POWELL, T. P. S., COWAN, W. M., & RAISMAN, G. The central
    olfactory connexions. Journal of Anatomy, 1965, 99, 791.

RAMON-MOLINER, E. A tungstate modification of the Golgi-Cox
    method. Stain Technology, 1958, 33, 19.

ROSE, M. Cytoarchitektonischer Atlas der Grosshirnrinde der Maus.
    Journal of Psychology and Neurology (Leipzig), 1929, 40, 1.

SHIMIZU, N., MORIKAWA, N., & OKADA, M. Histochemical studies of
    monoamine oxidase of the brain of rodents. Zeitschrift für
    Zellforschung und Mikroskopische Anatomie (Berlin), 1959,
    49, 389.

SHUTE, C. C. D., & LEWIS, P. R.   Cholinesterase-containing
    systems of the brain of the rat.   Nature, 1963, 199, 1160.

TIMM, F.   Zur Histochemie der Schwermetalle.   Das Sulfid-Silber-
    verfahren.   Deutsche Zeitschrift für die Gesamte Gericht-
    liche Medizin (Berlin), 1958, 46, 706.

UCHIDA, Y.   A contribution to the comparative anatomy of the
    amygdaloid nuclei in mammals, especially in rodents.
    Part I.   Rat and mouse.   Folia Psychiatrica et Neurologica
    Japonica (Niigata), 1950a, 4, 25.

UCHIDA, Y.   A contribution to the comparative anatomy of the
    amygdaloid nuclei in mammals, especially in rodents.
    Part II.   Guinea pig, rabbit and squirrel.   Folia
    Psychiatrica et Neurologica Japonica (Niigata), 1950b, 4, 91.

VALVERDE, F.   Intrinsic organization of the amygdaloid complex.
    A Golgi study in the mouse.   Trabajos del Instituto Cajal de
    Investigaciones Biologicas (Madrid), 1962, 54, 291.

VALVERDE, F.   Studies on the forebrain of the mouse.   Golgi
    observations.   Journal of Anatomy, 1963, 97, 157.

VALVERDE, F.   Studies on the Piriform Lobe.   Cambridge: Harvard
    University Press, 1965.

WAKEFIELD, C.   Thesis in preparation.   University of Ottawa, 1971.

WHITLOCK, D. G., & NAUTA, W. J. H.   Subcortical projections from
    the temporal neocortex in Macaca mulatta.   Journal of
    Comparative Neurology, 1956, 106, 183.

YOUNG, M. W.   The nuclear pattern and fiber connections of the
    non-cortical centers of the telencephalon of the rabbit
    (Lepus cuniculus).   Journal of Comparative Neurology, 1936,
    65, 295.

YU, H.   The amygdaloid complex in the rat.   Thesis, University of
    Ottawa, 1969.

Abbreviations used in figures:

AD	area dentata
AF	amygdaloid fissure
C	central nucleus
Co	cortical nucleus
D	dendrite
DB	degenerating bouton
dL	dorsal part of the lateral nucleus
I	intercalated nucleus
L	lateral nucleus
lB	large-celled part of the basal nucleus
lC	lateral part of the central nucleus
M	medial nucleus
mC	medial part of the central nucleus
P	putamen
sB	small-celled part of the basal nucleus
T	cortico-amygdaloid transitional area
tB	transitional zone of the basal nucleus
vL	ventral part of the lateral nucleus
X	area X

THE NEURAL CONNECTIONS OF THE AMYGDALOID COMPLEX IN MAMMALS

Hubert J. Lammers

Department of Anatomy and Embryology
Medical Faculty, Catholic University
Nijmegen, The Netherlands

INTRODUCTION

There is an extensive literature regarding the neural connections of the amygdaloid complex. Reviews of the older literature, based mainly on descriptions from normal material or from experimental Marchi material, have been published previously (Pribram and Kruger, 1954; Thomalske, Klingler and Worringer, 1957; Gastaut and Lammers, 1961). In the present survey, I intend to concentrate on the work done since then which, owing to improved histological and experimental techniques, has provided us with more reliable and more detailed information.

From the old days, the amygdala has been thought to be related to the olfactory system. However, experimental-anatomical as well as physiological studies have given evidence that, on one hand, the direct connection of the amygdala with the olfactory system - certainly for mammals - is only a limited one but that, on the other hand, the amygdala, apart from a considerable indirect olfactory input, also must receive a not insignificant, aspecific sensory input. Hence, the amygdala has assumed importance as a subcortical center of co-ordination for olfactory impulses with other sensory influences. This non-olfactory input can be realized only through afferent projections, originating either from the diencephalon or from the adjacent neocortical areas of the telencephalon. Tracing the anatomical substratum of the olfactory projection as well as that of the non-olfactory projections to and from the amygdala continues to be a challenge to neuro-anatomists interested in this area.

123

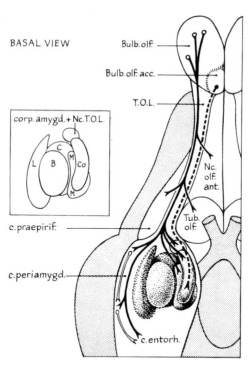

Fig. 1.   Secondary olfactory projection.
Black lines indicate the projection of the main olfactory bulb.
White lines indicate the projection from the periamygdaloid
cortex to the entorhinal cortex. The projection of the accessory
olfactory bulb (Winans and Scalia, 1970) is indicated by the
stippled line.

        I shall now deal with the neural connections of the
amygdala, classifying them as follows:

        1.   The connections with the secondary olfactory area.
        2.   The connections with the neocortex.
        3.   The connections with the lateral preoptic area and
                the hypothalamus.
        4.   The connections with the dorsal thalamus.

1.   Connections of the Amygdala with the Secondary Olfactory Area

        The secondary olfactory area is situated at the base of the
hemisphere of the telencephalon, medially to the rhinal sulcus
and extending rostro-caudally from the olfactory bulb as far as,
and including, the piriform lobe. It comprises the areas receiving a direct projection from the olfactory bulb. To this

Abbreviations

A.A.A.	area amygdaloidea anterior
A.H.A.	area hypothalamica anterior
A.P.O.	area praeoptica
B.	nucleus amygdalae basalis
Bednc.Str.t.	bed nucleus of the stria terminalis
Br.	diagonal band of Broca
Bulb.olf.	bulbus olfactorius
Bulb.olf.acc.	bulbus olfactorius accessorius
C.	nucleus amygdalae centralis
C.A.	commissura anterior
C.m.	corpus mammillare
Co.	nucleus
C.entorh.	cortex entorhinalis
C.periamygd.	cortex periamygdaloidea
C.praepirif.	cortex praepiriformis
F.Hipp.	formatio hippocampi
L.	nucleus amygdalae lateralis
M.	nucleus amygdalae medialis
M.F.B.	medial forebrain bundle
Nc.olf.ant.	nucleus olfactorius anterior
Nc.T.O.L.	nucleus of the tractus olfactorius lateralis
T.O.L.	tractus olfactorius lateralis
Tub.olf.	tuberculum olfactorium
V.M.	nucleus ventromedialis hypothalami
V.Pr.M.	regio premammillaris ventralis

category belong the retro-bulbar area (the anterior olfactory nucleus), the olfactory tubercle, the prepiriform cortex and the periamygdaloid cortex, as well as the nucleus of the lateral olfactory tract and the cortical nucleus of the amygdala (Fig. 1). According to White (1965) and Heimer (1968) this projection extends, in the rat, as far as the ventro-lateral part of the entorhinal cortex.

Scalia (1966) found in the rabbit a direct projection to the anterior continuation of the hippocampus as well, but this has not been reported by other workers. For an exhaustive survey of the secondary olfactory projection I refer to the studies by Scalia (1968) and Girgis (1970).

Concerning the amygdala, a large number of studies in a variety of mammalian species have made it clear that the direct olfactory projection to this nuclear complex is a much more

limited one than was thought after the early experimental-anatomical investigations done by Le Gros Clark and Meyer (1947) in
the rabbit (for a review of the older literature see Pribram and
Kruger, 1954; Gastaut and Lammers, 1961).

The majority of workers now have come to accept the view that
the olfactory projection to the amygdaloid complex is restricted
to the nucleus of the lateral olfactory tract and the cortical
nucleus of the amygdala, in particular its antero-lateral part.
Heimer and Lohman (personal communication) contend, however, that
in the rat there is not any degeneration in the nucleus of the
lateral olfactory tract after removal of the ipsilateral olfactory
bulb. In addition, a few authors report a projection to the
anterior amygdaloid area (White, 1965) and the anterior part of
the medial amygdaloid nucleus (Scalia, 1966).

Recently, Winans and Scalia (1970) have succeeded in demonstrating a clear difference between the projection to the cortical
amygdaloid nucleus from the main olfactory bulb and that from the
accessory one. The former projects only to the antero-lateral
part of the cortical nucleus and the latter projects only to its
postero-medial part. The two projection areas of this nucleus
are easily distinguished. It has not been possible yet to confirm
this remarkable finding by Winans and Scalia - the first indication of a topical differentiation in the olfactory projection
system - in other species than the rat. By the side of the
above-mentioned limited direct projection from the olfactory bulb
there is, however, an important indirect olfactory projection to
the amygdala by way of the piriform cortex.

Before discussing this projection in detail, however, I
would like to point to the current great confusion concerning
the use of the terms piriform, prepiriform, and periamygdaloid
cortex (see also Scalia, 1968). By 'piriform lobe' we should
understand the pair-shaped part of the basal telencephalon,
caudally to the olfactory tubercle. In some species, this is
marked on the surface by the olfactory incisure, in others such
a marking is absent. However, 'piriform lobe' is not to be used
in case of the primates. Here the (parahippocampal) uncus is the
homologue of the piriform lobe in the lower mammals. Strictly
speaking, the name of piriform cortex should only be applied to
the cortex of the corresponding lobe. Yet, Gray (1924) spoke of
a 'piriform cortex' which extended over the entire basal area;
he distinguished an anterior part, a medial part and a posterior
part. Gray's view is taken over by Valverde (1965), but the
latter designates these parts as 'prepiriform area,' 'piriform
area' and 'entorhinal area,' respectively. We (Gastaut and
Lammers, 1961), in denoting these areas, prefer to use the terms
'prepiriformal area,' 'periamygdaloid area' and 'entorhinal area.'
Other workers follow the terminology devised by Rose (1931), who

denoted the whole of the basal area from the olfactory bulb to
the caudal end of the amygdala as the 'prepiriform area' (regio
praepyriformis), and its adjoining caudal area as the 'entorhinal area' (regio entorhinalis). Within these areas there are
further subdivisions. The term 'periamygdaloid area' (regio
periamygdaloidea) - subdivided into six fields - was used by
Rose to indicate the medial surface area, which is in close
relation to the amygdala. A consideration of all the arguments
for or against these various terms would carry us too far, but
it seems desirable that workers in this field should agree upon
a common terminology in order to prevent any misunderstanding
in the interpretation of each other's findings, or at least
state clearly which set of terms they are going to use.

As regards the piriform-amygdaloid connections, various
investigations have led to the conclusion that from the entire
piriform cortex, with the exception of its posterior area (the
entorhinal cortex), fibers pass to the amygdaloid nuclei.
According to Cowan et al. (1965), in the rat, the anterior
piriform area projects by way of the longitudinal association
bundle to the lateral and basal nuclei. Valverde (1965), in
his study, reports on a single prepiriform lesion only (cat 8).
In this animal, he did not find any projection to the amygdala.
Degeneration of the cortical nucleus in this case must be
attributed to lesion of the lateral olfactory tract. In a
quite recent study of the rat, hamster and mouse, Scott and
Leonard (1971) have likewise failed to find any terminal degeneration in the amygdaloid nuclei after superficial lesion
of the prepiriform cortex. Here, too, terminal degeneration in
the cortical nucleus must be ascribed to an accidental lesion
of the lateral olfactory tract. Powell et al. in the above-
mentioned studies stress the fact that, in lesions of the more
caudal part of the piriform cortex, the majority of degenerating
fibers pass by way of the external capsule or through the
amygdala to the anterior amygdaloid area and the lateral preoptic area, but that there also is clear evidence of pre-terminal
degeneration in the basal and lateral amygdaloid nuclei.

A more detailed description of the relationship between the
periamygdaloid cortex (his 'area piriformis medialis') and the
amygdala is given by Valverde (1965) from his Golgi material.
It shows that the axons from this cortex project mainly to the
lateral amygdaloid nucleus. According to this author, in the
projection system of the animal species examined by him (mouse,
rat, cat), three parts should be distinguished: a medial part,
an intermediary part, and a lateral one. The medial part has
its origin in the most medial periamygdaloid cortex, laterally
adjacent to the cortical nucleus; its fibers traverse the basal
nucleus arc-like in a dorso-lateral direction, terminating in

the lateral nucleus or passing on into the ventral amygdalo-fugal
system. Collaterals of these axons branch off towards the basal
nucleus. Laterally to this medial system is the intermediary
part of the piriformo-amygdaloid projection. Its fibers arise
from a narrow cortical strip, laterally to the medial cortical
area and pass on dorsally in the fibrous layer between the
lateral and basal amygdaloid nuclei to join the medial fibers.
The lateral and greater part of the piriformo-amygdaloid projection has its origin in the periamygdaloid cortex, from its
intermediary part to the rhinal sulcus. The axons from this
cortical area run by way of the external capsule in a dorsal
direction, either turning off towards the lateral amygdaloid
nucleus or passing on medially, underneath the putamen, to join
the ventral amygdalo-fugal system. Valverde (1965) found that
the piriform-amygdaloid system also is joined by short axons,
which link the ventral part of the amygdala with its dorsal
part, thus forming an intrinsically amygdaloid relay-system.

    So far as is known at present, apart from the prepiriform
and periamygdaloid cortex, no other amygdalo-petal fibers arise
from the basal telencephalic area. In the older literature,
based chiefly on observations made from normal material, mention
is made of fibers passing from Broca's diagonal band to the
amygdala, but a recent experimental investigation in the rat by
Price and Powell (1970b) concerning projections to Broca's band
has failed to confirm this. It appears exceedingly difficult to
furnish experimental-anatomical proof of the existence of any
amygdalo-petal projection from the olfactory tubercle. Any
reports on this in the literature are negative.

    Unlike the connections of the amygdaloid nuclei to be dis-
cussed next, none or only a few reciprocal connections seem to
exist between these nuclei and the piriform cortex. Only
Valverde (1965) mentions the presence of axons passing from the
amygdaloid nuclei via the external capsule extending partly to
the deep plexus of the piriform cortex, partly continuing
caudally and terminating in the hippocampal formation. Such a
direct projection of the amygdala to the hippocampus has not
been found by other experimental workers.

2. Connections of the Amygdala with the Neocortex

    The amygdala receives fibers not only from the paleocortical
piriform cortex, but also from a number of neocortical areas.

    In view of physiological investigations (Kaada, 1951; Gastaut,
Naquet and Roger, 1952; Segundo, Naquet and Arana, 1955), the
existence of a direct temporo-amygdaloid projection was to be
expected, an assumption that was substantiated in 1956 by Nauta

and Whitlock through an experimental-anatomical investigation in
the monkey. They found a projection from the inferior temporal
area to the baso-lateral and central nuclei of the amygdala. We
(Lammers and Lohman, 1957) were able to confirm such a temporo-
amygdaloid projection in the cat, the projection being directed
chiefly from the temporal 'pole' towards the dorsal part of the
lateral nucleus and towards the central nucleus. Nauta (1961)
holds that, at least in the monkey, there is a reciprocal relationship between the amygdala and these neocortical areas, for
his examination has brought to light that fibers pass from the
amygdala to the rostral parts of the superior, inferior and
medial temporal gyri, as well as to the ventral insular area. A
projection to the amygdala has been reported not only from the
temporal area but also from the orbito-frontal cortex. Koikegami
(1963) found in the cat a projection from the anterior orbital
area to the lateral part of the amygdala, and from the orbito-
sylvian area to its medial part. Valverde (1965), likewise in
the cat, observed that from the orbital gyrus there is a projection to all nuclei of the amygdala, with the exception of the
central nucleus. According to him, this orbito-amygdaloid projection decreases from rostral to caudal and from medial to
lateral. On the other hand, Powell et al. (1965), after extensive
lesions of the neocortex above the rhinal sulcus, in the rat, did
not observe any projections either to the amygdaloid nuclei, the
piriform cortex, or the hypothalamus. In a more recent study of
the projections of the prefrontal cortex in the same animal,
Leonard (1968) concludes that from the sulcal cortex ('the cortex
forming the dorsal lip of the rhinal sulcus') there is no projection to the amygdala (there is one to the olfactory tubercle, the
substantia innominata and the most rostral part of the lateral
hypothalamic area). Nor did she find an amygdalo-petal projection
from the medial pregenual prefrontal cortex. The question arises
whether or not we have to contend here with a species-difference
(monkey and cat against rat). If it were indeed a matter of
species-difference, this could mean that with an increase of the
neocortex in the cat and the monkey as compared with that of the
rat the amygdala, alongside its function as a subcortical center
for the paleocortex, is becoming more and more important also as
a subcortical center for the neocortex, in particular for the
latter's temporal and orbital areas. In this respect, it would
be worthwhile to verify in higher mammals whether or not the
temporal projection is indeed directed mainly towards the laterobasal and central nuclei of the amygdala, and the orbital projection mainly towards the medial amygdaloid area.

For the neocortical projection to the amygdala, I would like
to refer also to a recent study by van Alphen (1969), who, after
lesions in the parietal, occipital as well as temporal areas of
the rabbit found a limited number of degenerative fibers, which,

via the posterior limb of the anterior commisure, pass to the
heterolateral (pre)piriform cortex, as well as to the anterior
part of the lateral nucleus of the amygdala.

3. Connections of the Amygdala with the Preoptic Area and the
    Hypothalamus

The amygdala is connected with the preoptic area and the
hypothalamus along two pathways, viz a dorsal pathway (the stria
terminalis) and ventral pathway (the ventral amygdalo-fugal
system). Whereas the view used to be held that these two systems
were only efferent, amygdalo-fugal, more recent studies have shown
that both of them contain amygdalo-fugal as well as amygdalo-petal
fibers.

The efferent fibers of the stria terminalis (Fig. 2) have
their origin in the cortico-medial and basal nuclei of the amygdala, but the lateral nucleus, also, contributes to such fibers.

The fibers of the stria terminalis end in an area extending
caudally from the ventral or precommissural area of the septum as
far as the ventro-medial nucleus of the hypothalamus and, in part,
even further caudally into the ventral premammillary area. As a
result of numerous investigations with normal material (Johnston,
1923; Berkelbach van der Sprenkel, 1926; Young, 1936; Humphrey,
1936; Fox, 1940; Fukuchi, 1952; Lammers and Magnus, 1955),
various components came to be distinguished in the stria terminalis.
Depending on their relationship with the anterior commissure, a distinction was made between a supra- or precommissural component, a
postcommissural component, and a commissural component. In addition, some workers also introduced a stria-medullaris component.
Later experimental investigations have not confirmed this. Nor has
the commissural component, which by Johnston (1923) and Humphrey
(1936) was described as a commissural connection between the two
nuclei of the lateral olfactory bulb, ever been found in this form
by others. Fox (1943), Lammers and Lohman (1957), Ban and Omukai
(1959), and Nauta (1959), were unable to follow this component
beyond the bed nucleus of the anterior commissure and some way
along its posterior limb. More recent studies (van Alphen, 1969;
Heimer and Nauta, 1969) have described a projection from the
lateral olfactory bulb to the heterolateral bed nucleus of the
stria terminalis.

Nevertheless, in view of our own (unpublished) observations
in the rabbit (Fink-Heimer technique, 3 days' survival) we feel
justified in assuming a genuine commissural component, linking
the two cortical nuclei. After lesions of the stria terminalis
in its mid-course, severe terminal degeneration presented itself
in the molecular layer of both cortical nuclei, attended by a

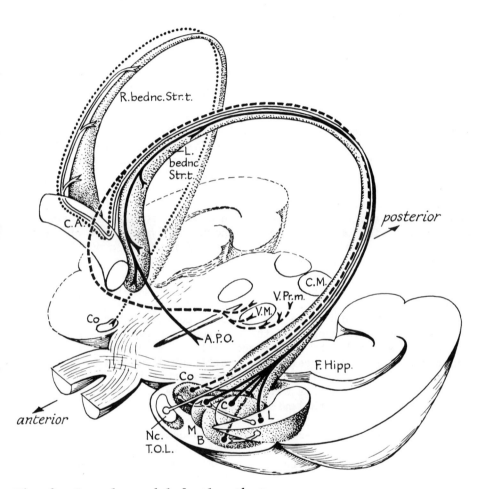

Fig. 2.   Dorsal amygdalofugal pathway.

diffuse distribution, only at the ipsilateral side, of degenerating terminals in the adjoining medial and basal nuclei. In the
cortical nucleus itself, too, we have seen degeneration of
terminals, but here it is more diffuse and, although more difficult to assess, considerably less than in the molecular layer of
this nucleus. In these observations, we are supported by de
Olmos' findings in the rat. We may refer the reader to the
latter's interesting contributions to the present conference.

As regards the pre- and postcommissural components of the
stria terminalis, some new findings, by means of the Fink-Heimer
technique, have been reported recently. In 1969, Heimer and
Nauta found that, in the rat, the precommissural component not
only terminates in the medial preoptic area and in the anterior
hypothalamus, but also continues caudally to end in a shell-like
cell-poor zone around the ventromedial hypothalamic nucleus and
in the ventral premammillary region. In its caudal course, this
component traverses the postcommissural component, which terminates in the bed nucleus of the stria terminalis and in the
anterior hypothalamus. Although the precommissural component
also has a termination inside the ventromedial hypothalamic
nucleus, Heimer and Nauta assume this component to have its main
site of termination on the outlying dendrites of the cells of
this nucleus. For this aspect, also, we refer to the contributions
by Raisman and de Olmos to this conference.

Quite recently, the course and termination of this component
has been confirmed by Leonard and Scott (1971), who also have
shown that, in the rat, mouse and hamster, the component has its
origin in the posterior part of the cortical nucleus. Taken together with the finding by Winans and Scalia (1970) that this part
of the cortical nucleus receives a specific projection from the
accessory bulb, it might lead us to conclude that the vomero-nasal
organ (Jacobson) has in this way a more or less separate projection
to the ventromedial hypothalamic nucleus.

In regard to the post-commissural component, Leonard and
Scott hold that it is built up from various elements, which,
originating from various parts of the medial and baso-lateral
amygdaloid nuclei, terminate in separate fields in the bed nucleus
of the stria terminalis, adjoining rostrally the anterior thalamus.
De Olmos, also, has found that, in the rat, the stria terminalis
is built up of several components, each having a course and
terminal area of its own. One of the components is shown to continue rostrally into the posterior and medial parts of the anterior
olfactory nucleus, and even as far as the deep layer of the accessory olfactory bulb (see de Olmos, this volume). The fact that
the stria terminalis does not only contain amygdalo-fugal but also
amygdalo-petal fibers was reported already by Gurdjian in 1928

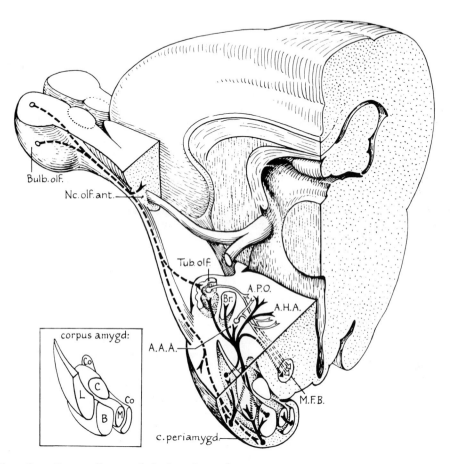

Fig. 3. Ventral amygdalofugal pathway.
Stippled line indicates secondary olfactory projection.

from normal rat material, but not confirmed experimentally until
a much later date. The origin of these afferent fibers passing
to the amygdala is in the preoptic area, the anterior hypothalamic
area and in the bed nucleus of the stria terminalis (Nauta, 1961;
Cowan et al., 1965; van Alphen, 1969). Valverde (1965) holds
that, at least in the cat, these fibers are joined by a limited
number of fibers that arise in the midline and rostral nuclei of
the dorsal thalamus. The afferent fibers terminate both in the
cortico-medial and in the baso-lateral nuclei of the amygdala.

The ventral amygdalo-fugal pathway (Fig. 3) has been
described in detail by numerous workers, from normal as well as
from experimental material. In the older literature (Crosby,
1917; Johnston, 1923; Howe, 1923; Loo, 1931), this system has
received widely divergent names (see Gastaut and Lammers, 1961).
Currently, the term 'ventral amygdalo-fugal pathway' is much in
use. As in this pathway, there also occur amygdalo-petal fibers;
this designation, too, seems inadequate. It was pointed out by
Johnston ('amygdalo-piriform association bundle,' 1923), by Fox
(1940) and by Fukuchi (1952), working with normal material, and
by Fox (1943) and us (Lammers and Magnus, 1955; Lammers and
Lohman, 1957), working with experimental material, that this
bundle not only contains fibers from the amygdala but also from
the piriform cortex.

Powell et al., in their studies on the connections of the
piriform cortex and the amygdala in the rat (Powell, Cowan, and
Raisman, 1965; Cowan, Raisman, and Powell, 1965), emphasize the
fact that "it is not possible to determine with certainty whether
the ventral pathway receives fibers from the amygdala in addition
to those which arise in the pyriform cortex. If there is an amygdaloid projection to this pathway, however, it must arise solely
from the baso-lateral group ..."

In this, they receive support from Leonard (1970), who states
that there is no convincing proof of a long-axon ventral pathway
originating in the rat's amygdala. According to her, long fibers
continuing into the preoptic area and the hypothalamic area,
beyond the amygdaloid area, have their origin solely in the periamygdaloid cortex. Leonard and Scott (1971) have found that
after small lesions in the basal area of the amygdala no more
than sparse terminal degeneration can be observed in the anterior
amygdaloid area and in the lateral preoptic area. Degeneration
in the medial-forebrain bundle and in the lateral hypothalamus is
seen only after lesion of the periamygdaloid cortex. Valverde
(1965) from his Golgi material also concludes that, in the rat
and mouse, the longitudinal association bundle, apart from long
fibers from the periamygdaloid cortex, only contains short axons
from the amygdaloid nuclei. In the cat, however, there also are
long amygdaloid axons, while in the monkey (Nauta, 1961) a con-

siderable contribution, continuing into the hypothalamus, from
the amygdaloid nuclei to this ventral system may likewise be
assumed. As an indirect argument for this, it also could be
argued that, in the primates, where the periamygdaloid piriform
cortex in proportion to the basal and lateral nuclei of the
amygdala is only of limited extent, the ventral projection system
nevertheless is highly developed. Klingler and Gloor (1960) hold
that in man the longitudinal association bundle is of about the
same size as the anterior commissure.

The terminal area of the piriformo-amygdalo-fugal system as
a whole is an oblong area, extending, as numerous studies have
shown, from the medio-frontal cortex (regio infra-radiata), the
accumbent nucleus, the olfactory tubercle and Broca's diagonal
band*, along the lateral preoptic and anterior hypothalamic areas,
into the hypothalamus (mid-tuberal area, Cowan et al., 1965). In
the lateral preoptic and hypothalamic area, these fibers join the
medial forebrain bundle. Part of the fibers of the ventral pathway turn dorsally to terminate in the dorso-medial thalamic
nucleus (see below). In comparing the terminal site of the
ventral pathways with that of the stria terminalis, it appears
that the termination of the ventral pathway in the preoptic area
and the rostral part of the lateral hypothalamus is lateral,
that of the stria terminalis more medial.

From investigations by Nauta (1961) in the monkey, and by
Cowan et al. (1965) in the rat, it has become evident that not
only the stria terminalis, but also the ventral projection-system,
contains amygdalo-petal fibers. These fibers have their origin
in the preoptic area, and in the rostral part of the hypothalamus.
According to Cowan et al. (1965) these amygdalo-petal fibers
spread over all the nuclei of the amygdala, with the exception
perhaps of the central nucleus.

4. Connections of the Amygdala with the Dorsal Thalamus

Fox (1943) was the first to demonstrate experimentally, in
the cat, that there is a projection from the amygdala to the
dorso-medial thalamic nucleus. His finding, based on Marchi preparations, was afterwards confirmed and added to by other workers
using the Nauta and the Nauta-Gygax silver-impregnation techniques
(Nauta and Valenstein, 1958, and Nauta, 1961, in the monkey;
Sanders-Woudstra, 1961, and Powell, Cowan and Raisman, 1963, in
the rat; Valverde, 1965, in the cat). The relevant fibers, together with the ventral amygdalo-fugal system, pass via the

---

* According to a recent study by Price and Powell (1970b),
  Broca's band does not receive any fibers from the piriform
  cortex and the amygdala.

lateral preoptic area, and from there turn dorso-medially to
reach finally the dorso-medial nucleus by way of the inferior
thalamic peduncle. A limited number of these fibers terminate
in the rostral midline region and in the paracentral intra-
laminar nucleus. According to Powell et al. (1965), this projection is bilateral. Besides, they found a small number of
fibers which continued into the lateral habenular nucleus of
both sides.

Nauta suggests that these fibers originate from the basal
and lateral amygdaloid nuclei. Valverde feels that they arise
chiefly from the anterior amygdaloid area, but he does not rule
out a contribution from the basal or other amygdaloid nuclei.
Powell et al. (1965) hold that the fibers do not originate from
the amygdala but from the piriform cortex. According to Nauta
(1962), in the monkey, they are also joined by fibers that arise
from the temporal cortex, but Valverde (1965), using the cat,
was unable to confirm this finding of Nauta. Other authors,
also, are inclined to locate the origins of these fibers mainly
in the piriform cortex. Leonard and Scott (1971) are convinced
that in the species studied by them (rat, hamster, mouse) only
short fibers pass from the amygdaloid nuclei to the anterior
amygdaloid area, and that from there or from the anterior periamygdaloid cortex fibers pass to the dorso-medial thalamic
nucleus via the stria medullaris, and not by way of the inferior
thalamic peduncle, as described by the other workers.

Here, also, it remains difficult to determine the share of
the amygdala itself in this projection to the thalamus. We must
account for differences in species, in the sense that in the
monkey and the cat a direct amygdalo-thalamic projection has
developed, whereas in lower mammals, such as the mouse, rat and
hamster, there is a multi-synaptic amygdalo-thalamic relationship. Regarding the termination in the medio-dorsal thalamic
nucleus, there is no complete agreement among workers, either.
According to Nauta (1961, monkey) and Valverde (1965, cat) the
amygdalo-fugal fibers end in the medial or magnocellular part
of the medio-dorsal nucleus, which has a reciprocal relationship
with the orbito-frontal cortex (Nauta, 1962). Powell et al.
(1965) found with lesions limited strictly to the piriform cortex
of the rat a terminal degeneration in the central part of the
medio-dorsal nucleus, throughout its antero-posterior extent,
but especially in its caudal third part. This might suggest that
the piriform cortex and amygdala could have a differentiated projection to the dorso-medial thalamic nucleus. On the other hand,
we still have no clear insight into the comparative anatomy of
this nucleus and its different parts.

The relationship between the amygdala and the dorso-medial

thalamic nucleus is a reciprocal one. The thalamo-amygdaloid
fibers reach the baso-lateral nuclei via the inferior thalamic
peduncle. Their number turns out to be a great deal less than
that of the amygdalo-thalamic fibers (Nauta, 1961). However,
so long as there remains so much uncertainty about the latter
fibers, a pronouncement on the quantitative correlation between
the efferent and afferent components of the amygdalo-thalamic
interrelation would appear little appropriate.

We already have seen that Valverde (1965) holds that in the
cat there is also a projection from the rostral thalamic area,
in particular from the lateral central nucleus, to the bed nucleus
of the stria terminalis. A small part of these fibers, together
with the stria, passes to the amygdala and terminates in the
medial nucleus, but the greater part of these fibers run in a
rostral direction via the internal capsule and end in the part
of the caudate nucleus adjacent to the capsule and in that part
of the bed nucleus of the stria terminalis which is situated
above the anterior commissure.

DISCUSSION

Historically seen, the amygdaloid complex is a descriptive-
anatomical concept. It is the subcortical grisea of the piriform
lobe, the piriform lobe being the caudal part of the basal telencephalic area. In it, we distinguish various parts, on account
of descriptive-topographical and cyto-architectonic data, supple-
mented by hodological ones, at first obtained from normal and
later from experimental material. In the literature, we find a
great many descriptions and classifications of the amygdaloid
complex in diverse animal species, higher as well as lower
vertebrates. The contributions by Tryphena Humphrey and by
Elizabeth Hall to the present seminar consider these classifications and allied problems at greater length, so that, while
referring the reader to these papers, I shall restrict myself
to a few hodological notes.

In these classifications and groupings, it is especially
the following nuclei or areas that appear to pose a problem:
the central nucleus, the cortical nucleus, the nucleus of the
lateral olfactory tract, the bed nucleus of the stria terminalis,
and the anterior amygdaloid area. The central nucleus, situated
between amygdala and putamen, is often difficult to distinguish,
in particular as regards its lateral part, from the ventral part
of the putamen (see also Hall's contribution). Berkelbach van
der Sprenkel (1926) held this nucleus to be a continuum with the
bed nucleus of the stria terminalis. Hodologically, the position
of the central nucleus is not easy to determine because of the

numerous fibers passing through and near the nucleus.

Although the <u>cortical</u> <u>nucleus</u> and the <u>nucleus</u> <u>of</u> <u>the</u> <u>lateral</u> <u>olfactory</u> <u>tract</u> with regard to structure and localisation cannot simply be linked with the amygdala, it is especially because of their efferent and afferent relations with the stria-terminalis system that they may be considered as belonging to the amygdaloid complex, thus ranking with the nuclei of the amygdala proper. In this respect, both structures are distinguished clearly from the surrounding piriform cortex, as part of which Valverde (1965) prefers to see them; not only for their cortical structure but also because of the direct afferent olfactory projection which they have in common with the piriform cortex and in which they are distinguished from the nuclei of the amygdala proper. Against Valverde's view may be held the fact that, unlike the piriform cortex, neither of these two nuclei contributes to the ventral projection system. Assessing hodological and cyto-architectonical data on their strength in an argument for or against a certain view must remain a ticklish job. Should, however, Heimer and Lohman's recent finding in the rat also be confirmed for other animal species, viz that the nucleus of the lateral olfactory tract does not receive any terminals coming from the ipsilateral olfactory bulb, then not much support would be left for Valverde's view.

The <u>stria</u> <u>terminalis</u> together with its accompanying bed nucleus form a highly complex system, consisting of various components. Short and long axons effect a cascade-like reciprocal relationship between the amygdaloid complex and the ventral-septal area, the medial preoptic area, the medial anterior hypthalamus, the ventro-medial hypothalamic nucleus and the ventral premammillary region. If de Olmos's findings were confirmed for other species than the rat, the projection area of the stria terminalis may be assumed to extend rostrally even into the accessory olfactory bulb. In addition to these unilateral projections, the stria-terminalis system brings about a crossed relationship between the nucleus of the lateral olfactory tract and the bed nucleus, and likewise effects a genuinely commissural connection between the two cortical nuclei of the amygdaloid complex. Whether and, if so, how extensively the central nucleus is part of this stria-terminalis system, as has already been pointed out here, cannot be decided, as yet.

Together with the preoptic area the <u>anterior</u> <u>amygdaloid</u> <u>area</u> constitutes "a wide area extending from the midline to the deep plexus of the cortex piriformis" (Valverde, 1965, p. 63). Cowan <u>et</u> <u>al</u>. (1965) hold that this area may be considered as the bed nucleus of the lateral extension of the medial forebrain bundle. For the piriform cortex as well as for the amygdaloid complex,

the area constitutes an important nodal point. It is the intermediary for reciprocal relationships between the amygdalo-piriform complex on the one hand and the preoptic area together with
the lateral hypothalamus (via the medial forebrain bundle), the
dorso-medial and rostral mid-line nuclei of the thalamus (via
the inferior thalamic peduncle), and the habenula (via the stria
medullaris) on the other hand.

The share of the amygdaloid nuclei proper in these projections is difficult to distinguish from that of the surrounding
piriform cortex. It might be concluded from the relevant literature that in such lower mammals as the mouse, rat and hamster
the efferent ventral projection system of the amygdala is built
up mainly of short axons, it having for this reason a multi-
synaptic character, whereas in higher mammals (cat, primates)
the system by the side of short axons also has long ones, which
realize a monosynaptic relationship between the baso-lateral
part of the amygdala and the above-mentioned areas of the diencephalon.

In view of these findings, it would not be correct, I think,
to consider this area as belonging to the amygdaloid complex.
The term 'anterior amygdaloid area' (area amygdaloidea anterior)
may be retained merely as a topographic indication.

The piriform cortex and the amygdaloid complex appear to
stand in close relationship to each other. This cortex forms
the secondary olfactory projection field. How extensively the
entorhinal cortex, in particular its ventro-lateral rostral part,
belongs to this is still a point under discussion. At any rate,
the entorhinal cortex receives olfactory impulses, either
direct or indirect, via a multi-synaptic conduction (Cragg, 1961;
Powell, Cowan and Raisman, 1965) or a zigzag conduction (Valverde,
1965) coming from the prepiriform and periamygdaloid cortex.
The significance of the piriform cortex, then, is that it has
its effect on the amygdaloid nuclei as well as on the hippocampal
formation, and through these structures indirectly 'works' upon
the hypothalamus as a whole. In addition, the prepiriform and
periamygdaloid cortex send out fibers directly to the lateral
preoptic and hypothalamic areas, and also to the dorsal thalamus.
Such relationships, that are in part reciprocal, appear in these
areas as having reversely a direct or indirect projection to the
piriform cortex. From the investigations of van Alphen (1969),
we may conclude that Rose's prepiriform cortex (in our terminology
the prepiriform and periamygdaloid cortex) by way of the posterior
limb of the anterior commissure has a projection to the contralateral prepiriform cortex, the putamen, the anterior part of
the lateral amygdaloid nucleus, the accumbent nucleus and perhaps
also to the olfactory tubercle. Van Alphen also postulates that

these areas, hence the prepiriform and the periamygdaloid cortex included, are reached, if only to a limited extent, by fibers from neocortical areas via the posterior limb of the anterior commissure. The relevant piriform cortical areas not only emit fibers to the hypothalamus and the dorsal thalamus. Heimer (1968) sees especially in the anterior prepiriform cortex the origin of the fibers passing to the olfactory bulb. In this, however, Price and Powell (1970a) are not in agreement with him. Their view is that these bulbo-petal fibers originate in the horizontal limb of Broca's diagonal band.

The question whether and how extensively the data available to us on the afferent and efferent relationships of the amygdaloid complex as presented in this paper allow for any functional conclusions is a matter for physiologists to decide. But, clearly, hodological research at light-microscopic level will not yield sufficient information to bridge the gap between the structural and functional relationships of a nuclear mass in the central nervous system. Such work, however, will provide us with a framework within which a well-directed and more detailed quantitative as well as qualitative histological and cytological research may be realized. In this respect, recent electron-microscopic research investigating the termination of projection systems at submicroscopic levels is of great importance. In this connection, I may refer to the important contribution made by Raisman to the present seminar. Here, we still have a vast field ripe for further exploration.

ACKNOWLEDGEMENTS

I wish to express my gratitude to my colleagues, Dr. A. H. M. Lohman and Dr. R. Nieuwenhuys, for their valuable advice and assistance in the preparation of this survey, to Mr. C. van Huyzen for the illustrations, to Mr. L. Grooten for the translation, and to Miss E. Maassen for the preparation of the manuscript.

REFERENCES

ALPHEN, H. A. M. VAN. The anterior commissure of the rabbit. Acta Anatomica, 1969, 74, Supplement 57, 9-111.

BAN, T., & OMUKAI, F. Experimental studies on the fiber connections of the amygdaloid nuclei in the rabbit. Journal of Comparative Neurology, 1959, 113, 245-280.

BERKELBACH VAN DER SPRENKEL, H. Stria terminalis and amygdala in
the brain of the opossum (Didelphys virginiana). Journal of
Comparative Neurology, 1926, 42, 211-254.

CROSBY, E. C. The forebrain of Alligator mississippiensis.
Journal of Comparative Neurology, 1917, 27, 325-402.

COWAN, W. M., RAISMAN, G., & POWELL, T. S. The connexions of
the amygdala. Journal of Neurology, Neurosurgery, and
Psychiatry, 1965, 28, 137-151.

CROSBY, E., & HUMPHREY, T. Studies of the vertebrate telencephalon. II. The nuclear pattern of the anterior olfactory
nucleus, tuberculum olfactorium and the amygdaloid complex in
adult man. Journal of Comparative Neurology, 1941, 74,
309-352.

FOX, C. A. Certain basal telencephalic centers in the cat.
Journal of Comparative Neurology, 1940, 72, 1-62.

FOX, C. A. The stria terminalis, longitudinal association
bundle and precommissural fornix fibres in the cat. Journal
of Comparative Neurology, 1943, 79, 277-295.

FUKUCHI, S. Comparative-anatomical studies on the amygdaloid
complex in mammals, especially in Ungulata. Folia Psychiatrica et Neurologica Japonica (Niigata), 1952, 5, 241-262.

GASTAUT, H., & LAMMERS, H. G. Anatomie du Rhinencéphale. Les
grandes activités du Rhinencéphale. Paris: Masson & Cie,
1961.

GASTAUT, H., NAQUET, R., & ROGER, A. Etude des post-décharges
électriques provoquées par stimulation du complexe nucléaire
amygdalien chez le chat. Review of Neurology, 1952, 2,
224-231.

GIRGIS, M. The rhinencephalon. Acta Anatomica, 1970, 76, 157-199.

GRAY, P. A. The cortical lamination pattern of the opossum,
Didelphys virginiana. Journal of Comparative Neurology, 1924,
37, 221-263.

GURDJIAN, E. S. Corpus striatum of the rat. Studies on the brain
of the rat, No. 3. Journal of Comparative Neurology, 1928,
45, 249-281.

HEIMER, L.   Synaptic distribution of centripetal and centrifugal
    nerve fibres in the olfactory system of the rat.   An experi-
    mental anatomical study.   Journal of Anatomy, 1968, 103,
    413-432.

HEIMER, L., & NAUTA, W. J. H.   The hypothalamic distribution of
    the stria terminalis in the rat.   Brain Research, 1969,
    13, 284-297.

HUMPHREY, T.   The telencephalon of the bat.   I. The non-cortical
    nuclear masses and certain pertinent fibre connections.
    Journal of Comparative Neurology, 1936, 65, 603-711.

JOHNSTON, J. B.   Further contributions to the study of the
    evolution of the forebrain.   Journal of Comparative Neurology,
    1923, 36, 143-192.

KAADA, B. R.   Somato-motor, autonomic and electrocorticographic
    responses to electrical stimulation of rhinencephalic and
    other structures in primates, cat and dog.   Acta Physio-
    logica Scandinavica, 1951, 24, Supplement 83.   285 pp.

KOIKEGAMI, H.   Amygdala and other related limbic structures;
    experimental studies on the anatomy and function.   I.
    Anatomical researches with some neurophysiological observa-
    tions.   Acta Medica et Biologica, 1963, 10, 161-277.

LAMMERS, H. J., & LOHMAN, A. H. M.   Experimenteel anatomisch
    onderzoek naar de verbindingen van piriforme cortex en
    amygdalakernen bij de kat.   Nederlands Tijdschrift voor
    Geneeskunde, 1957, 101, 1-2.

LAMMERS, H. J., & MAGNUS, O.   Etude expérimentale de la région
    du noyau amygdalien du chat.   Comptes Rendus de la Associa-
    tion Anatomistes XLIIe Réunion, 1955, 840-844.

LE GROS CLARK, W. E., & MEYER, M.   The terminal connexions of
    the olfactory tract in the rabbit.   Brain, 1947, 70,
    304-328.

LEONARD, C. M.   Origin of the amygdalofugal pathways in the rat.
    Anatomical Record, 1970, 166, 337.

LEONARD, C. M., & SCOTT, J. W.   Origin and distribution of the
    amygdalofugal pathways in the rat: an experimental neuro-
    anatomical study.   Journal of Comparative Neurology, 1971,
    144, 313-330.

LOO, Y. T.   The forebrain of the opossum, Didelphis virginiana,
    II.   Journal of Comparative Neurology, 1931, 52, 1-148.

NAUTA, W. J. H. Fibre degeneration following lesions of the amygdaloid complex in the monkey. Journal of Anatomy, 1961, 95, 515-531.

NAUTA, W. J. H. Neural associations of the amygdaloid complex in the monkey. Brain, 1962, 85, 505-519.

NAUTA, W. J. H., & HAYMAKER, W. Hypothalamic nuclei and fiber connections. In W. Haymaker, E. Anderson and W. J. H. Nauta (Eds.), The Hypothalamus. Springfield, Illinois: Charles C. Thomas, 1969. Pp. 136-209.

NAUTA, W. J. H., & VALENSTEIN, E. Some projections of the amygdaloid complex in the monkey. Anatomical Record, 1958, 130, 346.

POWELL, T. P. S., COWAN, W. M., & RAISMAN, G. The central olfactory connexions. Journal of Anatomy, 1965, 99, 791-813.

PRIBRAM, K. H., & KRUGER, L. Functions of the "olfactory brain." Annals of the New York Academy of Science, 1954, 58, 109-138.

PRICE, J. L., & POWELL, T. P. S. An experimental study of the origin and the course of the centrifugal fibres to the olfactory bulb in the rat. Journal of Anatomy, 1970a, 107, 215-237.

PRICE, J. L., & POWELL, T. P. S. The afferent connexions of the nucleus of the horizontal limb of the diagonal band. Journal of Anatomy, 1970b, 107, 239-256.

ROSE, M. Cytoarchitektonischer Atlas der Grosshirnrinde des Kaninchens. Journal of Psychology and Neurology (Leipzig), 1931, 43, 353.

SANDERS-WOUDSTRA, J. A. Experimenteel anatomisch onderzoek over de verbindingen van enkele basale telencephale hersengebienden bij de albinorat. Thesis, Groningen University, 1961.

SCALIA, F. Some olfactory pathways in the rabbit brain. Journal of Comparative Neurology, 1966, 126, 285-310.

SCALIA, F. A review of recent experimental studies on the distribution of the olfactory tracts in mammals. Brain Behavior and Evolution, 1968, 1, 101-123.

SCOTT, J. W., & LEONARD, C. M. The olfactory connections of the lateral hypothalamus in the rat, mouse and hamster. Journal of Comparative Neurology, 1971, 141, 331-344.

SEGUNDO, J. P., NAQUET, R., & ARANA, R.   Subcortical connections
    from temporal cortex of monkey.   A. M. A. Archives of
    Neurology and Psychiatry, 1955, 73, 5515-5524.

THOMALSKE, G., KLINGLER, J., & WORRINGER, E.   Ueber das Rhinen-
    cephalon.   Physiologischer und anatomischer Ueberblick.
    Acta Anatomica, 1957, 30, 865-902.

VALVERDE, F.   Studies on the Piriform Lobe.   Cambridge: Harvard
    University Press, 1965.

VÖLSCH, M.   Zür vergleichenden Anatomie des Mandelkernes und
    seiner Nachbargebilde.   Teil I, Archiv fur mikroscopische
    Anatomie, 1906, 68, 573-683; Teil II, ibid, 76, 373-523.

WHITE, L. E.   Olfactory bulb projections of the rat.   Anatomical
    Record, 1965, 152, 465-480.

WHITLOCK, D. G., & NAUTA, W. J. H.   Subcortical projections from
    the temporal neocortex in Macaca mulatta.   Journal of
    Comparative Neurology, 1956, 106, 183-213.

WINANS, S. S., & SCALIA, F.   Amygdaloid nucleus: new afferent
    input from the vomeronasal organ.   Science, 1970, 170, 330-332.

THE AMYGDALOID PROJECTION FIELD IN THE RAT

AS STUDIED WITH THE CUPRIC-SILVER METHOD

José S. de Olmos

Department of Anatomy

The University of Iowa, Iowa City, Iowa

INTRODUCTION

Much has been written about the connections of the amygdala, first of descriptions of normal material as stained with Weigert or Bielschowski type techniques, later of experimental material impregnated with Glees (1946) and Nauta-Gygax (1954) silver procedures. Reviews on the subject can be found in publications by Gloor (1955), Valverde (1965), Nauta and Haymaker (1969) and, in the present meeting, by Professor Lammers (1971). However, despite the pioneer value of such works, it was not until very recently that more reliable information has been produced, specifically by Heimer and Nauta (1969) who utilized the Fink-Heimer (1967) and electron microscopic methods. Further elaboration of their findings has been presented by Leonard and Scott (1971), who also employed the Fink-Heimer technique, and these contributions find strong support in additional electron microscopic observations by Raisman (1970, 1971). However, perhaps due to differences in experimental and/or technical approaches, it has been possible for the present writer not only to confirm at the optic microscope level some of the observations of the above-mentioned authors, but also to demonstrate a wider diencephalic and telencephalic terminal distribution of some of the fiber contingents of the stria terminalis, as well as a spatial organization of the efferent components in the latter, according to their origin in the amygdala. On a much reduced scale, additional information also has been obtained with regard to some of the connections within the so-called ventral amygdalofugal systems. This has been accomplished mainly by using the cupric-silver method originally developed in this laboratory (de Olmos, 1969), and a more recent modification

145

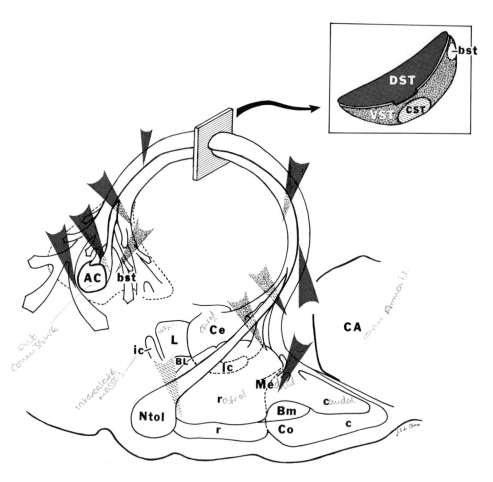

Fig. 1.   Schematic representation of the localization of the
lesions transecting the stria terminalis either totally or partially at different levels of its course.   The stria of the right
side is shown as seen from its medial aspect.   The different
components of the stria, the anterior commissure, the cornu
ammonis and portions or all of the amygdaloid nuclei except the
anterior amygdaloid area are outlined in continuous line.   The bed
nucleus of the stria terminalis is indicated by broken lines as
are the caudal intercalate masses, the lower boundary of the
central amygdaloid nucleus and the divisory line separating the
rostral and caudal portions of the medial, basomedial and cortical
amygdaloid nuclei.   At the upper corner there is a cross-sectional
schema of the components of the stria and the shade code representing them.   The arrows represent the level and extent of
transections of the stria terminalis and the type of shading indicates which of its 3 components were involved separately or in
combination.

(de Olmos and Ingram, 1971), which stains the Wallerian degeneration produced by suitably placed experimental lesions. Preliminary reports of some of these findings were presented at the
American Association of Anatomy meetings held in 1968 (de Olmos,
1968) and 1970 (de Olmos, 1970) and a more comprehensive description and analysis will be published very shortly (de Olmos
and Ingram, 1971b).

The following account will deal with the efferent connections
of the amygdalopiriform region as embodied by the stria terminalis and the so-called ventral amygdalofugal pathways, plus some
additional observations on the intraamygdaloid connections.

MATERIAL AND METHODS

A total of 60 rats ranging in weight between 47 and 140 grams
was used. Several surgical approaches were utilized, but in most
cases the lesions in the stria terminalis, amygdaloid nuclei or
the bed nucleus of the stria terminalis (Figs. 1, 2) were placed
stereotactically using usually Bernardis's (1967) coordinates,
or sometimes those of De Groot (1959) and König and Klippel (1963).
In several cases, the electrode was inserted at an angle of 25-30°
to a paramedian sagittal plane passing through the amygdaloid
complex in order to avoid track injury to the limbic cortex and
hippocampus. Electrolytic lesions were produced by passage of
d.c. anodal current of 0.8 to 1.2 mA for 6 to 10 seconds. The
electrodes were stainless steel wires less than 0.25 mm in
diameter and insulated, except for a few fractions of a millimeter
at the tip. The lesions in the piriform cortex were made either
by aspiration or by electrolysis.

For control purposes, the hippocampus, overlying neocortex and
the fimbria fornicis were aspirated with varying degrees of involvement of the cingulum fibers. Other controls were provided
by making electrolytic lesions in the dorsal hippocampus and/or
the fimbria fornicis as well as the ventral hippocampus-subiculum
formation. Also, in the control series, there were included
brains with lesions in the lateral preoptic and hypothalamic areas
or in the ventral striatum. After survival periods, varying from
30 hours to 4 days (the best results being obtained with 36-38
hours of survival), the animals were anesthetized and the brains
perfused and processed according to the methods outlined elsewhere
(de Olmos, 1969; de Olmos and Ingram, 1971a).

In the following account, Brodal's (1947) criteria for identifying the amygdaloid nuclei are most often adopted. However, as
here used the term anterior amygdaloid area refers to the diffusely
outlined gray mass which extends laterocaudally from the nucleus
of the horizontal limb of the diagonal band (see Price and Powell,

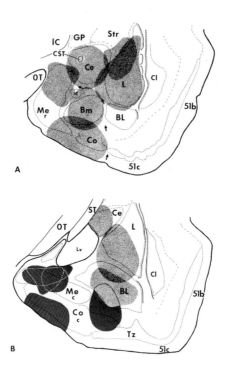

Fig. 2. Diagrammatic representation of two frontal sections
through the lesions in the rat amygdala of 14 of the most representative cases ranging from small to large coagulations in
different amygdaloid nuclei. The lesions are indicated according
to the shading code on Fig. 1, which represents the patterns of
degeneration which are produced by the various lesions. In Fig.
2A those lesions placed in the amygdaloid nuclei situated medially
to the row of arrows produced degeneration in the medial long-
projecting division of the ventral strial component. Coagulations
in the nuclei lateral to this row of arrows produced degeneration
in the lateral short-projecting division of the ventral strial
component. The involvement of the "commissural" component is also
represented with the shading code. In Fig. 2B three of the five
lesions associated with degeneration in the dorsal strial component
overlap, and their limits are indicated by broken white lines.

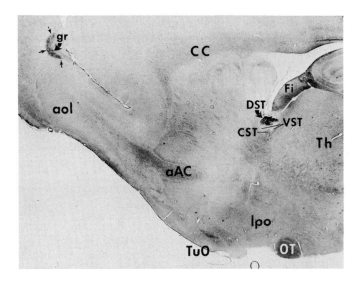

Fig. 3. Photomicrograph of a sagittal section through the stria
terminalis at a level just before its entrance in the rostral
expansion of its bed nucleus. The degenerating dorsal strial
component (DST) stands out markedly among the remainder of normal
ventral (VST) and commissural (CST) components of the stria because of its much heavier staining with silver. Modified cupric-
silver method. x10. Note the terminal degeneration in the internal granular layer (gr) of the accessory olfactory bulb derived
from the parolfactory radiation of the DST. In this case the
lesions interrupted the DST just after its outset from the caudal
corticomedial amygdala. Small arrows indicate the boundaries of
the terminal degeneration fields.

1970a) and which is bordered laterally by the prepiriform and
piriform cortices, caudally by the rostral pole of the amygdaloid
complex proper and dorsomedially by the sublenticular portions of
the substantia innominata. The band-like gray formation which
extends diagonally along the ventral aspect of the globus pallidus and dorsal to the nucleus of the horizontal limb of the
diagonal band and which merges rostromedially with the bed
nucleus of the stria terminalis and laterocaudally with the cephalic end of the central amygdaloid nucleus, constitutes the sublenticular part of the substantia innominata. The latter is considered to be a gray entity which is separate from the anterior
amygdaloid and lateral preoptic areas within which it has very
often been included. A main reason for such a separation lies
in the differential impregnation of its normal neuropil by the
cupric-silver methods, a reaction which characterizes it together
with the bed nucleus of the stria terminalis and the central
amygdaloid nucleus, with both of which it is continuous. Furthermore, this sublenticular gray mass appears to form a bed nucleus
for one of the branches of the ventral amygdalofugal system by
which it is abundantly supplied.

RESULTS

The results of the experimental material to be presented may
be discussed conveniently in three sections, dealing first with
the stria terminalis, second with the so-called ventral amygdalofugal pathways, and third with the intraamygdaloid connections.

I.   THE STRIA TERMINALIS

Since complete correlation between the various descriptions of
the components of the stria terminalis given in the literature
and the findings reported here is beset with difficulties, the
following account will refer to these components according to
their relative positions within the supracapsular portion of the
looped course of the stria. Thus, the stria may be considered as
comprised of three parts: (1) a dorsal or subventricular component, (2) a ventral or juxtacapsular component and (3) a
commissural component.

A.   The Dorsal Strial Component

Lesions damaging the superficial or subventricular portion
of the stria terminalis, or the nuclear group formed by the medial
and cortical amygdaloid nuclei cause Wallerian degeneration of
fibers which travel rostrally in the bundle just beneath the
ventricle (Fig. 3). At the level of the anterior commissure, this
contingent of degenerating fibers divides into three streams
arranged loosely which, because of their topographical relation-

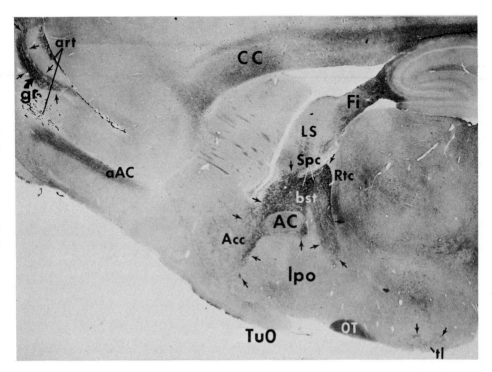

Fig. 4. The same case as Fig. 3 but the section passes now
through the bed nucleus of the stria terminalis (bst) showing the
distribution of the supracommissural (Spc) and retrocommissural
(Rtc) divisions of the DST as indicated by the terminal degeneration in the bst. Modified cupric-silver method. x10. The
terminal degeneration in the basal part of the lateral septal
nucleus (LS), in the nucleus accumbens septi (Acc) and the
internal granular layer (gr) of the accessory olfactory bulb
belongs to the parolfactory radiation of Spc. The terminal degeneration indicated at Diepen's nucleus tuberis lateralis (tl)
belongs to the hypothalamic radiation of Spc.

Fig. 5. (Opposite page). Parasagittal section through the
forebrain of a young rat to show the distribution pattern of
terminal degeneration after lesions confined to the dorsal strial
component. Modified cupric-silver method. x 11. The terminal
degeneration in the pars medialis of the anterior olfactory
nucleus (aom) and in the pars medialis of the olfactory tubercle
(TuOm) comes from the parolfactory radiation of the supracommissural division. The terminal degeneration in the medial
preoptic-hypothalamic junction area (mph), in the capsule surrounding the ventromedial hypothalamic nucleus (vm) and in the dorsal
premammillary nucleus (dpm) marks the distribution of its
hypothalamic radiation (hr).

Fig. 8 (Opposite page). A higher magnification of the terminal
degeneration (arrows) found in the internal granular layer of
the accessory olfactory bulb as shown in a frontal section of a
brain in which the dorsal strial component was damaged. Modified
cupric-silver method. x 230.

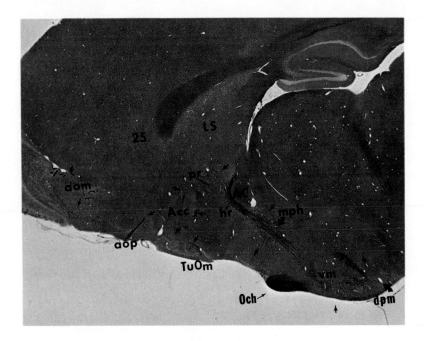

Fig. 5.

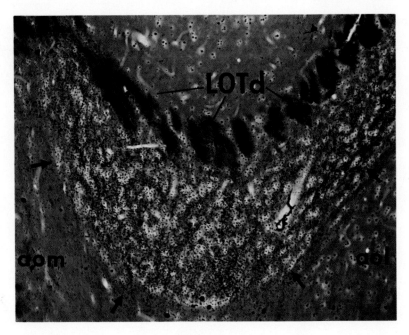

Fig. 8.

ship to the commissure are defined as the supracommissural, retrocommissural and commissural divisions of the dorsal componet
(Fig. 4). These three divisions account for the heavy terminal
degeneration in those portions of the bed nuclei of the stria
terminalis (Fig. 4, bst) and of the anterior commissure through
which they pass.

The supracommissural division, which is the most voluminous,
divides further into parolfactory (or precommissural proper) and
hypothalamic radiations (Fig. 5) (cf. Johnston, 1923). Through
its parolfactory radiation the supracommissural division supplies
the laterobasal septum (Figs. 4, 5, 18) (cf. Cajal, 1911; Fox,
1943; Knook, 1965; Valverde, 1965; Ishikawa et al., 1969), the
posteromedial aspect of the nucleus accumbens septi (Figs. 5, 6)
(cf. Fox, 1943; Gloor, 1955; Cowan et al., 1965; Knook, 1965, etc.),
and after traversing the fiber stratum (diagnonal band) intervening
between the nucleus accumbens and the olfactory tubercle, enters
the posteromedial portion of the latter paleocortical formation
(Figs. 5, 6). Here, the terminal degeneration is disseminated
profusely among cell bodies of its pyramidal layer, and invades
also the deepest portion (Ib) of its external plexiform layer.
The small islands of Calleja, within the polymorph layer, and
this layer itself are free of argyrophilic granules (cf. Beccari,
1910; Hilpert, 1928; Marburg, 1948; Klingler and Gloor, 1960).

Other areas in which termianl degeneration occurs are: Rose's
(1912) cortical area praegenualis 25 in the medial frontal cortex,
the partes posterior and medialis of the nucleus olfactorius anterior in the olfactory peduncle (Figs. 5, 7) (cf. Marburg, 1948),
and the internal granular layer of the accessory olfactory bulb
(Figs. 3, 4, 8). The bundles reaching these structures branch off
from the strial parolfactory radiation and appear to run diffusely
along the sulcus limitans septi and the olfactory ventricular
cleft (ov).

The degenerative changes occurring at the level of the
cortical area praegenualis 25 are limited to its posterior
portion and to its layer V sparing almost completely the remaining
layers I, II-III and VI. Similarly, the degeneration in the pars
medialis of the anterior olfactory nucleus is confined to its
dorsal portion (area praepiriformis 51e of Rose, 1912), and
shows a layered pattern of distribution in such a manner that only
the deep polymorph-celled layer III and the inner half (or sublamina tangentialis Ib) of the external plexiform layer are
recipients of the stria parolfactory projection. This pattern of
distribution appears to favor the view sustaining a cortical
nature of the gray formations in the olfactory peduncle, and
which places them among the paleocortical structures.

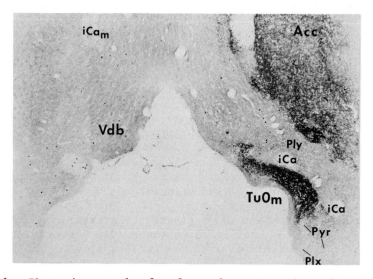

Fig. 6. Photomicrograph of a frontal section through the parolfactory region to show the distribution pattern of the terminal
degeneration after a lesion confined to the dorsal strial component. Modified cupric-silver method. x 45. The degenerating
terminals are located in the pars medialis of the olfactory
tubercle (TuOm) and in the nearby nucleus accumbens septi (acc).
Note that the polymorph (Ply) and plexiform (Plx) layers of the
olfactory tubercle as well as the small (iCa) and medial (iCam)
islands of Calleja are free of such terminals.

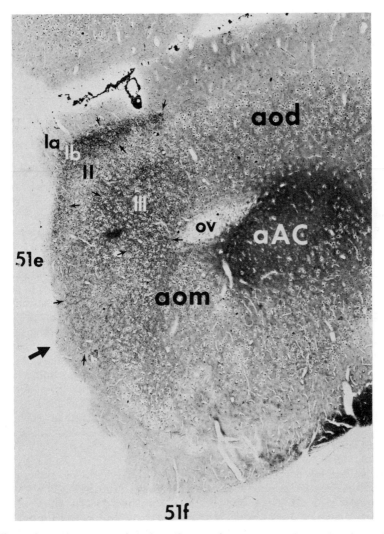

Fig. 7. Photomicrograph of a frontal section through the olfactory peduncle to show the distribution pattern of the terminal degeneration in this region after a lesion in the dorsal strial component. Modified cupric-silver method. x 92. Degenerating terminals fill the deep polymorph-celled layer III and the inner sublamina Ib of the external plexiform layer of the dorsal portion of the anterior olfactory nucleus, pars medialis (aom). The big arrow marks the approximate location of the so-called medial olfactory tract, as indicated in König and Klippel's atlas of the rat brain, the boundary between the dorsal (51e) and ventral (51f) portions of the nucleus and the location of the superficial offshoots of the parolfactory radiation.

(Fig. 4), while from the caudoventral end of the above
mentioned capsule another group of fibers distribute to the
dorsal (Fig. 5) and ventral premammillary nuclei (Fig. 10). No
other hypothalamic nuclei seem to receive an afferent supply from
this fiber system (cf. Ban and Omukai, 1959; Lundberg, 1960;
Hall, 1963; Knook, 1965; Ishikawa, 1969; Heimer and Nauta, 1969;
Leonard and Scott, 1971).

The retrocommissural division of the dorsal strial component
(Fig. 4), on the other hand, accounts for a massive but diffuse
terminal degeneration in the postcommissural part of the bed
nucleus of the stria terminalis. Fibers from this division appear
to reach also the medial preoptic-hypothalamic junction area where
their terminal arborizations overlap to some extent those of the
supracommissural-hypothalamic system. No contribution from this
division is traceable to other hypothalamic nuclei (cf. Heimer
and Nauta, 1969; Leonard and Scott, 1971a).

Finally, the small commissural division of the dorsal strial
component, after crossing the midline in the dorsalmost stratum
of the anterior commissure, and distributing some degenerating
terminals to the contralateral bed nuclei of the anterior
commissure and of the stria, swing dorsolaterally, enter the
contralateral stria and reach the amygdala where it ends mostly
in the caudal one-third of the cortical amygdaloid nucleus. A
small band of terminal degeneration also is seen along that
portion of the medial amygdaloid nucleus which is traversed by
this fiber system. Interestingly, the degeneration in this
commissural connection is reinforced in other experiments in
which there were lesions in the bed nucleus of the stria terminalis as is illustrated in Fig. 11 (cf. Lammers, 1971).

Data provided by comparison of the effects of total transection of the dorsal stria component with those obtained from
partial interruption of this bundle, or, further, with those
acquired by variously localized lesions within the caudal portions
of the corticomedial nuclear group of the amygdala, suggest the
existence of a mediolateral organization within the system.

Thus, cases with lesions involving medially located fiber
contingents show very abundant terminal degeneration in the accessory olfactory bulb and in the more medial hypothalamic nuclei.
This contrasts with the very sparse degeneration which can be
detected in the remaining structures listed previously as
recipients of the dorsal stria component. Conversely, the richness of the terminal degenerative changes in the nucleus
accumbens septi, olfactory tubercle and Diepen's nucleus tuberis
lateralis with lesions affecting the more laterally coursing
fibers point to the above gray formations as the chief points of

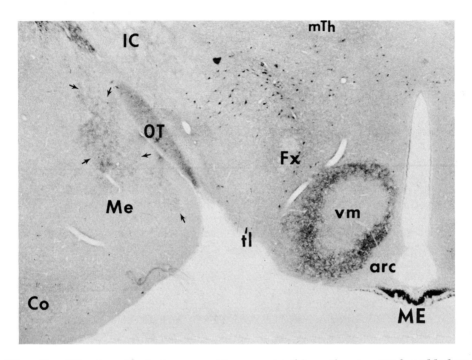

Fig. 9. The terminal degeneration encircling the central cellular
core of the ventromedial nucleus (vm) as seen in a frontal section
through the mid-tuberal level of the hypothalamus. Original
cupric-silver method. x 36. Note that the terminal degeneration
does not invade the area of Diepen's nucleus tuberis lateralis
(tl). For the terminal degeneration in the rostral portion of the
medial amygdaloid nucleus refer to the description of Fig. 10.

    The hypothalamic radiation of the dorsal stria component passes
caudally and ventrally toward the basal tuberal region of the
hypothalamus. Along its journey it sends terminal projections to
an ovid area in the central sector of the boundary zone between
the medial preoptic and anterior hypothalamic nuclei (to be called
subsequently the medial preoptic-hypothalamic junction area, mph)
(Figs. 5, 18) and, in a much reduced number, to the retrochiasmatic area of the hypothalamus. Most of the fibers, however,
proceed farther caudally and terminate in the cell-poor capsule
encircling the ventromedial hypothalamic nucleus (Figs. 5, 9).
Other more laterally coursing fiber contingents appear to account
for terminal degeneration in the parvocellular tuberal gray
beneath the fornix [Diepen's (1962) nucleus tuberis lateralis, tl]

termination of the intermediate portions of the dorsal strial
component.   Brains with lesions placed even more laterally and
which cut the lateral margin of the bundle, exhibit also degeneration of a delicate bundle of fibers terminating in a small
area in the ventrolateral aspect of the postcommissural portion
of the bed nucleus of the stria terminalis.   This bundle perhaps
represents in the rat brain the Johnston (1923) infracommissural
component or bundle 3.

In cases with lesions confined to the caudal or laminar
portion of the medial amygdaloid nucleus, the pattern of degeneration (Figs. 9, 10) resembles closely that described after interruption of the medial sectors of the dorsal strial component by
a lesion such as that shown in Fig 12a.   Furthermore, lesions
involving the caudomedial one-third of the cortical amygdaloid
nucleus (Fig. 12b) lead to the same sort of picture as the one described in the instance of lesions affecting the intermediate portions of the component under discussion except for the presence of
a considerable amount of terminal degeneration in the accessory
olfactory bulb.   Finally, in one brain in which the lesion is
lateral to that in Fig 12b (see also Fig. 2b) a different pattern
of degeneration is present in which no terminal degeneration can
be detected in the accessory olfactory bulb nor in the medial
hypothalamic nuclei.

Other experimental material demonstrates in addition that the
rostral portions of the corticomedial nuclei of the amygdala do
not contribute to the formation of the dorsal strial component.

From the above series of experiments, it appears that the caudal portion of the cortical and medial amygdaloid nuclei contribute to the formation of the dorsal strial component (cf. Ban and
Omukai, 1959; Lundberg, 1960; Sanders-Woudstra, 1961; Hall, 1963;
Knook, 1965; Valverde, 1965; Ishikawa et al., 1969; Leonard and
Scott, 1971).   Another finding is that this posterior part of the
cortical nucleus sends stronger projections to the rostral paleocortical formations than to the diencephalon and that a reverse
pattern appears to be true for the medial amygdaloid nucleus.   The
writer is inclined to such a viewpoint.   However, one must consider that lesions in the posterior subventricular portion of the
medial amygdaloid nucleus may cause coincidental damage to axons
which traverse it after originating in the posterior part of the
cortical nucleus.

Furthermore, it must be considered that the posterior subventricular portion of the medial amygdaloid nucleus covers, like
a cup, that part of the cortical nucleus supposedly concerned in
the formation of the dorsal strial component.   This laminated or
cup-like extension of the medial amygdaloid nucleus can be

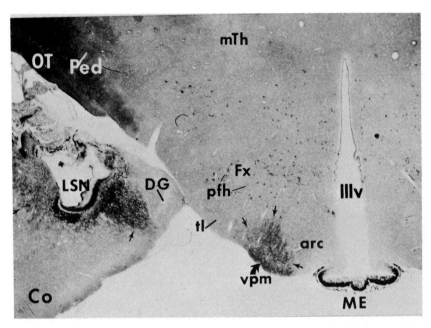

Fig. 10. Photomicrograph of a frontal section through the caudal
tuberal region of the hypothalamus of the same case as that in
Fig. 9, to show terminal degeneration in the ventral premammillary
nucleus (vpm). Original cupric-silver method. x 28. In this
case the Wallerian degeneration of the dorsal strial component was
produced by a lesion (LSN) in the caudal portion of the medial
amygdaloid nucleus. The terminal degeneration in the hypothalamus as in Fig. 9 does not extend into the area of Diepen's nucleus
tuberis lateralis. On the other hand, the terminal degenerative
changes in the amygdala are confined to the remaining portions of
the caudal medial amygdaloid nucleus and extend also to the rostral
portions of this nucleus as shown in Fig. 9. In both Figs. 9 and
10 the dark spots diffusely scattered through that level of the
hypothalamus are granular argyrophilic neurons (de Olmos, 1969).

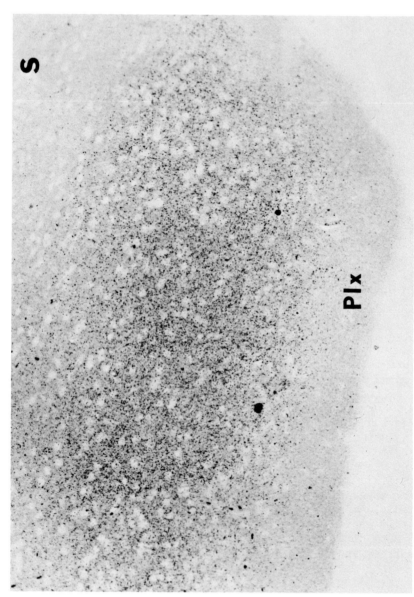

Fig. 11. Photomicrograph of a frontal section through the caudal pole of the amygdala to show terminal degeneration in the caudal portion of the cortical amygdaloid nucleus contralateral to a lesion in the bed nucleus of the stria terminalis. Modified cupric-silver method. x 221. Note that the terminal degenerative changes are confined to the cellular portion of the nucleus, sparing its plexiform layer (Plx).

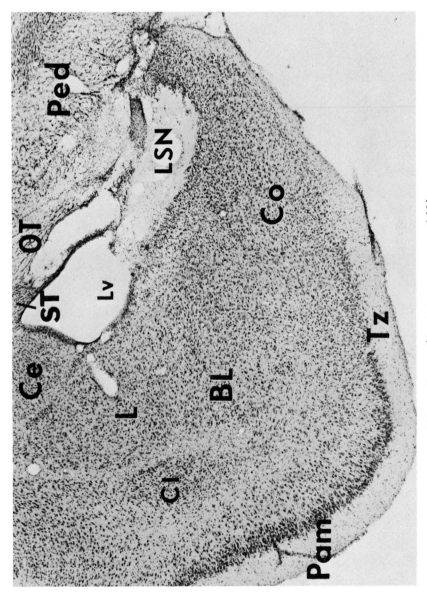

Fig. 12a. (Caption on page 163).

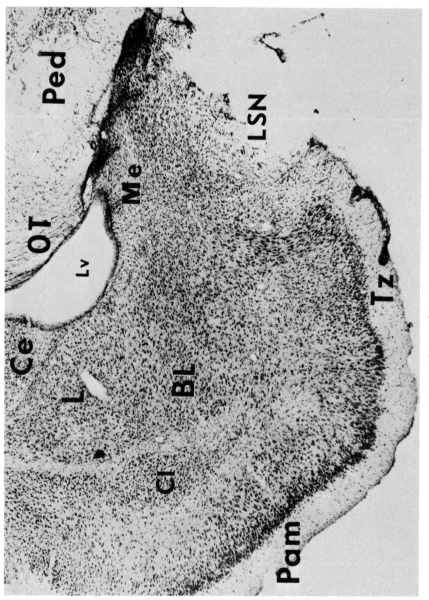

Fig. 12b. (Caption on page 163).

affected either by too extensive coagulation of the underlying
target nucleus or simply by the electrode track. This consideration is, of course, without significance if one considers,
as Uchida (1950) does, that the caudal part of the medial amygdaloid nucleus is part of the cortical nucleus.

B. The Ventral Strial Component

After lesions injuring the ventral portion of the stria
terminalis (sometimes called the preoptic or postcommissural component or bundle), either as it leaves the amygdala or, more
rostrally, in the dorsal portion of the retrocommissural part of
its bed nucleus, a degeneration picture has been found which
differs in some important features from those described by other
workers, perhaps because these fibers are rather feebly stainable.

The bulk of this fiber system radiates downward and caudally
through the retrocommissural portion of the bed nucleus of the
stria terminalis, to which it contributes abundantly in its
lateral portions (Fig. 13). Farther on part of it reaches and
terminates in the medial preoptic hypothalamic junction area
(mph). Other contingents enter the basal tuberal region of the
hypothalamus where they account for terminal degeneration which

---

Figs. 12a and 12b. (pp. 161, 162) Photomicrographs of Nissal
frontal sections from lesions placed in the caudal portions of the
medial (12a) and cortical (12b), amygdaloid nuclei, respectively,
which evoke patterns of terminal degeneration within the projection field of the DST suggesting a differential contribution
and distribution of the fibers arising from either amygdaloid gray
masses. x22. In Fig. 12a, the lesion (LSN) (cf. Fig. 2a) is
mostly limited to the caudal laminar portion of the medial
amygdaloid nucleus, but also involves the caudomedial part of the
stria. The picture of terminal degeneration in the hypothalamus
following this lesion is similar to that in the case illustrated
by Figs. 9 and 10. In the rostral telencephalon of both cases
the internal granular layer of the accessory olfactory bulbs
shows abundant degenerating terminals but this is minimal in the
nucleus accumbens septi and olfactory tubercle. In Fig. 12b
the rostromedial half of the caudal portion of the cortical
amygdaloid nucleus has been destroyed (LSN) by a medial oblique
approach, being almost totally limited to that structure (cf.
Fig. 2b). The distribution of the terminal degeneration in the
above listed areas follows a reverse pattern than in the case
illustrated in Figs. 9 and 10 and in 12a.

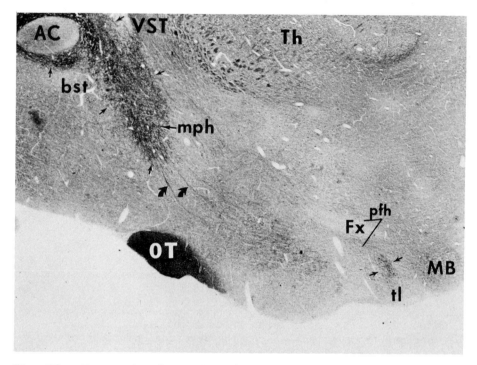

Fig. 13. Parasagittal section through the perifornical hypothalamic area (pfh) to show the arrival of the degenerating fiber
contingents and distribution of the terminal degeneration of the
ventral strial component. Modified cupric-silver method. x32.
The thick small arrows point to fine degenerating fibers from
this component leaving the area of terminal degeneration in the
bed nucleus of the stria terminalis (bst) and medial preoptic
hypothalamic junction area (mph) to reach eventually the parvocellular lateral tuberal area or Diepen's nucleus tuberis
lateralis (tl) where terminal degeneration is also present.

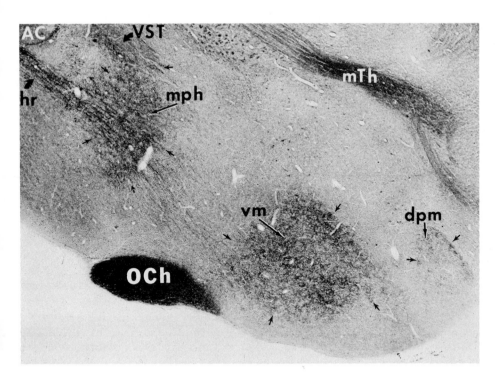

Fig. 14. A parasagittal section through the hypothalamus of a
young rat to show the distribution pattern of terminal degeneration (small arrows) in different gray formations after a total
transection of the stria terminalis. Modified cupric-silver
method. x25. The filling of the whole of the ventromedial
hypothalamic area marks the effects of the involving of the ventral
strial component (VST). Compare with Fig. 5.

fills Diepen's nucleus tuberis lateralis (Fig. 13, tl), part of
the retrochiasmatic area, and the dorsal and ventral premammillary
nuclei (Figs. 10 and 14). The most remarkable finding is that
terminal degeneration is spread throughout the whole area of the
ventromedial hypothalamic nucleus (Figs. 14, 15, 16) (cf. Lammers,
1971). Furthermore, another much reduced field of termination
extends along the anterolateral portion of the bed nucleus of the
stria. No other terminal degeneration can be found in the hypothalamus nor do any degenerated fibers of this component become
incorporated in the medial forebrain bundle as some authors
have suggested (cf. Cajal, 1911; Johnston, 1923; Sanders-Woudstra,
1961; Valverde, 1965; Millhouse, 1969), unless those fibers which
reach Diepen's nucleus tuberis lateralis can be considered as

part of this bundle. Terminal degeneration seen in other brain areas after such lesions may be considered to be due to incidental damage of fiber systems which pass near the sites of the lesions.

Since the interruption of this component in the extraamygdaloid portion of its course necessarily causes coincidental damage to other components, lesions were placed within the amygdala which involved in varying degree, together or singly, different amygdaloid nuclei and portions of the ventral component. From such experiments, the following conclusions were drawn:

a) The ventral component is composed of two divisions, medial and lateral, separated one from the other by the so-called commissural component of the stria.

b) Damage to the medial division accounts for the terminal degeneration in the ventromedial hypothalamic nucleus, the pre-mammillary nuclei, etc., while the lateral contingent or division seems to end almost exclusively in the lateral portion of the bed nucleus of the stria terminalis.

c) The anterior amygdaloid area does not appear to participate in the formation of this nor of any other component of the stria terminalis (cf. Fox, 1943; Adey and Meyer, 1952; Nauta, 1961; Valverde, 1965).

d) The group formed by the lateral and basal lateral amygdaloid nuclei together with the central amygdaloid nucleus give origin to the lateral division of the ventral strial component (cf. Ban and Omukai, 1959; Ishikawa et al., 1969; Leonard and Scott, 1971) (Fig. 17).

However, involvement of the central amygdaloid nucleus poses some problems since lesions in this nucleus usually produce coincidental damage to fibers passing from the laterobasal complex to this portion of the stria terminalis. Thus, in one case with a lesion well localized to this nucleus the pattern of degeneration along the lateral portions of the bed nucleus of the stria matches very closely that seen after lesions placed in the basolateral cell group, except for the additional presence of terminal degeneration in the bed nucleus of the anterior commissure homo- and contra-laterally as well as in other contralateral structures. The latter could be due to unavoidable damage to the so-called "commissural component" in such lesions. A very similar case is described by Valverde (1965, case 3), who, in addition, in his Golgi studies of the rat and cat amygdalae was able to trace the axons of the neurons in the central amygdaloid nucleus toward the stria, in contrast with the negative experimental results of Fox (1943) and Lammers and Lohman (1957).

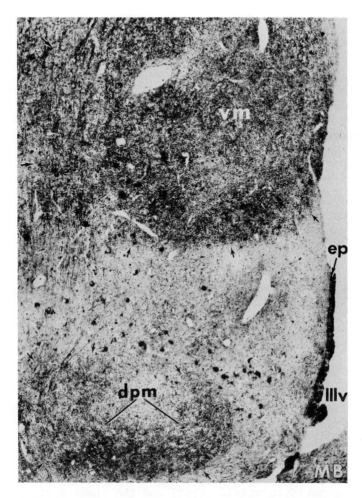

Fig. 15. A horizontal section through the left hypothalamus to
show the distribution of the terminal degeneration in the ventromedial hypothalamic nucleus (vm) and in the dorsal premammillary
nucleus (dpm) 36 hours after a total transection of the ipsilateral
stria terminalis. Original cupric-silver method. x86. As in
Fig. 14 the filling of the central cellular core of the ventromedial hypothalamic nucleus marks the effects of the involvement
of the ventral strial component. The dark gross spots scattered
in the area between the ventromedial and the dorsal premammillary
nuclei are granular argyrophilic neurons.

Fig. 16. (Opposite page.) Photomicrograph to compare results
obtained with modifications of the Nauta-Gygax (1954) (N) and
Fink-Heimer (1967) (FH) techniques adapted for the material and
that with the original cupric-silver method (CU) in staining the
terminal degeneration in the cellular core of the ventromedial
hypothalamic nucleus after a small lesion which destroyed part of
the rostral portion of the medial amygdaloid nucleus with the consequent degeneration of the ventral strial component. The
density of the terminal degenerative changes correlates well with
the volume of tissue damaged in the amygdala and this was used as
a means of checking the staining capabilities of the three
methods. x 360.

Fig. 18. (Opposite page.) Frontal section through the caudal
portion of the anterior commissure of a young rat to show the
distribution of terminal degeneration 36 hours after production
of a lesion which damaged both dorsal and "commissural" components
of the stria on the right side. Original cupric-silver method. x 25.
The terminal degeneration filling the rostral part of the bed
nucleus of the stria terminalis (bst) and the basal part of the
lateral septal nucleus (LS) must be due to degeneration of the
right DST, whose hypothalamic radiation (hr) has also degenerated
as seen ventral to the anterior commissure. The thick arrows
point to the decussating degenerating axons composing the so-
called "commissural" component (CST), in which the small arrows on
the left mark the areas of terminal degeneration in the contralateral bed nucleus of the stria terminalis.

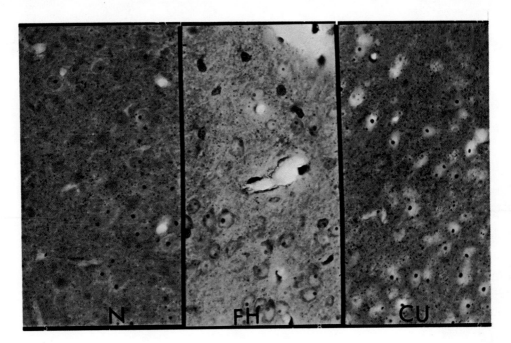

Fig. 16.

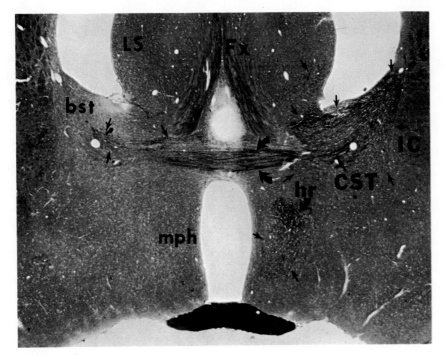

Fig. 18.

e) The rostral two-thirds of the cortical and medial amygdaloid nuclei and perhaps also the basal medial nucleus seem to contribute to the formation of the medial segment of the ventral strial component.

The present experimental series provides substantial evidence for the above statement (cf. Leonard and Scott, 1971). The other implications, i.e., the arrival and eventual termination of the medial division of the ventral component in the nuclei of the medial tuberal hypothalamus, need some consideration here particularly in view of the negative results obtained by authors who employed the Fink-Heimer technique (Heimer and Nauta, 1969; Leonard and Scott, 1971).

In the first place, the possibility exists that the new findings reported here may be a consequence of a major affinity of the products of Wallerian degeneration for the staining procedure used here because of such factors as the type or size of the lesions, age of the animals, survival periods, and pH of the fixing solutions.

In the second place, one must consider the possibility that lesions of the stria, whether total or involving only the ventral component may produce degeneration of axons which originate in the slender strans of bed nuclear cells which are distributed marginally along the supracapsular portion of the stria. In this connection, one must propose that the destruction, never complete, of such a small rim of neurons cannot alone account for the terminal degeneration in the projection field of the ventral component as described and illustrated here.

A third possibility exists in cases of ventral strial transections or lesions in the rostral portions of the medial and cortical amygdaloid nuclei, wherein degeneration appearing in the ventromedial hypothalamic nucleus may be due to coincidental damage of amygdalo-hypothalamic pathways which run ventrally to reach that nucleus (Szentágothai et al., 1962; Ishikawa et al., 1969).

Regarding this third hypothesis the following considerations argue against its relevance: 1) Golgi studies by Valverde (1965) and Millhouse (1969) do not offer evidence for the existence, at least in the rat brain, of a direct ventral amygdalo-hypothalamic pathway as has been reported for the cat by some authors (Szentágothai et al., 1962; Ishikawa et al., 1969); and 2) in the present experiments lesions placed directly between the amygdala and the hypothalamus failed to reproduce the pattern of terminal degeneration which characteristically follows damage to the ventral strial component or its nuclei of origin in the

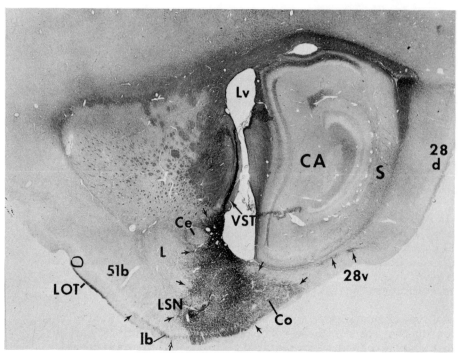

Fig. 17. Photomicrograph of a sagittal section through the
lateral portion of the stria terminalis and the amygdala to show
the anterograde degeneration of the lateral division of the ventral strial component (VST) following a lesion (LSN) which destroyed the basal lateral nucleus, basal medial ventromedial
portion of the caudal parvocellular part of the lateral nucleus
(L). The destruction also encroaches upon the deepest layers of
the periamygdaloid area 51d. Modified cupric-silver method. x10.
The area of terminal degeneration (arrows) covers the caudal
portions of the basolateral nuclear complex and of the central
amygdaloid nucleus (Ce) and all the posterior portion of the
cortical nucleus. Ventral to the lesion the terminal degeneration
also fills the remaining layers of the periamygdaloid cortical
field affected by the damage. Rostral to it, terminal degeneration is seen in the sublaminatangentialis (Ib) of the external
plexiform layer of the periamygdaloid field extending between the
latter and the area praeperiformis 51b. Caudally the small
arrows point to the degeneration in the deep layer of the ventromedial portion of the entorhinal cortex (28v). Terminal degeneration is also present in its layer 1 but cannot be visualized at
this magnification. The rostral sectors of the central amygdaloid nucleus and the anterior magnocellular portion of the
laternal amygdanucleus (L) contain degenerating terminals but
these are not visible at the magnification used, in marked contrast with the remainder of the amygdala.

amygdala (cf. Heimer and Nauta, 1961; Chi, 1970; Eager et al.,
1971; Leonard and Scott, 1971).

C. The "Commissural" Component

Lesions involving the nucleus of the lateral olfactory tract,
or the region through which the compact fascicle which originates
in it passes, provoke Wallerian degeneration of Johnston's (1923)
commissural component or bundle 1. Before and after its decussation in the posteroventral aspect of the anterior commissure, the
bundle sends off a dense terminal projection to the ipsi- and
contralateral bed nuclei of the anterior commissure (Fig. 18)
(cf. Lammers and Lohman, 1957; Valverde, 1965; Millhouse, 1969;
Leonard and Scott, 1971). Subsequently, it splits into two contingents. One of these, the dorsal division, just before swing-
ing dorsolaterally into the contralateral stria terminalis, distributes some terminals to a small area of its bed nucleus (Fig.
18) (cf. Valverde, 1965; Heimer and Nauta, 1969; van Alphen,
1969; Leonard and Scott, 1971). Next, this component reaches the
amygdaloid region where its terminals especially are concentrated
in the caudolmedial extreme of the magnocellular portion of the
lateral amygdaloid nucleus.

The other contingent or ventral division of the "commissural"
bundle joins the posterior limb of the anterior commissure (cf.
Cajal, 1911; Berkelbach van der Sprenkel, 1926; Gurdjian, 1929;
Knook, 1965; Morgan, 1968) and its terminals can be identified in
the cell masses surrounding this bundle (here called interstitial
nucleus of the posterior limb of the anterior commissure, ipac,
Figs. 22, 31) and in the anterior magnocellular portion of the
lateral amygdaloid nucleus as well as in the most medial portion of
the area prepiriformis 51a at the periphery of its pyramidal celled
layer II (Fig. 19) and in the convoluted foldings of the pyramidal
layer of the anterolateral portion of the olfactory tubercle. No
degenerating terminals could be found in the contralateral nucleus
of the lateral olfactory tract. From these descriptions, it is
apparent that the so-called "commissural" component of the stria
terminalis is a decussation rather than a commissure.

With respect to the sources of this component, the massive
degeneration undergone by the bundle after lesions which destroyed
the nucleus of the lateral olfactory tract but spared all of the
other amygdaloid nuclei not only confirms the traditional view,
but leaves little basis for consideration of additional sources.
Thus, although it is true that extensive lesions of the central
amygdaloid nucleus result also in degeneration of the commissural
bundle (cf. Valverde, 1965), such effects can be attributed to
the unavoidable interruption of the bundle in question which this
type of lesion involves. Support for this proposal may be found
in the fact that lesions which encroached upon the caudolateral
portion of the central nucleus failed to evoke the kind of degenerative picture under discussion. Likewise, lesions of other

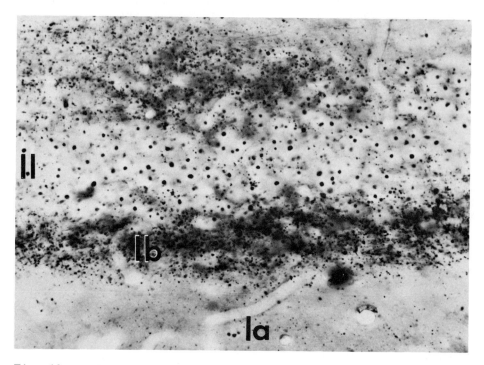

Fig. 19. A horizontal section showing the terminal degeneration
in the area praepiriformis 51a after a lesion of the contralateral
"commissural" component of the stria. Modified cupric-silver
method. x 644. The terminal degeneration encapsulates the superficial pyramidal-celled layer II, invading the sublamina tangentialis Ib of the external plexiform layer of this olfactory area.
Very dense terminal degeneration is also seen on the deep (upper)
side of layer II while fewer silver granules are spread among the
cell bodies of the neurons forming this layer.

amygdaloid nuclei which spared the nucleus of the lateral olfactory tract or the compact bundle of fibers emerging from it,
were equally ineffective.

II. THE VENTRAL "AMYGDALOFUGAL" SYSTEM

It now is accepted generally that in the rat brain the so-
called ventral amygdalofugal systems originate mostly, if not
entirely, from the periamygdaloid cortex (Cowan et al., 1965;
Leonard and Scott, 1971), and evidence for participation of the
basolateral amygdala in their formation has been adduced only in
higher species (Nauta, 1961; Valverde, 1965). Therefore, in the
following account data obtained after lesions of the periamygdaloid
cortex with varying degrees of involvement of the basolateral
amygdaloid nuclei, and from experiments in which the latter were
destroyed together with adjacent portions of the external capsule
will be used to show the total projection field of these structures.
Coincidental consideration will be given to the degenerative
effects of smaller lesions along the intraamygdaloid courses of
these pathways as well as those due to injury of nuclei which
might contribute to them.

A. Telencephalic Projections

From the sites of the lesions very fine degenerating fibers
and terminals form a continuous band along the sublamina tangentialis Ib or inner half of the external plexiform layer at the level
of the ventrolateral part of the entorhinal area 28, the remaining
portions of the periamygdaloid areas 51b (Fig. 17) and 51c, and
very densely, in the amygdalo-piriform transitional area 51d. This
pattern of degeneration does not extend rostrally very far from
the lesions. This short superficial associational pathway probably
represents at least part of Kreiner's (1949) accessory association
tract (see also Cragg, 1961; Powell et al., 1965; Valverde, 1965).

Degenerating elements of a longer association pathway are
traceable along the deep fibrillar plexus of the periform cortex
(Cajal's 1911 "voie sagittale d'association"; see also Kreiner,
1949; Sanders-Woudstra, 1961; Powell et al., 1965; Valverde, 1965).
These deeply running fiber contingents not only supply the deep
cell layers of the region, but also its sublamina tangentialis Ib.
This pattern of distribution is found rostrally as far as the
caudal portions of the prepiriform areas 51a and b, and caudally
are limited to the ventrolateral part of the area entorhinalis 28
(cf. Powell et al., 1965; Raisman et al., 1965).

Other degenerating elements are incorporated into the ventral
part of the external capsule and reach the claustrum, all parts of
the olfactory tubercle, the cortical area praegenualis 25 in the
medial frontal cortex, and the partes posterior, dorsalis and
medialis of the anterior olfactory nucleus (cf. Lammers and Lohman,
1957; Sanders-Woudstra, 1961; Nauta, 1961; Powell et al., 1965;
Valverde, 1965). Other fibers of this capsular system enter the

posterior limb of the anterior commissure, cross the midline and distribute to the pars anterior of the lateral amygdaloid nucleus and the interstitial nucleus of the posterior limb of the anterior commissure.

The distribution pattern of terminal degeneration in the above listed areas is largely as follows:

Degenerating terminals are distributed diffusely among the cell bodies of the claustrum and pars posterior of the anterior olfactory nucleus. The pyramidal layer of the olfactory tubercle presents a similar picture, although the argyrophilic granules which mark the presence of terminal degeneration appear to be more concentrated at the periphery of this layer. There is little invasion of the plexiform and other polymorphic layers of this paleocortical formation. In the area praegenualis 25, on the other hand, terminal degeneration is present throughout layers I, II-III and V but not in layer VI of this juxtallocortical field. Finally, the projection to the partes dorsalis and medialis of the anterior olfactory nucleus shows the same layered pattern already described for those piriform areas receiving afferent projections from the deep associational pathway, i.e., the terminal degeneration is confined to both the sublamina tangentialis Ib or inner half of the external plexiform layer, and the deep polymorph-celled layer III. Obviously, all of the above described connections contribute to the piriform-prepiriform association system already described by Cajal (1911).

At this point it is interesting to mention that lesions encroaching upon the anterior amygdaloid area and/or the caudal end of the area praepiriformis 51a and the nucleus of the horizontal limb of the diagonal band, respectively, evoke terminal degeneration in all of the divisions of the anterior olfactory nucleus although maintaining the layered pattern which has been indicated. However, only in those cases with involvement of the nucleus of the horizontal limb of the diagonal band are degenerating terminals found in the main olfactory bulb (cf. Sanders-Woudstra, 1961; Heimer, 1968; Price and Powell, 1970a).

On the other hand, lesions involving the rostral and dorsal portions of the medial amygdaloid nucleus and/or the basal medial nucleus and/or the medial aspect of the basolateral amygdaloid complex, alternatively, cause degeneration of a caudally oriented fantail system of fibers which supplies, besides other portions of the amygdaloid complex, the ventrolateral portion of the area entorhinalis 28 and also its ventromedial part (Fig. 17, 28), where its terminals are more heavily concentrated in the layers I and IV. Furthermore, some medial fiber contingents appear to reach also the molecular and cellular layers of the ventral subiculum

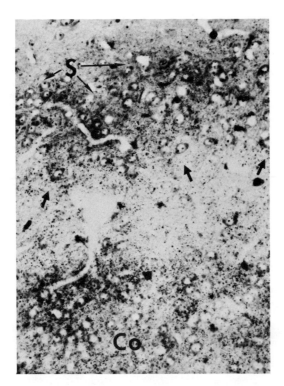

Fig. 20. Photomicrograph of a cross section through the caudal end
of the cortical amygdaloid nucleus (Co) showing the terminal degeneration in this nucleus and in the rostral end of the ventral
subiculum (S) after a lesion in the rostral portion of the corticomedial amygdala probably causing interruption of fibers coming
from extraamygdaloid sources, perhaps the lateral preoptic area
(see text). Original cupric-silver method. x270. The thick
arrows mark the lower boundary of the terminal degeneration within
the subiculum.

(Fig. 20) (cf. Nauta, 1959; Cragg, 1961; Cowan et al., 1965;
Valverde, 1965; Shute and Lewis, 1967). A careful examination of
cases with smaller and more caudal lesions reveals a lateromedial
arrangement within this fantail system whereby, for instance, the
more laterally placed lesions provoke terminal degeneration only in
the lateral portion of the field, i.e., the ventrolateral part of
the area entorhinalis 28. The medial lesions have a contrasting
medial distribution. With regard to the possibility that this
caudal fantail system has at least a partial origin within the
amygdala the present experiments do not allow definitive conclusions since the size and location of the lesions necessarily

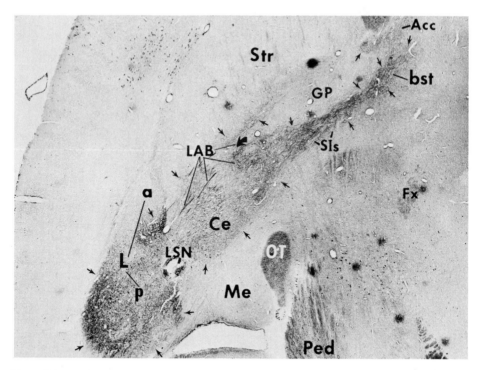

Fig. 21. Photomicrographs of a horizontal section through the sublenticular portion of the substantia innominata (SIs) to show the
distribution pattern of the terminal degeneration following a
lesion confined to the ventral portion of the posterior parvocellular division of the lateral amygdaloid nucleus (Lp) and the
whole of the basal lateral nucleus. Modified cupric-silver
method. x24. At the level of the section only a very small part
of the lesion (LSN) is shown. Terminal degeneration is profuse
in the remainder of the lateral amygdaloid nucleus except at its
rostral magnocellular pole. The central amygdaloid nucleus (Ce),
the subventricular portion of the substantia innominata (SIs),
and the caudoventral pole of the bed nucleus of the stria terminalis (bst) form a continuous field of terminal degeneration.
Degenerating terminals are also present in the nucleus accumbens
septi (Acc). The dorsal and rostral portions of the medial amygdaloid nucleus in contrast do not contain them. In the lateral
margin of the field of degeneration outlining the central amygdaloid nucleus, the most dorsal elements of the longitudinal association bundle (LAB) are seen and also the point (thick arrow)
at which the medial and lateral (or Gurdjian's tractus A) streams
split off.

caused the interruption of some fibers from extraamygdaloid sources
(cf. Cragg, 1961; Valverde, 1965; Shute and Lewis, 1967). However,
in cases bearing lesions in the anterior amygdaloid area, alone or
together with the sublenticular portion of the substantia innominata as well as in the central amygdaloid nucleus, no terminal
degeneration is discernible in the subiculum or ventromedial
entorhinal formation. It does occur in the ventrolateral field of
the area entorhinalis 28.

Superior to those portions of the lesions which encroached
upon the basolateral amygdaloid complex, the degenerating fibers
which follow an intraamygdalar course form in the rat amygdala a
rostral fantail system similar to that described in the cat by
Valverde (1965). This fiber system is formed by axons coming
not only from the piriform lobe but also from the basolateral cell
group of the amygdala, in agreement with Valverde's description.
At the level of the caudoventral end of the central amygdaloid
nucleus the more medially located fibers leave the main bundle,
enter the stria terminalis and form the lateral portion of the
ventral strial component with end stations in the lateral part of
the bed nucleus of the stria terminalis. The remaining fiber
contingent of this rostral fantail system continues rostromedially between the central and lateral amygdaloid nuclei and
forms a considerable part if not all of the "compact" portion of
the rostral projection system of the amygdalo-piriform region
[Johnston's (1923) longitudinal association bundle]. Once it
has reached the level of the cephalic end of the central amygdaloid nucleus, the bundle sends off a diffuse, medially directed
system of fibers which traverses the sublenticular portion of the
substantia innominata, to which it contributes abundantly, and
finally ends in the ventral portion of the bed nucleus of the
stria terminalis (Figs. 21, 24). The remaining main contingent of
the longitudinal association bundle (LAB) proceeds rostromedially
close to the ventrocaudal aspect of the posterior limb of the
anterior commissure and constitutes the tractus A of Gurdjian
(1928) (Fig. 22). This "lateral" division of the longitudinal
association bundle accounts for the terminal degeneration visible
in the ventral part of the caudate-putamen (Figs. 22, 24), the
interstitial nucleus of the posterior limb of the anterior
commissure, the posteroventral part of the nucleus accumbens septi
(Figs. 21, 24) and in the medial part of the ofactory tubercle.
Finally, remaining contingents of this system which now run
scattered through the ventral aspect of the rostromedial striatum
contribute to the terminal degeneration in the partes posterior and
medialis of the anterior olfactory nucleus and in the cortical
area praegenualis 25 (cf. Lammers and Lohman, 1957; Sanders-Woudstra, 1961; Nauta, 1961; Hall, 1963; Cowan et al., 1965; Valverde,
1965; Morgan, 1968; Ishikawa et al., 1969; Price and Powell, 1970).
The pattern of distribution within the latter areas is similar to
that described for the capsular system.

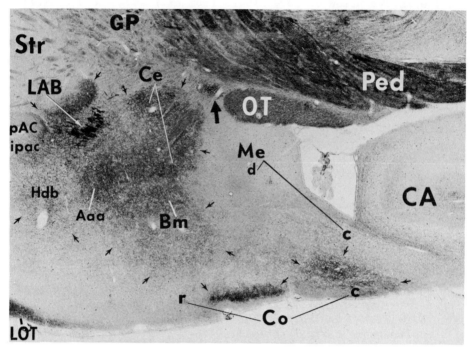

Fig. 22. Photomicrograph of a sagittal section through the caudal
and dorsal portions of the medial amygdaloid nucleus (Me, d and
c, respectively) to show the distribution of terminal degeneration in a case with a lesion (not shown) which encroaches upon
the total extent of the basolateral nuclear complex of the amygdala and the deep layers of the posterior part of the periamygdaloid fields 51b and 51c. Modified cupric-silver method. x24.
The terminal degeneration in the cortical (Co, r and c), basal
medial (Bm) and central (Ce) amygdaloid nuclei might be accounted
for at least in part by the involvement of a diffuse fine-fibered
intraamygdaloid association system. The area of degeneration below the postlenticular portion of the internal capsule (thick
arrow) might also be assigned to the fine-fibered diffuse component of the ventral amygdalofugal pathways. Moreover, the terminal degeneration in the nucleus of the horizontal limb of the diagonal band (Hdb) might belong to the thick-fibered component of the
above diffuse system, while that in the anterior amygdaloid area
(Aaa), the interstitial nucleus of the posterior limb of the
anterior commissure (ipac) and in the ventral part of the striatum is derived from the degenerating longitudinal association
bundle (LAB) or "compact" division of the rostral projection
system of the amygdalo-piriform region. The medial amygdaloid
nucleus, or more specifically its dorsal (Me d) and caudal (Me c)
portions, is almost free of degenerating terminals.

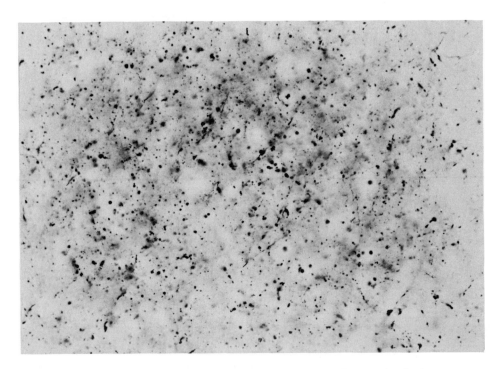

Fig. 23. A higher magnification of the area of terminal degeneration in the nucleus of the horizontal limb of the diagonal band
shown in Fig. 22.   x490.

In addition to the fiber degeneration patterns described above,
the type of lesion constituting the basis for the present description also produces a diffusely arranged fiber system which
follows a more ventral intraamygdalar course than the "compact"
system.   This diffuse amygdalofugal pathway is composed of both
fine and thick fibers.   Its fine-fibered portion, part periform
and part amygdaloid in origin, appears to be concerned mostly
with intraamygdaloid connections, though it might also participate
in the formation of the rat representative of the medial amygdalo-
hypothalamic tract (Valverde, 1965).   The thick-fibered part, on
the other hand, very likely originates to great extent in the
piriform cortex and contributes to the bundles which reach the
diencephalon.   It appears to be responsible for the terminal degeneration seen in the lateral portion of the nucleus of the
horizontal limb of the diagonal band (Fig. 23), and its fibers also
form the pathway which runs through the diagonal band to reach the
precommissural (Fig. 25) and supracommissural hippocampus (cf.
Price and Powell, 1970a).   Finally, a comparison of these data with

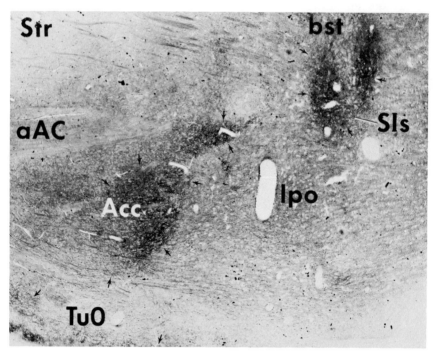

Fig. 24. Photomicrograph of a sagittal section through the area of
junction between the bed nucleus of the stria terminalis (bst) and
the sublenticular portion of the substantia innominata (SIs) in
a case with a lesion affecting the anterior amygdaloid area, part
of the anterior magnocellular portion of the lateral amygdaloid
nucleus and of the lateral part of the central amygdaloid nucleus.
Original cupric-silver method. x 57. While the terminal degeneration in the olfactory tubercle (TuO), the ventral portion of the
nucleus accumbens septi (Acc) are assignable to the interruption
of the more lateral fibers of the "compact" division of the so-
called ventral amygdalofugal pathways, that which extends continuously between the rostromedial end of the sublenticular
portion of the substantia innominata (SIs) the bed nucleus of the
stria terminalis (bst) is very probably due to the lesions in the
central amygdaloid nucleus.

those resulting from more deeply placed intraamygdaloid lesions
allows the following conclusions to be made:

a) Only the periamygdaloid cortex projects to the pre-
commissural and supracommissural hippocampus, and this via a
thick-fibered and diffuse "amygdalofugal" system which becomes
incorporated in the diagonal band.

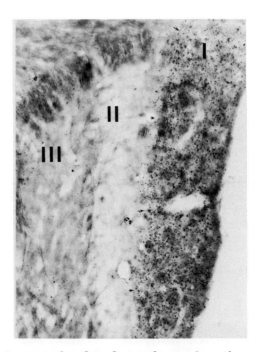

Fig. 25. Photomicrograph of a frontal section through the pre-
commissural hippocampus to show the distribution of the terminal
degeneration in this archicortical formation following lesions
which affected posterior portions of the periamygdaloid cortex.
Modified cupric-silver method. x 260. Note that the terminal
degeneration is entirely confined to the external plexiform layer
I of this formation.

     b)   The "compact" component or longitudinal association bundle
of Johnston supplies the more medial and caudal portions of the
general telencephalic projection field as revealed by lesions of
the periamygdaloid cortex, the basolateral amygdala and the external capsule, combined or individual.

     c)   None of the components of the amygdalo-piriform rostral
projection system reach the medial nuclei of the preoptic-hypothalamic region but are limited to supply its more laterally located
cell aggregations.

     d)   The medial and lateral septal nuclei do not receive
afferent projections via the fiber systems under consideration.
The latter nucleus, however, is a recipient of fibers of the
strial parolfactory radiation, whose source is in the caudomedial

amygdala.

e)   The caudal one-third of the medial and cortical amygdaloid
nuclei do not contribute to the formation of the systems here
considered.   (Figs. 9, 10).

f)   The lateral amygdaloid nucleus does not appear to be the
source of capsular elements which cross the midline in the
posterior limb of the anterior commissure (cf. Brodal, 1948;
Sanders-Woudstra, 1961; van Alphen, 1961).

B.   Diencephalic Projections

Both the "compact" and diffuse divisions of the rostral projection system of the amygdalo-piriform complex participate in the
afferent supply of the diencephalon.

Fibers from the compact groups leave the main bundle during
its journey through the sublenticular region.   This occurs inter-
mittently in such a manner that they become diffusely scattered
through the caudal dorsolateral aspect of the lateral preoptic
area.   Some of them turn medially and end shortly in that part of
the lateral preoptic region which is in contact with the ventrocaudal portion of the bed nucleus of the stria terminalis where
the postcommissural fibers of the stria terminalis end.   Others
run ventrally to end in a rim of small cells close to the dorsomedial margin of the nucleus of the horizontal limb of the diagonal
band.   Still other fibers pass caudally and become incorporated
into the medial forebrain bundle where they can be followed for
only a short distance.   They probably contribute, in part at
least, to the terminal degeneration in the rostral part of the
lateral hypothalamus.

Undoubtedly, the thick-fibered component of the diffuse
division constitutes the major affluent of the diffuse fiber
stream passing from the amygdalo-piriform complex to the diencephalon via the medial forebrain bundle.   On the other hand, the
characteristic fragmentation undergone by its fibers, after
experimental transection, permits verification not only of its
participation in the formation of the inferior thalamic peduncle
and of the stria medullaris thalami but also its parenthood
relative to the terminal degeneration found in the whole extent of
the lateral hypothalamic areas, in the nuclei gemini (Fig. 26), the
dorsomedial and ventromedial thalamic nuclei and the lateral
habenular nucleus (cf. Lundberg, 1960, 1962; Sanders-Woudstra,
1961; Nauta, 1961; Hall, 1963; Powell et al., 1965; Cowan et al.,
1965; Knook, 1965; Valverde, 1965; Morgan, 1968; Ishikawa et al.,
1969; Price and Powell, 1970; Scott and Leonard, 1971).

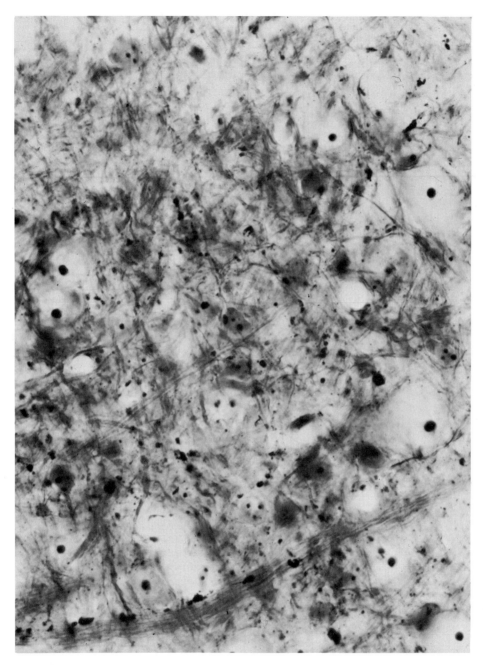

Fig. 26. Photomicrograph to show the terminal degeneration in the nuclei gemini following a lesion interrupting the so-called ventral amygdalofugal pathways. Original cupric-silver method. x 1000.

Finally, a ventromedially directed stream of very fine degenerating elements can be seen to extend continuously from the dorsal portion of the degeneration field within the amygdala to the ventrolateral hypothalamus through the subcapsular extension of the gray area beneath the ansa lenticularis (Fig. 22, thick arrow). This flattened subcapsular region which lies compressed between the retrolenticular portion of the internal capsule above and the dorsal supraoptic commissure and optic tract below, contains among its cell population neurons of the same type as those in the lateral hypothalamic area. Around these cells and those in the ventrolateral aspect of the lateral hypothalamic area, there are many argyrophilic granules which represent the terminals of the rat brain counterpart of the medial amygdalohypothalamic tract of higher species (cf. Nauta, 1961; Hall, 1963; Valverde, 1965; Morgan, 1968; Ishikawa et al., 1969).

A comparison with brain which bear smaller and more medially placed lesions shows:

a) The thick-fibered component of the amygdalo-piriform rostral projection system degenerates only after lesions which damage the periamygdaloid cortex.

b) Lesions confined to the central amygdaloid nucleus with slight involvement of the caudal part of the sublenticular portion of the substantia innominata as here described, evoke dramatic fiber degeneration in the medial forebrain bundle with, consequently, dense terminal degeneration in the entire extent of the lateral hypothalamic area (Fig. 27) and in minor scale in the nuclei gemini (cf. Escolar, 1965; Cowan et al., 1965; Valverde, 1965; Scott and Leonard, 1971). Such a picture is never so intense even with very extensive combined lesions of the periamygdaloid cortex and basolateral amygdala, or of the anterior amygdaloid area.

c) Similarly, lesions involving the central amygdaloid nucleus or the rostral part of the medial nucleus evoke more marked degenerative changes in the medial amygdalo-hypothalamic system than combined destruction of the piriform-basolateral amygdaloid region. The field of terminal degeneration extends even farther medially in the retrochiasmatic hypothalamic area, but there is also involvement of the thick-fibered component already discussed (cf. Golgi studies by Valverde, 1965; Millhouse, 1969).

d) Lesions of the caudal portions of the cortical and medial amygdaloid nuclei or which in addition encroached to some extent upon the caudal aspect of the basal amygdaloid complex do not show degeneration in any of the pathways under discussion (cf. Sanders-Woudstra, 1961; Cowan et al., 1965; Leonard and Scott, 1971).

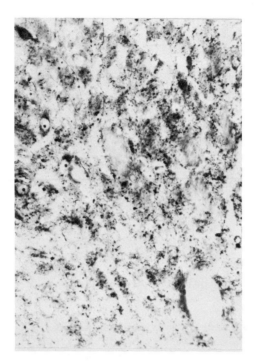

Fig. 27. Photomicrograph to show the terminal degeneration in the
lateral hypothalamic area as well as fiber degeneration in the
medial forebrain bundle after a lesion which involved the whole
extent of the central amygdaloid nucleus and the laterocaudal pole
of the sublenticular portion of the substantia innominata.
Original cupric-silver method. x 350.

III.   INTRAAMYGDALOID CONNECTIONS

In relation to this subject, any critical evaluation of distribution patterns of stem and terminal axon degeneration resulting
from lesions within the amygdala must take into account that even
small injuries to its nuclei may affect not only fibers originated
in them but also axons coming from extraamygdaloid structures, most
specifically those coming from the periamygdaloid cortex and from
diencephalon (see Cragg, 1961; Powell et al., 1965; Valverde, 1965;
Shute and Lewis, 1967). Notwithstanding these difficulties, some
conclusions can be reached from the present material.

a) Apparently, no part of the periamygdaloid cortex sends projections to the medial amygdaloid nucleus, neither does, at least

in appreciable volume, the basolateral nuclear complex (Figs. 21, 22).

b) The caudal laminar or subventricular portion (Me c) of the medial amygdaloid nucleus as well as its dorsal portion (Me d) stand out because of the paucity of their intraamygdaloid connections, both afferent and efferent (Figs. 22, 28). Moreover, it seems to send and receive projections mostly to or from the rostral portion of the medial nucleus (Fig. 9), that is, an intrinsic type of connection. Adjacent parts of the cortical amygdaloid nucleus and, on a much reduced scale, the basal medial nucleus appear to be the other sources of significant afferent supply to the caudal part of the medial nucleus.

c) The rostral part of the medial amygdaloid nucleus is distinguished from the caudal portion by its richer afferent supply, which comes mainly from the basal medial and cortical amygdaloid nuclei (Fig. 28). On the other hand, the material at hand does not allow any definitive conclusion with respect to the efferent intraamygdaloid connection of this cell group due to the probable involvement of axons from other sources which pass diffusely through or close to it. However, it is possible to verify its possession of reciprocal connections with the caudal portion of the medial nucleus and also its lack of relationship with the anterior magnocellular portion of the lateral amygdaloid nucleus or the anterior amygdaloid area.

d) The caudal one-third of the cortical amygdaloid nucleus is conspicuous for the richness of its afferent supply which comes particularly from the area of the basolateral complex and the ventral portion of the periamygdaloid cortex (areas 51c and 51d) (Figs. 17, 22, 28). The basal medial nucleus may also contribute to this afferent stream.

The remaining rostral portion of the cortical nucleus, on the other hand, does not offer the same picture, being more diffusely and sparsely innervated (Fig. 22). Both portions of the cortical nucleus, however, are the sources of most of the intraamygdaloid afferents of the medial amygdaloid nucleus, the anterior and posterior portions equally. The cortical nucleus seems also to send axons to the basal medial and perhaps even to the central amygdaloid nuclei.

e) Establishing the afferent and efferent intraamygdaloid connections of the basolateral complex and of the central amygdaloid nucleus is difficult because these nuclei in particular lie in the route of pathways of extraamygdaloid origin which send terminals to them as they do to other rostromedial nuclei. However, although they evidently do not contribute an afferent supply to the medial amygdaloid nucleus, it is possible to verify that

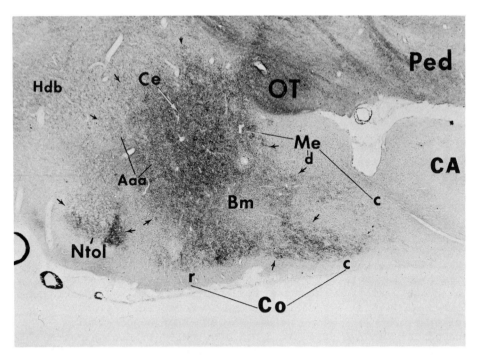

Fig. 28. Photomicrograph of a sagittal section passing through the
nucleus of the lateral olfactory tract (Ntol) to show the distribution pattern of the terminal degeneration following a lesion
which destroyed the lateral portion of the basal medial nucleus and
the ventromedial portion of the basal lateral amygdaloid nucleus.
Modified cupric-silver method. x 35. Note that the terminal degeneration invades the rostral part of the medial amygdaloid
nucleus (Me r), in less degree its caudal part (Me c) while it
spares almost totally its dorsal part (Me d). This and the terminal degeneration in the remaining portion of the basal medial
nucleus (Bm), the rostral and caudal portions of the cortical
nucleus (Co r and Co c) as well as in the central amygdaloid
nucleus (Ce) might be derived at least in part from the interruption of the diffuse fine-fibered intraamygdaloid association
system (see text).

interestingly enough the anterior magnocellular portion of the
lateral amygdaloid nucleus and the posterior parvocellular one are
rather poorly interconnected (Figs. 17, 21), and that the basal
lateral nucleus sends very few axons dorsally to the lateral nucleus.

On the other hand, cases bearing medial lesions, which encroach only upon parts of the central amygdaloid nucleus show, in
contrast with those in which this nucleus is totally destroyed,
that the terminal degeneration related to axonic processes of
central amygdaloid cells is restricted to the surviving portions of
the nucleus, although it extends also to the subventricular
portion of the substantia innominata via the longitudinal association bundle.

f) The nucleus of the lateral olfactory tract appears to receive afferent intraamygdaloid connections from all the laterally
placed structures of the amygdala (Fig. 28) as well as from the
periamygdaloid cortex. However, it seems that it is not interconnected with the medial amygdaloid nucleus and is the recipient
of an apparently restricted afferent supply from the cortical
nucleus.

g) Finally, the anterior amygdaloid area as here defined
receives a strong afferent supply from the periamygdaloid cortex,
very probably also from the basolateral nuclear complex and perhaps
from the basal medial amygdaloid nucleus. Moreover, lesions of
the latter areas which also include the anterior magnocellular
portion of the lateral amygdaloid nucleus, while producing sparse
terminal degeneration in the posterior parvocellular part of the
latter nucleus does so more abundantly in the nucleus of the
lateral olfactory tract, the basal complex and in the lateral half
of the posterior portion of the cortical amygdaloid nucleus.
Little if any terminal degeneration is detectable in the remaining
portions of the corticomedial amygdala.

COMMENT

From the above experimental anatomical observations, summarized in part in Figs. 29a, 29b, 30 and 31, it becomes evident
that the stria terminalis constitutes, at least in the rat brain,
not only the major efferent pathway linking the amygdala, or more
properly its corticomedial nuclear group, directly with the ipsilateral medial hypothalamus but also, and what is more striking,
with telencephalic formations in both hemispheres. Notable among
the long list of strial efferent connections are those established
with the ipsilateral accessory olfactory bulb, the pars medialis
of the anterior olfactory nucleus, the ventromedial hypothalamic
nucleus and with the contralateral olfactory tubercle and prepiriform cortex.

The inclusion of the accessory olfactory bulb within the projection field of the stria terminalis attains major relevance in
the light of the recent report by Winans and Scalia (1970) according to which the accessory olfactory bulb projects directly to the

posterior portion of the corticomedial amygdala, which is the
source of the axons which traverse the dorsal strial component to
end in the internal granular layer of this olfactory formation.
Furthermore, the same group of amygdaloid neurons projects to the
pars medialis of the anterior olfactory nucleus and to the cell-
poor capsule around the ventromedial hypothalamic nucleus, as well
as to other terminal areas. Contrariwise, the rostral portion of
the corticomedial amygdala, which receives a direct afferent
supply from the main olfactory bulb (Heimer, 1968; Scalia, 1969)
and projects via the ventral strial component directly into the
cellular core of the ventromedial hypothalamic nucleus does not re-
ciprocate at least directly in such an olfactory connection. Interestingly, the nucleus of the horizontal limb of the diagonal
band which has been shown by Price and Powell (1970a) to be the
site of origin of centrifugal fibers to the main olfactory bulb,
shows sparse signs of terminal degeneration after lesions in the
rostral portions of the corticomedial amygdala.

Such differences in the patterns of projection in the longitudinal axis of the brain are also present in the transverse axis
as is evidenced by segregation of the axons passing from the basolateral nuclear complex and the central amygdaloid nucleus via the
lateral division of the ventral strial component to terminate
eventually in the lateral portions of the bed nucleus of the stria
terminalis (cf. Ban and Omukai, 1959; Hall, 1963; Morgan, 1968;
Ishikawa et al., 1969; Leonard and Scott, 1971). The dorsal strial
component seems to be organized in a similar fashion as shown by
experiments in which lesions affected alternatively separate segments of its mediolateral organization. In view of the results described it appears logical to assume that the above arrangement reflects the way by which the posterior portions of both the medial
and cortical amygdaloid nuclei contribute to the formation of the
dorsal strial component. In this arrangement the medially located
gray mass sends its axons chiefly into the medial sectors of the
bundle while the more laterally located one contributes to the
lateral portions. Of course, due to the special topographical relationship between these two nuclei it is not possible to discern
sharp boundaries between their fiber groups within the dorsal
strial component and their projection fields. However, the strong
projection traced from the cortical nucleus to the accumbens septi,
the medial portion of the olfactory tubercle and Diepen's nucleus
tuberis lateralis which contrast markedly with the very weak one, if
any, from the medial nucleus, may have a special meaning if considered together with other morphological characteristics distinguishing the nuclei under discussion.

Thus, it can be noted from the present descriptions that the
posterior part of the cortical nucleus receives a strong afferent
supply from the lateral olfactory tract as well as from the ventral

Figs. 29a and 29b. (See opposite page.)

Serial schematic representation in a medial view of the right
amygdala of the general origin of the three componets of the stria
terminalis analyzed in the present report: dorsal, ventral, and
commissural. Inserted in the left upper corner of Fig. 29a is a
color code representing each of the above-mentioned three components of the stria. Furthermore, in the right corner of both
Figs. 29a and 29b is also inserted a cross-sectional schema of
the components of the stria colored in agreement with the segment
respectively involved in each representation.

Fig. 29a.

Diagrammatic representation of the nuclei of origin of the
long-projecting amygdalofugal fibers of the stria and their relative position within the latter as well as their topographical relationships with relation to the anterior commissure. Their
arrowed endings on the other hand only indicate the general direction toward which they are oriented. The terminal distribution of
these far-reaching projections are illustrated in Figs. 30 and 31.

Fig. 29b.

This schema shows the nuclei of origin of the lateral short-
projecting amygdalofugal axons of the stria and their relative
position within the latter as well as their topographical relationship in relation to the anterior commissure. Their shorter character is illustrated by their restriction within the limits of the
bed nucleus of the stria terminalis which is represented in
broken lines.

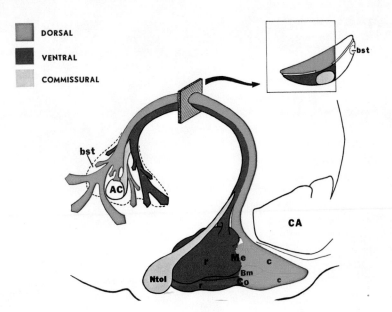

Fig. 29a.

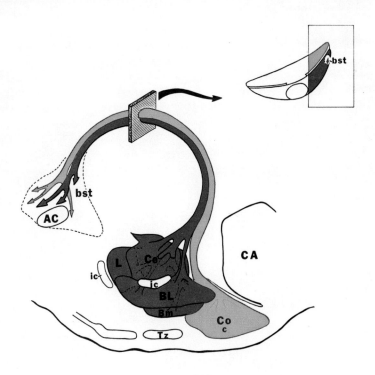

Fig. 29b.

periamygdaloid cortex and the basolateral nuclear complex. Moreover, it also possesses efferent intraamygdaloid connections with
almost all of the other amygdaloid nuclei; features which are very
little, if at all, developed in the medial amygdaloid neurons.

These data added to those of Valverde (1965), obtained in
his Golgi studies of the amygdala, seem to sustain his view that
the cortical nucleus should be classified as a cortical formation--
related to the periamygdaloid cortex--rather than a nucleus of the
amygdala. Equivalent consideration can be applied to the nucleus of
the lateral olfactory tract (cf. Valverde, 1965), which has been
shown here to be predominantly connected with paleocortical formations of the contralateral hemisphere.

Apart from the enumeration of the connections established by
the so-called ventral amygdalofugal pathways, and pointing out that
the "compact" division of this system probably represents the
counterpart for the rat brain of Johnston's (1923) longitudinal
association bundle, very little can be added here to what has already been said about these pathways by other researchers who
studied the rat brain (Sanders-Woudstra, 1961; Cowan et al., 1965;
Powell et al., 1965; Leonard and Scott, 1971a, b). However,
among the anatomical observations presented in this report, there
are two which might have some interesting implications. According
to one of these, the central amygdaloid nucleus appears to emit
fibers which become incorporated into the "compact" division of
the ventral amygdalofugal pathways and form a continuous field of
terminals along the nucleus itself which extends along the subventricular portion of the substantia innominata as far as the
ventral postcommissural portion of the bed nucleus of the stria
terminalis. The other observation is that lesions damaging the
central amygdaloid nucleus in its total extent but which encroached
upon the caudolateral end of the sublenticular portion of the substantia innominata were associated with abundant fiber degeneration
in the medial forebrain bundle and consequent heavy terminal degeneration in the lateral hypothalamic area and nuclei gemini.
Such degenerative changes in the MFB and its terminal field were
never so pronounced after extensive lesions of the periamygdaloid
cortex or of the anterior amygdaloid area as here defined. However,
these observations do not provide complete evidence that the central amygdaloid nucleus is a source of part at least of the ventral
pathways which end in the lateral hypothalamus since, as has been
pointed out, this nucleus is traversed by fiber streams of extraamygdaloid origin which when interrupted in an area of concentration might explain the density of the changes described in
the lateral portions of the hypothalamus. This point deserves
further study.

The same exception can be adduced with relation to the path-

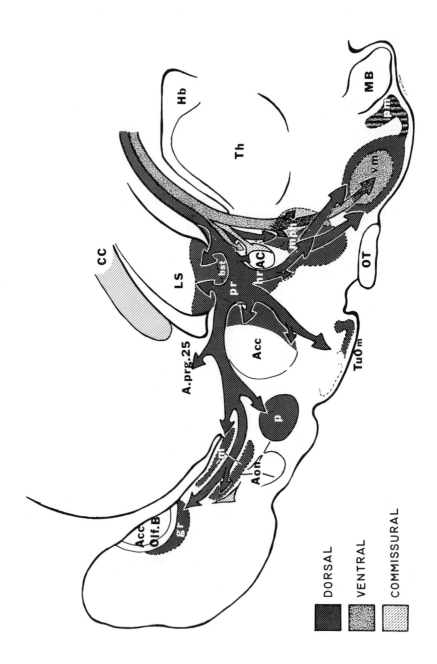

Fig. 30. (See opposite page.) Schematic representation of the distribution of the 3 components of the stria terminalis: dorsal, ventral, and commissural. The diagram shows the stria of the right as seen from its medial side. The gray groups supplied from it are represented in broken lines and filled with the type of shading representing that particular component of the stria. The shading code is in the lower left corner of the figure and represents each of the three components illustrated. The dorsal strial component supplies the bed nucleus of the stria terminalis, the basal part of the lateral septal nucleus, the posteromedial part of the nucleus accumbens septi and olfactory tubercle, the partes posterior and medialis of the anterior olfactory nucleus, the internal granular layer of the accessory olfactory bulb, the medial preoptic hypothalamic junction area, the capsule encircling the cellular core of the ventromedial hypothalamic nucleus, and the premammillary area. Not represented in the figure are the retrochiasmatic area and Diepen's nucleus tuberis lateralis which are also recipients of fibers from this bundle. The ventral strial component distributes to the bed nucleus of the stria terminalis, the medial preoptic-hypothalamic junction area, the central core of the ventromedial hypothalamic nucleus, and the premammillary area. Again, the retrochiasmatic area and the Diepen's nucleus tuberis lateralis are not represented although they also receive fibers from this component. Finally, the "commissural" component is seen to enter the anterior commissure as does also the commissural division of the dorsal strial component.

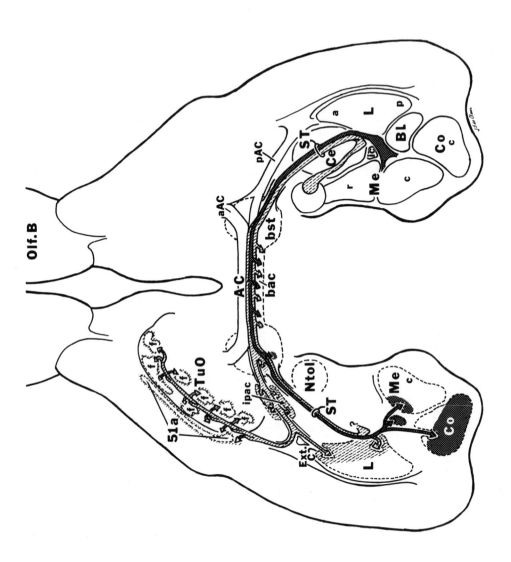

Fig. 31. (See opposite page.) Diagram, as viewed from above,
illustrating the sources, courses and terminations of the commissural division of the dorsal component and of the so-called "commissural" component. Both fiber systems are identified according to
the shading code used in Fig. 30, but they can also be recognized
by their pathways of departure from the right amygdala, i.e., the
first bundle from the caudal portions of the corticomedial nuclear
complex, and in the second instance, from the nucleus of the
lateral olfactory tract. On the left hemisphere, the outlines of
the gray formation where these two fiber systems terminate are
represented in broken lines which contrast with the continuously
lined outlines of the amygdaloid nuclei in the right hemisphere.
Their respective relationships are further stressed by the shad-
ings which match those of the fiber contingents supplying them.
Thus, the commissural division of the dorsal component is shown to
supply the bed nuclei of the anterior commissure and of the stria
terminalis, and after joining the contralateral stria terminalis,
restricted areas of the caudal portions of the medial and cortical amygdaloid nuclei. The "commissural" component, on the other
hand, connects with the bed nuclei, and, after bifurcating and
joining thereafter the contralateral stria terminalis and the
posterior limb of the anterior commissure, terminates in restricted portions of the lateral amygdaloid nucleus, in cell masses
surrounding the posterior limb of the anterior commissure, the
area praepiriformis 51a and the convolutions in the anterolateral
portions of the olfactory tubercle.

ways which were traced to the entorhinal cortex and subiculum of
the hippocampus after lesions in various portions of the amygdala
(cf. Cragg, 1961; Valverde, 1965; Shute and Lewis, 1965). As in
the case of the former problem, it will be the subject of future
studies.

ACKNOWLEDGMENTS

The work reported here was supported by NINDS grants NS05249
and NS08166 (W. R. Ingram, principal investigator). The author
thanks Drs. Basil E. Eleftheriou and Geoffrey Raisman for the
privilege of participating in this conference. Advice received
from Dr. Raisman is also greatly appreciated. Aid from the
Departments of Anatomy and Neurology, The University of Iowa,
for publication expenses is gratefully acknowledged.

Figures 1, 5, 13, 30, and 31 are taken from another paper by
de Olmos and Ingram which has been submitted for publication to a
journal.

LIST OF ABBREVIATIONS

Aaa, anterior amygdaloid area
aAC, anterior limb of the anterior commissure
AC, anterior commissure
Acc, nucleus accumbens septi
Acc. Olf. B., accessory olfactory bulb
Aon, anterior olfactory nucleus
aod, anterior olfactory nucleus, pars dorsalis
aol, anterior olfactory nucleus, pars lateralis
aom, anterior olfactory nucleus, pars medialis
aop, anterior olfactory nucleus, pars posterior
A. prg. 25, area corticalis praegenualis 25
arc, arcuate nucleus
art, artifact
BL, basal lateral amygdaloid nucleus
Bm, basal medial amygdaloid nucleus
bac, bed nucleus of the anterior commissure
bst, bed nucleus of the stria terminalis
CA, cornu Ammonis
Ce, central amygdaloid nucleus
CC, corpus callosum

Cl, claustrum
Co, cortical amygdaloid nucleus
Co c, caudal portion
Co r, rostral portion
CST, commissural component of the stria terminalis
CU, original cupric-silver technique
DG, dentate gyrus
dpm, dorsal premammillary nucleus
DST, dorsal strial component
ep, ependyma
Ext. C, external capsule
f, convolutions of olfactory tubercle
FH, Fink-Heimer (1967) technique modified
Fi, fimbria fornicis
Fx, columna fornicis
GP, globus pallidus
gr, internal granular layer of the accessory olfactory bulb
Hb, habenula
Hdb, nucleus of the horizontal limb of the diagonal band
hr, hypothalamic radiation of the supracommissural division of
    the dorsal strial component
IC, internal capsule
ic, intercalate masses
iCa, islands of Calleja
iCam, medial island of Calleja
ipac, interstitial nucleus of the posterior limb of anterior
    commissure
L, lateral amygdaloid nucleus
La, lateral amygdaloid nucleus, anterior magnocellular portion
LAB, longitudinal association bundle
Lp, lateral amygdaloid nucleus, posterior parvocellular portion
lpo, lateral preoptic area
LS, lateral septal nucleus
LSN, lesion
LOT, lateral olfactory tract
LOTd, lateral olfactory tract, dorsal peduncle
Lv, lateral ventricle
m, medial part of the anterior olfactory nucleus (Aon)
MB, mammillary body
ME, median eminence
Me, medial amygdaloid nucleus
Me c, medial amygdaloid nucleus, caudal portion
Me d, medial amygdaloid nucleus, dorsal portion
Me r, medial amygdaloid nucleus, rostral portion
mph, medial preoptic-hypothalamic junction area
mTh, mammillo-thalamic tract
N, Nauta-Gygax (1954) technique modified
Ntol, nucleus of the lateral olfactory tract
OCh, optic chiasma
Olf. B, main olfactory bulb

OT, optic tract
ov, olfactory ventricular cleft
p, posterior part of anterior olfactory nucleus (Aon)
pAC, posterior limb of the anterior commissure
Pam, periamygdaloid cortex
Ped, pedunculus cerebri
pfh, perifornical area
pm, premammillary area
pr, parolfactory radiation of the supracommissural division of
    the dorsal strial component
Rtc, retrocommissural division of the dorsal strial component
S, subiculum
SIs, substantia innominata, sublenticular portion
Spc, supracommissural division of the dorsal component
ST, stria terminalis
Str, striatum
Th, thalamus
tl, nucleus tuberis lateralis (Diepen, 1962)
TuO, olfactory tubercle
TuOm, olfactory tubercle, medial
Tz, amygdalo-piriform transitional zone or 51d
V db nucleus of the vertical limb of diagonal band
vm, ventromedial hypothalamic nucleus
vpm, ventral premammillary nucleus
VST, ventral strial component
IIIv, third ventricle
25, area praegenualis 25

Structures of the prepiriform cortex and olfactory tubercle:

 I, external plexiform layer of lamina zonalis or molecular
    layer
 Ia, sublamina supratangentialis of the plexiform layer
 Ib, sublamina tangentialis of the plexiform layer
 II, superficial pyramidal-celled layer
III, deep polymorph-celled layer
Plx, external plexiform layer
Ply, polymorph layer
Pyr, pyramidal layer
51a, area praepiriformis 51a
51b, area praepiriformis 51b
51e, area praepiriformis 51e
51f, area praepiriformis 51f
28d, area entorhinalis 28, pars dorsalis
28v, area entorhinalis 28, pars ventralis

REFERENCES

ADEY, W. R., & MEYER, M. Hippocampal and hypothalamic connexions of the temporal lobe in the monkey. Brain, 1952, 75, 358-383.

BAN, T., & OMUKAI, F. Experimental studies on the fiber connections of the amygdaloid nuclei in the rabbit. Journal of Comparative Neurology, 1959, 113, 245-280.

BECCARI, N. Il lobo paraolfattorio nei mammiferi. Archivio Italiano di Anatomia e di Embriologia, 1910, 9, 173-220.

BERKELBACH VAN DER SPRENKEL, H. Stria terminalis and amygdala in the brain of the opossum (Didelphis virginiana). Journal of Comparative Neurology, 1926, 42, 211-254.

BERNARDIS, L. L. Stereotaxic localization of amygdaloid nuclei in rats from weaning to adulthood. Experientia, 1967, 23, 158-160.

BRODAL, A. The amygdaloid nucleus in the cat. Journal of Comparative Neurology, 1947, 87, 1-6.

BRODAL, A. The origin of the fibres of the anterior commissure in the rat; experimental studies. Journal of Comparative Neurology, 1948, 88, 157-205.

CAJAL, S. RAMON Y. Histologie du System Nerveus de l'Homme et des Vertébrés. Paris: Maloine, 1911. Tome II.

COWAN, W. M., RAISMAN, G., & POWELL, T. P. S. The connexions of the amygdala. Journal of Neurology, Neurosurgery and Psychiatry, 1965, 28, 137-151.

CHI, C. C. Afferent connections to the ventromedial nucleus of the hypothalamus in the rat. Brain Research, 1970, 17, 439.

CRAGG, B. G. Olfactory and other afferent connections of the hippocampus in the rabbit, rat and cat. Experimental Neurology, 1961, 3, 588-600.

DE GROOT, J. The rat forebrain in stereotaxic coordinates. Verhandl Koninklinic Nederlands Wetenschels, Section II, 1959, 1-40.

DE OLMOS, J. S. The stria terminalis: its projection field in the rat. Anatomical Record, 1968, 160, 339 (Abstract).

DE OLMOS, J. S. A cupric-silver method for impregnation of
    terminal axon degeneration and its further use in staining
    granular argyrophilic neurons. Brain, Behavior and
    Evolution, 1969, 2, 213-237.

DE OLMOS, J. S. The amygdaloid projection field in the rat brain
    as studied by different silver procedures. Anatomical
    Record, 1970, 166, 298 (Abstract).

DE OLMOS, J. S., & INGRAM, W. R. An improved cupric-silver method
    for impregnation of axonal and terminal degeneration. Brain
    Research, 1971a, in press.

DE OLMOS, J. S., & INGRAM, W. R. The projection field of the
    stria terminalis in the rat brain. An experimental study
    (submitted for publication) 1971b.

DIEPEN, R. Hypothalamus. In Handbuch der Mikroskopische
    Anatomie des Menschen. Berlin-Gottingen-Heidelberg:
    Springer-Verlag, 1962. Vol. 4, VII.

EAGER, R. P., CHI, C. C., & WOLF, G. Lateral hypothalamic pro-
    jections to the hypothalamic ventromedial nucleus in the
    albino rat: demonstration by means of a simplified
    ammoniacal silver degeneration method. Brain Research,
    1971, 29, 128-132.

ESCOLAR, J. Apport á l'organization du complexe amygdalien (les
    connexions du supraamygdaleum). Comptes Rendus de la
    Association des Anatomistes, XLII Réunion, 1955, 496-505.

FINK, R. P., & HEIMER, L. Two methods for selective silver
    impregnation of degenerating axons and their synaptic end-
    ings in the central nervous system. Brain Research, 1967,
    4, 369-374.

FOX, C. A. The stria terminalis, longitudinal association bundle
    and precommissural fornix fibers in the cat. Journal of
    Comparative Neurology, 1943, 79, 277-295.

GLEES, P. Terminal degeneration within the central nervous
    system as studied by a new silver method. Journal of Neuro-
    pathology and Experimental Neurology, 1946, 5, 54-59.

GLOOR, P. Electrophysiological studies on the connections of the
    amygdaloid nucleus in the cat. Part I. The neuronal organi-
    zation of the amygdaloid projection system. Electroencephalo-
    graphy and Clinical Neurophysiology, 1955, 7, 223-242.

GURDJIAN, E. S. Olfactory connections of the albino rat, with
    special reference to stria medullaris and anterior commissure.
    Journal of Comparative Neurology, 1925, 38, 127-163.

GURDJIAN, E. S. The corpus striatum of the rat. Journal of
    Comparative Neurology, 1928, 45, 249-281.

HALL, E. Efferent connections of the basal and lateral nuclei of
    the amygdala in the cat. American Journal of Anatomy, 1963,
    113, 139-151.

HEIMER, L. Synaptic distribution of centripetal and centrifugal
    nerve fibres in the olfactory system of the rat. An experi-
    mental study. Journal of Anatomy, 1968, 103, 413-432.

HEIMER, L., & NAUTA, W. J. H. The hypothalamic distribution of
    the stria terminalis in the rat. Brain Research, 1969, 13,
    284-297.

HILPERT, P. Der Mandelkerne des Menschen. I. Cytoarchitektonik
    und Faserverbindungen. Journal of Psychology and
    Neurology (Leipzig), 1928, 36, 44-74.

ISHIKAWA, L., KAWAMURA, S., & TANAKA, O. An experimental study
    on the efferent connections of the amygdaloid complex in the
    cat. Acta Medica Okayama, 1969, 23, 519-539.

JOHNSTON, J. B. Further contributions to the study of the evolu-
    tion of the forebrain. Journal of Comparative Neurology,
    1923, 35, 337-481.

KLINGLER, J., & GLOOR, P. The connections of the amygdala and of
    the anterior temporal cortex in the human brain. Journal of
    Comparative Neurology, 1960, 115, 333-369.

KNOOK, H. L. The Fibre-Connections of the Forebrain. Royal
    Vangoreum. Philadelphia: Davis Co., 1965.

KONIG, J. F. R., & KLIPPEL, R. A. The Rat Brain: A Stereotaxic
    Atlas of the Forebrain and Lower Parts of the Brain Stem.
    Baltimore: Williams and Wilkins, 1963.

KREINER, J. Myeloarchitectonics of the lateral olfactory tract
    and of the piriform cortex of the albino rat. Journal of
    Comparative Neurology, 1949, 91, 103-127.

LAMMERS, H. J. The neural connections of the amygdaloid complex
    in mammals. In this volume, 1972.

LAMMERS, H. J., & LOHMAN, A. H. Experimental anatomisch onder-
    zoek naar de verbindingen van piriform cortex en amygdala-
    kernen bij de kat. Nederlands Tijdschrift voor Geneeskunde,
    1957, 101, 1-2.

LEONARD, C. M., & SCOTT, W. S. Origin and distribution of the
    amygdalofugal pathways in the rat: an experimental neuro-
    anatomical study. Journal of Comparative Neurology, 1971,
    141, 313-330.

LUNDBERG, P. O. Cortico-hypothalamic connexions in the rabbit.
    Acta Physiologica Scandinavica, 1960, 49, 171, 1-80.

LUNDBERG, P. O. The nuclei gemini. Two hitherto undescribed
    nerve cell collections in the hypothalamus of the rabbit.
    Journal of Comparative Neurology, 119, 311-316.

MARBURG, O. The amygdaloid complex. Confinia Neurologia, 1948,
    9, 211-216.

MILLHOUSE, O. E. A Golgi study of the descending medial fore-
    brain bundle. Brain Research, 1969, 15, 341-363.

MORGAN, M. V. Some efferent fiber projections of the amygdala in
    the cat. Ph.D. Thesis, 1968, Duke University.

NAUTA, W. J. H. Hippocampal projections and related neural path-
    ways to the midbrain in the cat. Brain, 1958, 81, 319-340.

NAUTA, W. J. H. Fibre degeneration following lesions of the
    amygdaloid complex in the monkey. Journal of Anatomy, 1961,
    95, 515-531.

NAUTA, W. J. H., & GYGAX, P. A. Silver impregnation of degen-
    erating axons in the central nervous system: A modified
    technique. Stain Technology, 1954, 29, 91-93.

NAUTA, W. J. H., & HAYMAKER, W. Hypothalamic nuclei and fiber
    connections. In W. Haymaker, E. Anderson, and W. J. H.
    Nauta (Eds.), The Hypothalamus. Springfield, Illinois:
    Charles C. Thomas, 1969.

POWELL, T. P. S., COWAN, W. M., & RAISMAN, G. The central
    olfactory connections. Journal of Anatomy, 1965, 99,
    791-813.

PRICE, J. L., & POWELL, T. P. S. An experimental study of the
    origin and the course of the centrifugal fibers to the olfac-
    tory bulb in the rat. Journal of Anatomy, 1970a, 107, 215-
    237.

PRICE, J. O., & POWELL, T. P. S. The afferent connexions of the
nucleus of the horizontal limb of the diagonal band.
Journal of Anatomy, 1970b, 107, 239-256.

RAISMAN, G. An evaluation of the basic pattern of connections
between the limbic system and the hypothalamus. American
Journal of Anatomy, 1970, 129, 197-202.

RAISMAN, G. Some anatomical projections of the stria terminalis.
In this volume, 1972.

RAISMAN, G., COWAN, W. M., & POWELL, T. P. S. An experimental
analysis of the efferent projection of the hippocampus.
Brain, 1966, 89, 83-108.

ROSE, M. Histologische Localisation der Grosshirnrinde beim
kleinin Saügetieren (Rodentia, Insectivor, Chiroptera).
Journal of Psychology and Neurology (Leipzig), 1912, 19,
389-479.

SANDERS-WOUDSTRA, J. A. R. Experimenteel anatomisch onderzoek
over de verbindingen van enkele basale telencefale hersengebienden bij de albino rat. Thesis, 1961, Groningen
University.

SCALIA, F. A review of recent experimental studies on the distribution of the olfactory tracts in mammals. Brain, Behavior,
and Evolution, 1969, 1, 101-123.

SCOTT, J. W., & LEONARD, C. M. The olfactory connections of the
lateral hypothalamus in the rat, mouse and hamster. Journal
of Comparative Neurology, 1971, 141, 331-344.

SHUTE, C. C. D., & LEWIS, P. R. The ascending cholinergic
reticular system: Neocortical, olfactory and subcortical
projections. Brain, 1967, 90, 497-520.

Szentágothai, J., Flerkó, B., Mess, B., & Halász, B. Hypothalamic
Control of the Anterior Pituitary. An Experimental-Morpho-
logical Study, 3rd Edition. Budapest: Akademiai Kiadó, 1968.

UCHIDA, Y. A contribution to the comparative anatomy of the
amygdaloid nuclei in mammals, especially in rodents. Part I.
Rat and mouse. Folia Psychiatrica et Neurologica Japonica
(Niigata), 1950, 4, 25-42.

VALVERDE, F. Studies on the Piriform Lobe. Cambridge: Harvard
University Press, 1951.

VAN ALPHEN, H. A. M. The anterior commissure of the rabbit.
    Acta Anatomica, 1969, Supplement 57, 74, 1-112.

WINANS, S. S., & SCALIA, F. Amygdaloid nucleus: new afferent
    input from the vomeronasal organ. Science, 1970, 170,
    330-332.

STIMULATION AND REGIONAL ABLATION OF THE AMYGDALOID COMPLEX

WITH REFERENCE TO FUNCTIONAL REPRESENTATIONS

Birger R. Kaada

Institute of Neurophysiology, University of Oslo

Oslo, Norway

CONTENTS

I. INTRODUCTION

Stimulation and ablation of the amygdaloid nuclear complex result in a variety of somatic, visceral, endocrine and behavioral effects. Since this brain area is a very heterogeneous structure, with a number of rather distinct subdivisions, and since differences in structure obviously implies functional differences, it is reasonable that attempts have been made to correlate structure and patterns of responses.

The aim of the present communication has been to summarize and discuss available evidences for a functional representation within the amygdaloid complex. Twenty years have passed since the first stimulations and attempts at localization were made (Kaada, 1951, Fig. 12), and a decade more since the first restricted ablation study was published (Spiegel et al., 1940). Since then 500-600 articles dealing with functional aspects of the amygdaloid complex have been published. However, only a restricted number of these allow any conclusion with respect to the question of whether the various effects produced by stimulation or ablation are related to specific nuclei or regions within this complex, and only these reports will be considered in this survey. Several of these studies have resulted in apparently contradictory reports and these have led some investigators to conclude that the amygdala acts in a less specific way than the structures to which it projects, or even that in this brain area there is no functional localization.

However, careful review of the experimental data from these apparently conflicting reports, and consideration of these in the light of more recent experimental findings, the opinion of the author is that most of the controversies can be resolved. Although there are still several conflicting results in the literature, it appears that some main features of the functional organization within this complex structure can be visualized, at least with respect to the main anatomical subdivisions. A failure to recognize existing functional representations will delay scientific progress, as otherwise lesions are likely to be made across functional borders with corresponding difficulties in interpretations of the results.

Several factors have to be taken into account before ascribing a certain function to a specific area:

(1) _Interference with traversing fiber tracts_. The methods of electrical stimulation and ablation have their limitations when applied to a relatively compact and complex structure like the amygdala. Although nuclear masses are stimulated or lesioned, so are the afferent and efferent projections along with the

intra-amygdaloid association fibers. In particular, activities
specifically related to the lateral part of the amygdala may be
interfered with by stimulating or ablating its dorsomedial part.
From the schematic drawing of Fig. 1, it is seen that fibers
of the ventral amygdalofugal path, which forms the main efferent
projection from the lateral and basal amygdaloid nuclei, and which
also contains fibers from the periamygdaloid and piriform cortex,
spread medially through the region of the central nucleus. At
this frontal plane, fibers of the stria terminalis, the main
efferent projection from the cortical, medial and central nuclei,
cross the ventral path at right angles and may be similarly interfered. This difficulty can be reduced by careful mapping at
threshold intensities. The ambiguity of experimental findings
also can be reduced by the use of chemical rather than electrical
stimulation techniques, as the former does not affect transmission along nerve fibers but excites cell bodies and dendrites
only.

(2) Experimental variables. The direction of the response
to stimulation may be influenced by the nature and depth of
anesthesia as exemplified by the effects on respiration (Ursin
and Kaada, 1960a). Further, the response may be reversed by
changing the stimulus frequency, as has been shown for the
effects on the cardiovascular system (Koikegami et al., 1957).
Finally, accompanying electrical afterdischarges may favor the
appearance of a particular cardiovascular response (Reis and
Oliphant, 1964). Therefore, more emphasis should be given to
results obtained in the non-anesthetized animal and to experiments in which the stimulus parameters have been controlled and
simultaneous records of the amygdaloid electrical activity have
been obtained.

(3) Definition of behavior patterns. The terms used to
describe the complex emotional behavior changes resulting from
brain stimulation and ablation vary considerably. This has led
to some confusion, creating unnecessary conflicting reports,
particularly with respect to topical localization. Also, the
term 'avoidance' has been employed for behavior performance under
various experimental conditions without a clear distinction for
example, between active and passive avoidance. These are known
to be selectively interfered by brain lesions, indicating different underlying physiological mechanisms (McCleary, 1961; Ursin,
1965b).

II.   ANATOMICAL SUBDIVISIONS AND FIBER CONNECTIONS

In most functional studies the effects of stimulation and
ablation have been related to either the basolateral or to the
corticomedial groups of nuclei (Fig. 2, left). This traditional

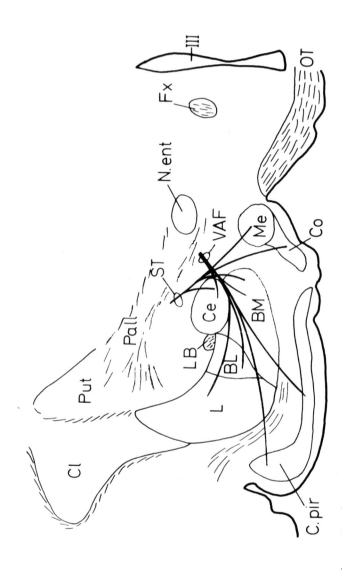

Fig. 1. Schematic drawing through the amygdaloid complex of cat to illustrate how stimulation and ablation of its dorsomedial portion (in the region of the central nucleus) may influence the functions of other parts of this nuclear complex by interference with the stria terminalis and the ventral amygdalofugal path which both pass through this area. BL - n. basalis pars magnocellularis amygdalae; BM - n. basalis pars parvocellularis amygdalae; Ce - n. centralis amygdalae; Cl - claustrum; Co - n. corticalis amygdalae; C. pir. - cortex piriformis; Fx - fornix; L - n. lateralis amygdalae; LB - longitudinal association bundle; Me - n. medialis amygdalae; N. ent. - n. entopeduncularis; OT - tractus opticus; Pall - pallidum; Put - putamen; ST - stria terminalis; VA - ventral amygdalofugar path.

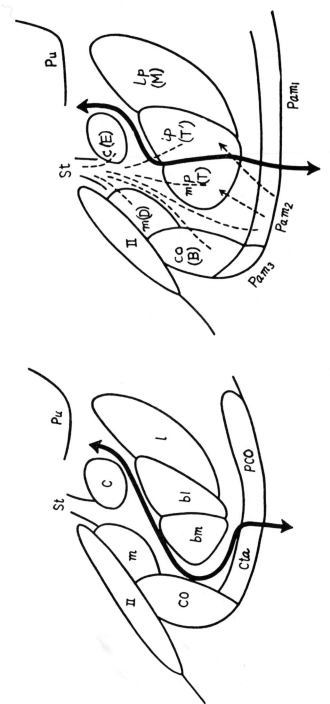

Fig. 2. Diagram showing two main divisions of the amygdaloid nuclei. The traditional anatomical division of the amygdala into a basolateral and a corticomedial division (left) contrasted with a proposed functional division (right). (From Koikegami, 1963).

subdivision of the amygdaloid nuclei is based on embryological
(Holmgren, 1925) and comparative anatomical studies (Johnston,
1923). The phylogenetically old corticomedial group was composed
of the central, medial and cortical nuclei and the nucleus of
the lateral olfactory tract. The phylogenetically younger basolateral group included the lateral and basal nuclei. Within the
latter, a lateral magnocellular part and a medial parvocellular
part could be distinguished. Later investigators (e.g., Gurdjian,
1928) separated from the central nucleus a more ill-defined anterior amygdaloid area. Together with the lateral olfactory tract
the latter has, by some authors, been termed the anterior group
of nuclei.

More recently, several Japanese investigators (Uchida, 1950;
Koikegami, 1963) have described more subdivisions within the
amygdaloid complex and have adopted terms from the studies of the
early German anatomists. Based on fiber projections (Omukai, 1958)
and functional studies (Koikegami, 1963), a main division of the
amygdaloid complex into a lateral and a medial part has been proposed, the important difference from the division proposed by
Johnston (1923) being that the medial small-celled part of the
basal nucleus (medial principle nucleus of Koikegami) was included
in the medial group (Fig. 2, right). In general, the medial group
was found to project its fibers to the medial sympathetic zone of
the hypothalamus while the lateral group projected mainly to the
lateral parasympathetic hypothalamic zone of Kurotsu. Some of
the published experimental data provide some support for a functional division in favor of Koikegami's viewpoint (cf. also Egger
and Flynn, 1967).

Macchi (1951) stated that the ventral claustrum is related
anatomically to the lateral amygdaloid nucleus. This view also
finds physiological support in stimulation as well as ablation
studies (Wood, 1958).

Using the Golgi technique and studying the chemoarchitecture
(acetylcholinesterase, monoamine oxidase, dithizone and silver
sulphide stain) and neocortical afferents, one may, with each
method, reach at a different grouping of the amygdaloid nuclei.
This applies both to a further subdivision of the various nuclei
as determined in Nissl preparation and grouping of nuclei in a
uniform manner. The results of these studies have been summarized
by Hall (1971), and will not be dealt with further in this review
since at the present state of our knowledge no definite correlation can be made between effects of stimulation and ablation and
chemoarchitecture. The studies indicate a heterogeneity of the
amygdaloid complex which is poorly understood. It is of particular interest, however, that several of these studies demonstrate
that the lateral nucleus as well as the medial part of the basal

nucleus can be divided into a ventral and a dorsal part (Hall,
1971). Several functional studies to be reported similarly indicate such a separation, particularly of the lateral nucleus.

Fiber connections. The main efferent projections from the
amygdala are the stria terminalis and the ventral amygdalofugal
pathway. The projection areas of these two fiber systems have
been mapped in detail in the monkey (Nauta, 1961). For more
information, the reader is referred to the chapters by Dreifuss,
Egger, Gloor, Hall, Lammers, Murphy and Raisman and de Olmos in
this volume.

The stria terminalis originates mostly in the caudal one-
half of the amygdaloid complex, mainly from the nuclei of the
corticomedial division and distributes its fibers to the hypothalamus and preoptic area, the bed nucleus of the stria terminalis, a.o. The stria terminalis also contains important
afferent connections to the amygdala (Hilton and Zbrozyna, 1963;
Powell et al., 1963).

The ventral amygdalofugal path is the main projection from
the lateral and basal nuclei. The fibers spread medially and
forward through the region of the central nucleus, ventral to the
internal capsule and pallidum, and form a direct connection to
thalamic, septal, lateral hypothalamic and preoptic areas,
olfactory tubercle, gyrus subcallosus, a.o. Just ventral to the
central nucleus, the fibers from the two efferent systems cross
at right angles (Fig. 1). In the cat, the basal nucleus sends
fibers through both the stria terminalis and the ventral pathway,
while the lateral nucleus projects only through the ventral
amygdalofugal path (Hall, 1963). The latter has exactly the same
course and distribution as the fibers from the piriform cortex
(Powell et al., 1963). Like the stria terminalis, the ventral
pathway also contains afferent fibers which can be traced to the
lateral, basal and medial nuclei. This ventral path generally
has not been recognized as a source of afferents to the amygdala,
but is probably more important from a functional point of view
since it contains considerably more fibers than the stria terminalis afferents.

According to Powell et al. (1963, p. 711), this "reciprocal
relationship between the amygdaloid region and the hypothalamus
suggests that the ventral pathway is essentially a laterally
directed extension of the medial forebrain bundle, and that the
stria terminalis should be regarded as a dorsal component of
this bundle which has become separated from the main part of the
medial forebrain bundle by the development of the internal capsule."

Gloor (1960) has given an extensive review of the wide
amygdaloid projection fields, mainly based on his own electro-

physiological studies, to which the reader is referred. Of particular interest, for localization studies within the amygdala, is the demonstration of convergence of stria terminalis and ventral amygdalofugal impulses upon single neurons of the ventromedial hypothalamic nuclei, the former being inhibitory and the latter excitatory followed by inhibition (Dreifuss, Murphy et al., 1968).

Among the afferent connections, the fibers of the olfactory tract are mainly confined to the corticomedial portion and the anterior amygdaloid area (Le Gros Clark and Meyer, 1947; Cowan et al., 1965). The lateral and basal nuclei receive olfactory impulses indirectly from the piriform cortex.

However, sensory input is not confined to the olfactory system. The amygdala, in particular its basolateral part, also receives input from all the various sensory modalities (Bonvallet et al., 1952; Machne and Segundo, 1956; Gloor, 1960; Wendt and Albe-Fessard, 1962; Sawa and Delgado, 1968; O'Keefe and Bouma, 1969). There is often a convergence of impulses from several modalities to the same amygdaloid cell.

Fiber connections with surrounding neocortical and rhinencephalic structures, brain stem, thalamus a.o. have also been demonstrated (Gloor, 1960; Klinger and Gloor, 1960; Nauta, 1961). The fibers of the inferior temporal gyrus of monkeys are distributed primarily to the basolateral complex (Whitlock and Nauta, 1956). In the cat the anterior and posterior sylvian gyri (Lescault, 1969, 1967; Druga, 1969) as well as the anterior and posterior ectosylvian gyri (Lescault, 1971) project mainly to dorsal part of the lateral nucleus, with some fibers to the large-celled part of the basal nucleus and the small-celled lateral part of the central nucleus, while the orbital gyrus projects only to the more ventromedial segment of the lateral nucleus (Lescault, 1971). No neocortical projections to the small-celled part of the basal nucleus or to the cortical nucleus were observed by these authors. From the preoptic region, mainly its lateral part, fibers project to the anterior, central medial and basal amygdaloid nuclei, but not to the lateral or the cortical nucleus (Nauta, 1958; Cowan et al., 1965).

III. FUNCTIONAL SUBDIVISIONS OF THE AMYGDALOID COMPLEX

Functions related to the amygdaloid complex, as demonstrated by stimulation and ablation, include: (A) general arousal, orienting reaction and sleep, (B) agonistic behavior (flight, defense and predatory attack), active and passive avoidance behavior, (C) feeding activities, (D) sexual activities, and (E) reward and punishment in self-stimulation experiments.

Included in this survey are studies which mainly contribute
to the problems of localizing functions within the amygdaloid
nuclear complex. This survey does not claim to be complete in
this respect.

A. General arousal, orienting reaction and sleep

(1) The orienting reaction is the most common response
elicited by amygdaloid stimulation in the unanesthetized animal
(Ursin and Kaada, 1960a, 1960b). The initial phase consists of
an almost immediate arrest of all spontaneous ongoing activities,
such as licking, walking and side-to-side movements of the tail.
The animal then exhibits signs of arousal: it becomes alert,
the facial expression and the whole attitude of the animal
changes to one of attention. From most areas, the arousal is
followed by movements of an orienting nature. The animal looks
around with glancing or searching movements in an inquisitive
manner, usually towards the contralateral side. There is retraction of the nictitating membranes with opening of the eyes
and slow pupillary dilatation. The searching is accompanied
frequently by sniffing, swallowing, chewing and by twitching
of the ipsilateral facial musculature. During the stimulation,
the animal still responds appropriately to various environmental
stimuli. These effects were obtained with no afterdischarges
in the amygdala.

This response was first described by Kaada (1951, pp.106-
110) as "arrest" reaction and later the term "attention" or
"searching" response has been used (Andersen, Jansen and Kaada,
1952; Gastaut et al., 1952; Kaada, Andersen and Jansen, 1954;
Magnus and Lammers, 1956; Shealy and Peele, 1957; Ursin and
Kaada, 1960a). The response usually is indistinguishable from
the arousal or orienting reaction induced by brain stem reticular activation and is associated with cortical desynchronization (Ursin and Kaada, 1960a; Feindel and Gloor, 1954). However,
contrary to the reticular induced orienting response, the
amygdaloid evoked responses habituate rapidly on repeated stimulation (Ursin et al., 1967), and is more susceptible to chlorpromazine (Kaada and Bruland, 1960).

Electrode sites from which the orienting response has been
induced are scattered over widespread but restricted areas of
the amygdala (Fig. 3). This seems reasonable as the orienting
response is the initial phase of the flight and defense responses
as well as of the feeding and sexual activities which are
elicited from various parts of this nuclear complex. It is
noteworthy that almost all electrode sites in the periamygdaloid
cortex and ventral part of the corticomedial nuclear group are

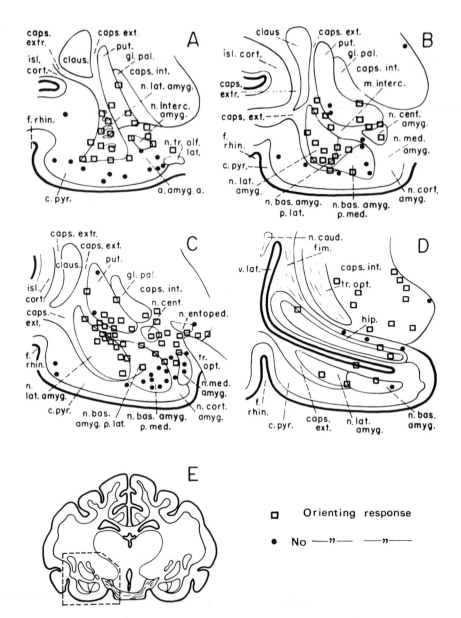

Fig. 3. Frontal sections through the amygdala in cats indicating points (open squares) from which behavior orienting responses have been produced on electrical stimulation. Dots, no response. Section C corresponds to the frontal plane shown in E. A rostral, and D caudal end of amygdala. (From Ursin and Kaada (1960a). Courtesy of Elsevier Publishing Company, Amsterdam.)

negative with respect to producing orienting reactions. As will be discussed below, stimulation of this zone produces the opposite effect, neocortical synchronization and sleep.

Arousal, associated with searching or orienting movements, is found in the anterior amygdaloid area, in the lateral and in the magnocellular part of the basolateral nucleus and in an area extending medially through the region of the central nucleus and into the internal capsule just dorsal to the optic tract, in the region of the entopeduncular nucleus. This medial extension corresponds to the course of the diffuse ventral amygdalofugal path. From this area Hassler (1956) produced similar contro-versive searching movements in cats by stimulation of a rather narrow zone continuing through the internal capsule, the zona incerta and the subthalamic nucleus to the mesencephalon. Using combined stimulation-ablation techniques, it has been shown that the arousal response is mediated also by fibers of the stria terminalis (Ursin and Kaada, 1960b).

About one-half of the searching responses were accompanied by sniffing, but sniffing also was elicited with searching be-haviour absent. Comparison of the map for the orienting res-ponse (Fig. 3) and sniffing (Fig. 4) reveals that there is a considerable overlap. The points yielding sniffing behavior are grouped in the anterior and dorsal part of the lateral nucleus, the magnocellular part of the basal nucleus as well as in the area extending medially through the region of the central nucleus, corresponding to the ventral amygdalofugal fibers.

It is somewhat surprising to find sniffing responses so far laterally since, as mentioned previously, only the anterior and corticomedial nuclear groups appear to receive direct olfactory projections. However, the basolateral division receives ol-factory impulses indirectly and may be involved in the efferent link in olfactory reflexes.

The amygdala appears to be essential for some components of the orienting reaction. Removal of this structure in monkeys causes a depression of the galvanic skin response (Bagshaw et al., 1965), heart-rate, and respiratory-rate components of the orient-ing reaction (Bagshaw and Benzies, 1968), while EEG activation and ear movement-orienting responses remain essentially intact but fails to habituate (Schwartzbaum et al., 1961; Bagshaw and Benzies, 1968). Thus, the orienting reaction can be fraction-ated into two major components by amygdalectomy. It was suggested that the autonomic indicators signify some sort of registrational process; the significance of the locomotor and EEG-activation remains to be explored.

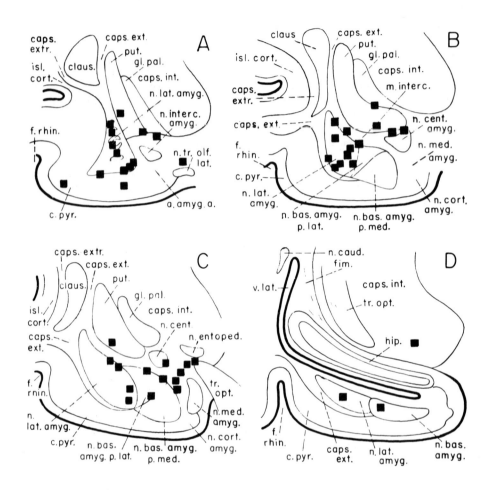

■   sniffing

Fig. 4.   Frontal sections in rostro-caudal direction through the amygdaloid nuclear complex indicating electrode sites yielding sniffing by electrical stimulation.   Cf. Fig. 3.   (From Ursin and Kaada (1960a).   Courtesy of Elsevier Publishing Company, Amsterdam.)

Kreindler and Steriade (1964) have studied the EEG responses
to amygdala stimulation in the "encéphale isolé and cerveau
isolé cat" (Fig. 5). Cortical desynchronization was obtained in
essentially the same amygdaloid areas as outlined above for the
orienting response, i.e., mainly from dorsal levels of the amygdaloid nuclear complex. The effective zone included the dorsal
parts of the anterior amygdaloid area and lateral nucleus, the
magnocellular part of the basal nucleus as well as the region
of the central nucleus. From ventral amygdaloid levels the
opposite effect, a neocortical synchronization, was produced (cf.
below).

Pagano and Gault (1964) have correlated recordings taken of
the spontaneous fast electrical activity from the basolateral
division of the amygdala with behavioral and neocortical measures
of arousal. The amygdala records show an increasing amount of
large amplitude, fast activity as the subject passes from the
"sleep" to the "aroused" state, and this measure is quite sensitive throughout the arousal continuum and is a better predictor
of behavioral state in the higher arousal regions than is the
neocortical activity.

(2) Sleep. Increased neocortical spindle activity induced
from the periamygdaloid cortex and olfactory tubercle in the cat
was described by Kaada (1951, p. 234) (Fig. 6). Stimulation of
the same area produces a number of inhibitory effects on somato-
motor reflexes, respiration and blood pressure, a.o. In the
conscious patient, the respiratory arrest evoked from this region
was found to be associated with impaired consciousness, with a
tendency to close the eyes, and with a feeling of tiredness and
sleepiness, without epileptic afterdischarges (Kaada, 1951, p.
62-64; Kaada and Jasper, 1952).

More recently, Hernández-Peón et al. (1967) has confirmed
these observations and reported that local chemical stimulation
of the prepiriform and periamygdaloid cortex, the olfactory tubercle and other structures with acetylcholine, successively induced behavioral and electrographic manifestations of sleep,
indistinguishable from spontaneous physiological sleep, when
applied to the prepiriform and periamygdaloid cortex, the olfactory tubercle and other structures. There was evidence that
the hypnogenic effects were mediated through the medial forebrain bundle, chemical stimulation of which is also followed by
sleep.

Kreindler and Steriade (1964) observed that electrical stimulation of ventral amygdaloid levels,in particular the ventral
part of the anterior amygdaloid area, the ventral part of the
lateral nucleus as well as the parvocellular part of the basal

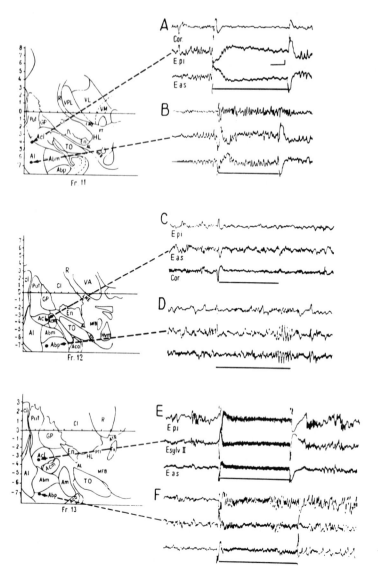

Fig. 5.   Different patterns of cortical electrographic reactions
obtained by stimulating dorsal and ventral amygdaloid portions.
Three different experiments, A and B, C and D, E, F. Dorsal and
ventral points within the amygdaloid complex, which were stimulated in each experiment, are indicated by black points in the
figure at left.   In each experiment, stimulation of dorsal and
ventral points is performed at the same rate and intensity.   In
A and B, C and D: 200/sec, 1 msec, 0.5 mA.   In E and F: 150/sec,
1 msec, 1 mA. (From Kreindler and Steriade, 1964.)

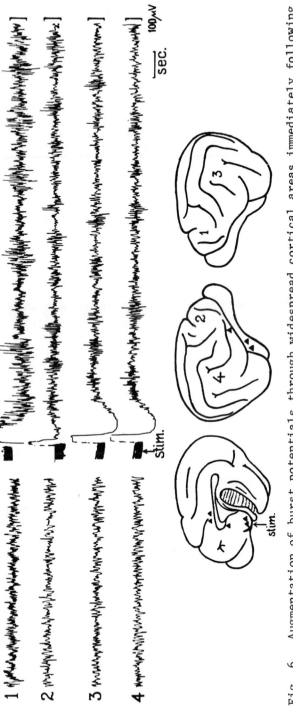

Fig. 6. Augmentation of burst potentials through widespread cortical areas immediately following electrical stimulation of the olfactory tubercle in cat (at arrow). The numbers on the brain indicate the placement of the four pairs of recording electrodes. Solid triangles indicate points found to yield such effect in the ECG. Dial anesthesia. (From Kaada, 1951, Fig. 60. Courtesy of Acta Physiologica Scandinavica.)

nucleus, produced neocortical synchronization (Fig. 5). Similar
effects have been reported by Russek and Hernández-Peón (1961),
Sterman and Clemente (1962) and Caruthers (1969). The neocortical
synchronization was not dependent on structures of the lower
brain stem since it persisted in the "cerveau isolé" preparation
(Kreindler and Steriade, 1964).

This inhibitory or 'anti-arousal' area of the lateral and
ventral amygdala and surrounding paleocortex is possibly related
to the inhibitory zone for flight and defense reactions as well
as to the inhibitory zone for feeding and sexual activities to be
discussed in the following sections.

B. Agonistic behavior and avoidance learning

(1) Flight, defense and predatory attack. Two types of
emotional responses have been induced by stimulation of the amygdala, flight and defense. The first is necessary for a particular part of "fear" behavior, the latter represents a particular
type of "aggression." The terms used to describe the complex
behavior patterns observed on brain stimulation vary considerably,
and this has led to some confusion, creating unnecessary conflicting reports, particularly with respect to topical localization.

Some ethologists have used the term agonistic behavior
to describe various kinds of adaptation which occur during conflict or right. It has been found useful to distinguish the
following three patterns of agonistic behavior, at least in the
cat: flight, defense and attack. Placed in front of a superior
enemy, the animal will display either flight or defense behavior,
depending on whether flight is possible. Faced with an inferior
enemy, the cat may adopt a threatening posture, termed the
"attack" response by Leyhausen (1956).

In the flight (Flucht) response, the animal first appears
restless and looks in all directions; it then withdraws or es-
capes without growling or hissing. There are signs of sympathetic outburst with pupillary dilatation and sometimes piloerection. Micturition may occur.

In the defense reaction (affektive Abwehr) the cat initially
exhibits the orienting behavior described above. It then re-
tracts its head (possibly to protect the neck) and crouches
(Fig. 7A-B). The ears are flattened to a posterior position,
the animal growls or hisses, the pupils are dilated, and there
is piloerection. On stronger stimulation the animal may raise
a forepaw, ready to strike with protruded claws (Fig. 7B).

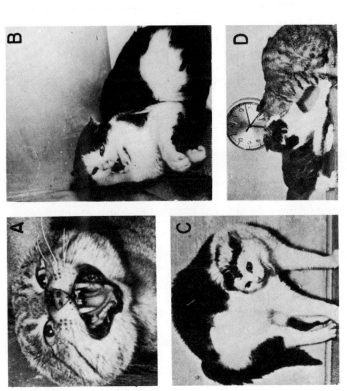

Fig. 7. Defense responses to (A) external stimulation (after Leyhausen, 1956) and (B) electrical stimulation of the amygdala. Responses to hypothalamic stimulation (after Brown and Hunsperger, 1963): (C) the characteristic humped back position, interpreted by Leyhausen (1956) as a super-imposition of attack on a high-intensity defense response; (D) attack on a stuffed dummy cat. (From Kaada, 1967. Courtesy of the University of California Press; reprinted with permission of the Regents of the University of California.)

In <u>attack</u> (Angriff) the cats start their bouts by a characteristic threatening posture with stretching of their legs and straightening their backs and necks (Fig. 7C). The cat may crouch close to the ground with its back arched. Slowly, it approaches the enemy uttering a series of growls. The animal always aims at biting the adversary at its head and neck (Fig. 7D), but also may fight the aggressor with its claws rather than bite.

Several investigators have suggested that attack is not a unitary concept. In the cat, Egger and Flynn (1963, 1967) and Flynn <u>et al</u>. (1970) distinguished between predatory or quiet attack and "affective" attack. Predatory attacks were characterized by the absence of concomitant autonomic signs. The cat does not growl or hiss, but makes a quiet deadly attack on the rat, biting viciously at its head and neck. The attach pattern closely resembles that of normal cats hunting or stalking prey (cf. also Leyhausen, 1956).

The "affective" attack is characterized by growling and hissing and the full complement of sympathetic signs indicative of feline rage. The cat strikes with its paw with claws unsheathed, in a series of swift, accurate blows. If the stimulus is continued, the cat will bite savagely the rat, but the initial part of the attack clearly is with its claws (Flynn <u>et al</u>., 1970). The two types of attack were elicited from different electrode locations in the hypothalamus and midbrain.

According to Hutchinson and Renfrew (1966), the quiet, biting attack or prey-killing in cats is a form of food acquisition that is used when smaller animals such as rats are the food. Eating and biting attacks invariably could be elicited from the same lateral hypothalamic electrode sites suggesting that the area concerned is responsible for the mediation of appetitive behavior. However, King and Hoebel (1968) and Flynn <u>et al</u>. (1970) observed hypothalamic locations from which prey-killing, and not eating, is elicitable. Hunger, probably therefore, is not always the motive for killing. From the hypothalamus, prey-killing, flight and defense are evoked from three discrete zones with no overlapping. The results of the various authors have been summarized by Kaada (1967).

From the amygdala, only flight and defense responses have been elicited, and there has been no report on directed attack behavior. However, hypothalamically elicited predatory attacks, as well as the spontaneous mouse-killing behavior in rats, may be <u>facilitated</u> or <u>suppressed</u> by amygdaloid stimulation and ablation (cf. below).

Moyer (1968) reviewed the physiological basis of aggressive behavior, and listed tentatively seven kinds of aggression which may be differentiated on the basis of the stimulus situation and which have different neural and possibly endocrine mechanisms: predatory, fear-induced, inter-male, irritable, territorial, maternal and instrumental. As it is not yet possible to identify all these types of aggression with confidence, and in particular not with reference to their respective neuronal substrates, they will not all be considered here. Only the three types of agonistic behavior defined above, flight, defense and predatory attack, will be discussed in this context.

(a) Stimulation

Flight and defense. In the amygdala, flight and defense responses have been elicited from separate zones as indicated from the map in Fig. 8 (Ursin and Kaada, 1960a). Flight and defense were obtained from 50 electrode sites, whereas about 150 points negative for these effects were recorded in the surroundings. The two zones run approximately parallel in a medio-dorso-caudal direction. Flight resulted from excitation of a rather restricted area extending from the rostral part of the lateral nucleus, and the preamygdaloid area through the region of the central nucleus and into the ventral part of the internal capsule.

Defense responses, on the other hand, resulted from stimulation of more posterior and medial parts of the amygdaloid complex, i.e. from the region of the central nucleus (Fig. 8C) and the adjacent dorsal portion of the lateral and basal nuclei (Fig. 8D-E).

The defense response produced by stimulation of the amygdala builds up gradually in the course of 20-40 secs and outlasts the stimulation period for 20-120 secs, in contrast to the defense response elicited from the hypothalamus which appears and disappears promptly (Hilton and Zbrozyna, 1963; Hunsperger and Buchner, 1967; Zbrozyna, 1971).

Separate flight and defense zones also appear to exist in the primate brain. Ursin (1971) has had the opportunity to analyze temporal lobe points in Bryan Robinson's and Mort Mishkin's large material collected from brain stimulation in the rhesus monkey. The responses were classified as "fear-like" and "defense-like" behavior. As seen from Fig. 9 there is again a localization into two zones, a rostral and lateral zone yielding flight, or fear, and a more caudal and medial zone for defense-like behavior.

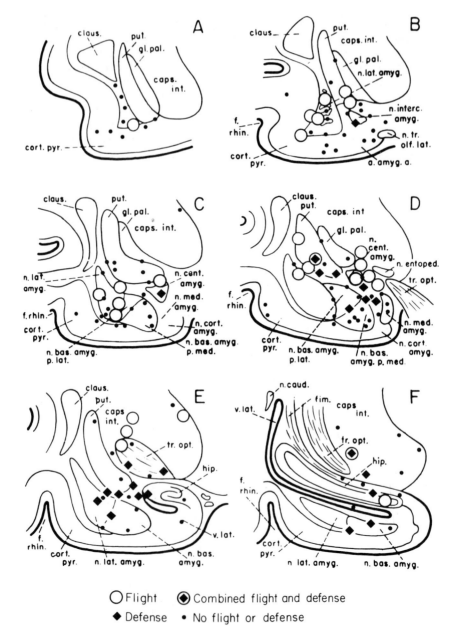

Fig. 8. Serial frontal sections in rostro-caudal direction through the amygdaloid nuclear complex indicating electrode sites from which flight and defense and a combination of these were obtained. (Modified from Ursin and Kaada, 1960a. Courtesy of Elsevier Publishing Company, Amsterdam.)

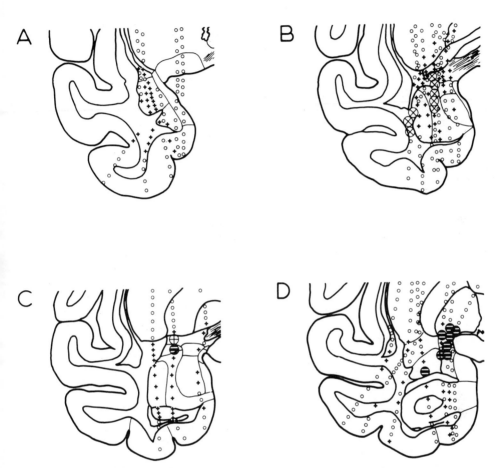

Fig. 9. Frontal sections through the temporal lobe in the rhesus monkey. A rostral, D caudal, medial to the right. Localization of points yielding 'fear-like' (circles with X) and 'defense-like' (black circles with horizontal bars) behavior in the rhesus monkey. Small circles: no response, small pluses indicate orienting behavior. (Analysis by Ursin, 1971 based on unpublished data from Robinson and Mishkin.)

     In other studies, a clear distinction has not been made be-
tween the flight (fear) and defense (threat, rage) patterns of
behavior and, consequently, these studies have not attempted a
topical differentiation of the two responses. In studies where
such a distinction has been made, or where it appears from the
description that one is dealing with the flight or defense responses as defined above, the localization of the effective areas
is in fairly good agreement with the results of Ursin and Kaada
(1960a). Thus, the map published by Wood et al. (1958) shows
components of flight and defense in the basal and central nuclei
in the rat. The amygdaloid defense area in the cat, as determined by Hilton and Zbrozyna (1963) and Zbrozyna (1971) included
part of the anterior amygdala, the basal nucleus (mainly its
lateral magnocellular part) and the central nucleus. Fonberg
(1968) similarly found that the defense area occupied mainly the
dorsomedial portion of the amygdaloid complex and extended to
the piriform cortex (Fig. 11).

     An apparent discrepancy from these studies is that some
investigators include the medial amygdaloid nucleus in the defense
area. Thus, Shealy and Peele (1957) evoked escape responses
from "the basal and lateral components of the amygdala, although
the central area was sometimes involved," and rage (defense)
reactions from the central and medial nuclei. Magnus and Lammers
(1956) similarly induced fear responses from the preamygdaloid
area, the parvocellular part of the basal nucleus and the central
and medial nuclei, whereas growling (defense) was elicited from
the medial nucleus and from the ventral part of the basal nucleus.
Since stimulation of the medial nucleus yields high self-stimulation rates (cf. below), one would not expect that this nucleus is
included in the flight and defense zone, unless it is composed
of subdivisions which differ functionally.

     As seen from Fig. 10, the responsive field for defense
(threat) and flight as delimited by Fernandez-deMolina and
Hunsperger (1959), mainly occupies the dorsomedial portion of
the amygdaloid complex in the region of the central nucleus but
with positive electrode sites scattered also in the adjoining
portions of the basal and medial nuclei. Points yielding flight
are intermingled with points yielding growling and hissing (defense). This is not in agreement with the findings of Ursin and
Kaada (1960a) and Ursin (1971) or with the differential effects
on selective ablation of the flight and defense zones (cf. below).

     Since positive sites were found along the course of the
stria terminalis (Fig. 10), and since such responses could be
traced to the bed nucleus of the stria, it was suggested by
Fernandez-deMolina and Hunsperger (1959, 1962) that, in the cat,
the emotional responses were mediated via the stria terminalis

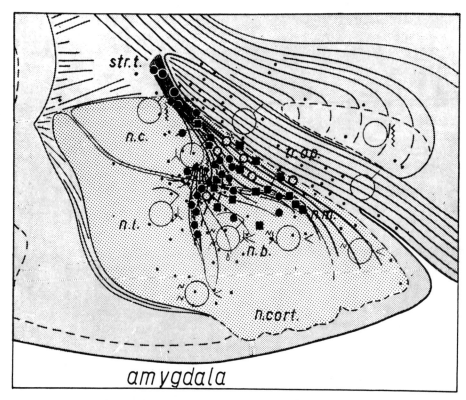

Fig. 10. Semi-schematic frontal section through the middle portion of the amygdala. Level of greatest extension of responsive
field for threat and flight as deliminated by Fernandez-deMolina
and Hunsperger (1959). Filled circles = threat with growling;
filled squares = threat with growling followed by hissing; small
open circles = flight of unpredicted character; dots 2 negative
points with regard to affective reactions; large circles = areas
yielding the effects indicated by the symbols in their vicinity
(cf. original publication).

    N.B. - n. basalis; n.c. - n. centralis; n. cort. - n. corti-
calis; n.l. - n. lateralis; n.m. - n. medialis; str.t. - stria
terminalis; tr. op. - tractus opticus.   (Courtesy of Journal
of Physiology.)

to corresponding areas in the preoptic region, hypothalamus and
midbrain.   However, it was later shown that the stria terminalis
contains afferent fibers to the amygdaloid area for the defense
reaction (Zbrozyna, 1960, 1971; Hilton and Zbrozyna, 1963).
These are likely cholinergic fibers (Lewis and Schute, 1963).
Hilton and Zbrozyna observed that defense responses, including
active muscle cholinergic vasodilatation (which also is part of
the defense reaction evoked from the hypothalamus) also were
elicited along the course of the ventral amygdalofugal pathway
to the preoptic and hypothalamic area for defense.   Lesions
severing this ventral connecting band abolished the response
from the amygdala provided the lesions extended to the most anterior and posterior extremes of the band.   On the other hand,
complete bilateral section of the stria terminalis did not reduce the defense reaction elicited from the amygdala (Zbrozyna,
1960, 1963, 1971; Hilton and Zbrozyna, 1963).

In a recent study, Hunsperger and Bucher (1967) have re-
explored the area between the optic tract and pallidum, an area
corresponding to the ventral, diffuse fibers described by Johnston (1923), Fox (1940) and Nauta (1961).   They could not find
support for the contention that the responsive fields for flight
and defense in the amygdala and hypothalamus are connected by
way of a direct ventral route.   On the other hand, Zbrozyna
(1971) has given further evidence for a ventral amygdaloid route.
Thus, this important problem of the efferent pathways for flight
and defense seems to require further investigation before definite
conclusions can be drawn.

Using local acetylcholine stimulation of the amygdala in
cats, Hernández-Peón et al. (1967) produced flight and rage
responses, without attacks.   Since acetylcholine excites cell
bodies and dendrites and presumably not axons, it is of importance in this connection that the positive sites included several
points in the magnocellular part of the basal nucleus, and also
in the prepiriform and piriform cortex.   If, in an excited cat,
the cannula was lowered into an inhibitory or hypnogenic point in
the piriform cortex, the animal sank into a sleep very quickly.
Unfortunately, the region of the central nucleus was not stimulated chemically, as this experiment would have solved the problem
of whether the positive effects obtained from electrical stimulation of the central region is merely due to traversing fibers,
or to the central nucleus participation in the response as is the
case with cells of the magnocellular part of the basal nucleus
and surrounding cortex.   In the experiments of Desci et al.
(1969), local chemical stimulation in the area of the central
amygdaloid nucleus with carbachol surprisingly inhibited the rage
reactions resulting from injection of the same drug into the
hypothalamus.

In man, stimulation of the amygdaloid region similarly has resulted in either feelings of fear (Chapman et al., 1954) or rage (Heath et al., 1955; Delgado, 1960; Feindel, 1961; Delgado et al., 1968; Stevens et al., 1969), again indicating a segregation of the two types of emotional responses. Stimulation of the anterior and inferior surface of the temporal lobe in epileptic patients produced fear, but never anger or rage (Penfield and Jasper, 1954; Mullan and Penfield, 1959). There was a high incidence of abdominal, thoracic and other bodily sensations accompanying the less intensive manifestations of fear. Fear also was induced from deep electrodes in the anterior temporal region.

As shown by Fonberg (1963, 1968), electrical stimulation of part of the basolateral division of the amygdala effectively inhibited fear reactions produced by external nociceptive or direct hypothalamic stimulation, as well as conditioned classical defense responses. Fig. 11 shows the inhibitory points for fear (black circles) in the lateral and basal nuclei with the points yielding defense responses more medially. This inhibition was not a mere 'arrest' reaction; the animals were able to walk and play during stimulation. Neither was it due to some nonspecific distraction effect since external distracting stimuli, such as a loud sound, did not produce the effects (Fonberg and Delgado, 1961; Egger and Flynn, 1963).

Predatory attack. As mentioned previously, directed attack responses have not been elicited by amygdaloid stimulation, but such stimulation may either facilitate or suppress hypothalamically elicited predatory attack behavior in cats (Egger and Flynn, 1962, 1963, 1967). Naturally, elicited attacks on mice similarly were blocked (Egger and Flynn, 1962). In general, suppression was elicited most consistently in the lateral portion of the basal nucleus, and in the anterior and medial portions of the lateral nucleus of the amygdala. Facilitation was elicited in the dorsolateral portion of the posterior part of the lateral nucleus. Data from trials during which electrical afterdischarges occurred were excluded. Further, Vergnes and Karli (1969), Karli et al. (1969, 1971) observed that the spontaneous aggressive behavior displayed by mouse-killing and mouse-eating rats was suppressed by electrical amygdaloid stimulation. Facilitation of the mouse-killing behavior was never seen.

(b) Ablation studies

Flight and defense. Evidence in support of the localization hypothesis may be derived also from selective ablation studies. It would be expected that small bilateral lesions restricted to the amygdaloid flight or defense zones specifically would reduce these behavior patterns, whereas a lesion restricted to the area

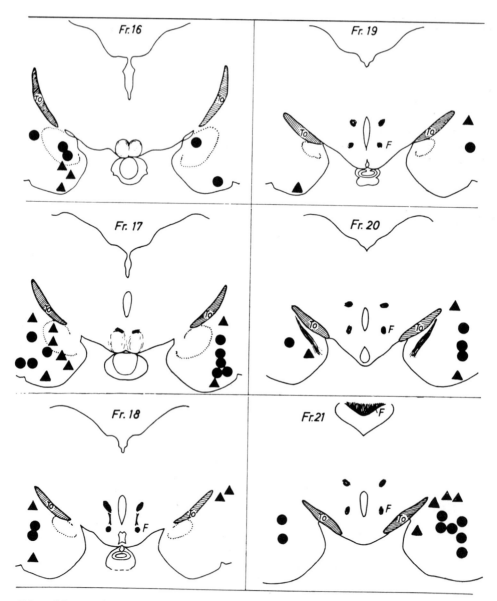

Fig. 11.   Schematic diagram of the localization of stimulated
points in fifteen dogs influencing the defense reaction.   The
defense field (triangles) is situated in the dorsomedial portion
(mostly nucleus centralis) and also in the piriform cortex.   The
inhibitory points (circles) are found mostly in the lateral nucleus, extending to the basal nucleus.   (From Fonger, 1968.
Courtesy of Elsevier Publishing Company, Amsterdam.)

inhibiting these responses would result in increased aggressiveness and fear by release of hypothalamic defense and flight
mechanisms.

It is a well established fact that total bilateral removal
of the amygdala, or the anterior portion of the temporal lobe
results in increased tameness with postoperative reduction of
fear, escape behavior or aggressiveness. Such placidity has been
observed in all species investigated, including humans. For
references cf. Gloor (1960), Koikegami (1964, p. 207), Goddard
(1964b), Ursin (1965a) and Moyer (1968). However, only very few
of these studies contribute a more exact localization of the
effective areas. It is almost beyond doubt, by the great number
of control lesions in areas outside of the amygdala, that the
placidity is primarily due to amygdala ablation.

On the other hand, a number of studies, all in cats, have
shown the opposite effect, i.e., removal of the amygdala leads to
increased aggressiveness (Bard and Rioch, 1937; Spiegel et al.,
1940; Bard and Mountcastle, 1947; Green et al., 1957; Wood, 1958).
Since it was not possible to correlate these observations with
damage to any specific area of the amygdaloid complex, and since
it has been difficult to reproduce the results, various other
hypotheses were put forward to explain the apparent discrepancy
between these studies and those reporting increased tameness.
The main non-localizing theory for the increased aggressiveness
has been that of Green et al. (1957) who observed that the cats
displaying postoperative rage all developed epileptic seizures,
and who therefore suggested that the savage behavior could be
due to a discharging focus in the periphery of the lesion.
Summers and Kaelber (1962) postulated that the hostile tendencies
following incomplete bilateral removal of the amygdalae and piriform cortex in two cats were due to additional injury to the
pallido-hypothalamic fascicle and ventromedial hypothalamic
nucleus on one side.

Ursin (1965b) placed small bilateral lesions restricted to
the amygdaloid flight zone, and could reduce specifically flight
behavior in wild cats with no effect on defense behavior. The
latter could similarly be specifically eliminated or reduced by
small amygdala lesions, again indicating that separate neural
mechanisms are involved. The relatively small number of cats in
which such a specific reduction of defense behavior was obtained,
did not allow any definite conclusion concerning the exact determination of the effective area.

It has been suggested that taming, following amygdaloid
lesions in cats, may be due only to adaptation to the laboratory
environment (Morgane and Kosman, 1957). However, small or mis-

placed lesions outside the flight and defense zones did not produce any significant taming effect (Urwin, 1965a).   Second, the
often observed late postoperative recovery of affective behavior
is contrary to the adaptation hypothesis.   Third, wild stray cats
have been kept in the laboratory for as long as 6 months without
showing any significant change in affective behavior (Ursin, 1964).
Finally, the dramatic taming of the wild Norway rat after bilateral amygdalectomy (Woods, 1956) cannot be explained by adaptation.

Fonberg (1965) produced increased tameness in dogs by
lesions restricted to the medial part of the amygdala, an area
shown by various investigators to participate in the defense
reaction.   However, a lesion placed in the dorsomedial part
of the amygdala (Fig. 12) unexpectedly resulted in increased
defensive behavior.   This increase appeared on the day after
the operation.   The most tempting explanation would be that these
lesions involved an inhibitory system, either by damaging the adjacent basolateral division or (in the cases when such damage was
not found) the efferent pathway from a basolateral inhibitory
system (Fonberg, 1965), releasing the brain stem defense system
through disinhibition.

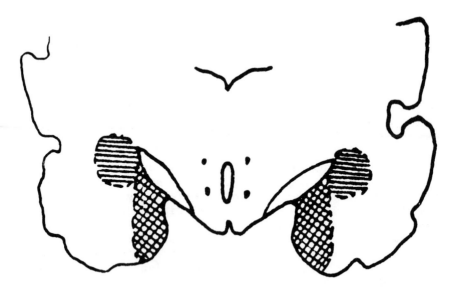

Fig. 12.   Amygdaloid lesions in dogs yielding increased tameness
(cross-hatched medial field) and increased defensive behavior
(stripped dorsal field).   (From Fonberg, 1965. Courtesy of
Polska Akademia Nauk, Warsaw.)

In reviewing the literature, some support may be found for this notion. Thus, Wood (1958) reported that in the cat small bilateral lesions in the amygdala produced increased aggressiveness only when the basal or central nuclei were destroyed. As mentioned previously, a possible participation of neurons in the region of the central nucleus in an inhibitory system with respect to rage also was indicated by the effects of carbachol stimulation (Desci et al., 1969). Masserman et al. (1958) observed that lesions in the lateral portions of the amygdaloid nuclei in cats produced a moderate hypersexuality and developed markedly lower thresholds for startle and fear. Further, Lewinska (1967), in a study of the hyperphagia resulting from lesions in the basolateral area, mentions that some cats with lesions of the posterior part of the parvocellular basal nucleus were more agitated than before surgery. When part of the magnocellular basolateral nucleus also was involved, aggressiveness became manifest. On the other hand, animals with lesions in the corticomedial area, and which developed aphagia, generally were more quiet after the operation.

Spiegel et al. (1940) observed increased aggressiveness after amygdaloid lesions, whereas superficial lesions of the piriform lobes evoked slight and transient symptoms of rage.

In conclusion, it seems that the increased aggressiveness following amygdaloid lesions previously reported by some investigators is caused by removal of areas exerting inhibitory influences on aggression. There appears to be no data which are against this assumption, but only further experimental work can resolve the problem and more accurately delimit the responsible structures.

Prey-killing. A lesion in the region of the central nuclei of the amygdaloid complex reduced the spontaneous aggressive behavior displayed by mouse-killing and mouse-eating rats (Vergnes and Karli, 1964; Karli and Vergnes, 1965; Karli et al., 1971). The lesions encroached more or less upon the medial nucleus. Such lesions did not affect eating-behavior, demonstrating a differential effect upon eating and mouse-killing behavior (Karli and Vergnes, 1964, 1965). Interruption of the diffuse ventral amygdalofugal path had the same effect, whereas section of the stria terminalis was without any influence (Vergnes and Karli, 1964). Also, lesions sparing the centro-medial region and involving either the cortical or basal and lateral nuclei had no significant effect on the mouse-killing behavior (Karli and Vergnes, 1965), whereas bilateral injection of antidepressant drugs into the centro-medial region produced an immediate inhibition of the mouse-killing behavior for 1-2 hours (Horowitz and Leaf, 1967).

Removal of the olfactory bulb converts rats that are not natural, spontaneous killers to mouse-killers (Vergnes and Karli, 1963). This inhibitory influence exerted through the olfactory system is most likely funneled through the amygdala, but the mechanism and inhibitory amygdaloid region is not known. The prepiriform cortex probably acts as a relay station in the inhibitory pathway (Vergnes and Karli, 1963, 1965).

(2) Active and passive avoidance behavior. A great number of studies deal with the role of the amygdaloid complex in acquisition and retention of various types of avoidance behavior. In the majority of the earlier studies, the lesions involve all or extensive parts of the amygdaloid complex with surrounding cortex. Therefore, these studies are of limited value with respect to the present problem of topical localization. A review of this literature has been given by Gloor (1960) and Goddard (1964a,b, 1969). Further, in several of these reports no distinction has been made between various types of avoidance behavior with different underlying mechanisms and, consequently, seemingly conflicting reports have appeared.

A distinction between active and passive avoidance behavior (Mowrer, 1960) has proved fruitful in studies of the functional significance of the cingulate-septal region. Thus, lesions in the supracallosal cingulate cortex, an area facilitating various motor and visceral activities (Kaada, 1951, 1960), disrupts performance in an active avoidance task (McCleary, 1961). Lesions in the subcallosal-septal region, an inhibitory area for such activities (Kaada, 1951, 1960), disrupts an animal's ability to inhibit its response in a passive avoidance situation, resulting in repetition or perseveration of the learned response.

By analogy, one would expect that lesions of the amygdala, involving the facilitatory flight, would reduce the animal's ability to exhibit escape responses and to solve a problem that requires active avoidance. Since lesions in these zones in a tame cat have little effect on overt behavior, one might by the use of active avoidance tests reveal changes in performance in such animals. By contrast, passive avoidance behavior would be impaired by lesions in the inhibitory basolateral amygdaloid region. Recent studies have verified these predictions.

Lesions restricted to the flight zone in the rostral part of the lateral nucleus or damage of the ventral amygdalofugal pathway in cats resulted in an impaired active avoidance behavior, whereas passive avoidance was not influenced by these lesions (Ursin, 1965b). Horwath (1963) reported similar findings with amygdaloid lesions which, from the data presented, appear to involve the flight zone in the region of the central nucleus and

adjacent dorsal part of the basolateral complex. A significant
active avoidance defect was observed only for the conventional,
double-grill box, whereas the performance was not substantially
impaired for the simple one-way active avoidance and passive
avoidance tests. Horvath concluded that the basolateral amygdaloid nuclei subserve an integrative function in the acquisition
of avoidance response in problem-solving situations of a high
order of complexity. An alternate explanation would be that the
active avoidance component in a double-grill box is more fear-
motivated than in a one-way situation and, therefore, would be
impaired more seriously by removal of a fear zone.

On the other hand, Pellegrino (1968), in an extensive and
careful experimental analysis, observed a clear-cut passive
avoidance deficit in rats with basolateral amygdaloid lesions
(Fig. 13). Measures indicated that the deficit could not be
attributed to an increased motivation for food, or water due to
the lesions, or to an increase in general activity level. Perseveration of previously learned responses after amygdalectomy
could occur also in tasks where no shock was employed (DRL-20
performance), and in spatial alternations without cues, but not
in visual or spatial alternations, or "go, no-go" visual discriminations and reversals. The results suggested that rats with
basolateral lesions are unable to inhibit established responses
when they must depend on the information provided by internal
cues, but can inhibit responses when there is a visual cue to
guide their behavior.

With respect to localization, Pellegrino stated that these
results were in apparent conflict with the report by Ursin (1965b)
who found that lesions of the medial amygdaloid nucleus or in the
region of the stria terminalis caused a passive avoidance deficit,
whereas a lesion in the rostral part of the lateral nucleus did
not affect passive avoidance but disrupted the acquisition of an
active avoidance. However, some passive avoidance deficit after
corticomedial lesions was recorded also by Pellegrino (1968).
Thus, both these studies suggest that within this ventromedial
quadrant of the amygdala there are some inhibitory structures
(as also indicated by the production of sleep, inhibition of
micturition and adrenocortical output a.o.). With respect to
the basolateral lesions yielding deficits in passive avoidance
in Pellegrino's experiments with rats, these do not seem to
correspond to the ineffective lesions in Ursin's experiments
with cats. The former were confined to the lateral nucleus,
whereas the latter included relatively large parts of the basolateral complex (Ursin, 1964b, Fig. 3, cats 19 and 22). Attention also should be paid to the fact that the lateral lesions
in Pellegrino's experiment were situated close to the capsula
extrema containing projections from the adjacent insular cortex,

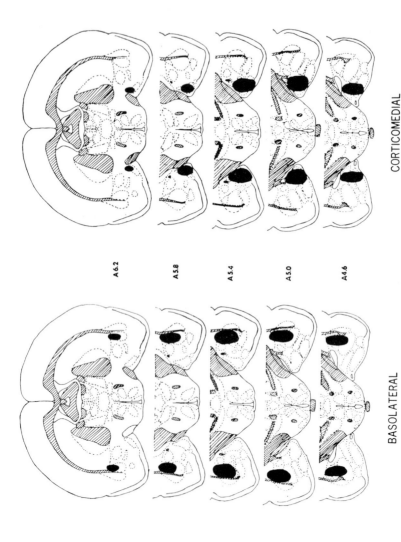

BASOLATERAL

CORTICOMEDIAL

Fig. 13. Lesions influencing avoidance behavior in rats (cf. text). Reconstructions of typical basolateral and corticomedial lesions on sections drawn from the de Groot (1959) stereotaxic atlas. Numbers in center refer to anterior-posterior coordinates in the de Groot atlas. (From Pellegrino, 1968. Courtesy of American Psychological Association.)

an area which, by removal, similarly causes a deficit in passive
avoidance behavior (Kaada et al., 1962). Pellegrino (1965) found
that low-level stimulation of the basolateral region in rats produced much greater passive avoidance deficits than stimulation
of the corticomedial area. It may either be a question of quantitative differences, or that different mechanisms are interfered with in lateral and medial ablation and stimulation, the
latter zone being a rewarding one (cf. later).

Thus, the apparent disagreement regarding the effects of
basolateral lesions may be due to the fact that this division
of the amygdala is looked upon as an entity, whereas functionally
it should be further subdivided. A passive avoidance deficit
by basolateral ablations would be expected in view of the inhibitory influence exerted by the lateral, and ventral portions
of this area on a number of other activities.

Differential effects on active and passive avoidance behavior
also were obtained using cholinergic stimulation of the amygdala
(Goddard, 1969). Passive avoidance and conditional emotional
responses were impaired severely whereas simple active avoidance
was not. However, the passive avoidance deficit was apparent
only in situations requiring long-term retention but not in
short-term memory. Generally, the site of injection was limited
to the posterior half of the amygdaloid complex, but a precise
description of the affected area was not warranted because of
the probable diffusion of the drug.

(3) Adrenocortical response. It would be expected that
stimulation of the area of the amygdala, which produces arousal
and emotional responses, would influence also the adrenocortical
system as integrated parts of these responses. Conversely, stimulation of other parts of the basolateral amygdala and piriform
cortex would be predicted to yield a decrease in adrenocortical
output. This is what the experiments show, again indicating a
functional localization. Mason (1958, 1959) reported elevation
of plasma 17-hydroxycorticosteroids (17-OHCS) level after electrical stimulation of the amygdala in conscious rhesus monkeys
through electrodes chronically implanted. The effective electrode sites were, as might be expected, distributed widely
within most parts of the amygdala (Mason, 1959). A similar wide
distribution was reported by Ishihara et al. (1964) and Kawakami
et al. (1968). Setekleiv et al. (1961), working with cats
lightly anesthetized found that the most effective area corresponds to the flight and defense zone. Three electrode sites
in the piriform cortex were negative.

Slusher and Hyde (1961) found an increase in corticosteroid
output from the adrenal vein by stimulation of the medial part
of the basal nucleus of the amygdala. A significant decrease

was observed after stimulation in the uncus (periamygdaloid cortex) and lateral amygdala as well as from the preoptic region, diagonal band of Broca, or the septum.

Rubin et al. (1966) stimulated the anterior temporal region prior to unilateral temporal lobectomy in four patients. The sites of stimulation could be determined through analysis of the removed tissue. An increase in plasma and urine corticosteroid levels followed stimulation through electrodes in the basolateral amygdala whereas hippocampal stimulation led to decreases in corticosteroid output. Mandell et al. (1963) made similar observations in man. The stimulus intensity used was too low to produce any detectable behavioral or subjective changes.

McHugh and Smith (1967) have shown recently that the plasma 17-OHCS response to amygdaloid stimulation in rhesus monkeys occur only in connection with local afterdischarges. Comparable stimuli, which did not evoke afterdischarges, did not produce significant changes in plasma 17-OHCS. Afterdischarges induced from the frontal lobe did not result in any effect, whereas direct hypothalamic stimulation could produce a 17-OHCS response with no afterdischarge.

Eleftheriou et al. (1966) observed that bilateral lesions restricted to the medial amygdaloid nucleus in the deermouse caused a significant increase in plasma and pituitary ACTH and remained significantly higher than control, unlesioned animals. There was also an increase of plasma and adrenal corticosterone. Thus, this experiment lends support to the view that the medial nucleus, or a portion of it, in some respect possibly is part of an inhibitory amygdaloid area.

(4) Cardiovascular, respiratory, pupillary and bladder responses. In view of the profound influence on emotional behavior exerted by the amygdala, one might anticipate alteration in various visceral activities, particularly following amygdaloid stimulation. Four such activities integrated in emotional behavior will be discussed in this connection, with reference to topical localization.

Cardiovascular responses. It would be hard to predict any specific cardiovascular response pattern related to any particular amygdaloid area. Even in the flight and defense area such predictions would be hazardous due to the complex cardiovascular adjustment associated with natural fighting (summarized by Adams et al., 1969) or with the defense reaction elicited from the hypothalamus (reviewed by Lisander, 1970).

The cardiovascular pattern preparatory for normal fighting consists of a fall in cardiac output and arterial pressure, either a bradycardia or no significant change in heart rate, and vasoconstriction of visceral organs (Adams et al., 1969). During the fighting itself there is tachycardia, increased cardiac output, visceral vasoconstriction and vasodilatation in contracting muscles. The mechanisms contributing to the latter response are under dispute. In this connection it is relevant to state that the defense reaction elicited by amygdaloid stimulation is associated with similar muscle vasodilatation through cholinergic sympathetic fibers (Hilton and Zbrozyna, 1963) to that induced by hypothalamic stimulation (for references see Adams et al., 1969; Lisander, 1970).

In general, pressor responses appear to dominate following amygdaloid stimulation in the unanesthetized animal (Koikegami et al., 1953; Reis and Oliphant, 1964) with the strongest effects elicited from the areas yielding flight and defense behavior (Morin et al., 1962; Koikegami et al., 1953, 1964), including an area corresponding to the course of the ventral amygdalofugal path (Morin et al., 1952) (Fig. 14). From these areas pressor responses are also present under anesthesia. However, low-frequency and low-intensity stimulation favor depressor responses, whereas high-frequency and higher-intensity stimulation favor pressor effects (Kaada, 1951; Koikegami et al., 1957). Morin et al. (1952) obtained a fall in blood pressure on stimulating the lateral nucleus, the responsive area being continuous with the depressor area of the claustrum (Fig. 14). Similar effects were obtained by Wood et al. (1958), but the strongest depressor points were located in the corticomedial division of the amygdala (Andy et al., 1959).

The most extensive study dealing with heart rate changes is that of Reis and Oliphant (1964). Responsive loci for bradycardia were found in all subdivisions of the amygdala but particularly in its basomedial part. Tachycardia, on the other hand, followed stimulation of points which tended to cluster in areas largely encircling the area from which bradycardia was elicited. These were often concentrated in the white matter of external and internal capsule but tachycardia was also at times elicited from all amygdaloid subdivisions and the surrounding paleocortex. Bradycardia was favored by low-frequency stimulation (6 cycles/sec) and reduced by stimulation at 100 cycles/sec. Stimulation at 30 cycles/sec. indirectly favored the appearance of bradycardia because of the propensity of this stimulus to provoke afterdischarges. Tachycardia did not seem to be influenced by the frequency of stimulation. Bradycardia could be totally blocked by atropine and bilateral vagotomy, whereas vagotomy did not influence the tachycardia due primarily to excitation of sympathetic cardio-acceleratory discharge.

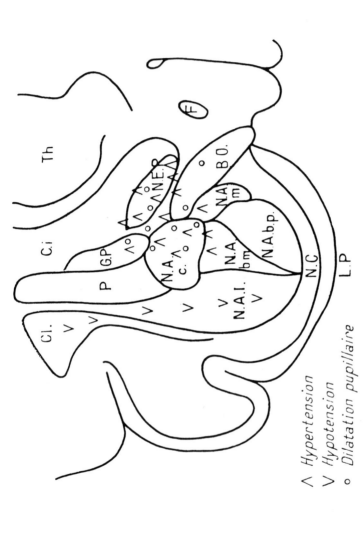

∧ *Hypertension*
∨ *Hypotension*
○ *Dilatation pupillaire*

Fig. 14. Localization of stimulated points within the amygdaloid region yielding changes in blood pressure and pupillary dilatation in cat.

B.o. - tractus opticus; C.i. - capsula interna; Cl. - claustrum; F. - fornix; G.P. - globus pallidus; L.P. - lobus piriformis; N.A.b.m. - n. basalis pars magnocellularis amygdalae; N.A.b.p. - n. basalis pars parvocellularis amygdalae; N.A.c. - n. centralis amygdalae; N.A.l. - n. lateralis amygdalae; N.A.m. - n. medialis amygdalae; N.C. - n. corticalis amygdalae; N.E.P. - n. entopeduncularis; P. - putamen; Th. - thalamus. (From Morin et al., 1952. Courtesy of Masson et Cie, Paris.)

Respiratory movements. The most common respiratory response obtained by amygdaloid stimulation in the unanesthetized animal is a lowering of the amplitude associated with acceleration of breathing (Kaada, 1951; Magnus and Lammers, 1956; Ursin and Kaada, 1960a). This pattern is usually observed in association with the general arousal and sniffing behavior, and the distribution of the active points therefore largely coincides with those yielding such responses, i.e., in parts of the basolateral amygdala and in an area extending dorso-medio-caudally through the region of the central nucleus, with some parts also located in the rostral piriform cortex (Magnus and Lammers, 1956; Ursin and Kaada, 1960a). This response is also present under urethane anesthesia (Koikegami and Fuse, 1952); it is favored by chloralose but is converted to one of pure inhibition (in expiration) by barbiturates (Kaada, 1951). This is possibly the reason why inhibition in expiration has been the most common response under barbiturate anesthesia from the amygdala (Kaada, 1951; Kaada, Andersen and Jansen, 1954; MacLean and Delgado, 1953) as well as the rostral piriform cortex (Kaada, Pribram and Epstein, 1949; Kaada, 1951; Wall and Davis, 1951; Glusman et al., 1953). Pure inhibition of breathing has also been elicited in unanesthetized animals (Kaada, 1951; Shealy and Peele, 1957) and in humans (Liberson et al., 1951; Kaada and Jasper, 1952), particularly from the ventral part of the amygdaloid region.

Strong respiratory inhibition has been obtained from the medial group of amygdaloid nuclei (Kaada, 1951; Kaada, Andersen and Jansen, 1954; Baldwin et al., 1954; Wood et al., 1958). This is possibly related to inhibitory olfactory reflexes elicited by vaporous stimuli (Allen, 1922a, 1922b; Frankenhauser and Lundervold, 1949; Andersen, 1954), and which appears to be mediated via the rostral piriform cortex, medial group of amygdaloid nuclei and stria terminalis (Kaada, 1951; Wood et al., 1958).

Pupillary dilatation is invariably associated with the general arousal and orienting response. The distribution of the effective electrode sites for pupillo-dilatation therefore closely corresponds to those yielding these behavior patterns and increases in blood pressure (cf. Fig. 14). Maximal effects are found in the areas for flight and defense reactions (Koikegami and Yoshida, 1953; Kaada, Andersen and Jansen, 1954; Magnus and Lammers, 1956). The pupillary response can be abolished by section of fibers corresponding to the M.V. and M.D. bundles of Fuchuchi (1952) of which the M.V. bundle, which appears to correspond to part of the ventral amygdalofugal fibers (Koikegami and Yoshida, 1953), is the most important one.

Micturition was elicited from electrodes situated in the anterior amygdaloid area, in the anteromedial group of nuclei and

in the basal nucleus, whereas the lateral nucleus was practically
unresponsive (Magnus and Lammers, 1956; Ursin and Kaada, 1960a).

Micturition was usually associated with fear (Magnus and
Lammers, 1956). A somewhat wider distribution of positive sites
within the amygdala was reported by Shealy and Peele (1957) and
Gjone (1966). Using intravesical bladder recording the latter
described an excitatory basolateral zone and an inhibitory anteromedial zone within the amygdala. Koikegami et al. (1957) similarly demonstrated excitatory effects on stimulating the lateral
magnocellular basal nucleus and inhibitory effects from the medial
parvocellular basal nucleus and the cortical nucleus. It was assumed by Gjone (1966) that the area yielding bladder contractions
corresponded to the flight zone and that yielding bladder inhibition corresponded to the defense zone. Attention was paid
to the common experience that fear frequently is accompanied
by an increase in the need for micturition. Selective removal
of the excitatory and inhibitory zones caused an increase and
a decrease, respectively, of the micturition threshold, indicating that the amygdala exerts a tonic influence on the
urinary bladder (Edvardsen and Ursin, 1968).

In conclusion, flight and defense responses have been
elicited from two separate amygdaloid zones: flight from an
area extending from the rostral part of the lateral nucleus,
and overlying piriform cortex, through the region of the
central nucleus and into the ventral amygdalofugal path, and
defense from the central region and adjacent portions of the
lateral and basal nuclei. Some authors include the medial
nucleus in the defense area. Flight and defense behavior can
be reduced selectively by restricted amygdaloid lesions. Parts
of the lateral and basal nuclei appear to inhibit fear reactions
and aggression, as suggested from the results of stimulation and
ablation. The amygdala also exerts a facilitatory and suppressing influence on prey-killing behavior. The adrenocortical and
visceral responses, presumably integrated in these various
behavior patterns, in the whole have been elicited from areas
from which such responses would be expected. Similarly,
active avoidance learning is impaired by removal of the
facilitatory flight zone, whereas a passive avoidance deficit
results from lesioning the inhibitory basolateral region and
possibly the medial nucleus.

C. Feeding activities

(1) Food and water intake. It now appears well documented
that the amygdala plays an important role in food and water intake.
In a number of species, bilateral ablation of the amygdala has
resulted in striking hyperphagia or in hypophagia; electrical or
chemical stimulation has increased or decreased food and water
intake. Several of the reports throw no or little light on the
question concerning the crucial amygdaloid areas responsible for
the effects. Only those studies which serve to elucidate the
problem of topical representation are included in this survey.

Most results of stimulation indicate that anterior and medial
parts of the amygdala exert a positive, excitatory effect on alimentary reactions, whereas neurons within the basolateral division
play an inhibitory role. In some respects, these two areas duplicate the "feeding center" of the lateral hypothalamus and the
inhibitory or "satiation center" of the ventromedial hypothalamus.
Conversely, removal of the excitatory anterior and medial zone
results in aphagia, whereas removal of the lateral inhibitory zone
results in hyperphagia, as would be anticipated from the results
of stimulation.

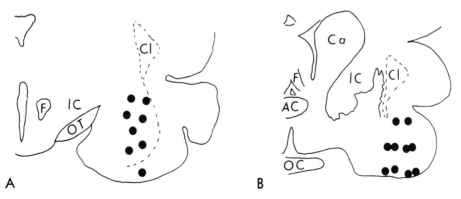

Fig. 15. Composite diagram showing location of contacts which in
different animals inhibited food intake at low thresholds of
stimulation. Coronal sections are at approximately (A) 12 and
(B) 13-15 mm. anterior to interaural plane. (From Fonberg and
Delgado, 1961. Courtesy of American Physiological Society.)

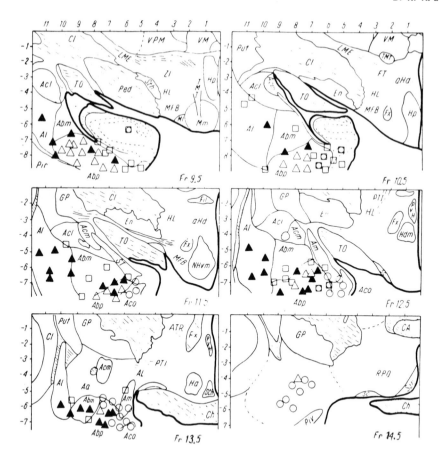

○ *Facilitation*
▲ *Intrinsic inhibition type a*
△ *Intrinsic inhibition type b*
□ *Extrinsic inhibition*

Fig. 16. Series of frontal sections in caudo-rostral direction
through the amygdala in cats. Sites of electrodes are indicated
from which facilitation and inhibition of alimentary reflexes were
obtained. Note, that stimulation of some points gives antagonistic
effects.

Abp - nucleus amygdaloideus basalis pars parvocellularis;
Abm - n. amygdaloideus basalis pars magnocellularis; Aco - n. amygdaloideus corticalis; Am - n. amygdaloideus medialis; Al - n. amygdaloideus lateralis; Acl - n. amygdaloideus centralis (pars
lateralis); Acm - n. amygdaloideus centralis (pars medialis);
Aa - area amygdaloidea anterior; Pir - lobus piriformis; ST - stria
terminalis. (From Lewińska, 1968a. Courtesy of Acta Biologiae
Experimentalis, Warsaw.)

Stimulation. It was shown by Fonberg and Delgado (1961) in cats and by Fonberg (1963) in dogs that electrical excitation of basolateral and anterior parts of the amygdaloid complex inhibited spontaneous food intake in hungry animals at intensities which did not modify playful behavior or produce fear or anxiety (Fig. 15). Previously learned instrumental alimentary reactions were similarly inhibited. The inhibition outlasted stimulation from ten seconds to several hours or even days. Chronic stimulation of points in the basolateral amygdala by remote radiocontrol reduced daily food intake (Fonberg and Delgado, 1961).

This inhibitory effect on stimulating within the basolateral division of the amygdala was confirmed by Lewińska (1968a) using an alimentary instrumental conditioning response in cats (Fig. 16). Three types of inhibition were described. In one type (a) the animal was completely indifferent towards food, as in satiated cats, and as in the experiments of Fonberg and Delgado (1961). Stimulation of these points decreased daily food intake, and lesions at the same electrode sites led to hyperphagia. In another form of inhibition (type b) the animal showed aversion towards food, refusing to eat it. In these cases a state of satiety might be excluded as judged from sniffing at the food and attempts to eat it. Nausea and retching were often evoked from these points at higher stimulus intensities. A third type of alimentary inhibition (c) occurred when stimulating the flight and defense areas. At intensities subthreshold for these emotional effects, there was no inhibition of food intake, but inhibition was seen when such effects were manifest.

Points responding with ejection of food and vomiting were also mapped by Robinson and Mishkin (1968a). The effective field in the amygdaloid region appears to be located in the basolateral division and prepiriform cortex (their Fig. 5, Section A 15 AP).

A facilitatory area for food intake comprised the cortical and medial nuclei, the adjoining part of the parvocellular area of the basal nucleus and the anterior amygdaloid area (Lewińska, 1968a). Searching, sniffing, licking, chewing and contractions of the ipsilateral face were obtained from some points. Facilitation of food responses to amygdaloid stimulation has also been observed by Gastaut (1952) and Robinson and Mishkin (1962). However, these authors did not localize their sites of stimulation. In a recent and more comprehensive study, Robinson and Mishkin (1968a) observed increased food intake in monkeys when stimulating the stria terminalis and the bed nuclei.

In chronic stimulation of the facilitatory area in cats, Lewińska (1968b) recorded an increase in food and milk intake, in particular such food which under normal conditions was most attractive to them (raw meat and milk). In a few cases in which the electrodes were situated at the border between the facilitatory and inhibitory zones, there was an increase in only one kind of food and a decrease in another kind.

These observations are in close agreement with those of Grossman and Grossman (1963) who obtained inhibition of feeding and drinking behavior in rats by stimulating posterior points of the ventral amygdala, whereas stimulation of anterior points increased water consumption but inhibited food intake. Small lesions placed in these areas produced opposite effects which appeared to be permanent.

It is possible that the inhibitory effect of basolateral stimulation is secondary to interference with some positive emotional state (Fonberg, 1968). There is some evidence suggesting that stimulation of parts of the basolateral part is positively reinforcing. Basolateral stimulation may produce a highly pleasant rewarding state and, therefore, interfere with both hunger drive and with defensive and aggressive behavior, and this may account for the lack of response.

Grossman (1964) has studied food and water intake using adrenergic and cholinergic stimulation of the ventral amygdala. Previous studies by the same author (Grossman, 1962) had shown that anatomically overlapping neuronal systems in the lateral hypothalamus, involved in the control of food and water intake, could be separated on a neurochemical basis. Hypothalamic feeding mechanisms are found to be selectively sensitive to adrenergic stimulation and inhibition, whereas regulation of water intake appeared to be mediated by cholinergic stimulation.

These findings were duplicated in the amygdala. Adrenergic stimulation, with the tip of the electrode in the cortical nucleus or close to uncus, increased food intake and presumably hunger but reduced water consumption. On the other hand, cholinergic stimulation of the same area increased thirst but reduced food intake. This pattern of results was reversed following local application of an adrenergic or a cholinergic blocking agent, respectively. Various control substances failed to duplicate these effects and demonstrated their specificity. Gamma-aminobutyric acid (GABA) produced effects similar to those following cholinergic stimulation, and hydroxylamine produced the opposite effect. Chemical stimulation lateral to the cortical nucleus produced only weak responses.

Thus, we are faced with the possibility that functional specificity may not be determined entirely by the anatomical locus but also neuropharmacologically. Specific functional systems may be characterized by similar identical, chemical properties.

It is very probable that the amygdala exerts its influence by modulating the activity of hypothalamic mechanisms. In agreement with what might be expected, the deficits following amygdaloid damage are much less severe than those seen after small hypothalamic lesions (Grossman, 1964). Also, compatible with this view, certain differences in the effects of amygdaloid and hypothalamic stimulation were present. Stimulation of the hypothalamus elicited feeding and drinking in the _sated_ animals, whereas comparable stimulation of the amygdaloid complex had little effect on sated subjects, but produced a significant increase in the food and water intake of _deprived_ subjects.

For a discussion of the various mechanisms that might be involved in the alimentary effects resulting from chemical stimulation, the article by Singer and Montgomery (1968) should be consulted.

Ablation. The results of electrical and chemical stimulation are in essential agreement with those observed in ablation studies both with respect to the direction of the response and the amygdaloid location. A large number of reports deal with alimentary changes, but contribute less to the precise localization (for references cf. Gloor, 1960; Goddard, 1964b).

Comprehensive selective ablation studies have been made recently in cats by Lewińska (1967) and by Fonberg (1966) and Fonberg and Sychowa (1968). The experiments of Grossman and Grossman (1963) have been mentioned already.

Aphagia was produced in dogs when the medial and central nuclei were destroyed (Fonberg and Sychowa, 1968). According to Lewińska (1967), the effective area (Fig. 17) extends more ventrally and also includes the cortical nucleus, the rostral part of the parvocellular basal nucleus and the anterior amygdaloid area.

The severity of the anorexia after bilateral amygdalectomy is greatest in the rat, less so in the cat, and is not present in the monkey (Kling and Schwartz, 1961b). Eleftheriou (personal communication) has indicated that, in the deermouse, there is a 40% mortality resulting from aphagia in animals lesioned in the cortical (ACO) amygdaloid nuclear group. Thus, it may be that the feeding center is located within or near this nuclear complex. Further, there is little or no defect in feeding in the infant as opposed to the older animal after removal of the amygdala. In rats, the aphagia and adipsia are associated with lack of grooming (Schwartz and Kling, 1964).

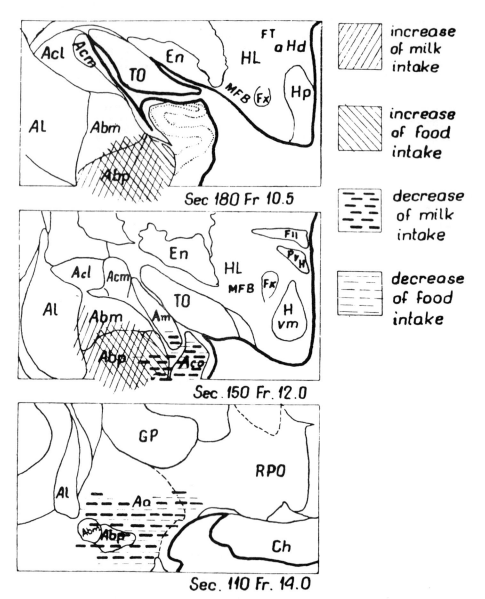

Fig. 17. Localization of lesions responsible for changes in food
and milk intake in cats projected on three representative frontal
planes (according to the atlas of Jasper and Ajmone-Marsan, 1954).
Abbreviations as in Fig. 16. (From Lewińska, 1967. Courtesy of
Polska Akademia Nauk, Warsaw.)

Hyperphagia and hyperdipsia were obtained in cats after
electrocoagulation of the greater part of the parvocellular
portion of the basal nucleus and the adjacent part of its magnocellular part as seen in Figure 17 (Lewińska, 1967). By removal
of basolateral parts of the amygdala, Fonberg (1968) similarly
produced hyperphagia, as well as disinhibition of conditioned
inhibitory alimentary reactions, whereas positive conditioned
responses were not changed. This, again, indicates that the
basolateral division contains inhibitory mechanisms with respect
to feeding. Brutkowski (1968) made similar observations in dogs.
The increased impairment of the inhibitory performance was thought
to be due to an increase in drive functions.

When reviewing previous reports, there appears to be some
agreement with these more recent observations. Thus, Green et al.
(1957) observed hyperphagia with marked weight increases in cats
most consistently when the lesions were placed in the basolateral
division of the amygdaloid complex, near the junction of the
lateral and basal nuclei. They did not observe excessive oral
investigation by their lesioned cats. In some animals, there also
was hypersexuality. Morgane and Kosman (1959) observed a significant hyperphagia by extensive removal of the amygdala and piriform
cortex in the cat. From the data presented, it appears that the
most effective lesions included mainly the basolateral division,
sparing most of the medial portion.

Wood (1958) described a marked hyperphagia after a lesion in
the central nucleus in the cat. This does not fit with the
hypothesis presented, but a possible explanation is that fibers
from the lateral inhibitory area, which run in a medial direction
through the region of the central nucleus, may have been damaged.

In this connection, it is an important problem whether the
amygdaloid area inhibiting feeding reactions corresponds, or is
related, to that suppressing defense and attack responses. If
this is true, hyperphagia and increased aggressiveness should be
expected to accompany each other. This appears to be the case
for some of the hyperphagic animals in the experiments of Wood
(1958) and Lewińska (1967). However, the hyperphagia resulting
from lesions of the basolateral amygdala in the experiments of
Morgane and Kosman (1959) was not accompanied by increased
aggressiveness and there was no hypersexuality. Further, aphagic
dogs with lesions in the dorsomedial amygdala showed the aggressive-
defense syndrome (Fonberg, 1966). This dissociation possibly
may be due to interference with the presumed inhibitory path
for defense and attack from the basolateral region which courses
through the lesioned area. As mentioned above, a differential
effect upon eating and mouse-killing behavior was also observed
by Karli and Vergnes (1964, 1965) after a lesion in the region
of the central nuclei.

(2) <u>Chewing</u>, <u>licking</u> and <u>salivation</u>. It has been known that superficial electrical stimulation of the pre- and periamygdaloid cortex produces masticatory movements and sniffing (Rioch and Brenner, 1938; Kaada, 1951; Takahashi, 1951). From the amygdala itself, such motor effects and licking have been elicited from rather diverse points in the amygdala, but still from only restricted parts of the nuclear complex.

Licking and chewing usually were elicited from more ventral parts of the amygdala than was sniffing, but to some extent there was overlapping with the sniffing area (Ursin and Kaada, 1960a). The most marked responses were evoked from the anterior part of the amygdala and positive sites are found in the periamygdaloid cortex, in the lateral and basal nuclei as well as in the region of the central nucleus. Swallowing, retching and salivation frequently were evoked from the same electrodes as licking and chewing.

These findings are in essential agreement with those of MacLean and Delgado (1953), Magnus and Lammers (1956), Shealy and Peele (1957), Wood (1958), Fernandez-deMolina and Hunsperger (1959), Hilton and Zbrozyna (1963), Koikegami (1964), and others.

(3) <u>Gastrointestinal</u> <u>secretion</u> and <u>motility</u>. Since changes in gastrointestinal activities are integrated in feeding as well as in emotional responses, such changes might be expected to be elicited from rather widespread amygdaloid areas. A facilitatory effect would be anticipated mainly from the anterior and centromedial parts which receive the majority of olfactory fibers and which exerts a facilitatory influence on food intake. Stimulation of the olfactory bulb and tract increases gastric motility (Eliasson, 1952). Inhibition would be expected from parts eliciting emotional responses.

Sen and Anand (1957) observed an increase in <u>gastric secretion</u> (volume and acidity) on stimulating the anteromedial as well as the basolateral amygdaloid nuclei in cats. Similar effects were produced by Shealy and Peele (1957) on stimulating the central, basal and part of the lateral nuclei of the amygdala in cats under anesthesia. Zawoiski (1967), working with unanesthetized cats, obtained consistent increases in gastric acid on stimulating the anterior amygdaloid area and the dorsal part of the lateral amygdala without concomitant behavioral or motor changes. It was suggested that the positive effects were mediated via the anteromedial part of the hypothalamus. If the stimulus intensity was raised to a level which produced overt behavioral or motor responses, there was a reduction of the gastric secretory response.

Similarly, Smith and McHugh (1967) observed significant inhibition of gastric acid outputs in conscious macaques equipped with gastric fistulas, when stimulating electrically the areas of the amygdala and hypothalamus which produced the defense reaction and increased plasma 17-OHCS. It was suggested that the inhibitory gastric response was an integrative visceral component of defense behavior. It is of interest that the gastric response, like the vascular adjustments which occur during the defense reaction (Abrahams et al., 1960), appears at maximal intensity during arousal which, as mentioned, is the initial stage of the defense response. The gastric response did not increase as the defense reaction became more vivid.

Local chemical stimulation of the amygdala suggests that acetylcholine is a possible neurotransmitter for gastric secretion while serotonin is an inhibitory transmitter in the amygdaloid complex (Lee et al., 1969).

Only a few of the reports dealing with the amygdaloid control of gastrointestinal motility are of value concerning their contribution to the problem of functional localization.

Increased gastric motility was obtained from the central and medial nuclei in anesthetized cats (Shealy and Peele, 1957). Such effects have been elicited also in anesthetized cat from the magnocellular basal nucleus (Koikegami, 1964) and, in unanesthetized rats, from the dorsal part of the lateral nucleus (Fennegan and Puigarri, 1966). Alterations in the stimulus parameters did not reverse the response.

Inhibition of gastrointestinal motility was obtained by Koikegami et al. (1952, 1953) and Koikegami (1964) on stimulating the parvocellular basal nucleus in anesthetized cats, an area producing maximal sympathetic outflow, according to the same authors. Inhibition was produced also in anesthetized cats from the magnocellular basal nucleus and adjoining parts of the lateral and central nuclei (Shealy and Peele, 1957) and, in unanesthetized rats from the anterior amygdaloid area (Fennegan and Puigarri, 1966). In the latter case the gastric inhibition was associated with minor alerting, chewing and licking.

Thus, regarding the localization of gastrointestinal influences, the results are not as unequivocal as might be expected. However, the majority of the observations support the postulate advanced in the introduction of this section.

Vagotomy has been shown to abolish the excitatory (Eliasson, 1952) as well as the inhibitory response (Fennegan and Puigarri, 1966) elicited from the amygdala.

In conclusion, most experiments indicate that feeding activities and associated autonomic effects appear to have a main representation in the medial and cortical nuclei and possibly in the parvocellular part of the basal nucleus with the overlying piriform cortex. The motor effects have a wider representation in the amygdala. The lateral nucleus and the magnocellular part of the basal nucleus, particularly their ventral portions, appear to exert an inhibitory influence on feeding activities.

D. Sexual activities

(1) Mating behavior. Little work has been done on the effects of amygdaloid stimulation on sexual behavior. Some sexual hyperactivity occurs during or immediately following stimulation in the poststimulation period (Gastaut, 1952; Alonzo-deFlorida and Delgado, 1958; Lissák and Endröczi, 1961). Knowledge of a localizing value comes almost entirely from ablation studies during which an increase as well as a decrease in mating behavior has been observed.

Schreiner and Kling (1953, 1954, 1956) and Green et al. (1957) produced a state of chronic hypersexuality with marked increases in copulatory behavior in which male cats also attempted copulation with members of other species and inanimate objects. The hypersexuality was abolished by castration and restored by androgen hormones (Schreiner and Kling, 1954; Green et al., 1957), and could also be counteracted by septal lesions (Kling et al., 1960). Hypersexuality resulting from anterior temporal lesions has also been recorded in man (Sawa et al., 1954; Terzian and Ore, 1955).

The effective lesions were primarily restricted to the amygdaloid complex and the overlying piriform cortex. Green et al. (1957) found that piriform removals alone could account for the hypersexual behavior. One would perhaps expect, in analogy to the increased emotional and feeding activities, that the area inhibiting sexual drive would include also the lateral and ventral parts of the basolateral amygdaloid division. This appears to be the case. Wood (1958) observed that lesions restricted to the lateral nucleus alone caused hypersexuality in males and females. Further, Eleftheriou and Zolovick (1966) produced a hypersexed state in the female deermouse by basolateral amygdaloid lesions, evaluated by excessive mating and reception of the male during dioestrus.

Kaada et al. (1968) used a quantitative method for assessing changes in sexual behavior following various brain lesions in rats.

In principle, the animals had to cross an electrified grid in an obstruction box to approach a rat of the opposite sex. There was a high correlation between number of crossings and number of mountings. Lesions restricted to the piriform and adjacent temporal cortex, as well as the septal region, habenula and stria medullaris, resulted in a persistent increase in number of crossings.

A decrease in sexual behavior is difficult to assess unless a measure of the strength of the sexual activity is made, as in the aforementioned experiments of Kaada et al. In these studies, removal of amygdaloid nuclei, the olfactory bulb and the genital region of the somatosensory cortex reduced the number of crossings. Spontaneous motor activity, as measured by a wheel-running test, sensitivity to electrified grids, estrous cyclicity and copulatory behavior, were not impaired.

Beach (1943, 1951) distinguished two principle processes in mating behavior: (i) an arousal mechanism constituting the active exploration, stimulation and pursuit of the sexual object by the animal playing the male role and (ii) an intromission and ejaculatory mechanism. Since the latter was retained in these forebrain lesions, it is the former that might have been interfered with in the obstruction test.

It is of considerable interest that the decreased sexual activity resulting from any of the effective forebrain lesions could be counteracted by bilateral removal of any area shown to increase sexual behavior. Thus, the two sets of regions serve to facilitate or inhibit, respectively, the excitability of subcortical structures that are essential for arousal instigated by sexual stimuli, and that removal of these forebrain structures interferes with this arousal mechanism.

With this quantitative test, the changes in sexual behavior were shown to be permanent. In previous work, a decrease has been apparent only in the first week or two after the operation, when the animal grooms little, and it also has been associated with hypophagia (Thompson and Walker, 1950, 1951; Walker et al., 1953; Kling and Schwartz, 1961a). Aphagic male and female rats with corticomedial amygdaloid lesions made in the prepuberal period were maintained for five months by forced feeding (Schwartz and Kling, 1964). These rats did not mate in spite of ongoing spermatogenesis and normal testis and seminal vesicle weight. Anterior pituitary and reproductive functions were normal. Similarly, lesions placed in the medial amygdaloid nuclei of adult female deermice abolished mating whereas the oestrous cyclic activity was normal (Eleftheriou and Zolovick, 1966).

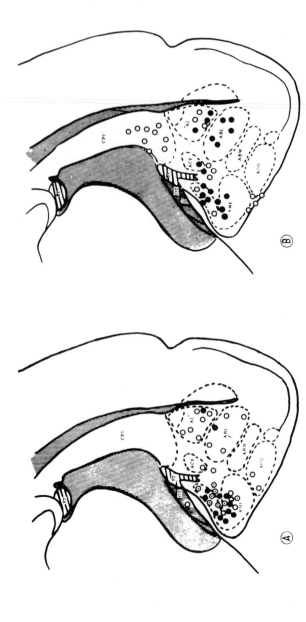

Fig. 18. Ovulatory responses in persistent estrous rats under continuous illumination induced by electrochemical (A) and chemical (B) stimulation. Each point located in a frontal section of the amygdaloid complex of the rat (adapted from de Groot's atlas, 1959) includes projections from an anterior posterior distance of 0.4 mm. Black points indicate positive ovulatory responses; white points, negative responses in animals in which the stria terminalis was damaged by the electrode track.

ABM - nucleus basalis amygdalae; ABL - nucleus lateralis amygdalae; ACE - nucleus centralis amygdalae; ACO - nucleus corticalis amygdalae; AL - nucleus lateralis amygdalae. (From Velasco and Taleisnik, 1969. Courtesy of Endocrinology.)

(2) Genital organs and sexual hormones.

Ovulation and estrogen uptake. Many investigators have
induced ovulation by stimulation of various nuclei in the corticomedial division of the amygdala (Koikegami et al., 1954; Shealy
and Peele, 1957; Everett, 1959; Velasco and Taleisnik, 1969, and
others). These include the cortical, medial, central and parvocellular part of the basal nuclei and the stria terminalis. The
latter authors obtained some effect from the lateral and magnocellular basal nuclei of rats as well (Fig. 18). Transection of
the stria terminalis abolished the response, whereas severing
of the ventral amygdalofugal pathway did not prevent the ovulation
after stimulation of the basolateral complex (Velasco and Taleisnik, 1969). These authors also found an increase in plasma
luteinizing (LH) and follicle-stimulating hormone (FSH) in females
by carbachol stimulation of the magnocellular basal amygdaloid,
whereas in males there was no increase in LH. Since preoptic
stimulation is effective in provoking LH-release in both females
and males, this finding has to be considered as a characteristic
feature of the amygdala.

This result apparently contradicts that of Eleftheriou and
Zolovick (1967) who observed that small lesions in the basolateral
amygdaloid complex induced a continuous release of LH from the
hypophysis of female deermouse, suggesting an inhibitory effect
on the pituitary secretion of LH by the basolateral complex.
Similar effects were obtained in the male deermouse (Eleftheriou
et al., 1967) where the effective region also included the cortical
nucleus (Eleftheriou et al., 1970). Small lesions in the latter
or in the basolateral area caused a significant increase of
pituitary and plasma LH concentrations while the hypothalamic
factor for this hormone also increased. Velasco and Taleisnik
(1969) interpreted the release of LH as the result of stimulation
by iron deposition, rather than inhibition of the amygdala by a
destructive lesion. On the other hand, Elwers and Critchlow
(1960, 1961) observed that small lesions in the medial portion
of the amygdala, including the stria terminalis, were associated
with precocious ovarian stimulation. Lesions of the cortical
amygdaloid nucleus and stria terminalis, in ovariectomized adult
rats, increased synthesis and release of LH, indicating that the
amygdala appears to exert a tonic inhibitory effect on LH secretion (Lawton and Sawyer, 1970).

Stumpf (1970, 1971) and Stumpf and Sar (1971), using autoradiography with labeled estradiol, have provided a precise
anatomical definition of estradiol areas within the amygdala,
hypothalamus and other brain structures. So-called estrogen-
neurons were found continuous throughout different nuclei, including neurons of the n. medialis, n. corticalis, n. basalis

parvocellularis, n. centralis and the anterior dorsal part of
n. lateralis. Most other neurons and the piriform cortex were
unlabeled. (Cf. chapter by Stumpf in this volume). There are
indications that these neurons are target cells for estrogens
and respond to the hormone in a specific manner. The various
nuclei are not labeled throughout their entire course, suggesting
that there are neurons subserving different and specific functions
within each nucleus commonly described as an entity. Similar
results were obtained by Pfaff (1968) and Pfaff and Keiner (1971,
this volume).

Hayward et al. (1964) found that hypothalamic and medial
amygdala stimulation at parameters which subsequently induced
ovulation produced an immediate increase in ovarian progestin
output. Stimulation of closely adjacent portions produced neither
a rise in progestin nor ovulation.

Total bilateral amygdalectomy causing aphagia reduces gonadotropin production in male rats (Yamada and Green, 1960).

Lesions placed only in the medial amygdaloid nuclei significantly increased both the hypophyseal growth-hormone activity and
hypothalamic growth-hormone releasing factor potency (Eleftheriou
et al., 1969).

Lactogenetic response. Bilateral estrogen implantation in
the amygdaloid complex in rabbits that were made pseudopregnant by
gonadotropin induced lactogenesis when the implants were placed in
the medial, central and medial part of the basal nuclei as well as
in the stria terminalis (Tindal et al., 1967). It was suggested
that the lactogenetic responses were caused by estrogen-sensitive
neurons in the amygdala acting via the stria terminalis on the
preoptic and/or basal hypothalamus to cause the release of prolactin.

Bilateral lesions in the dorsal and medial parts of the
amygdala in rats abolished the milk reflex (réflexe d'éjection de
lait) (Stutinsky and Terminn, 1965).

Uterine movements. Increased uterine contractions were
recorded by Koikegami et al. (1954) in cats, dogs and rabbits
on stimulating the parvocellular basal nucleus whereas all other
amygdaloid areas failed to yield any such response. Shealy and
Peele (1957) and Setekleiv (1964) made similar observations
stimulating the central nucleus and medial area.

Penile erection. Robinson and Mishkin (1968b) observed penile
erection in Macaca mulatta following stimulation of three loci in
the corticomedial area as well as in eight loci in the stria

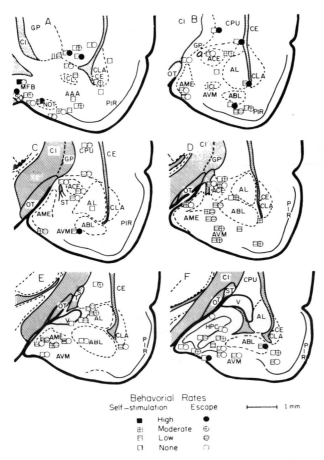

Fig. 19. Anatomical location of 79 electrodes in the amygdala and surrounding structures. (There are four types of squares to indicate the self-stimulation rate obtained from a point, and four types of circles indicate the escape rate obtained from the same point. An unpaired square or circle indicates that only one test was performed.) Sections are modified from de Groot, 1959, and are arranged in an anterior-to-posterior order.

Abbreviations: AAA - anterior amygdaloid area; ABL - amygdaloid lateral basal nucleus; ACE - amygdaloid central nucleus; AL - amygdaloid lateral nucleus; AME - amygdaloid medial nucleus; AVM - amygdaloid ventromedial area; CE - external capsule; CI - internal capsule; CLA - claustrum; CPU - caudate/putamen; GP - globus pallidus; HPC - hippocampus; ICL - intercalated nucleus; MFB - medial forebrain bundle; NOT - nucleus of the olfactory tract; OT - optic tract; PIR - piriform cortex; SO - supraoptic nucleus; ST - stria terminalis; V - lateral ventricle. From Wurtz and Olds, 1963. Courtesy of American Psychological Association.)

terminalis and its bed nucleus. Using autoradiography, Sar and Stumpf (1971) have demonstrated accumulation of labeled androgen in n. medialis amygdalae and the n. interstitialis striae terminalis.

In conclusion, a maximal facilitatory amygdaloid area for sexual functions appears to be the corticomedial nuclei and possibly the parvocellular basal nucleus. These areas exhibit significant estrogen uptake; stimulation induces ovulation, lactogenetic responses, uterine movements and penile erection, and removal reduces mating behavior. The stria terminalis seems to be an important projection system for the influence on the hypothalamus. The ventral part of the basolateral complex, including part of the piriform cortex, appears to exert a tonic "inhibitory" influence on mating behavior.

E. Reward and punishment in self-stimulation

It seems natural that a brain area like the amygdala which, on stimulation, evokes fear and aggression and which subserves feeding and sexual activities would be negative as well as positive in operant reinforcement. Further, on the basis of the foregoing discussion, one would anticipate positive reinforcement (reward) of behavior in self-stimulation experiments from the corticomedial area involved in positive feeding and sexual activities, whereas self-stimulation in the region yielding fear and aggression would be negatively reinforcing. It is hard to predict the result of self-stimulation in the basolateral inhibitory zone. This might inhibit a reward as well as a punishment system. In analogy to the self-stimulation effects of the ventromedial hypothalamus, which has similar inhibitory effects and which yields negative reinforcement (Olds and Olds, 1963), the same result would be expected from the amygdaloid inhibitory area.

Strong positive, negative and ambivalent reactions have been recorded in monkeys (Bursten and Delgado, 1958; Porter et al., 1959; Brady, 1961), rats (Oldd, 1956; Olds and Olds, 1963; Wurtz, 1962; Wurtz and Olds, 1963) and rabbits (Bruner, 1967). Figure 19 summarizes the results of self-stimulation in rats by Wurtz and Olds (1963). Most electrodes yielding self-stimulation per se or the highest self-stimulation rates were in the central, medial, cortical and parvocellular basal nuclei, with the strongest effects along the medial edge of the corticomedial group. Most electrodes yielding escape alone and the highest escape rates were located in the magnocellular basal nucleus and the lateral nuclei. However, some approach behavior was also evoked from electrode sites in the basolateral division and some escape behavior elicited from the corticomedial division. Frequently, both effects were evoked from the same electrode. Compared with the brain stem, self-stimulation rates in the amygdala were moderate but escape rates high.

The elicitation of strong positive reinforcement from points in the central region is somewhat surprising since stimulation of this area also yields flight and defense responses. However, there is the possibility that the positive reinforcement is due to stimulation of the stria terminalis fibers coursing through the central nucleus (Fig. 1).

Total bilateral amygdalectomy did not influence self-stimulation behavior in response to basal tegmental stimuli in rats (Ward, 1961). The amygdala is thus not essential for the mediation of the self-stimulation phenomenon.

The reward and punishment systems in the amygdala are dependent on different transmitter substances (Margules, 1968). Neurons that produce reward are innervated by cholinergic synapses, whereas neurons that excite the punishment system are under noradrenergic inhibitory control. These noradrenergic synapses may attenuate the behavior-suppressant effect of punishment when rewarding stimuli occur, and may thus be part of the neurochemical basis of reciprocal inhibition between punishment and reward.

IV.   CONCLUDING REMARKS

A close analysis of published data on amygdaloid stimulation and ablation reveals that several of the previous reports that were in apparent contradiction to each other are in fact in basic agreement with respect to topical localization. A denial of the existence of any functional representation within the amygdaloid complex finds no support in the wealth of experimental data available. Even if these data are based on the relatively crude methods of stimulation and ablation, they, nevertheless, have provided some of the main features of a correlation between function and structure. In brief, these are as follows:

The corticomedial nuclei, and possibly the adjoining part of the basal nucleus and the anterior amygdaloid area, appear to represent a facilitatory area for feeding and sexual activities.

Emotional responses (flight and defense) are obtained - in addition to the anterior amygdala and prepiriform cortex - from dorsal parts of the amygdaloid complex, at the level of its middle and posterior parts.

At these levels the more ventral part of the amygdaloid nuclear complex, in particular its lateral portion and the overlying piriform cortex, appear to exert a tonic, inhibitory influ-

ence on various activities, such as flight and defense, adrenocortical and cardiovascular responses, feeding and mating behavior,
and the release of luteinizing hormones. The mechanisms behind
the suppressing effect on stimulation, and the release of several
of these activities on ablation (increased aggressiveness, hyperphagia and hypersexuality), are not known. Also, it remains to
be shown whether there is one common inhibitory area within this
lateral, ventral region, or whether there are several inhibitory
zones different for each of the functions concerned.

The recent use of chemoarchitectural methods has demonstrated
a further subdivision of the amygdaloid nuclei, as well as a
different nuclear grouping, than had the traditional anatomical
methods. This new information may serve as a guide in future
attempts to correlate function and structure, in providing better
defined physiological and behavioral experiments with a more
rational placement of stimulating electrodes and lesions. This
will necessitate a more accurate and detailed description of the
lesion and the electrode placement than has often been the case
up to now.

The amygdaloid nuclei (and hippocampus) have a lower threshold for seizure discharges than almost any other brain area.
Therefore, electrical afterdischarges are a frequent accompani-
ment to the various types of responses to amygdaloid stimulation.
If the seizure discharge spreads widely across natural, functional
borders, this, of course, represents a serious limitation in
studies of the topical representation of the various behavioral,
motor, visceral and endocrine responses. Most stimulation studies
described in this survey have been performed without obtaining
simultaneous records of the electrical activity at the site of
stimulation and at distant areas. However, where such controls
have been made, the behavioral and other responses usually have
been elicited also without electrical afterdischarges. Further,
at relatively low-intensity stimulation, the afterdischarges are
fairly restricted and confined to the neurons at the site of
stimulation and to their efferent projection fields. In this
case, the local afterdischarges probably do not represent a more
unphysiological situation than does the artificial stimulation by
itself. The fact that various types of responses, associated with
electrical afterdischarges, may be obtained with consistency from
different electrode sites, sometimes only a fraction of a millimeter apart, strongly indicates that such responses do not need to
be excluded in an analysis of the topical representation of a given
function. The observation that ablation of the same areas, and not
of other parts, usually produces the opposite effect of that obtained by stimulation, similarly indicates that the presence of
electrical afterdischarges does not invalidate the results obtained
under such conditions.

The present survey does not discuss the mechanisms by which the amygdaloid nuclei exert their effect on behavior or motor, visceral and endocrine functions, nor to any extent the pathways mediating the effects. There is accumulating experimental evidence that the amygdala mainly acts by its influence on preoptic and hypothalamic mechanisms. This influence appears not to be a controlling one, but rather by modulation or adaptation of the response. The effects of stimulation and ablation of the amygdala appear to be smaller than those following hypothalamic stimulation and lesioning. The amygdala adds plasticity to the basic inborn and more fixed reflex mechanisms of the brain stem, possibly by incorporating past experiences with the present stimulus situation, thus determining the final response pattern. In the ontogenetic development, the functional maturation of the hypothalamus for autonomic and behavioral responses precedes the amygdala. Thus, in the kitten the defense reactions were obtained from the hypothalamus at 12 days and from the amygdala by 3 weeks of age (Kling and Coustan, 1964).

REFERENCES

ABRAHAMS, V. C., HILTON, S. M., & ZBROZYNA, A. Active muscle
vasodilatation produced by stimulation of the brain stem.
Its significance in the defense reaction. Journal of
Physiology (London), 1960, 154, 491-513.

ADAMS, D. B., BACCELLI, G., MANICA, G., & ZANCHETTI, A. Cardiovascular changes during naturally elicited fighting behavior
in the cat. American Journal of Physiology, 1969, 216, 1226-
1235.

ALFONSO-DEFLORIDA, F., & DELGADO, J. M. R. Lasting behavioral and
EEG changes in cats induced by prolonged stimulation of amygdala. American Journal of Physiology, 1958, 193, 223-229.

ALLEN, W. F. Effects on respiration blood pressure and carotid
pulse of various inhaled and insufflated vapors when stimulating one cranial nerve and various combinations of cranial
nerves. III. Olfactory and trigeminals stimulated. American
Journal of Physiology, 1929a, 88, 117-129.

ALLEN, W. F. Effect of various inhaled vapors on respiration and
blood pressure in anesthetized, unanesthetized, sleeping and
anosmic subjects. American Journal of Physiology, 1929b, 88,
620-632.

ANDERSEN, P. Inhibitory reflexes elicited from the trigeminal and
olfactory nerves in rabbit. Acta Physiologica Scandinavica,
1954, 30, 137-148.

ANDERSEN, P., JANSEN, J. JR., & KAADA, B. R. Electrical stimulation of the amygdaloid nuclear complex in unanesthetized cats.
Acta Psychiatrica et Neurologica Scandinavica, 1952, 29, 55.

ANDY, O. J., BONN, P., CHINN, R. MCC., & ALLEN, M. Blood pressure
alterations secondary to amygdaloid and periamygdaloid afterdischarges. Journal of Neurophysiology, 1959, 22, 51-60.

BAGSHAW, M., & BENZIES, S. Multiple measures of the orienting reaction and their dissociation after amygdalectomy in monkeys.
Experimental Neurology, 1968, 20, 175-187.

BAGSHAW, M. H., KIMBLE, D. P., & PRIBRAM, K. H. The GSR of monkeys
during orienting and habituation after ablation of the amygdala, hippocampus and inferotemporal cortex. Neuropsychologia,
1965, 11, 111-119.

BALDWIN, M., FROST, L. L., & WOOD, C. D.   Investigation of the
   primate amygdala. Movements of the face and jaws, Neurology,
   1954, 4, 586-598.

BARD, P., & MOUNTCASTLE, V. B.   Some forebrain mechanisms involved
   in expression of rage with special reference to suppression
   of angry behavior. Research Publications, Association for
   Research in Nervous and Mental Disease, 1947, 27, 362-404.

BARD, P., & RIOCH, D. MCK.   A study of four cats deprived of neocortex and additional portions of the forebrain. Johns Hopkins
   Bulletin, 1937, 60, 73-148.

BEACH, F. A.   Effects of injury to the cerebral cortex upon display of masculine and feminine mating behavior by female rats.
   Journal of Comparative Psychology, 1943, 36, 169-199.

BEACH, F. A.   Instinctive behavior: reproductive activities.   In
   Handbook of Experimental Psychology. New York: Wiley, 1951.
   Pp. 387-434.

BONVALLET, M., DELL, P., & HUGELIN, A.   Projections olfactives,
   gustatives, viscerales, vagales, visuelles, et auditives, au
   niveau des formations grises du cerveau antérieur du chat.
   Journal de Physiologie et Pathologie Général, 1952, 44, 222-4.

BRADY, J. V. Motivational-emotional factors and intracranial
   self-stimulation. In D. E. Sheer (Ed.), Electrical Stimulation of the Brain. Austin: University of Texas Press.
   Pp. 413-430.

BROWN, J. L., & HUNSPERGER, R. W. Neuroethology and the motivation
   of agonistic behaviour. Animal Behaviour, 1963, 11, 439-448.

BRUNER, A. Self-stimulation in the rabbit: An anatomical map of
   stimulation effects. Journal of Comparative Neurology, 1967,
   131, 615-629.

BRUTKOWSKI, S. A cortical-subcortical system controlling differentiation ability. Progress Brain Research, 1968, 22, 265-272.

BURSTEIN, B., & DELGADO, J. M. R. Positive reinforcement induced
   by intracerebral stimulation in the monkey. Journal of
   Comparative and Physiological Psychology, 1958, 51, 6-10.

CARUTHERS, R. P. Temporal lobe recruitment systems. Electroencephalography and Clinical Neurophysiology, 1969, 26, 336.

CHAPMAN, W. P., SCHROEDER, H. R., GEYER, G., BRAZIER, M. A. B.,
    FAGER, C., POPPEN, T. L., SOLOMAN, H. C., & YAKOVLEV, P. I.
    Physiological evidence concerning importance of the amygda-
    loid nuclear region in the integration of circulatory function
    and emotion in man. Science, 1954, 949-950.

CLARK, W. E. LE GROS, & MEYER, M. The terminal connexions of the
    olfactory tract in the rabbit. Brain, 1947, 70, 304-328.

COWAN, W. M., RAISMAN, G., & POWELL, T. P. S. The connexions of
    the amygdala. Journal of Neurology, Neurosurgery and
    Psychiatry, 1965, 28, 137-151.

DELGADO, J. M. R. Emotional behavior in animals and humans.
    Psychiatric Research Report, 1960, 12, 259-266.

DELGADO, J. M. R., MARK, V., SWEET, W., ERVIN, F., WEISS, G.,
    BACH-Y-RITA, G., & HAGIWARA, R. Intracerebral radio stimula-
    tion and recording in completely free patients. Journal of
    Nervous and Mental Disease, 1968, 147, 329-340.

DECSI, L., VÁRSZEGI, M. K., & MÉHES, J. Direct chemical stimula-
    tion of various subcortical brain areas in unrestrained cats.
    In K. Lissák (Ed.), Recent Developments of Neurobiology in
    Hungary. II. Results in Neurophysiology, Neuropharmacology
    and Behaviour. Budapest: Akadémiai Kiadó, 211 pp.

DREIFUSS, J. J., MURPHY, J. T., & GLOOR, P. Contrasting effects of
    two identified amygdaloid efferent pathways on single hypotha-
    lamic neurons. Journal of Neurophysiology, 1968, 237-248.

DRUGA, R. Neocortical projections to the amygdala. (An experi-
    mental study with the Nauta method). Journal Hirnforschung,
    1969, 11, 467.

EDVARDSEN, P., & URSIN, H. Micturition threshold in cats with amyg-
    dala lesions. Experimental Neurology, 1968, 21, 495-501.

EGGER, M. D., & FLYNN, J. P. Amygdaloid suppression of hypothalam-
    ically elicited attack behavior. Science, 1962, 136 (3510), 43.

EGGER, M. D., & FLYNN, J. P. Effects of electrical stimulation of
    the amygdala on hypothalamically elicited attack behavior in
    cats. Journal of Neurophysiology, 1963, 26, 705-720.

EGGER, M. D., & FLYNN, J. P. Further studies on the effects of
    amygdaloid stimulation and ablation on hypothalamically
    elicited attack behavior in cats. Progress in Brain Research,
    1967, 27, 165-182.

ELEFTHERIOU, B. E., & ZOLOVICK, A. J. Effect of amygdaloid lesions on oestrous behaviour in the deermouse. Journal of Reproduction and Fertility, 1966, 11, 451-453.

ELEFTHERIOU, B. E., & ZOLOVICK, A. J. Effect of amygdaloid lesions on plasma and pituitary levels of luteinizing hormone. Journal of Reproduction and Fertility, 1967, 14, 33-37.

ELEFTHERIOU, B. E., ZOLOVICK, A. J., & PEARSE, R. Effect of amygdaloid lesions on pituitary-adrenal axis in the deermouse. Proceedings of the Society for Experimental Biology and Medicine, 1966, 122, 1259-1262.

ELEFTHERIOU, B. E., ZOLOVICK, A. J., & NORMAN, R. L. Effects of amygdaloid lesions on plasma and pituitary levels of luteinizing hormone in the male deermouse. Journal of Endocrinology, 1967, 38, 469-474.

ELEFTHERIOU, B. E., DESJARDINS, C., PATTISON, M. L., NORMAN, R. L., & ZOLOVICK, A. J. Effects of amygdaloid lesions on hypothalamic-hypophysial growth-hormone activity. Neuroendocrinology, 1969, 5, 132-139.

ELEFTHERIOU, B. E., DESJARDINS, C., & ZOLOVICK, A. J. Effects of amygdaloid lesions on hypothalamic-hypophysial luteinizing hormone activity. Journal of Reproduction and Fertility, 1970, 21, 249-254.

ELIASSON, S. Cerebral influence on gastric motility in the cat. Acta Physiologica Scandinavica, 1952, 26 (Supplement 95), 70 pp.

ELWERS, M., & CRITCHLOW, V. Precocious ovarian stimulation following hypothalamic and amygdaloid lesions in rats. American Journal of Physiology, 1960, 198, 381-385.

ELWERS, M., & CRITCHLOW, V. Precocious ovarian stimulation following interruption of stria terminalis. American Journal of Physiology, 1961, 201, 281-284.

EVERETT, J. W. Neuroendocrine mechanisms in control of the mammalian ovary. In A. Gorbman (Ed.), Comparative Endocrinology. New York and London: J. Wiley & Sons, 1959. Pp. 174-168.

FEINDEL, W. Response patterns elicited from the amygdala and deep temporoinsular cortex. In D. E. Sheer (Ed.), Electrical Stimulation of the Brain. Austin: University of Texas Press, 1961, pp. 519-533.

FEINDEL, W., & GLOOR, P.   Comparison of electrographic effects of
    stimulation of the amygdala and brain stem reticular forma-
    tion in cats.   Electroencephalography and Clinical Neuro-
    physiology, 1954, 6, 389-402.

FENNEGAN, F. M., & PUIGGARI, M. J.   Hypothalamic and amygdaloid
    influence on gastric motility in dogs.   Journal of Neuro-
    surgery, 1966, 24, 497-504.

FERNANDEZ-DEMOLINA, A., & HUNSPERGER, R. W.   Central representa-
    tion of affective reactions in forebrain and brain stem:
    Electrical stimulation of amygdala, stria terminalis, and
    adjacent structures.   Journal of Physiology (London), 1959,
    145, 251-265.

FERNANDEZ-DEMOLINA, A., & HUNSPERGER, R. W.   Organization of the
    subcortical system governing defense and flight reactions in
    the cat.   Journal of Physiology (London), 1962, 160, 200-213.

FLYNN, J. P., VANEGAS, H., FOOTE, W., & EDWARDS, S.   Neural
    mechanisms involved in a cat's attack on a rat.   In R. Wholer
    (Ed.), The Neural Control of Behavior.   New York: Academic
    Press, Inc., 1970, pp. 135-173.

FONBERG, E.   The inhibitory role of amygdala stimulation.   Acta
    Biologiae Experimentalis (Warsaw), 1963, 23, 171-180.

FONBERG, E.   Effect of partial destruction of the amygdaloid
    complex on the emotional-defensive behaviour of dogs.
    Bulletin of the Polish Academy of Science (Biology), 1965,
    13, 429-432.

FONBERG, E.   Aphagia, produced by destruction of the dorsomedial
    amygdala in dogs.   Bulletin of the Polish Academy of Science
    (Biology), 1966, 14, 719-722.

FONBERG, E.   The role of the amygdaloid nucleus in animal behav-
    iour.   Progress in Brain Research, 1968, 22, 273-281.

FONBERG, E., & DELGADO, J. M. R.   Avoidance and alimentary re-
    actions during amygdala stimulation.   Journal of Neuro-
    physiology, 1961, 24, 651-664.

FONBERG, E., & SYCHOWA, B.   Effects of partial lesions of the
    amygdala in dogs.   I.   Aphagia.   Acta Biologiae Experimentalis
    (Warsaw), 1968, 28, 35-46.

FOX, C. A.   Certain basal telencephalic centers in the cat.
    Journal of Comparative Neurology, 1940, 72, 1-62.

FRANKENHAEUSER, B., & LUNDERVOLD, A. A note on an inhibitory
    reflex from the nose of the rabbit. Acta Physiologica
    Scandinavica, 1949, 18, 238-242.

FUKUCHI, S. Comparative anatomical studies on the amygdaloid
    complex in mammals, especially in Ungulata. Folia Psychiat-
    rica et Neurologica Japonica (Niigata), 1952, 5, 241-262.

GASTAUT, H. Corrélations entre le système nerveux végétatif et
    le système de la vie de relation dans le rhinencéphale.
    Journal de Physiologie (Paris), 1952, 44, 431-470.

GASTAUT, H., NAQUET, R., VIGOUROUX, R., & CORRIOL, J. Provocation
    de comportements emotionnels divers per stimulation rhinen-
    céphalique chez le chat avec électrodes à demeure. Review of
    Neurology (Paris), 1952, 86, 319.

GJONE, R. Excitatory and inhibitory bladder responses to stimula-
    tion of 'limbic', diencephalic and mesencephalic structures
    in the cat. Acta Physiologica Scandinavica, 1966, 66, 91-102.

GLOOR, P. Amygdala. In J. Field (Ed.), Handbook of Physiology,
    Vol. 2. Washington: American Physiological Society, 1960,
    pp. 1395-1420.

GLUSMAN, M., RANSOHOFF, J., POOL, J. L., & SLOAN, N. Electrical
    excitability of human uncus. Journal of Neurophysiology,
    1953, 16, 528-536.

GODDARD, G. V. Amygdaloid stimulation and learning in the rat.
    Journal of Comparative and Physiological Psychology, 1964a,
    58, 23-30.

GODDARD, G. V. Functions of the amygdala. Psychological Bulletin,
    1964b, 62, 89-109.

GODDARD, G. V. Analysis of avoidance conditioning following
    cholinergic stimulation of amygdala in rats. Journal of
    Comparative and Physiological Psychology, 1969, 68 (No. 2,
    Part 2), 1-8.

GREEN, J. D., CLEMENTE, C. D., & DE GROOT, J. Rhinencephalic
    lesions and behavior in cats. Journal of Comparative Neurology,
    1957, 108, 505-546.

GROOT, J. DE. The rat forebrain in stereotaxic coordinates.
    Verhandelingen der Koninklijke Nederlandse Akademie van
    Wetgenshappen, Afd. Naturrkunde (Amsterdam), 1959, 2, 1-40.

GROSSMAN, S. P. Direct adrenergic and cholinergic stimulation of hypothalamic mechanisms. American Journal of Physiology, 1962, 202, 872-882.

GROSSMAN, S. P. Behavioral effects of chemical stimulation of the ventral amygdala. Journal of Comparative and Physiological Psychology, 1964, 57, 29-36.

GROSSMAN, S. P., AND GROSSMAN, L. Food and water intake following lesions or electrical stimulation of the amygdala. American Journal of Physiology, 1963, 205, 761-765.

GURDJIAN, E. S. The corpus striatum of the rat. Journal of Comparative Neurology, 1928, 45, 249.

HALL, E. Efferent connections of the basal and lateral nuclei of the amygdala in cat. American Journal of Anatomy, 1963, 113, 139-145.

HALL, E. Some aspects of the structural organization of the amygdala. In B. Eleftheriou (Ed.), The Neurobiology of the Amygdala. New York: Plenum Press, in press.

HASSLER, R. Die zentralen Apparate der Wendebewegungen. Archiv für Psychiatrie und Nervenkrankheiten, 1956, 194, 481-516.

HAYWARD, J. N., HILLIARD, J., & SAWYER, C. H. Time of release of pituitary gonadotropin induced by electrical stimulation of the rabbit brain. Endocrinology, 1964, 74, 108-113.

HEATH, R. G., MONROE, R. R., & MICKLE, W. A. Stimulation of the amygdaloid nucleus in a schizophrenic patient. American Journal of Psychiatry, 1955, 111, 862-863.

HERNÁNDEZ-PEÓN, R., O'FLAHERTY, J. J., & MAZZUCHELLI-O'FLAHERTY, A. L. Sleep and other behavioural effects induced by acetyl-cholinic stimulation of basal temporal cortex and striate structures. Brain Research, 1967, 4, 243-267.

HILTON, S. M., & ZBROZYNA, A. W. Amygdaloid region for defense reactions and its efferent pathway to the brain stem. Journal of Physiology (London), 1963, 165, 160-173.

HOLMGREN, N. Points of views concerning forebrain morphology in higher vertebrates. Acta Zoologica (Stockholm), 1925, 6, 414-477.

HOROVITZ, Z. P., & LEAF, R. The effects of direct injections of psychotropic drugs into the amygdala of rats, and its relationship to antidepressant site of action. In H. Brill, J. O.

Cole, P. Deniker, H. Hippins, and P. B. Bradley (Eds.),
Neuropharmacology, Proceedings of the Vth International
Congress , Collegium Internationale Neuropsychopharmacologi-
cum, 1967. Amsterdam: Excerpta Medica ICS 129, pp. 1042.

HORVATH, F. E. Effects of basolateral amygdalectomy on three
types of avoidance behavior in cats. Journal of Comparative
and Physiological Psychology, 1963, 56, 380-389.

HUNSPERGER, R. W., & BUCHER, V. M. Affective behaviour produced
by electrical stimulation in the forebrain and brain stem of
the cat. In W. R. Adey and T. Tokizane (Eds.), Structure and
Function of the Limbic System. Progress in Brain Research,
1967, 27, 103-127.

HUTCHINSON, R. R., & RENFREW, J. W. Stalking attack and eating
behavior elicited from the same sites in the hypothalamus.
Journal of Comparative and Physiological Psychology, 1966,
61, 300-367.

ISHIHARA, I., KOMORI, Y., & MARUYAMA, T. Amygdala and adrenocortical response. Ann. Res. Rep. Inst. Environ. Med.
Nagoya University, 1964, 12, 9-17.

JASPER, H. H., & AJMONE MARSAN, C. A Stereotaxic Atlas of the
Diencephalon of the Cat. Ottawa: National Research Council
of Canada, 1954.

JOHNSTON, J. B. Further contributions to the study of the evolution of the forebrain. Journal of Comparative Neurology,
1923, 35, 337-481.

KAADA, B. R. Somato-motor, autonomic and electrocorticographic
responses to electrical stimulation of "rhinencephalic" and
other structures in primates, cat and dog: A study of
responses from the limbic, subcallosal, orbito-insular,
piriform and temporal cortex, hippocampus-fornix and
amygdala. Acta Physiologica Scandinavica 24 (Supplement No.
83), 1951, 285 pp.

KAADA, B. R. Cingulate, posterior orbital, anterior insular and
temporal pole cortex. In J. Field, A. W. Magoun, and V. E.
Hall (Eds.), Handbook of Physiology, Section I: Neurophysiology, Vol. 2. Baltimore: Williams and Wilkins Co.,
1960. Pp. 1345-1372.

KAADA, B. R. Brain mechanisms related to aggressive behavior. In
D. C. Clemente and D. B. Lindsley (Eds.), Aggression and
Defense. Neural Mechanisms and Social Patterns. Berkeley

and Los Angeles: University of California Press (UCLA Forum in Medical Science), 1967. Pp. 95-133.

KAADA, B. R., & BRULAND, H. Blocking of the cortically induced behavioral attention response by chlorpromazine. Psychopharmacologia, 1960, 1, 372-388.

KAADA, B. R., & JASPER, H. Respiratory responses to stimulation of temporal pole, insular, and hippocampal and limbic gyri in man. A.M.A. Archives of Neurology and Psychiatry, 1952, 68, 609-619.

KAADA, B. R., PRIBRAM, K. H., & EPSTEIN, J. A. Respiratory and vascular responses in monkeys from temporal pole, insula, orbital surface and cingulate gyrus. Journal of Neurophysiology, 1949, 12, 347-356.

KAADA, B. R., ANDERSEN, P., & JANSEN, J. Stimulation of the amygdaloid nuclear complex in unanesthetized cats. Neurology, 1954, 4, 48-64.

KAADA, B. R., RASMUSSEN, E. W., & KVEIM, O. Impaired acquisition of passive avoidance behavior by subcallosal, septal, hypothalamic and insular lesions in rats. Journal of Comparative and Physiological Psychology, 1962, 55, 661-670.

KAADA, B. R., RASMUSSEN, E. W., & BRULAND, H. Approach behavior towards a sex incentive following forebrain lesions in rats. International Journal of Neurology, 1968, 6, 306-323.

KARLI, P., & VERGNES, M. Nouvelles données sur les bases neurophysiologiques du comportement d'aggression interspécifique rat-souris. Journal de Physiologie (Paris), 1964, 56, 384.

KARLI, P., & VERGNES, M. Rôle des différentes composantes du complexe nucléaire amygdalien dans la facilitation de l'agressivité interspécifique du Rat. Comptes Rendus des Seances de la Société de Biologie, 1965, 159, 754.

KARLI, P., VERGNES, M., & DIDIERGEORGES, F. Rat-mouse interspecific aggressive behaviour and its manipulation by brain ablation and by brain stimulation. In S. Garattini and E. B. Sigg (Eds.), Aggressive Behaviour. Proceedings of the Symposium on the Biology of Aggressive Behaviour, Milan, May, 1968. Amsterdam: Excerpta Medica, 1969.

KARLI, F., VERGNES, M., ECLANCHER, F., SCHMITT, P., & CHAURAND, J. P. Role of the amygdala in the control of "mouse-killing" behavior in the rat. In B. E. Eleftheriou (Ed.), The Neurobiology of the Amygdala. New York: Plenum Press, in press.

KAWAHAMI, M., SETO, K., TERASAWA, E., YOSHIDA, K., MIYAMATO, T., SEKIGUCHI, M., & HATTORI, Y. Influence on electrical stimulation and lesion in limbic structure upon biosynthesis of adrenocorticoid in the rabbit. Neuroendocrinology, 1968, 3, 337-348.

KING, M. B., & HOEBEL, B. G. Killing elicited by brain stimulation in rats. Communications in Behavioral Biology, Part A, 1968, a, 173-177.

KLING, A., & COUSTAN, D. Electrical stimulation of the amygdala and hypothalamus in the kitten. Experimental Neurology, 10, 81-89.

KLING, A., & SCHWARTZ, N. B. Effects of amygdalectomy on sexual behaviour and reproductive capacity in the male rat. Federation Proceedings, 1961a, 20, 335.

KLING, A., & SCHWARTZ, N. B. Effects of amygdalectomy on feeding in infant and adult animals. Federation Proceedings, 1961b, 20, 335.

KLING, A., ORBACH, J., SCHWARTZ, N. B., & TOWNE, J. C. Injury to the limbic system and associated structures in cats. Archives of General Psychiatry, 1960, 3, 391-420.

KLINGER, J., & GLOOR, P. The connections of the amygdala and of the anterior temporal cortex in the human brain. Journal of Comparative Neurology, 1960, 115, 333-369.

KOIKEGAMI, H. Amygdala and other related limbic structures; experimental studies on the anatomy and function. I. Anatomical researches with some neurophysiological observations. Acta Medica Biologica (Niigata), 1963, 10, 161-277.

KOIKEGAMI, H. Amygdala and other related limbic structures; experimental studies on the anatomy and function. II. Functional experiments. Acta Medica Biologica (Niigata), 1964, 12, 73-266.

KOIKEGAMI, H., & FUSE, S. Studies on the functions and fiber connections of the amygdaloid nuclei and periamygdaloid cortex. Experiments on respiratory movements. Folia Psychiatrica et Neurologica Japonica, 1952, 5, 188-197; 6, 94-103.

KOIKEGAMI, H., & YOSHIDA, K. Pupillary dilatation induced by stimulation of amygdaloid nuclei. Folia Psychiatrica Neurologica Japonica, 1953, 7, 109-126.

KOIKEGAMI, H., DODO, T., MOCHIDA, Y., & TAKAHASHI, H. Stimulation experiments on the amygdaloid nuclear complex and related structures: Effects upon the renal volume, urinary secretion, movements of the urinary bladder, blood pressure and respiratory movements. Folia Psychiatrica Neurologica Japonica, 1957, 11, 157-207.

KOIKEGAMI, H., FUSE, S., YOKOYAMA, T., WATANABE, T., & WATANABE, H. Contributions to the comparative anatomy of the amygdaloid nuclei of mammals with some experiments of their desctruction or stimulation. Folia Psychiatrica Neurologica Japonica, 1955, 8, 336-370.

KOIKEGAMI, H., KIMOTO, A., & KIDO, C. Studies on the amygdaloid nuclei and periamygdaloid cortex: Experiments on the influence of their stimulation upon motility of small intestine and blood pressure. Folia Psychiatrica Neurologica Japonica, 1953, 7, 86-108.

KOIKEGAMI, H., KUSHIRO, H., & KIMOTO, A. Studies on the functions and fiber connections of the amygdaloid nuclei and periamygdaloid cortex: Experiments on gastro-intestinal motility and body temperature in cats. Folia Psychiatrica Neurologica Japonica, 1952, 6, 76-93.

KREINDLER, A., & STERIADE, M. EEG patterns of arousal and sleep induced by stimulating various amygdaloid levels in the cat. Archivio Italiano Biologia, 1964, 102, 576-586.

LAWTON, I. E., & SAWYER, C. H. Role of amygdala in regulating LH secretion in the adult female rat. American Journal of Physiology, 1970, 218, 622-626.

LEE, Y. H., THOMPSON, J. H., & MCNEW, J. J. Possible role of amygdala in regulation of gastric secretion in chronic fistula rats. American Journal of Physiology, 1969, 217, 505-510.

LESCAULT, H. Some neocortico-amygdaloid connections in the cat. Proceedings of the Canadian Federation of Biological Societies, 1969, 12, 24.

LESCAULT, H. Some neocortico-amygdaloid connections in the cat. Thesis, University of Ottawa, 1971.

LEWIŃSKA, M. K. Changes in eating and drinking produced by partial amygdalar lesions in cat. Bulletin of the Polish Academy of Science (Biology), 1967, 15, 301-305.

LEWIŃSKA, M. K. Inhibition and facilitation of alimentary behavior elicited by stimulation of amygdala in the cat. Acta Biologiae Experimentalis (Warsaw), 1968a, 28, 23-34.

LEWIŃSKA, M. K. The effect of amygdaloid stimulation on daily food intake in cats. Acta Biologiae Experimentalis (Warsaw), 1968b, 28, 71-81.

LEWIS, P. R., & SCHUTE, C. C. D. Tracing presumed cholinergic fibres in the rat forebrain. Proceedings of the Physiological Society, 1963 (May), 5-6.

LEYHAUSEN, P. Verhaltensstudien an Katzen. Berlin: Parey, 1956, 120 pp.

LIBERSON, W. T., SCOVILLE, W. B., & DUNSMORE, R. H. Stimulation studies of the prefrontal lobe and uncus in man. Electroencephalography and Clinical Neurophysiology, 1951, 3, 1-8.

LISANDER, B. Factors influencing the autonomic component of the defense reaction. Acta Physiologica Scandinavica, 1970, Supplement 351, 42 pp.

LISSAK, K., & ENDRÖCZI, E. Neurohumoral factors in the control of animal behavior. In J. F. Delafresnaye (Ed.), Brain Mechanisms and Learning. Toronto: Ryerson Press, 1961.

MCCLEARY, R. A. Response specificity in the behavioral effects of limbic system lesions in the cat. Journal of Comparative and Physiological Psychology, 1961, 54, 605-613.

MCHUGH, P. R., & SMITH, G. S. Plasma 17-OHCS response to amygdaloid stimulation with and without afterdischarges. American Journal of Physiology, 1967, 212, 619-622.

MACLEAN, P. D., & DELGADO, J. M. R. Electrical and chemical stimulation of fronto-temporal portion of limbic system in the waking animal. Electroencephalography and Clinical Neurophysiology, 1953, 5, 91-100.

MACCHI, G. The ontogenetic development of the olfactory telencephalon in man. Journal of Comparative Neurology, 1951, 95, 245.

MACHNE, X., & SEGUNDO, J. P. Unitary responses to afferent volleys in amygdaloid complex. Journal of Neurophysiology, 1956, 19, 232-240.

MAGNUS, O., & LAMMERS, H. J.   The amygdaloid-nuclear complex:
    Part 1. Folia Psychiatrica, Neurologica et Neurochirurgia
    Neerlandica, 1956, 59, 555-581.

MANDELL, A. J., CHAPMAN, L. F., RAND, R. W., & WALTER, R. D.
    Plasma corticosteroid: Changes in concentration after stimu-
    lation of hippocampus and amygdala. Science, 1963, 139, 1212.

MARGULES, D. L. Noradrenergic basis of inhibition between reward
    and punishment in amygdala. Journal of Comparative and Physi-
    ological Psychology, 1968, 66, 329-334.

MASON, J. W. The central nervous system regulation of ACTH secre-
    tion. In H. H. Jasper et al. (Eds.), Reticular Formation of
    the Brain. Boston: Little, Brown, 1958. Pp. 645-670.

MASON, J. W. Plasma 17-hydroxycorticosteroid levels during elec-
    trical stimulation in the amygdaloid complex in conscious
    monkeys. American Journal of Physiology, 1959, 196, 44-48.

MASSERMAN, J. H., LEVITT, M., MCAVOV, T., KLING, A., & PECHTEL, C.
    The amygdalae and behaviour. American Journal of Psychiatry.
    1958, 115, 14-17.

MORGANE, P. J., & KOSMAN, A. J. Alterations in feline behaviour
    following bilateral amygdalectomy. Nature, 1957, 180, 598-
    600.

MORGANE, P. J., & KOSMAN, A. J. A rhinencephalic feeding center
    in the cat. American Journal of Physiology, 1959, 197,
    158-162.

MORIN, G., NAQUET, R., & BADIER, M. Stimulation électrique de la
    région amygdalienne et pression artérielle chez le Chat.
    Journal de Physiologie (Paris), 1952, 44, 303-305.

MOWRER, O. H. Learning Theory and Behavior. New York: Wiley,
    1960. 555 pp.

MOYER, K. E. Kinds of aggression and their physiological basis.
    Communications in Behavioral Biology, Part A, 1968, 2, 65-87.

MULLAN, S., & PENFIELD, W. Illusions of comparative interpreta-
    tion and emotion: Production by epileptic discharge and by
    electrical stimulation in the temporal cortex. A.M.A.
    Archives of Neurology and Psychiatry, 1959, 81, 269-284.

NAUTA, W. J. H. Hippocampal projections and related neural path-
    ways to the mid-brain in the cat. Brain, 1958, 81, 319.

NAUTA, W. J. H.   Fibre degeneration following lesions of the
    amygdaloid complex in the monkey.   Journal of Anatomy
    (London), 1961, 95, 515-531.

O'KEEFE, J., & BOUMA, H.   Complex sensory properties of certain
    amygdala units in the freely moving cat.   Experimental
    Neurology, 1969, 23, 384-398.

OLDS, J.   A preliminary mapping of electrical reinforcing effects
    in the rat brain.   Journal of Comparative and Physiological
    Psychology, 1956, 49, 281-285.

OLDS, M. E., & OLDS, J.   Approach-avoidance analysis of rat dien-
    cephalon.   Journal of Comparative Neurology, 1963, 120,
    259-295.

OMUKAI, F.   Experimental studies on fiber connection of the
    amygdaloid complex in rabbit (Japanese text with English
    summary).   Kaibogaku Zasshi (Acta Anatomica Nippon), 1958,
    33, 499-522.

PAGANO, R. R., & GAULT, F. P.   Amygdala activity: A central
    measure of arousal.   Electroencephalography and Clinical
    Neurophysiology, 1964, 17, 255-260.

PELLEGRINO, L.   The effects of amygdaloid stimulation or passive
    avoidance.   Psychonomic Science, 1965, 2, 189-190.

PELLEGRINO, L.   Amygdaloid lesions and behavioral inhibition in
    the rat.   Journal of Comparative and Physiological Psychology,
    1968, 65, 483-491.

PENFIELD, W., & JASPER, H.   Epilepsy and the Functional Anatomy of
    the Human Brain.   Boston: Little, Brown and Company, 1954.

PFAFF, D. W.   Uptake of ^{3}H-estradiol by the female rat brain.   An
    autoradiographic study.   Endocrinology, 1968, 82, 1149-1155.

PFAFF, D. W., & KEINER, M.   Estradiol-concentrating cells in the
    rat amygdala as part of a limbic-hypothalamic hormone-sensi-
    tive system.   In B. E. Eleftheriou (Ed.), The Neurobiology
    of the Amygdala.   New York: Plenum Press, in press.

PORTER, R. W., CONRAD, D. G., & BRADY, J. V.   Some neural and
    behavioral correlates of electrical self-stimulation of the
    limbic system.   Journal of Experimental Analysis of Behavior,
    1959, 2, 43-55.

POWELL, T. P. S., COWAN, V. M., & RAISMAN, G. Olfactory relationships of the diencephalon. Nature (London), 1963, 199,
710-712.

REIS, D. J., & OLIPHANT, M. C. Bradycardia and tachycardia
following electrical stimulation of the amygdaloid region
in monkey. Journal of Neurophysiology, 1964, 27, 893-912.

RIOCH, D. MCK., & BRENNER, C. Experiments on the striatum and
rhinencephalon. Journal of Comparative Neurology, 1938,
68, 491-507.

ROBINSON, B. W., & MISHKIN, M. Alimentary responses evoked from
forebrain structures in Macaca mulatta. Science, 1962,
136, 260-261.

ROBINSON, B. W., & MISHKIN, M. Alimentary responses to forebrain
stimulation in monkeys. Experimental Brain Research, 1968a,
4, 330-366.

ROBINSON, B. W., & MISHKIN, M. Penile erection evoked from forebrain structures in Macaca mulatta. Archives of Neurology,
1968b, 19, 184-198.

RUBIN, R. T., MANDELL, A. J., & CRANDALL, P. H. Corticosteroid
responses to limbic stimulation in man: Localization of
stimulus sites. Science, 1966, 153, 767-768.

RUSSEK, M., & HERNÁNDEZ-PEÓN, R. Olfactory bulb activity during
sleep induced by stimulation of limbic structures. Acta
Neurologica Latinoamericana, 1961, 7, 299-302.

SAWA, M., & DELGADO, M. R. Amygdala unitary activity in the unrestrained cat. Electroencephalography and Clinical Neurophysiology, 1963, 15, 637-650.

SAWA, M., UEKI, Y., ARITA, M., & HARADA, T. Preliminary report
on the amygdaloidectomy on psychotic patients, with interpretation of oral-emotional manifestation in schizophrenics.
Folia Psychiatrica et Neurologica Japonica, 1954, 7, 309-329.

SCHREINER, L., & KLING, A. Behavioral changes following rhinencephalic injury in cats. Journal of Neurophysiology, 1953,
16, 643-659.

SCHREINER, L., & KLING, A. Effects of castration on hypersexual
behavior induced by rhinencephalic injury in cats. A.M.A.
Archives of Neurology and Psychiatry, 1954, 72, 180-186.

SCHREINER, L., & KLING, A. Rhinencephalon and behavior. American Journal of Physiology, 1956, 184, 486-490.

SCHWARTZ, N. B., & KLING, A. The effect of amygdaloid lesions on feeding, grooming and reproduction in rats. Acta Neurovegetativa (Vienna), 1964, 26, 12-34.

SCHWARTZBAUM, J. S., WILSON, W. A., JR., & MORRISSETTE, R. The effects of amygdalectomy on locomotor activity in monkeys. Journal of Comparative and Physiological Psychology, 1961, 54, 334-336.

SEN, R. N., & ANAND, B. K. Effect of electrical stimulation of the limbic system of brain ("visceral brain") on gastric secretory activity and ulceration. Indian Journal of Medical Research, 1957, 45, 515-521.

SETEKLEIV, J. Uterine motility of the estrogenized rabbit. V. Response to brain stimulation. Acta Physiologica Scandinavica, 1964, 62, 313-322.

SETEKLEIV, J., SKAUG, O. E., & KAADA, B. R. Increase of plasma 17-hydroxycorticosteroids by cerebral cortical and amygdaloid stimulation in the cat. Journal of Endocrinology, 1961, 22, 119-127.

SHEALY, C. N., & PEELE, T. L. Studies on amygdaloid nucleus of the cat. Journal of Neurophysiology, 1957, 20, 125-139.

SINGER, G., & MONTGOMERY, R. B. Neurohumoral interaction in the rat amygdala after central chemical stimulation. Science, 1968, 160, 1017-1018.

SLUSHER, M., & HYDE, J. E. Effect of limbic stimulation on release of corticosteroids into the adrenal venous effluent of the cat. Endocrinology, 1961, 69, 1080-1084.

SMITH, G. P., AND MCHUGH, P. R. Gastric secretory response to amygdaloid or hypothalamic stimulation in monkeys. American Journal of Physiology, 1967, 213, 640-644.

SPIEGEL, E. A., MILLER, H. R., & OPPENHEIMER, M. J. Forebrain and rage reactions. Journal of Neurophysiology, 1940, 3, 538-548.

STERMAN, M., & CLEMENTE, C. D. Forebrain inhibitory mechanisms: Cortical synchronization induced by basal forebrain stimulation. Experimental Neurology, 1962, 6, 91-102.

STEVENS, J. R., MARK, V. H., ERWIN, F., PACHERO, P., & SUEMATSU, K.
Deep temporal stimulation in man. Long latency, long lasting
psychological changes. Archives of Neurology, 1969, 21,
157-169.

STUTINSKY, F., & TERMINN, Y. Effets des lésions du complexe amygdalien sur le réflexe d'éjection de lait chez la Ratte.
Journal de Physiologie (Paris), 1965, 57, 279-280.

STUMPF, W. E. Estrogen-neurons and estrogen-neuron systems in the
periventricular brain. American Journal of Anatomy, 1970,
129, 207-218.

STUMPF, W. E. Probable sites for estrogen receptors in brain and
pituitary. Journal of Neuro-Visceral Relations, 1971,
Supplement X, 51-64.

STUMPF, W. E., & SAR, M. Estradiol concentrating neurons in the
amygdala. Proceedings of the Society for Experimental
Biology and Medicine, 1971, 136, 102-106.

SUMMERS, T. B., & KAELBER, W. W. Amygdalectomy: effects in cats
and a survey of its present status. American Journal of
Physiology, 1962, 203, 1117-1119.

TAKAHASHI, K. Experiments on the periamygdaloid cortex of the
cat and dog. Folia Psychiatrica et Neurologica Japonica,
1951, 5, 147-154.

TERZIAN, H., & ORE, G. D. Syndrome of Klüver and Bucy reproduced
in man by bilateral removal of the temporal lobes. Neurology,
1955, 5, 373-380.

THOMPSON, A. F., & WALKER, A. E. Behavioral alterations following
lesions of the medial surface of the temporal lobe. Folia
Psychiatrica, Neurologica et Neurochirurgia, 1950, 53,
444-452.

THOMPSON, A. F., & WALKER, A. E. Behavioral alterations following
lesions of the medial surface of the temporal lobe. Archives
of Neurology and Psychiatry, 1951, 65, 251-252.

TINDAL, J. S., KNAGGS, G. S., & TURVEY, A. Central nervous control of prolactin secretion in the rabbit: Effect of local
oestrogen implants in the amygdaloid complex. Journal of
Endocrinology, 1967, 37, 279-287.

UCHIDA, Y. A contribution to the comparative anatomy of the
amygdaloid nuclei in mammals, especially in rodents. Folia

Psychiatrica et Neurologica Japonica (Niigata), 1950, 4,
25 and 91.

URSIN, H. Flight and defense behavior in cats. Journal of
Comparative and Physiological Psychology, 1964, 58, 180-186.

URSIN, H. The effect of amygdaloid lesions on flight and defense
behavior in cats. Experimental Neurology, 1965a, 11, 61-79.

URSIN, H. Effect of amygdaloid lesions on avoidance behavior and
vision discrimination in cats. Experimental Neurology,
1965b, 11, 298-317.

URSIN, H. Limbic control of emotional behavior. In E. R.
Hitchock and K. Varnet (Eds.), Proceedings, 2nd International
Conference on Psychosurgery (Copenhagen 24th - 26th August,
1970). Springfield: C. C. Thomas, 1971, in press.

URSIN, H., & KAADA, B. R. Functional localization within the
amygdaloid complex in the cat. Electroencephalography and
Clinical Neurophysiology, 1960a, 12, 1-20.

URSIN, H., & KAADA, B. R. Subcortical structures mediating the
attention response induced by amygdala stimulation. Experimental Neurology, 1960b, 2, 109-122.

URSIN, H., WESTER, K., & URSIN, R. Habituation to electrical
stimulation of the brain in unanesthetized cats. Electroencephalography and Clinical Neurophysiology, 1967, 23, 41-49.

VELASCO, M. E., & TALEISNIK, S. Release of gonadotropins induced
by amygdaloid stimulation in the rat. Endocrinology, 1969,
84, 132-139.

VERGNES, M., & KARLI, P. Déclenchement du comportement d'aggression interspécifique Rat-Souris par ablation bilatérale des
bulbes olfactifs. Action de l'hydroxyzine sur cette agressivité provoquée. Comptes Rendus des Seances de la Société
de Biologie, 1963, 157, 1061.

VERGNES, M., & KARLI, P. Etude des voies nerveuses de l'influence
facilitatrice exercée par les noyaux amygdaliens sur le comportement d'agression interspécifique Rat-Souris. Comptes
Rendus des Seances de la Société de Biologie, 1964, 158(I),
856-858.

VERGNES, M., & KARLI, P. Etude des voies nerveuses d'une influence inhibitatrice s'exercant sur l'agressivité interspécifique du Rat. Comptes Rendus des Seances de la Société de
Biologie, 1965, 159, 972.

VERGNES, M., & KARLI, P. Comportement d'agression interspécifique Rat-Souris: effets de la stimulation électrique de l'hypothalamus latéral, de l'amygdale et de l'hippocampe. Journal de Physiologie (Paris), 1969, 61 (supplement 2); 425.

WALKER, A. E., THOMPSON, A. F., & MCQUEEN, J. D. Behavior and the temporal rhinencephalon in the monkey. Johns Hopkins Bulletin, 1953, 93, 65-93.

WALL, P. D., & DAVIS, G. D. Three cerebral cortical systems affecting autonomic function. Journal of Neurophysiology, 1951, 14, 507-517.

WARD, H. P. Tegmental self-stimulation after amygdaloid ablation. Archives of Neurology, 1961, 4, 657-659.

WENDT, R., & ALBE-FESSARD, D. Sensory responses of the amygdala with special reference to somatic afferent pathways. In Physiologie de l'Hippocampe. Paris: Centre National de la Recherche Scientifique, 1962. Pp. 171-200.

WHITLOCK, D. G., & NAUTA, W. J. H. Subcortical projections from the temporal neocortex in Macaca mulatta. Journal of Comparative Neurology, 1956, 106, 183.

WOOD, C. D. Behavioral changes following discrete lesions of temporal lobe structures. Neurology, 1958, 8, 215-220.

WOOD, C. D., SCHOTTELIUS, B., FROST, L. L., & BALDWIN, M. Localization within the amygdaloid complex of anesthetized animals. Neurology, 1958, 8, 477-480.

WOODS, J. W. Loss of aggressiveness in wild rats following lesions in the rhinencephalon. International Physiological Congress Abstract, 1956, 20, 978-979.

WURTZ, R. H. Self-Stimulation and Escape Behavior in Response to Stimulation of the Rat Amygdala. Unpublished doctoral dissertation. Ann Arbor: University of Michigan, 1962 (University Microfilms, Inc., Ann Arbor, 63-481).

WURTZ, R. H., & OLDS, J. Amygdaloid stimulation and operant reinforcement in the rat. Journal of Comparative and Physiological Psychology, 1963, 56, 941-949.

YAMADA, T., & GREER, M. A. The effect of bilateral ablation of the amygdala on endocrine function in the rat. Endocrinology, 1960, 66, 565-574.

ZAWOISKI, E. J. Gastric secretory response of the unrestrained
cat following electrical stimulation of the hypothalamus,
amygdala, and basal ganglia. Experimental Neurology, 1967,
17, 128-139.

ZBROZYNA, A. W. Defense reactions from the amygdala and the stria
terminalis. Journal of Physiology (London), 1960, 153, 27-28.

ZBROZYNA, A. W. The anatomical basis of patterns of autonomic and
behavioural response effected via the amygdala. In W. Bargmann and J. P. Schade (Eds.), Progress in Brain Research,
Vol. 13. Amsterdam: Elsevier, 1963. Pp. 50-70.

ZBROZYNA, A. W. The organization of the defence reaction elicited
from amygdala and its connections. This volume p. 597.

THE DISTRIBUTION OF ACETYLCHOLINESTERASE ENZYME IN

THE AMYGDALA AND ITS ROLE IN AGGRESSIVE BEHAVIOR

Makram Girgis

Missouri Institute of Psychiatry
University of Missouri School of Medicine
Columbia, Missouri

The terminology and classification of the amygdaloid nuclei is based mainly on the work of Johnston (1923). Johnston postulated that many of the amygdaloid nuclei were formed by an ingrowth from the lower border of the pyriform cortex in the region of the endorhinal sulcus (usually referred to as the amygdaloid fissure at this level), the cells of which remained superficial medial to the sulcus forming a cortical amygdaloid nucleus. He thinks that this nucleus and the nucleus of the lateral olfactory tract, also superficial in position, and the more deeply situated central and medial nuclei are phylogenetically older than two other nuclei which he referred to as basal and lateral. Later workers (Crosby and Humphrey, 1941) refer to these nuclear groups as cortico-medial and baso-lateral, respectively. However, the question of their relative phylogenetic age is doubtful because it is difficult to establish the homologies of the individual nuclei in sub-mammalian vertebrates.

Among the mammals, the arrangement of the amygdaloid nuclei is remarkably uniform and no clear phylogenetic trends can be recognized. Crosby and Humphrey (1941), in their study of the human amygdala, were able to identify the baso-lateral and cortico-medial group and to use the same terminology as for sub-primate mammals. It also is usual to include in the amygdala a third or anterior group (Fox, 1940) consisting of the nucleus of the lateral olfactory tract, an ill-defined anterior amygdaloid area, and certain intercalated masses of cells. It should be emphasized, however, that the exact anatomical structures and fiber connections of these large nuclear groups are not yet known thoroughly and that these nuclei are highly developed in primates, including man, in direct contrast with their reduced olfaction.

As mentioned above, the conventional classification of the amygdaloid nuclei is based on Johnston's embryological and comparative anatomical observations, but recently Koikegami (1963) has pointed out that this may be unsatisfactory from a functional point of view. He agrees that the large-celled part of the basal nucleus should be grouped with the lateral nucleus because it has autonomic and extrapyramidal functions. The small-celled medial part, on the other hand, he considers, should be classed with the medial and cortical nuclei which carry out autonomic functions only. In his view, the baso-lateral amygdaloid nuclei (the striatal part of the amygdala of Holmgren (1925) form the amygdala proper. He accepts, however, Holmgren's definition of the cortico-medial nuclei (but including the small-celled part of the basal nucleus) as a sub-pallial part of the amygdala.

Our own cytoarchitectonic studies (Girgis, 1968a; 1969a; 1970) have revealed the major subdivisions into cortico-medial and baso-lateral groups of nuclei. The finer subdivisions, however, run up against the difficulty that many of the described nuclei are extremely ill-defined and often blend with surrounding structures. It has been suggested (Girgis, 1968a) that the cortical nucleus and the cortico-amygdaloid transition area are more naturally classed with other paleocortical structures of the basal rhinencephalon, and, possibly, the same can be said of the nucleus of the lateral olfactory tract. It is of interest to note here that we have found in our experimental material (Girgis and Goldby, 1967) that these are the only parts of the amygdala that receive afferents from the olfactory bulb.

Many reports have appeared in the electrophysiological literature concerning the functional significance of the amygdala. These are very varied and have been very well summarized by Gloor (1960). Physiologically, two different portions have been distinguished. Kaada (1951) and Kaada et al. (1954) distinguish an antero-medial division (phylogenetically old) which responds with autonomic and somatomotor effects to stimulation, and a lateral division (phylogenetically younger) which on stimulation yields behavioral changes which indicate anxiety and sometimes fear and anger. Ursin (1965) suggested that the medial amygdaloid nucleus may play a role in the general suppression of learned responses and pointed out the functionally heterogeneous nature of the amygdaloid nuclei. Hall, Haug and Ursin (1969) emphasized this point by stating that "on comparing the structure and function of the cellular groups within each subdivision, it is obvious that they are not entirely homogenous, and that in some instances there may be variations even within an individual nucleus."

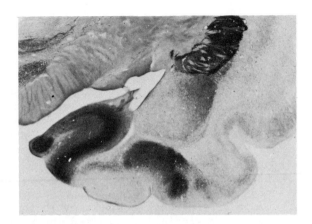

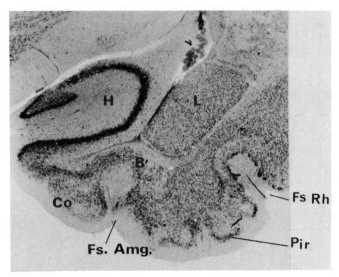

Fig. 1. A frontal section of the brain of the coypu (Myocaster
coypus) stained by Koelle's thiocholine method to show the
amygdaloid nuclei. x 10.

Fig. 2. A frontal section of the same brain but stained with the
Nissl method for comparison. x 12.

Abbreviations for Figs. 1 to 4:

B	=	n. amygdaloid basalis	Fs. Amg.	= fissura amygdala
BA	=	n. amygdaloid basalis accessorius	Fs. Rh. H	= fissura retrinalis = hippocampus
C	=	n. amygdaloid centralis	L	= n. amygdaloid
Co	=	n. amygdaloid corticulis		lateralis

Acetylcholinesterase (AChE) Distribution in the Amygdala

In an attempt to help unravel the structural and functional
complexities of this heterogeneous structure, we have undertaken a
histochemical investigation of the amygdala using Lewis' (1961)
modification of the thiocholine technique. It has been suggested
that not only are AChE-containing neurons better developed in
older parts of the brain (Gerebtzoff, 1959), but also that they
are more deeply stained in the brains of simpler animals (Shen,
Greenfield and Boell, 1956). There exists, however, some doubt
concerning the existence of a phylogenetic trend in the distribution of AChE in the brains of animals. Because of these uncertainties, we have decided to investigate the distribution of the
enzyme in different mammals, progressing up the evolutionary scale.
In this series of comparative studies, we studied the rodent, galago and monkey brains. The results of our earlier observations
have been reported elsewhere (Girgis, 1967; 1968b; 1968c; 1969b).
In these investigations, we have studied the limbic structures
with particular reference to the amygdala.

Although there are some differences, there seems to be a
general similarity between these brains examined insofar as the
distribution of cholinesterase is concerned. The intensity of
staining is on the whole similar. In all brains examined, however, the individual amygdaloid nuclei show marked variations in
the enzyme content varying from negligible to mild to moderate
and to even very intense. The structurally and functionally
heterogeneous nature of the amygdaloid nuclei, referred to above,
is well demonstrated in these histochemical studies.

Baso-Lateral Group. The most intensely stained and best
circumscribed is the basal amygdaloid nuclei (lateral magnocellular
part), in the perikaria as well as the surrounding neuropil (Figs. 1
and 3). This has been the case in all brains examined by us and
other investigators, including the human brain (see Ishii and
Friede, 1967). In our series, it is only in the coypu brain that
the lateral amygdaloid nucleus shows definite AChE activity. In
this animal it exhibits its typical quadrilateral shape which is
so prominent a feature in Nissl preparation of this brain (Figs. 1
and 2). Here, the lateral nucleus is outlined clearly by the
unstained external capsule laterally and by the more deeply stained
nucleus ventrally.

Cortico-Medial Group. Only the coypu brain exhibits AChE
activity in most of the nuclei which are included in this group.
The cortical nucleus shows slight staining in the superficial part
of the molecular layer but it is otherwise negative. The principal
nuclei in this group (medial and central) both show moderately
diffuse staining. The bed nucleus of the stria terminalis stains
heavily, and many of the fibers of the stria itself show quite

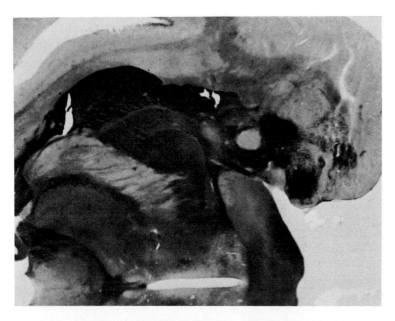

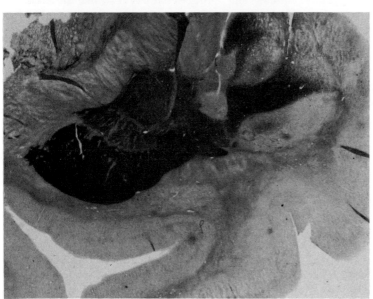

Fig. 3. Photograph of a frontal section of the cebus monkey brain showing intense acetylcholinesterase activity in the amygdala. Koelle's thiochcline technique x6.

Fig. 4. Photograph of a frontal section of the monkey brain showing cholinesterase activity in the thalamic region and in the internal and external medullary lamina. Also note the staining in the amygdala. Koelle's method x6.

definite activity. A commissural, hypothalamic and lateral optic
component can be individually recognized rostrally on account of
their positive staining. The stria terminalis does not stain in
its caudal part since it can not be identified in its usual
position medial to the baso-lateral amygdaloid nuclei.

Anterior Group. The anterior amygdaloid area of the coypu
and galago brains shows selective staining which is mainly intracellular. Also in these two animals the nucleus of the lateral
olfactory tract gives an intense reaction. It appears as two
small rounded masses, deeply stained, at the caudo-lateral border
of the olfactory tubercle. In the monkey brain, all the nuclei
included in this group do not stain at all.

Cholinergic System of the Amygdala and Aggression

Recently, many investigators attempted to determine the
chemical basis of the amygdaloid mediation of aggression, and to
evaluate the role of the amygdala in the overall regulation of
aggression by direct stimulation by different cholinergic agents.
Igic, Stern and Basagic (1970) found that local application of
amitone (a cholinesterase inhibitor) in the baso-lateral amygdala
cause an increased aggressiveness and hyper-reactivity and in some
cases muricide in the rat. This increased reactivity was depressed
by atropine. These facts indicate that behavioral changes may be
elicited by direct stimulation of cholinergic mechanisms and that
anticholinergic drugs would modify these behavioral changes. It
is to be recalled that our histochemical studies mentioned above
reveal an intense cholinesterase activity in mainly the basolateral amygdala.

In order to compare the effects of electrical and chemical
stimulation of the brain, Baxter (1967) applied "chemitrodes"
which permit either crystalline chemical compound or electric
current to be applied at the same site in the basal amygdala.
In this study, stimulation of the amygdala showed that sites from
which no emotional behavior could be elicited electrically produced emotional behavior upon carbachol injection. It was not
surprising that such behavior was elicited from the amygdala since
earlier workers have reported anatomical pathways from the amygdala
to hypothalamus, emotional responses after electrical or carbachol
stimulation (Grossman, 1963), and that amygdaloid stimulation
modulates hypothalamically elicited emotional behavior (Egger and
Flynn, 1963). Baxter suggested that his amygdaloid chemitrodes
were so located that they did not stimulate the axons of fibers
mediating emotional responses upon electrical activation but
could activate synaptic membrane so involved via carbachol. He
also proposed the hypothesis that "ventricular diffusion"

could help explain the basically different patterns of behavior produced by carbachol and electrical stimulation; the selective depression of carbachol "receptors" without depression of electrical "receptors," and the production of emotional behavior after carbachol from sites which yielded no such behavior upon electrical stimulation.

It is pertinent to mention here that facilitation of aggressive behavior was produced in rats by direct cholinergic (carbachol) stimulation of the lateral hypothalamus (Bandler, 1969; Smith, King and Hoebel, 1970). As additional evidence for a cholinoceptive mechanism, Smith et al. (1970) showed that neostigmine (cholinesterase inhibitor) elicited killing and methyl atropine blocked it.

It is of interest to note here that our histochemical studies indicate intense AChE activity and the existence of cholinergic pathways between these lateral hypothalamic areas and the amygdaloid sites from which aggressive behavior has been elicited by carbachol injections. The presence of the enzyme in abundance at these sites appears to act as a protective mechanism to cope with excess acetylcholine release. An imbalance between the acetylcholine-cholinesterase system may lead in one way or the other to emotional imbalance.

Finally, all of the above mentioned studies and findings raise the practical possibility that pharmacological manipulation of the cholinergic limbic system could be used in the treatment of pathological aggressive behavior.

REFERENCES

BANDLER, R. J., Jr. Facilitation of aggressive behavior in rat by direct cholinergic stimulation of the hypothalamus. Nature, 1969, 224, 1035-1036.

BAXTER, B. L. Comparison of the behavioral effects of electrical or chemical stimulation applied at the same brain loci. Experimental Neurology, 1967, 19, 412-432.

CROSBY, E. C., & HUMPHREY, T. Studies on the vertebrate telencephalon. II. The nuclear pattern of the anterior olfactory nucleus, tuberculum olfactorium and amygdaloid complex in adult man. Journal of Comparative Neurology, 1941, 74, 309-352.

EGGER, M. D., & FLYNN, J. P. Effect of electrical stimulation of the amygdala in hypothalamically elicited attack behavior in cats. Journal of Neurophysiology, 1963, 26, 705-720.

FOX, C. A. Certain basal telencephalic centers in the cat. Journal of Comparative Neurology, 1940, 72, 1-62.

GEREBTZOFF, M. A. Cholinesterases. In P. Alexander and Z. M. Bacq (Eds.), International series in Monographs on Pure and Applied Biology. (Division: Modern Trends in Physiological Sciences). New York: Pergamon Press, 1959.

GIRGIS, M. Distribution of cholinesterase in the basal rhinencephalic structures of the coypu (Myocaster coypus). Journal of Comparative Neurology, 1967, 129, 85-96.

GIRGIS, M. Some features of the basal rhinencephalic structures in the coypu. Acta Anatomica, 1968a, 70, 352-381.

GIRGIS, M. Histochemical localization of cholinesterase in a rodent, a subprimate and a primate brain. Histochemie, 1968b, 16, 307-314.

GIRGIS, M. Distribution of cholinesterase in the basal rhinencephalic structures of the grivet monkey (Cercopithecus aethiops). Acta Anatomica, 1968c, 70, 568-576.

GIRGIS, M. The amygdala and the sense of smell. Acta Anatomica, 1969a, 72, 502-519.

GIRGIS, M. Distribution of cholinesterase in the basal rhinencephalic structures of the Senegal bush baby (Galago senegalensis). Acta Anatomica, 1969b, 72, 94-100.

GIRGIS, M. The rhinencephalon. Acta Anatomica, 1970, 76, 157-199.

GIRGIS, M., & GOLDBY, F. Secondary olfactory connexions and the anterior commissure in the coypu (Myocastor coypus). Journal of Anatomy (London), 1967, 101, 33-44.

GLOOR, P. The Amygdala. In Field, Magoun & Hall (Eds.) Handbook of Physiology, Section 1, Neurophysiology, II, 1395. Washington, D.C.: American Physiological Society, 1960.

GROSSMAN, S. P. Chemically induced epileptiform seizures in the cat. Science, 1963, 142, 409-411.

HALL, E., HAUG, F. M. S., & URSIN, H. Dithizone and sulphide
    silver staining of the amygdala in the cat. Zeitschrift fur
    Zellforschung und Mikroskopische Anatomie, 1969, 102, 40-48.

HOLMGREN, N. Points of view concerning forebrain morphology in
    higher vertebrates. Acta Zoologica, 1925, 6, 414-477.

IGIC, R., STERN, P., & BASAGIC, E. Changes in emotional behavior
    after application of cholinesterase inhibiter in the septal
    and amygdala region. Neuropharmacology, 1970, 9, 73-75.

ISHII, T., & FRIEDE, R. L. A comparative histochemical mapping
    of the distribution of acetylcholinesterase and nicotinimide
    adenine dinucleotide-diaphorase activities in the human
    brain. International Review of Neurobiology, 1967, 10, 231.

JOHNSTON, J. B. Further contributions to the study of the evolu-
    tion of the forebrain. Journal of Comparative Neurology,
    1923, 35, 337-481.

KAADA, B. R. Somato-motor, autonomic and electrocorticographic
    responses to electrical stimulation of "rhinencephalic" and
    other structures in primates, cat, and dog. A study of
    responses from the limbic, subcallosal, orbitoinsular, piri-
    form and temporal cortex, hippocampus-fornix and amygdala.
    Acta Physiologica Scand., Suppl. 83, 1951, 24, 1-285.

KAADA, B. R., ANDERSEN, P., & JANSEN, J. Stimulation of the
    amygdaloid nuclear complex in unanesthetized cat.
    Neurology (Minneap.), 1954, 4, 48-64.

KOIKEGAMI, H. Amygdala and other related limbic structures;
    experimental studies on the anatomy and function. 1. Anatom-
    ical researches with some neurophysiological observations.
    Acta Medica Biologica Niigata, 1963, 10, 161-277.

LEWIS, P. R. The effect of varying the conditions in the Koelle
    technique. (Histochemistry of Cholinesterase, Symposium,
    Basel, 1960). Bibliographica Anastatica, 1961, 2, 11-20.

SHEN, S. C., GREENFIELD, P., & BOELL, E. J. Localization of
    acetylcholinesterase in chick retina during histogenesis.
    Journal of Comparative Neurology, 1956, 106, 433-461.

SMITH, D. E., KRIEG, M. B., & HOEBEL, B. G. Lateral hypothalamic
    control of killing: Evidence for a cholinoceptive mechanism.
    Science, 1970, 167, 900-901.

URSIN, H. Effect of amygdaloid lesions on avoidance behavior
and visual discrimination in cats. Experimental Neurology,
1965, 11, 298-317.

ELECTROPHYSIOLOGY–NEUROPHYSIOLOGY

EFFECTS OF ELECTRICAL STIMULATION OF THE AMYGDALOID
COMPLEX ON THE VENTROMEDIAL HYPOTHALAMUS*

J. J. Dreifuss

Department of Physiology
University of Geneva Medical School
Geneva, Switzerland

The basal hypothalamus serves as a primary control region for
a multitude of visceral regulatory mechanisms essential for the
organism, such as the regulation of temperature (Hammel, 1968),
feeding behavior (Stevenson, 1969), cardiovascular function
(Zanchetti, 1970), and many forms of emotion (Kaada, 1967). In
addition, as a consequence of its intimate relationship with the
pituitary, and of the arrangement of the vascular system in the
infundibular region, the tuberal hypothalamus has the unique role
of being a neuroendocrine organ; this field has been reviewed
extensively (McCann et al., 1968; Beyer and Sawyer, 1969;
Martini et al., 1970).

This wide variety of functions subserved by the hypothalamus
has prompted numerous investigations of surrounding brain regions,
in order to find which of these might influence the activity of
hypothalamic neurones. Both from anatomical (Nauta, 1961) and
electrophysiological evidence (Sawa et al., 1959; Wendt, 1961;
Tsubokawa and Sutin, 1963; Stuart et al., 1964; Egger, 1967;
Fernandez de Molina and Ruiz Marcos, 1967; Murphy et al., 1968a;
Van Atta and Sutin, 1971), it is readily apparent that most of
the important ascending and descending fibre systems supplying
the basal hypothalamus originate in limbic structures.

This review will deal solely with the organization of the
efferent projection systems of the amygdaloid complex to the
mammalian hypothalamus, and especially with those connections
directed towards the hypothalamic ventromedial nucleus.

In mammals, the amygdaloid complex projects to a wide subcortical region, which extends from the septum to the midbrain

tegmentum (Gloor, 1955; Hall, 1963); it reaches these areas via
two separate fibre systems: the stria terminalis, which takes a
semi-circular course dorsal to the basal ganglia and internal
capsule, and serves as the efferent pathway for amygdaloid nuclear
masses lying medially; and a ventral fibre system, which takes a
sublenticular and subcapsular course and through which more
laterally situated amygdaloid nuclei project to the septum and
hypothalamus. Several studies suggest that these two amygdaloid
efferent fibre tracts, and hence the regions from which they
originate, may subserve different functions. Thus, Hilton and
Zbrozyna (1963) showed that the ventral, but not the stria, system
mediates the defense reaction in cats. Vergnes and Karli (1964)
demonstrated that bilateral interruption of the ventral fibre
tracts abolishes the "mouse-killing" behavior observed in certain
rats, while bilateral section of the stria terminalis had no
effect on this behavior. The ovulatory response obtained following stimulation of the amygdala is blocked by transection of the
stria, but not by interruption of the ventral amygdalofugal,
fibres (Velasco and Taleisnik, 1969).

   An Inhibitory Action of Impulses Travelling along the
   Stria Terminalis on Hypothalamic Ventromedial Neurones

   In an investigation designed to define the influences exerted
by amygdaloid stimulation, Dreifuss et al. (1968) used microelectrodes to record compound and unit responses in the tuberal hypothalamus following electrical stimulation of discrete amygdaloid
areas in "cerveau isolé" cats. The most conspicuous evoked
responses were obtained in the ventromedial hypothalamus, suggesting that this zone might represent a focal point within the amygdaloid projection field.

   Thus, when stimuli were applied to the corticomedial complex
of the amygdala, a large positive compound action potential was
recorded in the ventromedial nucleus. It had a long onset latency
and the peak of the positive transient followed stimulation by
30-40 msec. It was confined virtually to the ventromedial nucleus,
with very little spread into surrounding areas. Figure 1 illustrates the compound potentials obtained in such an experiment at
various recording locations in frontal plane F 11.5, according to
the atlas of Jasper and Ajmone Marsan (1954).

   This positive compound potential was obtained only when the
stimulating electrodes were located in a fairly restricted area
of the amygdaloid complex (corresponding to the cortical and
central nuclei), or when stria terminalis fibres were stimulated
directly near their origin within the amygdala. Trace 1 of
Figure 2 shows a positive compound potential obtained in another
cat, and illustrates how the potential recorded in the ventro-

medial hypothalamus was altered in its form when the stimulating
electrode was lowered in vertical steps through the amygdala at
stereotaxic plane F 11.5. It should be noted that the form and
latency of the evoked response changed when the electrode left
the area crossed by stria fibres and penetrated the basal nucleus
of the amygdala; but then it remained essentially unchanged with
progressive, 1 mm by 1 mm displacement throughout the basal
nucleus (Fig. 2, traces 2-4).

Fig. 1. Mapping of the compound potentials recorded in the cat
hypothalamus (at stereotaxic plane F 11.5) in response to single
stimuli applied to the ipsilateral stria terminalis at the level
of its origin within the amygdaloid complex. Note that long
latency, positive potentials are found in the region of the
ventromedial nucleus. In this and subsequent figures, upward
deflections represent negative polarity. (From Driefuss et al.,
1968).

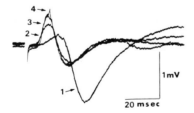

Fig. 2. Compound potentials recorded in the ventromedial nucleus
of the hypothalamus during vertical displacement of a concentric
bipolar stimulating electrode through the ipsilateral amygdala.
Trace 1 was obtained when the stimulating electrode was located
in the amygdaloid area crossed by stria nerve fibres (cf. Fig. 1);
traces 2-4 were recorded after the stimulating electrode had been
lowered 1, 2 and 3 mm respectively into the basal nucleus.

In our study, an estimated conduction velocity of less than
1 m/sec was obtained for the stria fibres that reach the ventromedial nucleus. The existence of slow and fast conducting nerve
fibres in the stria has been described by Fernandez de Molina and
Garcia Sanches (1967), with peak conduction velocities of 0.6-0.8
m/sec, and 1.8-2.3 m/sec respectively. It is very likely that
the stria fibres which reach the ventromedial hypothalamus
directly belong to the slow conducting ones.

Experimental Neuroanatomical Studies of the Stria terminalis

The stria terminalis is the best studied route for amygdaloid
control of the hypothalamus, because it forms a compact bundle
which can be lesioned easily without producing any damage to the
hypothalamus itself. However, the tracing of fibre terminations
within the hypothalamus has until recently been complicated due
to the small size of the axons. Thus, in early neuroanatomical
studies using the Marchi or Nauta-Gygax methods, no clearcut
evidence of axon degeneration was found in the hypothalamic ventromedial nucleus of either side after lesioning of the stria in rats
(Cowan et al., 1965), cats (Szentágothai et al., 1962; but cf.
Ishikawa et al., 1969) or monkeys (Nauta, 1961; but cf. Adey and
Meyer, 1952). These earlier findings suggested that the stria
distributes its fibres mainly to hypothalamic areas anterior to
this nucleus. The development of silver impregnation methods
which afford better visualization of small diameter degenerating
axon terminals (Fink and Heimer, 1967; De Olmos, 1969) has led
to a reappraisal of this concept, in showing that hypothalamic
terminations of the stria do actually reach the ventromedial
nucleus. Degenerating boutons are found mainly in three circum-
script hypothalamic regions. Post-commissural fibres of the
stria terminate in the stria's bed nucleus (Fig. 3A); in addition, other post-commissural fibres of the stria terminate in
the anterior hypothalamus (Fig. 3 B-E). Other stria fibres form
the so-called pre-commissural component, which follows a nearly
sagittal course through the medial hypothalamus. The most
posterior area of synaptic termination of these axons is in
the core of the ventral pre-mammillary nucleus (Fig. 3H); moreover, fibres of the pre-commissural component of the stria, upon
entering the tuberal hypothalamus, disperse into a zone of dense
termination which surrounds the ventromedial nucleus (Fig. 3 F,
G). The terminals actually lie in a shell surrounding this
nucleus, where they establish synaptic contacts with dendrites
which radiate out of the nucleus into the relatively cell-free
zone around it.

The mode of termination of the strial amygdalo-ventromedial
pathway has been investigated recently by means of electron
microscopy of degenerating axon terminals following section of

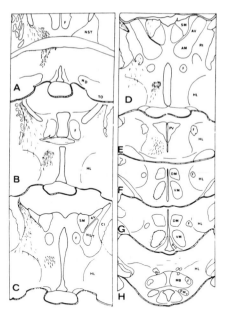

Fig. 3. Degeneration patterns elicited by stria terminalis section in the rat. NST, bed nucleus of stria terminalis; VM, ventromedial nucleus; PMv, ventral premammillary nucleus. (From Heimer and Nauta, 1969)

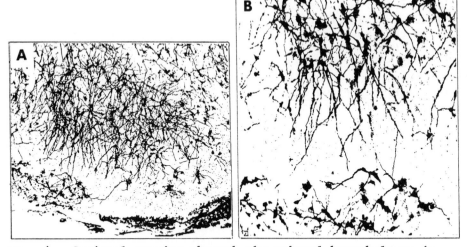

Fig. 4. Sagittal section through the tuberal hypothalamus in the rat. Golgi-Cox stain. Note the large number of dendrites radiating from the ventromedial nucleus into the surrounding, cell-poor zone. B, detail of A at higher magnification; near lower margin are cells of the arcuate nucleus. (From Heimer and Nauta, 1969)

the stria (Raisman, 1970; and this symposium). These studies
confirm the importance of the stria terminalis projection onto
ventromedial neurones, since 20 per cent of all the synapses
impinging upon these neurones degenerate after interruption of
the stria. The neuropil of the outer part of the ventromedial
nucleus, in rats, consists largely of axo-dendritic synapses;
synapses upon dendritic shafts outnumber those contacting
dendritic spines in a proportion of 5 to 1. However, in animals
whose stria had been transected, degenerating terminals were
found almost exclusively upon dendritic spines. Synaptic
terminals of stria fibres account for more than half of all
spine synapses. They are of the symmetrical type first described by Gray (1959), and might thus be expected to be excitatory in nature. This obviously raises the question whether the
positive compound potential recorded in the ventromedial nucleus
following stimulation of stria fibres in the amygdala, which
will be later shown to be associated with a pause in firing of
ventromedial neurones, is not mediated through a synaptic relay
station.

The field of origin of the stria terminalis fibres within
the amygdala has been investigated thoroughly by Leonard and
Scott (1971) and by De Olmos (personal communication). The
post-commissural component of the stria, which projects to the
bed nucleus and to the anterior hypothalamic area, originates
in a widespread zone within the amygdaloid complex, comprising
the basal, medial and central nuclei. In contrast, a surprisingly restricted field of origin was found by Leonard and Scott
(1971) for the pre-commissural component of the stria, which
projects to the ventromedial hypothalamus: only a lesion in the
cortical nucleus of the amygdala produced degeneration of this
pathway. According to De Olmos, the stria fibres which reach
the ventromedial nucleus arise in the medial nucleus and in the
posterior part of the cortical nucleus.

The origin, course and termination of the pathway that leads
from the corticomedial amygdala to the outer zone of the hypothalamic ventromedial nucleus is thus well established. However,
consideration of this alone would not enable full comprehension
of its possible functions. For such comprehension, a knowledge
of the type of synaptic transfer, namely excitation and/or
inhibition, exerted by this fibre tract must be available.

An analysis of the discharge patterns of single ventromedial
neurones following stimulation of the tract is a way of obtaining
this information. Dreifuss et al. (1968) have conducted a microelectrode study which suggests that excitation of the stria
terminalis in the cat produces an inhibition of firing of
ventromedial neurones. Formvar-coated, tungsten electrodes

were inserted stereotaxically into the tuberal hypothalamus of
"cerveau isole" animals. Action potentials from ventromedial
neurones were biphasic, a majority being positive-negative.
An inflection on the rising phase was often seen, and was most
clearly evident when two spikes occurred in close succession
(Fig. 5). This was taken as proof that the recording was from
the neurone soma or proximal dendrites, but not from an axon.

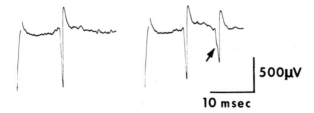

500μV

10 msec

Fig. 5. Extracellularly recorded action potentials from a cat
hypothalamic neurone evoked by stimulation of the ipsilateral
basal nucleus of the amygdala. The first transient in both
traces is the stimulus artifact. Note the inflection (arrow)
of the initial phase of the action potential seen when two
spikes occur in close succession. (From Murphy et al., 1968a)

A signal averager was used to generate for each ventromedial
neurone: (a) an algebraically summed compound potential response
and (b) a post-stimulus time histogram of unit activity. Figure
6 shows the compound potential (A) and post-stimulus time histogram (B) recorded simultaneously through the same microelectrode
located near a ventromedial neurone, during cortical amygdaloid
stimulation. It is apparent that the probability of discharge
of this spontaneously active cell was reduced during the positive
transient of the stria terminalis mediated compound potential,
and that the time course of the two phenomena was almost the same.
Another example which shows the inhibitory action of impulses
travelling along the stria on another ventromedial neurone is
illustrated in Figure 11B. Actually, inhibitory responses were
the most common, excitatory responses of very long latency being
only observed very rarely following stimulation of the corticomedial complex of the amygdala.

Excitation of Hypothalamic Ventromedial Neurones by
Basal Amygdaloid Stimulation

Figure 6 shows that the envelope of a post-stimulus time
histogram may reproduce rather faithfully the form of the compound potential recorded at the same site, a decrease of cell
firing being observed during a positive transient of the slow

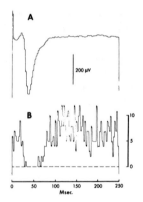

Fig. 6. Compound potential (A), and post-stimulus time histogram
(B) recorded simultaneously, from a ventromedial neurone following stimulation of the ipsilateral corticomedial amygdala.   In
B, 50 responses were summated; vertical scale, number of action
potentials per address.

potential.   In view of this observation, it was of interest to see
whether the negative-positive compound potential recorded from the
ventromedial hypothalamus during basal amygdaloid stimulation was
associated with predictable changes in the probability of cell
firing.   Figure 7 illustrates that this is indeed the case, there
being again a good correspondence between the phases of the
compound potential and cell firing.   In fact, for a total of
nearly 100 ventromedial neurones studied, negative phases of slow
potentials consistently coincided with accelerations of neurone
discharges.

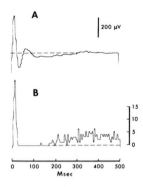

Fig. 7. Compound potential (A) and post-stimulus time histogram
(B) from a ventromedial neurone following stimulation of the
basal nucleus of the amygdala.   Same format as Fig. 6.
(From Dreifuss et al., 1968).

As the positive response obtained following electrical stimulation of the corticomedial amygdala, this negative-positive compound potential was also most prominent in the ventromedial hypothalamus. Figure 8 illustrates the potentials recorded at various levels in the tuberal hypothalamus during a vertical

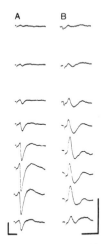

Fig. 8. Compound potentials recorded in the tuberal hypothalamus at F 11.5, L 1.0, following single stimuli applied to the cortical (A) and basal (B) amygdaloid nuclei. The first records (from above downwards) were obtained when the recording microelectrode was lowered in vertical steps of 1 mm, starting at H-1.0; from H-3.0 to H-6.5, records were obtained every 0.5 mm. Note that both responses are of maximum amplitude in the ventromedial nucleus. Calibrations: 1 mV, 50 msec.

penetration of a recording microelectrode at frontal plane 11.5, 1 mm lateral to the midline, in response to stimulation of the cortical (A) and basal (B) amygdaloid nuclei. The first records (from above downwards) were obtained when the electrode was lowered through the hypothalamus in steps of 1 mm, starting at H-1.0; when reaching the ventromedial nucleus, records were obtained every 0.5 mm. It may be seen that both responses are of maximum amplitude at the level of the nucleus. Similar results were obtained in frontal (Fig. 1) and parasagittal (Fig. 9) planes. Figure 9 shows isopotential fields at various frontal and horizontal levels in the ventromedial hypothalamus, 1 mm lateral to the midline, drawn for the negative phase of the potential following stimulation of the basal amygdala; the two-dimensional shape obtained corresponds to the actual location of the ventromedial nucleus.

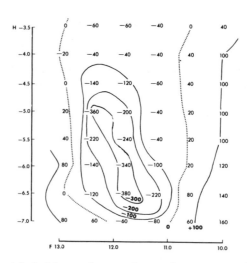

Fig. 9.   Isopotential lines drawn from the compound potentials
obtained from 6 vertical recording electrode penetrations, at
stereotaxic plane L 1.0, during stimulation of the ipsilateral
basal nucleus of the amygdala.   Each number represents the
amplitude of the evoked compound potential at this particular
point in the plane, expressed in μV.

        These observations indicate that the projection fields of
the responses evoked by corticomedial and basal amygdaloid stimulation overlap in the ventromedial hypothalamus.   Experiments
during which stimuli to these structures were applied at various
time intervals, and the potentials shown to interact, as illustrated in Figure 10, provide further evidence for this overlap.

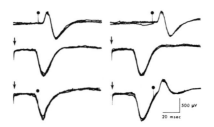

Fig. 10.   Compound potentials evoked by amygdaloid stimulation,
and their interaction in the ipsilateral ventromedial nucleus.
Arrows and filled circles mark artifacts of stimuli applied to
the cortical and basal amygdaloid nuclei respectively.   Note
reduced amplitude of the VAF-responses when they occur during
(lower left corner) or at the end (lower right corner) of an
ST-response.

In Figure 11, oscilloscope traces obtained from a ventromedial neurone during stimulation of the basal nucleus (A), and of the origin of the stria within the amygdala (B) have been photographed. While this neurone was excited by basal amygdaloid stimulation, its firing was reduced for a period of approximately 0.2 sec by stimuli applied to the stria terminalis.

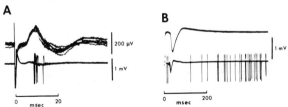

Fig. 11. Oscilloscope traces of a ventromedial neurone during 1/sec stimulation of the basal (A) and cortical (B) nuclei of the ipsilateral amygdala with 0.2-250 Hz (upper traces) and 0.2-1 kHz (lower traces) band-passes. Stimulus artifacts occur at time 0. A, 10, B, 25 superimposed sweeps.

Identification of the Pathways that Mediate Amygdaloid Influences to the Ventromedial Hypothalamus

Electrophysiological data suggest that the observations reported previously can be explained by assuming that stimulation of the corticomedial amygdala generates impulses which travel to the ventromedial hypothalamus via the stria terminalis (ST), whereas the biphasic response which follows basal amygdaloid stimulation is mediated over the ventral amygdalofugal fibre (VAF) system. Evidence in favor of these assumptions was obtained as follows:

(a) Stimulation of the stria terminalis in the floor of the lateral ventricle, where it runs in a groove between the caudate nucleus and the thalamus, elicited a monophasic, positive response in the ventromedial hypothalamus. Its shape was similar to the response recorded in the same animal with cortical amygdaloid stimulation, except that its peak latency was reduced by 11 msec to 24 msec. In the same experiment, the stimulating electrode then was lowered vertically by approximately 8 mm so as to lie immediately dorsally to the optic tract. When stimuli were applied, a biphasic, negative-positive potential was recorded from the ventromedial hypothalamus; it resembled the response obtained in the same animal with basal amygdaloid stimulation, but its peak latency was reduced from 12 to 7 msec.

(b) The ST-response could be abolished selectively after

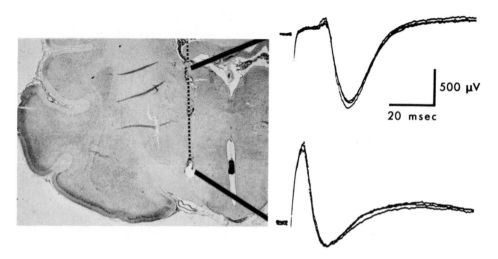

Fig. 12.   Compound potentials recorded in the ventromedial nucleus
following stimulation of the stria terminalis and of ventral amygdalofugal nerve fibres, at the sites indicated on the cresyl
violet stained histological section. See text. (From Dreifuss
et al., 1968)

unilateral section of the stria in the floor of the lateral
ventricle; such a section left the VAF-response unaltered. Conversely, section along a parasagittal plane, midway between the
amygdala and the hypothalamus, eliminated the VAF-potential,
while leaving the ST-response intact (Fig. 13A, 14B). If, subsequent to this procedure, the stria was interrupted at the level
of the caudo-thalamic groove, responses to amygdaloid stimulation
could no longer be recorded (Fig. 14C). It should be stressed
that the VAF-response was eliminated only when a sagittal cut,
such as shown in Fig. 13A, extended from a position lateral to
the septum to the posterior hypothalamus, i.e. over a wide
antero-posterior region.

                            DISCUSSION

     Electrophysiological data suggest that a ventral amygdalo-
fugal fibre system mediates excitatory influences onto ventromedial
hypothalamic neurones. However, neuroanatomical studies have shown
that the VAF system terminates almost exclusively in the lateral

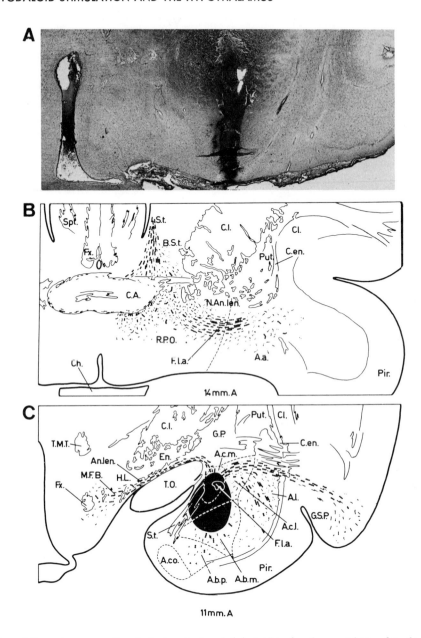

Fig. 13. A, cresyl violet stained histological section showing cut that abolished VAF-response without altering shape of ST-response. Same experiment as Fig. 14. B, C, degeneration patterns seen in a cat following a lesion made in the medial region of the amygdaloid complex, showing the course of ventral amygdalofugal nerve fibres (B and C, from Valverde, 1965).

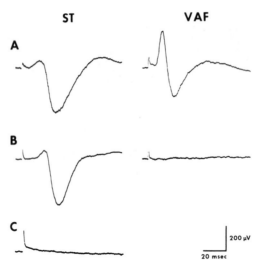

Fig. 14. Compound potentials recorded in the ventromedial nucleus upon stimulation of the cortico-medial (ST) and basal (VAF) areas of the amygdala. B, responses recorded in the same animal with identical stimulation, but after longitudinal section illustrated in preceding figure. Note that the VAF-response is abolished, whereas the ST-response is essentially unchanged; C, after section of the stria terminalis at the level of the caudo-thalamic groove.

hypothalamus (Heimer and Nauta, 1969; De Olmos, this symposium), with few fibres reaching more medial hypothalamic zones. Millhouse (1969, Fig. 11) has published a Golgi picture in which collaterals of a VAF-fibre establish synaptic contact with distal dendrites of a ventromedial neurone. Nevertheless, it is likely that most VAF fibres reach the ventromedial nucleus across a synaptic relay station.

The origin of afferent connections to the ventromedial nucleus has been recently reinvestigated by Chi (1970) in rats after placement of lesions at a short distance from it. Massive terminal degeneration in and around the ventromedial nucleus was observed after lesions of the anterior hypothalamic area. This observation points to the existence in the rodent of an intra-hypothalamic connection from the anterior hypothalamus to the core of the ventromedial nucleus. Since the post-commissural component of the stria terminalis ends massively in the anterior hypothalamus, this intra-hypothalamic connection may provide a trans-synaptic conduction route for impulses travelling along the stria to the ventromedial hypothalamus. This postulated pathway parallels the more direct, pre-commissural component of the stria which reaches the outer zone of the ventromedial nucleus.

If one considers that according to the electron microscope study of Raisman (1970, and this symposium) the terminal boutons of stria terminalis fibres that reach the ventromedial nucleus directly are likely to be excitatory in nature, then the possibility exists that the inhibitory responses of ventromedial neurones following corticomedial amygdaloid stimulation may well travel along the pathway postulated by Chi (1970). However, this latter pathway cannot be involved in mediating the VAF-responses obtained with basal amygdaloid stimulation, since interruption of the stria terminalis at the level of the caudothalamic groove did not alter the shape of these responses.

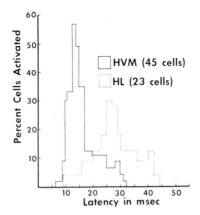

Fig. 15. Latency distribution of single unit responses to stimulation of the basal amygdaloid nucleus in the region of the hypothalamic ventromedial nucleus (HVM) and in the lateral hypothalamus (HL).

According to Chi (1970), no other lesions around the ventromedial nucleus produce significant degeneration in this nucleus. Thus, lesions of the posterior hypothalamus produced fibre degeneration in the lateral hypothalamus, but not medially. Lesions in the lateral hypothalamus caused degeneration in dorsal and dorsomedial hypothalamic areas; the question whether axons arising in the lateral hypothalamus reach the ventromedial nucleus has long been open to controversy (Wolf and Sutin, 1967; Eager et al., 1971), but seems somewhat academic, since there is no clear separation of the dendritic fields of lateral and ventromedial hypothalamic neurones (Millhouse, 1969). Functional connections from the lateral hypothalamus to the ventromedial nucleus are in all probability few in numbers, as Murphy and Renaud (1969) found that electrical stimulation in the lateral hypothalamus failed to modify the discharges recorded extra-

cellularly from ventromedial neurones in cats. Also, Murphy et
al. (1968a) showed that following amygdaloid stimulation, lateral
hypothalamic neurones tend to fire at longer latencies and less
regularly than neurones lying more medially (Fig. 15), an observation which seems to rule out the lateral hypothalamus as a
likely relay station for amygdalofugal impulses that reach the
ventromedial nucleus.

    In view of the excellent correlation found between the
probability of unit discharges in the ventromedial hypothalamus,
and the deflections of the compound potentials following amygdaloid stimulation, we have postulated that the latter are indicative of membrane potential changes of synaptic origin (Dreifuss
et al., 1968). Oomura et al. (1970) have published electrophysiological data recently, including intracellular recordings (Fig. 16)
obtained from lateral hypothalamic neurones in rats in response
to electrical stimulation of the amygdaloid complex. In view of

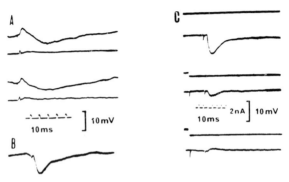

Fig. 16. Intracellular recordings of synaptic potentials in rat
lateral hypothalamic neurones following stimulation of the amygdala. A is from two different neurones; the upper trace in each
pair shows an EPSP-IPSP sequence; the lower traces are extracellular recordings obtained after withdrawal of the recording
electrodes from the cells. B and C show IPSPs not preceded by
EPSPs, obtained from other neurones. In C, inward currents
were applied through the recording micropipette and lead to a
reversal of polarity. (From Oomura et al., 1970)

the overlap of the dendritic fields of lateral and ventromedial
hypothalamic neurones, it is of interest that their results are
very similar to those that Dreifuss et al. (1968) obtained from
ventromedial neurones. Oomura et al. (1970) observed positive
as well as negative-positive compound potentials, positive
transients being associated with a reduction, negative ones with

an increase in neuronal firing probability; since transection of
the stria terminalis resulted in the disappearance of the positive
potentials, they confirm that inhibitory effects of amygdaloid
stimulation may be due to long-lasting inhibitory post-synaptic
potentials mediated through the stria.

Oomura et al. (1970) also suggest that excitatory responses
could be due to impulses travelling along the VAF fibre system.
However, unequivocal anatomical evidence for a direct or oligo-
synaptic connection from the amygdala to the ventromedial hypothalamus through a ventral fibre system is still lacking.   In
this respect, a further complication arises from the fact that
fibres which form the VAF system originate not only in the basolateral complex of the amygdala, but also in the pyriform cortex.

CONCLUSIONS AND SUMMARY

Electrophysiological studies demonstrate that the basal and
the corticomedial amygdaloid nuclei project through two powerful
pathways to the ventromedial nucleus of the hypothalamus, establishing opposite control of the firing of these neurones.   The
hypothalamic ventromedial nucleus is implied in the regulation of
food intake (Stevenson, 1969) and of growth hormone secretion
(Frohman et al., 1968) and thus plays an essential role in homeo-

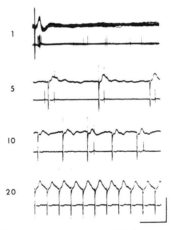

Fig. 17.   Responses of a cat ventromedial neurone to stimulation
of the basal nucleus of the amygdala.   The upper trace in each
pair is the "unfiltered," the lower one the "filtered," response
(cf. Fig. 11).   Uppermost pair, superimposed sweeps showing
excitation of the neurone coinciding with the negative component
of the evoked potential (1/sec).   Other tracings are single,
continuous records obtained when the stimulation frequency was
5/sec, 10/sec and 20/sec respectively.   Note that the neuron is
no longer "driven" during high frequency stimulation.   Calibrations:
100 msec, 0.5 mV.   (From Murphy et al., 1968b).

stasis for the organism. It appears that the dorsal amygdaloid
projection system, through the stria terminalis, exerts an
inhibitory action on ventromedial neurones, whereas a ventral
pathway is excitatory, later followed by inhibition. These observations favor the concept that limbic structures may exert a
modulatory influence on hypothalamic functions. Whether this
functional dichotomy of two descending amygdaloid projection
systems also applies to other projection zones of the amygdala
in the basal forebrain remains at present unknown.

The question evidently arises whether the functional identity
which each of these amygdalofugal pathways discloses in electrophysiological studies can be translated into specific differences
in their biological significance for the organism. Although it
certainly is premature to make definite statements with regard to
such a possibility, there are a certain number of observations,
reviewed by Egger and Flynn (1967), Gloor et al. (1971) and
Kaada (this symposium) which suggest a functional dichotomy within the amygdala.

We do not know, however, to what extent differences which
have been observed in behavioral studies reflect activities
mediated over the dorsal and ventral amygdalofugal systems. The
only exceptions to this statement known to this author are the
defense reaction in the cat (Hilton and Zbrozyna, 1963) and the
rat's "mouse-killing" behavior (Vergnes and Karli, 1964), both
of which are abolished by interruption of the VAF system, but unaltered by stria section. In contrast, the stria terminalis,
and the amygdaloid nuclei from which this tract originates, appear
to be involved in reproductive functions (Elwers and Critchlow,
1961; Velasco and Taleisnik, 1969).

It must be kept in mind that the elicitation of patterned
behavioral or endocrine responses following stimulation of the
amygdaloid complex necessitates, as a rule, high frequency
trains of stimuli. It seems legitimate to assume that the rate
of impulse transmission may be another important factor in determining whether their trans-synaptic effects will be excitatory
or inhibitory (Murphy et al., 1968b). Observations such as those
illustrated in Figure 17, which show that single hypothalamic
neurones can be affected in at times opposite ways by increases
in the stimulation frequency, suggest that a division of the
amygdaloid complex into functional compartments based on data
obtained following electrical stimulation of the amygdala may
be premature.

ACKNOWLEDGMENTS

*Supported by grants from the Swiss National Science Foundation.

REFERENCES

ADEY, W. R., & MEYER, M. Hippocampal and hypothalamic connections
     of the temporal lobe in the monkey.   Brain, 1952, 75, 358-
     384.

BEYER, C., & SAWYER, C. H. Hypothalamic unit activity related
     to control of the pituitary gland. In W. F. Ganong and L.
     Martini (Eds.), Frontiers in Neuroendocrinology. New York
     and London: Oxford University Press, 1969. Pp. 255-288.

CHI, C. C. Afferent connections to the ventromedial nucleus of
     the hypothalamus in the rat. Brain Research, 1970, 17, 439-
     445.

COWAN, W. M., RAISMAN, G., & POWELL, T. S. P.   The connections
     of the amygdala. Journal of Neurology, Neurosurgery and
     Psychiatry, 1965, 28, 137-151.

De OLMOS, J. S. A cupric silver method for impregnation of ter-
     minal axon degeneration and its further use in staining
     granular argyrophilic neurons. Brain Evolution and Behavior,
     1969, 2, 213-237.

DREIFUSS, J. J., MURPHY, J. T., & GLOOR, P. Contrasting effects
     of two identified amygdaloid efferent pathways on single
     hypothalamic neurons. Journal of Neurophysiology, 1968, 31,
     237-248.

EAGER, R. P., CHI, C. C., & WOLF, G. Lateral hypothalamic pro-
     jections to the hypothalamic ventromedial nucleus in the al-
     bino rat: demonstration by means of a simplified ammoniacal
     silver degeneration method. Brain Research, 1971, 29, 128-
     132.

EGGER, M. D. Responses of hypothalamic neurons to electrical
     stimulation in the amygdala and the hypothalamus. Electro-
     encephalography and Clinical Neurophysiology, 1967, 23, 6-15.

EGGER, M. D., & FLYNN, J. P. Further studies on the effects of
     amygdaloid stimulation and ablation on hypothalamically
     elicited attack behavior in cats. In W. R. Adey and T.
     Tokizane (Eds.) Progress in Brain Research, Vol. 27,
     Structure and Function of the Limbic System. Amsterdam:
     Elsevier, 1967. Pp. 165-182.

ELWERS, M., & CRITCHLOW, V. Precocious ovarian stimulation follow-
    ing interruption of the stria terminalis. American Journal of
    Physiology, 1961, 201, 281-284.

FERNANDEZ de MOLINA, A., & GARCIA SANCHEZ, J. L. The properties of
    the stria terminalis fibers. Physiology & Behavior, 2, 225-227.

FERNANDEZ de MOLINA, A., & RUIZ MARCOS, A. A study on the neuronal
    activity in the amygdaloid projection field. Trabajos del
    Laboratorio de Investigaciones biologicas de la Universidad
    de Madrid, 1967, 59, 137-151.

FINK, R. P., & HEIMER, L. Two methods for selective silver impreg-
    nation of degenerating axons and their synaptic endings in the
    central nervous system. Brain Research, 1967, 4, 369-374.

FROHMAN, L. A., BERNHARDIS, L. L., & KANT, K. J. Hypothalamic
    stimulation of growth hormone secretion. Science, 1968, 162,
    580-582.

GLOOR, P. Electrophysiological studies on the connections of the
    amygdaloid nucleus in the cat. I. The neuronal organization
    of the amygdaloid projection system. Electroencephalography
    and Clinical Neurophysiology, 1955, 7, 223-242.

GLOOR, P., MURPHY, J. T., & DREIFUSS, J. J. Anatomical and physio-
    logical characteristics of the two amygdaloid projection
    systems to the ventromedial hypothalamus. In C. Hockman (Ed.),
    Limbic System Influences on Autonomic Function. Springfield,
    Illinois: C. C. Thomas, 1971, in press.

GRAY, E. G. Axosomatic and axodendritic synapses in the cerebral
    cortex: an electron microscope study. Journal of Anatomy
    (London), 1959, 93, 420-433.

HALL, E. Efferent connections of the basal and lateral nuclei of
    the amygdala in the cat. American Journal of Anatomy, 1963,
    113, 139-151.

HAMMEL, H. T. Regulation of internal body temperature. Annual
    Review of Physiology, 1968, 30, 641-710.

HEIMER, L., & NAUTA, W. J. H. The hypothalamic distribution of the
    stria terminalis in the rat. Brain Research, 1969, 13, 284-297.

HILTON, S. M., & ZBROZYNA, A. W. Amygdaloid region for defence
    reactions and its efferent pathway to the brain stem. Journal
    of Physiology (London), 1963, 196, 160-173.

ISHIKAWA, I., KAWAMURA, S., & TANAKA, O. An experimental study
    on the efferent connections of the amygdaloid complex in
    the cat. Acta medica Okayama, 1969, 23, 519-539.

JASPER, H. H., & AJMONE MARSAN, C. A Stereotaxic Atlas of the
    Diencephalon of the Cat. Ottawa: National Research Council
    Canada, 1954.

KAADA, B. Brain mechanisms related to aggressive behavior. In
    UCLA Forum of Medical Sciences, Vol. 7, Aggression and
    Defense, Neural Mechanisms and Social Pattern. Washington,
    D.C.: American Institute of Biological Sciences, 1967.
    Pp. 95-133.

KAADA, B. Stimulation and regional ablation of the amygdaloid
    complex with reference to functional representation. In
    this volume p. 205.

LEONARD, C. M., & SCOTT, J. W. Origin and distribution of the
    amygdalofugal pathways in the rat: an experimental neuro-
    anatomical study. Journal of Comparative Neurology, 1971,
    141, 313-330.

MARTINI, L., MOTTA, M., & FRASCHINI, F. The Hypothalamus.
    New York and London: Academic Press, 1970.

McCANN, S. McD., DHARIWAL, A. P. S., & PORTER, J. C. Regulation
    of the adenohypophysis. Annual Review of Physiology, 1968,
    30, 589-640.

MILLHOUSE, O. E. A Golgi study of the descending medial fore-
    brain bundle. Brain Research, 1969, 15, 341-363.

MURPHY, J. T., & RENAUD, L. Mechanisms of inhibition in the
    ventromedial nucleus of the hypothalamus. Journal of
    Neurophysiology, 1969, 32, 85-102.

MURPHY, J. T., DREIFUSS, J. J., & GLOOR, P. Topographical diff-
    erences in the responses of single hypothalamic neurons to
    limbic stimulation. American Journal of Physiology, 1968a,
    214, 1443-1453.

MURPHY, J. T., DREIFUSS, J. J., & GLOOR, P. Responses of hypo-
    thalamic neurons to repetitive amygdaloid stimulation.
    Brain Research, 1968b, 8, 153-166.

NAUTA, W. J. H. Fibre degeneration following lesions of the amygdaloid complex in the monkey. Journal of Anatomy (London), 1961, 95, 515-531.

OOMURA, Y., ONO, T., & OOYAMA, H. Inhibitory action of the amygdala on the lateral hypothalamic area in rats. Nature (London), 1970, 228, 1108-1110.

RAISMAN, G. An evaluation of the basic pattern of connections between the limbic system and the hypothalamus. American Journal of Anatomy, 1970, 129, 197-202.

SAWA, M., MARUYAMA, N., HANAI, T., & KAJI, S. Regulating influence of amygdaloid nuclei upon the unitary activity in the ventromedial nucleus of the hypothalamus. Folia Psychiatrica et Neurologica Japonica (Niigata), 1959, 13, 235-256.

STEVENSON, J. A. F. Neural control of food and water intake. In W. Haymaker et al. (Eds.) The Hypothalamus. Springfield, Ill.: C. C. Thomas, 1969. Pp. 524-621.

STUART, D. G., PORTER, R. W., & ADEY, W. R. Hypothalamic unit activity. II. Central and peripheral influences. Electroencephalography and Clinical Neurophysiology, 1964, 16, 248-258.

SZENTAGOTHAI, J., FLERKO, B., MESS, B., & HALASZ, B. Hypothalamic Control of the Anterior Pituitary. Budapest: Akad. Kiado, 1962.

TSUBOKAWA, T., & SUTIN, J. Mesencephalic influence upon the ventromedial hypothalamic nucleus. Electroencephalography and Clinical Neurophysiology, 1963, 15, 804-810.

VALVERDE, F. Studies on the Piriform Lobe. Cambridge, Mass.: Harvard University Press, 1965.

VAN ATTA, L., & SUTIN, J. The response of single lateral hypothalamic neurons to ventromedial nucleus and limbic stimulation. Physiology & Behavior, 1971, 6, 523-536.

VELASCO, M. E., & TALEISNIK, S. Release of gonadotrophins induced by amygdaloid stimulation in the rat. Endocrinology, 1969, 84, 132-139.

VERGNES, M., & KARLI, P. Etude des voies nerveuses de l'influence facilitatrice exercee par les moyaux amygdaliens sur le comportement d'aggression interspécifique Rat-Souris. Comptes Rendus des Seances de la Society de Biologie, 1964, 158, 856-858.

WENDT, R. H. Amygdaloid and Peripheral Influences upon the Activity of Hypothalamic Neurons in the Cat. Ph.D. Thesis, UCLA, 1961.

WOLF, G., & SUTIN, J. Fiber degeneration after hypothalamic lesions in the rat. Journal of Comparative Neurology, 1967, 127, 137-156.

ZANCHETTI, A. Control of the cardiovascular system. In L. Martini et al. (Eds.), The Hypothalamus. New York and London: Academic Press, 1970. Pp. 233-244.

AMYGDALOID-HYPOTHALAMIC NEUROPHYSIOLOGICAL INTERRELATIONSHIPS

M. David Egger
Department of Anatomy
Yale University School of Medicine
New Haven, Connecticut 06510

I. ANATOMY

Anatomy is logically prior to physiology, and as a member of an Anatomy Department, I feel some obligation at least to allude to what is known about the anatomy of amygdala-hypothalamic interconnections. In experimental studies, the anatomy of the amygdala has been studied extensively in cat, rat, rabbit and monkey. Although strong similarities between species exist, it is best to keep very clear which species one is talking about, otherwise, one gets into a muddle, making impossible an already very difficult subject. Unless otherwise noted, I intend to concentrate on cats, for both the anatomy and electrophysiology.

It is agreed generally that there are two main pathways from the amygdala to the hypothalamus, the stria terminalis (ST) and the ventral amygdalofugal system (VAF). The former is a more well defined, compact bundle. These two main pathways can be fractionated further, but that is not necessary for our purposes.

The stria terminalis (ST) in the cat arises primarily from the corticomedial nuclei, and, perhaps to some extent, from the basal and lateral nuclei (see, for instance, Valverde, 1965). The ventral amygdalofugal fiber system (VAF) arises most probably from the basal and lateral nuclei, though the possibility has been raised that these fibers arise almost solely from the pyriform cortex and only pass through the basolateral nuclei (see, for instance, with reference to the rat, Cowan, Raisman, and Powell, 1965).

The stria terminalis (ST) projections from the amygdala include the septum, rostral preoptic region, and the dorsomedial hypothalamus. In rats, the ST has been shown to project to the cell free zone around the ventromedial nucleus of the hypothalamus (HVM), which contains dendrites of HVM neurons (Heimer and Nauta, 1969). Szentágothai et al. (1968) state that in the cat most ST fibers terminate in the bed nucleus of ST, while only a few make it to the anterior hypothalamus. But the staining method employed by Szentágothai et al. (1968) may have made it impossible for them to trace the finer fibers.

The VAF projections from the amygdala include the rostral preoptic region and the lateral hypothalamus, overlapping in many parts of the hypothalamus with the stria terminalis projection (Valverde, 1965). A direct projection via the VAF to the HVM is doubtful, but negative evidence in neuroanatomy must be taken with a grain of salt, especially since we are now in a period when new techniques are being developed to reveal connections mediated by fine fibers that we were unable to detect a few years ago (e.g., Heimer and Nauta, 1969; Eager, Chi, and Wolf, 1971).

The ST fibers are fine, the VAF fibers probably even finer, so that in large part we are dealing with comparatively slow-conducting, lightly myelinated systems (Gloor, 1955a; Fernandez de Molina and Garcia-Sanchez, 1967).

Now for some complications. (1) Both of these amygdalo-fugal systems conveying fibers to the hypothalamus also are amygdalopetal, conveying fibers to the amygdala from hypothalamus and related regions (e.g., Nauta, 1958; Valverde, 1965). The amygdalofugal pathways have been much more studied, in part because behavioral effects of amygdaloid stimulation or lesions depend on an intact hypothalamus, but, apparently, not the other way around (e.g., Kling and Hutt, 1958). (2) The amygdaloid nuclei are richly interconnected with one another. (3) Indirect pathways from the amygdala to the hypothalamus have been demonstrated. For instance, in cats, there may be a pathway from the amygdala to the hippocampus via the pyriform cortex, and then to the hypothalamus via the fornix (Valverde, 1965). In addition, in the monkey at least (Nauta, 1962), there is a pathway from the amygdala to the dorsomedial nucleus of the thalamus, to the orbito-frontal cortex, then back to the hypothalamus. The roles of these indirect pathways with respect to the responses of hypo-thalamic neurons following electrical stimulation in the amygdala are unknown.

Because the VAF and ST project, in part, to different portions of the hypothalamus, in order to interpret the patterns of responses of hypothalamic neurons to stimulation in the amygdala, we need to know something about the interconnections within the hypothalamus. Much functional evidence points to reciprocal interactions between the medial and the lateral hypothalamus (HL), especially between HVM and HL, but the anatomical evidence has been hard to come by. Now, using the newer anatomical techniques, it recently has been possible to demonstrate such reciprocal connections, from HL to HVM in the rat (Eager, Chi, and Wolf, 1971), and from HVM, or its lateral shell, to HL in the cat (Sutin and Eager, 1969), and in the mouse (Arees and Mayer, 1967).

One moral of all this is that one may assume as a first approximation at least, that if a relationship is demonstrated rather convincingly physiologically, sooner or later it may be demonstrated anatomically, too.

II.   ELECTROPHYSIOLOGY

A. Evoked potentials.

The basic electrophysiology of the efferent connections of the amygdala was established elegantly by Gloor (1955a,b). He was concerned with evoked potentials following 1 Hz and higher frequency stimulation of the amygdala in various subcortical portions of the cat's brain, including the hypothalamus.

In brief summary, Gloor (1955a) established that in cats the subcortical projections from both the basolateral and corti-comedial amygdala overlap widely. Short latency (9 msec or less) responses from the basolateral nuclei occurred in the rostral preoptic area, at the base of the septal area, and in the an-terior lateral hypothalamus in the region of medial forebrain bundle. From here, responses presumably were relayed to, among other regions, most of the rest of the hypothalamus. Short latency responses from the corticomedial nuclei were found to extend more caudally into the hypothalamus --including the HVM-- than for the responses elicited from the basolateral nuclei.

Gloor (1955a) demonstrated further, by recording in ST, that the short latency projections from the basolateral nuclei do not run in ST, and estimated that the conduction velocities tended to be slow, comparable to what in the periphery would be in the C-fiber range.

The hypothalamus, like the amygdala, is a heterogeneous region. It might be possible to make more sense of recent neuro-physiological findings by looking at amygdaloid-hypothalamic interconnections a portion at a time. Rather arbitrarily, I would like to divide the hypothalamus and preoptic regions into five zones (Fig. 1), not corresponding precisely to the known

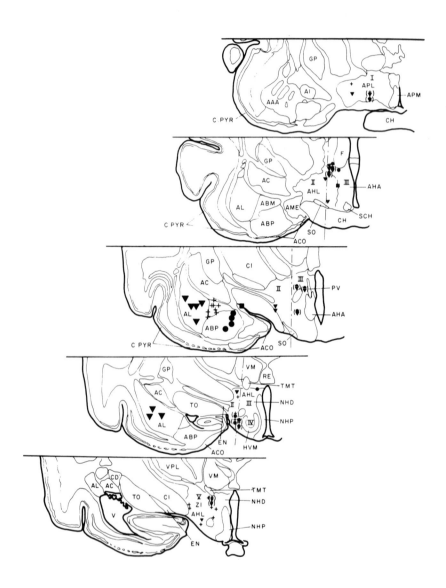

Fig. 1. Location of units whose action potentials were
driven by 1 Hz stimulation in various nuclei or subnuclei of
the amygdala. The symbols in the hypothalamus and preoptic
region show the locations of units whose action potentials were
driven by stimulation through electrodes indicated by larger,
corresponding symbols in the amygdala. When a unit was driven
by two amygdaloid stimulating electrodes, both symbols are indicated, contained within parentheses, with the symbol on top
corresponding to the electrode producing driving at the shortest
latency. Triangles (▼) refer to stimulation in the lateral
nucleus (AL). Crosses (✛) refer to stimulation in the magnocellular portion of the basal nucleus (ABM). Dots (●) refer
to stimulation in the parvocellular portion of the basal nucleus
(ABP). Squares (■) refer to stimulation in the medial nucleus
(AME). Roman numerals I - V indicate the five hypothalamic zones
discussed in the text. The center lines indicate the division
between zones II and III.

The drawing of frontal sections of the cat's brain are
based on the atlas in Bures̆ et al. (1962). Additional anatomical abbreviations are as follows: AAA = area amygdalaris anterior; AC = nucleus centralis amygdalae; ACO = nucleus corticalis
amygdalae; AHA = area hypothalamica anterior; AHL = area hypothalamica lateralis; AI = nucleus intercalatus amygdalae; APL =
area praeoptica lateralis; APM = area praeoptica medialis; CD
= nucleus caudatus; CH = chiasma opticum; CI = capsula interna;
C PYR = cortex pyriformis; EN = nucleus entopeduncularis; F =
fornix; GP = globus pallidus; HVM = nucleus ventro-medialis
hypothalami; NHD = nucleus hypothalamicus dorsalis; NHP =
nucleus hypothalamicus posterior; PV = nucleus paraventricularis;
RE = nucleus reuniens; SCH = nucleus suprachiasmaticus; SO =
nucleus supraopticus; TMT = tractus mammillo-thalamicus; TO =
tractus opticus; V = ventriculus lateralis; VM = nucleus ventralis medialis; VPL = nucleus ventralis postero-lateralis; ZI =
zona incerta.

anatomical regions and subdivisions: the rostral preoptic
(RPO, I), the anterior lateral hypothalamus ("AHL", II), the
medial and dorsal hypothalamus exclusive of the HVM ("AHD", III),
the HVM (IV) and the hypothalamus caudal to the HVM ("PH", V).
In zone (I), the rostral preoptic (RPO), Gloor (1955a) found
short latency responses ($\leq$ 9 msec) mediated via both VAF and ST.
In zone (II), "AHL", there were short latency responses mediated
via VAF, but few or none via ST. In zone (III), "AHD", there
were some short latency responses via ST, but few or none via
VAF. In the HVM (IV), there were short latency responses via
ST, but the latencies of responses via VAF were slightly longer,
10-14 msec. In zone (V), "PH", there were a few short latency
responses via ST, while those via VAF were at least 15 msec or,
more typically, longer than 25 msec. Thus, these data suggest
the possibility that the ST effects into the hypothalamus go in
large part directly to the "AHD" and HVM, from which they are
perhaps relayed to other structures, including other parts of
hypothalamus and preoptic region, whereas the VAF projects, in
part, directly to RPO and rostral AHL, from which further projections may occur more or less along the course of the medial
forebrain bundle and medial to it.

Studying effects of 1 Hz and 50 Hz stimulation on the
response patterns evoked in various parts of the amygdaloid projection field, Gloor (1955b) found that most areas receiving
evoked potentials at 1 Hz showed potentiation following 10 or
50 Hz stimulation in the hypothalamus. This was especially the
case with HVM, whereas the lateral preoptic region was unusual
in showing little or no potentiation. Gloor concluded that
monosynaptic projections of the amygdala tend not to potentiate,
whereas relayed impulses do, contrary to what is observed (albeit
with stimulation of higher frequency) in the spinal cord. Could
it be rather that, at least as far as the hypothalamus is concerned, the direct ST connections tend to produce greater potentiation than the direct VAF connections do?

B. Single units.

This brings us to the unit era in amygdala-hypothalamus
studies. One of the earliest studies was by Sawa and co-workers
(1959), who found that high frequency stimulation of the basolateral amygdala in cats affected firing patterns of HVM, usually
inhibiting spontaneous firing.

Wendt (1961) in his doctoral dissertation found that units
in many parts of the hypothalamus responded to stimulation in the
amygdala, but that responsive neurons tended to be concentrated
in HVM (IV) and the anterior hypothalamic area (mostly in "AHD"

(III) rather than in "AHL" (II)). Wendt found that for low frequency stimulation, the predominant response patterns of hypothalamic units was driving, i.e., a response of the unit following each electrical stimulation, or at least acceleration of the
spontaneous response rate.   Wendt also noted that some of the
hypothalamic units affected by amygdaloid stimulation were also
affected by stimulation of the sciatic nerve. He found that cutting the ST did not affect the evoked potentials in HVM following stimulation of basolateral amygdala, though interrupting the
VAF did.

    In summary, looking at the hypothalamus on the basis of the
five zones defined above, most of Wendt's driven units were in
"AHD" (III) or HVM (IV), with relatively more inhibited units in
"PH" (V).

    Wendt found that the shortest latencies for driving hypo-
thalamic units by amygdaloid stimulation were 10-12 msec. These
also were the modal latencies.

    Tsubokawa and Sutin (1963) studied units in the HVM in cats.
Of the 272 units they studied, 31% were fired by amygdaloid stimulation; a third of these also were fired by septal stimulation.
Latency for driving ranged from 4-33 msec, with a mode at 10-12
msec, as in Wendt's study. Tsubokawa and Sutin noted that their
latency distribution was similar to the form of evoked potentials
in the HVM following amygdaloid stimulation. They also noted
that high frequency stimulation of the dorsomedial mesencephalic
tegmentum decreased the amygdaloid-HVM evoked response, whereas
stimulation of the lateral mesencephalic reticular formation increased the amygdaloid-HVM evoked response. High frequency medial
mesencephalic reticular formation stimulation inhibited most of
the units driven by amygdaloid stimulation.

    Stuart et al. (1964) compared the effects on hypothalamic
units of sciatic stimulation or bladder distention to that of
stimulation in various forebrain regions, including the amygdala.
They found, generally, that stimulation within the amygdala had
a similar effect on hypothalamic firing patterns whether added
to peripheral somatic stimulation or to visceral stimulation.

    This brings us to work published, or in progress, during the
last five years. I have concentrated on the following studies in
cats: Egger, 1967, and unpublished; Dreifuss, Murphy, and Gloor,
1968; Murphy, Dreifuss, and Gloor, 1968a, 1968b; Dreifuss and
Murphy, 1968; Gloor, Murphy, and Dreifuss, 1969; Murphy and
Renaud, 1969; Van Atta and Sutin, 1971, and unpublished. These
data are in general agreement with, but add much detail to the
earlier work. Where the results are comparable, the agreement

among these various studies is substantial.

C. The HVM (Zone IV).

Gloor et al. (1969) emphasized that the HVM is a preferred reception site for evoked potentials in the hypothalamus. In most parts of the hypothalamus, evoked potentials following stimulation in the amygdala were monophasic, but in the HVM some of the evoked potentials were biphasic, and very large. Furthermore, it appeared that the form of the evoked potential in HVM depended on the conduction pathway into the hypothalamus, in agreement with Wendt (1961).

Evoked potentials in HVM followed stimulation frequencies up to about 8-10 Hz. However, evoked potentials following stimulation in the lateral nucleus of the amygdala (AL) did not follow at 8-10 Hz; rather, they showed augmentation, then fatigue. Potentiation of evoked potentials was marked following AL stimulation, much less so following stimulation in the basal nucleus of the amygdala (AB). It is possible that the VAF input to HVM may be direct from AB, but relayed from AL (perhaps through AB), leading to different electrophysiological properties of the response in the HVM. As mentioned above, Gloor (1955b) found that the direct VAF response showed little potentiation in RPO and HL. Perhaps AL potentiation actually occurred in relay from AL to AB, i.e., within the amygdala itself.

Gloor et al. (1969) pointed out that "It seems legitimate to assume that the rate of impulse transmission from the amygdala to the hypothalamus may be an important factor in determining whether the net effect of a stream of amygdaloid impulses arriving at the hypothalamus will be of an excitatory or an inhibitory nature," emphasizing a cautionary note well to remember in trying to extrapolate from electrophysiological to behavioral studies.

Dreifuss et al. (1968) analyzed the evoked potential and unit responses in HVM in great detail. Stimulation in the cortico-medial nuclei, mediated over ST, elicited an evoked potential (typically positive, monophasic) in the HVM with a latency of 11 msec and a maximum positive wave at 30-35 msec. Stimulation in AB, mediated via the VAF, elicited a biphasic wave in the HVM, with a latency of 7 msec, a maximum negative response at 12-14 msec, and a maximum positive response at about 25 msec. Stimulation in AL elicited a response similar to that from AB, but smaller, or, sometimes, just a negative wave. The ST and VAF responses were mutually inhibitory.

An analysis of unit responses in HVM led Dreifuss et al. (1968) to the observation that the negative wave in the evoked potential reflected increased probability of unit firing, and the positive wave reflected decreased probability of unit firing in the HVM. This correlation also was true of spontaneous fluctuations of potential in the HVM. Latencies of VAF unit driving in HVM were 10-15 msec, typically followed by inhibition for 150-200 msec. Some driving occurred from ST at a latency of 45-90 msec, but the principal effect was inhibition. The shortest driving latency was 24 msec.

The evoked potential latencies and some of the known features of the anatomy were used to calculate approximate conduction velocities of the two pathways. Dreifuss et al. (1968) estimated 1-1.5 m/sec in VAF, assuming one synapse. This is in essential agreement with Gloor (1955a). The ST conduction velocity, assuming monosynaptic connection to HVM, was 0.6 - 1.0 m/sec. This is slower than the 4.2 m/sec estimated earlier by Gloor (1955a). Fernandez de Molina and Garcia-Sanchez (1967) more recently measured the conduction velocities in ST, and found two conduction bands, one about 1.8-2.3 m/sec and the other at 0.6-0.75 m/sec, which suggests that it may be the slower fibers in the ST that are conducting impulses to the HVM.

Murphy et al. (1968a) extended their study of amygdaloid influence on HVM to the effects of repetitive stimulation, especially with reference to differences in the effects of AL vs AB stimulation (chiefly the magnocellular portion of AB, ABm). They also looked at the DC potentials in the HVM, and how they were affected by repetitive amygdaloid stimulation. Murphy et al. (1968a) found a negative shift in HVM with ABm stimulation, maximum at about 32 Hz. A similar, but less marked shift occurred with AL stimulation.

The magnitude of the evoked potentials from both ABm and AL began to drop off at 8-10 Hz, and decreased about 50% at 20 Hz. Potentiation occurred following 10 Hz stimulation. As noted earlier, the potentiation was greater following stimulation in AL than AB.

Using a double shock technique with ABm stimulation, Murphy et al. (1968a) found facilitation at an intershock interval of about 15-30 msec and maximum inhibition at an interval of about 40 msec. This might imply that facilitory build-up would occur at about 35-60 Hz, with inhibition predominating at about 25 Hz, which might have implications for how different stimulation frequencies in the amygdala might produce different effects in behavior studies.

Murphy et al. (1968a) studied units in "AHD" (III) as well as in HVM (IV). Some units driven at 1 Hz with a latency of

12-20 msec were inhibited for up to 200 msec following stimulation.
These units showed an increase in firing rate at 5 Hz, decreased
firing at 10 Hz, and shut off completely at 20 Hz. Apparently, the
inhibition summated. Some units were only inhibited (for 150-200
msec) at 1 Hz; these stopped firing completely at 10 Hz. Some
units were driven at 1 Hz, but showed no after-inhibition: these
continued to increase in firing rates as the stimulation frequency
increased.

In summary, ABm seemed to have more powerful input than AL to
"AHD" and HVM, as judged by the amplitudes of evoked potential and
DC shifts, but AL showed more potentiation following high frequency
stimulation, and more frequent after-discharges.

Finally, to focus again on HVM, Murphy and Renaud (1969)
demonstrated the existence of two cell types in the HVM, strongly
implicating the smaller cells in an inhibitory role with respect
to the large HVM neurons. The small cells presumably mediate both
the inhibition via ST and the inhibition following activation via
VAF.

In summary, the HVM is an extremely important recipient of
signals from the amygdala, via both ST and VAF. The ST and the VAF
appear to have different effects on large HVM neurons, the ST input
being primarily inhibitory, the VAF producing activation, or activation-inhibition. Furthermore, within VAF, there is some indication that effects may be different, depending on whether activation
is originally in ABm or in AL. Interaction studies indicate that
some individual neurons receive inputs via both VAF and ST.

D. The remainder of the hypothalamus (Zones I, II, III, V).

As Gloor (1955a) showed, much of the hypothalamus, especially
the more anterior regions, appear to receive powerful inputs from
the amygdala, at least as judged by evoked potentials. Wendt
(1961) showed that many hypothalamic units, in addition to those
in HVM, respond to amygdaloid stimulation.

In my own investigations of hypothalamic unit responses to
amygdaloid stimulation (Egger, 1967, and unpublished), several
suggestive findings turned up.

As an aside, the spontaneous firing frequencies of hypothalamic neurons appear to be different in different parts of the
hypothalamus, with medial hypothalamic neurons firing on the
average more slowly than lateral hypothalamic neurons (Egger,
1967; Murphy, Dreifuss, and Gloor, 1968b; Van Atta and Sutin,
1971), at least in unanesthetized cats with bilateral lesions in
the brain stem reticular formation. This is in substantial agree-

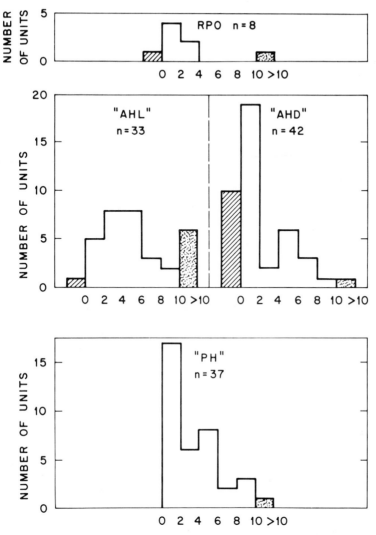

Fig. 2. Histograms for spontaneous firing frequencies of units in four of the five hypothalamic zones depicted in Figure 1. The abscissae indicate the number of firings per sec. The ordinates indicate the number of units at each average rate. The striped bars to the left of zero indicate the number of units studied that were not spontaneously active, detected only during amygdaloid stimulation. The stippled bars at right indicate the number of units firing at greater than 10/sec. For each of the four histograms, n = the total number of units, each studied for at least ten minutes. RPO corresponds to Zone I; "AHL" to Zone II; "AHD" to Zone III; and "PH" to Zone V.

ment with Oomura et al. (1967), for deeply anesthetized, and
Oomura et al. (1969), for sleeping cats. Furthermore, Murphy et
al. (1968b) noted that spontaneous rates in HVM (IV) and "AHD"
(III) were the same.

In analyzing in more detail the anatomical distribution of
108 spontaneously active units within the various hypothalamic
regions, I found that in most of the hypothalamus (Zones I, III,
IV, V), the modal spontaneous frequency was about 2/sec. Only in
"AHL" (II) were the spontaneous frequencies significantly higher
($P < 0.001$), with the mode at 5/sec (Fig. 2).

The various hypothalamic zones also seemed to receive slightly
different mixes of inputs from the various amygdaloid nuclei
(Fig. 3). The following analysis is based on data from 120 units.
For a description of methods, see Egger (1967).

For instance, in "PH" (V), a long latency region for both ST
and VAF evoked potentials (Gloor, 1955a), a high percentage of
units were driven by stimulation in AL and in the magnocellular
portion of AB (ABm) (18% and 27% respectively) versus no driving
by stimulation in the parvocellular portion of AB (ABp) or in the
medial nucleus (AME). The number (n) of units studied was 21
during stimulation in ABp; n = 11 during stimulation in AME. In
"AHD" (III) the driving was distributed much more evenly, though,
oppositely to "PH" (V), it tended to be more frequent from the
more medial nuclei, with 23% from AL, 33% from ABm, and 43% from
ABp. (The 50% from AME represents observations on only two units,
so it is unreliable.) In "AHL" (II), driving by stimulation in
AL, ABm and ABp occurred in 20-26% of the units studied, but was
absent following stimulation in AME (n = 9). In RPO (I) there
were, unfortunately, few units, but those few were markedly
driven by stimulation in AL and ABm. No observations were made
in RPO of effects of stimulation in ABp. Unfortunately, no
recordings were made within HVM (IV).

The most remarkable finding with respect to the basal nuclei
is the fact that, although ABp was a very potent activator of
units in "AHL" (II) (25%) and "AHD" (III) (43%), it did not drive
any (n = 21) units in "PH" (V) ($P < 0.001$). In contrast, ABm
drove 26% of the units in "AHL" (II); 33% in "AHD" (III); and 27%
in "PH" (V). Stimulation in AL seemed to have effects similar to
stimulation in ABm, with ABm being more potent, as noted by
Murphy et al. (1968a).

An analysis of latencies of driving bears out these generalizations (Fig. 4). Most of the short latency responses in "AHD"
(III) were elicited by stimulation in ABp, whereas ABm stimulation
was followed most often by short latency driving of units in "AHL"
(II) and "PH" (V) ($P < .05$). Although some suppression at 1 Hz

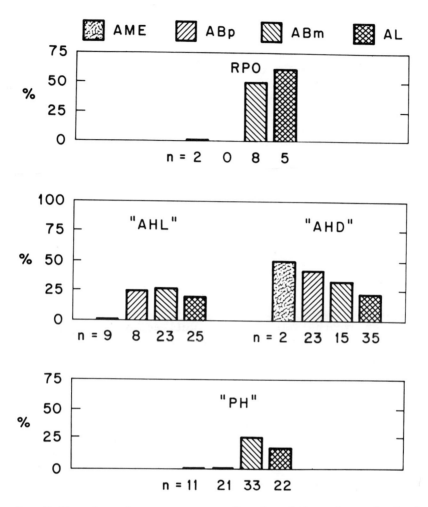

Fig. 3. Indicating the percentage of units driven by stimulation
in various portions of the amygdala. The pattern of shading in
each bar indicates the location of the stimulating electrodes
eliciting driving. Under each bar, n = total number of units
observed during stimulation of an electrode in the indicated amygdaloid regions. The ordinates indicate the percentages of these
units driven by 1 Hz stimulation in the indicated regions. Data
from some units contributed to more than one bar. AME refers to
stimulation in the medial nucleus; ABp refers to stimulation in
the parvocellular portion of the basal nucleus; ABm refers to
stimulation in the magnocellular portion of the basal nucleus;
and AL refers to stimulation in the lateral nucleus. RPO
corresponds to Zone I; "AHL" to Zone II; "AHD" to Zone III; and
"PH" to Zone V.

was seen in each of the zones, the only marked effect was in "AHL" (II) following stimulation in AME (22%) (n = 9). Effects follow-ing 60 Hz stimulation were also somewhat scattered. For instance, in "AHD" (III), statistically significant increases in firing following 5 sec of stimulation occurred most frequently following stimulation in ABp (22%), whereas no increases occurred following stimulation in ABm, and only in 6% of units following stimulation in AL. Interestingly, 22% of units in "PH" (V) (n = 18) increased in firing after 60 Hz stimulation in the central amygdaloid nucleus (AC). Inhibition after 60 Hz stimulation in AB and AL was most marked in "PH" (V). A few cases of convergent, but opposite, effects on single units from different portions of the amygdala occurred, both at 1 Hz and at 60 Hz, but these were rare occur-rences, less than 10%. Much more common was a similarity of effects (on a single unit), from all the stimulation electrodes in the amygdala, though often with one placement being more potent than the others. The duration of some of the statistical signi-ficant changes in spontaneous firing rates following the end of 5 sec of 60 Hz stimulation was 5-10 sec, or even longer, at stimula-tion intensities below threshold for after-discharges (Egger, 1967).

Dreifuss and Murphy (1968) looked at the effects of amygdaloid stimulation on hypothalamic units, as well as convergent effects of stimulation in the septum, hippocampus and the midbrain tegmentum. 64.6% of units in their sample were affected by amygdaloid stimula-tion. Septal stimulation affected 56.4%, with a high degree of convergent effects between septum and basolateral amygdala. The amygdaloid stimulations more often produced activation than did the septal stimulation.

Dreifuss and Murphy (1968) specifically looked at effects of stimulation in ABp versus stimulation in AL. In 42/44 neurons affected by stimulation in these two regions (presumably at or near 1 Hz), effects were in the same direction. That is, opposite effects were seen only in 2/44, or 4.5%. ABp was generally a more effective site of stimulation than AL. Of the 49 units affected by stimulation in the amygdala and septum, 83.6% showed the same direction of effects, and 16.4% showed opposite effects. While amygdaloid and septal stimulation each affected about 60% of units, hippocampus and midbrain stimulation only affected about 20% each.

Murphy et al. (1968b) examined the effects of 1 Hz stimulation in AB (ABp and ABm were not differentiated), AL, septum and mid-brain tegmentum on 454 hypothalamic units. They divided the hypo-thalamus into three regions, corresponding roughly to zones II ("AHL"), III ("AHD"), and IV (HVM). Murphy et al. (1968b) found "AHD" (III) neurons less responsive (63%) to limbic (not just amygdaloid) stimulation than those of "AHL" (II) (77%) or HVM (IV) (84%) (P <0.005). They found stimulation in AL activated fewer units at 1 Hz than did stimulation in AB (P <0.05).

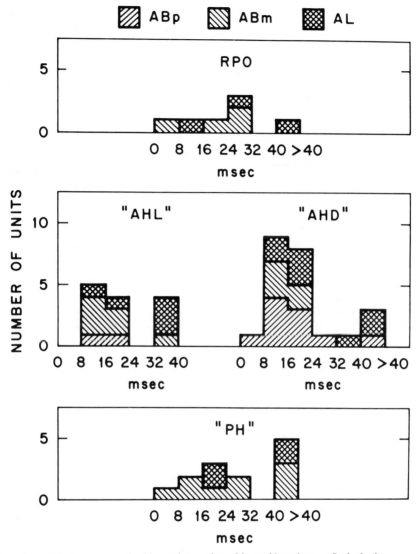

Fig. 4. Histograms indicating the distribution of driving latencies following amygdaloid stimulation at 1 Hz. The pattern of shading indicates the location of the stimulating electrodes eliciting driving at the latencies indicated by the abscissae. The ordinates indicate the number of units at each latency. ABp refers to the stimulation in the parvocellular portion of the basal nucleus; ABm refers to stimulation in the magnocellular portion of the basal nucleus; AL refers to stimulation in the lateral nucleus. RPO corresponds to Zone I; "AHL" to Zone II; "AHD" to Zone III; and "PH" to Zone V.

From AB and AL, activation, or activation-inhibition was much more common in all three of their zones than pure inhibition (about 4/1 for the population of influenced neurons). In my data, the ratio of activation to inhibition was even higher (8/1). However, there is a suggestion in my data that inhibition/activation ratios from a given stimulation site in the amygdala are not distributed randomly across various regions of the hypothalamus, e. g., there was no inhibition in "AHD" (III) following stimulation in ABp, although 10/23 units were driven. An extremely interesting finding by Murphy et al. (1968b) was that HVM neurons responded, as a group, at shorter latencies than neurons in HL or AHD. HVM was also the chief recipient of short latency driving from the septum. A latency analysis suggested that some of the impulses in HL were relayed through HVM.

Van Atta and Sutin (1971) looked at responses of HL units to stimulation both in limbic structures and in HVM. Driving at 1 Hz was seen in 14.7% of the units during stimulation in ABm, 10.7% for AL, and 22.7% for the ST. These percentages are somewhat lower for ABm and AL than those I found. On the other hand, Van Atta and Sutin (1971) observed suppression in 13.0% of hypothalamic units following stimulation in ABm, in 17.0% following AL, and in 26.5% following stimulation activating chiefly ST. There was much convergence between the amygdaloid and septal inputs, with 93% of these convergent responses in the same direction.

A very large portion of the HL units sampled by Van Atta and Sutin (1971) were also driven (51.6%) or suppressed (31.4%) by HVM stimulation, which data add further strength to the suggestion that HVM is an important relay station from amygdala on the way to HL, at least for many HL units.

High frequency (50 Hz) effects (Van Atta and Sutin, unpublished) were observed more frequently with ST or AL stimulation than with stimulation in ABp. Sometimes ST or AL stimulation produced long periods of inhibition following stimulation, without signs of afterdischarges. Stimulation in HVM also produced arrest of firing in some units during and for many seconds following stimulation. Effects of high frequency stimulation in the amygdala lasted in some cases for several seconds, e.g., 6-7 sec., even when no changes in firing rates occurred during stimulation.

Oomura et al. (1970) have begun to record intracellularly in hypothalamic neurons in rats during stimulation in the amygdala. Further studies in this direction should provide clearer evidence about the nature of amygdaloid influences on hypothalamic units.

E. Amygdalopetal influences.

Briefly, what about the connections from the hypothalamus to

the amygdala? Caruthers et al. (1964) investigated in cats
whether interactions of neocortical (posterior ectosylvian gyrus)
and hypothalamic (anterior hypothalamus) evoked potentials
occurred in the amygdala. They did.

Happel and Bach (1970) briefly reported that in the amygdala
stimulation of ST inhibits, and stimulation of VAF facilitates,
activity in AB. AB recordings showed summation of VAF and ST
effects, though the cells of origin of VAF appeared to receive
fibers from VAF, but not from ST.

What sorts of impulses, in addition to hypothalamic, get
into the amygdala? Complex sensory stimuli affect units, as well
as simple light flashes, tones, and tactile stimulation (Machne
and Segundo, 1956; Sawa and Delgado, 1963; O'Keefe and Bouma,
1969). Nasal air puffs affect electrical activity in the
amygdala (e.g., McLennan and Graystone, 1965). Brain regions
known to affect amygdaloid activity in cats, in addition to the
hypothalamus, include neocortex, especially temporal (ectosylvian)
(Niemer and Goodfellow, 1966), thalamus, especially the anterior
reticular, anterior and ventral nuclei (Niemer et al., 1970), and
the magnocellular portion of medial geniculate (Wepsic and Sutin,
1964).

F. Indirect pathways to the hypothalamus.

As mentioned above, there are indirect pathways into the
hypothalamus from the amygdala. These, in fact, may be of crucial
importance in mediating behavioral or neuroendocrine influences
of the amygdala on the hypothalamus, but we know nothing about
the electrophysiology of these pathways.

III. DISCUSSION

On the basis of what we know now, it appears that different
portions of the amygdala, that is, different amygdaloid nuclei,
or subnuclei, have different patterns of anatomical projection
to the hypothalamus, and that the hypothalamus itself has various
patterns of anatomical organization within itself (e.g.,
Szentágothai et al., 1968). On the basis of phylogenetic considerations, it has been traditional to divide the amygdala into
a phylogenetically older corticomedial division (primarily associated with ST), rather intimately associated with the olfactory
system, and a phylogenetically more recent basolateral division
(primarily associated with the VAF). On the basis of behavioral
and other studies, Koikegami (1963) proposed an alternative
division, with the ABp included with the corticomedial division
of the amygdala. Egger and Flynn (1967), on the basis of our
own behavioral data, plus a review of the literature on the
effects of stimulation and ablation in the amygdala in cats,

provided support for Koikegami's suggestion.

I think we may now say that the electrophysiological data provide some evidence that stimulation in ABp produces effects different from stimulation in AL and ABm, and such stimulation affects differentially various parts of the hypothalamus and preoptic region. But stimulation in ABp is not equivalent to corticomedial (or ST) stimulation.

Hall (1972, this symposium) reviewed anatomical and histochemical data consistent with the idea that the ABp cannot be classified as a part either of the corticomedial or of the basolateral nuclei, but in some sense might be considered a transition zone between these two regions.

Also, different frequencies of stimulation in the amygdala would be expected to produce different patterns of effects, behaviorally and physiologically, on the hypothalamus. At 1 Hz, ABm and AL stimulation generally tend to produce similar patterns of effects, though ABm appears to have more potent effects on hypothalamic units than does AL. At higher frequencies, the AL effects are more indirect and variable, in agreement with the behavioral observations of Egger and Flynn (1967).

It appears, on the basis of physiological studies, that there is at least a rudimentary somatotopic mapping of the amygdala into the hypothalamus (Fig. 1). The medial amygdala may relate more directly to medial and ventral hypothalamus (Zones III and IV), via ST, with ABm and AL relating more directly to RPO (I), and "AHL" (II). It is "AHL" (II) from which many behavioral and somatomotor acts are most easily elicited by electrical stimulation (Sutin, 1966). The posterior hypothalamus (V), to which ABp input seems to be relatively slight, receives much of the polysensory input into the hypothalamus (Dafny and Feldman, 1970). The more medial hypothalamic regions, (III) and (IV), may be more directly concerned with neuroendocrine effects (e.g., Szentágothai, 1964; Szentágothai et al., 1968).

Finally, the HVM (IV) appears to be an important node in the input to the hypothalamus from the amygdala. HVM lesions may reverse or abolish the behavioral effects of amygdaloid lesions (Kling and Hutt, 1958; Sclafani et al., 1970) or amygdaloid stimulation (White and Fisher, 1969).

IV. SPECULATIONS

The amygdala has been implicated in the control of endocrine function, and there is some evidence that at least a portion of the HVM neurons have a neurosecretory function (Kaelber and

Leeson, 1967). The long time courses of some of the behavioral
effects of lesions in the amygdala (Bard and Mountcastle, 1948)
and HVM (Wheatley, 1944) suggest that perhaps some of the
behavioral effects of lesions and stimulation here may be
secondary to neuroendocrine changes (see, for instance, Elwers
and Critchlow, 1961; Elwers Bar-Sela and Critchlow, 1966).

Perhaps the VAF is predominantly excitatory, at least in so
far as it is concerned with behavior, and the ST, predominantly
inhibitory, at least in so far as it is concerned with neuro-
endocrinological timing. Perhaps projections from ABp represent
a transition or link between these two systems.

Both the hypothalamus and the amygdala exhibit hints of a
rough functional organization, medial to lateral, with the most
medial and ventral portions of both amygdala and hypothalamus
perhaps related primarily to neuroendocrine systems, the intermediate regions more related to "internal" behavior, e.g., rage,
alarm (Folkow and Rubinstein, 1966) and sympathetic arousal.
Finally, the more lateral regions of both amygdala and hypothalamus may be more concerned with behavioral alerting, outward looking, food seeking, reward, and prey catching (e.g., Sutin, 1966;
Egger and Flynn, 1967; Stein, 1968; and Flynn et al., 1970).

Within the amygdala, this sort of rough functional localization is consistent with many behavioral studies (e.g., Kaada,
1972, this symposium).

V.   CONCLUSIONS

1. The amygdala and the hypothalamus are connected
reciprocally.

2. The HVM is a very important focus of amygdaloid input.

3. There may exist at least the rudiments of an organized
topographical projection of the amygdala onto the hypothalamus.

4. The amygdala probably acts as a biaser, rather than a
controller, influencing hypothalamic neurons along with many
other limbic and nonlimbic structures. The amygdala, perhaps
acting as an intermediate gray region between cortical regions
and the hypothalamus, modulates and times some important functions
over which the hypothalamus presumably exerts a controlling
integration.

ACKNOWLEDGMENTS

I would like to thank Miss E. Clark and J. Bishop for technical assistance during the preparation of this paper, and Mrs. V. Simon for the illustrations. Work reported in this paper was aided in part by NIH grant NS-06297, and Research Scientist Development grant 5-K02-MH-11,952 from NIMH.

REFERENCES

AREES, E. A., & MAYER, J.   Anatomical connections between medial and lateral regions of the hypothalamus concerned with food intake. Science, 1967, 157, 1574.

BARD, P., & MOUNTCASTLE, V. B.   Some forebrain mechanisms involved in expression of rage with special reference to suppression of angry behavior. Research Publications of the Association for Research in Nervous and Mental Disease, 1948, 27, 362.

BUREŠ, J., PETRÁŇ, M., & ZACHAR, J.   Electrophysiological Methods in Biological Research. New York: Academic Press, 1962.

CARUTHERS, R., MÜLLER, A. K., MULLER, H. F., & GLOOR, P.   Interaction of evoked potentials of neocortical and hypothalamic origin in the amygdala. Science, 1964, 144, 422.

COWAN, W. M., RAISMAN, G., & POWELL, T. P. S.   The connexions of the amygdala. Journal of Neurology, Neurosurgery, and Psychiatry, 1965, 28, 137.

DAFNY, N., & FELDMAN, S.   Unit responses and convergence of sensory stimuli in the hypothalamus. Brain Research, 1970, 17, 243.

DREIFUSS, J. J., & MURPHY, J. T.   Convergence of impulses upon single hypothalamic neurons. Brain Research, 1968, 8, 167.

DREIFUSS, J. J., MURPHY, J. T., & GLOOR, P.   Contrasting effects of two identified amygdaloid efferent pathways on single hypothalamic neurons. Journal of Neurophysiology, 1968, 31, 237.

EAGER, R. P., CHI, C. C., & WOLF, G.   Lateral hypothalamic projections to the hypothalamic ventromedial nucleus in the albino rat: demonstration by means of a simplified ammoniacal silver degeneration method. Brain Research, 1971, 29, 128.

EGGER, M. D. Responses of hypothalamic neurons to electrical
    stimulation in the amygdala and the hypothalamus. Electro-
    encephalography and Clinical Neurophysiology, 1967, 23, 6.

EGGER, M. D., & FLYNN, J. P. Further studies on the effects of
    amygdaloid stimulation and ablation on hypothalamically
    elicited attack behavior in cats. In W. R. Adey and
    T. Tokizane (Eds.) Progress in Brain Research, Vol. 27,
    Structure and Function of the Limbic System. Amsterdam:
    Elsevier Publishing Co., 1967. Pp. 165-182.

ELWERS, M., & CRITCHLOW, V. Precocious ovarian stimulation
    following interruption of stria terminalis. American Journal
    of Physiology, 1961, 201, 281.

ELWERS BAR-SELA, M., & CRITCHLOW, V. Delayed puberty following
    electrical stimulation of amygdala in female rats. American
    Journal of Physiology, 1966, 211, 1103.

FERNANDEZ DE MOLINA, A., & GARCIA-SANCHEZ, J. L. The properties of
    the stria terminalis fibres. Physiology and Behavior,
    1967, 2, 225.

FLYNN, J. P., VANEGAS, H., FOOTE, W., & EDWARDS, S. Neural
    mechanisms involved in a cat's attack on a rat. In R. Whalen
    (Ed.) The Neural Control of Behavior. New York: Academic
    Press Inc., 1970. Pp. 135-173.

FOLKOW, B., & RUBINSTEIN, E. The functional role of some autonomic
    and behavioral patterns evoked from the lateral hypothalamus
    of the cat. Acta Physiologica Scandinavica, 1966, 66, 182.

GLOOR, P. Electrophysiological studies on the connections of the
    amygdaloid nucleus in the cat. I. The neuronal organization
    of the amygdaloid projection system. Electroencephalography
    and Clinical Neurophysiology, 1955a, 7, 223.

GLOOR, P. Electrophysiological studies on the connections of the
    amygdaloid nucleus in the cat. II. The electrophysiological
    properties of the amygdaloid projection system. Electro-
    encephalography and Clinical Neurophysiology, 1955b, 7, 243.

GLOOR, P., MURPHY, J. T., & DREIFUSS, J. J. Electrophysiological
    studies of amygdalo-hypothalamic connections. Annals of the
    New York Academy of Science, 1969, 157, 629.

HALL, E. Some aspects of the structural organization of the
    amygdala. In B. E. Eleftheriou (Ed.) The Neurobiology of the
    Amygdala. New York: Plenum Press, 1972, in press.

HAPPEL, L. T., & BACH, L. M. N.   Amygdalopetal fiber influences
    upon excitability of amygdaloid nuclei.   Federation
    Proceedings, 1970, 29, No. 829 (abstract).

HEIMER, L., & NAUTA, W. J. H.   The hypothalamic distribution of
    the stria terminalis in the rat.   Brain Research, 1969,
    13, 284.

KAADA, B.   Electrical and chemical stimulation of the amygdala
    with reference to topical and functional representations.
    In B. E. Eleftheriou (Ed.) The Neurobiology of the Amygdala.
    New York: Plenum Press, 1972, in press.

KAELBER, W. W., & LEESON, C. R.   A degeneration and electron
    microscopic study of the nucleus hypothalamicus ventro-
    medialis of the cat.   Journal of Anatomy, 1967, 101, 209.

KLING, A., & HUTT, P. J.   Effect of hypothalamic lesions on the
    amygdala syndrome in the cat.   A.M.A. Archives of Neurology
    and Psychiatry, 1958, 79, 511.

KOIKEGAMI, H.   Amygdala and other related limbic structures;
    experimental studies on the anatomy and function.   I. Anato-
    mical researches with some neurophysiological observations.
    Acta Medica et Biologica, 1963, 10, 161.

MACHNE, X., & SEGUNDO, J. P.   Unitary responses to afferent volleys
    in amygdaloid complex.   Journal of Neurophysiology, 1956,
    19, 232.

McLENNAN, H., & GRAYSTONE, P.   The electrical activity of the
    amygdala, and its relationship to that of the olfactory bulb.
    Canadian Journal of Physiology and Pharmacology, 1965,
    43, 1009.

MURPHY, J. T., DREIFUSS, J. J., & GLOOR, P.   Responses of hypo-
    thalamic neurons to repetitive amygdaloid stimulation.   Brain
    Research, 1968a, 8, 153.

MURPHY, J. T., DREIFUSS, J. J., & GLOOR, P.   Topographical differ-
    ences in the responses of single hypothalamic neurons to
    limbic stimulation.   American Journal of Physiology, 1968b,
    214, 1443.

MURPHY, J. T., & RENAUD, L. P.   Mechanisms of inhibition in the
    ventromedial nucleus of the hypothalamus.   Journal of Neuro-
    physiology, 1969, 32, 85.

NAUTA, W. J. H.   Hippocampal projections and related neural
    pathways to the mid-brain in the cat.   Brain, 1958, 81, 319.

NAUTA, W. J. H. Neural associations of the amygdaloid complex in
the monkey. Brain, 1962, 85, 505.

NIEMER, W. T., & GOODFELLOW, E. F. Neocortical influence on the
amygdala. Electroencephalography and Clinical Neurophysiology, 1966, 21, 429.

NIEMER, W. T., GOODFELLOW, E. F., BERTUCCINI, T. V., & SCHNEIDER,
G. T. Thalamo-amygdalar relationships. An evoked potential
study. Brain Research, 1970, 24, 191.

O'KEEFE, J., & BOUMA, H. Complex sensory properties of certain
amygdala units in the freely moving cat. Experimental
Neurology, 1969, 23, 384.

OOMURA, Y., OOYAMA, H., YAMAMOTO, T., NAKA, F., KOBAYASHI, N., &
ONO, T. Neuronal mechanism of feeding. In W. R. Adey and
T. Tokizane (Eds.) Progress in Brain Research, Vol. 27,
Structure and Function of the Limbic System. Amsterdam:
Elsevier Publishing Co., 1967. Pp. 1-33.

OOMURA, Y., OOYAMA, H., NAKA, F., YAMAMOTO, T., ONO, T., &
KOBAYASHI, N. Some stochastical patterns of single unit
discharges in the cat hypothalamus under chronic conditions.
Annals of the New York Academy of Science, 1969, 157, 666.

OOMURA, Y., ONO, T., & OOYAMA, H. Inhibitory action of the amygdala
on the lateral hypothalamic area in rats. Nature, 1970, 228,
1108.

SAWA, M., MARUYAMA, N., HANAI, T., & KAJI, S. Regulatory influence
of amygdaloid nuclei upon the unitary activity in ventromedial
nucleus of hypothalamus. Folia Psychiatrica et Neurologica
Japonica (Niigata), 1959, 13, 235.

SAWA, M., & DELGADO, J. M. R. Amygdala unitary activity in the unrestrained cat. Electroencephalography and Clinical Neurophysiology, 1963, 15, 637.

SCLAFANI, A., BELLUZZI, J. D., & GROSSMAN, S. P. Effects of lesions
in the hypothalamus and amygdala on feeding behavior in the
rat. Journal of Comparative and Physiological Psychology,
1970, 72, 394.

STEIN, L. Chemistry of reward and punishment. In D. H. Efron
(Ed.) Psychopharmacology; A Review of Progress. Washington:
U. S. Government Printing Office, 1968.

STUART, D. G., PORTER, R. W., ADEY, W. R., & KAMIKAWA, Y. Hypothalamic unit activity. I. Visceral and somatic influences.
Electroencephalography and Clinical Neurophysiology,

1964, 16, 237.

SUTIN, J.   The periventricular stratum of the hypothalamus.   In
    Carl   Pfeiffer and John R. Smythies (Eds.) International
    Review of Neurobiology, Vol. 9. New York: Academic Press
    I$_{nc}$, 1966. Pp. 263-300.

SUTIN, J., & EAGER, R. P.   Fiber degeneration following lesions
    in the hypothalamic ventromedial nucleus. Annals of the
    New York Academy of Science, 1969, 157, 610.

SZENTÁGOTHAI, J.   The parvicellular neurosecretory system.   In
    W. Bargmann and J. P. Schade (Eds.) Progress in Brain Research,
    Vol. 5, Lectures on the Diencephalon. Amsterdam: Elsevier
    Publishing Co., 1964. Pp. 135-146.

SZENTÁGOTHAI, J., FLERKÓ, B., MESS, B., & HALÁSZ, B.   Hypothalamic
    Control of the Anterior Pituitary. An Experimental-Morpho-
    logical Study, 3rd Edition. Budapest: Akadémiai Kiadó, 1968.

TSUBOKAWA, T., & SUTIN, J.   Mesencephalic influence upon the hypo-
    thalamic ventromedial nucleus. Electroencephalography and
    Clinical Neurophysiology, 1963, 15, 804.

VALVERDE, F.   Studies on the Piriform Lobe. Cambridge: Harvard
    University Press, 1965.

VAN ATTA, L., & SUTIN, J.   The response of single lateral hypo-
    thalamic neurons to ventromedial nucleus and limbic
    stimulation. Physiology and Behavior, 1971, 6, 523.

WENDT, R. H.   Amygdaloid and peripheral influences upon the
    activity of hypothalamic neurons in the cat. Unpublished
    Doctoral Dissertation. Los Angeles: University of
    California, 1961. Pp. 1-115.

WEPSIC, J. G., & SUTIN, J.   Posterior thalamic and septal influence
    upon pallidal and amygdaloid slow-wave and unitary activity.
    Experimental Neurology, 1964, 10, 67.

WHEATLEY, M. D.   The hypothalamus and affective behavior in cats;
    A study of the effects of experimental lesions, with anatomic
    correlations. A.M.A. Archives of Neurology and Psychiatry,
    1944, 52, 1.

WHITE, N. M., & FISHER, A. E.   Relationship between amygdala and
    hypothalamus in the control of eating behavior. Physiology
    and Behavior, 1969, 4, 199.

RELATIONSHIPS AMONG AMYGDALOID AND OTHER LIMBIC STRUCTURES IN INFLUENCING ACTIVITY OF LATERAL HYPOTHALAMIC NEURONS

Loche Van Atta, Department of Psychology, Oberlin

College, and Jerome Sutin, Department of Anatomy,

Emory University School of Medicine, Atlanta, Georgia

The term, "limbic system," has come into vogue during the past decade or two, but it is difficult to find either anatomical or physiological justification for lumping a diverse, multi-functional collection of cortical areas and subcortical structures together as the limbic system, and this designation appears not to have sufficient descriptive value to justify its continued use (Livingston and Escobar, 1971). However, there seems to be little doubt that certain limbic structures are important in the initiation and control of motivational and emotional processes, and this has been made particularly clear through studies of brain and behavioral correlates in a wide range of mammalian species from rodents to primates. Furthermore, anatomical study of limbic structures and their fiber projection systems has demonstrated rich and intricate interconnections among them which appear to be increasingly complex as newer techniques are developed for detecting finer fiber systems and degenerated axon terminals.

One of the most remarkable common features of limbic structures is the fact that they all project in greater or lesser degree onto hypothalamic structures. Nauta (1963) described the hypothalamus as a nodal point in the interrelation between limbic forebrain structures and the midbrain, and went on to emphasize that "the functional state of the hypothalamus is inseparably related to the patterns of neural activity in (this) circuit as a whole" (p. 14).

Since Hess (1928) demonstrated that both skeletal and autonomic motor patterns associated with rage could be elicited by electrical stimulation of the cat hypothalamus, an extensive

experimental literature has accumulated showing that direct
electrical or chemical stimulation of the hypothalamus is capable
of producing a wide variety of behavioral patterns, such as feeding, drinking, flight and defensive reactions, and aggression in
a variety of mammals. Experiments of this sort demonstrate convincingly that the hypothalamus exerts important selective influences on patterns of neural activity leading to diverse overt
behavior. However, it also is clear that hypothalamic functioning
is biased by sensory activity, most probably by way of midbrain
reticular formation and non-specific thalamic nuclear relays
(Nauta, 1963). Flynn (1967) and his co-workers have demonstrated
most elegantly that the form and direction of behavioral patterns
elicited by direct electrical stimulation of the hypothalamus
depend upon sensory cues from the environment, showing that elec-
trically-aroused patterns of central neural activity are integrated with normally perceived sensory information in order to
produce specific coordinated and goal directed motor patterns.
Most recently, Bandler and Flynn (1970) have shown that electrical
stimulation of a point in the lateral hypothalamus of the cat
which normally elicits an attack on a rat selectively facilitates
attack responses directed toward a mouse presented to the eye
contralateral to the site of stimulation. This finding is consistent with similar results reported by MacDonnell and Flynn
(1966), showing that stimulation of a lateral hypothalamic attack
point produces a sensory field on the muzzle and lip from which
a reflex pattern of head-turning and mouth-opening can be elicited
by light touch when the hypothalamus is being stimulated. These
effects were detected on the lip and muzzle areas contralateral
to the site of brain stimulation at lower stimulation intensities
than those required to elicit the reflex patterns on the ipsilateral side. These lines of evidence, taken together, suggest
that the hypothalamus receives and integrates information from
the sensory systems and also is capable, in turn, of modifying
activity in these sensory systems.

The conceptual scheme which emerges from such considerations
is one in which the hypothalamus, together with its midbrain
connections, functions to alter the motor system so as to facilitate particular patterns of activity, biases sensory systems for
reception of particular inputs, and is subjected to the modulatory effects of activity in a variety of limbic structures. This
view is the outgrowth largely of the work of Nauta (1958, 1963)
and Flynn (1967), but has been more or less implicit in the
writing of many manuscripts. The important point to be emphasized is that a model such as this implies clearly that it is not
possible to understand hypothalamic functions without also examining limbic functions. Furthermore, if Nauta's (1963) characterization of the hypothalamus as a "nodal point" in limbic-fore-
brain and limbic-midbrain interactions is to be defined in more

specific terms, it requires that we obtain more data about the
functional nature of limbic influences on hypothalamic activity.
Unfortunately, large gaps exist in our knowledge of how activity
in limbic structures influences hypothalamic activity, and the
manner in which hypothalamic activity acts selectively on sensory
and motor processes. Partly, this is due to insufficient information concerning the functional organization of the hypothalamus
itself. We know next to nothing about the nature of synaptic
linkages in the hypothalamus owing to the lack of an orderly
cytoarchitecture and to the relatively small size of cells in
this region. Intracellular recordings in this region are extremely difficult. Extracellular recordings of single cell activity are easily accomplished, however, and since the first
single cell studies of hypothalamic neurons by Sawa et al.
(1959) and Cross and Green (1959), considerable information has
accumulated concerning the changes in activity of single hypothalamic cells which may be brought about by hypo- and hyper-
glycemia, visceral stimulation, vaginal stimulation in estrous
and anestrous animals, and stimulation of various peripheral sensory systems. For the most part, such studies have succeeded in
showing changes in firing frequency or changes in the form of
interspike interval histograms. Fixed-latency responses, time-
locked to the application of peripheral sensory stimulation,have
rarely been reported outside the posterior hypothalamus. Because
of this, investigators more recently have employed intracranial
stimulation of regions of the brain which project upon hypothalamic cells in order to determine features of the functional organization of the hypothalamus. Examples of this approach to the
study of hypothalamic organization and afferent connections include the work of Gloor (1955), Tsubokawa and Sutin (1963), Egger
(1967), Murphy et al. (1968), Dreifuss and Murphy (1968), Murphy
and Renaud (1969), and Van Atta and Sutin (1971). This latter
work, which will be summarized here, began with the assumption
that since limbic structures project upon the hypothalamus, we
may obtain a statistical description of the excitatory and inhibitory actions and their spatial distribution by placing stimulating electrodes in several limbic structures and systematically
examining single cell responses in the lateral hypothalamus (LH).
We restricted the recording area to the lateral hypothalamus, for
we wished also to investigate projections via the efferent system
of the ventromedial nucleus of the hypothalamus into LH.

METHODS AND MATERIALS

The experiments were performed on 32 acutely prepared, un- ·
anesthetized adult male and female cats. The details of preparation of the animals, procedures used to insure the comfort of the
animals, and methods of recording and stimulation are detailed
elsewhere (1971). Each cat had from three to seven stainless

steel concentric bipolar stimulating electrodes inserted stereotaxically into a variety of limbic structures, including the
corticomedial amygdala and stria terminalis, the magnocellular
portion of the basomedial amygdala, lateral amygdaloid nucleus,
septum, preoptic region in the vicinity of the origin of the
medial forebrain bundle, bed nucleus of stria terminalis, dorsal
hippocampus, ventral hippocampus, and hypothalamic ventromedial
nucleus.

For the most part, we used single pulse stimuli applied at
a 1/sec rate in order to evaluate response properties of LH
neurons, since this procedure makes it relatively easy to relate
responses of single cells to the time of stimulation, and the
characteristically low discharge rate of LH neurons requires
summation techniques for the detection of changes in firing
rates. By using 1/sec repetition rates, we could summate the
effects of many test trials, looking for consistencies in response
characteristics on successive applications of the electrical
stimulation.

On the other hand, very few behavioral effects elicited by
hypothalamic or limbic stimulation can be obtained by very
low-frequency stimulation, and we are not aware of any that have
been elicited by 1/sec stimulation. Most commonly, repetition
rates ranging from 50 to 100 Hz have been used to produce effects
on feeding, drinking, flight, or aggressive behavioral patterns.
Therefore, we tested some LH units for response to stimulation
frequencies ranging from 10 to 100 Hz, including some units
which were totally unresponsive to 1/sec stimulation of any test
stimulation site and some units which did respond to 1/sec stimulation of one or more limbic sites. In the latter case, we
applied high-frequency stimulation trains to the test sites only
after the single-pulse analysis had been completed, since it was
not uncommon for units to be injured, killed, or otherwise lost
when a high-frequency pulse train was applied to the brain.
Therefore, the population of units tested with high-frequency
stimulation is considerably smaller than the population analyzed
with single-pulse stimulation.

RESULTS

Since the findings of many of these experiments have been
published elsewhere (Van Atta and Sutin, 1971), we will present
only a summary of some of the more interesting of those results
together with data not previously published. We examined a
total of 302 LH units which were affected by stimulation of one
or more brain structures and classified them according to the
discharge pattern produced by single-pulse stimulation of the
various limbic test sites. Sixty-two per cent of our population

of LH units discharged spontaneously, the remainder showing
action potentials only in response to brain stimulation.

Unit responses to brain stimulation were classified on the
basis of an excitatory effect or a suppression of spontaneous
firing to single-pulse stimulation. Those cells which showed
action potentials driven with a constant latency following stimulation were designated "D" type units, while those cells which
showed a suppression of spontaneous firing for periods ranging
from thirty to several hundred milliseconds were classified as
"S" type units. Both D and S categories were further divided
into two subcategories, depending upon whether the cells were
affected by stimulation of only one structure (single-site effects)
or responded to stimulation of two or more brain structures
(convergent effects). Finally, the convergent category was
further subdivided into synergistic effects, if all effective
stimulus sites affected the unit in the same manner, or antagonistic if some sites excited the cell while others suppressed
its activity.

Suppression of firing. Eighty-five LH cells exhibited
suppression of spontaneous firing in response to brain stimulation, of which 39 were in the single-site category and 46 were
classified as convergently suppressed from stimulation of two
to five different brain sites. Since we could detect a suppression effect by means of extracellular recording only in those
neurons which maintained a spontaneous firing rate, we have
taken the 186 spontaneously active units as our sample base, so
that 46 per cent of LH neurons available for test demonstrated
suppression effects. If this estimate is biased, it probably
represents an underestimation of inhibitory effects of limbic
structures and HVM on LH cells, inasmuch as there may have been
cells not spontaneously active in which IPSPs developed which
cannot be detected in extracellular recordings.

Figure 1 will serve to illustrate a summation technique
for recording suppression effects, our criteria for judging such
effects, three principal types of suppression effects observed,
and some other properties of cells exhibiting suppression of
firing in response to remote brain stimulation. These records
all were made by superimposing multiple, successive traces on
the storage oscilloscope screen, since the spontaneous firing
rates of most LH neurons are so low that single-trace analysis
ordinarily will not reveal a suppression effect of brain stimulation. However, superimposing 5, 10, 20, or more successive
traces on the storage screen clearly reveals those periods in
time during which the unit does not fire and whether or not such
periods are time-locked to the occurrence of the brain stimulus.
A control record is also available by superimposing the same

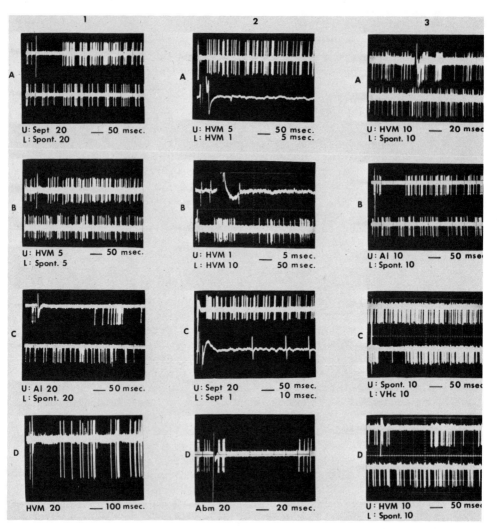

Fig. 1. Oscillographic records illustrating three types of suppression effects exerted on spontaneously active LH neurons by HVM and limbic structure stimulation. The legend beneath each record indicates the content of the upper (U) and lower (L) traces, designated by Spont. (spontaneous, unstimulated firing record), and site of stimulation, designated as follows: Sept. = septum; HVM = ventromedial nucleus of the hypothalamus; Al = lateral amygdaloid nucleus; Abm = magnocellular portion of the basal amygdaloid nucleus; VHc = ventral hippocampus. The numbers following the U and L trace key indicate the number of superimposed, stored traces recorded. See Text.

number of sweeps without the stimulus, recording only the summated spontaneous activity of the unit. In judging whether or
not stimulation of a particular brain site produces suppression of
the spontaneous firing pattern of a LH unit, we look simply for
gaps in the summated unit firing record which are: (1) time-
locked to the occurrence of the stimulus, and (2) are longer than
those appearing in the control record. As the number of such
superimposed traces of stimulated and unstimulated unit activity
increases, the reliability of these estimates is improved, and
simple statistical tests are available for judging whether or not
the probability of observing periods of time without a unit action
potential as long as that measured by means of n number of summated, superimposed sweeps exceeds some critical chance level
(Van Atta, unpublished).

Figure 1 illustrates three types of suppression effects
which we observed. In column 1 there are shown units whose spontaneous firing stopped abruptly, coincident with the time of
stimulation. Column 2 shows cells in which arrest of firing
followed the occurrence of a driven action potential. In records
A, B, and C of column 2, traces recorded at an expanded sweep
speed are shown in order to indicate more clearly the driven
action potentials masked by the stimulus artifact in the summated
records taken at a slower sweep speed. Column 3 shows the third
type of suppression effect in which a brief period following the
stimulus artifact occurred during which the unit continued to
discharge, succeeded by a second period during which unit firing
was suppressed. The action potentials following the stimulus
artifact were not driven at a fixed latency, although in record
3-A the probability of discharge rises above the spontaneous
firing rate immediately after the stimulus artifact, suggesting
a facilitation followed by inhibitory pause. This particular
record was made with a relatively low-impedance microelectrode
and the time-constant of the preamplifier adjusted to pass slow
potential changes. In this record negative is down, so that a
negative slow potential is seen to follow the stimulus artifact,
with unit action potentials densely concentrated in the period
occupied by this evoked potential, followed by a brief inhibitory
pause, a second band of densely concentrated firing, and a second
inhibitory pause. Most of our records were made under conditions
which did not permit simultaneous recording of slow potentials
and unit action potentials; however, when this occurred it was
the case that unit firing tended to occur during the negative
phase of the slow potential, consistent with the findings described by Dreifuss elsewhere in this volume.

There was no relationship between the type of inhibitory
effect produced and the site of brain stimulation. However, the
suppression variant shown in column 1 of Fig. 1 was related
definitely to the mean spontaneous firing rate of the units

showing this effect.[1]   The mean interspike interval during 1
minute or longer samples of spontaneous firing was determined for
a sample of units, including those shown in Figure 1, exhibiting
each of the three variants of suppression effects. The mean frequency was calculated from these data for each unit, and the
medians for the three distributions of mean firing rates determined. For those showing the type of suppression effect illustrated in column 1, the median of the mean firing rates was 4.3
spikes per second (range 3.1 - 7.3); that for the effect shown
in column 2 was 15.2 spikes per second (range 9.0 - 20.2), while
that for the column 3 effect was 14.5 spikes per second (range
5.6 - 24.4). Therefore, the inhibitory pause in firing beginning immediately after the stimulus artifact, shown in column 1,
Figure 1, reflects the low probability of a spike occurring before the onset of inhibition in cells with low spontaneous firing
rates. The possibility of defacilitation resulting from inactivation of a tonic excitatory influence by cells at the stimulus site must also be considered.

Figure 2 shows the relationship between the type of inhibitory effect observed and the interspike interval distributions
for three of the neurons shown in Fig. 1, representing one of
each of the three types of suppression depicted there. These
ISI distributions are typical of those obtained from recordings
of spontaneous firing trains of neurons in each of these categories, and Fig. 2 illustrates quite clearly the fact that units
such as those found in column 1, Fig. 1 are characterized by very
low rates of spontaneous firing and tend to have a high degree
of variability in interspike intervals, while those found in
columns 2 and 3 of Fig. 1 have much higher spontaneous firing
rates and lower interspike interval variability.

The effect shown in column 1, Fig. 1 was by far the most
frequently observed of the three variants, in keeping with the
fact that about 70 per cent of all spontaneously active LH units
recorded showed firing rates of fewer than 10 spikes per second
and yielded heavily skewed interspike interval distributions.
The effect shown in column 2 was least frequently observed. We
found only 8 units which showed driven action potentials followed
by suppression of firing with stimulation of all effective brain
sites. However, these effects may be of more than passing
interest, inasmuch as the pattern of excitation followed by

[1]The analysis which follows was not included in the original publication of these data (Van Atta and Sutin, 1971), but was suggested to the first author by Dr. Pierre Gloor during the discussion period following the presentation of these results at this
conference. We are indebted to Dr. Gloor for his suggestion.

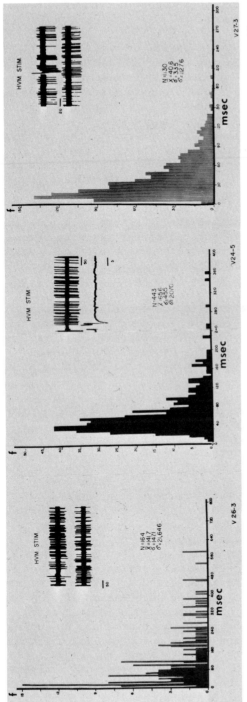

Fig. 2. Interspike interval distributions obtained from computer analysis of spontaneous spike trains for each of the types of suppression effect shown in Fig. 1. The inset oscillographic records for these three units from left to right are identical with those shown in Fig. 1 as Records 1-B, 2-A, and 3-A, respectively. In each distribution, the ordinate represents simple spike frequency, and the abscissa is scaled in five msec. addresses. The figure illustrates differences in mean spontaneous firing frequency and variability of interspike intervals around that mean for each of the three types of suppression effect observed. Distribution statistics: N = number of single spikes counted in the spontaneous firing sample; X = mean interspike interal; σ = standard deviation; σ^2 = variance.

inhibition is similar to that associated with recurrent collateral inhibition involving interneurons.

Quite commonly, we observed a period of variable duration following the suppression interval during which the unit firing rate was facilitated--a kind of "rebound" effect following an inhibitory pause in firing. Figure 3 provides a clear example of this phenomenon. The inset shows 10 superimposed sweeps of spontaneous firing on the top trace, and 10 sweeps following 1/sec stimulation of the magnocellular portion of the basomedial amygdaloid nucleus on the bottom line. The poststimulus firing probability distribution computed for forty stimulus presentations also is shown. This cell had a near-zero probability of firing for a period of 140 msec after the stimulus, followed by a greatly increased probability of discharge between about 160 and 350 msec, followed by a rather abrupt return to a baseline spontaneous firing rate. This rebound effect was seen frequently following stimulation of all brain sites which produced inhibitory pauses in firing of LH cell and was not specific to any particular site. Although it might seem reasonable, a priori, to assume that there could be some relationship between duration of inhibitory effects and variability in spontaneous firing or mean spontaneous firing rates, we found no relationship between such parameters and duration of suppression effects. Nor was a relationship found between the pattern of spontaneous activity of LH cells and S or D type responses to limbic or HVM stimulation.

When we compared the suppression effects with the site of brain stimulation, it was clear that HVM and the various limbic structures tested were not equally effective in producing suppression of the spontaneous activity of LH units. The data were analyzed in terms of percentages of units spontaneously active available for test with each stimulation site which were suppressed by stimulation of each structure, percentages of such units which were suppressed from a single site of stimulation, and percentages of these units which were suppressed from more than one site. HVM, the lateral amygdaloid nucleus, the stria terminalis--corticomedial amygdala, septum, and bed nucleus of the stria terminalis were all relatively effective sites of stimulation in producing suppression effects. HVM was the most effective site of stimulation, showing suppression of 31.4 per cent of 169 LH units tested. The two least effective sites of stimulation were dorsal hippocampus, which produced no suppression effects in 25 units tested, and the preoptic region, which suppressed 5 per cent of 40 units tested. Among the amygdaloid placements, stimulation of the basomedial amygdaloid nucleus was least effective, with more lateral (lateral amygdaloid nucleus) and more medial (stria terminalis--corticomedial nucleus) sites

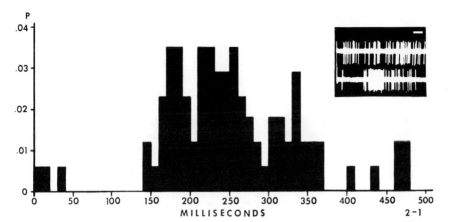

Fig. 3. Poststimulus probability distribution of a LH neuron during 1/sec stimulation of the basomedial amygdaloid nucleus (Abm). The inset record shows fifteen superimposed sweeps of spontaneous activity (top) and response to stimulation (bottom). Calibration for inset = 50 msec. Note the pronounced post-inhibitory rebound. (Reproduced with permission of Pergamon Press).

sites producing most of the suppression effects observed.

Single-site suppression effects were rarely obtained from
stimulation of structures other than HVM, which was the single
effective site for suppression of 10.1 per cent of 169 units
tested.   By comparison, stimulation of the stria terminalis--
corticomedial amygdala resulted in suppression of 26.5 per cent
of 102 units tested, only 2.9 per cent of these being single-
site effects, and stimulation of the septum produced suppression
effects in 24.2 per cent of 149 units, 4.7 per cent of these
being single-site effects.   On the other hand, convergent suppression effects were quite common, both with HVM stimulation
and stimulation of other brain sites, most prominently the
lateral amygdala, stria terminalis--corticomedial amygdala,
septum, ventral hippocampus, and the bed nucleus of the stria
terminalis.   While convergent suppression effects involving
various pairs of these sites were very common, inhibitory effects
upon a single LH neuron from four or five different sites of
stimulation were also frequently seen.   Analysis of the convergent suppression data according to pairs of effective sites indicated that HVM was the structure most commonly involved, showing peaks of interaction with stria terminalis--corticomedial
amygdala, septum, and ventral hippocampal stimulation.   Of 12
units suppressed by ventral hippocampal and one or more other
sites of stimulation, 100 per cent were suppressed by stimulation of stria terminalis--corticomedial amygdala, and 92 per
cent were also suppressed by HVM stimulation, suggesting a heavy
convergence of inhibitory effects on LH units from these three
structures.   Of the three amygdaloid sites of stimulation, the
lateral amygdaloid nucleus and the stria terminalis--cortico-
medial amygdala were the most effective in producing convergent
inhibitory effects on LH neurons.   These two sites showed relatively heavy convergent inhibitory effects between them, affecting 58 per cent of 24 LH neurons jointly and both being involved
in high percentages of convergent effects with HVM stimulation.
The lowest convergence found in this analysis was obtained between each of these two sites and stimulation of the septum,
both showing an identical 34 per cent convergence of suppression
effects involving 29 LH cells also suppressed by septal stimulation.

Driving of neuronal discharge.   There were 229 D-type cells,
of which 140 (61%) were driven from single sites and 89 from

multiple sites.[2]   When we examined the D-type effects by
effective site of stimulation, we found considerable variation
in the relative effectiveness of the several brain structures in
driving LH unit discharge.   In this regard the data on D-type
effects are similar to those on S-type effects.   Stimulation of
HVM drove 51.6 per cent of 285 units tested and was by far the
most effective site of stimulation for driving LH unit action
potentials.   At the other extreme, dorsal hippocampal stimulation drove only 2 LH units out of a total of 65 neurons tested.
Between these two extremes, the probability of obtaining excitation in LH cells was relatively high with activation of stria
terminalis--corticomedial amygdala (22.7%), septum (27.6%),
bed nucleus of the stria teminalis (23.5%), and the preoptic
region (36.1%), while stimulation of the dorsal or ventral hippocampus, basomedial amygdala, or lateral amygdala was relatively
less effective, ranging from 10.7 to 16.9 per cent of the units
tested.

We compared the single-site and convergent driving effects
and found a picture very similar to that found with suppression
effects.   Among the structures tested, HVM is by far the most
effective in single-site driving capabilities.   The other site
which exhibited an appreciably large proportion of single-site
driving effects was the septum.   However, several sites in
addition to HVM exhibited relatively high proportions of convergent excitatory effects, principally stria terminalis--corti-
comedial amygdala, septum, ventral hippocampus, bed nucleus of
the stria terminalis, and preoptic region.

Comparing single-site and convergent <u>driving</u> capabilities
with single-site and convergent <u>suppression</u> capabilities revealed
a difference which might have some significance with regard to
the functional organization of LH.   Of 229 D-type units, 61 per
cent were driven from single sites and 39 per cent driven convergently.   Of 85 S-type cells, 46 per cent were suppressed
from single sites, while a majority (54%) were suppressed convergently.   Casting these proportions into a Chi-square analysis
and computing expected proportions from marginal totals yielded

---

[2]Some units which we recorded were affected in different ways by
different sites of stimulation.   In such cases, we have included
S-type effects obtained from stimulation of particular structures
in our analysis of suppression effects, and for these same units
the D-type effects obtained from stimulation of other structures
appear in the present analysis.   Therefore, the sum of D-type
effects plus S-type effects is greater than the total population
of 302 recorded units.   See Van Atta and Sutin (1971) for a full
description of the breakdown of the population for statistical
analysis.

a Chi-square value of 5.94 with 1 df, significant beyond the .02 level of confidence. Therefore, it is likely that single-site driving and suppression effects are distributed disproportion-ately with respect to chance expectations. The heavy contribu-tion of HVM stimulation to single-site driving effects in large measure accounts for this difference; however, except for the lateral amygdala and the bed nucleus of the stria terminalis, the proportion of single-site driving effects exceeds the proportion of single-site suppression effects at all stimulation sites tested. Our observations indicate, therefore, that a larger popu-lation of LH cells receive excitatory inputs than the population of cells which receive inhibitory inputs.

Analysis of the convergent driving data by pairs of effective sites indicated that HVM was the structure most commonly involved in convergent driving of LH cells and showed peaks of interaction with the lateral amygdala, stria terminalis--corticomedial amyg-dala, septum, ventral hippocampus, and the preoptic region. Of 33 cells driven convergently by stria terminalis--corticomedial amygdala, HVM stimulation also drove 70 per cent of these units. Of particular interest is the fact that of 21 cells driven con-vergently by preoptic stimulation, HVM drove 100 per cent of these, while stria terminalis--corticomedial amygdala stimulation drove no units in common with preoptic stimulation. Raisman (this volume) reported that, following section of the stria ter-minalis in the rat, terminal degeneration was found on one in twenty preoptic region cells, while one cell in five showed ter-minal degeneration in the cell-poor region just ventral and medial to the ventromedial nucleus. If Raisman's anatomical data for the rat are approximately descriptive of the same system in the cat, there seems to be an excellent correspondence between his anatomical data and our electrophysiological findings. The heavy convergence of driving effects between HVM and stria-terminalis--corticomedial amygdala stimulation may reflect the projections of the postcommissural portions of the stria upon HVM. However, stimulation of the stria terminalis--corticomedial amygdala drove 0 per cent of 21 cells driven by preoptic stimulation. If stria terminalis fibers terminate on a much smaller proportion of preoptic region cells than HVM cells, the likelihood of finding many cells driven convergently by these two sites would be very small and consistent with our findings.

Synergistic and antagonistic effects. A total 135 LH cells showed convergent inputs, and 111 (82%) of these were syner-gistically affected. Eighty-two cells (74%) showed synergistic driving, while 29 (26%) were synergistically suppressed. This preponderance of excitatory over inhibitory events, especially regarding those synergistic responses in which HVM is one of the stimulated sites, is quite similar to the situation regarding single-site responses: we found many more single-site excitatory effects than inhibitory effects, and HVM was the prime site for obtaining single-site effects of both classes.

Where our data on proportions of excitatory and inhibitory effects and synergistic vs. antagonistic convergent effects are directly comparable to those reported by other investigators, there is a very good degree of correspondence, in spite of some differences in procedure and criteria for judging effects of stimulation on LH neurons. For example, Dreifuss and Murphy (1968) found that 84 per cent of the hypothalamic cells which they recorded were synergistically affected by stimulation of the septum and amygdala, while we found 93 per cent of our convergent stimulation effects to be synergistic from activation of these same structures. Dreifuss and Murphy (1968) also examined stimulation of dorsal hippocampus for convergence with other structures stimulated. They found nine hypothalamic cells whose location was not specified which could be affected by dorsal hippocampal stimulation, but they reported no clear pattern of antagonistic or synergistic changes. We also were unable to find any convergent effects in the 65 LH cells which we tested with dorsal hippocampal stimulation, but were able to obtain a considerable number of synergistic effects between ventral hippocampal and HVM, amygdala, and septal stimulation.

Where differences exist, they are not extreme. For example, Murphy et al. (1968) reported 29 per cent of 35 LH cells affected by stimulation of the lateral amygdala were suppressed, while we found 17 per cent of 182 LH cells which responded in this fashion when the same structure was electrically activated. However, the ratio of inhibited LH cells to excited cells obtained with stimulation of the basomedial amygdala or septum are in good agreement with the results produced by Murphy et al. (1968). The differences which exist may reflect the more restrictive criteria we have employed in our classification scheme. Only neurons which were affected in a time-locked fashion following the stimulation of our test structures were regarded as "excited." Of 241 cells which were "excited" by stimulation of our test structures, we found only 12 in which 1/sec stimulation produced a long lasting increase in the discharge rate.

Higher Frequency Stimulation Effects

We recorded a total of 166 LH units tested with stimulation frequencies ranging between 10 and 100 Hz. Of these, 52 units were totally unresponsive to stimulation of any brain site at any frequency tested. An additional six units were driven by 1/sec stimulation, but failed to respond to stimulation frequencies between 10 and 100 Hz., leaving a total of 108 units which were

responsive in some manner to higher frequency stimulation of one
or several brain sites.

We wish here to describe only some of the effects we have
seen; 28 units have been identified which were not responsive in
any manner to 1/sec stimulation of any brain site but which were
affected when tested at higher frequencies of stimulation. Among
these, there was not a single case in which driven action potentials followed at higher frequencies of stimulation. Of the 28
units referred to above, all higher frequency stimulation effects
were confined either to suppression of unit discharge (18 units)
or to facilitation of the spontaneous firing rate (10 units).

The sort of facilitation effect most commonly observed was
simply an increase in the unit firing rate over the spontaneous
rate. Figure 4 illustrates one such unit response. The top
trace represents the spontaneous firing rate which averaged 0.14
spikes per second. B shows the effect of HVM stimulation at
10/sec for 4-1/2 seconds, producing a negligible change in discharge rate. C shows the effect obtained with 50/sec HVM stimulation for five seconds. A marked increase in firing rate occurred.
D is a continuation of C, with a ten-second interval between
traces, showing that the unit continued to fire at a rate greatly
exceeding the spontaneous rate for at least 1 minute 35 seconds
after the termination of the stimulus train. When higher stimulation frequencies were used, the electrical activity in the
vicinity of the tip of the stimulating electrode was monitored
in order to detect afterdischarges. When such effects were found,
these tests were not included in our data analysis. The long
time-course of the higher frequency stimulation effects is of
greatest interest, but their interpretation is difficult without
concomitant EEG and blood pressure recordings.

Not all facilitation effects persisted beyond the period
of stimulation, however. Figure 5 illustrates a unit in which a
facilitatory effect of higher frequency stimulation was confined
to the period of stimulation. The top record shows the spontaneous firing. B shows a sequence of 1/sec stimuli to the ventral
hippocampus, followed by a stimulus train at 10/sec which continues through record C. 1/sec stimulation of the ventral hippocampus inhibited the spontaneous firing of the unit completely;
however, when 10 Hz. stimulation was applied, the unit promptly
began firing again. Trace D shows the shift from 1/sec stimulation to a train of 50/sec stimulation. The unit firing rate increased abruptly with the onset of the 50/sec train and terminated just as abruptly at the end of the high frequency stimulation.
A five-sec . train of 50/sec stimulation delivered without preceding or succeeding 1/sec stimulation had the idential effect--
prompt facilitation of the firing rate confined strictly to the

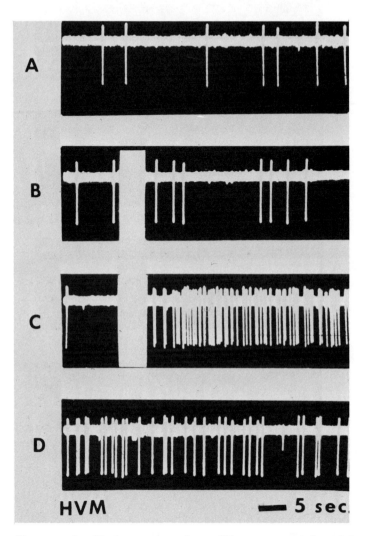

Fig. 4. Changes in firing rate of an LH neuron produced by
high frequency stimulation of HVM. Trace A, spontaneous activity
sample; B, 10 Hz. stimulation effect; C, 50 Hz. stimulation effect; D, continuation of trace C. See text.

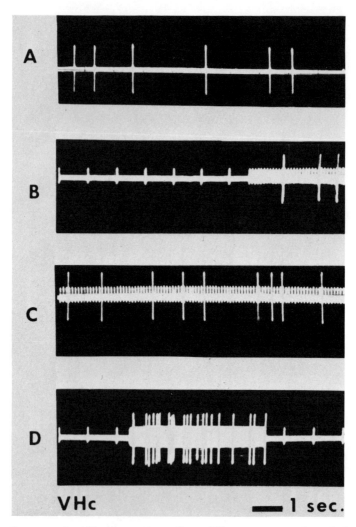

Fig. 5. Changes in firing rate of an LH neuron produced by high
frequency stimulation of the ventral hippocampus. Trace A, spontaneous activity sample; B, 1/sec stimulation, followed by a 10 Hz.
stimulus train, which is continued through trace C. Note suppression of firing during 1/sec stimulation. Trace D, 50 Hz. stimulus
train preceded and succeeded by 1/sec stimulation, producing a
suppression of firing during 1/sec and facilitation of firing
during 50 Hz. stimulation. See text.

duration of the stimulus train. The more commonly observed
effect among units whose spontaneous firing rate was suppressed
by 1/sec stimulation was a much more prolonged and complete
suppression of spontaneous firing when high frequency stimulation
was applied.

Of the 108 units which responded in some manner to higher
frequency stimulation of one or several sites,   66 (61%) responded to HVM stimulation; 48 responded to amygdala stimulation,
with the very large majority of these responding either to
lateral amygdala or stria terminalis--corticomedial amygdala
stimulation. Only 8 units responded to higher frequency stimulation of the basomedial amygdala, and only 14 units responded
to stimulation of the septum, a rather surprising outcome in
view of the large number of LH neurons which responded to 1/sec
stimulation of the septum. There were no units responding
either to preoptic region or dorsal hippocampal higher frequency
stimulation. Of the 66 units which responded to HVM stimulation,
26 were single-site effects and 40 were convergently affected.
Virtually all of the HVM single-site effects occurred in LH
neurons which responded in this same fashion to 1/sec stimulation
of HVM. The only exceptions were units which did not respond
at all to any site of stimulation at 1/sec but did respond to
higher frequency stimulation of HVM.

Inhibitory effects of higher frequency stimulation. The
inhibitory properties of higher frequency stimulus trains revealed some differences from the types of suppression effects
seen with 1/sec stimulation. Some cells exhibited suppression
of unit spontaneous firing for the duration of the stimulus
train, with prompt recovery of the spontaneous rate when the
higher frequency stimulus was terminated, but other effects were
complex, and Figure 6 shows some of these properties. Record 1
illustrates a unit which was unresponsive to stimulation of any
site at 1/sec. The top trace is a 20 sec. sample of spontaneous
firing, and the bottom trace shows the effect of 2 sec. of
50/sec stimulation of the stria terminalis--corticomedial amygdala. The firing rate of this unit did not change appreciably
during stimulation when examined at faster sweep rates. However,
it continued to fire for approximately 400 msec. following the
end of the stimulus train, then stopped firing for 6.8 seconds.
What is more interesting here is the post-stimulus time-course
of the inhibitory effect--much longer than anything seen with
single-pulse stimulation analysis and within the range of behavioral effects observed following application of brief trains
of stimulation to limbic and hypothalamic structures.

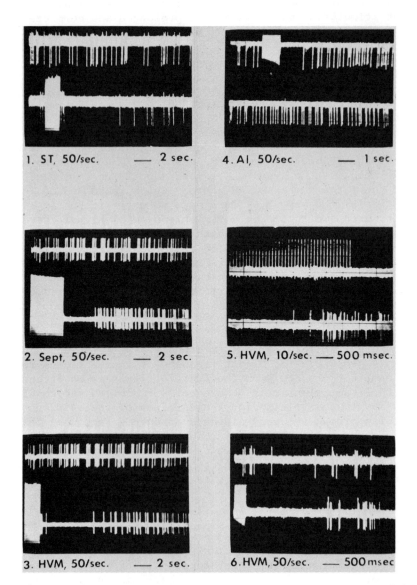

Fig. 6. Oscillographic records showing various effects of high frequency stimulation on LH cells. Sites Stimulated: ST = stria terminalis--corticomedial amygdala; Sept = septum; A1 = lateral amygdaloid nucleus; HVM = ventromedial nucleus. See text.

Records 2 and 3 show different effects produced in the activity of a single unit by 50/sec stimulation of two different brain sites. In record 2, the unit was suppressed completely during stimulation of the septum, and the inhibitory effect continued for about four seconds after termination of the stimulus train. The post-suppression firing rate was somewhat greater than the spontaneous rate shown in the upper trace of record 2, illustrating a mild rebound effect very commonly observed with both 1/sec and higher frequency stimulation. This same unit was also suppressed by 50/sec stimulation of HVM, shown in record 3. However, in this case, the firing rate increased during stimulation and continued at this rate for 380 msec. before the onset of a 5.6 second period of arrest of firing. This particular LH cell was inhibited convergently and synergistically by 1/sec stimulation of both septum and HVM, for 62 and 80 msec., respectively. The remaining records in Figure 6 are simply further illustrations of the relatively long time-course of inhibitory effects obtained with stimulus trains.

Among the units which exhibited relatively long time-courses of post-stimulus suppression of firing, stimulation of HVM, stria terminalis--corticomedial amygdala, and lateral amygdala were by far the most effective in producing this effect. Among these units which exhibited an inhibitory pause in firing when tested with 1/sec stimulus trains, higher frequency stimulation produced post-stimulus suppression of firing. The chief differences between effects produced with 1/sec and train stimuli were (a) train stimulation in some cases produced a facilitation of firing during stimulation, followed by inhibition, and (b) the time-course of the inhibitory pause was generally greatly increased by higher frequency stimulation.

Localization of Unit Responses

Sutin and Eager (1969) found that discrete lesions confined to the ventromedial nucleus of the hypothalamus resulted in degeneration only within the nucleus. However, if the lesion were larger and included tissues immediately surrounding HVM, degeneration occurred dorsally and laterally in the hypothalamus, beyond the confines of the HVM nucleus itself. Figure 7 is a drawing representing the pattern of degenerating fibers in the lateral hypothalamus following the larger tuberal lesions, depicted on a single Horsely-Clarke frontal plane. Figure 8 is a localization plot of 225 of the 302 units recorded by Van Atta and Sutin (1971). There are two important generalizations derived from comparison of these two figures. First, Figure 7 indicates heavy degeneration in the perifornical region, and Figure 8 shows that this was a region in which many single-

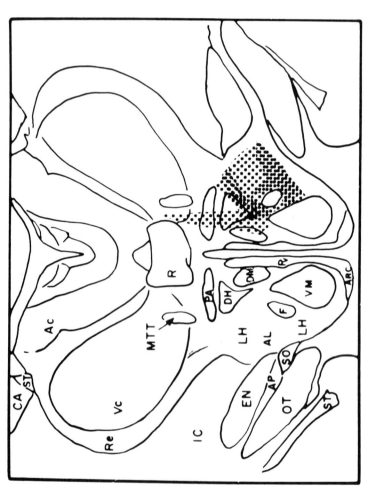

Fig. 7. Scheme of the pattern of degeneration found in the lateral hypothalamus following destruction of the hypothalamic ventromedial nucleus and immediately surrounding tissue, based upon the studies of Sutin and Eager (1969). While degeneration is shown on a single Horsley-Clarke frontal plane, it actually extends from the level of the optic chiasm to the posterior hypothalamus. Relative density of stippling represents density of degenerating fibers. Abbreviations: LH = lateral hypothalamic area; DM = dorsomedial nucleus; DH = dorsal hypothalamic area; F = fornix; VM = ventromedial nucleus. (Reproduced by permission of Pergamon Press.)

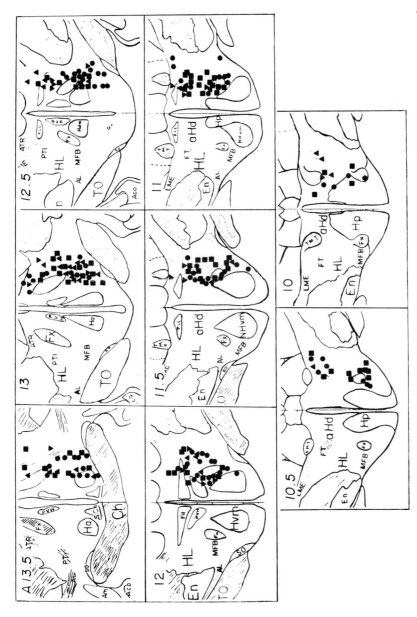

Fig. 8. Diagrams showing the locations of 225 hypothalamic units. Circles = convergently driven units; squares = units driven from single stimulation sites; upright triangles = convergently suppressed units; inverted triangles = units suppressed from single stimulation sites. Horsley-Clarke planes A13.5--A10 are represented. (Reproduced by permission of Pergamon Press.)

unit responses were concentrated; furthermore, in the mid-tuberal region (frontal planes A 12 to A 11, Fig. 8), only driven action potentials occurred in the perifornical zone, while at levels more rostral and caudal to these planes there was a rather uniform admixture of excitatory and inhibitory effects of stimulation. Secondly, Figure 7 shows a region of very sparse degeneration lateral and dorsal to the fornix, as though the fornix were casting a "shadow" in the LH degeneration pattern; likewise, Figure 8 shows a similar zone at planes A 12 and A 11.5 in which a few responsive units were located, regardless of the site of stimulation and despite repeated microelectrode sampling of this region. While many of the units located in the perifornical region showing driven action potentials responded either uniquely or convergently to HVM stimulation, other sites of stimulation were also effective.

With regard especially to the sparse degeneration and absence of single-unit responses dorsal and lateral to the fornix, the concordance of these two sets of observations (one anatomical and the other electrophysiological), suggests that many of the effects of limbic stimulation upon LH cells may be mediated through HVM. This possibility was put forth earlier by Dreifuss and Murphy (1968) and is reiterated elsewhere in this volume separately by Dreifuss, Egger, Murphy, and Raisman. The data we have obtained, using a wider variety of limbic structures for test stimulation than elsewhere reported, supports this point of view, which has been developed independently by several investigators and may provide new insight into limbic--hypothalamic functional relationships. Heimer and Nauta (1969) and Raisman (this volume) have shown that the stria terminalis projects to HVM. Sutin (1963) and Chi (1970) have demonstrated projections from the preoptic region and anterior hypothalamus into HVM. Also, there is electrophysiological evidence reported by Dreifuss and Murphy (1968), Murphy and Renaud (1969), Sutin (1963), and Tsubokawa and Sutin (1963) showing amygdaloid and septal pathways to HVM. The major involvement of HVM in convergent stimulation effects on LH neurons and the scarcity of single-site driving and suppression effects obtained with stimulation of structures other than HVM (Van Atta and Sutin, 1971) is wholly consistent with the hypothesis that HVM is an important relay in the pathways from various limbic structures to the lateral hypothalamus. It is well known that there are direct pathways to LH from the septum, preoptic region, and amygdala; therefore, it seems important at this time to determine the extent to which limbic influences on the lateral hypothalamus are relayed via HVM and studies have been initiated in the senior author's laboratory to obtain information on this question.

Concluding Remarks

The manner in which data gathered by anatomical, electro-physiological, and neurochemical methods relating to amygdaloid functions and relationships among the amygdala, other limbic structures, and the hypothalamus are in general agreement is most striking. It is clear that, in order to understand the effects of amygdaloid activity on the hypothalamus, it is necessary to study the interactions between the amygdala and other limbic structures. Without a detailed understanding of the functional interrelationships among limbic and hypothalamic structures, it is not likely that we will be able to answer global questions like, "How does limbic activity modify motivational and emotional states?", or "How does the hypothalamus affect sexual behavior?"

It is quite evident that the influences exerted by amygdaloid activity are not independent of the activities of other structures. The high proportions of convergent influences reported by Dreifuss and Murphy (1968) and those which we have found using a larger number of stimulation test sites, indicate the multiple possibilities for interactive effects exerted by two, three, or more limbic structures upon domains of cells in the hypothalamus. Perhaps, the most pressing problem is unveiling the principle by which limbic and hypothalamic neuronal networks operate so as to select one or another output pattern.

ACKNOWLEDGMENTS

Aided by General Research Support Grant No. FR5364 to Emory University School of Medicine.

The data reported herein were gathered while the senior author was a National Science Foundation Science Faculty Fellow (No. 66303), in the Department of Anatomy, Emory University, Atlanta, Georgia.

REFERENCES

BANDLER, R., JR. & FLYNN, J. P. Visual patterned reflex present during hypothalamically elicited attack. Science, 1971, 171, 817-818.

CHI, C. C. Afferent connections to the ventromedial nucleus of the hypothalamus in the rat. Brain Research, 1970, 17, 439-445.

CROSS, B. A., & GREEN, J. D. Activity of single neurones in the hypothalamus: effect of osmotic and other stimuli. Journal of Physiology, 1959, 148, 554-569.

DREIFUSS, J. J., & MURPHY, J. T. Convergence of impulses upon single hypothalamic neurons. Brain Research, 1968, 8, 167-176.

EGGER, M. D. Responses of hypothalamic neurons to electrical stimulation in the amygdala and the hypothalamus. Electroencephalography and Clinical Neurophysiology, 1967, 23, 6-15.

FLYNN, J. P. The neural basis of aggression in cats. In D. C. Glass (Ed.) Neurophysiology and Emotion. New York: Rockefeller University Press, 1967, Pp. 40-60.

GLOOR, P. Electrophysiological studies on the connections of the amygdaloid nucleus in the cat. Part I: The neuronal organization of the amygdaloid projection system. Electroencephalography and Clinical Neurophysiology, 1955, 7, 223-242.

HEIMER, L., & NAUTA, W. J. H. The hypothalamic distribution of the stria terminalis in the rat. Brain Research, 1969, 13, 284-297.

HESS, W. R. Stammganglien-Reizversuche, 10. Tagung der Deutschen Physiologischen Gesellschaft, Frankfurt am Main, Ber. ges. Physiol., 1928, 42, 554-555.

LIVINGSTON, K. E., & ESCOBAR, A. Anatomical bias of the limbic system concept. A proposed reorientation. Archives of Neurology, 1971, 24, 17-21.

MacDONNEL, M. F., & FLYNN, J. P. Control of sensory fields by stimulation of hypothalamus. Science, 1966, 152, 1406-1408.

MURPHY, J. T., DREIFUSS, J. J., & GLOOR, P. Topographical differences in the response of single hypothalamic neurons to limbic stimulation. American Journal of Physiology, 1968, 214, 1443-1453.

MURPHY, J. T., & RENAUD, L. P. Mechanisms of inhibition in the ventromedial nucleus of the hypothalamus. Journal of Neurophysiology, 1969, 32, 85-102.

NAUTA, W. J. H. Hippocampal projections and related neural pathways to the midbrain in the cat. Brain, 1958, 81, 319-340.

NAUTA, W. J. H. Central nervous organization and the endocrine motor system. In A. V. Nalbandov (Ed.) Advances in Neuro-endocrinology. Urbana: University of Illinois Press, 1963. Pp. 5-21.

SAWA, M., MARUYAMA, N., HANAI, T., & KAJI, S. Regulatory influence of amygdaloid nuclei upon the unitary activity in ventromedial nucleus of hypothalamus. Folia Psychiatrica et Neurologica Japonica (Niigata) 1959, 13, 235-256.

SUTIN, J. An electrophysiological study of the hypothalamic ventromedial nucleus in the cat. Electroencephalography and Clinical Neurophysiology, 1963, 15, 786-795.

SUTIN, J., & EAGER, R. P. Fiber degeneration following lesions in the hypothalamic ventromedial nucleus. Annals of the New York Academy of Science, 1969, 157, 610-628.

TSUBOKAWA, T., & SUTIN, J. Mesencephalic influence upon the hypothalamic ventromedial nucleus. Electroencephalography and Clinical Neurophysiology, 1963, 15, 804-810.

VAN ATTA, L. Unpublished technique for judging statistical significance of inhibitory intervals in single-unit spike trains following brain stimulation. 1971.

VAN ATTA, L., & SUTIN, J. The response of single lateral hypothalamic neurons to ventromedial nucleus and limbic stimulation. Physiology and Behavior, 1971, 6, 523-536.

THE ROLE OF THE AMYGDALA IN CONTROLLING

HYPOTHALAMIC OUTPUT

John T. Murphy

Department of Physiology, University of Toronto School
of Medicine, Toronto, Ontario, Canada

INTRODUCTION

There are many fascinating accounts of the role which the
amygdala plays in determining the precise form of the output of
the hypothalamus. Most of the pertinent studies have utilized as
measurable indicators of output various behavioural, physiological
or biochemical parameters. This approach has increased knowledge
not only about amygdalo-hypothalamic interrelationships, but also
about the underlying role of each structure in the organism's
adaptations to environmental changes. It also has been of considerable heuristic benefit in encouraging investigation into the
cellular basis underlying such adaptations.

One of the results of such a back-and-forth interplay between
different experimental approaches is that positive feedback may
occur. For example, enhanced understanding of cellular basis
further increases insight into actual behavioural or physiological
function and may suggest further experimentation with consequent
refinement of our conceptual framework regarding the system's
normal operation. The work to be described in this paper will, I
believe, serve to illustrate the need for greater utilization of
this principle. In doing so, it will offer the general idea that
biological science can best progress by referring the results of
a particular approach, say that of organ function analysis, back
upon the results of another approach, say cellular function
analysis. Other approaches, for example molecular, organismal or
societal, are, of course, equally valid and must eventually be
incorporated if one is to obtain a truly general theory. While
these ideas are undoubtedly merely truisms for many here, they are,

perhaps, worth restating at times to prevent our views from being too constrained by our particular technological orientation (often shaped in large measure by the graduate training of each investigator).

With this preamble, I would like to turn to a further consideration of the dual pathway from the amygdala to the hypothalamus upon which Dr. Dreifuss has elaborated in this volume (1971). The original neurophysiological investigation of these two pathways by Dr. Gloor (1955a, b) was preceded by the important studies of Kaada (1951) which showed very clearly how the amygdala could influence functions which were believed to be primarily controlled by the hypothalamus. The more recent electrophysiological analysis to which reference was made by Dr. Dreifuss has confirmed the presence of these two pathways and defined additionally their anatomical sphere of influence. These findings help to explain some of the conflicts among various investigators as to whether the amygdala should be considered facilitatory or inhibitory in its effects on various autonomic (cardiovascular, respiratory), endocrine, or behavioural (feeding and drinking, sexual, defense and aggression) activities. We may conclude from the work outlined by Dr. Dreifuss that it can be either, at least insofar as the hypothalamic ventromedial nucleus plays a role in these activities.

One may observe that, in fact, the demonstration of these two unique pathways provides a satisfactory enough resolution of some of the conceptual schizms to which reference was made above. However, there remained uncertainties which prompted further study of these pathways in collaboration with Dr. L. Renaud (Murphy and Renaud, 1968, 1969). We knew that the demonstration of an amygdaloid outflow system sufficiently complex to explain dichotomies between the results of different investigations[1] did not a priori ensure that, in fact, the system did not have additional as yet undemonstrated capability. Moreover, there remained as yet unexplained complexities at the hypothalamic end of the system. The puzzling role of the ventromedial nucleus and its immediate environs in manifestations of anger or other hyperemotional states has been discussed by MacLean (1969).

An additional difficulty was raised by our consideration of an extremely important and, at times, overlooked general principle of nervous system cytoarchitectonics, namely, that discrete cellular regions in the CNS are composed of diverse types of neurons. The

---

[1] An early and interesting example of such a dichotomy is provided by comparing the temporal lobe extirpation required to tame the monkeys of Klüver and Bucy (1939) with the temporal lobe sparing required to make the cats of Bard and Mountcastle (1948) placid.

converse idea of a regional homogeneous set of neurons, which is implicit to some extent in the conceptual formulation of functional 'centres'[2] so often used to explain the results of stimulation or lesioning experiments in the CNS, is not in general supported by morphologic studies. This principle of diverse neuron types, which is obvious with respect to cortical structures, is no less valid in the case of the phylogenetically most primitive core of nervous tissue in mammals which includes the central grey of the spinal cord and, more rostrally, the reticular formation and hypothalamus. Golgi studies at the turn of the century by Cajal (1911) revealed the presence of more than one cell type in these areas. There are many classical anatomical studies of the spinal cord which extend Cajal's impression (Rexed, 1964), and the Scheibels (1958) have added elegant Golgi studies of the reticular formation which also support this multiple neuron hypothesis. With regard to the ventromedial nucleus of the hypothalamus, Szentagothai and co-workers (1968) have confirmed the presence of more than a single type of neuron within the confines of the nucleus.

For the above reasons, we were led to search for a second cell type at the receiving end of the two amygdaloid efferent pathways, and to try to elaborate its function.

EXPERIMENTAL OBSERVATIONS

The initial approach was to stain cell bodies of the ventromedial region with thionin for purposes of comparison with the Golgi material of Szentagothai and Cajal. Examinations of sections in various places revealed primarily two main cell types with this staining procedure. The two types are illustrated in the coronal sections of Figure 1. The form of the soma and proximal dendrites could be detected using light microscopy at a magnification of 690 as shown in these sections. One type is demonstrated in the upper section of Figure 1. This type was found in heaviest concentration at the periphery of the nucleus, especially at the lateral border, and has bipolar dendrites which are usually oriented radially with respect to the centre of the nucleus. This feature would be especially suitable for the development of synaptic connections with axons both from outside the nucleus, as in the case of the two amygdaloid efferent pathways, and from within the nucleus. Other distinguishing features of these neurons include a somewhat ovoid-shaped nucleus and dark staining cytoplasmic particles. Moreover, the cell bodies are quite small, usually being about 7 μ at their widest diameter.

[2] See for example discussion by Arees and Meyer (1967) and by Oomura et al. (1967) concerning the highly publicized 'satiety' and 'feeding' centre concept. The literature on this subject has been critically reviewed by Hoebel (1971) and by Stevenson (1969).

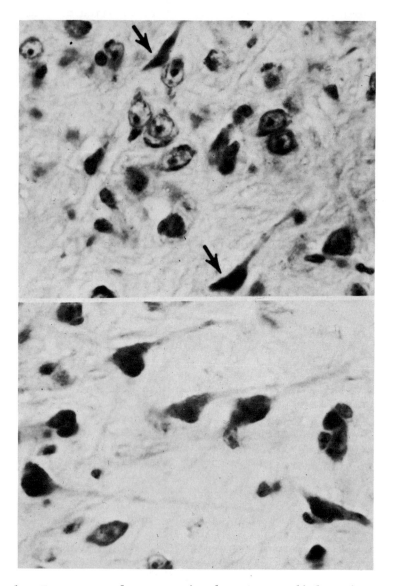

Fig. 1. Two types of neurons in the ventromedial nucleus of
the hypothalamus as seen in thionin stained coronal sections.
The upper section is from the lateral region of the nucleus
and shows bipolar neurons. The lower section, taken from
the ventral portion of the nucleus, shows several multipolar
neurons and two nearly bipolar neurons (arrows). (From
Murphy and Renaud, 1968; courtesy of Brain Research.)

A second type of neuron is shown in the lower section of Figure 1. This type was more commonly found in the central and ventral portions of the nucleus. It has multipolar dendrites, often four or five in number, which extend in all directions away from the soma. The latter is significantly larger than in the first type, having a maximum diameter of about 12 μ. The nucleus usually is more rounded and there are fewer darkly stained particles in the cytoplasm which, consequently, has a more pale appearance than that of the first type. Bipolar cells were, at times, seen in the same microscopic field as the multi-polar cells; two bipolar cells are indicated by arrows in the lower section of Figure 1.

The axons of these neurons are not demonstrated with this staining technique. However, clues as to their course are pro-vided by the Golgi material of Cajal (1911; cf. Fig. 313) and of Szentagothai (1963; cf. Figs. 24, 25). The peripheral bipolar neurons send axons into the central regions of the nucleus where they branch, and are distributed widely throughout the nucleus. These axons usually terminate as pericellular nests about other larger cells in the nucleus; the latter are probably equivalent to the multipolar neurons seen in our thionin sections. The axons of the multipolar cells apparently course in a primarily dorsal direction. They give off collaterals within the nucleus whose terminations are unknown, but which may be assumed, in the absence of specific information, to contact the inward facing dendrites of the bipolar neurons. The parent axon then leaves the nucleus. Thus, the multipolar neurons may be considered as the efferent portion of this nuclear system. The final route(s) of this efferent axonal pathway remain(s) a matter of conjecture. Some axons terminate within the hypothalamus (Arees and Meyer, 1967; Szentagothai et al., 1968); others probably are assimilated into the medial forebrain bundle, the dorsal longitudinal fasciculus and the tuberohypophysial tract (Crosby and Showers, 1969; Haymaker, 1969; Nauta and Haymaker, 1969). Further neuroanatom-ical investigation of this region is required to answer some of the questions raised by the existing observations.

The morphologic evidence suggested experiments to discover the neurophysiological presence and functional role of the bipolar cells. The experimental arrangements are outlined in Figure 2. In addition to the two separately controlled concentric stimulating electrodes in the amygdala, which could activate the stria termi-nalis (ST) and ventral amygdalo-fugal (VAF) pathways as described previously (Dreifuss, 1971), a very fine bore concentric stainless steel stimulating electrode (25 gauge o.d.; interpolar distance 0.1 mm) was placed stereotaxically at the lateral edge of the nucleus. We hoped that this would enable selective stimulating (LOC) of the presynaptic fibres to the bipolar cells and/or of the soma of these cells. Insulated tungsten microelectrodes

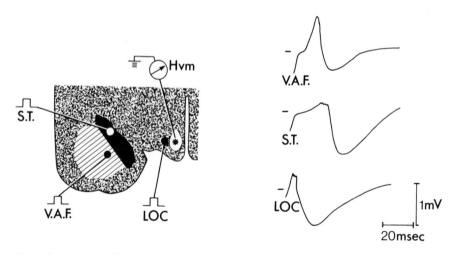

Fig. 2. A. A diagram of experimental arrangements. Full
explanation in text.
     B. Field potentials in the ventromedial nucleus evoked
by stimulation in the amygdala (ST and VAF) and at the lateral
edge of the nucleus itself (LOC). Downward deflection indicates positivity in all illustrations.

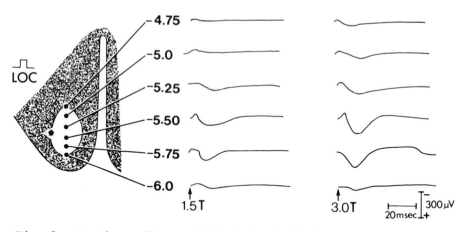

Fig. 3. Depth profiles of LOC induced field potentials in the
ventromedial nucleus. Horizontal (H) recording positions
relative to the inter-aural line are given in millimeters.
Stimulus intensities are relative to threshold (T) for eliciting
response in the centre of the nucleus. Asterisk indicates point
at which LOC stimulation was applied.

with tip diameters less than 1 μ were used for extracellular recording of both field potentials and single neuron spike trains within the nucleus. The positions of recording and stimulating electrodes were verified by histological examination.

A comparison, of the effects in the ventromedial nucleus of LOC stimulation with those of VAF and ST stimulation provides important preliminary evidence about the bipolar cells. Typical field potentials elicited by activation of each pathway are illustrated in Figure 2. It may be seen that the LOC-induced field potential resembles the ST induced potential, and, similarly, the second phase of the VAF potential. It is of a primarily positive polarity and its duration varies with stimulation intensity. This latter phenomenon may be taken as evidence of spatial summation of postsynaptic potentials (PSPs) within the nucleus, as there is now evidence that field potentials recorded in a volume conductor of neuronal tissue result mainly from PSP activity, with action potentials being to a large extent filtered out in such recording situations (Humphrey, 1968). From the analysis of the effects of the two amygdaloid efferent pathways upon single hypothalamic neurons (Dreifuss, 1971), it may be presumed that these potentials are, in fact, manifestations of inhibitory PSPs, since inhibition of cell firing is found during the positive phase of both the ST and VAF induced field potentials.

When the depth profile of the LOC-induced field potentials is examined with respect to extent, the potentials are found to be confined to the ventromedial nucleus (Fig. 3). Similar results are observed when the recording microelectrode is advanced in other planes indicating that almost no current flows outside the very roughly spherical nucleus in this experimental situation (Rall and Shepherd, 1968). Moreover, the point of maximum current flow in the experiment illustrated in Figure 3 is at the approximate geometric center of the nucleus. When we positioned the recording electrode at this point and applied stimulations at various locations in the lateral and medial hypothalamic regions, we found that the lateral edge of the nucleus requires the lowest stimulation intensity to evoke the characteristic field potential. This must mean that the greatest density of neural elements responsible for the LOC effect, namely presynaptic fibres and interneurons, are in this region.

The implication of these field potential studies is that the ST and VAF pathways traverse at least in part the region influenced by LOC stimulation. This premise may be tested directly by interacting LOC stimulation with either VAF or ST stimulation. An interesting example of such an experiment is shown in Figure 4. When just-threshold ST and sub-threshold LOC stimulations (Figure 4A) are interacted at suitable intervals, spatial

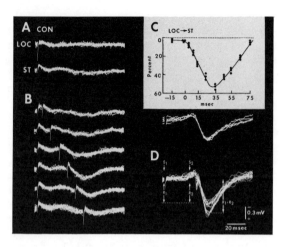

Fig. 4. Identity of synaptic terminals activated by stria
terminalis (ST) and local stimulation (LOC). A-C: facilitation
of response to stria terminalis stimulation by subthreshold local
stimulation. A: control responses in the ventromedial nucleus
to subthreshold local and just threshold stria terminalis stimulation. B: responses to the two stimuli at various intervals with
same stimulation intensities as in A. The first stimulation artifact in each tracing represents stria terminalis (test), the
second local (conditioning) stimulation. C: results of three
experiments in which the amplitudes of the positive wave,
illustrated in B, are plotted on the vertical axis as percent of
positive wave amplitude produced by 3 x threshold stria terminalis
stimulation alone (illustrated below graph). Horizontal axis
represents intervals, measured as real time, between stria
terminalis and local stimulation. Negative values refer to
instances in which local stimulation preceded stria terminalis
stimulation. The record depicting the response to 3 x threshold
stria terminalis stimulation, shown below the graph, has been
shifted to the left on the horizontal axis by an amount equal to
the differences in conduction time for the two pathways. D: interaction of 2 x threshold stria terminalis (S_1) and local (S_2)
stimulations at an interval of 27 msec demonstrating occlusion.
Amplitudes of the positive wave, measured from base line to peak,
of the responses to stria terminalis and local stimulation alone
are indicated by the upper dotted line (R_1 and R_2 respectively).
Similarly the amplitude of the response produced by the interaction of the two stimuli, at the same intensity, is given by the
lower dotted line ($R_1 + R_2$). Note that although R_1 and R_2 is only
approximately the same size, the response to the interaction of
the two stimuli ($R_1 + R_2$) is only slightly greater than either R_1
or R_2 alone. Vertical and horizontal calibrations refer to all
tracings in A-D. (From Murphy and Renaud, 1969; courtesy of
Journal of Neurophysiology).

facilitation occurs (Figure 4B). The time course of this
facilitation mirrors that of the ST evoked field potential
(Figure 4C) indicating that LOC and ST stimulation each result
in excitation of the same elements, namely, the bipolar cells
at the lateral periphery of the nucleus. This interpretation is
substantiated by the experiment illustrated in Figure 4D which
demonstrates spatial occlusion as a result of supra-threshold
LOC and ST stimulations interacted at an appropriate interval
determined by the response latency for each pathway. Thus, the
electrophysiological evidence indicates that the stria terminalis
excites synaptically the bipolar cells which, in turn, inhibit
multipolar cells within the nucleus. Similar phenomena are
observed when LOC stimulation is interacted with VAF stimulation.

It had been assumed previously that the ST did not reach the
medial hypothalamic regions (Nauta, 1961). Subsequent to the
neurophysiological experiments described herein, Heimer and Nauta
(1969) reinvestigated the ST projections with the aid of the
Fink-Heimer modification of the Nauta-Gygax staining technique
(Fink and Heimer, 1967). This technique is believed to stain the
terminal boutons and arborizations which exhibit Wallerian
degeneration after sectioning of the parent axons. Their results
showed not only that ST axons terminate in the medial hypothalamus,
but also that they are clustered about the periphery of the ventromedial nucleus (Figure 5). While these experiments in the rat
have not yet been confirmed in higher mammals, they appear to
support to some extent the electrophysiological data concerning
the stria terminalis, as the latter has a remarkably uniform distribution in different species (Klinger and Gloor, 1960).

Our interpretation of these results is that both the ST and
VAF pathways to the ventromedial nucleus exert their inhibitory
effect by activating synaptically bipolar interneurons located
mainly at the lateral margin of the nucleus. As these neurons
are relatively small, they would be expected to generate action
potentials of low amplitude. This is in fact the case. Figure 6
illustrates this and other aspects of bipolar neuron firing
patterns in response to VAF stimulation. These neurons respond,
usually with a high frequency train of spikes, in contrast to the
multipolar neurons which characteristically fire once in response
to VAF stimulation (Dreifuss et al., 1968). Similar trains also
occur with ST or LOC stimulation. Their onset coincides with the
negative phase of the VAF induced field potential as seen in the
lower trace of Figure 6 or with the small negativity which usually
precedes the ST or LOC induced field potentials (Figure 2). This
suggests that the action potentials are the result of excitatory
PSPs by virtue of the theoretical considerations discussed
previously and of the experimental evidence that negative field
potentials in the ventromedial nucleus are associated with
excitatory events (Dreifuss, 1971).

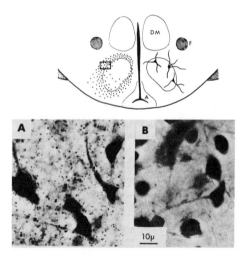

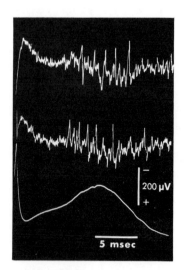

Fig. 5. Photograph A shows terminal degeneration near the
lateral border of the ventromedial nucleus in a rat sacrificed
4 days after ipsilateral amygdalectomy. Approximate position of
the field shown is indicated by rectangular frame in upper drawing. B shows the corresponding field on the opposite side.
Fink-Heimer stain, procedure 1. (From Heimer and Nauta, 1969;
courtesy of Brain Research.)

Fig. 6. Firing patterns of bipolar neuron in response to activation of the VAF pathway by stimulation in the basomedial amygdala.
Breaks in tracings indicate stimulation artifacts. Slow potentials are filtered out in upper two traces, spikes in lowest
trace.

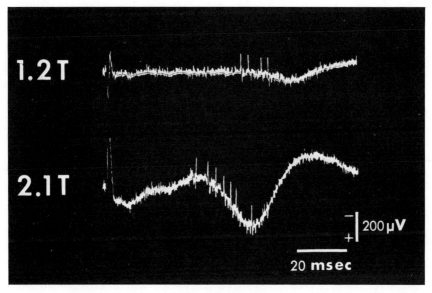

Fig. 7. Relationship between positive wave and bipolar neuron response to stria terminalis stimulation at different intensities of stimulation relative to threshold (T) for evoking the positive wave.

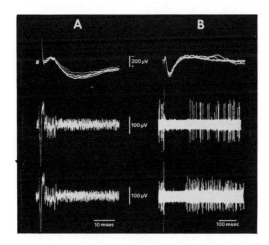

Fig. 8. Fast sweep speeds show burst response of bipolar neuron associated with negative phase of field potential (A). Slow sweep speeds show inhibition of multipolar neuron firing during positive phase (B). See text for additional explanation. The bandpass is 0.2-300 Hz for the field potential recordings, and 2-5 kHz for the unit recordings.

Further evidence that these unit potentials are a result of synaptic events is given in Figure 7. Here, it can be observed that the number of spikes in the train grades with the intensity of stimulation applied to the ST; in addition, the latency of the first spike is diminished with higher strength stimulation due to the greater number of synapses activated. This experiment also shows quite clearly the relationship of the bipolar spike train to the inhibitory events within the nucleus. The amplitude and latency of the positive wave, which is the indicator of inhibition, parallels the response of the bipolar neuron. Similarly, graded spike trains are characteristic of local interneuron behaviour elsewhere in the central nervous system (Eccles, 1969; Marco et al., 1967).

Multipolar neurons, which have significantly larger spike amplitudes, may, at times, be recorded at or near the same location as are the bipolar neurons. An example of this phenomenon, elicited by LOC stimulation, is shown in Figure 8. The relative constancy of the high frequency burst from a bipolar neuron in association with the initial negative phase of the field potential can be seen in Figure 8A. The firing of a multipolar neuron, recorded with the microelectrode in the same location, is inhibited during the positive phase of the field potential (Figure 8B, middle trace). A second multipolar neuron, similarly inhibited, is found after a 100 μ advancement of the microelectrode (Figure 8B, lowest trace). Rarely electrode placements allowed a demonstration that afferents excited by all three stimulations converge on the same bipolar neuron (Figure 9A) which, in turn, apparently contributes to the inhibition of a multipolar neuron found in the same recording location (Figure 9B). The high frequency burst from the bipolar neuron appears at a different latency, as expected, with each of the three stimulations in Figure 9. The fact that the duration of inhibition of the multipolar neurons is identical approximately in each case further suggests a common inhibitory mechanism for each pathway; we presume that the bipolar neuron is a significant part of this mechanism.

The stria terminalis, which is a discrete fibre bundle in its early course away from the amygdala, is relatively accessible for neuroanatomical study and has been well described both in terms of fibre composition (Fernandez de Molina and Garcia-Sanches, 1967) and terminal projections (Heimer and Nauta, 1969). The ventral amygdalo-fugal pathway is less accessible and no truly comparable neuroanatomical studies exist. Sectioning experiments indicate that it is extremely diffuse in its ventral passage from the amygdala to the ventromedial nucleus (Dreifuss et al., 1968), and the relatively long response latency suggests the possibility of one or more intervening synaptic delays. As the lateral hypo-

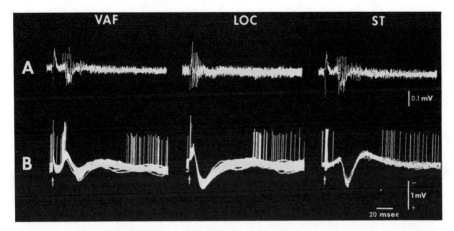

Fig. 9.   A: convergence of fibre systems activated by ventral
amygdalofugal, local and stria terminalis stimulation on the same
interneuron.   Each tracing represents a single response photographed at high gain after passage through a high-pass filter.
Barrage responses produced at a different latency by each pathway
are associated in each case with a negative wave (B).
B: a second unit of larger amplitude recorded at same time as
the interneuron in A and photographed at low gain.   In addition
to being inhibited for about the same duration by all 3 pathways,
this unit was also activated in association with the initial
negative wave by VAF stimulation.   Each example shows 100 superimposed responses.   Arrows refer to onset of stimulation artifact.
(From Murphy and Renaud, 1969; courtesy of J. Neurophysiol.)

thalamic area, an ill-defined collection of neurons and axons, lies directly between the amygdala and the ventromedial nucleus, it has been assumed logically to act as a relay for the VAF pathway (Heimer and Nauta, 1969). We have been unable to demonstrate this with lateral hypothalamic stimulation. This must suggest one of three possibilities: (1) that the VAF pathway does not traverse the lateral hypothalamus; (2) that the stimulating current is too weak or is ineffective for some other technical reason; or (3) that the concerned fibres or cell bodies are separated so widely in this area that insufficient numbers of neuronal elements are recruited by the stimulus. The latter possibility seems most tenable. In view of the postulated relationship between the lateral and ventromedial hypothalamus in the control of feeding and drinking behaviour, this particular problem requires further investigations. In this regard, negative (i.e. inhibitory) cross correlations between the firing patterns of lateral and ventromedial neurons have been reported (Oomura et al., 1964, 1967).

Whatever its intermediate course, the VAF pathway must, in part, enter the ventromedial nucleus in the regions delimited by the influence of LOC stimulation, at least, with respect to the part of the pathway which excites bipolar cells (Figures 6 and 9). The excitatory components of this pathway also enter, in part, at or near the lateral aspect of the nucleus, since LOC stimulation at times produces synaptic excitation of multipolar neurons (Figure 10A). The evidence that the activation is synaptic rather than antidromic or direct includes the relatively long latency which must include a synaptic delay, the slight jitter in the latency (Figure 10B), and the prolongation of latency and spike attenuation (Figure 10C) as well as eventual failure of excitation (Figure 10D) at high frequencies of stimulation. Inhibitory periods, similar to those produced with VAF stimulation, invariably follow the excitation.

DISCUSSION

Undoubtedly, it is obvious to all who read the literature on this subject how little we know either about how the hypothalamus functions or how the amygdala controls its operation. The studies described in this paper, despite its title, add very little to our knowledge on these large questions. Thus, any profitable discussion would perhaps best be directed towards future research strategies. The amygdalo-ventromedial nucleus system must serve as a prototype as we have virtually no detailed knowledge of circuitry at the cellular level in other levels of limbic-hypothalamic interaction. Thus, our consideration will be limited to this one system. As information becomes available at other levels it may be treated similarly. Five questions may help focus attention on some pertinent issues raised by our current knowledge concerning this

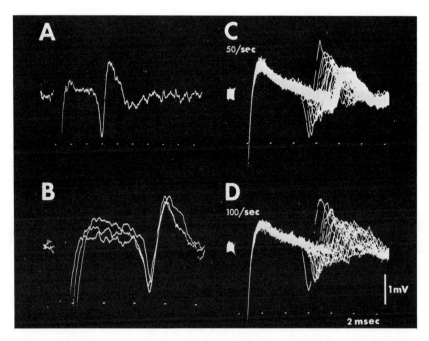

Fig. 10. Synaptic excitation of a multipolar neuron in the ventromedial nucleus by LOC stimulation. Break in each tracing denotes
stimulation artifact. A: Single response showing latency of
approximately 5.0 msec. B: Responses to stimulation at 1, 2,
and 4 x threshold intensity (lower, middle, and upper tracings,
respectively, after stimulation artifact) showing the relative
invariance of latency with increasing intensity. C: Super-
imposition of responses to 50/sec stimulation. Film was exposed
for about 3 sec beginning with onset of stimulation. Note final
stabilization of action potential amplitude at longer latency.
D: Same procedure as in C, but at higher stimulation frequency.
Note final inability of cell to follow this stimulation frequency
after gradual decline in amplitude and prolongation of latency.

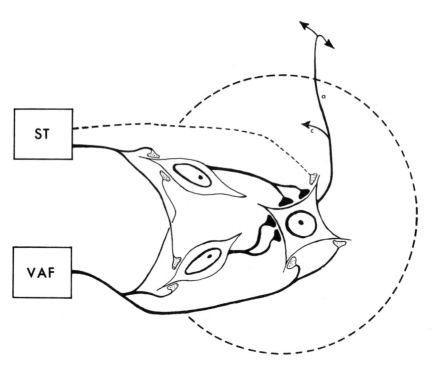

Fig. 11. Schematic interpretation of experimental results.
Light-coloured knobs represent excitatory, dark-coloured inhibitory presynaptic endings. Cells with bipolar dendritic processes
are inhibitory interneurons. Those with multipolar processes are
the recipients of this inhibition and are presumably the major
effector cells in the ventromedial nucleus. Note manner in which
the ventral amygdalofugal pathway might produce excitation-inhibi-
tion sequences. The rather ill-defined excitatory effects of the
stria terminalis pathway (Dreifuss et al., 1968) are represented
by the dotted line. The major axonal outflow (a) from multipolar
neurons courses in a primarily dorsal direction. Axon collaterals
within the nucleus exist (c) but their termination on inhibitory
interneurons is not yet certain. The apparent concentration of
inhibitory interneurons at the lateral edge of the nucleus places
these neurons in an excellent position to act as transmission
cells for the amygdala's inhibitory afferent pathways. (From
Murphy and Renaud, 1969; courtesy of J. Neurophysiol.)

system.

What insight to fundamental cellular mechanisms has been
gained?

Several conclusions may be drawn from the schematized interpretation illustrated in Figure 11.

1. Both efferent pathways from the amygdala pharmacologically
    must be excitatory, if Eccles' modification of Dale's
    principle (Eccles, 1964) is not to be contravened. The inter-
    vening bipolar interneuron provides the inhibitory function
    for each pathway. This implication may be taken into con-
    sideration in microinjection or microiontophoresis experiments
    on this system. It also may be of importance in our eventual
    understanding of how certain drugs are able to alter mood;

2. An inversion module is incorporated for both amygdaloid output
    pathways. Stated another way, a positive signal into the
    amygdala may, in the case of either pathway, result in a
    negative signal into the ventromedial nucleus. This conclusion
    is somewhat oversimplified in that in the course of the normal
    operation of the system it probably applies only to the ST
    pathway;

3. Direct transmission lines are present. We have no evidence
    that the VAF pathway can, in fact, produce inversion without
    an immediately preceding non-inverted signal. This positive
    (i.e. excitatory) signal is the shortest in latency and,
    thus, provides a relatively quick replication of amygdaloid
    state to the ventromedial nucleus;

4. Delay lines are available. The presence of an interposed
    neuron in a pathway provides the opportunity for delay or
    complete block of transmission, depending on the overall
    synaptic input to the interneuron. The bipolar interneurons
    may act in this capacity for both inhibitory pathways. The
    probability of a synaptic delay in the VAF excitatory pathway
    has been previously discussed (Dreifuss et al., 1968; Heimer
    and Nauta, 1969);

5. Computational error is reduced through redundancy (Von Neumann,
    1956). The redundancy takes two forms. First, there are
    multiple excitatory fibres in the VAF pathway and multiple
    inhibitory fibres from the interneurons; in either case, the
    fibres act in a uniform manner, thus yielding spatial redun-
    dancy. Second, there is temporal redundancy in the inhibitory
    actions; if the signals in the basomedial and corticomedial
    complexes of the amygdala simultaneously have the same sign,
    the inhibition produced by the former (VAF pathway) will be

enhanced profoundly by the latter (ST) about 15-30 msec later.
The critical question here is how does the sign at the two
amygdaloid regions in the normal operating state vary in time
with respect to a given function.

What additional information would be most helpful?

Most of the neurophysiological evidence which has been brought
forth in our series of studies bears on the forward flow of information from the amygdala to the hypothalamus. To understand fully
the normal operation of this or any other system, the amount of
feedback must also be determined. The presence of feedback imparts
to a linear system several important features (Horowitz, 1963)
including: (1) increased accuracy in response to input;
(2) reduced sensitivity to variations in system characteristics
of the ratio of output to input; this feature would be useful
especially in shielding a system which functions as this one does
essentially to provide homeostatis for the organism in the face
of ordinary physiological perturbations such as alterations in
circulation, electrolytes, or pH at the local tissue level;
(3) reduction in the effects of non-linearities or distortions
at the input, in this case at the amygdala; (4) increased band-
width or range of frequencies over which the system will respond
satisfactorily. There also are other more subtle ramifications
of feedback such as the tendency toward oscillation or instability
at certain frequencies of input.

Implicit in the above discussion is the requirement of
linearity in the neuronal interrelationships of the system. It
appears that CNS systems are indeed linear over physiologically
meaningful ranges (Werner and Mountcastle, 1965) and that any non-
linearities in a physiological situation are most likely to occur
at the level of peripheral sensory transducers (Stevens, 1970).
At any rate, some idea of the range of linearity must be obtained
for the amygdalo-ventromedial nucleus system in the context of
natural input, as opposed to electrical stimulation, both in
the feed-forward and feedback of information. Feedback from
multipolar neurons to either bipolar neurons or to neurons in the
basomedial or corticomedial amygdala is poorly understood at
present. There is two-way transmission in the stria terminalis
(Hilton and Zbrozyna, 1963; Crosby and Showers, 1969), but a
ventromedial origin of such fibres is not known (Valverde, 1965).
Amygdalopetal fibres of the VAF pathway appear to arise exclusively
in the lateral hypothalamus and course through the substantia
innominata (Crosby and Showers, 1969). The possibility of
collaterals of multipolar axons recurrently feeding back upon the
bipolar interneurons has been discussed previously (cf. Figure 11).
Some of these feedback questions may be answered by the converse
of the experiments described herein, i.e. restricted stimulation

in the ventromedial nucleus while recording from amygdaloid neurons or bipolar ventromedial neurons. Attempts at recording from the latter in such an experiment have yielded equivocal results (Murphy and Renaud, 1969).

How might the above information be used?

This is, of course, the most fundamental question of those raised thus far. A flurry of excitement occurred undoubtedly when man was able to alter dramatically an organism's behaviour by selective lesioning or stimulation in the central nervous system. Many of the most profound alterations, at least insofar as the limits of the observer's perceptions are concerned, have centred about hypothalamic function (Bard, 1928). While a great deal of descriptive information has been amassed in the past century about the effects of lesioning or stimulation, the early anticipation has certainly not been realized in that we are really very little further along in developing a general neuronal theory, analogous to the 'laws' of physics or chemistry, which explains behaviour[3]. What has often emerged, rather, is a confusing picture in which the variability of results is more impressive than the actual behavioural alterations incurred by the lesions or stimulations[4]. Clearly a different or modified approach is needed.

One such approach, as referred to in the INTRODUCTION, might be to overlay and interact different methodologies. However, for this approach to succeed, a common denominator must be found which can act as a unifying reference point. I believe that mathematical models incorporating information such as that in the preceding section would serve as a common denominator. Predictions could be derived as the model was refined which would be directly testable by each of the various methodologies. Stated more directly, a model might predict, as an example, the cellular, endocrine and behavioural result of an input action. These predictions might be tested simultaneously, or separately, and not necessarily by the same investigator in the latter case. Obviously, a team approach is to be encouraged and would be necessary in

---

[3] A few notable exceptions exist in the realm of invertebrate behaviour. Interesting examples of studies of central nervous system organization underlying behaviour include those on crayfish motor reflexes (Kennedy, 1968) and locust flight motor patterns (Wilson and Waldron, 1968).

[4] Some of the pertinent literature may be found in the collection of original papers edited by Isaacson (1964) and in the recent reviews of MacLean (1969), Stevenson (1969) and Hoebel (1971).

most instances.

Has any unifying concept emerged as a starting point for a model of this particular neural system?

The answer to this question must, of necessity, be equivocal. However, a hint of such a unifying concept exists. The neuro-physiological evidence about this system presented by Dr. Dreifuss (1971), and extended herein, shows clearly that the basomedial and corticomedial amygdala influences strongly only the ventro-medial nucleus among the various hypothalamic regions investigated (Dreifuss et al., 1968; Murphy et al., 1968). Moreover, it is likely that this nucleus plays some role in many homeostatic behavioural functions, rather than having a single function. These functions include feeding behaviour, sexual and mating behaviour, defense and aggressive behaviour and pituitary endocrine control[5]. It should also be noted that according to Szentagothai and co-workers (1968), the most abundant intra-hypothalamic connections originate from the ventromedial nucleus and spread from it to lateral, anterior and suprachiasmatic cell groups of the same and (via the supraoptic commissure) opposite sides.

These observations lead us to a reconsideration of the possi-bility that the ventromedial nucleus may indeed be a 'centre.' In this consideration, the emphasis in the term differs from that rejected previously in the INTRODUCTION. Thus, the ventromedial nucleus in this conception may be thought of as an adaptive centre, rather than a rigid one, capable of switching and sorting out among many homeostatic functions, and aided in this endeavour by, among other things, input from the amygdala. This concept is in direct conflict with that of Hess (1957), who derived from the classical lesioning and stimulation experiments the interpretation that 'specific autonomic functions are correlated with certain circumscribed regions of the diencephalon...' Hess went on to state that '...there is overlapping and intermingling of (these) various systems...' although the intended meaning of this apparent contradiction is unclear. It may be noted that there is some experimental support in the studies of Valenstein and coworkers (1970), albeit limited in extent to variations on oral activity, for the present idea of the ventromedial nucleus being an adaptive or 'plastic' centre.

---

[5] No attempt is made to review the literature concerning possible functions of the ventromedial nucleus. References on this subject may be found in Crosby et al. (1962); Murphy and Renaud (1969); Hoebel (1971) and in the compendium edited by Haymaker et al. (1969).

What problems may impede progress?

Two major difficulties, one conceptual and one procedural, must be faced. The conceptual problem relates to the probability that the amygdalo-ventromedial nucleus system, despite its attractiveness as a prototype system for understanding the neuronal basis of homeostatic behaviour in mammals, is, at most, only a minute slice of the neural substrate underlying this behaviour. This difficulty probably is not unsuperable, however, in that a larger comprehension may be attained gradually in modular fashion from an understanding of individual isolated systems. An incursive first step in expanding neurophysiological knowledge of this system would be to study effects of input systems to the ventromedial nucleus which are contained in the lateral and medial forebrain bundles, the fornix and the fronto-hypothalamic pathway (Nauta and Haymaker, 1969).

The second problem relates mainly to neurophysiological experiments and concerns the use of electrical stimulation of brain structures. The results of such stimulations are un-physiological for two reasons. First, they provide highly synchronous volleys rather than the more temporally spaced ones which occur naturally; second, they excite a variety of neurons, parts of neurons, and passing fibres alike in an indiscriminate fashion. Such experiments provide insight to neuronal circuitry, but are not helpful in elaborating the operations of this circuitry in a behavioural context. The use of chronically implanted microelectrodes to monitor neural activity during natural inputs together with the application of powerful stochastic analysis techniques may in large measure surmount this problem.

SUMMARY

Details about the pathways from the amygdala to the ventromedial nucleus of the hypothalamus as elaborated in neuroanatomical and neurophysiological experiments are presented. A consideration of their potential significance in the initiation of future research is provided. Some problems involved in using this system as an initial vehicle to develop a conceptual base concerning the neuronal organization underlying mammalian homeostatic behaviour are discussed.

ACKNOWLEDGMENTS

The author is supported by grants from the Medical Research Council of Canada (MA 4140) and from the Playfair Foundation of Toronto.

REFERENCES

AREES, E. A., & MEYER, J. Anatomical connections between medial
    and lateral regions of the hypothalamus concerned with food
    intake. Science , 1967, 157, 1574.

BARD, P. A diencephalic mechanism for the expression of rage
    with special reference to the sympathetic nervous system.
    American Journal of Physiology, 1928, 84, 490.

BARD, P., & MOUNTCASTLE, V. B. Some forebrain mechanisms involved
    in expression of rage with special reference to suppression
    of angry behaviour. Research Publications of the Association
    for Research in Nervous and Mental Disease, 1948, 27, 362.

CAJAL, S. R. Histologie du systeme nerveux de l'Homme et des
    Vertebres, Vol. II. Paris: Maloine, 1911.

CROSBY, E. C., & SHOWERS, M. J. C. Comparative anatomy of the
    preoptic and hypothalamic areas. In W. Haymaker, E.
    Anderson, and W. J. H. Nauta (Eds.), The Hypothalamus.
    Springfield: Charles C. Thomas, 1969. Pp. 61-135.

CROSBY, E. C., HUMPHREY, T., & LAUER, E. W. The Correlative
    Anatomy of the Nervous System. New York: MacMillan, 1962.

DREIFUSS, J. J. Effects of amygdaloid stimulation and functional
    subdivision of the amygdala. In B. Eleftheriou (Ed.), The
    Neurobiology of the Amygdala. New York: Plenum Press, 1971.

DREIFUSS, J. J., MURPHY, J. T., & GLOOR, P. Contrasting effects
    of two identified amygdaloid efferent pathways on single
    hypothalamic neurons. Journal of Neurophysiology, 1968,
    31, 237.

ECCLES, J. C. The Physiology of Synapses. New York: Academic
    Press, 1964.

ECCLES, J. C. The Inhibitory Pathways of the Central Nervous
    System. Springfield: Charles C. Thomas, 1969.

FERNANDEZ DE MOLINA, A., & GARCIA-SANCHEZ, J. L. The properties
    of stria terminalis fibers. Physiology & Behavior, 1967,
    2, 225.

FINK, R. P., & HEIMER, L. Two methods for selective silver
    impregnation of degenerating axons and their synaptic endings
    in the central nervous system. Brain Research, 1967, 4, 369.

GLOOR, P. Electrophysiological studies on the connections of the amygdaloid nucleus in the cat. I. The neuronal organization of the amygdaloid projection system. Electroencephalography and Clinical Neurophysiology, 1955a, 7, 223.

GLOOR, P. Electrophysiological studies on the connections of the amygdaloid nucleus in the cat. II. The electrophysiological properties of the amygdaloid projection system. Electroencephalography and Clinical Neurophysiology, 1955b, 7, 243.

HAYMAKER, W. Hypothalamo-pituitary neural pathways and the circulation system of the pituitary. In W. Haymaker, E. Anderson, and W. J. H. Nauta (Eds.), The Hypothalamus. Springfield: Charles C. Thomas, 1969. Pp. 219-250.

HAYMAKER, W., ANDERSON, E., & NAUTA, W. J. H. (Eds.) The Hypothalamus. Springfield: Charles C. Thomas, 1969.

HEIMER, L., & NAUTA, W. J. H. The hypothalamic distribution of the stria terminalis in the rat. Brain Research, 1969, 13, 284.

HESS, W. R. In J. R. Hughes (Ed.), The Functional Organization of the Diencephalon. New York: Grune and Stratton, 1957.

HILTON, S. M., & ZBROZYNA, A. W. Amygdaloid region for defense reactions and its efferent pathways to the brain stem. Journal of Physiology, 1963, 165, 160.

HOEBEL, B. G. Feeding: neural control of intake. Annual Review of Physiology, 1971, 33, 533.

HOROWITZ, I. M. Synthesis of Feedback Systems. New York: Academic Press, 1963.

HUMPHREY, D. R. Re-analysis of the antidromic cortical response. II. On the contribution of cell discharge and PSPs to the evoked potentials. Electroencephalography and Clinical Neurophysiology, 1968, 25, 421.

ISAACSON, R. L. (Ed.). Basic Readings in Neuropsychology. New York: Harper and Row, 1964.

KAADA, B. R. Somatomotor, autonomic and electrocorticographic responses to electrical stimulation of 'rhinencephalic' and other structures in primates, cat and dog. Acta Physiologica Scandinavica, 1951, 24 (Suppl. 83), 1.

KENNEDY, D.   Input and output connections of single arthropod
    neurons.   In F. D. Carlson (Ed.), Physiological and Bio-
    chemical Aspects of Nervous Integration.   Englewood Cliffs:
    Prentice-Hall, 1968.   Pp. 285-306.

KLINGLER, J., & GLOOR, P.   The connections of the amygdala and of
    the anterior temporal cortex in the human brain.   Journal of
    Comparative Neurology, 1960, 115, 333.

KLÜVER, H., & BUCY, P. D.   Preliminary analysis of the functions
    of the temporal lobes in monkeys.   Archives of Neurology
    and Psychiatry, 1939, 42, 979.

MacLEAN, P. D.   The hypothalamus and emotional behavior.   In
    W. Haymaker, E. Anderson, and W. J. H. Nauta (Eds.), The
    Hypothalamus.   Springfield: Charles C. Thomas, 1969,
    Pp. 659-678.

MARCO, L. A., BROWN, T. S., & ROUSE, M. E.   Unitary responses in
    ventrolateral thalamus upon intranuclear stimulation.
    Journal of Neurophysiology, 1967, 30, 482.

MURPHY, J. T., & RENAUD, L. P.   Inhibitory interneurons in the
    ventromedial nucleus of the hypothalamus.   Brain Research,
    1968, 9, 385.

MURPHY, J. T., & RENAUD, L. P.   Mechanisms of inhibition in the
    ventromedial nucleus of the hypothalamus.   Journal of Neuro-
    physiology, 1969, 32, 85.

MURPHY, J. T., DREIFUSS, J. T., & GLOOR, P.   Topographical differ-
    ences in the responses of single hypothalamic neurons to
    limbic stimulation.   American Journal of Physiology, 1968,
    214, 1443.

NAUTA, W. J. H.   Fibre degeneration following lesions of the
    amygdaloid complex in the monkey.   Journal of Anatomy, 1961,
    95, 515.

NAUTA, W. J. H., & HAYMAKER, W.   Hypothalamic nuclei and fiber
    connections.   In The Hypothalamus.   Springfield: Charles C.
    Thomas, 1969.   Pp. 136-269.

OOMURA, Y., KIMURA, K., OOYAMA, H., MAENO, T., IKI, M., &
    KUNIYOSHI, M.   Reciprocal activities of the ventromedial and
    lateral hypothalamic areas of cats.   Science, 1964, 143, 484.

OOMURA, Y., OOYAMA, H., YAMAMOTO, T., & NAKA, F. Neural mechanism of feeding. In W. R. Adey and T. Tokizane (Eds.), Progress in Brain Research, Vol. 27, Structure and Function of the Limbic System. Amsterdam: Elsevier, 1967. Pp. 1-33.

RALL, W., & SHEPHERD, G. M. Theoretical reconstruction of field potentials and dendrodendritic synaptic interactions in olfactory bulb. Journal of Neurophysiology, 1968, 31, 884.

REXED, B. Some aspects of the cytoarchitectonics and synaptology of the spinal cord. In J. C. Eccles and J. P. Schade (Eds.), Progress in Brain Research, Vol. 11, Organization of the Spinal Cord. Amsterdam: Elsevier, 1964. Pp. 58-92.

SCHEIBEL, M. E., & SCHEIBEL, A. B. Structural substrates for integrative patterns in the brain stem reticular core. In H. H. Jasper et al. (Eds.), Reticular Formation of the Brain. Boston: Little, Brown and Co., 1958. Pp. 31-55.

STEVENS, S. S. Neural events and the psychophysical law. Science, 1970, 170, 1043.

STEVENSON, J. A. F. Neural control of food and water intake. In W. Haymaker, E. Anderson, and W. J. H. Nauta (Eds.), The Hypothalamus. Springfield: Charles C. Thomas, 1969. Pp. 524-621.

SZENTAGOTHAI, J., FLERKO, B., MESS, B., & HALASZ, B. Hypothalamic Control of the Anterior Pituitary. An Experimental Morphological Study. Budapest: Akademiai Kiado, 1968.

VALENSTEIN, E. S., COX, V. C., & KAKOLEWSKI, J. W. Re-examination of the role of the hypothalamus in motivation. Psychological Review, 1970, 77, 16.

VALVERDE, F. Studies on the Piriform Lobe. Cambridge: Harvard University Press, 1965.

VON NEUMANN, J. Probabilistic logic and the synthesis of reliable organisms from unreliable components. In C. E. Shannon and J. McCarthy (Eds.), Automata Studies. Princeton Univ., 1956.

WERNER, G., & MOUNTCASTLE, V. B. Neural activity in mechanoreceptive cutaneous afferents: stimulus-response relations, Weber functions, and information transmission. Journal of Neurophysiology, 1965, 28, 359.

WILSON, D. M., & WALDRON, I. Models for the generation of the motor output pattern in flying locusts. Proceedings IEEE, 1968, 56, 1058.

THE HUMAN AMYGDALA: ELECTROPHYSIOLOGICAL STUDIES

Mary A. B. Brazier

Brain Research Institute, University of California

Los Angeles, California

INTRODUCTION

This report will be restricted to electrophysiological studies
of the amygdala and related structures in man and will not present
material from parallel experiments in this laboratory on cats
because of the extreme species differences found. In fact, this
is one of the most striking results that has emerged from running
similar studies in the two species. It will, however, be necessary
to refer from time to time to the anatomical connections of the
amygdala in lower animals, and especially in Macaque, for the
sources for human material are more limited.

The goal has been to search for electrophysiological evidence
of the activity in known connections of the amygdala when these are
in functional use in the various conditions that the living brain
experiences. The anatomists have established the morphology of the
fiber tracts; this electrophysiological effort attempts to define
which of these pathways functions during life in such varying
states as sleep and alertness, or under the influence of various
drugs including anesthetic agents, as well as during the trauma of
elipeptic seizures. The circuitry is known, but when is it used?

The principal functional relationships of the amygdala which
will be described here have suggested themselves for exploration
by the following anatomical findings established by others: connections with the septum, with the dorsal medial nucleus of the
thalamus, and the controversial connection with the hippocampus.
There has been no clinical, and hence no ethical, reason for
electrode placements in the hypothalamus or brain stem of the

patients studied, and hence no report can be given of electrical
activity in those pathways.

PATIENT SERIES

The human subjects on whom this report is based are mostly
patients with temporal lobe epilepsy being studied under a joint
clinical and neurophysiological research program at the Brain
Research Institute at the University of California Los Angeles
in which Dr. Paul Crandall of the Division of Neurosurgery and
Dr. Richard Walter of the Department of Neurology are the key
figures.

The program, which is supported by NINDS*, is well named for
two main streams of information have been derived from it -- the
one clinical and the other neurophysiological. It is the latter
information that will be reported here in as far as it is germane
to the amygdala.

The patients in this series, who now number over 50, are
cases of epilepsy whose seizures have proved uncontrollable by
medication and whose EEGs, as recorded from the scalp or from
sphenoidal leads, give insufficiently clear lateralizing signs,
either when they are awake, or asleep, with any of the usual
activating techniques used in clinical electroencephalography.
These patients are, therefore, candidates for therapeutic surgery
by Dr. Crandall. In the cases of temporal lobe epilepsy, who
form the majority of cases studied so far, lateralization of the
more impaired hemisphere is of importance, bilateral temporal resection being ruled out owing to the severe impairment of memory
that ensues (Scoville and Milner, 1957; Milner, 1959).

In addition to the epileptic patients, there has been the
opportunity to record from 3 non-epileptic chronic psychotic
patients in whom indwelling electrodes had been placed in the
course of a research project in the Department of Psychiatry.
These, as non-epileptic brains, form our only available "controls"
for the major series of patients.

ELECTRODE PLACEMENTS

The patients, by their own consent and by that of their next
of kin, have electrodes inserted through twist drill holes into
various structures as indicated by the clinical signs and the need
for refinement of the diagnostic information.

---

* This Clinical Neurophysiology Program is supported by Grant
NS 02808 from the National Institutes of Health.

Dr. Crandall implants the electodes according to stereotactic coordinates determined from internal landmarks viewed by X-ray using contrast media (Crandall, 1963). Placements vary according to the patient's symptoms. In the largest series, the temporal lobe epileptics in whom lateralization and possible localization is sought, the usual placements include 3 bipolar electrodes inserted into the hippocampus of each hemisphere; three in each hippocampal gyrus; and at least one pair in each amygdala. In those patients whose symptoms include some centrencephalic signs, electrodes also have been placed in the centre median and anterior nucleus of the thalamus and, occasionally, in the ventrolateral nucleus. Some patients, whose symptoms so indicate, have had electrodes placed in the dorsal medial nucleus of the thalamus and others in the cingulate gyrus. In every case, the placements have been determined by the clinical problem presented by the case. In addition to the deep electrodes, all patients have a minimum of 12 cortical electrodes inserted through the skull at operation so that they just reach the dura.

The important feature of these depth implantations is that the electrodes are left in place for 4 to 6 weeks so that recordings may be made during many varying states of the patient's behavior: waking, sleeping, actively engaged in various situations including interviews, and during different forms of sensory stimulation: photic, auditory or somato-sensory, as well as under the influence of various drugs.

Electrode placement, decided initially by stereotactic measurement, is checked by X-ray for both hemispheres and later by histology in the removed lobe. The close agreement of the histological check with the previous X-ray has led to considerable confidence in the placements in the unoperated hemisphere.

TESTING PROCEDURES

The electrodes are insulated bipolar stainless steel with tips one above the other, each bared of insulation for 1 mm, thus giving an ovoid stimulus zone, if used for stimulation. Either point can be used also as a unipole with reference to an electrode elsewhere. This reference electrode must not, of course, be on the ear or mastoid, for these act as recorders from the temporal convexity. This error of interpretation crept early into clinical electroencephalography although well exposed by Hill (1950). Recordings are made simultaneously on a 16-channel ink writer and on either a 14-channel or a 32-channel

F.M. tape recorder.   All records receive computer analysis.*

The analyses most frequently used have been: frequency
specta of the on-going activity; coherence calculation of frequencies common to any pair of recording sites; phase relations
of any wave-trains common to a pair; and averaging of evoked
potentials.**

On-line, while the patient is still in the laboratory,
frequency spectra of one channel at a time can be monitored on
the laboratory's PDP 12 computer and coherences between a pair
of sites calculated. This facility is used only for monitoring
and editing, for it is too slow for processing coherences
between all possible pairs that are calculable between the
32 or more electrodes that many patients have (including the
superficial ones).

The computer program used allows the investigator to choose
the parameters for analysis. In the tests reported here, frequency bands of 2 cps generally have been chosen, analysis being
made of 30-second epochs through an EEG recording of at least
7 1/2 min. in real time. The parameters chosen affect the
degrees of freedom and hence of the level of coherence needed
for significance. Coherence depends only on frequencies
common to two loci and is independent of amplitude.

In the section of this multifaceted program that is reported
here, results of studies of the on-going wave activity will be
emphasized. This approach has been developed as ancillary to
the usual search for spike potentials. The latter have been the
specific sign of epileptogenic tissue sought classically by
electroencephalographers ever since the first demonstration of
experimental epilepsy (Kaufman, 1912).

However, studies of the potential fields around a spike show
them to involve an extremely restricted region so that the high
probability of not having an electrode in that field may hide
this activity from the electroencephalographer. This is a well-

---

* For computing, other than that on the PDP 12 in this investigator's own laboratory, the facilities have been used of the Data
Processing Laboratory, in the Brain Research Institute (supported
by NS 02501) and the Health Sciences Computer Facility in the
School of Medicine (FR-3) from the U. S. Public Health Service.

** The programs used are available from the BMD (Biomedical
Computer Programs) publication of the Health Sciences Computer
Facility at UCLA. (The Fast Fourier Transform X92 and the Aver
Program of the Data Processing Laboratory).

known hazard in scalp electroencephalography, but becomes a
serious handicap when using a very restricted number of inserted
electrodes.

Wave activity on the other hand has more extensive fields
(Brazier, 1949, 1951, 1961) and, when recorded from neuronal
aggregates, is found to cover more extensive regions. Thus, in
the section of this research devoted to refining lateralizing
and localizing techniques, for diagnostic purposes, one line
has been the development of more intensive study (using computer
analyses) of the on-going wave activity in the two hemispheres
of these patients with a comparison of the activity from homologous contralateral recording points. In the course of analysing
records from 50 of these patients and from 3 non-epileptic cases,
it is now possible to estimate the probability of normality or
otherwise of the on-going activity found. The clinical develop-
ments of these analyses will not be reported here and have, in
part, been published (Brazier, 1965, 1966a, 1966b, 1967a, 1967b,
1968a, 1968b, 1969). The criteria that have a high probability
for classification of normality in activity of limbic circuits
are described below.

RESULTS ON THE AWAKE STATE

A. <u>Electrogenesis of the amygdaloid complex.</u>

In comparison with other nuclear structures in the human
brain, the amygdaloid complex has poor electrogenesis and what
there is proves to be largely in the theta band. This observation holds for all our patients and for the unoperated, and
hence deemed less abnormal, hemisphere. It also holds for the
non-epileptic patients in the series.

Even when strong alpha activity is present, not only at the
occiput, but in leads from the temporal scalp, the amygdala shows
a quite different frequency spectrum (Figure 1).

The recordings shown in Figure 1 were made from the left (unoperated) hemisphere of a wide-awake patient whose abnormal activity had been found in the homologous placement on the right. The
amygdaloid leads, which matched position by X-ray bilaterally,
were found by histology* on the right in the basolateral nucleus,
the most prominent division of the amygdaloid complex in man
(Crosby and Humphrey, 1941).

---

* The author is indebted to Dr. P. H. Crandall for the histological reports on the tissue removed in these patients.

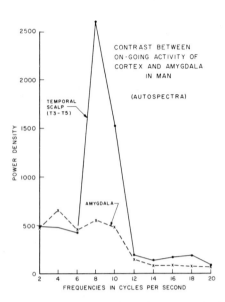

Fig. 1. Frequency spectra recorded from an indwelling bipolar electrode in the basolateral part of the amygdala in man and simultaneously from electrodes inserted through the skull over the temporal convexity. Computer averaged activity over a 7.5 minute recording period. Subject awake.

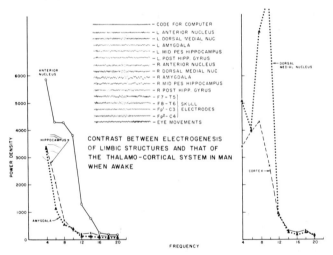

Fig. 2. Frequency spectra recorded from indwelling electrodes in a non-epileptic patient. Note the comparatively poor electrogenesis in the amygdala and hippocampus and the lower frequencies dominant in these leads when contrasted with the thalamo-cortical recordings. Computer averages. Subject awake.

Figure 2 is from a non-epileptic patient whose electrode placements were checked by X-ray since no operation followed. On the left, the limbic structures, amygdala and hippocampus, are seen to exhibit minimal activity as fast as 8 cps whereas this is the dominant frequency in the overlying cortex (F8 - T6) while the dorsal medial nucleus has maximal electrogenesis at 8 and 10 cps. The clinical electroencephalographer must not expect scalp recordings to give him any information about the on-going electrical activity of the amygdala.

B.   Study of on-going activity in limbic connections.

In an attempt to find an alternative to the classical, but extremely unphysiological, technique of electrical stimulation in one site and recording, either of evoked potentials or after discharges in another, a variation of a technique developed by Petsche and his colleagues in Vienna has been adopted (Petsche, 1960, 1962, 1965, 1970; Gogolák, 1967, 1968; Brücke, 1959).

Although useful in clinical electroencephalography, the evocation of afterdischarges cannot give information about physiological use of neuronal connections, for electrical stimulation has the effect of synchronizing all discharge in the stimulus site thus impinging a totally abnormally concentrated barrage of impulses at the next synapses, a barrage condensed in temporal terms and probably excessive in threshold.

The technique of Petsche and his group is to follow the ongoing activity of wave trains from one locus to another in the brain. These workers have used an ingenious method to compute the phase shift between peaks of the dominant theta waves in the two locations. Their studies were initially made in the rabbit in whom theta activity is almost monorhythmic and can be evoked by any arousing stimulus. This is not so in man where theta activity is unrelated to arousal and is accompanied by many other frequency components.

Petsche's initial observation was that electrodes implanted in other regions of the brain, thalamus, hypothalamus and septum, all showed the same rhythm though not exactly synchronously. There was always some detectable phase difference when the peaks were compared. To cut a long story short, it was found that the waves in the septum of the rabbit led all the rest in phase and appeared therefore to be a pacemaker. Moreover lesions made in the medial portion of the septum abolished theta rhythm in all regions of the brain (Mayer and Stumpf, 1958). Moving then to unit recording in the septum (Petsche, 1965), they were able to define a class of septal cell units which discharged in bursts, each burst being locked to the same phase of a theta wave.

C. Adoption of this line of exploration to man.

In our human subjects, the proliferation of theta waves in
limbic structures is striking, although they do not show an
arousal characteristic as in the rabbit. It seemed then of interest to explore whether or not similar coherences of ave trains
existed in man and, if so, whether or not they exhibited the
locked phase relations found in the rabbit.

Similar studies have therefore been made in our patients
using the electrode placements already available. Such analyses
require computer aid since in man the theta activity is not monorhythmic and needs selection by filtering before any single
frequency band can be studied. For this purpose a program available in our Health Sciences Computing Facility that calculates
the coherence between wave trains has been used.* This computer
analysis detects the presence of frequencies common to any two
recording sites and, for each frequency, calculates the percentage
of the activity that occurs in both places and the phase lag existing between any two wave trains. From this phase displacement one
can calculate the time displacement in the pathway.

1. Amygdala and septum

The term "septum" is used by some workers for a very large
region in the human brain, but we owe to Andy and Stephan (1968)
an intensive study in which they differentiate a restricted zone,
the septum verum, from the septum pellucidum. Andy and Stephan
restrict this latter term to the dorsal part in which they found
fiber tracts and glia but no cell bodies. Klinger and Gloor
(1960) regard the septum pellucidum in man as the homolog of the
lateral septal nuclei in lower animals.

It is not known whether the results reported below will prove
to be the basis for generalizations, for these recordings have
been possible only in the very small number of psychotic patients
studied, there being no clinical reason for septal placements in
the epileptic patients. There is also, of course, no histological
check available since these cases do not come to operation, but
X-ray check places the recording points outside the restricted
zone of the septum verum. However, as abundant wave trains of
theta activity were recorded, it is assumed that these probably
emanate from the potential field of the nucleus of the diagonal
band of Broca. Figure 3 illustrates the results from one of our
patients; when he was awake, both sites, amygdala and septum, had
dominant wave activity at 6 cps and at this frequency (and this
frequency only) there was a consistent coherence in all fifteen
30-second epochs analyzed throughout the 7 1/2 minute recording

---

* X92 BMD

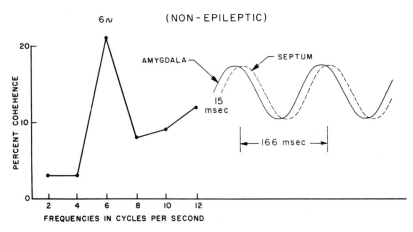

Fig. 3. On the left, the graph found for coherences between
frequency bands present in amygdala and septum of a non-epileptic
patient. On the right a schematic representation of the lag of
the septal waves behind those of the amygdala. Computer averages
of 7 1/2 minutes' recordings. Subject awake.

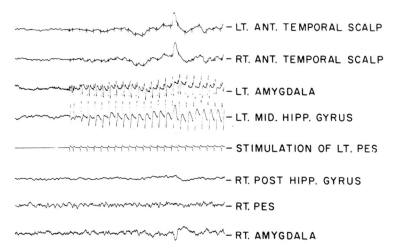

Fig. 4. "Recruitment" evoked in the ipsilateral amygdala of man
by repetitive pulsed stimuli in the hippocampus. Note that only
the ipsilateral amygdala and hippocampal gyrus respond. There is
no contralateral effect and the surface electrodes pick up only
the stimulus artifact.

in real time. In every 30-second sample of recording the amygdaloid waves led consistently the septal waves. The average
phase displacement was 33° which, for a frequency of 6 cps
(i.e. of waves 166 msec in duration), indicates a lead-time of
15 milliseconds between peaks.

From this extremely consistent result one may infer that a
pacemaking influence for this theta rhythm probably is exerted
on the septal activity by the amygdala in the waking state.
Anatomical connections running in this direction via the stria
terminalis have been found in man by Klinger and Gloor (1960)
and it may be presumed that these are the structural substrate
of this electrophysiological phenomenon in the living brain.

2. Amygdala and dorsal medial nucleus of the thalamus

Moving to the other strong anatomical link found by the
anatomists (Angevine et al., 1961; Klinger and Gloor, 1960),
namely, that between the amygdala and the dorsal medial nucleus
of the thalamus, one may expect some electrophysiological interaction from the results of several workers who have reported
that electrical stimulation of the amygdala evokes responses in
the dorsal medial nucleus of lower animals.

Using the less abnormal electrophysiological evidence of
relations between the on-going wave trains in the two loci, i.e.
by computing the coherence between amygdala and thalamus, this
was again found only in the 6 cycle theta although the dominant
frequency of this thalamic nucleus was 10 cps. In this case,
the phase displacement showed the thalamic site to lag behind
the amygdala by 15°. For a 6 cycle wave train, this means a
time displacement of 7 msec between peaks with the amygdala
leading. This is the direction found in the electrical stimulation experiments of Ajmone Marsan and Stoll (1951) and of Gloor
(1955) in the cat.

The anatomical substrate for this connection found by Vogt
(1898), by Fox (1949) and since confirmed by Nauta (1961, 1962)
and by Nauta and Valenstein (1958) in the monkey has been identified by Klinger and Gloor (1960) in man. Our tests in man shed
no light on any theta-pacemaking influence passing in the opposite
direction although doubtless some fibers exist.

3. Septum and hippocampus

If the septum is indeed the pacemaker for hippocampal
activity, as seems proven in lower animals, and provided one
respects species differences, one might well expect some correlation between septal and hippocampal activity. However, in spite

of some comparative studies (Andy et al., 1962) almost all the reports describing work on lower animals are on the dorsal hippocampus which has so scant an analogous structure in the human brain--just the small strip adjacent to the corpus callosum, the hippocampal rudiment or induseum griseum.

It is possibly owing to this marked species difference that, although strong coherences were found between septum and hippocampus, no consistently constant phase relationship was present. Such phase relations as were found indicated a long time lag of the hippocampal wave trains behind those in the septum but were so variable as to make a definitive statement about this interrelationship unwise, especially in the light of the small number of opportunities to look for data on this point.

4. Hippocampus to amygdala

For hippocampal connections to the amygdala there is scant anatomical evidence. Yet, several categories of electrophysiological observations imply such connections, e.g. responses evoked in the amygdala by stimulation of the hippocampus in lower animals (Green and Adey, 1956; Green, 1964; Gloor, 1955, 1959).

Pulsed electrical stimuli to the hippocampus in man evoke a response in the ipsilateral amygdala (Brazier, 1964). This is of variable latency and appears to travel by some polysynaptic route. Such a conclusion is strengthened by the fact that recruitment* can be obtained in the amygdala by repetitive stimulation in the hippocampus of man (Fig. 4).

5. Amygdala to hippocampus

In this direction, there is more suggestion of anatomical connections though some controversy still exists (Gloor, 1959). In our hands, pulsed stimuli to the amygdala of man evokes responses in the ipsilateral hippocampus. These are of variable latency and appear to be polysynaptic in pathway, possibly via the pyriform cortex or the long circuit through the septum. They

---

* This term was given to this electrophysiological phenomenon by Morison and Dempsey (1942) in their classic work on repetitive stimulation of the non-specific thalamic projection system. This was before knowledge had been obtained of excitatory postsynaptic potentials and their ability to summate on repetitive stimulation. The earlier explanation (that more and more cells were drawn in) led to the term "recruitment," which can therefore be misleading, though firmly embedded in neurophysiological texts. The term is used here for that reason.

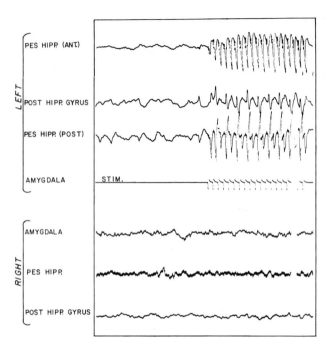

Fig. 5. "Recruitment" evoked in the ipsilateral hippocampus of man by stimulation in the amygdala. Note lack of effect in contralateral hemisphere. Subject awake.

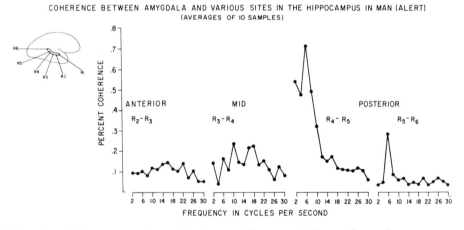

Fig. 6. Coherences found between the amygdala and various zones of the ipsilateral hippocampus as recorded by linkages along its anterior-posterior extension. Electrode placements were 6 mm apart. Computer averages. Subject awake.

also, with suitable stimulus rates, exhibit the phenomenon of recruitment (Fig. 5).

Avoiding the unphysiological technique of electrical stimulation and studying the on-going wave activity in awake man, high coherences are found between some (but not all) regions of the ipsilateral hippocampus. Elsewhere data on the regional differences of various zones in the human hippocampus have been published (Brazier, 1970) and it is by study of these hippocampal zones that the incidence of coherence with the amygdala has emerged.

In the patient whose results are shown in Figure 6, six unipolar electrodes had been placed in each hippocampus. All tests led to the implication that the left hemisphere was the abnormal side, so only results from the right are illustrated here. When the hippocampal leads were linked in pairs, as indicated in the diagram, the only significant coherence found with the ipsilateral amygdala was with the linkages labelled R_4 - R_5, with a trace in R_5 - R_6. It will be noticed that this coherence was maximal for the 6 cps band.

Although these high coherences were found, the phase displacement of the waves was far less consistent, and usually very long. One would infer some polysynaptic and, possibly, complex interaction. However, the inconsistency of the phase relations makes it unlikely that this is a pacemaker reaction. It seems more likely that this merely reflects the preferred frequency of the neurons in these limbic structures.

6. Amygdala and centre median

In 1965, a paper was published by Sommer-Smith and his associates describing responses evoked in the centre median of the thalamus by electrical stimulation of the amygdala and septum in cats. Jasper (1958) has published a schematic diagram showing a connection between amygdala and centre median in man, but the accompanying text does not give the data on which this is based.

The type of response Sommer-Smith (1965) was studying was a change in the firing pattern of single units in the centre median. He found stimulation of the septum to be more effective than the amygdala in provoking changes in firing rate.

There is no patient in our series who had both septal and centre median placements, and thus no report on this can be given. Several cases have had both amygdala and centre median placements, but all tests for electrophysiological interrelationships were negative. There was no coherence found between the two sites in

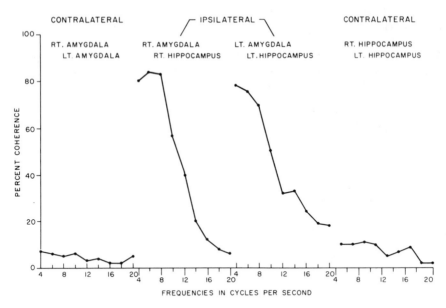

Fig. 7.   Coherence levels found in man ipsilaterally and contra-
laterally in various pairings of amygdala and hippocampus.   See
text.   Subject awake.

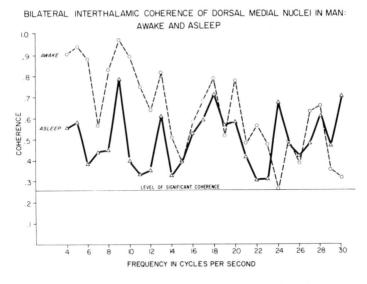

Fig. 8.   Contralateral coherences of wave activity between electrode
sites in the left and right dorsal medial thalamic nuclei at all
frequencies from 4 to 30 cps.   Non-epileptic patient.   Broken curve
when awake, solid curve when asleep.   (Reproduced from Brazier,
1968c).

any patient, let alone any phase relationships between their on-
going activities.

7. Contralateral electrophysiological interaction

Of more interest, perhaps, in a report bearing on the amygdala, is that these high coherences are entirely restricted to
the ipsilateral hemisphere. No correlations were found between
the wave trains of one amygdala with those of its opposite
homolog, and the same holds for the activity of the two hippo-
campi. Figure 7 illustrates these points from one of our patients.
The result cannot be explained as a feature of the epileptic brain
for similar results were found in the non-epileptic patients.
Possibly only the cortico-medial nuclei have commissural connections.

One of the strongest coherences in limbic structures is that
between certain zones of the hippocampus and the hippocampal gyrus.
These zones have been explored and defined in a previous publication (Brazier, 1970), and will not be repeated in this symposium
devoted to the amygdala. It may be noted, however, this, too, is
an ipsilateral relationship only.

The same lack of contralateral effect was found by pulsed
stimulation of either the amygdala or the hippocampus (Brazier,
1964). One explanation of the lack of evoked responses could be
that the contralateral recording electrode must lie in the exact
receiving zone of the afferent axons of the homologous region.
As these are gross electrodes this seems an unlikely requisite.
Moreover, using the less abnormal test for coherence of on-going
wave activity (which has a wide potential field) the exact
position of the receiving neurons is less exacting.

Possibly the weak stimuli*, coupled with computer averaging
to emphasize small responses used in this laboratory in the search
for evoked potentials may be the clue. It is suggested that if
the stimulus is insufficiently strong to propagate from the
amygdala to the thalamus, impulses will not cross to the opposite
hemisphere in man. Excessive strengths of stimulation (sufficient
to evoke afterdischarge) may force this route when spreading to
the opposite hemisphere.

Interhemispheric coherence of wave activity between dorsal
medial thalamic nuclei is strong in man and not restricted to the
theta band (Fig. 8). A similar observation has been made in
respect to the anterior nuclei of thalamus.

---

* Biphasic pulses, 10 microseconds in duration of each phase, 2 to
3 m.amps, delivered approximately once per second.

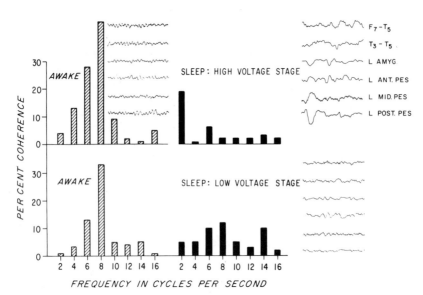

Fig. 9. Coherence between on-going activity of amygdala and
hippocampus plotted as histograms with samples of the original
EEG recordings. The strong coherences present in the waking
state were lost at both stages of sleep. Each plot represents
the analysis of a 7.5 minute recording in real time.
(Reproduced from Brazier, 1968b).

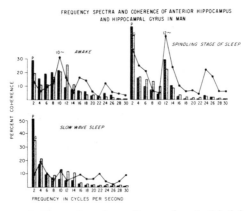

Fig. 10. Sleep spindles develop throughout limbic structures of
the same hemisphere at the same time and result in emergence of
high values for coherence in their frequency band. Slashed
columns give the frequency spectra of the activity in the pes
hippocampus, the white columns that in the ipsilateral hippocampal gyrus. Black dots represent the coherence at each
frequency.
(Reproduced from Brazier, 1968a)

The species difference in this respect is striking. The technique of observing afterdischarges evoked by electrical stimulation of limbic sites has been used by many workers, notably by Gloor and his colleagues (1961) and by Andy (1967, 1968). Andy has reported that, even in the lower animals with which they worked, there was a marked reduction in contralateral spread in the monkey compared with that in the cat or the dog. It is even more rare in man, a finding of some importance in clinical work where the spread of epileptic activity from one hemisphere to the other (usually after a time lag) is a question of interest in temporal lobe epilepsy.

RESULTS DURING NORMAL SLEEP

As reported previously, high coherences found between limbic structures in the awake state fall to insignificant levels during the slow wave periods of sleep as well as when low voltage fast activity and rapid eye movements are present (Brazier, 1966a, 1967b, 1968 a, b, c). An example of this change in the case of amygdala-hippocampal relations is seen in Figure 9. In marked contrast is the stability of coherence between contralateral thalamic nuclei (as seen in Fig. 8). A striking difference is found in the so-called "spindling" stage of sleep when 12-16 cps waves become conspicuous. At this stage there is a marked increase in coherence with stronger coherences in this frequency band than are present in the waking stage. Figure 10 illustrates this for the pes hippocampus and the ipsilateral hippocampal gyrus. Similar increases in coherences in the 12-16 cps frequency band have been found between amygdala and septum and between septum and hippocampus. The implications of the increased traffic over these limbic connections at this particular level of sleep are intriguing and have yet to receive their explanation.

RESULTS DURING BARBITURATE ANESTHESIA

There is growing evidence from other work that the brain mechanisms operating during loss of consciousness due to normal sleep differ markedly from that artificially produced by barbiturate anesthesia. Among the evidence is that from the limbic coherences, as studied when the patient is under thiopental anesthesia.

During the prenarcotic stage, when fast activity (18 to 30 cps) is prominent in the ink-written EEG record, high coherences between many sites are found in these frequencies. High coherences are even found between the fast frequencies of limbic structures and scalp (Brazier, 1969). Moreover, at anesthetic levels, when delta waves occupy the traces, the coherences remain high (Brazier, 1967a, 1969). One example is given in Figure 11 at two levels of thiopental infusion. This illustration is especially drawn from the same patient whose data when in normal sleep are shown in

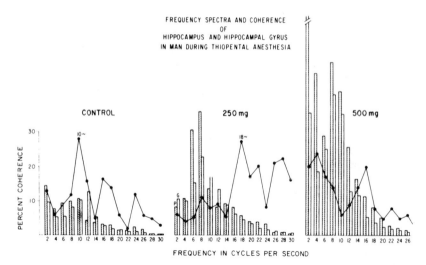

Fig. 11. Slashed columns to the left of each pair represent the
frequency spectra of the pes hippocampus, stippled columns those
of the ipsilateral hippocampal gyrus. The black dots indicate the
coherence at each frequency. At the prenarcotic dose (250 mg) of
thiopental the coherences rise in the fast frequencies. With loss
of consciousness after 500 mg, coherences are not lost as they are
in normal sleep. These are recordings (made on different days)
from the same patient as in Figure 10.

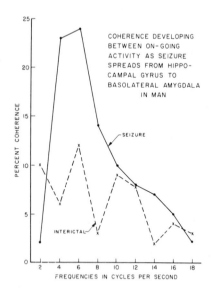

Fig. 12. Coherences are plotted with the broken line for the
interictal periods analyzed. The solid line shows the great
increase when electrical seizure activity develops in depth.
There was no clinical seizure.

Figure 10.

RESULTS DURING EPILEPTIC SEIZURE DISCHARGES

Coherences develop which are not present in the normal state. They have been found between deep structures not normally related, and between hippocampus and scalp as well as between amygdala and scalp (Brazier, 1969). Loci showing some degree of coherence during the normal waking state exhibit striking increases and transhemispheric coherences may emerge. An example of the greatly enhanced coherence of the amygdala with the ipsilateral hippocampal gyrus is seen in Figure 12.

These electrophysiological effects take place in the absence of overt clinical seizures and clearly indicate traffic forcing synaptic transmission in routes not normally operative.

SUMMARY

Studies are reported of the electrophysiological evidence for functioning of anatomical connections between amygdala and other limbic structures during various states studied in man with indwelling electrodes.

Emphasis is laid on the relationships of the naturally ongoing wave activity rather than on electrical stimulation which may force a path through synapses not normally transmissive.

Differences are reported between the apparent usages made of these various pathways in the varied conditions of wakefulness, sleep, thiopental anesthesia and spontaneously occurring electrical seizure activity.

ACKNOWLEDGMENTS

The work of this investigator is supported by Career Award #5 K6-18608 and Grant NS 09774 from the National Institutes of Health.

REFERENCES

AJMONE MARSAN, C., & STOLL, J. Subcortical connections of the temporal pole in relation to temporal lobe seizures. Archives of Neurology and Psychiatry, 1951, 66, 669.

ANDY, O. J., & KOSHINO, K. Duration and frequency patterns of
    the after-discharge from septum and amygdala. Electroen-
    cephalography and Clinical Neurophysiology, 1967, 22, 167.

ANDY, O. J., MUKAWA, J., & MELVIN, J. After-discharge propagation
    between the amygdalae in the cat. Journal of Nervous and
    Mental Diseases, 1968, 147, 85.

ANDY, O. J., & STEPHAN, H. The septum in the human brain.
    Journal of Comparative Neurology, 1968, 133, 383.

ANDY, O. J., WEBSTER, C. L., MUKAWA, J., & BONN, P. Electro-
    physiological comparisons of dorsal and ventral hippocampus.
    In Physiologie de l'Hippocampe. Paris: Centre Nationale de
    Recherche Scientifique, 1962. Pp. 411-427.

ANGEVINE, J. B., LOCKE, S., & YAKOVLEV, P. I. Limbic nuclei of
    thalamus and connections of limbic cortex. Archives of
    Neurology, 1961, 4, 355.

BRAZIER, M. A. B. The electrical fields at the surface of the
    head during sleep. Electroencephalography and Clinical
    Neurophysiology, 1949, 1, 195.

BRAZIER, M. A. B. A study of the electrical fields at the surface
    of the head. Electroencephalography and Clinical Neuro-
    physiology, 1951, Suppl. 2, 38.

BRAZIER, M. A. B. Recording from large electrodes. B. Electrical
    fields in a conducting medium. In J. H. Quastell (Ed.),
    Electrophysiological Techniques for Medical Research.
    Chicago: Yearbook Medical Publishers, 1961. Pp. 416-432.

BRAZIER, M. A. B. Evoked responses recorded from the depths of
    the human brain. Annals of the New York Academy of Science,
    1964, 112, 33.

BRAZIER, M. A. B. Electrophysiological studies of the hippocampus
    in man with averaged response computations. Acta Physio-
    logica, 1965, 26, 107.

BRAZIER, M. A. B. Electroencephalographic studies of sleep in
    man. In J. B. Dillon and C. M. Ballinger (Eds.), Anesthesi-
    ology and the Nervous System. Salt Lake City: University
    of Utah Press, 1966a. Pp. 106-128.

BRAZIER, M. A. B. The contribution of anesthesiology to electro-
encephalography. In J. B. Dillon and C. M. Ballinger (Eds.),
Anesthesiology and the Nervous System. Salt Lake City:
University of Utah Press, 1966b. Pp. 165-187.

BRAZIER, M. A. B. Thiopental: effects on subcortical mechanisms
in temporal lobe epilepsy. Anesthesiology, 1967a, 28, 192.

BRAZIER, M. A. B. Electrophysiological studies of the thalamus
and hippocampus in man. Sechenov Physiological Journal of
the USSR, 1967b, 53, 10.

BRAZIER, M. A. B. Absence of dreaming or failure to recall? In
C. D. Clemente (Ed.), Physiological Correlates of Dreaming,
Experimental Neurology Suppl. 4. New York: Academic Press,
1967c. Pp. 91-98.

BRAZIER, M. A. B. Varieties of computer analysis of electrophysio-
logical potentials. In W. A. Cobb and C. Morocutti (Eds.),
The Evoked Potentials. Electroencephalography and Clinical
Neurophysiology Suppl. 26, 1967d, 1-8. Amsterdam:
Elsevier.

BRAZIER, M. A. B. Studies of the EEG activity of limbic structures in man. Electroencephalography and Clinical Neurophysiology, 1968a, 25, 309.

BRAZIER, M. A. B. Étude electrophysiologique de l'hippocampe et
du thalamus chez l'homme. Actualités Neurophysiologiques,
1968b, 8, 149. Paris: Masson et Cie.

BRAZIER, M. A. B. Analysis of sleep activity as revealed by deep
recording in man. In H. Gastaut, E. Lugaresi, G. Berti
Ceroni, G. Coccagna (Eds.), The Abnormalities of Sleep in
Man. Bologna: Aulo Gaggi, 1968c. Pp. 35-43.

BRAZIER, M. A. B. Prenarcotic doses of barbiturates as an aid
in localizing diseased brain tissue. Anesthesiology, 1969,
31, 78.

BRAZIER, M. A. B. Regional activities within the human hippocampus
and hippocampal gyrus. Experimental Neurology, 1970, 26, 354.

BRÜCKE, F., PETSCHE, H., PILLAT, B., & DEISENHAMMER, E. Ein
Schrittmacher in der medialen Septumregion des Kaninchen-
gehirnes. Pflugers Archives fur die Gesamte Physiologie,
1959, 269, 135.

CRANDALL, P. H. Clinical applications of studies on stereotacticly implanted electrodes in temporal lobe epilepsy. Journal of Neurosurgery, 1963, 20, 827.

CROSBY, E. C., & HUMPHREY, T. Studies of the vertebrate telencephalon, II. The nuclear pattern of the anterior olfactory nucleus tuberculum olfactorium and the amygdaloid complex in adult man. Journal of Comparative Neurology, 1941, 74, 309.

FOX, C. Amygdalo-thalamic connections in Macaca mulatta. Anatomical Record, 1949, 103, 537.

GLOOR, P. Electrophysiological studies of the connections of the amygdaloid nucleus in the cat. Electroencephalography and Clinical Neurophysiology, 1955, 7, 223 and 243.

GLOOR, P. Amygdala, Handbook of Neurophysiology, 1959, 1, 1395.

GLOOR, P., SPERTI, L., & VERA, C. An analysis of hippocampal evoked responses and seizure discharges with extracellular microelectrode and DC recordings. In Physiologie de l'Hippocampe. Paris: Centre Nationale de Recherche Scientifique, 1961. Pp. 147-161.

GOGOLÁK, G., PETSCHE, H., STERC, J., & STUMPF, CH. Septum cell activity in the rabbit under reticular stimulation. Brain Research, 1967, 5, 508.

GOGOLÁK, G., STUMPF, CH., PETSCHE, H., & STERC, J. The firing pattern of septal neurons and the form of the hippocampal theta wave. Brain Research, 1968, 7, 201.

GREEN, J. D. The hippocampus. Physiological Review, 1964, 44, 561.

GREEN, J. D., & ADEY, W. R. Electrophysiological studies of hippocampal connections and excitability. Electroencephalography and Clinical Neurophysiology, 1956, 8, 245.

HILL, J. D. N., & PAR, G. Electroencephalography. London: MacDonald, 1950. 438 pp.

JASPER, H. H. Functional subdivisions of the temporal region. In M. Baldwin and P. Bailey (Eds.), Temporal Lobe Epilepsy. Springfield: Thomas, 1958. Pp. 40-51.

KAUFMAN, P. V. Electrical phenomena in the cerebral cortex. Obozrienie Psikhiatrii Nevrologii Eksperimental'noi Psikhologii, 1912, 7-8, 403, 9, 513 (in Russian).

KLINGER, J., & GLOOR, P. The connections of the amygdala and of the anterior temporal cortex in the human brain. Journal of Comparative Neurology, 1960, 115, 333.

MAYER, CH., & STUMPF, CH. Die Physostigminwirkung auf die Hippocampustätigkeit nach Septumläsionen. Naunyn-Schmiedeberg's Archiv fur Experimentelle Pathologie und Pharmakologie, 1958, 234, 490.

MILNER, B. The memory defect in bilateral hippocampal lesions. Psychiatric Research Report, 1959, 11, 43.

MORISON, R. S., & DEMPSEY, E. W. A study of thalamo-cortical relations. American Journal of Physiology, 1942, 135, 281.

NAUTA, W. J. H. Fiber degeneration following lesion of the amygdaloid complex in the monkey. Journal of Anatomy, 1961, 95, 515.

NAUTA, W. J. H. Neural associations of the amygdaloid complex in the monkey. Brain, 1962, 85, 505.

NAUTA, W. J. H., & VALENSTEIN, E. S. Some projections of the amygdaloid complex in the monkey. Anatomical Record, 1958, 130, 346.

PETSCHE, H., & STUMPF, CH. Topographic and toposcopic study of origin and spread of the regular synchronized arousal pattern in the rabbit. Electroencephalography and Clinical Neurophysiology, 1960, 12, 589.

PETSCHE, H., STUMPF, CH., & GOGOLÁK, G. Significance of the rabbit's septum as a relay station between the midbrain and the hippocampus. Electroencephalography and Clinical Neurophysiology, 1962, 14, 202.

PETSCHE, H., GOGOLÁK, G., & VAN ZWIETEN, P. A. Rhythmicity of septal cell discharges at various levels of reticular excitation. Electroencephalography and Clinical Neurophysiology, 1965, 19, 25.

PETSCHE, H. The quantitative analysis of EEG data. In J. P. Schadé and J. Smith (Eds.), Computers and Brains. Amsterdam: Elsevier, 1970. Pp. 63-86.

RAISMAN, G. The connexions of the septum. Brain, 1966, 89, 317.

SCOVILLE, W. B., & MILNER, B. Loss of recent memory after bilateral hippocampal lesions. Journal of Neurology, Neurosurgery and Psychiatry, 1957, 20, 11.

SOMMER-SMITH, J. A., POWARZYNSKI, J., STIRNER, A., & GRÜMBERG, V.
    Décharges cellulaires du noyeau centre-médian du thalamus
    induites par la stimulation amygdalienne et septale. Acta
    Neurologica Latinoamerica, 1965, 11, 360.

VALVERDE, F. Amygdaloid projection field. In W. Bargmann and
    J. P. Schadé (Eds.), Progress in Brain Research Vol. 3.
    The Rhinencephalon and Related Structures. Amsterdam:
    Elsevier, 1963. Pp. 20-30.

VOGT, M. Sur un faisceau septo-thalamique. Comptes Rendus des
    Seances de la Société de Biologie, 1898, 5, 206 (Series 10).

NEUROSURGERY

TEMPORAL LOBE EPILEPSY: ITS POSSIBLE CONTRIBUTION
TO THE UNDERSTANDING OF THE FUNCTIONAL SIGNIFICANCE
OF THE AMYGDALA AND OF ITS INTERACTION WITH
NEOCORTICAL-TEMPORAL MECHANISMS

P. Gloor

The Montreal Neurological Institute and the
Department of Neurology and Neurosurgery,
McGill University, Montreal, Canada

In the human brain, the amygdala forms a prominent subcortical
mass of grey matter located within the depths of the temporal lobe.
Its function usually has been discussed in terms of its connections
to the hypothalamus and to the various autonomic, endocrine and
motivational mechanisms represented there (Gloor, 1960).   Little
consideration has been given, up to now, to the nature and significance of the afferent input to the amygdala which in higher
mammals and man seems to be derived largely from the temporal neocortex (Segundo et al., 1955; Whitlock and Nauta, 1956; Niemer and
Goodfellow, 1966; Jones and Powell, 1970).   The view I would like
to put forward, in this essay, is that the amygdala and the
temporal neocortex of higher mammals and man can be regarded as
a functional system subserving complex motivated behavior patterns
dependent upon highly differentiated perceptual and cognitive
functions.   It is hoped that this holistic view of temporal lobe
function may further our understanding of the relationship between
neocortical and limbic physiology.

The material presented in this paper will include observations made on epileptic patients hospitalized at the Montreal
Neurological Institute and studied in detail by Dr. Penfield,
Dr. Rasmussen and others.   I will attempt to correlate these with
what is known about amygdaloid physiology from investigations in
experimental animals.   Furthermore, I shall discuss some aspects
of amygdaloid physiology in the light of recent studies on the
role of the sense of smell in subprimate mammalian behavior.

THE AMYGDALA AND ITS RELATIONSHIP TO FUNDAMENTAL
MOTIVATIONAL MECHANISMS

Amygdaloid neurons project to septal, preoptic and hypothalamic grey matter (Gloor, 1955; Nauta, 1962; Hall, 1963; Cowan
et al., 1965; Valverde, 1962; Hall, 1963; Dreifuss et al., 1968;
Murphy et al., 1968; Heimer and Nauta, 1969; Raisman, 1970;
for a review of the anatomical literature, see Gloor, 1960).
It is, therefore, not surprising that electrical stimulation of
the amygdala is capable of reproducing virtually the entire
range of response patterns obtained by hypothalamic stimulation
(see review of the literature in Gloor, 1960). This applies with
equal validity to autonomic, endocrine and motor responses as well
as to behavioral sequences related to basic drive mechanisms. It
is difficult to assess the proper functional significance of
these stimulation responses if they are looked upon in isolation
and without regard to the results of other studies involving,
among others, ablation techniques. The results of bilateral
amygdaloid ablations show that the basic homeostatic functions
integrated in the hypothalamus, such as temperature regulation,
electrolyte and water balance, and the autonomic control of the
cardiovascular and digestive systems, are not compromised
seriously by bilateral amygdaloid lesions. In contrast to this,
however, the behavior of a bilaterally amygdalectomized animal is
disturbed severely (Klüver and Bucy, 1937, 1938, 1939; Pribram
and Bagshaw, 1953; Schreiner and Kling, 1953; Weiskrantz, 1956;
for a review of the literature see Gloor, 1960), to the point
where survival in the animal's natural habitat is, in fact,
impossible (Dicks et al., 1969; Kling et al., 1969, 1970). The
behavioral disturbances which lead ultimately to the death of
the animal can be described as a lack of appropriate affective
responses to environmental cues, especially those important in
the organization of social and defensive behavior. This exposes
the animal to life threatening dangers against which the normal
animal is protected by its affective response patterns. Within
a short time these behavioral deficits lead to the animal's
demise.

Thus, of all the responses elicited by electrical stimulation
of the amygdala, those which involve the global behavior of the
animal probably are the most revealing with regard to its functional significance. It is, indeed, likely that the autonomic
and endocrine stimulation responses represent but partial aspects
of these global behavioral response patterns.

All the fundamental drive mechanisms which we find represented
at the hypothalamic level are re-represented at the amygdaloid
level. These include motivational mechanisms involved in feeding
and drinking, sexual behavior, and avoidance behavior as represented by rage and flight reactions. These behavior patterns

have autonomic, endocrine and motor concomitants which can be
reproduced by amygdaloid stimulation or interfered with by amyg-
daloid lesions (Fernandez de Molina and Hunsperger, 1959;
Fonberg and Delgado, 1961; Egger and Flynn, 1963; Grossman and
Grossman, 1963; Hilton and Zbrozyna, 1963; Wurtz and Olds, 1963;
Grossman, 1964; Eleftheriou and Zolovick, 1966; Fonberg, 1967;
Lewinska, 1967; Gentil et al., 1968; Russell et al., 1968;
Keating et al., 1970; Sclafani et al., 1970; Stokman and Glusman,
1970; see also Gloor, 1960, for a review of the literature.) The
subjective experiential concomitants of these behavior patterns,
which can only be communicated by man because of his power of
speech, are emotions or affective states.

As a first conclusion, one could, therefore, propose the
tentative hypothesis that the amygdala is related to motivational
drive mechanisms involved in feeding, drinking, reproductive and
avoidance behavior. The latter is not only important for the
defence of the individual or group against predators or rivals,
but also for the proper integration of the individual animal
within a social group of the same species. One would expect
observations on the human amygdala to be in agreement with the
postulates of this hypothesis. For instance, one would expect
stimulation of the amygdala in man to reproduce such elementary
affective drive states as thirst, hunger, libidinous feelings,
rage, fear and, perhaps, other emotions of both a pleasant or
unpleasant character. According to the Jacksonian hypothesis,
one would also expect naturally occurring epileptic discharge to
reproduce such emotions as part of the ictal event in the course
of an epileptic discharge arising in the human amygdala.

The study of human temporal lobe epilepsy indeed provides
evidence that emotional mechanisms are represented within the
temporal lobe with a preferred, although not exclusive, localization to deep temporal structures in the amygdala or its vicinity
(for review of the literature see Gloor and Feindel, 1963). The
emotional state which is most frequently elicited either by ictal
discharge in the temporal lobe or by electrical stimulation in
this area is that of fear (Mulder and Daly, 1952; Feindel and
Penfield, 1954; Penfield and Jasper, 1954; Gibbs, 1956; Williams,
1956; Daly, 1958; Bingley, 1958; Mullan and Penfield, 1959).
Other emotions are much more rarely thus elicited (Penfield and
Jasper, 1954). However, often electrical stimulation of the
human amygdala or discharge, originating in this area, fails to
produce an emotional state, but merely elicits autonomic changes,
masticatory or visceromotor responses, crude and ill defined
sensory experiences, or a period of amnesia with behavioral automatism (Feindel and Penfield, 1954; Jasper and Rasmussen, 1958).
The emotional responses are thus not produced consistently by
amygdaloid stimulation or amygdaloid ictal discharge in man.
Nevertheless, they provide valid evidence for the localization

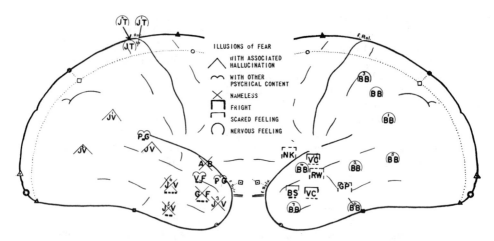

Fig. 1. Sites of stimulation producing illusion of fear and
nervousness. Deep stimulations indicated by horizontal dashes
and dots underneath symbols, each dash indicating 1 cm and each
dot 0.5 cm. Letters indicate initials of patients' names. Note
that all points producing fear are located within the temporal
lobe. Some points producing only a nervous feeling are also
found in the parietal lobe. (From Mullan and Penfield, 1959)

of affective mechanisms, for a topographical analysis of such
affective responses or ictal emotions reveals quite clearly that
they arise only in response to temporal lobe discharge, and not
as a consequence of stimulation of other areas of the cerebral
cortex (Fig. 1). Within the temporal lobe there is some clustering of these responses in deep mesial structures, presumably in
or near the amygdala (Mullan and Penfield, 1959).

     The following observation illustrates the elicitation of
fear by amygdaloid ictal discharge particularly well. This 29-
year-old epileptic patient, V. S., complained that she experienced short-lasting attacks of fear associated with palpitations.
She sometimes suddenly awoke at night with a fearful expression
on her face and called her husband to hold her. Such episodes
were sometimes followed by a motor seizure involving the left
side of the face, which at times evolved into a generalized tonic-
clonic convulsion. The patient was operated under local anesthesia for relief of her seizures by Dr. T. Rasmussen at the
Montreal Neurological Institute. During the operation, electro-
corticography and exploration of the mesial structures of the

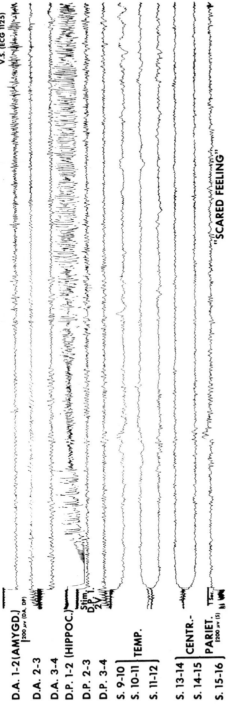

Fig. 2. Patient V. S. Electrocorticogram and recording from deep temporal structures after stimulation of hippocampus (through contact D. P. 1). The stimulation is followed by local afterdischarge in the hippocampus (D. P. 1 - 2), which after a few seconds spreads to involve the amygdala (D. A. 1 - 2), at which time the patient experienced a "scared feeling." Bipolar recording. D. A. - anterior depth electrodes with 4 contacts 1 cm apart (contact 1 is deepest). D. P. - posterior depth electrode (contact 1 is deepest). S - surface cortical electrodes. Amygd - amygdala. Hippoc - hippocampus. Temp - temporal cortex. Centr. Pariet - centro-parietal cortex.

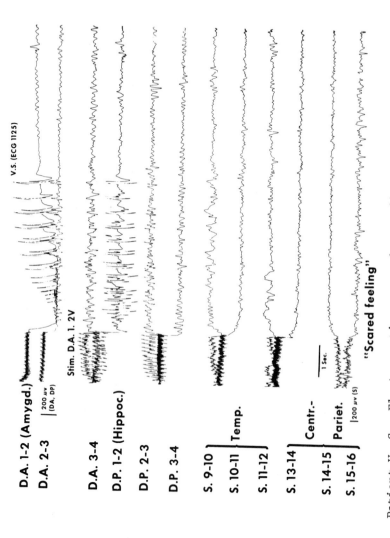

Fig. 3. Patient V. S. Electrocorticogram and recording from deep temporal structures a few minutes after record shown in Fig. 2. Stimulation of the amygdala (through contact D. A. 1) leads to immediate experience of fear ("scared feeling") associated with local afterdischarge in amygdala and hippocampus. For further explanations see legend to Fig. 2.

temporal lobe with depth electrodes was carried out. A small
focal seizure discharge was elicited by electrical stimulation of
the right hippocampus, as shown in Figure 2. The discharge first
remained confined to the hippocampus and, during this time, no
objective signs were noted, nor did the patient experience any
subjective symptoms. However, as soon as the elctrographic
record indicated that the epileptic discharge invaded the amygdala,
the patient said that she experienced a "scared feeling." Later,
as shown in Figure 3, the amygdala was stimulated directly and
this immediately elicited the feeling of fear, together with a
local afterdischarge involving the amygdala and the hippocampus.
Neither during the first nor the second stimulation was the
epileptic discharge conducted to the temporal neocortex. It is
quite clear that in this patient the evocation of fear was
dependent upon epileptic discharge involving the amygdaloid grey
matter. It was reproduced immediately by direct amygdaloid
stimulation. Hippocampal afterdischarge evoked fear only once
the ictal discharge invaded the amygdaloid grey matter.

A review of stimulation responses by Mullan and Penfield
(1959) showed that the largest number of the stimulation points
which evoked fear were located deep in the temporal lobe, presumably in the amygdaloid and periamygdaloid region. Nevertheless,
fear was evoked not infrequently also by stimulation applied to
the surface of the temporal neocortex. An analysis of deep
temporal stimulation responses by Jasper and Rasmussen (1958)
showed that the incidence of fear was even greater with stimulation
of depth electrode contacts located somewhere between the amygdala
and the deep neocortical grey matter which forms the transitional
fold separating the insula from the superior temporal cortex
buried within the Sylvian fissure.

These human observations show clearly that the emotion of
fear can be reproduced by electrical stimulation or by ictal discharge in temporal grey matter, either neocortical or amygdaloid,
but not from other regions of the cerebral cortex. A mechanism
activating this emotional response must, therefore, exist within
the temporal lobe. It is more difficult, however, to localize
this fear-producing neuronal mechanism more precisely within the
temporal lobe. Evidence suggests that deep structures are more
likely to be involved than superficial ones; in the light of
animal experiments demonstrating a neural substrate of escape
behavior in amygdaloid grey matter (Vigouroux et al., 1951;
Gastaut, 1952; MacLean and Delgado, 1953; Kaada et al, 1954;
Fernandez de Molina and Hunsperger, 1959; Wurtz and Olds, 1963;
Stokman and Glusman, 1970; Keating et al., 1970), the most
reasonable assumption is that the evocation of fear by temporal
lobe stimulation or discharge in man must involve the amygdala
directly or indirectly through fibers originating in temporal

neocortex and synapsing with amygdaloid neurons.

With regard to other emotional mechanisms, we are on much
shakier ground: rage, although it is the most common behavioral
response to amygdaloid stimulation in animals (Vigouroux et al.,
1951; Gastaut, 1952; MacLean and Delgado, 1953; Fernandez de
Molina and Hunsperger, 1959; Hilton and Zbrozyna, 1963), is an
exceedingly rare ictal event in man (Gastaut et al., 1955;
Williams, 1956; Ervin et al., 1969; Mark et al., 1969; Stevens et
al., 1969; Sweet et al., 1969). Outbursts of intense and frequently destructive rage, often triggered by very trivial events,
are reported to occur fairly frequently in temporal lobe epileptics (Roger and Dongier, 1950; Gastaut et al., 1955; Ervin et al.,
1969). These, however, rarely are seizures and in our experience
at the Montreal Neurological Institute rage has never been produced upon surface or deep temporal stimulation in man. However,
others (Heath et al., 1955; Ervin et al., 1969; Mark et al., 1969)
have, on a few occasions, elicited rage in man with amygdaloid
stimulation.

Other unpleasant emotional states are elicited sometimes by
temporal lobe discharge or temporal lobe stimulation, but are
rather uncommon; these are feelings of depression, sadness or
disgust (Penfield and Jasper, 1954; Weil, 1955, 1956, 1960;
Williams, 1956; Daly, 1958; Mullan and Penfield, 1959; Stevens
et al., 1969).

Pleasurable emotional experiences, feelings of joy, happiness
or mirth, are evoked only infrequently by temporal lobe epileptic
discharge (Mulder and Daly, 1952; Williams, 1956; Daley and Mulder,
1957; Daly, 1958; Gloor and Feindel, 1963). On a number of
occasions they have been elicited by deep mesial temporal stimulation, presumably in the amygdaloid region (Delgado, 1960; Sem-
Jacobsen, 1959; Sem-Jacobsen and Torkildsen, 1960; Stevens et al.,
1969). These reports have come from studies using chronic indwelling electrodes. It is of interest that, during neurosurgical
operations for the relief of temporal lobe epilepsy, when the
amygdala is stimulated through depth electrodes acutely inserted
into this structure in the course of a neurosurgical operation,
such feelings have never been evoked (Penfield and Jasper, 1954).
It is possible that pleasant emotions are difficult to elicit
under these rather unusual and stressful circumstances.

Sexual emotions very rarely are elicited by temporal lobe
ictal discharge (Bronstein, 1951; Bente and Kluge, 1953; Gastaut
and Collomb, 1954; Ajuriaguerra and Blanc, 1961); hunger and
thirst are even less common (Gastaut, 1955; Daly, 1958).

We may conclude from this survey of emotional responses,

produced by temporal lobe ictal discharge or electrical stimulation, that this part of the brain contains structures which when
excited can elaborate an emotional response, in man most often of
fear.   These observations thus support the evidence from animal
experimentation that the motivational mechanisms involved in
avoidance behavior have a representation in the amygdala and
perhaps, to some extent, also in the temporal neocortex. There
is less abundant evidence that pleasant emotional states including sexual feelings also are represented in deep temporal
structures, but in man evidence for representation of thirst and
hunger in the temporal lobe is very scanty.

     One may ask what the evolutionary pressures were that led to
a re-representation of basic motivational mechanisms in structures
of the telencephalon above the hypothalamic level, since all
these behavioral drive mechanisms, as animal experiments show,
are well represented in the hypothalamus (Hess, 1949; Anand and
Brobeck, 1951; Olds, 1956, 1958; Harris and Michael, 1964; Miller,
1965; Caggiula and Hoebel, 1966; Fitzsimons, 1966; Hoebel, 1969;
MacLean, 1969; Stevenson, 1969). Flight and aggression, the
outward expressions of the emotions of fear and rage with all
their behavioral, autonomic and endocrine concomitants, can be
reproduced by hypothalamic stimulation, and the same is true for
the drive mechanisms related to reproductive, feeding and drinking
behavior. What then is the purpose of the re-representation of
these mechanisms at a higher level? The answer to this question
may be guessed from a consideration of the evolutionary history
of the limbic system and of the amygdala in particular.

THE LEGACY OF THE OLFACTORY SENSE

     The primitive cerebral hemisphere of low vertebrates
primarily is an olfactory structure (Kappers, Huber and Crosby,
1936). Its main components are the olfactory bulb, the septum,
the hippocampus, the piriform cortex and the amygdala, which are
all parts of the limbic system of higher vertebrates. Thus, the
limbic system of mammals can be defined as the equivalent of the
primitive vertebrate hemisphere.

     Accordingly, the main afferent connections to the amygdala
in submammalian forms are olfactory. Herrick pointed out, in
1923, that a clearly identifiable amygdala makes its first appearance in tailless amphibians (Anura) in conjunction with that of
the vomeronasal organ from which it receives its main afferent
inflow by way of the accessory olfactory bulb. An identifiable
vomeronasal input into the amygdala still is clearly demonstrable
in mammals like the rabbit in which this organ has not become
vestigial (Winans and Scalia, 1970). Herrick (1923) speculates
that the new evolutionary pressures engendered when early
amphibians left their aquatic environment and started to live

on land may have brought about these changes in the organization of the nervous system. Unfortunately, little is known about the function of the vomeronasal organ. In snakes, where the paired openings of the organ admit neatly the tips of the animal's forked tongue, the vomeronasal organ has been shown to be important for tracking and recognition of prey (Burghardt and Hess, 1968). A possible role in sexual behavior has also been considered for some species (Winans and Scalia, 1970). Be this as it may, the vomeronasal organ, like the olfactory apparatus, must subserve some form of chemical sense and both contribute heavily to the afferent connections of the primitive amygdala. The mammalian amygdala, especially in the lower forms, still has a considerable olfactory input (Kappers, Huber and Crosby, 1936; for a review of the literature see Gloor, 1960).

In trying to understand the probable functional significance of the amygdala in lower animals, it is, therefore, useful to consider the role of olfaction in animal behavior. Obviously, one of the functions of the sense of smell is that of providing chemical clues leading to sources of food. It does much more than that, however. Recent investigations have shown clearly how important the sense of smell is for all kinds of complex behavioral patterns. Probably, for the majority of mammals below the level of primates, no sensory system has such a profound impact on behavior as olfaction. It provides the cues not only for the searching and ingestion of food, but also for reproductive behavior, maternal behavior, avoidance behavior and, most importantly, also for the integration of the individual within its social group as well as for individual recognition within the social group (Ralls, 1971; Schultze-Westrum, 1969; Pfaffmann, 1971). Even in fish, olfaction (and apparently no other sensory modality) provides the neural basis for plasticity of behavior making possible "sophisticated" forms of behavior which a priori one would not expect in these low forms. These include individual recognition, cooperative behavior and dominance (Nelson, 1964; Todd et al., 1967; Atema et al., 1969).

The importance of olfaction in mammalian behavior can be documented by many examples. Maternal behavior in mice, for instance, is dependent upon the integrity of the olfactory apparatus. Olfactory bulb removal eliminates maternal behavior in lactating mice (Gandelman et al., 1971).

Even more revealing of the profound influence of olfaction upon mammalian behavior are studies on scent marking. This subject has been reviewed recently by Ralls (1971). Most of the examples I shall cite are taken from her article. Many mammals possess special scent glands producing odorous substances which they deposit on the ground or rub on objects in their environment. Scent marking fulfills many functions: in slow loris it is used

for laying trails; in mice and rats, it serves as an alarm signal
alerting others to danger; in mice and deer, it subserves individual recognition; in the sugar glider, a marsupial, and in the
marmoset, a South American monkey, it subserves group recognition,
and in many other species the secretions of the scent glands
serve as a sexual attractant.

   Ralls (1971) points out that the most important aspect of
scent marking is its role in aggression and in the establishment
of dominance within a social group. Many mammals mark frequently
when they show an aggressive disposition. Scent marking often
precedes physical attack towards a member of the same or other
species. Johnston (1970) has shown, for instance, that, in the
hamster, scent marking is related to the aggressive drive of the
male, and that it increases when the animal is prone to display
aggressive behavior. It is stimulated by the smell of scent
deposited by another male hamster, but is reduced drastically by
the smell of the vaginal secretion of an estrous female. This
provides an interesting example of how, in this particular species,
two odors interact in complex behavioral patterns, one promoting
aggressive behavior and the other inhibiting it.

   In some social species, as for instance in the sugar glider,
the dominant male marks the territory occupied by the group and
also members of his group by rubbing his forehead which is
equipped with the scent gland on them. The scent deposited by
the dominant male inhibits marking by other members of the group
and, thus, provides a chemical signalling system which maintains
the hierarchical order within the group. The specific odor of
the dominant male which he has rubbed onto all other members of
his group also serves to identify an individual as belonging to
the group. The converse is of course also true; a member of the
same species, but of another group, will be recognized as a
stranger and will be attacked and driven off. Encountering the
smell of a strange individual of the same species enhances greatly
scent marking behavior (Schultze-Westrum, 1969; Ralls, 1971).

   Thus, in many animal species, motivational mechanisms involved
in aggressive, sexual and social behavior are activated by a rich
mosaic of signals provided by the olfactory sense. This provides
the organism with a highly differentiated set of cues which make
possible a high complexity of behavioral patterns. These depend
upon the recognition of very specific and learned, in contrast to
inborn, sets of signals. The sense of smell thus becomes very
important to the recall of the individual's past life experience,
in the light of which current behavior can be adapted presently to
existing needs. Thus, increasingly, as evolution progresses,
animals are freed from the stereotyped and genetically fixed mode
of operation characteristic of the hypothalamic level of organization of fundamental vertebrate drive mechanisms. The hypothalamic

neural substrate for motivational drive mechanisms does not by
itself possess sufficiently differentiated input systems which
could fulfill this role. The hypothalamus provides a neural
substrate for effector programs which are designed to redress
an impending homeostatic imbalance or to ward off a threat to
the organism's integrity in response to an actually injurious
stimulus. These mechanisms can be activated directly at the
hypothalamic level by an imbalance in homeostasis, for instance,
by an increase in body fluid osmolality which induces thirst
(Andersson and McCann, 1955; Fitzsimons, 1963), or by simple
nociceptive stimuli such as pain which elicit aggressive or
escape responses. Originally, in very early vertebrates,
these may well have been the only stimuli capable of triggering
these behavioral patterns. The same may be true for the early
stages of the development of the individual organism at higher
evolutionary levels. The newborn human infant, for instance,
as Ajuriaguerra and Blanc (1961) pointed out, is capable only of
displaying undifferentiated behavioral reactions induced by
nociceptive stimuli and non-satisfaction of basic needs. But, in
adult higher animals and man, these simple and basic triggering
mechanisms of motivational drives are used only as a method of
last resort. Aggression or escape, for instance, occurs before
the infliction of physical pain whenever sensory cues in the
light of past experience signal impending danger. In the course
of individual existence, some sets of originally neutral stimuli
acquire affective connotations by virtue of their association
with rewarding or punishing life situations. As evolution progresses, the refinements of sensory perception that can be put
to the service of basic motivational drives increase in proportion to the increased anatomical complexity of the brain,
especially with regard to the development of the neocortex. In
this evolutionary history, the olfactory sense seems to have
opened the way which freed animal behavior from the rigidity of
simple reflex mechanisms represented at the hypothalamic level.

It may be of interest to speculate why olfaction was selected
by evolution to subserve this role rather than other sensory
modalities. The answer probably lies in the fact that no other
sensory system could have fulfilled this function with the same
economy of means. Each specific odor activates a different set
of olfactory receptors and their central representation, but each
olfactory receptor is also sensitive to more than one, but never
to all, olfactory stimuli (Leveteau and MacLeod, 1966; Mathews,
1966). Thus, each specific odor can be represented centrally by
a specific matrix of excited neurons which is peculiar to it and
which can be recognized without equivocation (Fig. 4). To
differentiate one such matrix from any other is a relatively
simple computational task (Pfaffmann, 1969). A relatively simple
neuronal network is equal to it, in spite of the very large number
of different stimuli which such a system is able to handle. We

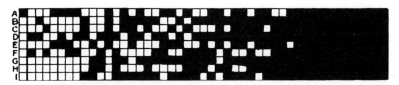

Fig. 4. Response matrix of 47 olfactory glomeruli (horizontal rows) of the rabbit to 9 different odorous stimuli (vertical columns A-I). Black squares indicate a glomerular response; white squares no response. (From Leveteau and MacLeod, 1966)

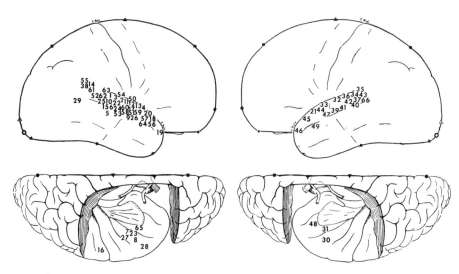

Fig. 5. Site of stimulation producing auditory experiential responses. (Numbers refer to patients.) Note the clustering of all the responses in the first temporal convolution and the superior temporal cortex. (From Penfield and Perot, 1963)

may contrast this, for instance, with the requirements an
advanced visual system must fulfill in order to perform this
type of function. It must be able to identify a specific
stimulus object with the same degree of certainty regardless of
its location in visual space, its apparent size, or its angular
orientation in visual space. This is a much more formidable
task which probably is beyond the powers of a simple brain.

We may now return to the previously asked question as to why
motivational mechanisms represented in the hypothalamus seem to
be re-represented at a higher level in the amygdala. The amygdala,
originally an olfactory structure of the primitive vertebrate
brain, must in lower forms at least be considered as part of a
system which processes olfactory signals, classifies them in the
light of past experience, and thus takes part in the programming
of motivated responses whose effector mechanisms are integrated
at the hypothalamic level. It may be regarded as part of a
neuronal system which, by establishing appropriate connections
between the store of information which an individual acquires in
his lifetime and the hypothalamic motivational effector mechanisms,
frees these fundamental drive mechanisms from their fixed dependence upon simple reflex-induced activation. While in lower forms
this store of information largely is olfactory, it involves increasingly the higher senses in higher animals.

AMYGDALOID FUNCTION IN PRIMATES AND MAN:
THE LESSON OF TEMPORAL LOBE EPILEPSY

In primates and especially man, the olfactory sense has lost
its dominant role in shaping motivated behavior. However, it has
not abdicated this function completely. There are few, if any,
emotionally neutral smells. Olfaction still provides man with
motivationally charged stimuli in the sphere of alimentary
behavior, sexual behavior, or in aversive situations - nothing,
for instance, produces a feeling of disgust more easily than a
bad smell.

However, in spite of this, the most important signals which
determine the emotional responses of primates and man, especially
with regard to social integration, have shifted from the olfactory sense to vision, and, in man, because of the evolvement of
speech, to hearing as well. Ethologists studying monkey behavior
have described many facial, postural and other bodily displays
which serve very much the same function in these species as
olfaction did in lower forms (Altmann, 1962; Hinde and Rowell,
1962; Van Hooff, 1962; Ploog, 1964, 1970; Marler, 1965; Kummer,
1968; Maurus and Ploog, 1971). They provoke fear or aggression,
maintain dominance, invite mating behavior, serve to form bonds
of affection between individuals and probably subserve individual
recognition. The signalling system here is almost entirely visual.

One would, therefore, surmise that visual perception and its cortical substrate (in man probably auditory perception, speech mechanisms and their cortical representation as well) must exert a powerful influence through anatomically definable routes of access to the motivational effector mechanisms represented in the hypothalamus. The visual perception of shapes and objects and the recognition of their meaning is, in the primate, a function of the inferior temporal neocortex (Klüver and Bucy, 1937, 1938; Milner, 1958, 1968; Kimura, 1963; Mishkin, 1966; Gross et al., 1969; Cowey and Gross, 1970). This cortex has no direct route of access to the hypothalamus. Whitlock and Nauta (1956) have shown that powerful connections to the amygdala originate from this inferior temporal neocortex. Recently, Jones and Powell (1970) have demonstrated that in the monkey each cortical sensory system exhibits an orderly sequence of cortical projections starting from the primary receiving areas to secondary and tertiary projection fields, with a final projection of all three sensory systems, auditory, visual and somatic, to the depths of the superior temporal sulcus as well as to the frontal pole and orbital cortex. The visual projection system, however, is unique in having a very powerful input into the amygdala.

Thus, we may surmise that visual input to the amygdala via temporal neocortex must be a very potent source of information for amygdaloid neural mechanisms. The important role of visual perception in primate amygdaloid physiology is demonstrated dramatically by the studies of Downer (1961). This investigator removed the amygdala on one side in a monkey in which previously all the forebrain commissures and the optic chiasm had been sectioned in the midline. This animal was aggressive towards human observers and unapproachable. It remained so when the eye ipsilateral to the amygdalectomy was occluded, and it thus was only able to process visual stimuli through the intact hemisphere. A dramatic change in behavior occurred when the eye on the side of the intact hemisphere was occluded. Now the animal was only able to process visual cues through the hemisphere from which the amygdala had been removed. Under these conditions, the monkey showed no signs of aggression or fear; it approached human observers readily and ate from their hands. This state of placidity, however, was only present in response to visual stimuli; touching or prodding the animal immediately produced an aggressive response.

It seems clear from this observation that when the animal was limited to processing visual information only through the hemisphere in which the amygdala had been removed, visual stimuli were no longer capable of inducing behavior motivated by fear or aggression. The expression of these motivational drives was now divorced from the animal's visual perceptive world, the cortex being prevented from processing visual information through an intact

amygdaloid outflow. For the amygdalectomized hemisphere visual
perception had lost its motivational significance. The capacity
to display aggressive behavior as such was not abolished as the
animal was capable of responding in an aggressive way to somato-
sensory stimuli.

Motivated behavior in primates and in man is guided by visual
and auditory cues which in the light of past experience have
assumed motivational significance. This presupposes not only an
accurate and detailed analysis of sensory stimuli and an ability
to form percepts, but also a matching of currently available
auditory and visual information with past experience. This is a
necessary prerequisite for the recognition of a set of stimuli as
familiar and meaningful in terms of present needs and past experience, or on the other hand as novel, unexpected and therefore
strange. Obviously, the primary sensory areas as represented in
the striate cortex and in Heschl's convolution only furnish the
raw material on the basis of which these higher perceptual and
mnemonic functions can be elaborated. Raw sensory data have to
be constructed into percepts and access has to be gained to de-
positories of past experience, both in the visual and auditory
spheres.

It is with regard to these functions that the analysis of
seizure patterns observed in temporal lobe epileptics, and of
some of the results of electrical stimulation of the temporal
lobe cortex in patients undergoing surgical treatment for temporal
lobe epilepsy, can provide valuable information concerning the
localization of these mechanisms within the human brain and their
relationship with emotional mechanisms represented in the amygdaloid and periamygdaloid grey matter.

Epileptic patients sometimes experience, at the beginning of
an attack, complex visual, auditory or combined visual and auditory
hallucinations. These are not simple elementary sensory hallucinations such as seeing light flashes or colors, or hearing a noise,
but like normal everyday experience they are structured experiences
and may be endowed with all the richness of detail of an actual
visual, auditory or combined auditory-visual experience. They may
consist of seeing a person or scene, frequently a familiar one,
hearing a familiar voice or music. Penfield and Perot (1963)
demonstrated that patients experiencing these complex hallucinations in the course of their seizures show electrographic or other
evidence that the discharge causing these seizures originated in
the temporal lobe. These hallucinations were observed in 10 per
cent of the patients operated upon for temporal lobe epilepsy.
Epileptic discharge involving other areas of the cerebral cortex
never gives rise to these experiential hallucinations. This
suggests that the temporal lobe may be the neural substrate for
the representation of the visual and auditory world, not at the

elementary, but at the higher perceptual level, and, since the perceptual data activated by temporal lobe epileptic discharge always, in one way or the other, derive from the patient's past experience, Penfield (1952, 1959, 1968, 1969) suggested that neuronal activity in temporal lobe cortex may reactivate fragments of the past stream of consciousness. The same visual and auditory experiential hallucinations are sometimes reproduced in these patients upon electrical stimulation of the temporal neocortex during neurosurgical operations for the relief of their seizures. Analyzing these results, Penfield and Perot (1963) showed that again the localization of these experiential responses to electrical stimulation of the cerebral cortex of man exclusively is temporal and never involves other cortical areas. The auditory type of experiential responses are all clustered along the first temporal convolution and on the superior surface of the temporal cortex buried within the Sylvian fissure (Fig. 5). They are found on both sides of the brain, with a slightly higher incidence on the nondominant side. The visual responses have a much more widely spread distribution including superior, lateral and inferior temporal cortex, as well as the cortex in the transitional area between occipital and temporal lobes (Fig. 6). These responses also are obtained more frequently from the nondominant than from the dominant hemisphere. These specific localization patterns, and the fact that these experiential responses can be reproduced even if the patient is unaware that his cortex is being stimulated, demonstrate that they represent genuine cerebral responses to electrical stimulation to the same extent as for instance finger movements occurring upon stimulation of the contralateral pre-central gyrus.

A few examples may illustrate these types of response: They are gleaned from observations made by Dr. Penfield and his associates at the Montreal Neurological Institute; they have been published in detail in Dr. Penfield's writings (Penfield and Perot, 1963).

The first illustrates an example of an auditory experiential response: This 22-year-old patient from South Africa, J. T., complained of seizures which began with a cephalic aura followed by automatism. At operation, a glioma was found in the right temporal lobe. When stimulation was applied to the cortex of the superior temporal surface in front of the transverse gyrus of Heschl, the patient, as soon as the current was turned on, exclaimed in great surprise, "Yes, doctor, yes, doctor! Now I hear people laughing - my friends in South Africa." He was asked whether he could recognize who these people were and he replied "Yes, they are my two cousins, Bessie and Anne Wheliaw." He said that he did not know why they were laughing, but that they must have been joking. The patient was able to recall this experience after the operation. There seems to be little doubt that electrical

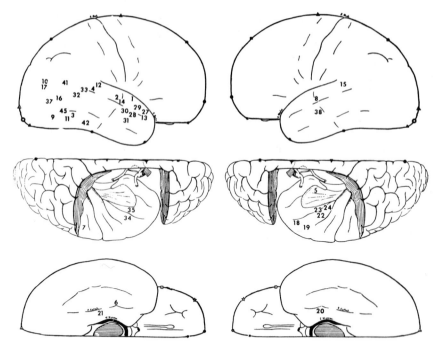

Fig. 6. Sites of stimulation producing visual experiential
responses. (Numbers refer to patients.) Note the distribution
of the responses over the entire temporal cortex. (From Penfield
and Perot, 1963)

stimulation had activated a complex auditory experience very
similar to a naturally occurring one. It most likely represented
reactivation of the auditory perceptual components of a true
experience the patient had had prior to his operation.

The second example illustrates a visual experiential
response: This is the case of a 12-year-old boy, R. W., who began
to have seizures at age 9. The attacks had the following pattern.
He first saw colored triangles, and this was followed by a visual
experiential hallucination; the patient saw a robber, a man with
a gun, moving towards him. The patient believed that the man was
someone he had seen in the movies or in comic strips. The figure
then moved to the left and the patient's head and eyes would turn
to the left, following which there was automatism and an occasional
generalized seizure. At operation, the right temporo-occipital
region was exposed. Stimulation of the occipital cortex produced
a visual sensation, the patient seeing triangles as at the

beginning of his attacks. Other elementary visual sensations were
also reproduced upon stimulation of the occipital cortex. The
patient described them as "lights, triangles, red, yellow, blue,
orange." When the stimulating electrode was applied to the
posterior temporal cortex, the nature of the responses changed.
Upon electrical stimulation of a point in the posterior part of
the second temporal convolution, the patient exclaimed "Oh gee!
Gosh! Robbers coming at me with guns." The robber seemed to
approach from the left side. This obviously was a reproduction of
the visual hallucination which characterized his spontaneous
seizures. Stimulation more inferiorly in the posterior temporal
neocortex induced the patient to say "Oh gosh! There they are,
my brother is there. He is aiming an air rifle at me." When
asked about what he had seen, he replied that his brother was
walking toward him and the gun was loaded. Auditory experiential
responses also were elicited in this patient upon stimulation of
the first temporal convolution. He had the feeling that he heard
his mother telling his aunt over the telephone to come up for a
visit. Another point on the first temporal convolution when
stimulated caused the patient to say "My mother is telling my
brother that he has got his coat on backwards. I can just hear
them." When asked whether he remembered this incident, he replied
"Oh yes, just before I came here." Sometimes auditory and visual
experiences occur together as part of a more complex experience,
just as in real life.

There are reasons to believe that these complex perceptual
experiences are fragmentary reactivations of past memories
(Penfield, 1952, 1959, 1968, 1969). For past memory material to
be of use in motivating present behavior, a mechanism must exist
whereby past experience can be matched with current one. The
result of this process will be a feeling of familiarity in the
case in which the present experience represents a close facsimile
of an earlier one or, on the contrary, it may be a feeling of unfamiliarity or strangeness when current experience is entirely
novel or differs from previous experience in some unexpected way.
One would have to postulate that some mechanism must exist in the
brain whereby this matching of past and current experience can be
achieved. Again, observations on temporal lobe epileptics suggest
that temporal neocortical grey matter may be involved in this
matching process. The neural code corresponding to this recognition of identity or similarity of experience, surprisingly enough,
can be activated again either by naturally occurring epileptic
discharge involving temporal neocortex or sometimes by artificial
electrical stimulation of the temporal lobe (Mullan and Penfield,
1959). This applies equally to the converse, the feeling of
strangeness or unfamiliarity. Temporal lobe epileptics sometimes
experience a feeling of "déjà vu" or inappropriate familiarity at
the beginning of a seizure, even if they may be in an unfamiliar

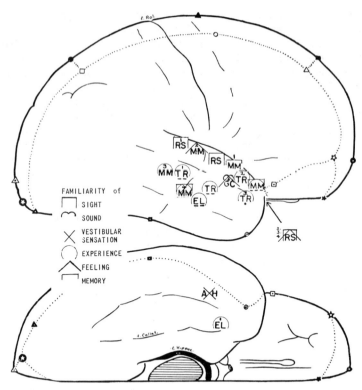

Fig. 7. Sites of stimulation producing an illusion of familiarity.
Deep stimulations indicated as in Fig. 1. Note the temporal localization of all the responses. No responses were obtained from the
right, nondominant hemisphere; left temporal stimulation never produced this response. (From Mullan and Penfield, 1959)

surrounding as is the case for instance when this feeling is produced by electrical stimulation of temporal cortex in the operating
theatre. Penfield (Mullan and Penfield, 1959) called this type of
response an interpretive illusion, because the patient inappropriately interprets current experience as either familiar or strange.
Only temporal lobe discharge or stimulation produces these interpretive illusions. Other parts of the cortex are incapable of
doing so (Fig. 7). The following example, described in detail in
the paper by Mullan and Penfield (1959), illustrates this.

This is the case of a 26-year-old woman, M. M., suffering from
temporal lobe attacks in which she experienced an illusion of
familiarity followed by a psychic hallucination, stiffening of the

body and automatic behavior with salivation. She described her
feeling of familiarity as a "feeling that I had lived through it
all before." At operation, stimulation of the temporal cortex in
several places evoked a typical feeling of familiarity. Stimulation for instance in the anterior part of the first temporal convolution caused her to say "I heard something familiar, I do not
know what it was." She later explained that this was a sound of
a mother calling her little boy, whom she had heard many years
before. Stimulation more posteriorly in the first temporal convolution made her say "A familiar feeling, very intense; I do not
know what it was." Stimulation of the second temporal convolution
caused her to say "I have a pain in my right eye, and the whole
operation now seems familiar." Stimulation at another point of
the temporal cortex caused her to say "As though I had been through
all this before, and I thought I knew exactly what you were going
to do next." Finally, stimulation at the undersurface of the
temporal lobe evoked what the patient described as "a familiar
memory, the place where I hang my coat up where I go to work."

The accurate detailed analysis of sensory cues leading to the
elaboration of a percept, the matching of a set of current stimuli
with past experience for recognition of identity, similarity or
unlikeness in normal life usually reactivates an affective response
appropriate to the past constellation of stimuli and this may
induce a motivated behavioral response which is appropriate to
present needs in the light of past circumstances of a similar
nature. The sequence from sensation to perception, to recognition
in the light of past experience inevitably leads, if the constellation of stimuli had in the past assumed motivational significance,
to an affective response with its autonomic, endocrine and behavioral concomitants. This signifies that the flow of information
finally has reached the basic motivational drive mechanisms at the
hypothalamic level. It is safe to assume in the light of anatomical, physiological and clinical observations that when affective or
motivational mechanisms are set in motion in response to appropriate
environmental stimuli, usually visual or auditory, or have been
activated by temporal lobe epileptic discharge, this signals the
activation of amygdaloid neurons. The amygdala, therefore, may be
conceived as an essential link in this information flow path from
temporal neocortex to hypothalamic motivational mechanisms. Without it, as in the amygdalectomized hemisphere of the split brain
monkey, visual cues provided by the external world and processed
through the hemisphere lacking an amygdala are no longer capable
of inducing appropriate motivated behavior.

Temporal lobe epilepsy, in exceptional circumstances, may
perform a particularly revealing experiment of nature, displaying
in a caricaturized form the natural flow of neural events which
in the light of this model we can follow through its various
stages as it involves neocortical temporal and associated amygdalo-

hypothalamic mechanisms. The following observation reported some
years ago by Dr. Penfield (Penfield and Perot, 1963) illustrates
this quite vividly.

J. V., a 32-year-old woman, had suffered from epileptic
seizures since the age of 11. At age 7, she had been frightened
by a man who came up behind her when she was walking through a
field of grass with her brothers in front of her. The man, who
was carrying a sack, approached her from behind and asked her
how she would like to get into his bag with the snakes and be
carried away. This was a very frightening experience for her and,
after that time, she had nightmares in which she re-experienced
this terrifying episode. Her attack pattern was as follows: she
had hallucinations of seeing herself as a little girl walking
through a field of grass. She then felt as though someone from
behind was going to smother her or hit her on the head, and she
became very frightened. The scene was almost always exactly the
same in each attack. This was followed by a short automatism and
sometimes by a generalized convulsive seizure beginning in the
left face and arm.

She was operated upon twice by Dr. Penfield. At the first
operation, the posterior temporo-occipital region was exposed and
a partial excision carried out in this area. Stimulation in the
middle part of the first temporal convolution made her say "I
imagine I hear a lot of people shouting at me." Stimulation at
a point nearby made her say "Oh, everybody is shouting at me
again, make them stop!" Then she added: "They are yelling at me
for doing something wrong. Everybody is yelling." Stimulation
again in the first temporal convolution, somewhat more posteriorly,
caused her to say "Oh, there it goes, everybody is yelling. Something dreadful is going to happen." Stimulation in the posterior
part of the second temporal convolution produced a visual hallucination and fear, the patient saying "I saw someone coming towards
me, as though he was going to hit me. Don't leave me." A cortical
excision in the posterior temporo-occipital region was carried out,
but the patient continued to have fits. The experiential hallucinations, however, were no longer part of her seizure pattern. The
attack now started with fear followed by a scream and automatism.
She was reoperated and the anterior part of the temporal lobe was
explored. Stimulation in the anterior temporal region near the
pole made her say "I had a terrible fear." Deep stimulation in
the medial temporal structures, presumably in or near the amygdala,
caused the patient to cry out whereupon she explained that she had
a strange feeling all over, starting with fear. Deep temporal
stimulation with the electrode directed toward the mesial temporal
structures made the patient seem disturbed; she wept, was terribly
afraid and looked afraid.

This case is of interest from many points of view. It illustrates how an actual childhood experience of a terrifying nature was built into the seizure pattern of this patient who suffered from temporal lobe epilepsy. The initial event was a visual hallucination consisting of a fairly accurate reenactment of the original terrifying event. Fragments of this experience were reproduced on posterior temporal neocortical stimulation and excision in this area eliminated the experiential hallucination, but not the fear, from the pattern of the seizures which still continued following the initial excision. The persisting epileptic mechanism for the evocation of fear was later localized to the anterior temporal region including its mesial part; it presumably involved the amygdala. It is probable that the march of the epileptic discharge in this patient retraced the path of the original neural events associated with the original frightening experience. Initial discharge in the occipital and posterior temporal neocortex spread forward and upon reaching the amygdaloid grey matter elicited the subjective experience of fear, which was associated with its appropriate behavioral concomitants, as for instance screaming.

It is probable that similar mechanisms linking the storehouse of information in the neocortex, particularly that of the temporal lobe, with the neuronal pools in the amygdala which project to the hypothalamus, may also be involved in other affective or motivational mechanisms such as those inducing angry behavior and aggression, or positive emotional feelings which cement affective bonds important for social cohesiveness, elicit mirth and, possibly, also those involved in feeding, drinking and sexual behavior. Evidence for the involvement of temporal lobe cortex and amygdala in these mechanisms, however, is only rarely provided by observations made in temporal lobe epileptics. This, of course, does not exclude the possibility that the human temporal lobe cortex and the amygdala are involved in other motivational mechanisms than fear.

It seems also likely that in man speech cortex may be closely linked to these temporo-limbic motivational mechanisms. The power of language as a means of communication and a motivator of human actions needs no emphasis. The fact that speech representation developed in the temporal lobe, therefore, may be no accident of evolution, but may be related to the necessity of providing speech mechanisms with a fairly direct access to the limbic structures, presumably the amygdala, through which language may gain control upon motivational mechanisms. Here, however, we enter the realm of pure speculation.

CONCLUSION AND SUMMARY

The view on the functional significance of the amygdala derived from these considerations can be briefly summarized as

follows. Every organism disposes of fundamental drives related
to feeding, drinking, reproduction, avoidance and approach, the
latter two including mechanisms involved in establishing and
maintaining social cohesiveness in animals living in social
groups. The neural substrates of these mechanisms are to be
found in the hypothalamus and limbic system. The basic effector
mechanisms which set into motion the global behavioral, autonomic
and endocrine responses characteristic of these fundamental drives
are integrated at the hypothalamic level. They can be activated
without the intervention of higher levels of the brain by disturbance of basic homeostatic balances, or, in the case of defence or
aggression, by nociceptive stimuli. The limbic system provides
important inputs into these fundamental hypothalamic drive mechanisms which render them more flexible and responsive to a great
variety of learned environmental signals. In lower animals, these
are provided mostly by olfactory stimuli. Phylogenetically, the
limbic system evolved in conjunction with the development of the
olfactory apparatus. In higher mammals, however, especially
primates, olfactory signals become less important for the initiation of motivated behavior. The importance of olfaction is super-
seded by that of other sensory mechanisms. Visual and auditory
perception, and in man undoubtedly language, provide the organism
with powerful triggers of motivational forces. Thus, the information processed by the neocortex becomes increasingly important
for the induction of motivated behavioral sequences. In the human
brain, higher level perceptual functions in the visual and auditory
sphere, including language, are represented in the temporal lobe.
It is there also that we find the main components of the human
limbic system, the hippocampus and the amygdala. Evidence derived
from the observations in temporal lobe epilepsy suggest that the
temporal neocortex is not only important for auditory and visual
perceptual functions, but also for the evocation of past memories
which involve these sensory modalities, and for the process of
matching present with past experience. This matching process is
of the foremost importance for the selection of behavioral
patterns which in the light of past experience satisfy present
needs. It is to be assumed that originally neutral perceptual
constellations experienced primarily through the visual and auditory
systems may have acquired in the past, for instance by association
with painful stimuli, an emotional connotation evoking in this
instance fear; upon recurrence of the same or a similar perceptual
constellation, the matching of the past record of the original
experience with present experience which involves the temporal
lobe, endows current experience with motivational significance.
This leads to the activation of the appropriate neural circuits
in the limbic system and its projections to the hypothalamus which
activate the mechanism for the expression and the subjective experience of fear. A similar mechanism may be involved in the
elaboration of other affective states and their behavioral con-

comitants. Therefore, it seems to be one of the main functions
of temporal lobe cortex and temporal lobe limbic structures,
especially the amygdala, to provide the link between the master
storehouse of information laid down in the neocortex and the
fundamental motivational drive mechanisms centered upon the
hypothalamus. Neural activity in these temporo-amygdaloid
motivational systems seems to represent the substrate for subjectively experienced emotions. They are the subjective counterpart of neural activity being directed towards the neuronal pools
in the hypothalamus which are in command of the fundamental drive
mechanisms of the organism.

This review leaves many questions unanswered. Thus, for
instance, the important problem of the mechanism of acquisition
of motivational significance has hardly been touched upon. To
convert an originally neutral stimulus into a motivationally
meaningful one, synaptic plasticity in some of the systems discussed here must be postulated. We may suspect that the amygdala
is the site of this plastic change, but proof is lacking.

Finally, there remains the much larger problem of how less
than 5 grams of tissue - the total weight of the human hypothalamus - can initiate the complex goal-directed behavioral patterns
characterizing motivated behavior in higher animals and man.
Obviously, these hypothalamic mechanisms have to call upon the
vast resources of the sensory, motor and associational systems
of the cerebral neocortex in order to perform these functions.
How this is achieved is, I believe, the greatest mystery of
neurophysiology.

ACKNOWLEDGMENT

I wish to thank Dr. W. Penfield for his permission to quote
his observations on patients J. T., R. W., M. M., and J. V.

REFERENCES

AJURIAGUERRA, J. DE & BLANC, C. Le rhinencéphale dans l'organisa-
    tion cérébrale. Neurobiologie du système limbique d'après
    les faits et les hypothèses. In Th. Alajouanine (Ed.),
    Les Grandes Activités du Rhinencéphale, Vol. II. Paris:
    Masson et Cie., 1961, 297-337.

ALTMANN, S. A. A field study of the sociobiology of rhesus monkeys
    (Macaca mulatta). Annals of the New York Academy of Science,
    1962, 102, 338-435.

ANAND, B. K., & BROBECK, J. R. Hypothalamic control of food intake.
    Yale Journal of Biology and Medicine, 1951, 24, 123-140.

ANDERSSON, B., & McCANN, S. M. Drinking, antidiuresis and milk
    ejection from electrical stimulation within the hypothalamus
    of the goat. Acta Physiologica Scandinavica, 1955, 35,
    191-201.

ATEMA, J., TODD, J. H., & BARDACH, J. E. Olfaction and behavioral
    sophistication in fish. In C. Pfaffmann (Ed.), Olfaction and
    Taste III, Proceedings of the Third International Symposium.
    New York: The Rockefeller University Press, 1969. Pp. 241-251.

BENTE, D., & KLUGE, E. Sexuelle Reizzustände im Rahmen des
    Uncinatus - Syndroms. Archives of Psychiatry, 1953, 190,
    357-376.

BINGLEY, T. Mental symptoms in temporal lobe epilepsy and tempor-
    al lobe gliomas. Acta Psychiatrica et Neurologica Scandina-
    vica, 1958, 33, Suppl. 120, 120-151.

BRONSTEIN, B. Zur Physiologie und Pathologie des Rhinencephalons.
    Schweizer Archiv fur Neurologie und Psychiatrie, 1951, 67,
    264-273.

BURGHARDT, G. M., & HESS, E. H. Factors influencing the chemical
    release of prey attack in newborn snakes. Journal of Com-
    parative and Physiological Psychology, 1968, 6, 289-295.

CAGGIULA, A. R., & HOEBEL, B. G. "Copulation-reward" site in the
    hypothalamus. Science, 1966, 153, 1284-1285.

COWAN, W. M., RAISMAN, G., & POWELL, T. P. S. The connexions of
    the amygdala. Journal of Neurology and Neurosurgery, 1965,
    28, 137-151.

COWEY, A., & GROSS, C. G. Effects of foveal prestriate and infero-
    temporal lesions on visual discrimination by rhesus monkeys.
    Experimental Brain Research, 1970, 11, 128-144.

DALY, D. Ictal affect. American Journal of Psychiatry, 1958,
    115, 97-108.

DALY, D. D., & MULDER, D. W. Gelastic epilepsy. Neurology, 1957,
    7, 189-192.

DELGADO, J. M. R. Emotional behavior in animals and humans.
    Psychiatric Research Report, 1960, 12, 259-266.

DICKS, D., MYERS, R. E., & KLING, A. Uncus and amygdala lesions:
    effects on social behavior in the free ranging rhesus monkey.
    Science, 1969, 165, 69-71.

DOWNER, J. L. de C. Changes in visual gnostic functions and emotional behavior following unilateral temporal pole damage in the "split brain" monkey. Nature, 1961, 191, 50-51.

DREIFUSS, J. J., MURPHY, J. T., & GLOOR, P. Contrasting effects of two identified amygdaloid efferent pathways on single hypothalamic neurons. Journal of Neurophysiology, 1968, 31, 237-248.

EGGER, M. D., & FLYNN, J. P. Effects of electrical stimulation of the amygdala upon hypothalamically elicited attack behavior in cats. Journal of Neurophysiology, 1963, 26, 705-720.

ELEFTHERIOU, B. E., & ZOLOVICK, A. J. Effect of amygdaloid lesions on oestrous behavior in the deermouse. Journal of Reproduction and Fertility, 1966, 11, 451-453.

ERVIN, F. R., DELGADO, J., MARK, V. H., & SWEET, W. H. Rage: A paraepileptic phenomenon? Epilepsia, 1969, 10, 417.

FEINDEL, W., & PENFIELD, W. Localization of discharge in temporal lobe automatism. Archives of Neurology and Psychiatry, 1954, 72, 605-630.

FERNANDEZ DE MOLINA, A., & HUNSPERGER, R. W. Central representation of affective reactions in forebrain and brainstem: electrical stimulation of amygdala, stria terminalis and adjacent structures. Journal of Physiology, 1959, 145, 251-265.

FITZSIMONS, J. T. The hypothalamus and drinking. British Medical Journal, 1966, 22, 232-237.

FONBERG, E. The role of the amygdaloid nucleus in animal behavior. Progress in Brain Research, 1967, 22, 273-281.

FONBERG, E., & DELGADO, J. M. R. Avoidance and alimentary reactions during amygdaloid stimulation. Journal of Neurophysiology, 1961, 24, 651-664.

GANDELMAN, R., ZARROW, M. X., DENENBERG, V. H., & MYERS, M. Olfactory bulb removal eliminates maternal behavior in the mouse. Science, 1971, 171, 210-211.

GASTAUT, H. Corrélations entre le système nerveux végétatif et le système de la vie de relation dans le rhinencéphale. Journal de Physiologie (Paris), 1952, 44, 431-470.

GASTAUT, H. Les troubles du comportement alimentaire chez les
    épileptiques psychomoteurs. Review of Neurology, 1955,
    92, 55-62.

GASTAUT, H., & COLLOMB, H. Etude du comportement sexuel chez les
    épileptiques psychomoteurs. Annales Médico Psychologiques,
    1954, 112, 657-696.

GASTAUT, H., MORIN, G., & LESÈVRE, N. Etude du comportement des
    épileptiques psychomoteurs dans l'intervalle de leur crises.
    Les troubles de l'activité globale et de la sociabilité.
    Annales Médico Psychologiques, 1955, 113, 1-27.

GENTIL, C. G., ANTUNES-RODRIGUES, J., NEGRO-VILAR, A., & COVIAN, M.
    Role of amygdaloid complex in sodium chloride and water in-
    take in the rat. Physiology & Behavior, 1968, 3, 981-985.

GIBBS, F. A. Abnormal electrical activity in the temporal regions
    and its relationship to abnormalities of behavior. Research
    Publications, Association of Nervous and Mental Disease,
    1956, 36, 278-294.

GLOOR, P. Electrophysiological studies on the connections of the
    amygdaloid nucleus in the cat. Part I: The neuronal
    organization of the amygdaloid projection system. Electro-
    encephalography and Clinical Neurophysiology, 1955, 7, 223-242.

GLOOR, P. Amygdala. In J. Field, H. W. Magoun and V. E.
    Hall (Eds.), Handbook of Physiology. Section I: Neuro-
    physiology, Vol. II. Washington, D. C.: American Physio-
    logical Society, 1960. Pp. 1395-1420.

GLOOR, P., & FEINDEL, W. Temporal lobe and affective behavior.
    In M. Monnier (Ed.), Physiologie des Vegetativen Nerven-
    systems, Vol. II. Stuttgart: Hippokrates Verlang, 1963.
    Pp. 685-716.

GROSS, C. G., BENDER, D. B., & ROCHA-MIRANDA, C. E. Visual
    receptive fields of neurons in inferotemporal cortex of the
    monkey. Science, 1969, 166, 1303-1306.

GROSSMAN, S. P. Behavioral effects of chemical stimulation of the
    ventral amygdala. Journal of Comparative and Physiological
    Psychology, 1964, 57, 29-36.

GROSSMAN, S. P., & GROSSMAN, L. Food and water intake following
    lesions or electrical stimulation of the amygdala. American
    Journal of Physiology, 1963, 205, 761-765.

HALL, E. A.   Efferent connections of the basal and lateral nuclei of the amygdala in the cat.   American Journal of Anatomy, 1963, 113, 139-145.

HARRIS, G. W., & MICHAEL, R. P.   The activation of sexual behavior by hypothalamic implants of oestrogen.   Journal of Physiology, 1964, 171, 275-301.

HEATH, R. G., MONROE, R. R., & MICKLE, W.   Stimulation of the amygdaloid nucleus in a schizophrenic patient.   American Journal of Psychiatry, 1955, 111, 862-863.

HEIMER, L., & NAUTA, W. J. H.   The hypothalamic distribution of the stria terminalis in the rat.   Brain Research, 1969, 13, 284-297.

HERRICK, C. J.   The connections of the vomeronasal nerve, accessory olfactory bulb and amygdala in amphibia.   Journal of Comparative Neurology, 1921, 33, 213-280.

HESS, W. R.   Das Zwischenhirn.   Syndrome, Lokalisationen, Funktionen.   Basel: Benno Schwabe & Co., Verlag, 1949.

HILTON, S. M., & ZBROZYNA, A. W.   Amygdaloid region for defense reactions and its efferent pathways to the brainstem. Journal of Physiology, 1963, 165, 160-173.

HINDE, R. A., & ROWELL, T. E.   Communications for postures and facial expressions in rhesus monkey (Macaca mulatta). Proceedings Zoological Society (London), 1962, 138 (I), 1-21.

HOEBEL, B. G.   Feeding and self-stimulation.   Annals New York Academy of Sciences, 1969, 157, 758-778.

JASPER, H. H., & RASMUSSEN, T.   Studies of clinical and electrical responses to deep temporal stimulation in man with some considerations of functional anatomy.   Research Publications, Association of Nervous and Mental Disease, 1958, 36, 316-334.

JOHNSTON, R. B.   Olfactory communication in the hamster.   Ph.D. Thesis.   Rockefeller University, 1970.

JONES, E. G., & POWELL, T. P. S.   An anatomical study of converging sensory pathways within the cerebral cortex of the monkey. Brain, 1970, 93, 793-820.

KAADA, B. R., ANDERSEN, P., & JANSEN, J.   Stimulation of the amygdaloid nucleus complex in unanesthetized cats.   Neurology, 1954, 4, 48-64.

KAPPERS, C. V. A., HUBER, G. C., & CROSBY, E. C.   The Comparative
     Anatomy of the Nervous System of Vertebrates, including Man.
     New York: The MacMillan Co., 1936.   (Reprinted in 1960 by
     Hafner Publishing Co.)

KEATING, E. G., KORMANN, L. A., & HOREL, J. A.   The behavioral
     effects of stimulating and ablating the reptilian amygdala
     (Caiman Sklerops).   Physiology & Behavior, 1970, 5, 55-59.

KIMURA, D.   Right temporal lobe damage.   Archives of Neurology,
     1963, 8, 264-271.

KLING, A., DICKS, D., & GUROWITZ, E. M.   Amygdalectomy and social
     behavior in a caged group of vertebrates (C. aethiops).
     Basel: Second International Congress on Primates, 1969,
     1, 232-241.

KLING, A., LANCASTER, J., & BENITONE, J.   Amygdalectomy in the
     free ranging vervet (Cercoptithecus aethiops).   Journal of
     Psychiatric Research, 1970, 7, 191-199.

KLÜVER, H., & BUCY, P.   "Psychic blindness" and other symptoms
     following bilateral temporal lobectomy in rhesus monkey.
     American Journal of Physiology, 1937, 119, 352-353.

KLÜVER, H., & BUCY, P.   An analysis of certain effects of
     bilateral temporal lobectomy in the rhesus monkey with
     special reference to "psychic blindness."   Journal of
     Psychology, 1938, 5, 33-54.

KLÜVER, H., & BUCY, P.   Preliminary analysis of functions of the
     temporal lobes in monkeys.   Archives of Neurology and
     Psychiatry, 1939, 42, 979-1000.

KUMMER, H.   Social Organization of Hamadryas Baboons.   Basel:
     S. Karger; Chicago and London: The University of Chicago
     Press; and Toronto: The University of Toronto Press, 1968.

LEVETEAU, J., & MacLEOD, P.   Olfactory discriminations in the
     rabbit olfactory glomerulus.   Science, 1966, 153, 175-176.

LEWINSKA, M. K.   Changes in eating and drinking produced by
     partial amygdala lesions in cat.   Academie Polonaise des
     Sciences, Bulletin Serie des Sciences Biologiques, 1967,
     15, 301-305.

MacLEAN, P. D.   The hypothalamus and emotional behavior.   In
     W. Haymaker, E. Anderson and W. J. H. Nauta (Eds.), The
     Hypothalamus.   Springfield, Illinois: Charles C. Thomas,
     1969.   Pp. 659-677.

MacLEAN, P., & DELGADO, J. M. R. Electrical and chemical
    stimulation of frontotemporal portion of limbic system in
    the waking animal. Electroencephalography and Clinical
    Neurophysiology, 1953, 5, 91-100.

MARK, V. H., ERVIN, F. R., SWEET, W. H., AND DELGADO, J. Remote
    telemeter stimulation and recording from implanted temporal
    lobe electrodes. Confinia Neurologica, 1969, 31, 86-93.

MARLER, P. Communication in monkeys and apes. In I. De Vore
    (Ed.), Primate Behavior: Field Studies of Monkeys and Apes.
    New York: Holt, Rinehart and Winston, 1965. Pp. 544-548.

MATHEWS, D. F. Response patterns of single units in the olfactory
    bulb of the unanesthetized, curarized cat to air and odor.
    Ph.D. Thesis, Brown University, 1966.

MILLER, N. E. Chemical coding of behavior in the brain. Science,
    1965, 148, 328-338.

MILNER, B. Psychological defects produced by temporal lobe
    excisions. Research Publications, Association for Research
    in Nervous and Mental Disease, 1958, 36, 244-257.

MILNER, B. Visual recognition and recall after right temporal
    lobe excisions in man. Neuropsychologia, 1968, 6, 191-210.

MISHKIN, M. Visual mechanisms beyond the striate cortex. In
    R. Russell (Ed.), Frontiers in Physiological Psychology.
    New York: Academic Press, 1966. Pp. 93-119.

MULDER, D. W., & DALY, D. Psychiatric symptoms associated with
    lesions of temporal lobe. Journal of the American Medical
    Association, 1952, 150, 173-176.

MULLAN, S., & PENFIELD, W. Illusions of comparative interpreta-
    tion and emotion. Archives of Neurology and Psychiatry,
    1959, 81, 269-284.

MURPHY, J. T., DREIFUSS, J. J., & GLOOR, P. Topographical
    differences in the responses of single hypothalamic neurons
    to limbic stimulation. American Journal of Physiology,
    1968, 214, 1443-1453.

NAUTA, W. J. H. Neural associations of the amygdaloid complex in
    the monkey. Brain, 1962, 85, 505-520.

NELSON, K. Behavior and morphology in the glandulocaudine fishes
    (Ostariophysi, Characidae). University of California Publica-
    tions in Zoology, 1964, 75(2), 59-152.

NIEMER, W. T., & GOODFELLOW, E. F. Neocortical influence on the
    amygdala. Electroencephalography and Clinical Neurophysio-
    logy, 1966, 21, 429-436.

OLDS, J. A preliminary mapping of electrical reinforcing effects
    in the cat brain. Journal of Comparative and Physiological
    Psychiatry, 1956, 49, 281-285.

OLDS, J. Self-stimulation experiments and differential reward
    systems. In H. Jasper and L. D. Proctor (Eds.), Reticular
    Formation of the Brain. Boston: Little Brown & Co. (Henry
    Ford Hospital International Symposium), 1958, Pp. 671-687.

PENFIELD, W. Memory mechanisms. Archives of Neurology and
    Psychiatry, 1952, 67, 178-191.

PENFIELD, W. The interpretive cortex. Science, 1959, 129,
    1719-1725.

PENFIELD, W. Engrams in the human brain. Proceedings, Royal
    Society of Medicine, 1968, 61, 831-840.

PENFIELD, W. Consciousness, memory and man's conditioned reflexes.
    In K. H. Pribram (Ed.), On the Biology of Learning. New
    York, Chicago, San Francisco, Atlanta: Harcourt, Brace and
    World Inc., 1969. Pp. 127-168.

PENFIELD, W., & JASPER, H. Epilepsy and the Functional Anatomy of
    the Human Brain. Boston: Little, Brown & Co., 1954.

PENFIELD, W., & PEROT, PH. The brain's record of auditory and
    visual experience - A final summary and discussion. Brain,
    1963, 86, 595-696.

PFAFFMANN, C. Summary of olfactory roundtable. In C. Pfaffmann
    (Ed.), Olfaction and Taste III, Proceedings of the Third
    International Symposium. New York: The Rockfeller Univ-
    ersity Press, 1969. Pp. 226-232.

PFAFFMANN, C. Recent advances in the study of olfaction. In
    P. Gloor and J. P. Cordeau (Eds.), Recent Contributions to
    Neurophysiology, Suppl. No. 30, Electroencephalography and
    Clinical Neurophysiology. Amsterdam: Elsevier Publishing
    Co., 1971, in press.

PLOOG, D. Verhaltensforschung und Psychiatrie. In H. W. Gruhle,
    R. Jung, W. Mayer-Gross, and M. Müller (Eds.), Psychiatrie
    der Gegenwart, Forschung und Praxis, Vol. I/1B Grundlagen-
    forschung der Psychiatrie, Part B. Berlin, Göttingen,
    Heidelberg: Springer Verlag, 1964. Pp. 291-443.

PLOOG, D. Social communication among animals. In F. O. Schmitt,
    G. C. Quarton, Th. Melnechuk and G. Adelman (Eds.), The
    Neurosciences Second Study Program. New York: The Rocke-
    feller University Press, 1970. Pp. 349-361.

PRIBRAM, K. H., & BAGSHAW, M. Further analysis of the temporal
    lobe syndrome utilizing fronto-temporal ablations. Journal
    of Comparative Neurology, 1953, 99, 347-375.

RAISMAN, G. An evaluation of the basic pattern of connections
    between the limbic system and the hypothalamus. American
    Journal of Anatomy, 1970, 129, 197-202.

RALLS, K. Mammalian scent marking. Science, 1971, 171, 443-449.

ROGER, A., & DONGIER, M. Corrélations électrocliniques chez 50
    épileptiques internés. Review of Neurology, 1950, 83,
    593-596.

RUSSELL, R. W., SINGER, G., FLANAGAN, F., STONE, M., & RUSSELL,
    J. W. Quantitative relations in amygdaloid modulation of
    drinking. Physiology & Behavior, 1968, 3, 871-875.

SCHREINER, L., & KLING, A. Behavioral changes following rhinen-
    cephalic injury in cat. Journal of Neurophysiology, 1953,
    16, 643-659.

SCHULTZE-WESTRUM, T. G. Social communication by chemical signals
    in flying phalangers (Petaurus breviceps papuanus). In
    C. Pfaffmann (Ed.), Olfaction and Taste III, Proceedings of
    the Third International Symposium. New York: The Rockefeller
    University Press, 1969. Pp. 269-277.

SCLAFANI, A., BELLUZZI, J. D., & GROSSMAN, S. P. Effects of
    lesions in the hypothalamus and amygdala on feeding behavior
    in the rat. Journal of Comparative and Physiological
    Psychology, 1970, 72, 394-403.

SEGUNDO, J. P., NAQUET, R., & ARANA, R. Subcortical connections
    from temporal cortex of monkey. Archives of Neurology and
    Psychiatry, 1955, 73, 515-524.

SEM-JACOBSEN, C. W. Depth-electrographic observations in
    psychotic patients. A system related to emotion and behavior.
    Acta Psychiatrica et Neurologica Scandinavica, 1959, 34
    (Suppl. 136), 412-416.

SEM-JACOBSEN, C. W., & TORKILDSEN, A. Depth recording and
     electrical stimulation in the human brain. In E. R. Ramey
     and D. S. O'Doherty (Eds.), Electrical Studies on the Un-
     anesthetized Brain. New York: Paul B. Hoeber, Inc.,
     Medical Division of Harper and Brothers, 1960. Pp. 275-290.

STEVENS, J. R., MARK, V. H., ERWIN, F., PACHECO, P., &
     SUEMATSU, K. Deep temporal stimulation in man. Long latency,
     long lasting psychological changes. Archives of Neurology,
     1969, 21, 157-169.

STEVENSON, J. A. F. Neural control of food and water intake.
     In W. Haymaker, E. Anderson and W. J. H. Nauta (Eds.), The
     Hypothalamus. Springfield, Illinois: Charles C. Thomas,
     1969. Pp. 524-621.

STOKMAN, C. L. J., & GLUSMAN, M. Amygdaloid modulation of hypo-
     thalamic flight in cats. Journal of Comparative and Physio-
     logical Psychology, 1970, 71, 365-375.

SWEET, W. H., ERVIN, F., & MARK, V. H. The relationship of
     violent behavior to focal cerebral disease. Aggressive
     Behavior (Excerpta Medica Foundation), 1969. Pp. 336-352.

TODD, J. H., ATEMA, J., & BARDACH, J. E. Chemical communication
     in social behavior of a fish, the yellow bullhead (Ictalurus
     natalis). Science, 1967, 158, 672-673.

VALVERDE, F. Studies on the Piriform Lobe. Cambridge: Harvard
     University Press, 1965.

VAN HOOFF, J. Facial expressions in higher primates. Symposium,
     Zoological Society (London), 1962, 8, 97-125.

VIGOUROUX, R., GASTAUT, H., & BADIER, M. Les formes expérimentales
     de l'épilepsie. Provocation des principales manifestations
     cliniques de l'épilepsie dite temporale par stimulation des
     structures rhinencéphaliques chez le chat non anesthésié.
     Review of Neurology, 1951, 85, 505-508.

WEIL, A. Depressive reactions associated with temporal lobe-
     uncinate seizures. Journal of Nervous and Mental Diseases,
     1955, 121, 505-510.

WEIL, A. Ictal depression and anxiety in temporal lobe disorders.
     American Journal of Psychology, 1956, 113, 149-157.

WEIL, A. Ictal emotion occurring in temporal lobe dysfunction.
     Archives of Neurology, 1960, 1, 101-111.

WEISKRANTZ, L. Behavioral changes associated with ablation of amygdaloid complex in monkeys. Journal of Comparative and Physiological Psychology, 1956, 49, 381-391.

WHITLOCK, D. G., & NAUTA, W. J. H. Subcortical projections from the temporal neocortex in Macaca mulatta. Journal of Comparative Neurology, 1956, 106, 183-212.

WILLIAMS, D. The structure of emotions reflected in epileptic experiences. Brain, 1956, 79, 29-67.

WINANS, S. S., & SCALIA, F. Amygdaloid nucleus: new afferent input from the vomeronasal organ. Science, 1970, 170, 330-332.

WURTZ, R. H., & OLDS, J. Amygdaloid stimulation and ope rant reinforcement in the rat. Journal of Comparative and Physiological Psychology, 1963, 56, 941-949.

STEREOTAXIC AMYGDALOTOMY

H. Narabayashi, Director of Neurological Clinic and
Professor of Neurology
Juntendo Medical School

Meguro, Tokyo, Japan

The human amygdaloid nucleus has not been investigated or
analysed extensively and the clinical importance of its function
has not been given due credit. It is included generally in the
important group of temporal lobe structures. A number of papers
on temporal lobe function, temporal lobe epilepsy and its surgical
treatment can be cited and discussed here, but, in these, the
localized cortical areas have been well analysed in detail, though
the deep-lying structures, especially the amygdaloid nucleus,
remain relatively obscure (Penfield and Flanigin, 1950; Penfield
and Jasper, 1954).

The commonest and the most constant symptoms in epileptics
of long history are in personality changes especially more preva-
lent in the emotional sphere than in any other neurological or
paroxysmal phenomenon. By the succession of repeated seizures of
varying intervals, the patients often become irritable, easily
excitable, explosive and sometimes violent. This tendency sometimes is nuanced by additional relative dementia, and these
patients become less controllable, more irritable, with shorter
concentration span and exhibit the so-called epileptic personality.
This hyperexcitability, irritability or poor concentration possibly
could be interpreted as positive symptomatology and could be related to the morphologically normal and still active, perhaps
hyperactive, structures, such as the amygdaloid complex. Toshima
(1961) reported that the amygdaloid nucleus remained almost unaffected in ten epileptic brains with long history, though the
neighbouring hippocampal structures were highly atrophied.

The amygdaloid nucleus is a well known structure in the
limbic-emotional circuit (Papez, 1937) and the electrical or

459

chemical stimulation of the nucleus often times causes rage reaction in animals. In patients, Chapman reported that aggression
and emotional excitation were exhibited in about one-half of his
cases following stimulation of the periamygdaloid area (1958).

Many other clinical data have suggested the close etiological
role of deep temporal structures for abnormal behavioral problems
in epileptics. The problem child also is accompanied very often
by the temporal lobe spikings (Aird and Yamamoto, 1966). Furthermore, temporal lobectomy produces marked calming effect on emotion
thus producing social adaptability of the patients, as well as
control of psychomotor fits, as reported by Milner (1958),
Penfield and his associates, Falconer (1963) and Taylor and
Falconer (1968).

I. OBSERVATIONS ON THE OPERATING TABLE

The surgical procedure for the violent and irritable cases
usually is performed under general anaesthesia, especially in
cases involving children. The patient is placed in the stereotaxic
frame in the supine position. The routine surgical procedure involves drilling a small burr-hole bilaterally at just behind the
frontal hair-line and 2.5 to 3.0 cm lateral from the midline.

The amygdaloid nucleus can be located and reached stereotaxically without much difficulty, using as a reference the outline of the tip of the temporal horn. However, since its extent is
large enough, more than 8 mm in diameter in adults, the more
detailed differentiation of various divisions within the nucleus is
not easy to perform. As in stereotaxy on the pallidum or on the
thalamus, both radiological and physiological devices in locating
the nucleus become important.

A. Radiological control: The amygdaloid nucleus lies on
the anterodorsal wall of the temporal horn, the ventral margin of
the nucleus being only a few millimeters above the tip of the horn
in the normal-sized ventricle. The air ventriculography is usually
enough to visualize the horn. In lateral distance from the midline,
the nucleus spreads between about 16 to 25 or 26 mm, in the adult
brain. Therefore, a little medial coordinate, such as 18 mm,
indicates medial nuclear group and the little lateral coordinate,
such as 22 or 24 mm may indicate the lateral nuclear group. However, this radiological measurement of the different intra-amygdaloid nuclear structures has not been so well established or
studied extensively, as in the thalamic nucleus, and the individual
variations due to size or pathology of each brain are less known.
However, the author's lateral coordinates for insertion to the
target are usually between 19 to 22 mm from the midline, which are
aimed at the central area of the amygdala, or the medial part of

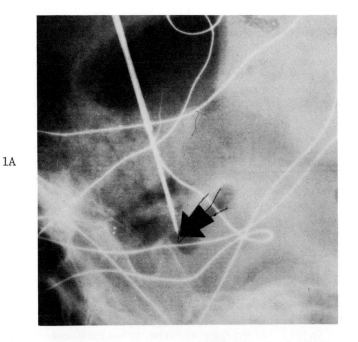

1A

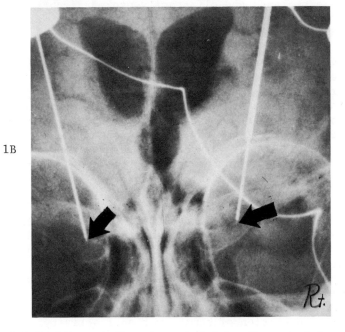

1B

Fig. 1 A & B. Two needles inserted into the amygdala of each
side referring to the figure of the temporal horn on X-ray.
(A) lateral view; (B) AP view.

the lateral nuclear group (Fig. 1 A & B). However, since no post-
mortem findings have been obtained to date, the further interpretation about localization of particular lesions must be postponed.
It should be stressed that the lesion produced does not include the
entire nucleus, but only a part of it, presumably the area within
the nucleus mentioned previously. This is the reason that the
term amygdalotomy, and not amygdalectomy, was used since the first
publication by the author.

B. Physiological devices: Several physiological findings,
which were described in the author's first report (1963 and 1964),
are used routinely since initiation of the procedure, and these
have been found quite convenient and practical. The needle of
insertion is a coated and insulated long and thin needle, covered
for protection by hole-needle, the outside diameter of the latter
being only 1 mm. Bipolar concentric electrodes are on the tip of
the insulated needle, the core electrode being of 200 μ diameter.
In some instances, for the purpose of more detailed information
at the cellular level, the core-needle is sharpened to 5 to 10 μ
(microelectrode).

Physiological criteria for detecting the target in the
amygdala are the injury discharges and the localized spontaneous
activity from the nucleus, the olfactorily-evoked discharges from
the nucleus and, finally, the somato-autonomic effects for high-
frequency stimulation of the nucleus.

The injury discharges are always evoked when the needle-tip
enters into the nucleus or when it is moved deeper within the
nucleus. Injury discharges can be detectable as high frequency
grouping phenomena of about 30 to 40 c/s. By using the microelectrode, these discharges are the continuous bursts of massive
small discharges of about 200 - 500 μV, in grouped fashion. In
all instances, when the injury discharges are obtained, they disappear usually within 20 or 30 seconds, and the spontaneous
activity of the nucleus tends to become manifest (Fig. 2). This
spontaneous activity is of the shape of very sharp spikings, and
these spikings are localized clearly within the nucleus. When
the needle-tip is located 1 - 2 mm outside the nucleus, they disappear totally (Fig. 3). These localized discharges are obtain-
able similarly from both medial and lateral parts of the nucleus,
but the general tendency is that they are a little greater in
number and larger in amplitude in the lateral part of the nucleus.
This spontaneous activity was thought to be most important not
only for the purpose of locating the nucleus, but also for
physiological interpretation of human emotional activity. The
first question that arises is whether or not this activity may,
even partially, reflect the biological basis of emotion. During
the course of our research, it once was postulated that the grade,

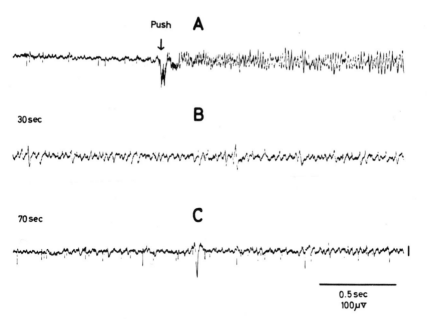

Fig. 2.   Injury discharges through the microelectrode, when the
needle reached the amygdala.   (A) immediately after insertion;
(B) 30 seconds later; (C) 70 seconds later.

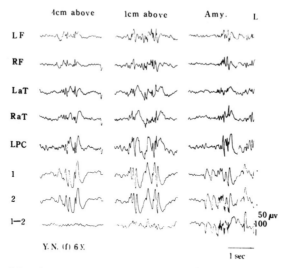

Fig. 3.   Localized spontaneous activity from the nucleus, which
cannot be detected when the needle is outside the nucleus.
Activities at 4 cm, 1 cm above, and inside of the amygdala;
1 is the core electrode of the needle-tip, 2 sheath electrode;
1-2 means bipolar lead at the tip.   F: frontal, aT: anterior
temporal, PC: parieto-central.

character, or number of these spontaneous spikings, which are sometimes larger than 500 μV, might be, at least partially, indicative
or suggestive of the grade of emotional irritability or unsteadiness in each clinical case. Actually, the number and grade of
spikings in the single operated schizophrenic patient in the awake
state were much fewer than in epileptics in the similar anaesthetic
condition, and, similarly, from the Parkinsonian amygdala in which
they were almost nonexistent. However, in contrast to this situation, under general anesthesia, especially when induced by intravenous barbiturate, there is an increase in the spikings from the
amygdala, and the number of spikings appears to differ depending
on the level of anesthesia. The number of cases involving severe
emotional or behavioral disturbances in which the amygdaloid recording in the awake state was accomplished satisfactorily is
still small, when we consider carefully the age, basic neurological
diseases and other conditions, and, therefore, no conclusive
summary can be drawn at the present stage. Through the implanted
electrodes in the amygdaloid area of two patients, the increase of
these spontaneous discharges during different stages of sleep was
reported by Yoshida (1964).

The second question of importance is whether or not these
spontaneous spikings are the origin of cortical spikings, i.e.,
whether they project themselves to the cortical recording or
vice versa. There exist a number of reports which suggest the
close relationship of cortical spiking activity with those of
deep structures. But what we could see is that most of the depth
discharges are not projected or represented to cortical spikings,
and, therefore, such depth activity rarely can be detectable on
cortical recording. Nashold has described beautifully the situation, that we are seeing only the surface of an ocean by surface
recording, and in the real depth of this ocean the very massive
and continuous activity is going on which cannot be observed or
even imagined from the surface activity (1970).

There are obviously some spikings in cortical leads which are
coincident in phase to those in depth and could be considered as
the projected ones from depth activity (Fig. 4). These projected
ones, however, in reality are not great in number when we observe
carefully the numerous depth spike activities, most of which are
not detectable at all by surface recording. On the other hand,
there also can be found the independent cortical spikes, which do
not reflect any of the depth activity. Figure 5 is the cortical
and depth EEG in the postencephalitic epileptic patients with focal
and generalized seizures. The large independent spikes are considered of localized cortical origin and have no correspondence to
the depth. These large spikes could not be modified at all by
amygdaloid surgery.

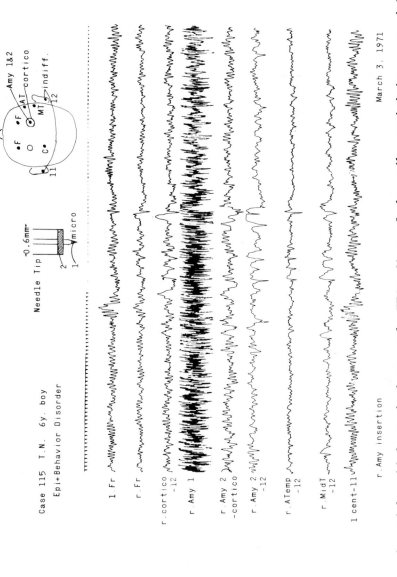

Fig. 4. Right-sided cortical or scalp EEG. Most of the spike activities, especially on anterior temporal leads, correspond to the amygdaloid discharges. r. Amy 1: core of depth electrode, r. Amy 2: sheath of depth electrode.

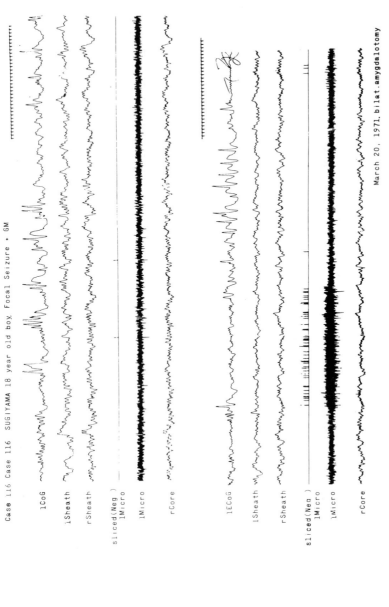

Fig. 5. Cortical spike activity, which is not dependent on the depth activity (case of left fronto-parietal focus). l ECoG; corticogram at the burr-hole made on left fronto-parietal area, left sheath, left core (micro needle) right sheath, right core are recordings from the amygdala of each side.

The third physiological criterion is the olfactorily-evoked response from the nucleus, which has been known in animals for some time (Imamura, Kawamura and Tokizane, 1957). This also can be demonstrated in human cases and, furthermore, it has specific importance in differentiating the medial nuclear group and the lateral nuclear group. In both situations, medial insertion and lateral insertion within the nucleus, it has the appearance of "spindle burst" composed of several spindles corresponding to the phases of inhalation. The olfactory stimulant, mainly ether, is used in our laboratory, is applied for ten to twenty seconds to the nose of the patient. In the medial insertion, the spindle usually is larger in amplitude and much sharper in resolution than in the lateral part. When the lateral part is inserted, its shape is blunt and smaller in amplitude (Fig. 6). These olfactorily-evoked responses are adapted quickly and then disappear after three or four bursts, even when ether is applied continuously.

Using different substances for olfactory stimulation, slight differences in pattern of response are observed. Figure 7 is a record of responses for ether, camphor and purin substance. Of extreme interest for future consideration is the question whether such differences in responses may be related to the physiological effects of the different odors or to the different odor sensations. On the other hand, it should be noted that our experiences indicate that even the bilateral amygdaloid destruction did not produce any marked loss or changes of smell function.

The three physiological indicators described would ease the difficulty of the radiological estimation.

By high frequency stimulation of the nucleus, the most commonly observable effects, even in the deeply anesthetized state, are the pupillar dilatation and arrest of respiration. Facial reddening, change of blood pressure or elevation of muscle tone was not observed in our anesthetized series. Pupillar dilatation occurs usually to about twice or three times larger in size in the anesthetized myotic state. Arrest of respiration occurs mostly in the inspiratory phase and not in the expiratory phase. Rhythmic respiratory movement stops suddenly in the inspiratory phase with initiation of stimulation and continues for about 30 to 40 seconds, even after cessation of stimulation, and then it reappears automatically. These autonomic responses would suggest that targeting of the needle was exactly to the aimed area, presumably to the medial part of the lateral nuclei, and that the destruction of this area might produce the marked calming effect. From our observations, we are confident that the clinical calming effects by surgery could be achieved better by destruction of the area which produced the most prominent autonomic responses as described above.

Olfactory Response

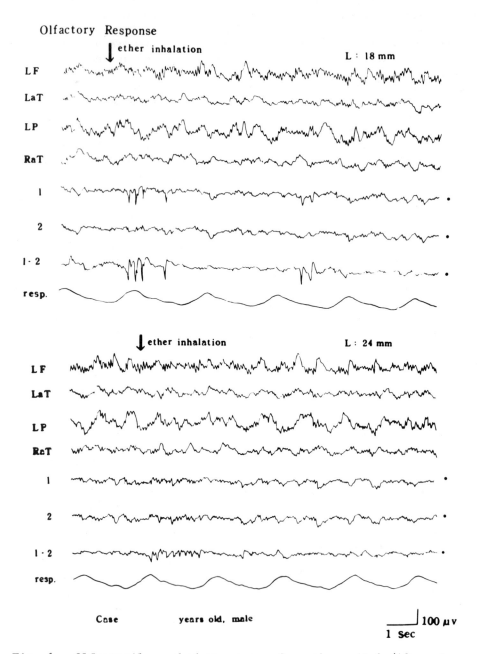

Fig. 6. Olfactorily-evoked responses from the medial (18 mm from the mid-line) and the lateral (24 mm) part of the amygdala for ether inhalation (cited from Archives of Neurology, 9, 6, 1963).

C. <u>Relationship</u> <u>between</u> <u>right</u> <u>and</u> <u>left</u> <u>amygdala</u>: Bilateral surgery in a single operation was not performed in this author's clinic until recently, though several others already have performed it without hesitation (Heimburger, 1966; Balasbramanian <u>et al.</u>, 1967). Diemath and Nievoll also performed unilateral amygdaloid destruction in combination with dorsomedial thalamic lesion of contralateral side (1966). This is due to the fact that the author was very much concerned about even the minimum possibility of producing the bilateral temporal lobe syndrome, Klüver-Bucy syndrome (Klüver and Bucy, 1939; Terzian, 1958). Until recently, in the author's institution each side was operated, usually with the interval of from two weeks to six months, in order to observe closely the clinical changes and effects, and possible presence of side-effects of unilateral destruction.

In bilateral surgery, through two depth electrodes, each of them being inserted simultaneously to either side, stimulation of one side of the amygdala produces the evoked biphasic responses on the other side (Fig. 8). Latency of positive peak is about 12 - 13 msec, which then is followed by slow large negative phase. Small initial negative phase sometimes precedes the positive peak. These evoked discharges quite commonly are observable, and, since they follow up to the frequency of about 20 c/s, the connection between the two nuclei must be assumed to be quite direct. Considering such a close internuclear relation, it can be assumed that even unilateral surgery would produce general calming effect to some extent. In our series of 25 epileptic cases (Table I), 14 cases were improved moderately or highly in behavioral sphere by unilateral lesion.

D. <u>Surgical</u> <u>procedure</u>: For producing lesions, in the initial 70 cases, the blocking of the nucleus was performed by Jordilax (Yoshitomi Co., Osaka; mixture of olive oil, bees wax and lipijodole) installation to produce mechanical lesion of about 5 to 8 mm in diameter. In subsequent cases up to the present, the controlled thermocoagulation using "Coagulador" was applied. Clinical results do not differ much by these two different devices of destruction. By controlled thermocoagulation, using 70°C for 30 seconds, a lesion of about 3 or 4 mm diameter is produced. For relatively larger amygdaloid lesions, several successive coagulations on the same needle tracks, but each differing in position by 3 mm, are usually necessary.

II. CLINICAL OBSERVATIONS ON THE EFFECTS BY SURGERY

Clinical effects on the emotional changes of these pathological cases and also on epileptic paroxysm, have been described in several papers published previously (Narabayashi <u>et al.</u>, 1963; Narabayashi, 1964, 1969, 1971; Narabayashi and Uno, 1966;

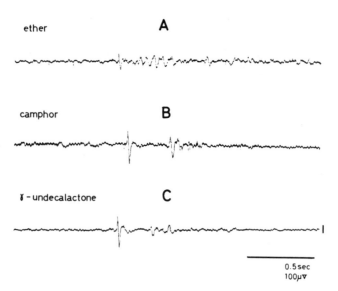

Fig. 7. Olfactory response from the amygdala through the micro-electrode for inhalation of: ether (A), camphor (B) and γ-unde-calacton (C).

Case 118

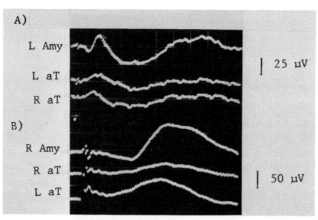

Fig. 8. Evoked amygdaloid responses for stimulation of the contra-lateral amygdala. (A) Right amygdaloid stimulation; recording from left amygdala, left anterior temporal and right anterior temporal lead...total sweep; 200 msec. (B) Left amygdaloid stimu-lation; recording from right amygdala, right anterior temporal, left anterior temporal...total sweep; 50 msec.

Stimulation parameter 10 V, 1 msec, 6 cps.

Table I*

Observations on long term results of 25 operated cases due to epileptic etiology. Effects on paroxysmal activity and on the behavioral sphere are classified.

	Effects on behavioral disorder**				
	A	B	C	D	Total
Seizure and spikes abolished	8 (4 cases: A′)	1	0	0	9
Seizure and spikes reduced	1	1	0	0	2
EEG spike reduced	1 (A′)	0	0	0	1
Seizure reduced	1	5	0	0	6
No change	1	3	3	0	7
Total cases	12	10	3	0	25

* Cited from Confinia Neurologica, 32, page 290, 1970.
** A = marked improvement
   A′ = dramatic improvement
   B = moderate improvement
   C = slight improvement
   D - no change or worsened

Narabayashi and Mizutani, 1970). Therefore, only a summary will be introduced here.

A. Description of patients: About two thirds of our surgery series are child cases under the age of 13, and more than one-half of the latter range in age between 5 and 8.

Irritability, unsteady mood, poor and short concentration span and even violence or assaultiveness in these child cases is not identical to the violent destructiveness, assaultiveness or antisocial aggressive behavior that is exhibited by adults. For instance, "explosive psychopathic personality in adult" is not conceived as the same as uncontrollable behavior or irritable state in children. As to the pathological emotional states in the child, the author has especially noticed the following three features: firstly, the poor and very short concentration span; secondly, the hyperactiveness, and, thirdly, the violence and destructiveness in beating people and throwing or damaging the inanimate things in the vicinity.

The short concentration span in the child is observed as a
rapidly changing focus of interest from one page of the book to the
next every second, changing the toy every few seconds from one hand
to the other, changing the TV channels quickly or never talking
continuously or calmly. Usually, only fragments of words are found
in these cases. Very often the child's interest in one subject continues only very shortly, one to five seconds. Hyperactivity,
sometimes called hyperkinesis, not in the sense of extrapyramidal
disorders, also is easily observable as always running or moving
around, as animals in the cage, and as never staying long and
quiet enough on a bed or on chairs, and touching everything or
sometimes even accompanied by climbing and jumping. When the
mother takes him to the department store, he is easily lost by
running away and pulling down many things and toys from show case,
with damaging results. Hair-cutting, EEG recording or injection
in the awake state is impossible.

Change or improvement of such uncontrollable behavior, after
surgery, can be observed quite vividly by family members, all
medical staff members and attending psychologists. The grade of
calming or taming, whether it is an almost complete one to the
normal level or is a relative one, can be determined without much
difficulty.

After the procedure, in the well-improved case, the child is
very quiet and obedient. Hair-cutting now is normally possible
with no difficulty, and even injections are accepted easily and
with no force. EEG without anaesthesia becomes often possible
although it was naturally quite impossible and unimaginable preoperatively. A visit to the department store or toy shop, taking
him to the party in a friend's house, or even on a trip by train
are now performed with no special difficulty.

The violent behavior and destructiveness in such child cases,
though not so seriously antisocial as in the adult cases because
of their age, also are much calmed. The child becomes attentive
to a parent's instruction and is capable of staying in bed or on
a chair quietly as instructed, playing with some toys, reading a
book or watching TV as he is interested. He may talk with his
mother or nurses in a more friendly and calm manner. Destructiveness disappears or is lessened significantly. Violent behavior
also disappears, with basic taming in mood.

In order to establish the more objective and quantitative
evaluation of the results, the author devised so-called behaviometry with Drs. Nagahata and Sumino, both of whom have been working in the field of pediatric neurology and psychiatry (Nagahata,
1968). They have designed the observation room for this purpose
(Fig. 9 A & B), which is a small room of 3.3 × 4.5 meters in size.

The observers observe, through a one-way mirror, the child's movements, play and general behavior. The floor is divided in twenty equal sections, each section numbered from 1 to 20, with several toy-boxes on one side. All observations of the patient's movements are recorded on tape, with emphasis placed on particular floor-section, toy, etc., and are accompanied by brief interpretation by the observer. At the center of the ceiling above, a 16 mm wide-angle camera is set to record the child's general behavior. Simultaneous to tape-recording by the observer, this film-picture is recorded by video-monitor. Change of interest on different toys can be checked in this manner.

The length of the routine observation period is for 20 minutes, and at least twice preoperatively and twice postoperatively. Taking the average of each two-stage observation periods, the data are analyzed statistically.

B. Behavioral observations and results: Regarding hyperactivity, the duration or the period of stay in each section (Blocks 1 to 20) is statistically treated, plotted and described in a graphic manner. Preoperatively, a significantly short stay in one block of less than 5 or 10 seconds is most frequent. But, postoperatively, much longer periods averaging more than 100 seconds become commoner, which is indicative that the child sits more steadily and quietly in one block and exhibits interest in playing with toys. Figure 10 shows the example of an eight-year-old boy patient. Postoperatively, about fifty days, he remains more frequently and with better concentration in one block, compared with the preoperative stage. Figure 11 also is the same case, which presents the steady improvement in hyperkinesis after two years and three months of observation.

The concentration span on particular toys can be demonstrated in the similar fashion. The period or length of span is measured and analyzed statistically, similarly as the movement on the floor-section. Figure 12 presents the results in nine cases which were operated relatively recently and were observed up to three years. Except cases 101 and 100, the other seven cases show the steady improvement in their attitude, and concentration in toy-play.

General observation indicates clinical behavioral changes, with improvement in hyperactivity and improvement in concentration span which are quite parallel in all the cases studied.

Results in adult cases are not so uniform and frequently are a little more difficult to evaluate as described previously (Narabayashi, 1966). One of the reasons is that aggression or violence in adults may have various different origins, even though it is based on similar biological and epileptic bases. In the

Playroom

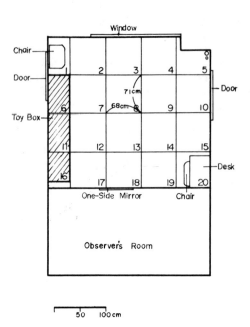

Fig. 9 A & B. Play room for observation and behaviometry of the behaviorally disturbed and hyperactive children cases.
A: play room.   B: toy box.

Toy   Box

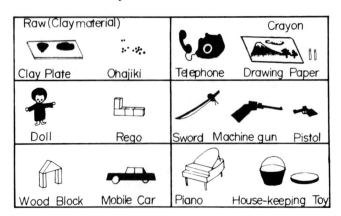

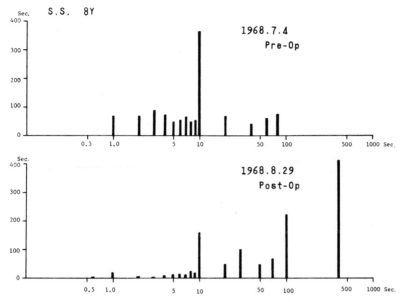

Fig. 10. Sum of periods of changing the blocks. Abscissa indicates the length of each stay in one block from 0.5 to 1000 sec. Ordinate indicates the sum of each stay of different length.

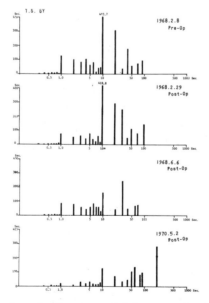

Fig. 11. Observation in 8-year-old boy patient. Explanation similar to Fig. 10. Preoperative; after right amygdaloid surgery; after left amygdaloid surgery; about two years later.

well-influenced adult cases, the improvement was quite satisfactory and gratifying, producing marked lessening of excitation and making the patient able to adjust to his environment, to return to his job or to keep a peaceful existence with his family.

Recent statistics in 25 cases of epileptic etiology that include both child and adult cases are shown in Table I.

C. <u>Long-term</u> <u>results</u>: In long-term postoperative period observations of two to twelve years, the late effect or late improvement also is quite marked. Late effects or changes are influenced naturally by or mixed with educational, familial or social conditions. However, this tendency of the late effect might be the most interesting and important one for evaluation of the procedure. Generally speaking, phenomenal elevation of IQ is not rare, when the child possessed some good latent intelligence-level such as of imbecility, debility or sub-normal IQ preoperatively. Sometimes, the preoperative attitude or aggression made psychological testing difficult or impossible, with diagnosis of erethistic, untestable idiots. When the postoperative test in the calmed state produces the relatively good IQ, the cases are considered as phenomenally improved in intelligence. This, however, might not be the real improvement in the intellectual sphere.

On the other hand, when the child becomes much quieter and follows instructions more easily, and concentration is better than preoperatively, his ability of gaining knowledge also may be increased and, as the result of this, his latent capacity will work better and the intellectual level later will be higher. This interpretation may mean that the improvement of intelligence is secondary to the improvement in attitude or behavior. However, it still can be postulated that the bombardment by the disturbing or inhibitory impulses from the preoperative amygdala are abolished and that in such new situations the delicately organized cortical activity will develop more normally.

Another very gratifying observation is that the speech begins to develop in a two- or three-week period after the surgery in preoperatively speechless cases. If it does not appear within this period, usually it does not appear at all. This also may suggest that the potential cortical capacity may develop quite rapidly when some strong disturbing factor is removed.

D. <u>Effect</u> <u>on</u> <u>epileptic</u> <u>seizures</u> <u>and</u> <u>EEG</u>: The effect on the clinical seizures and paroxysmal activity on EEG was described recently in this author's paper (Narabayashi, 1970). When the procedure was initiated, control or influence on grand-mal seizures was not imagined at all, but from our observation for more than a 10-years course after surgery the procedure seems to have definite

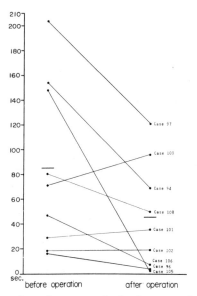

Fig. 12. Postoperative long-period observation of recent nine
cases concerning concentration span on the toy. Only short
concentration for less than 10 seconds is plotted. Ordinate
shows the total sum of this short concentration-span with a
20 minutes observation period.

influence on the seizure activity in these cases, especially on
the generalized seizures. In Table I, observations on 25 cases
of epileptic etiology are summarized, which were operated after
1963 and were followed carefully from one to six years. A indicates the grade of marked improvement in emotional and behavioral
aspects, such that the patient can adapt to his social milieu,
as kindergarten or school or getting a job, according to the
respective age and intellectual level. Five cases graded A´ in
the A group are of the especially dramatic and noticeable
improvement, becoming almost normalized in all aspects and
living as normal members of the society, getting standard school
results or doing a competent job in business or factory work.
B indicates moderate improvement, but the patient usually stays
at home with much less necessity of being watched or being taken
care of by the family. C indicates slightly improved cases and
D indicates no change at all.

On the left side of this table, the change or improvement in
paroxysmal seizure phenomena and in EEG pattern is listed. In
nine of the best improved cases, both clinical and electrical

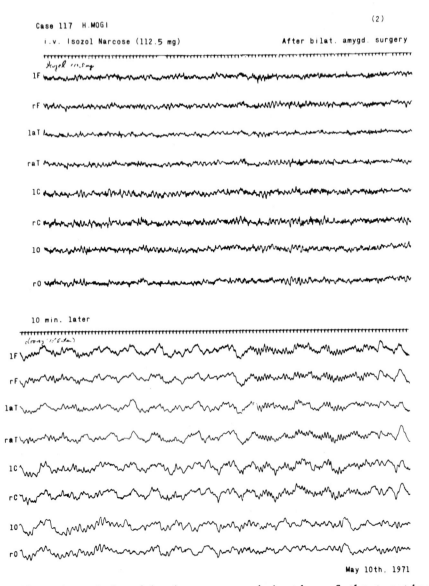

Fig. 13. Sleep induced by intravenous injection of short acting
barbiturate (Thiobarbitur Na, Isozol) in 7-year-old boy patient.
(A) In preoperative stage, dosage of 210 mg could induce the
slight drowsy state; immediately after injection and ten
minutes thereafter.

Case 117 H. MOGI 7 year old boy, severe behavior disorder

i.v. Isozol Narcose (210 mg) preoperative

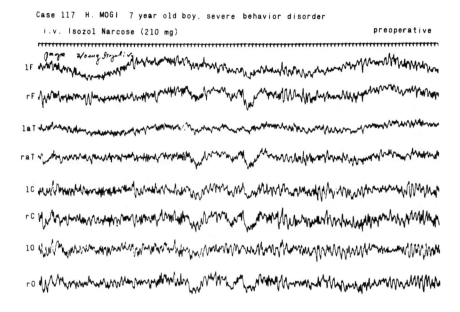

same (10 min. later --- drowsy)

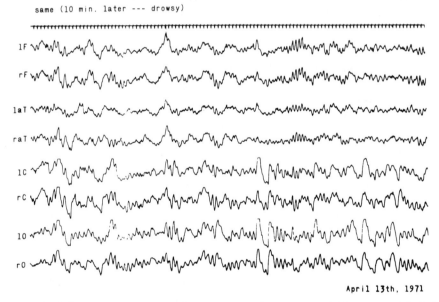

April 13th, 1971

Fig. 13 (B) In postoperative stage, dosage of 112 mg could induce better spindle phase sleep.

    1F, rF: left or right frontal lead

    1aT, raT: left or right anterior temporal lead

    1C, rC: left or right central temporal lead

    10, r0: left or right occipital lead

seizure activity disappeared completely (see for details, Narabayashi and Mizutani, 1970). In two other cases, both seizure and EEG spikes were much reduced. In one, mainly EEG improvement was observed, while clinical seizures were much reduced in six cases. Details in each of A´ and A groups are described in the above mentioned paper.

Concerning the paroxysmal phenomena, most influenced are the diffuse and synchronous EEG spikes and the accompanying GM seizures. Naturally, some tendency of asymmetry of these abnormal activities between hemispheres exists both electrically and clinically in the preoperative stage and even in such instances the effect also is similar. But the purely localized spikings of cortical origin and the clear focal seizures usually could not be controlled, except those of temporal lobe types in a few cases. As explained at the beginning of this paper, some of the focal spikes are of cortical origin and the amygdaloid lesion could produce no influence or change. It is quite noteworthy that the temporal lobe seizures could be dramatically improved immediately after surgery, with later recurrence in a high percentage, as has been described elsewhere (Narabayashi and Mizutani, 1970).

The mechanism of modifying or alleviating the paroxysmal activity, especially of generalized character, by amygdaloid surgery, is not known at all. In addition, there are several recent findings: (1) in most of the cases the calming effects in the behavioral sphere and improvement of generalized paroxysmal activity usually seem to be parallel and (2) disappearance or lessening of electrical paroxysmity usually occurs gradually within several weeks or months after surgery; usually it is not observable immediately after surgery, though clinical seizures disappear relatively earlier after surgery in the demonstrative cases; and (3), in the improved cases, necessary dosage of intravenous barbiturate to induce sleep changes markedly and becomes significantly reduced. Frequently, the necessary dosage becomes one-half of the original preoperative dose (Fig. 13A). This figure represents the sleep EEG of one case. Preoperatively even 210 mg of short-acting barbiturate (Isozol) did not induce enough sleep of spindle stage, but the same case is narcotized postoperatively more easily by about half dosage (112.5 mg) of the same substance (Fig. 13B).

All these would suggest some of the excitability threshold of whole brain might be changed to less excitable ones by way of neural or metabolic changes which could open new topics for the control of epileptic paroxysm.

In order to assess some of the metabolic bases, the value of steroids in serum and of growth hormone were measured in three cases which presented no marked changes between preoperative and postoperative samples.

E. <u>Side-effects</u>: Clinical side-effects by this procedure, even by bilateral surgery at the same sitting, usually are non-existent if the procedure is carried out correctly for destroying the central and basal part of the amygdaloid nucleus. Klüver-Bucy syndrome, which is widely known as the most serious damage for human mental activity by bilateral temporal lesions, was not observed in our series, except transiently observed in one case (#68; Narabayashi and Uno, 1965). This is the case operated bilaterally with relatively posteriorly located lesions, perhaps damaging a part of the hippocampus. Increased unsteadiness, polyphagia, oral tendency and hypersexuality with excitation were noted for about two months after surgery, and were difficult to control even by high-dosage of chlorpromazine, but these subsided gradually.

In other cases, in both adult and child cases, no marked or noticeable loss of memory, no rough change of taste or of smell, no hypersexual behavior was observed. Transient relative polyphagia for about two weeks postoperatively is not uncommon in about one fourth of these operated cases, but are not accompanied by an increase of body weight. This, however, usually subsides gradually. Slight transient capsular paresis did appear in two cases. Both were the cases in which the trephanation was made a little too posteriorly and perhaps the needle path damaged the capsule. Because of this possibility, we are careful to make burr-holes as frontal as possible.

Changes in the autonomic function by surgery, such as pulse rate, blood pressure, respiration, size of pupils, salivation, sweating or even in function of the digestive tract have not been noticed.

A 16 mm film of a well-improved 6-year-old boy was shown demonstrating the preoperative hyperactivity, very short concentration span on books or toys and violence with their postoperative improvement.

REFERENCES

AIRD, R. B., & YAMAMOTO, T. Behavior disorders of childhood. Electroencephalography and Clinical Neurophysiology, 1966, 21, 148.

BALASUBRAMANIAM, V., RAMAMURTHI, B., JAGANNATHAN, K., & KALYAMARAMAN, S. Steriotaxic amygdalotomy. Neurology (India), 1967, 15, 119.

CHAPMAN, W. P. Studies of the periamygdaloid area in relation to human behavior. Research Publications Association for Research in Nervous and Mental Disease, 1958, 36, 258.

DIEMATH, H. E., & NIEVOLL, A. Stereotaktische ausschaltungen im nucleus amygdalae und im gegenseitigen dorsomedialkern bei erethischen kindern. Confinia Neurologica, 1966, 27, 172.

FALCONER, M., & SERAFETINIDES, E. A follow-up study of surgery in temporal lobe epilepsy. Journal of Neurology, Neurosurgery and Psychiatry, 1963, 26, 154.

HEIMBURGER, R. F., WHITLOCK, C. C., & KALSBECK, J. E. Stereotaxic amygdalotomy for epilepsy with aggressive behavior. Journal of American Medical Association, 1966, 198, 741.

IMAMURA, G., KAWAMURA, H., & TOKIZANE, T. Electrophysiological study of archicortex. Speech at 121th Tokyo Physiological Society Meeting, September, 1957.

KLÜVER, H., & BUCY, P. Preliminary analysis of functions of the temporal lobe in monkeys. Archives of Neurology and Psychiatry, 1939, 42, 979.

MILNER, B. Psychological defects produced by temporal lobe excision. Research Publications Association for Research in Nervous and Mental Disease, 1958, 36, 244.

NAGAHATA, M. Behavior disorder and minor brain damage. Shonika Shinryo, 1968, 31, 1193. (In Japanese)

NARABAYASHI, H. Stereotaxic amygdalotomy. Brain and Nerve, 1964, 16, 400. (In Japanese)

NARABAYASHI, H. Stereotaxic amygdalotomy (its long-term results). Excerpta Medica International Congress Series, 1969, 193, 8.

NARABAYASHI, H. Stereotaxic operations for behavior disorders. W. Sweet (Ed.), being published.

NARABAYASHI, H., & UNO, M. Long range results of stereotaxic amygdalotomy for behavior disorders. Confinia Neurologica, 1966, 27, 168.

NARABAYASHI, H., & MIZUTANI, T. Epileptic seizures and the stereotaxic amygdalotomy. Confinia Neurologica, 1970, 32, 289.

NARABAYASHI, H., NAGAO, T., SAITO, Y., YOSHIDA, M., & NAGAHATA, M. Stereotaxic amygdalotomy for behavior disorders. Archives of Neurology, 1963, 9, 11.

NASHOLD, B. Personal communication, 1970.

PAPEZ, J. W. A proposed mechanism of emotion. Archives of
    Neurology and Psychiatry, 1937, 38, 725.

PENFIELD, W., & FLANIGIN, H. Surgical therapy of temporal lobe
    seizures. Archives of Neurology and Psychiatry, 1950,
    64, 491.

PENFIELD, W., & JASPER, H. Epilepsy and the Functional Anatomy
    of the Human Brain. Boston: Little Brown, 1954.

TAYLOR, D., & FALCONER, M. Clinical, socioeconomic, and psycho-
    logical changes after temporal lobectomy for epilepsy.
    British Journal of Psychiatry, 1968, 114, 1247.

TERZIAN, H. Observations on the clinical symptomatology of
    bilateral partial or total removal of the temporal lobes
    in man. Temporal Lobe Epilepsy. Springfield: Charles
    C. Thomas, 1958. Pp. 510-529.

TOSHIMA, Y. Histopathology of the amygdaloid nucleus.
    Psychiatria et Neurologia Japonica, 1961, 63, 1178.
    (In Japanese)

YOSHIDA, M. Correlation between spikes and seizure discharges
    in amygdala and emotion. Brain and Nerve, 1964, 16, 809.
    (In Japanese)

DEEP TEMPORAL LOBE STIMULATION IN MAN

Vernon H. Mark, Frank R. Ervin, and William H. Sweet

Neurosurgical Service, Boston City Hospital; Neurosurgical and Psychiatric Services, Massachusetts General
Hospital, Boston, Massachusetts

Before discussing deep temporal lobe stimulation with
chronically implanted stereotactic electrodes, it might be well
to mention some of the ethical considerations involved in this
kind of stereotactic surgery. First of all, the patients who
are candidates for stereotactic temporal lobe electrodes are all
temporal lobe epileptics with aggressive behavior who have had a
considerable trial period of anti-epileptic and ataractic drugs,
together with the various forms of psychotherapy. Almost all of
these patients would be considered as candidates for the more
traditional anterior temporal lobectomy, except that they had
multiple foci, which were usually independent, bilateral, and
non-synchronous in nature. All of our patients except one were
adults, and this one exception was a brain tumor suspect with a
pneumoencephalographic diagnosis of temporal lobe tumor.

Under the direction of a dean's committee at Harvard Medical
School, we have set up an independent review committee of physi-
cians not associated with the patient, for a consideration of
the suitability of surgical treatment in each case. Cognizance
is taken of past treatment, and the possibility of other kinds
of treatment, i.e., medical and psychiatric, before surgery is
recommended.

In a wider context, our group is surveying a number of
patients with focal brain disease and behavioral abnormalities
of the episodic variety (especially violent behavior). We are
not looking at these patients as surgical candidates, but are
seeking to apply other kinds of treatment, i.e., medical and
psychiatric, to their problem if a substantial link can be
proved between their brain abnormality and their behavioral

disorder. Our surgical approach has been restricted to episodi-
cally violent patients with intractable temporal lobe epilepsy
who are over the age of eighteen.

Deep temporal lobe stimulation in man is not novel; it has
been carried out for a number of years in temporal lobe epileptic
patients during their anterior temporal lobectomy. Many of these
patients were operated on under local anesthesia, and the effects
of stimulation were measured in the operating room. These effects
were mostly immediate in onset, coming on within one-half to five
seconds after stimulation, and subsiding at or shortly after the
time that the stimulation was terminated. These effects were
often associated with a seizure discharge or afterdischarge on
the electroencephalogram, and thus directly related to the ictal
episode.

The introduction of stereotactic surgery with chronic
implanted electrodes (Delgado et al., 1952) has taken the stimu-
lation-recording procedures out of the operating room. It has
enabled us to make a more thorough study of the patient's
responses under more natural conditions, without anesthetics or
sedation, and it has allowed us to evaluate a different kind of
response than could be evaluated in the operating room. The
response in question is one that may come on thirty seconds to
minutes after stimulation has subsided, and which may persist
for many minutes, or even hours, after the stimulation has
stopped.

The technique of stereotactic surgery with chronic electrode
implantation has been described in detail by our group (Heath and
Mickle, 1960). The procedure, in brief, involves the implantation of insulated multi-lead depth electrodes, with twelve recording points, directed to the lateral amygdala and medial
amygdala. The extracranial terminals of the electrodes were
fixed to the skull and soldered to the contacts of twenty-four-
connector plugs for subsequent stimulation and recording.
Recording and stimulation were then accomplished between the
various electrode points, while the patient was seated in a
comfortable chair, facing the examiner, but unable to visualize
the stimulus control panel. The EEG, from depth and scalp
derivations, heart rate, and respiration, were recorded on the
ten-channel polygraph. Stimulus current was monitored through
an oscilloscope; although various combinations of electrode
positions, current strength, pulse width, and frequencies were
employed, most stimulations were carried out between adjacent
electrode points in depth, with 60 cycles per second 1 msec
rectified square saves, at a current varying between .1 and 1 ma.
Individual stimuli lasted from 10 seconds to 2 minutes. A continuous written record of behavior was made before, during, and
following stimulation. Because stimulation and recording

techniques required the patient's electrodes to be physically connected to bulky pieces of electronic apparatus and required the patient to be in a confined and unnatural situation, we adopted the Delgado telestimulation and recording technique in some of our patients (Mark and Ervin, 1970).

Pertinent examples of the kinds of responses that were obtained through the stimulation of points on chronic inlying electrodes in the temporal lobe are illustrated by the following case reports:

CASE 1. The first patient, Julia P., exhibited some of the behavioral responses that can be obtained immediately after stimulation, associated with the ictal process. Julia was a 22-year-old girl with a history of brain disease that went back to the time when, before the age of 2, she had a severe attack of encephalitis. At the age of ten, she began to have epileptic seizures; occasionally these attacks were grand mal seizures. Most of the time, they consisted of brief lapses of consciousness, staring, lip-smacking, and chewing. Often, after such a seizure, she would be overcome by panic and would run off as fast as she could, without caring about destination. Her behavior between seizures was marked by severe temper tantrums, followed by extreme remorse. Four of these depressions ended in serious suicide attempts. On twelve occasions, Julia assaulted seriously other people without any apparent provocation. By far the most serious attack occurred when she was eighteen. She was at a movie with her parents when she felt a wave of terror pass over her body. She told her father she was going to have another one of her "racing spells", and agreed to wait for her parents in the ladies' lounge. As she went to it, she automatically took a small knife out of her handbag; she had gotten into the habit of carrying this knife for protection because her "racing spells" often took her into dangerous neighborhoods where she would come out of her fugue-like state to find herself helpless, alone and confused. When she got to the lounge, she looked in the mirror and perceived the left side of her face and trunk (including the left arm) as "shriveled, disfigured, and evil." At the same time, she noticed a drawing sensation in her face and hands. Just then another girl entered the lounge and inadvertently bumped against Julia's left arm and hand. Julia, in a panic, struck quickly with her knife, penetrating the other girl's heart, and then screamed loudly. Fortunately, help arrived in time to save the life of her victim.

The next serious attack occurred inside the mental hospital to which Julia had been sent. Her nurse was writing a report when Julia said, "I feel another spell coming on---please help

me." The nurse replied, "I'll be with you in just a moment."
Julia dragged a pair of scissors out of the nurse's pocket and
drove the point into the unfortunate woman's lungs. Luckily,
the nurse recovered.

Julia's case clearly illustrates the point that violent
behavior caused by brain dysfunction cannot be modified, except
by treating the dysfunction itself. She had had extensive
medical care and years of psychotherapy (including behavior
therapy). She had taken, consecutively and in combination, all
the known anti-seizure medications, as well as the entire range
of drugs used to help emotionally disturbed patients. She had
been treated in three of the major medical centers of North
America, with no signs of improvement. As a last resort, she
had been given over sixty electroshock treatments, without any
change in her seizures or in the pattern of her violence in
rage.

The neurological examinations showed Julia's ability to
assimilate newly-learned material was impaired, and she had a
severe deficiency in both recent and remote memory. Electroencephalographic examinations disclosed a typical epileptic
seizure pattern, with spikes in both temporal regions, in
addition to widespread abnormality over the rest of the brain.
Pneumoencephalograms disclosed central atrophy of the right
temporal lobe.

Electrodes were placed stereotactically into both temporal
lobes and, after she had recovered from the surgical procedures,
we recorded epileptic electrical activity from both amygdalas.
Electrical stimulation of either amygdala produced symptoms
characteristic of the beginning of her seizures. The symptoms
were more easily elicited by stimulating her left amygdala, a
RF lesion was placed in this amygdala, and all the electrodes
were withdrawn. However, her symptoms persisted and changed
to include signs that indicated a small portion of her brain
was firing abnormally, and that this area was related to the
movement of her left arm. This suggested that her persistent
seizures and attack behavior were initiated in her right
temporal lobe; therefore, we again placed electrodes in her
right amygdala. At the time of this second operation, Dr. Jose
Delgado's "stimo-ceiver" had become available to us, and we
attached one of these to Julia's right temporal lobe electrode.
This device made it possible to record the electrical activity
in her right amygdala and hippocampus from a hundred feet away,
while she moved around with others in her ward. We were thus
able to observe the interactions between brain stimulations and
environmental cues. We could also stimulate a selected target
in her brain from the same distance; because there were no

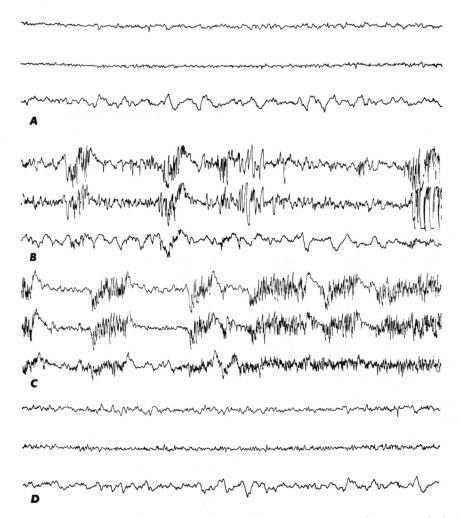

Fig. 1. Telemetered EEG tracings of spontaneous seizure activity
in patient Julia. Remotely recorded brain waves from Julia's
amygdaloid nucleus are sampled before electrical stimulation:
(A) resting record; (B) and (C), cascades of abnormal spikes are
seen; (D) resting record. During (B) and (C) the patient had
the behavioral change noted in the text.

wires involved, we could try to reproduce her violent symptoms, without fear that she might hurt herself by pulling out the electrodes, and we were also able to record the activity inside her brain continuously for up to 24-36 hours.

The following records were made from this patient in a hospital room, with the cooperation of Dr. Delgado. Both Julia and her parents knew that sometime during the day her brain was going to be recorded from and stimulated, but they had no idea just when this was going to happen. Before we had done any stimulating, but while we were recording, the electrical activity recorded from the leads in Julia's amygdaloid nucleus showed a typical epileptic seizure pattern, as seen in Figure 1. The behavior that accompanied this change in Julia's brain waves involved her getting up and running over to the wall of her bedroom; once there, she narrowed her eyes, bared her teeth, and clenched her fists; that is, she exhibited all the signs of being on the verge of making a physical attack.

The results of two episodes of electrical stimulation in her right amygdala can be seen in Figures 2 to 11. The stimulus was a 5-second train of 50 cycles per second biphasic square waves at 1 ma. Figure 2A shows the brain wave recordings in the resting state. There are 3 channels of recordings: the top represents the recordings from the electrode in the anterior amygdala, the second from the electrode in the posterior amygdala, and the third from the electrode in a part of the temporal lobe that is just behind the amygdala. In the first sequence, the electrical activity is near-normal; then there is an interference pattern, caused by the electrical stimulation (Fig. 2B). 130 seconds after the onset of the stimulation, an occasional abnormal spike can be seen; at that point, Julia was unresponsive to questioning. The next sequence, Figure 2C, shows the electrical activity gradually becoming seizure-like, then a characteristic electrical epileptic seizure of the amygdala occurs. Immediately afterwards, this patient exhibited rage behavior. The last EEG tracing was taken five minutes after stimulation, and disclosed a more normal record, with some persistent post-seizure activity. A motion picture was made of this patient's action while her brain was being stimulated, and the picture sequence had been abstracted to show the significant changes in behavior. Her initial behavior was placid (Fig. 3); after stimulation, she was out of contact; then she made a series of angry grimaces, which included lip retraction and baring of the teeth, the ancient "primate threat display" (Figs. 4A and 4B); her spring towards the wall was sudden and quite unexpected (Fig. 5). We were able to understand how victims of her attacks had not had time to defend themselves!

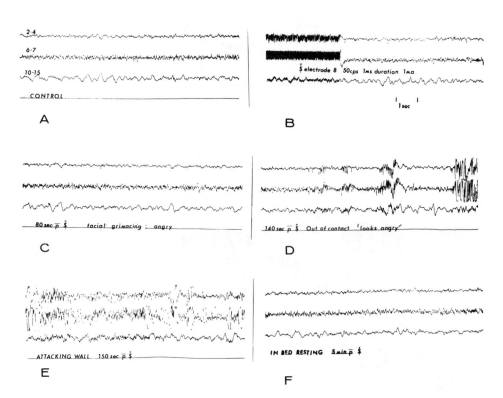

Fig. 2. Telemetered EEG tracings taken in Julia during episode in
which she attacked the wall after stimulation. Remotely recorded
brain waves from Julia's amygdala are seen in stimulus sequence:
(A) Control recording (before stimulation); (B) amygdala is remotely stimulated with weak current; (C) electrical correlates of facial
grimacing are not present; (D) cascades of spikes are followed by
frank amygdaloid seizures which precede her attack behavior. (E)
Her spring toward the wall occurs during amygdaloid seizure.
(F) Five minutes after the attack her brain wave record is similar
to control recording.

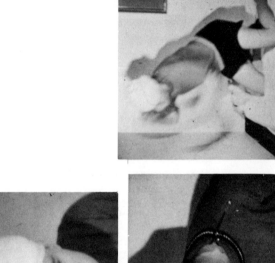

Fig. 3.

Fig. 4.

Fig. 5.

Fig. 3. Patient Julia in a pleasant mood. This photograph was taken from a motion picture sequence seen in Fig. 2. This shows her immediately before stimulation. (From LIFE Magazine, June 21, 1968.)

Fig. 4. (A) Julia in an angry mood at onset of rage attack after stimulation. This shows the change in facial expression immediately following stimulation. (B) Facial grimacing in this picture was followed by lip retraction and other signs of primate "threat display." (From LIFE Magazine, June 21, 1968.)

Fig. 5. Julia attacking wall after stimulation. Attack against the wall was both sudden and unexpected. We could see why it would be difficult to defend one's self against such an attack.

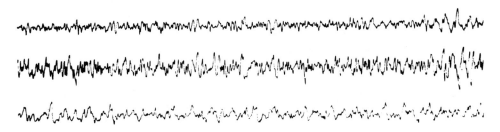

playing guitar   singing

Fig. 6. Telemetered EEG tracings taken as Julia played her guitar
and sang. Remote brain wave recordings were taken before another
stimulation sequence.

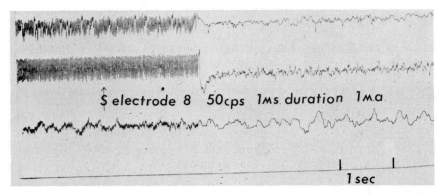

Fig. 7. Telemetered EEG tracings taken during and immediately
after stimulation of the amygdala in patient Julia. Stimulation
was carried out with immediate cessation of guitar playing and
singing.

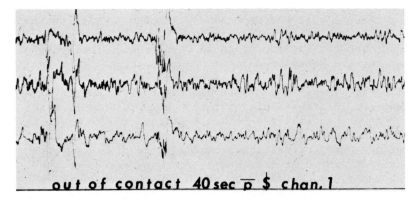

Fig. 8. Telemetered EEG tracings taken 40 seconds after amygdala
stimulation shown in Figure 7. Intermittent "abnormal spikes" in
brain wave record; patient would not answer questions.

The second stimulation sequence began while Julia was playing a guitar; Figures 6 and 7 show the brain wave recordings from the amygdala. The first recording taken while the patient was singing and playing was slightly abnormal (Fig. 6). The second recording shows the electrical artifact produced by the amygdala stimulation (Fig. 7). After five seconds of stimulation, Julia stopped singing and stared blankly ahead; during the next sequence, she slipped out of communication and was unable to answer the questions posed by the psychiatrist who was examining her. A cascade of abnormal spike-like epileptic brain waves from her amygdala was then recorded (Fig. 8). The posterior amygdala exhibited a constant abnormal electrical discharge, characteristic of seizure (Fig. 9). This was followed by a sudden and powerful swing of her guitar; she narrowly missed the head of the psychiatrist, and, instead, the guitar smashed against the wall (Fig. 10). Her resting record (Fig. 11) was taken later, and showed only post-seizure activity.

A short time after these sequences were recorded, we made a radiofrequency lesion in this patient's right amygdala. It is still too early to assess the results of the procedure, but the frequency of both the rage attacks and epileptic seizures have been markedly decreased since operation.

The following cases illustrate some of the long-lasting, long-latency behavioral responses, which can occur after electrical stimulation in and around the amygdala.

CASE 2. A 33-year-old engineer developed seizures at age 22, several months following a severe gastrointestinal hemorrhage resulting in vascular shock. Attacks began with loss of consciousness and stare, followed by salivation, licking, swallowing, head turning to the left, impulse to run, searching movements, and rapid speech, followed by a return to consciousness with a feeling of intense hurt and depression. Treatment with anticonvulsants successfully controlled the early motor portions of the seizure, but the patient began to have frequent episodes of violent aggressive behavior which commenced with the same feeling of hurt and hypersensitivity that previously had followed frank ictal episodes. These "attacks" would typically commence with the patient complaining to his wife that some relatively minor event was not to his liking. He would proceed to brood aloud, dwell upon, and increasingly elaborate on this single theme with mounting anger, verbal abuse, and irrational accusations over a period of three or four hours, reaching a crescendo of rage which was always climaxed by an outburst of physical aggression during which he threw his children against the walls, spit or kicked at his wife, and on one occasion pinned her down while burning her bared chest with a lighted cigarette. During

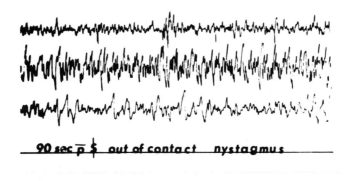

90 sec p̄ $ out of contact nystagmus

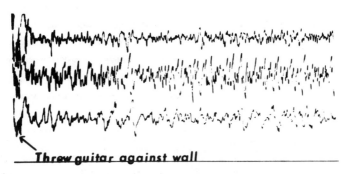

Threw guitar against wall

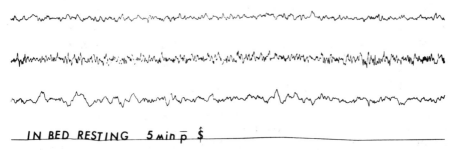

IN BED RESTING 5 min p̄ $

Fig. 9. (Top) Telemetered EEG tracings taken 90 seconds after stimulation shown in Figure 7. Seizure-like brain wave activity is seen in second brain wave channel.

Fig. 10. Telemetered EEG tracings taken as Julia violently smashes her guitar against wall, narrowly missing the head of the examining psychiatrist. Seizure-like brain wave activity continues in her posterior amygdala.

Fig. 11. Telemetered EEG tracings taken 5 minutes after stimulation shown in Figure 7. Brain wave recordings are similar to pre-stimulation record. Third lead shows slower electrical waves, which may be postseizure in character.

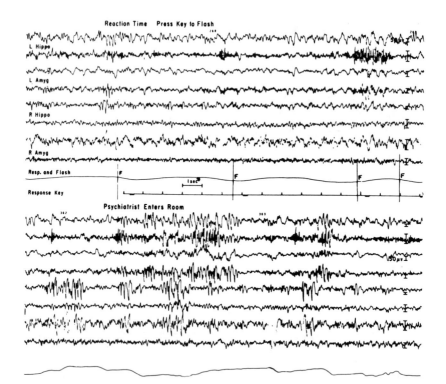

Fig. 12. <u>Top</u>, Control record. Patient pressing key immediately after flash. Shorter reaction time during high voltage hippocampal spindle was not a consistent finding. <u>Bottom</u>, Effect of psychiatrist entering room on depth EEG. Patient excited and distressed but neither hallucinated nor psychotic (case #2).

the latter part of these attacks, he appeared dazed and wild-eyed. As the anger spent itself in physical violence, he rather suddenly seemed to come to himself, wept violently, feeling hurt and broken. Nearly total amnesia was claimed for the portion of the attack between the early complaint and the arousal with weeping and remorse. Immediately following the attacks, he always felt exceptionally well and was considered so unusually creative in his field of design at these times that his partner often made a special effort to join him for work after a "seizure." Results of physical and neurological examination were unremarkable. The patient had many somatic complaints involving chest, abdomen, back, and limbs, and was extremely restless, irritable, depressed, and elated, by turns. Scalp EEGs demonstrated nonspecific sharp and ϱ-activity over both temporal regions. Depth electrode studies from amygdala and hippocampus disclosed sharp and spike activity bilaterally and high voltage, irregular slow and sharp activity, which was markedly activated by emotional stress (Fig. 12). The patient usually presented himself in the experimental laboratory with multiple pains, worry, dejection, and feelings of unbearable tension and anxiety. Bipolar stimulation of the most lateral points in the amygdalar complex on either side regularly gave him a "tremendous feeling of relaxation and relief" after some 10 to 30 seconds of current at 2 to 3 ma. These pleasurable sensations had a latency of 15 to 30 seconds and persisted for minutes to hours following cessation of the stimulation. A variety of other transient mood or sensory changes occurred (Fig. 13), most of which, in contrast to the pleasure and euphoric effect, were strictly limited to duration of the electrical stimulus. There were no sexual or gustatory sensations. Little or no EEG change was typically associated with the pleasure response to electrical stimulation. Blood pressure, pulse, and respiration were not affected. Although the remarkable relief of tension or somatic distress which the patient achieved following stimulation suggested the possibility that a purely psychological effect of the experimental situation might be responsible, on no occasion did repeated stimulation of other intracerebral sites yield similar results. Five separate trials of warningless stimulation of medial amygdala by remote telemetry induced typical relaxation and euphoria on a day when the patient was deeply depressed. No alteration in performance was measured following stimulation on digit symbol subtest of the Wechsler Adult Intelligence Scale (WAIS) and parts C and D of Raven Matrices. The patient became increasingly dependent and insistent upon the stimulation of the most medial pair of electrodes on either side, and, when stimulation was omitted for ten days, he became irritable and severely depressed. Placebo stimulations were ineffective in giving relief. At this time, somatic complaints and belligerent episodes returned, and spontaneous spike activity was marked from the depths of the temporal lobes bilaterally.

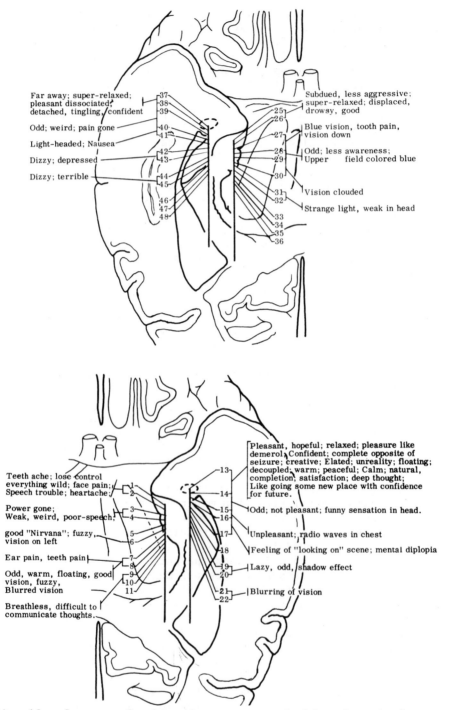

Fig. 13. Summary of subjective states evoked by electrical stimulation superimposed on diagrammatic reconstruction of depth recording stimulating points (case #2).

Little change in scalp EEG was evident. A lesion was made with
radio frequency current (two minutes, 10 ma) at the site of
maximum spike activity in right amygdala region. Although there
was no immediate effect, on the following day, the patient was
markedly depressed and remained so for nearly a week, following
which he returned to his previous tense, complaining, and
ruminating state. Spike activity in depth was only transiently
decreased. One month later, a second radio frequency lesion at
the point of maximum spike activity in the opposite amygdala
induced severe depression, which lasted several weeks and required treatment with a monoamine oxidase inhibitor for relief.
During this period, the patient held delusions (of remote brain
stimulation) and exhibited poor judgment and impulse control.
As his spirits improved, the delusional material faded, judgment
was somewhat improved, and attacks of anger did not reappear.
Three years later, rage attacks are still in abeyance.

     CASE 3. A 28-year-old veteran was discharged from the Navy
because of poorly controlled psychomotor automatisms and episodes
of violence and impulsivity. His first seizures occurred in infancy following encephalitis. He was then free of all attacks
from his third to 20th year without medication. Following a head
injury, spells recurred, commencing usually with arrest of speech
and on-going activity, after which the patient rose to his feet,
stared, uttered "uh uh uh," then began to vocalize in meaningless
gibberish, and engage in purposeful but inappropriate acts, such
as dealing out cigarettes like cards, bending a tin ashtray
double in his hands, or demonstrating parts of his shoes to
another patient. If interrupted or restrained, he might lash
out violently. Occasional attacks commenced with head turning
to the right or saying the same thing repeatedly. As the seizure
subsided, he commonly pulled at his nose three or four times.
Because of repeated episodes of verbal or physical aggression,
he was unemployable. Neurological examination revealed an alert,
cooperative man of above average intelligence. There was a left
lateral rectus palsy, mild nominal aphasia, occasional thought
blocking, and a slight droop to the right corner of the mouth.
A dilated left temporal horn was evident on the pneumoencephalo-
gram. Scalp EEG revealed bitemporal sharp and θ-activity. Preoperative pentylenetetrazol infusion induced left temporal spike
activity from scalp recording. Following stereotactic implantation of multilead electrodes in hippocampus and amygdala bilaterally, recordings regularly demonstrated high voltage brief one
to two second polyspike bursts from the left amygdala and hippocampal electrodes (Fig. 14, top). From the homologous regions
on the right, independent episodic paroxysms of rhythmic round
slow spikes appeared and occasionally endured for many minutes
or even hours without clinical change (Fig. 14, bottom). When
the patient experienced spontaneous feeling of "everything

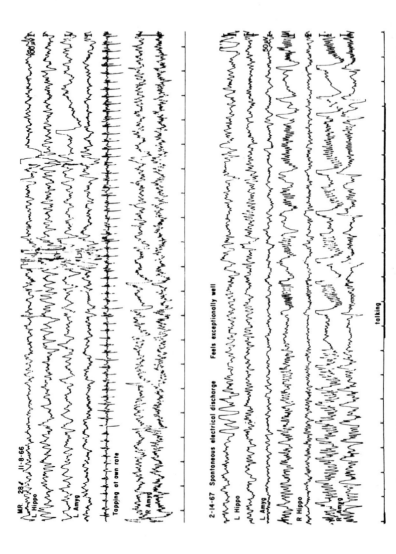

Fig. 14. Top, Patient told to tap key at own comfortable rate. Spontaneous high amplitude grouped spikes from left hippocampal and amygdalar regions have no effect on tapping rate. Random single spikes from right amygdala appear independent of or are diminished during left temporal paroxysms. Bottom, Spontaneous increase in rhythmic spike and spindle activity from right hippocampus-amygdala region. Left temporal spike now absent. Patient reports no distinct subjective change except that he feels "exceptionally well" (case #3).

suddenly strange," the slow spike-wave discharge in the right
amygdala ceased abruptly. Bipolar stimulation of the electrodes
in either amygdala complex occasionally led to afterdischarge
associated with lip smacking and clouded sensorium, but no dupli-
cation of the patient's characteristic automations. Pentylenetetrazol, 600 mg. administered intravenously, caused a prolonged
spike discharge restricted to contacts in the right amygdala
and accompanied by a sense of whirling, but no characteristic
clinical seizure. (As already noted, a previous pentylenetetrazol
activated EEG with scalp electrodes had shown a left temporal
spike focus.) Stimulation with a variety of parameters of the
48 depth points in amygdala and hippocampus induced numerous
unusual subjective changes, salient features of which are
summarized in Figure 15. Latency for these responses was 0.5
to 15 seconds, and the mood changes often outlasted the stimulus
by several minutes. Outstanding among the responses of this
patient were the detached peculiar distant feelings which
struck him with special force:

Stimulation of the right lateral depth electrode (directed
to amygdala): "I feel detached, it seemed hard to talk, everything is separated, the handles don't seem to belong to the doors
or the frames, I'm a little frightened, a feeling as though I
seem to be nowhere. I just don't exist. This is a completely new
mental feeling. I feel crazy, everything is very serious; things
are switching so fast I can't keep track, now I'm in a place all
by myself. Somebody else is controlling me and moving my arms
and legs (no movement visible). I feel morbid. I just want to
be alone."

Left lateral electrode (amygdala complex): "Something is
going to happen but it doesn't. Everything seems so distant: it
has no real connection with me or the present setup in the hospital; it doesn't feel like me talking; I seem to be concentrating
very deeply, thinking about something very hard, don't know what
it is; you're changing me into another world. I feel for a
moment as though I want to be alone, doing something having deep
thoughts about what is going to happen; I don't want to talk to
anyone, someone is saying to me I can't do it."

Stimulation between right and left amygdala produced the most
elaborate and subjectively composite mental states: "Several
things or moods are happening at the same time; I'm completely
alone in the room meditating but it's odd in here. I'm not part
of it, it is as though I'm talking to myself; something just
changed but nothing happened; I have a funny attitude as though
I'm in charge of the situation, it's up to me to make the next
decision; (patient suddenly smiled) I feel quite amused. I don't
know how to explain this, it is as though I can see myself."

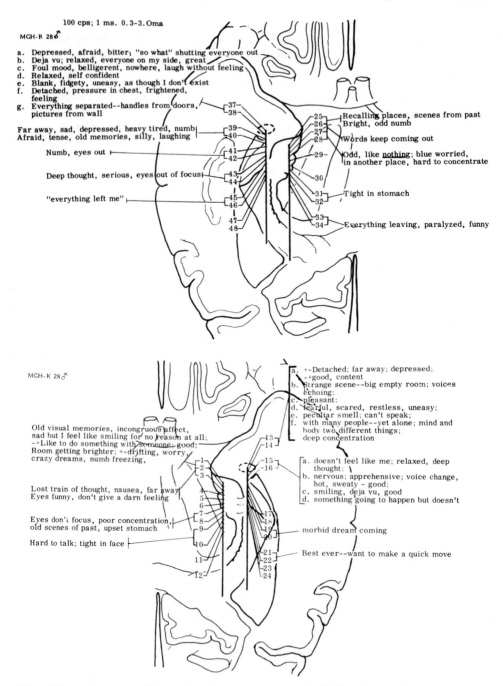

100 cps; 1 ms. 0.3-3.0ma

MGH-R 28♂

a. Depressed, afraid, bitter; "so what" shutting everyone out
b. Deja vu; relaxed, everyone on my side, great
c. Foul mood, belligerent, nowhere, laugh without feeling
d. Relaxed, self confident
e. Blank, fidgety, uneasy, as though I don't exist
f. Detached, pressure in chest, frightened, feeling
g. Everything separated--handles from doors, pictures from wall

Far away, sad, depressed, heavy tired, numb
Afraid, tense, old memories, silly, laughing

Numb, eyes out

Deep thought, serious, eyes out of focus

"everything left me"

—37—
—38—
—39—
—40—
—41—
—42—
—43—
—44—
—45—
—46—
47—
48—

—25— Recalling places, scenes from past
—26— Bright, odd numb
—27—
—28— Words keep coming out
—29— Odd, like nothing; blue worried, in another place, hard to concentrate
—30—
—31— Tight in stomach
—32—
—33—
—34— Everything leaving, paralyzed, funny

MGH-K 28♂

Old visual memories, incongruous affect, sad but I feel like smiling for no reason at all; -+Like to do something with someone, good; Room getting brighter; +-drifting, worry, crazy dreams, numb freezing,

Lost train of thought, nausea, far away
Eyes funny, don't give a darn feeling

Eyes don't focus, poor concentration, old scenes of past, upset stomach

Hard to talk; tight in face

—1—
—2—
—3—
4
5
6
—7—
—8—
—9—
—10—
11—
—12—

—13—
—14—
—15—
—16—
—17—
—18—
—19—
—20—
—21—
—22—
—23—
—24—

a. +-Detached; far away; depressed; -+good, content
b. Strange scene--big empty room; voices echoing;
c. pleasant;
d. fearful, scared, restless, uneasy;
e. peculiar smell; can't speak;
f. with many people--yet alone; mind and body two different things; deep concentration

a. doesn't feel like me; relaxed, deep thought;
b. nervous; apprehensive; voice change, hot, sweaty - good;
c. smiling, deja vu, good
d. something going to happen but doesn't

—— morbid dream coming

—— Best ever--want to make a quick move

Fig. 15. Summary of subjective states evoked by electrical stimulation superimposed on diagrammatic reconstruction of depth recording-stimulating points (case #3).

On another occasion: "I have a pleasant contented sensation, very relaxed, there doesn't seem to be any life to people around me, they look stuffed; deep concentration, there is no life to people around me, something real good just happened."

Like patient 2, this man felt changed by the stimulations for hours after the current was turned off, despite the fact that the vast majority of stimuli and sensations were not associated with discernible local or propagated electrical cahnges. In contrast to patient 2, his post-stimulus behavior was marked by confusion, excitability, press of speech, overactivity, restlessness, and difficulty with concentration and mental recollection of old situations which induced anger and excitement. He reported a sense of being "way out" with displaced thoughts, racing mind, and crazy mixed-up feelings for many hours following stimulation. Images and thoughts passed through his mind with extraordinary vividness and disjointedness, like a dream or something out of "fantastic features." Yet, in talking with him, no objective abnormalities could be detected by his physician. His disturbed subjective state led to considerable difficulty, irritability, and uncooperative behavior on the ward. Chlorpromazine, 200 mg. daily, was added to anticonvulsant medications and controlled these behavioral and subjective symptoms quite well. Speed and accuracy scores obtained on sections of the Raven Matrices and the digit symbol subtest of the WAIS were rarely changed following the stimulus. On two occasions, there was significant improvement in test performance beyond practice effect following stimulation of the anterior tip of amygdala electrode. Deterioration in test performance was never observed following stimulation of any of the depth points.

Because of the occurrence of separate spontaneous EEG spike discharges on left and right side and the alternate activation of left or right by pentylenetetrazol, a third method of activation was decided upon for this patient. Recalling the report of Eidelberg et al. (1963) of cocaine induction of seizure discharge in rat and cat, we introduced a relatively small amount (80 mg. in 4% solution) of this drug intranasally in our patient while recording from depth electrodes. Within two minutes, characteristic autonomic effects of mydriasis, conjunctival injection, lacrimation, and blood pressure depression (to 90/60 mm Hg from 130/80 mm Hg) occurred. After six minutes, blood pressure had returned to normal, and a slow high voltage (300 μv) spike-wave discharge appeared at the right lateral depth electrode tip. This discharge lasted some 20 minutes and was exaggerated by hyperventilation. The patient was moderately euphoric and overtalkative. At the time of this first cocain trial, the patient had been receiving the following drugs: chlorpromazine, 200 mg daily; diphenylhydantoin sodium (Dilantin) 300 mg daily; and

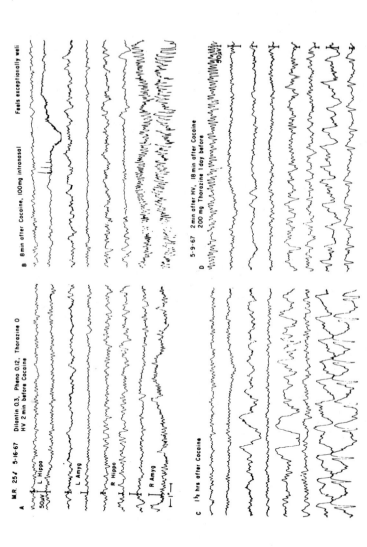

Fig. 16.   Effects of intranasal instillation of cocaine on depth EEG.   Top, left, Activity from right and left deep temporal region during hyperventilation prior to cocaine.   Top, right, Eight minutes following intranasal instillation of 100 mg. of cocaine: rhythmic high voltage spikes appeared from lateral distal pair of deep temporal electrodes.   This activity continued for several minutes during which patient felt exceptionally well.   Bottom, left, One and a half hours later slow spikes from same pair of electrodes, patient loquacious, euphoric.   Bottom, right, Slight effect of intranasal instillation of cocaine when patient receiving chlorpromazine, 200 mg. daily.

phenobarbital, 120 mg daily. One week later, off chlorpromazine
for eight days but still receiving the diphenylhydantoin sodium,
and phenobarbital, 100 mg of cocaine in 4% solution was introduced
intranasally. Two minutes after the administration of cocaine,
blood pressure decreased as before and autonomic changes similar
to those noted previously recurred. Six minutes after cocaine,
the high amplitude right amygdala rapid spike discharge appeared
and was exaggerated by hyperventilation (Fig. 16). There was no
effect on the seizure discharge by flickered light or odors. The
patient was euphoric and reported a sense of extreme well-being,
which he stated was totally different from that following electrical stimulation or associated with his usual seizures. The
right deep temporal spike discharge was momentarily arrested by
speaking to the patient. Performance on digit symbol and Raven
subtest was distinctly improved during the rhythmic 1 1/2 cps
spike activity. The patient completed the entire digit symbol
test in the allotted 90 seconds with no errors, while his usual
score (precocaine) was around 75%. On Raven matrices, part C,
he obtained 12 out of 12 correct, while his usual (precocaine)
score was eight out of 12 correct. A dose of 60 mg. of phenobarbital had no effect on the sustained seizure activity which
persisted for nearly four hours. Because of the decrease in
blood pressure following the cocaine, we were reluctant to use
adrenolytic agents such as chlorpromazine or dibenamine to
abolish the discharge (as had been shown by Eidelberg et al.
(1963) in the monkey). One week later, the cocaine given
intranasally was repeated with similar results. At this time,
nasopharyngeal and scalp leads were used simultaneously with
the depth electrodes to determine whether it would be possible
to use this method of activation in patients without intracerebral electrodes. Although a high amplitude seizure discharge was again induced in the right amygdala region as on
previous occasions, no change was evident at nasopharyngeal or
scalp leads. (We have since employed cocaine intranasally to
induce temporal spikes in a patient under anesthesia, prior to
temporal lobectomy. In this instance, spiking appeared only
from amygdalar probe and not from exposed lateral temporal cortex.)

                              DISCUSSION

     Previous reports of the effects of electrical stimulation in
deep temporal structures in man have emphasized the well-known
responses of short latency, which are limited in duration to
passage of the current or subsequent after-discharge, and which
so mimic brief epileptic seizures. The present report emphasizes
responses of longer latency which endure for minutes or hours and
often resemble the abrupt or chaotic mood and thought disturbances
of certain psychoses and interictal states. Our limited data from

man and studies in animals by others suggest that the long latency, long lasting effects of periamygdalar and hippocampal stimulations affect secretion of biologically active transmitters.

The behavioral changes following stimulation without detectable alteration in local EEG suggest that such activation may influence structures at a distance from the recording electrodes by prolonged release of the exciting neurotransmitters or exhaustion of antagonists. Extensive study of EEG from the depth and surface of the temporal lobe situations in normal animals and nonepileptic, nonpsychotic man does not suggest that electrical spike discharges are a part of the normal cerebral electrical repertoire, but they are, of course, a frequent and characteristic finding in the patient with temporal lobe epilepsy. Furthermore, data from our epileptic patients with behavior disorders indicates that certain interictal mood and behavioral aberrations may be associated with overactivity of a neuronal separate but physically proximate monoaminergic system which interacts reciprocally with the cholinergic systems.

SUMMARY

Experiences with electrical stimulation, recording, autonomic and behavioral observations from patients with chronic implanted, multicontact electrodes in amygdala and hippocampus are detailed. Long latency, long lasting effects of stimulation, unassociated with after-discharge, may be considered more representative of the usual activity of this region than ictal events precipitated by higher current and accompanied by after-discharge.

ACKNOWLEDGMENT

We should like to acknowledge the important collaborative efforts of Dr. Janet Stevens in accomplishing our investigative and therapeutic efforts in the patients reported in this paper.

REFERENCES

DELGADO, J. M. R., HAMLIN, H., & CHAPMAN, W. P. Technique of intracerebral electrode placement for recording and stimulation and its possible therapeutic value in psychotic patients. Confinia Neurologia, 1952, 315-319.

EIDELBERG, E., LESS, H., & GAULT, F. P. An Experimental Model of Temporal Lobe Epilepsy: Studies of the Convulsant Properties of Cocaine. In G. H. Glaser (Ed.), EEG and Behavior. New York: Basic Books, Inc., 1963. Pp. 272-283.

HEATH, R. G., & MICKLE, W. A.   Evaluation of Seven Years'
    Experience with Depth Electrode Studies in Human Patients.
    In E. R. Ramsey and D. S. Doherty (Eds.), Electrical
    Studies on the Unanesthetized Brain. New York: Hoeber,
    1960.

MARK, V. H., & ERVINE, F. R.   Relief of Pain by Stereotactic
    Surgery. In J. C. White and W. H. Sweet (Eds.), Pain and
    the Neurosurgeon: A Forty Year Experience. Springfield,
    Illinois: Thomas, 1969. Pp. 834-887.

MARK, V. H., & ERVIN, F. R.   Violence and the Brain. New York:
    Harper & Row, 1970, P. 111.

STEVENS, J. R., MARK, V. H., ERVIN, F. R., PACHECO, P., &
    SUEMATSU, K. Long latency, long lasting psychological
    changes induced by deep temporal lobe stimulation in man.
    A.M.A. Archives of Neurology and Psychiatry, 1969, 21,
    157-169.

PSYCHOLOGY–BEHAVIOR–PSYCHIATRY

EFFECTS OF AMYGDALECTOMY ON SOCIAL-AFFECTIVE BEHAVIOR IN NON-HUMAN PRIMATES

Arthur Kling

Department of Psychiatry, Rutgers Medical School

New Brunswick, New Jersey

The classic studies of Kluever and Bucy (1939), which by now have been reproduced repeatedly by many experimenters, have called attention to a syndrome of behavioral changes that occurs in the monkey after bilateral temporal lobectomy, or with lesions restricted to the amygdaloid nuclei. These changes can be briefly characterized as: 1) a decrease in belligerence and a reduction of fear toward normally fear-inducing objects including man; 2) a tendency to investigate orally and generally contact orally inedible objects including coprophagia and uriposia; 3) increased and inappropriate sexual behavior; 4) "hypermetamorphosis."

In the past three decades, a large body of evidence has been gathered on the influence of this nuclear group on affective and cognitive behavior as well as autonomic, endocrine and metabolic function. These results have been reviewed by a number of investigators (Gloor, 1960; Goddard, 1964; Kling, 1966; Kaada, 1951; Kluever, 1952; DeGroot, 1965; Pribram, 1967a; MacLean, 1949). More recently, increasing attention has been given to its pathophysiology in man, especially with regard to the convulsive states, and in the control of violent-aggressive behavior (Ervine et al., 1969; Chapman et al., 1950; Scoville et al., 1953; Narabayashi et al., 1963; Sawa et al., 1954; Falconer and Taylor, 1968; Penfield and Jasper, 1954; Gastaut et al., 1959).

A long neglected, but major, dimension of the behavioral repertoire of subhuman primate behavior is the behavioral interaction occurring within the social group. Both human and non-human primates are social animals and their survival depends, to a large extent, on the maintenance of social bonds and the integrity of the social group. Those behaviors which are

important in maintaining social bonds have been given increasing
attention in the past decade. These contributions have been included in recent volumes by DeVore (1965), Jay (1968) and Altmann
(1967). Concurrently, laboratory studies have focused on maternal-infant and early peer interactions, and their influence on
the development of social behavior as in the studies of Harlow
(1965), Jensen (1967), Rosenblum (1967) and Mason (1965). With
the advent of increasing interest in, and accumulation of knowledge of primate social bonds and its relevance for the evolution
of human behavior (Hamburg, 1968), a parallel interest is
developing in brain function and social behavior. Combining the
ethological approach of field workers with the techniques of
neurophysiology and related psychological disciplines offers
promising opportunities for research in this area.

This report is intended to review those studies relevant to
the influence of the amygdaloid nuclei on social behavior in
primates and to attempt to integrate these findings with current
hypotheses regarding its significance in the regulation of
affective behavior. While the contribution of environmental,
genetic, age and sex factors to the expression of affective
behavior has always been recognized, a fuller appreciation of
the influence of these variables becomes possible when behavior
is examined in the social context and, especially, under natural
field conditions. It is with this view in mind that the following
report has been prepared.

I. Maternal-Infant Behavior

a) <u>Amygdala</u> <u>lesioned</u> <u>mothers</u>: Amygdalectomy probably is
inconsistent with the expression even of the rudimentary elements
of maternal behavior. While no specific studies have been done
on this issue, some scattered observations have been made on
amygdalectomized females which have given birth in the laboratory.
In Dr. Jules Masserman's laboratory, I watched an amygdala
lesioned female mishandle, bite and kill her newborn shortly
after delivery. Several years later, a similar event occurred in
my laboratory at Michael Reese Hospital. In both cases, the
mothers behaved as though the infant was a strange object to be
mouthed, bitten and tossed around as though it were a rubber ball.
Lesions to related cortical structures also may disrupt the
maternal-infant bond. In the course of a field study on rhesus
colony of the Cayo Santiago, Dr. Michael Miller observed a multi-
parous mother, who had sustained a bilateral lesion of dorsolateral frontal cortex, reject and abandon her yearling and
subsequently leave the group. To the contrary, in the same
colony, he observed exemplary maternal behavior on the part of
another female who had undergone a bilateral lesion of superior
temporal neocortex.

Fig. 1. Infant M. speciosa several days after bilateral amygdalectomy. Mother is grooming the genital region.

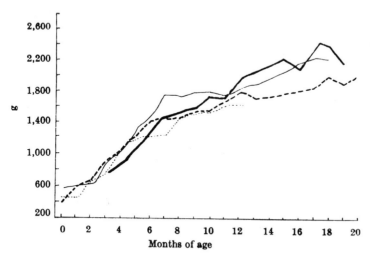

Fig. 2. Growth curves for eight infants from birth to 20 months. Each line represents two animals. ••••, Amygdala lesioned, maternal deprivation; ----, amygdala lesioned, maternal reared; ——, maternal deprivation; ▬▬ , maternal reared.

b) Infant-maternal behavior: In contrast to the behavior of
lesioned mothers toward their young, amygdalectomized infant
monkeys display grossly normal nipple orientation, sucking,
grasping and can be successfully reared maternally (Fig. 1).
Kling and Green (1967) reported on two such cases that were observed for the first year of life and which during this period
demonstrated normal somatic growth and grossly normal affective
behavior (Fig. 2). Infants sustaining cortical lesions (e.g.
dorsolateral frontal) also have been reared maternally in our
laboratory without difficulty. We separated them at 5 months of
age for formal cognitive testing at which time they were grossly
indistinguishable from unoperated monkeys of a similar age (Kling
and Tucker, 1968). In my experience, as long as the neonate can
suck and grasp adequately, most mothers will show adequate
maternal behavior toward their brain injured infants. Those
infants with large ablations or having gross neurological
deficits were incapable of being reared maternally (Kling, 1968a).

The intensity of the maternal-infant bond in non-human
primates is well known from both casual observations and formal
investigations (Rosenblum and Kaufman, 1967; Harlow, 1965; Jensen
et al., 1967; Mason, 1965). The apparent shattering of this
bond by lesions of the amygdala (and probably frontal cortex)
suggests the degree to which these areas may be involved in the
maintenance of social bonds. That the behavior of infants with
similar lesions toward their mothers is unaffected can be related
to incomplete neural maturation at the time of insult. As has
been demonstrated for cognitive tasks, the influence of these
structures on social behavior becomes more important at later
maturational states (Kling, 1968a). In this regard, it is not
precisely known when amygdalectomized infants would develop
eventually the characteristic aberrant behaviors. From previous
studies in the cat (Kling, 1965), I would guess that the onset of
puberty would be a critical period, and we would expect to see
at least some features of the syndrome at that time.

II. Juvenile-Peer Behavior

A number of studies now have been reported which deal directly
with the effects of amygdalar lesions and juvenile-juvenile interactions. The majority have been conducted on paired subjects in
a laboratory setting, although some observations on operated
juveniles have been made in semi- and free-ranging settings
within naturally composed social groups. Since the results of
observations from laboratory settings differ so completely from
field studies, they will be considered separately.

Using 2-3 year old male M. mulatta, I (Kling, 1968b) compared the interactions between normal and amygdala lesioned dyads.

The lesioned pairs (lesion-lesion interactions) showed less
aggressive interactions than the normals, but significantly more
rough and tumble play, attempted and appropriate mounts, and
grooming bouts. There were major qualitative differences as well.
In the normal pairs, one member, shortly after pairing, would be
clearly dominant and remain so in all subsequent test sessions.
Among the operates, there was frequent switching and no clearly
dominant subject for all interactions. In addition, the operates
displayed frequent mutual solicitation for mounting and grooming
with persistent erection. True aggressive acts such as biting
and hair pulling, with resulting cowering by the submissive
member, were common in the normals while mutual mouthing, nibbling
and rough tumble play without injuries or screaming was characteristic of the operates. Inappropriate and excessive oral behavior
was also characteristic of the operates. In the operated group,
exogenously administered testosterone had the effect of increasing rough and tumble play at the expense of grooming and other
behaviors. In the normal group, all social interactions were
increased, particularly grooming bouts (Fig. 3).

Thompson et al. (1969) recently have studied social fear
responses in infant female M. mulatta who were lesioned at 2.5
months of age and tested in three different pairing conditions:
lesioned-control, lesioned-lesioned and control-control pairings.
Age of testing varied from 3 months to 8 months of age. They concluded that the operated infants showed more social fear and performed less social exploration (grooming, sitting together,
clasping) than normals. The operates, however, showed less fear
toward novel situations, and appeared more intimidated by the
normals. These authors concluded that the amygdalar lesioned
infants showed heightened fear responses when in contact with
normal peers, while other behaviors were grossly normal.

In my laboratory, Miller (1968) observed operated juvenile
M. speciosa females paired with a normal male peer as well as
with each other. Post-operatively, she found a drop in the
frequency and duration of grooming in both conditions. With
respect to aggressive behavior prior to surgery, only a few instances of a female challenging the dominance of the male were
recorded. Post-operatively, there was an increase in aggression
by the male toward the females since the operates were inappropriately challenging the male over food and cage position. A
slight increase in sexual behavior occurred in the females after
operation. This included greater receptivity to the male and
more female-female mountings. While total social interactions
decreased in the females after operation, the responsible factor
clearly was the decrease in mutual grooming which was the most
frequent social interaction between subjects prior to surgery.

S^a	Aggression		Attempted mount		Mount		Groom	
	Pre	Post	Pre	Post	Pre	Post	Pre	Post
J1	200	180	94	24	7	0	35	130
J2	0	0	0	0	0	0	0	0
J3	30	78	4	65	0	16	3	80
J4	0	0	0	0	0	0	1	0
J5	0	0	0	0	2	0	0	0
J6	19	46	39	60	8	12	6	42
Total for ten 15-min. periods	249	304	117	149	18	28	45	252
AJ1	75[b]	160[b]	13	4	125	49	19	9
AJ2	75[b]	158[b]	39	14	173	196	11	1
AJ3	31	17	162	137	14	42	39	5
AJ4[c]	14	0	0	0	2	0	1	0
AJ5	6[b]	161[b]	32	14	223	136	79	37
AJ6	8[b]	4[b]	11	5	75	35	8	1
Total for ten 15-min. periods	209	500	257	174	612	458	157	53

[a] Ss 1 and 2, Ss 3 and 4, and Ss 5 and 6 were paired.
[b] Sham-biting, rough and tumble play; not true aggression.
[c] Inadequate lesion.

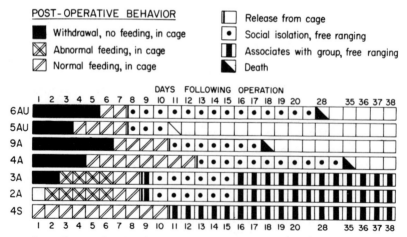

Figure 4. Chronological diagram of post-operative behavior for 6
operates (6AU, 5AU, 9A, 4A, 3A, 2A) and 1 control (4S). AU signifies removal of amygdala plus uncinate cortex, bilaterally; A
signifies removal of amygdala, bilaterally. Numerals signify age
of animal, in years.

Some observations of Rosvold et al. (1954) on a group of
juvenile male M. mulatta housed in a large cage, indicated that
amygdalectomy had the effect of reducing aggression and causing
a fall in rank in some, but not all, operated subjects; when
observed in individual cages, however, the operates appeared
less fearful and more aggressive.

In two laboratory studies of amygdalectomy in small but more
naturally composed social groups (C. aethiops, M. mulatta) containing adults and juveniles of both sexes (Kling and Cornell,
1971; Kling et al., 1968) we found also that operated juveniles
fell in rank and exhibited decreased social interactions. In a
group of C. aethiops (Kling et al., 1968), all operates would
huddle together in a corner of the cage apart from the normal
animals. In addition, they disregarded the previously existing
feeding order in the cage and occasionally were attacked for
their indiscretion.

In the semi-free-ranging rhesus colony of the Cayo Santiago
(Dicks et al., 1969) it was observed that two juvenile male
operates (2 and 3 years of age) with sub-total amygdalectomy,
after being isolated from their group for one week, returned to
their mothers and reengaged in grossly normal social activities.
Older animals, or those with lesions which included the uncus,
remained socially indifferent, isolated from the group, were
attacked by the normals and all eventually died (Fig. 4).

In a completely natural setting, along the Zambesi River in
Central Africa, Kling et al. (1970) observed the effects of
amygdalar lesions in 4 juveniles (3 males and 1 female) who were
trapped out of a completely free ranging social group of C.
aethiops. When released back into their own or a neighboring
group, they displayed withdrawal to positive social communications
by their peers; these included attempts to sit close, groom,
muzzle and play. In spite of repeated attempts by the normals to
socialize with them, they appeared fearful, withdrew, eventually
left the group and disappeared.

To summarize the effects of amygdaloid lesions on social
behavior between juveniles, the following trends seem to run
through the various studies reviewed. Paired operated subjects
exhibit heightened rough and tumble play, increased sexual and
oral behavior. In dyadic or caged small groups, the operated
subjects show much less social interaction than between equivalent
aged normals. As the amount of space is increased, social interaction decreases until, in more natural setting, social withdrawal and total isolation from the group ensues. That these
behaviors appear related to spatial and environmental factors is
supported by the observations on the caged, small social group of
C. aethiops, wherein the operated huddled together in a corner

TABLE I

SUMMARY OF STUDIES ON EFFECTS OF AMYGDALECTOMY ON SOCIAL BEHAVIOR

Authors	Species	Lesion	Group Composition	N	N Op.	Observation Conditions	Change in Rank	Social Interaction	Aggression	Orality	Sexual Behavior	Behavior to man
Kling 1968b	M. mulatta	amygdala	paired J♂	12	6	Cage, L-L, N-N	to indiscriminate →	←	(rough & tumble play) ←	←	←	tameness
Thompson et al. 1969	M. mulatta	amygdala	paired, infant op J♀	12	6	Cage, L-N, L-L, N-N	-	→	→	?	-	-
Miller 1968	M. specio-sa	amygdala	paired, J♀ op. J♂ (N) L-N, L-L	10	7	Cage, J♀ op. J♂ (N) L-N, L-L	-	→	→	←	slight ←	tameness
Rosvold et al. 1954	M. mulatta	amygdala	J♂	8	3	Group cage	2/3 →	-	→	?	?	?
Kling & Cornell 1971	M. specio-sa	amygdala	J-A, ♂ & ♀	6	3	Group cage	SA ♂ Ao A♀ →	→	(♂'s) → A♀ ←	←	slight ←	tameness
Kling et al. 1968	C. aethiops	amygdala	J-A, ♂ & ♀	6	5	Group cage	(all op's) →	→	→	← inappropriate feeding order		tameness
Plotnik 1968	S. sciureus	amygdala & ant. temp.	A♂	4	4	Group cage	(all op's) →	?	→	?	?	less avoidance
Dicks et al. 1969	M. mulatta	amygdala	natural social group	85	6	Semi-free field	→	adults, social, social isolation, 2J, partial lesions, resocialized	→	?	-	bizarre, avoided
Kling et al. 1968	C. aethiops	amygdala	natural social group	42	7	Free-ranging	→	social isolation	→	→	?	bizarre, avoided
Iwato & Ando 1970	M. fuscatta	temp. lobectomy	paired A♂	6	3	Cage, L-N, L-L	to indiscriminate →	←	(L-L) →	←	←	tameness

of the cage out of the main social hierarchy. In the field,
given the space to determine social distance, they became complete isolates. However, we have not yet had the opportunity to
observe a group of operates, released together in a field situation.

The location extent of the lesion also may be a critical
factor in determining the disruption in social behavior. As
previously noted, two subjects which sustained partial amygdaloid
lesions, sparing the uncal cortex, resocialized and eventually
seemed to behave quite normally for their age and sex. It is
not known, however, what would be their later behavior as they
matured.

III.   Social Bonding and Adult Behavior

Relationships between group members often are identifiable by
the spacing between them as well as the amount and kinds of activities occurring between them. These behaviors probably are only
studied meaningfully under free or semi-free ranging conditions
in naturally composed social groups since artificial restrictions
of the laboratory caging result in interactions which hardly
resemble the situation in nature. In all primates, each stage
of life is characterized by specific interactions with group
members which are based on its gender, age, rank within the
social group, environmental determinants and seasonal variables.
In spite of the variability between inter- and intra-species
behaviors, all primates seem to display these characteristic
spacing behaviors. Careful observations on the Cayo Santiago
colony have been most productive of this type of analysis since
all group members are well identified and have been studied
longitudinally over many years (Kaufman, 1967; Altmann, 1962).

Observations seem to suggest that: in the free field
condition, adult amygdalectomized monkeys become social isolates,
appear fearful and withdraw from any type of closeness with group
members. In the African field study by Kling et al. (1942), once
the operates were in the field, they tended to remain stationary
as long as they were left alone. At the approach of a peer, or
another group member of any size or sex, they would withdraw and,
if followed, would flee. An adult female, who was followed with
interest by the adult males and several juveniles, finally was
left sitting alone in the high branches of a tree after she
ignored repeatedly their attempts to interact with her. Even a
dominant male ignored and left his group after operation. While
we lost sight of him for several hours after he was released, he
turned up again two months later, when he was observed to be
sitting in a school yard in the nearby town of Livingstone. He
apparently had survived and lived as an isolate for at least two

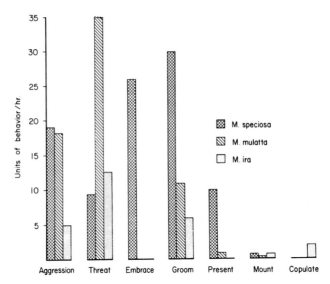

Fig. 5.   Comparison of normal social behavior of 3 species of macaques.

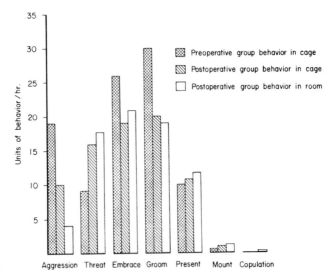

Fig. 6.   Normal and post lesion group behavior in M. speciosa

months. This behavioral withdrawal is even more significant considering that the operates were approached and greeted with non-
belligerent communications. While none of the adult rhesus
operates in the Cayo Santiago study with total amygdalectomy re-
joined a group, they had to contend with a more competitive and
spatially restricted field area in which they were attacked, and
driven off by normal group members and all eventually died of
wounds or inanition.

    The degree of disruption of social bonds after operation
also may vary depending upon species-specific behaviors. In a
study comparing the social behavior of small groups of 3 species
of macaques in the laboratory, Kling and Cornell (1971) found
that amygdalectomy had less of a disruptive effect in M. speciosa
than in either M. mulatta or M. ira. Pre-operative observations
indicated that M. speciosa displayed more intense positive social
behaviors of grooming and embracing than the other two species
(Fig. 5). Some degree of grooming, sitting close and embracing
were still present after the operation, especially in the adult
male and female. When this group was transferred from their cage
to a large room, the amount of aggression showed a further decline, while the other behaviors were not altered greatly (Fig. 6).

    On the basis of the few studies relevant to this issue, it
would appear that: (1) amygdalectomy results in the operates
placing increased social distance to social isolation from normals
in a field environment; (2) a tendency to remain in close physical
contact with each other in confined area; (3) an increase in
"social fear" with respect to normals; (4) an increase in aggressive assaults by normals, curiosity or indifference depending on
the species studied; (5) the disturbance tends to increase with
age; and (6) quantitative differences in the disruption of social
bonds may be related to pre-operative species specific patterns
of social bonding between group members.

                           Dominance Behavior

    Of all the identifiable social behaviors, one of the most well
studied and readily quantified is social rank. Clear evidence of
social rank in dyadic or multi-member groups may be reflected in
aggressive and submissive behaviors, spatial positions within a
cage, feeding order, and, in the males, access to females.

    In dyadic encounters or in small social groups, a fall in
rank of operated male subjects has been found consistently in all
studies. This is especially true if operates are compared with
normals. When observed with each other, the characteristic
elements of social rank appear to dissolve and become indeter-
minant. Among adult males, this has been observed by Iwato and
Indo in temporal lobectomized M. fuscata (1970) and by Plotnik

(1968) in the squirrel monkey; among juvenile M. mulatta by Kling
(1968), (Thompson et al., 1969; Rosvold et al., 1954).

The situation is less clear with regard to the females. Observations by Miller (1968) indicated that in her lesion-lesion
pairings, evidence of dominance behavior was observed. In M.
speciosa, an operated adult female did not lose rank after
operation as did operated C. aethiops females (Kling et al.,
1968). Much more work on female behavior is needed to clarify
this issue. While clear dominance relations among females exist
in free-ranging groups, they are not as noteworthy in laboratory
cages where they are forced into closer relations with males
than they would ordinarily maintain in the wild.

The fall in rank of the operated primates relative to normal
group members probably is related closely to the raised threshold
for the expression of threat and agonistic behaviors since social
rank, at least in males, is in part determined by these communications.

This deficiency in expression of aggressive behavior, and
the heightened fear of social interaction, would limit severely
their ability to maintain rank within the group. Their lack of
affective responses, even appropriate submissive gestures when
challenged by subordinate group members, would make them very
vulnerable to attack and limit their potential for survival.

IV. Oral and Ingestive Behavior

Of the many effects of amygdaloid lesions in primates, change
in orality and ingestive behavior is one of the most notable
symptoms of this lesion. These preparations have a tendency to
mouth and keep in their pouches both edible and inedible objects
and to persevere in examining with their mouth; items in their
environment not usually attended to, to eat in a distractable
fashion and in some cases to become hyperphagic and obese.
Coprophagia and uriposaia commonly are associated with the hyperorality.

While this aspect of the amygdala syndrome is seen regularly
after surgery in caged primates, it is noteworthy that it has not
yet been seen in the field studies. In the vervet (C. pygerythrus)
hyperorality was present soon after the animals recovered from
their post-operative stupor and anorexia, but lasted only while
they remained in captivity. Once in the field, we never observed
this symptom in spite of having several operates of various ages
under constant observation for at least 8 hours. The same was
true in the study of the rhesus on Cayo Santiago. In studies on
caged subjects, it seems more prominent in individually housed

operates than those housed in more naturally composed groups, and, when present, is more directed toward food items than inedible objects. The expression of this behavioral alteration then appears to be highly dependent on environmental factors.

V. Sexual Behavior

As with the hyperorality, the hypersexuality exhibited by amygdalectomized monkeys seems highly dependent on environmental factors. Observations of subjects caged individually have called particular attention to their excessive auto-erotic behavior. Not only do these operates display bizarre sexual activity but, depending on the settings, attempts at copulation are increased quantitatively as well (Kling, 1968b; Iwata and Ando, 1970).

In those studies in which operated subjects were studied in a group with normals, there is less evidence of hypersexual behavior than when conducted on paired lesioned subjects, and no observations of sexual behavior have been reported from the field studies, although the amount of observation time on operated subjects up to now has been so minimal that the absence of positive observations may not be significant.

In several studies that involve small social groups housed in the laboratory, we did not see gross evidence of hypersexuality or an increase in auto-erotic behavior. These include studies in three species, C. aethiops, M. speciosa, and M. mulatta. In the M. speciosa study (Kling and Cornell, 1971), while the adult male showed an increase in mounting over his pre-operative status, it was not a dramatic change. In the C. aethiops study (Kling et al., 1968), no pre- or post-operative copulations were observed during formal observation periods while in the rhesus the adult male operate was attacked and killed by normal group members.

Iwata and Ando (1970) reported hypersexual behavior in paired, adult male, temporal lobectomized M. fuscata which is comparable to that observed in my paired juvenile rhesus operates.

While Miller (1968) did not find a significant increase in mounting between paired female juvenile M. speciosa, I observed an adult female rhesus who would regularly mount and thrust on any monkey, normal or operated, with whom she was paired. She would rarely submit to being mounted, but rather preferred the dorsal position. As with dominance behavior, the effect of amygdalectomy on female sexual activity needs more systematic investigation. Several variables may be responsible for the differences in sexual behavior in paired operates vs. larger groups containing normal subjects: (1) suppression of the overt behavior by the normals, (2) the need of the operates to be alert and avoid attack by the normals, and (3) their increased fear of

the normals and a tendency to be social isolates might also preclude the expression of sexual behavior. This does not explain
the absence of auto-erotic behavior which could be manifested in
a group cage or in the field without involving other group
members.

In general then, dyadic encounters between male operates
results in hypersexual behavior while only a slight increase in
frequency of copulation, if at all, occurs in more naturally
composed groups caged in the laboratory. No sexual behavior has
as yet been observed in the field studies.

VI. Relation to Human Studies

By now, there have been a significant number of patients
who have been subjected to either bilateral amygdalectomy (or
otomy) or more extensive temporal lobectomy. These procedures
have been carried out largely on temporal lobe epileptics or
chronic schizophrenics with or without mental retardation.

A comparison of the results of human studies with those in
non-human primates presents certain difficulties in view of the
pre-operative neuro- or psychopathology existing in these cases,
and in evaluating the effects of the operation in chronically
psychotic individuals. Further complications arise when we consider the variability of environments and cultural determinants
on human behavior.

Nevertheless, such a comparison seems indicated to further
elucidate some aspects of amygdaloid function in the most adaptive of the living primates.

Scoville et al. (1965), reporting on 5 cases of bilateral
uncotomy in adult female schizophrenics, noted that 4 of the 5
showed an initial post-operative apathy and decrease in emotional
expression. Three showed a more persistent withdrawal and decrease in social interaction along with diminished aggressiveness. In 7 cases of bilateral medial temporal lobectomy, 4 showed
a decrease in social interaction and became seclusive while 2
clearly became emotionally unresponsive. One patient was described as becoming more amorous.

The classic case of Terzian and Ore (1955) involved an
18-year-old epileptic male who sustained bitemporal lobectomy in
two stages. Post-operatively, he was described as having a flat
affect, a decrease in social interaction and diminished aggressiveness. Hypersexual behavior was persistent and consisted
mostly of excessive masterbation. The only symptom relative to
the Kluever-Bucy preparation not present was the increase in oral
behavior.

Perhaps the largest series of cases has been reported by Narabayashi et al. (1954) who reported on 21 bilateral-stereotaxic amygdalotomies. Fifteen were mental retardates who pre-operatively were excitable and aggressive. The major effect of the operation was a diminution in aggressiveness, easier patient management, and improvement in trainability. They state that there was no deterioration in intellectual ability, no lack of emotional expression and no evidence of the Klüever-Bucy syndrome.

Others (Sawa et al. (1954) have reported on 5 chronic schizophrenics with impulsive and destructive behavior who were subjected to bilateral amygdalectomy by aspiration through the temporal lobe. It is not clear from the description of the operative procedure how much of the nucleus was removed, or whether the uncus was lesioned as well, but it may be assumed that the procedure was a subtotal resection. These patients displayed increased friendliness and were more tractable after operation. Particular emphasis was placed on the change in "oral" behavior in which the patients are described as displaying "phagomania", "heterophagia" and excessive requests for water. In this regard, the authors noted that the patients always asked for things in sight, while they were incapable of asking for things when their eyes were closed or from "pure imagination." Only small increments in sexual activity were noted. More recently Ervin et al. (1969) have reported the usefulness of subtotal amygdalotomy for the relief of violent and aggressive behavior characteristic of the interictal phase in selected temporal lobe epileptics.

From the studies reviewed, there is general agreement, especially those dealing with epileptics, that no significant deterioration in cognition has been observed after bilateral lesions limited to the amygdaloid nuclei (Anderson, 1970). This is in essential agreement with non-human primate studies, although some deficits in complex discrimination tasks have been found consistently (Douglas and Pribram, 1969; Schwartzbaum, 1965; Schwartzbaum and Pulas, 1965; Weiskrantz, 1956). These reports also indicate that bilateral lesions of the amygdala do not alter significantly the schizophrenic process. Rather, the beneficial effects seem related to the decrease in aggressive and violent behavior in the epileptic, psychotic and mentally retarded patients.

As in the non-human primate studies, a decrease in aggressive behavior is among the most consistent lesion effects in man. Similarly, the initial apathy, lack of emotional expression and social withdrawal may be related primarily to lesion size. Only in Sawa's (1954) series was there a clear increase in and inappropriate oral behavior. Of particular interest is the fact that it was directly related to visual input.

As in the non-human primate studies, the expression of overt hypersexual behavior was the least consistent finding and only rarely observed. Suppression of overt sexual activity by the presence of staff, family or other patients may be an important factor to be considered. If this were so, a significant degree of reality testing and appropriate responsiveness to social pressure must remain after operation.

DISCUSSION

The results of the studies reviewed in this report reveal some general relationships between environmental factors and the behavioral changes resulting from lesions of the amygdaloid nuclei and related temporal lobe structures.

The more natural the group composition, environmental space and complexity, the more the operates tend to avoid social interactions and to become isolates. The characteristic increase in sexual and oral behavior seen under caged conditions is not evident in more natural surroundings and tends to be suppressed when the operates are housed with normal con-specifics. Conversely, reducing environmental complexity or restricting the interactions of the operated subjects to each other allows for maximal expression of characteristic effects of amygdalectomy including the increase in oral and sexual behavior.

The deficit in emotional expression, especially those associated with threat and associated reduction in aggressive behavior, does not seem as related to environmental factors, since it occurs consistently in most species, including man, in a variety of settings and group compositions. These alterations in behavior may be more influenced by age, sex and species-specific behavior. In this regard, the ablation studies are consistent with the effects of electrical stimulation in that, in primates and man, stimulation of the amygdala usually results in heightened fear and anxiety along with the physiological concomitants of these emotional states (Chapman et al. 1950; Kaada, 1967; Ursin, 1965; Anand and Dua, 1956; Mason, 1959; Delgado, 1967; Heath and Mickle, 1960). Both stimulation and ablation studies have related these effects to its influence on the hypothalamic and brain stem substrate for these emotional states and accompanying physiological responses. This is further supported by repeated observations that amygdalectomy does not abolish emotional expression, but rather alters the threshold for its elaboration.

While the influence of lesion locus and size has not been specifically dealt with in this report, it is obviously crucial to the behavioral alterations seen in operated subjects. Unfortunately, there have been too few studies in which this variable has been well controlled or attempts to separate out,

especially in primates, the influence of the anatomically distinct
nuclear and cortical elements on the individual elements of the
syndrome resulting from the gross ablation. While good evidence
for distinct flight, defense and, perhaps, aggressive reactions
has been determined for specific loci within the amygdala for
cat, the situation is less clear for monkey (Kaada, 1967; Ursin,
1965).

In this regard, there is some evidence from the study of
Dicks et al. (1969) and from human stereotaxic ablation work
(Narabayashi et al., 1963; Ervin et al., 1969) that amygdaloid
lesions sparing the uncal cortex may reduce aggressive behavior
without producing excessive withdrawal from social interactions.

The tendency of the lesioned monkeys to withdraw from and
avoid social interactions, especially under free ranging conditions, suggests that the operates are inappropriately "fearful"
rather than indifferent to normal group members. Otherwise, it
would be expected that a certain amount of at least passive interactions would be tolerated. Nor does it seem likely that their
avoidance of normals is based solely on the reduction of aggressive or threat behavior, since communication of even appropriate
submissive gestures would allow for resocialization even though
at a lower rank. This kind of adaptation has been observed in
our laboratory in a social group of monkeys treated with α-methyl-
p-tyrosine (Redmond et al., 1971).

If amygdalectomy does in fact result in heightened fear of
social contact, how does one explain the well known tameness
toward man and their tendency to approach and explore objects
which were fear inducing and avoided prior to the ablation (Green
and Kling, 1966). A similar question with regard to their cognitive function was posed by Klüever some twenty years ago
(Kluever, 1952).

"The instrumental use of objects by bilateral temporal monkeys
raises the puzzling question as to why monkeys which behave as if
they cannot distinguish edible and inedible, dangerous and harm-
less objects can solve problems which supposedly require the
highest form of animal intelligence."

By now, a number of hypotheses have been advanced to explain
the defects exhibited by amygdalectomized preparations. The
theoretical positions of Papez (1937) and MacLean (1949) have
focused on rhinencephalic or limbic structures as being concerned
with the integration of visceral sensations and affective tone
with exteroceptive input. More recently, Pribram (1967b) has
proposed that the frontal and medial basal portions of the forebrain act as an efferent system that normally inhibits afferent
inhibitory processes. Thus, lesions of this system result in a

lack of habituation to novelty, lack of "self-inhibition" and
inefficiency in reinforcement. He notes that while lesioned
monkeys are alert to stimuli they have difficulty in focusing on
the alerting event. Other experimenters agree that these preparations have no defect in alerting, but rather in integrating,
informational input with past experience (Schwartzbaum and Pulas,
1965). This is demonstrated by their inefficiency in solving
problems related to transfer of learning and discrimination
reversal. At a more phenomological level, Williams (1968) pro-
poses that temporal lobe structures are involved with sensory
perceptional integration and the "I am" experiences. Absence or
disturbed function of temporal lobe structures results in depersonalization, inability to utilize memory (which is intact) and
a lack of adaptive capacity to environmental change.

If the lesioned monkeys are beset by an inability to inhibit
selectivity visual information from their surroundings, the
response characteristics of these preparations should be related
to the amount and complexity of their visual world. At another
level, this lack of selective afferent inhibition would result in
the subjects being flooded with perceptions from their visual
world resulting in distractability, hyperalertness, inability to
sort out visual communications, perhaps resulting in a state of
"depersonalization." In such a state, the affected monkey would
be expected to withdraw from complex stimuli or "freeze" in a
protective place. Such was the case when our operates were released into a social group in a natural setting. It will be
recalled that in the African study (Kling et al., 1970) they
either climbed to the highest branches of nearby trees and sat
immobile or hid in dense thickets. When approached by normals
or man, they withdrew repeatedly. This behavior appropriately is
self protective and can be observed when normal vervets are
pursued by predators or when they were mildly obtunded after
being fed barbituates. Humans in depersonalized states also display a withdrawal from social contacts and, in extreme cases such
as catatonic states, may remain immobile and outwardly unresponsive while actually being hyperattentive and hyperalert to
surrounding sensory stimuli.

Further support for the relationship between the amygdala
and visual input is suggested by the elegant experiment of Downer
(1962) and confirmed by Barrett (1969). They demonstrated in the
split brain monkey with a unilateral amygdala ablation that
closing the eye on the intact side resulted in a number of features
of the amygdala syndrome which were reversed when the eye was
opened. No effect was demonstrated when the contralateral eye
was closed. It will be recalled also that Sawa et al. (1954)
reported that their patients were "hyperoral" only when their
eyes were open.

It may be, then, that one of the effects of amygdalectomy is
a reduction of control over visual input resulting in a "deper-
sonalized-like" state in which the subject is hyperattentive,
distractable and confused resulting in withdrawal from those
communications which cannot be sorted out. As the environmental
complexity is reduced, the operated subject would show increased
attention and responsiveness to those limited objects in its
visual world permitting the expression of the characteristic
hyperorality, hypersexuality and approach behavior toward
normally avoided objects and man. Their deficiency in habituation to novelty would add to the perserverative nature of this
behavior.

The ability of these preparations to perform adequately on
a variety of cognitive tasks is not so surprising since such
procedures are carried out under conditions which limit distraction and to maximize the focus of the subject on the stimulus
object. When the task complexity reaches certain limits, however, especially with regard to a visually dependent task, their
inability to maintain sequential acts results in a failure to
solve the problem.

Some major questions raised by this hypothesis are: (1) how
specific are the changes in social behavior with respect to the
amygdala and related temporal lobe structures: (2) Would
equivalent lesions of frontal or other medial basal forebrain
structures result in a similar disruption in social behavior, or
do the various anatomically distinct structures have varying
degrees of influence, and, if so, in what direction and to what
extent? Some recent observations on the effects of lesions of
frontal cortex, cingulate gyrus and temporal neocortex on the
rhesus of the Cayo Santiago colony strongly suggest that lesions
of these areas also may result in a lack of resocialization, poor
defensive responses and eventual death (Meyers and Swett, 1970;
Meyers, 1970). Several laboratory studies (Brody and Rosvold,
1952; Mirsky et al., 1957; Ward, 1948; Deets et al., 1970) also
indicate that lesions of frontal and cingulate cortex may in
monkeys result in varying degrees of disruption in established
social hierarchy, but they do not show other similarities with
amygdala lesioned preparations. Unfortunately, there have been
no systematic studies done with more naturally composed groups,
or of the field studies, only on the rhesus colony of the Cayo
Santiago. On this island, the high population density, naturally
belligerent nature of the rhesus and the competition for food and
space may result in behaviors not seen with other species or in
less competitive surroundings.

It also would be of importance to document in greater detail
the influence of lesion locus, and size within the amygdaloid

nucleus itself as well as to separate the effects of uncal cortex
lesions from those restricted to the sub-cortical nuclei.

Returning to the major hypothesis, it has to be demonstrated
that altering the visual input perhaps by drugs or through environmental manipulation will alter substantially the defect in social
and affective behavior. Other considerations must be given as to
whether by repeated and forced interaction with normals operated
subjects may be capable of "relearning" appropriate social responses and eventually reintegrate within the social group. In
this regard, long term studies are needed on both adult animals
and infant operates to determine the sequence in which behavioral
impairments related to the lesion may appear or disappear with
time or maturation.

It is recognized that this report has not dealt with a large
body of experimental data relevant to amygdala function, but has
only attempted to review and integrate information related to
social behavior in non-human primates. It is hoped that this
meeting will go a long way to achieving this more ambitious goal.

REFERENCES

ALTMANN, S. A field study of the sociobiology of rhesus monkeys,
    Macaca mulatta. In H. E. Whipple (Ed.) Annals of the N.Y.
    Academy of Science, 1962. Pp. 338-435.

ALTMANN, S. Social communication among primates, Chicago:
    University of Chicago Press, 1967.

ANAND, B. K., & DUA, S. Electrical stimulation of the limbic
    system of the brain ("visceral brain") in the waking animals.
    Indiana Journal of Medical Research, 1956, 44, 107.

ANDERSON, R. Psychological difference after amygdalotomy.
    Acta Neurologica Scand. Suppl. 43, 1970, 46, 94.

BARRETT, T. W. Studies of the function of the amygdaloid com-
    plex in M. mulatta. Neurophologia, 1969, 1-12.

BRODY, E. B., & ROSVOLD, E. H. Influence of prefrontal lobotomy
    on social interaction in a monkey group. Psychosomatic
    Medicine, 1952, 14, 406.

CHAPMAN, W. P., LIVINGSTON, R., & LIVINGSTON, K. E. Effect of
    frontal lobotomy and of electrical stimulation of the orbital
    surface of the frontal lobes and tip of temporal lobes upon
    respirations and blood pressure in man. In M. Greenblatt,
    R. Arnot, and H. C. Solomon (Eds.), Studies in Lobotomy.
    New York: Grune and Stratton, 1950.

DEETS, A. C., HARLOW, H. F., SINGH, S. D., & BLOOMQUIST, A. V.
    Effects of bilateral lesions of the frontal granular cortex
    on the social behavior of rhesus monkeys. Journal of Com-
    parative Physiological Psychology, 1970, 72, 452.

DeGROOT, J. The influence of limbic structures on pituitary
    functions related to reproduction. In F. A. Beach (Ed.),
    Sex and Behavior. New York: John Wiley & Sons, Inc., 1965.
    Pp. 496-511.

DELGADO, J. M. R. Aggression and defense under cerebral radio
    control. In C. D. Clemente and D. B. Lindsley (Eds.),
    Brain Function (V. 5), 1967. Pp. 171-193.

DeVORE, I. Field studies of monkeys and apes. In I. DeVore (Ed.)
    Primate Behavior. New York: Holt, Rinehart & Winston, 1965.

DICKS, D., MEYERS, R. E., & KLING, A. Uncus and amygdala
    lesions: Effects on social behavior in the free-ranging
    rhesus monkey. Science, 1969, 165, 69.

DOUGLAS, R. J., & PRIBRAM, K. H. Distraction and habituation in
    monkeys with limbic lesions. Journal of Comparative Physio-
    logical Psychology, 1969, 3, 473.

DOWNER, C. J. L. Interhemispheric integration in the visual
    system. In V. B. Mountcarlle (Ed.), Interhemispheric Rela-
    tions and Cerebral Dominance. Baltimore: Johns Hopkins
    Press, 1962.

ERVIN, F. R., MARK, V. H., & STEVENS, J. Behavioral and affect-
    ive responses to brain stimulation in man. In J. Zubin and
    Shagass (Eds.), Neurobiological Aspects of Psychopathology.
    New York: Grune and Stratton, 1969. Pp. 54-65.

ERVIN, F. R., MARK, V. H., & SWETT, W. Focal brain disease and
    assaultive behaviour. Proceedings of the Symposium on the
    Biology of Aggressive Behaviour, Milan, May 1968. Excerpta
    Medica, Amsterdam, 1969.

FALCONER, M. A., & TAYLOR, D. C. Surgical treatment of drug
    resistant epilepsy due to mesial temporal scloerosis.
    Archives of Neurology, 1968, 19, 353.

GASTAUT, H., TOGA, M., ROBER, J., & GIBSON, W. C. A correlation
    of clinical, electroencephalographic and anatomical findings
    in nine autopsied cases of "temporal lobe epilepsy."
    Epilepsia, 1959, 1, 56.

GLOOR, P. Amygdala. In J. Field, H. W. Magoun, and V. E. Hall
    (Eds.) American Physiology Society Handbook of Physiology,
    Section I: Neurophysiology, V. II. 1960. Pp. 1395-1416.

GODDARD, G. V. Functions of the amygdala. Psychological Bulle-
    tin, 1964, 62.

GREEN, P. C., & KLING, A. Effects of amygdalectomy on affective
    behavior in juvenile and adult macaque monkeys. APA Pro-
    ceedings, 1966, 93-94.

HAMBURG, D. A. Evolution of emotional responses: Evidence from
    recent research on non-human primates. In Science and Psycho-
    analysis, Vol. 12. New York: Grune and Stratton, 1968.
    Pp. 39-54.

HARLOW, H. F., & HARLOW, M. K. The affectional systems. In
    A. Schrier, H. F. Harlow & Stollnitz (Eds.), Behavior of
    Non-Human Primates. New York: Academic Press, 1965.
    Pp. 287-333.

HEATH, R. G., & MICKLE, W. A. Evaluation of seven years experi-
    ence with depth electrode studies in human patients. In
    E. R. Ramey & D. S. O'Doherty (Eds.), Electrical Studies on
    the Unanesthetized Brain. New York: Hoeber, 1960.
    Pp. 214-247.

HEIMBURGER, R. F., WHITLOCK, C. C., & KALSBECK, J. E. Stereo-
    taxic amygdalotomy for epilepsy with aggressive behavior.
    Journal American Medical Association, 1966, 198, 165.

IWATA, K., & ANDO, Y. Socio-agonistic behavior in temporal
    lobectomized monkeys. (pre-print).

JAY, P. Primates: Studies in adaptation and variability.
    New York: Holt, Rinehart & Winston, 1968.

JENSEN, G. D., BABBITT, R. A., & GORDON, B. N. The development of mutual independence in mother-infant pigtailed monkeys. In S. Altmann (Ed.), Social Communication Among Primates, 1967. Pp. 43-53.

KAADA, B. R. Somato-motor, autonomic and electrocorticographic responses to electrical stimulation of "rhiencephalic" and other structures in primates, cat and dog. Acta Physiologica Scand., 1951, 24(1).

KAADA, B. Brain mechanisms related to aggressive behavior. In C. D. Clemente & D. B. Lindsley (Eds.), Aggression and Defense. Univ. California Press, 1967. Pp. 195-234.

KAUFMAN, J. H. Social relations of adult males in a free-ranging band of rhesus monkeys. In S. A. Altmann (Ed.), Communication Among Primates. Chicago: Univ. Chicago Press, 1967. Pp. 73-98.

KLING, A. Behavioral and somatic development following lesions of the amygdala in cat. Journal of Psychiatric Research, 1965, 3, 263.

KLING, A. Ontogenetic and phylogenetic studies on the amygdaloid nuclei. Psychosomatic Medicine V. XXVIII, No. 2 (March-April 1966).

KLING, A. The effect of cerebral ablation in infant monkeys on motor and cognitive function. In C. R. Angle and E. A. Bering, Jr. (Eds.), Physical Trauma, 1968a. Pp. 197-205.

KLING, A. Effects of amygdalectomy and testosterone on sexual behavior of male juvenile macaques. Journal of Physiological Psychology, 1968b, 65, 466.

KLING, A., & CORNELL, R. Amygdalectomy and social behavior in the caged stump-tailed macaque (M. speciosa). Folia Promatology (In press) 1971.

KLING, A., DICKS, D., & GUROWITZ, E. M. Amygdalectomy and social behavior in a caged group of vervets (C. aethiops). Proceedings 2nd International Congress of Primates, Atlanta, Georgia, v. 1, pp. 232-241, New York: (Karger, Basel), 1968.

KLING, A., & GREEN, P. C. Effects of amygdalectomy in the maternally reared and maternally deprived neonatal and juvenile macaque. Nature, 1967, 213, 742.

KLING, A., LANCASTER, J., & BENITONE, J. Amygdalectomy in the
    free-ranging vervet. Journal of Psychiatric Research, 1970,
    7, 191.

KLING, A., & TUCKER, T.   Sparing of function following localized
    brain lesions in neonatal monkeys. In R. Isaacson (Ed.),
    The Neuropsychology of Development. New York: John R. Wiley
    & Sons, 1968. Pp. 121-145.

KLÜEVER, H. Brain mechanisms and behavior with special reference
    to the rhiencephalon. Lancet, 1952, 72, 567.

KLÜEVER, H., & BUCY, P. Preliminary analysis of functions of the
    temporal lobes in monkeys. Archives of Neurology and Psy-
    chiatry, 1939, 42, 979.

MacLEAN, P. D. Psychosomatic disease and the "visceral brain"
    recent developments on the Papez theory of emotion. Psycho-
    somatic Medicine, 1949, 11, 338.

MASON, J. W. Plasma 17-hydroxycorticosteroid levels during
    electrical stimulation of the amygdaloid complex in conscious
    monkeys. American Journal of Physiology, 1959, 196, 44.

MASON, W. A. The social development of monkeys and apes. In
    I. DeVore (Ed.), Primate Behavior. New York: Holt, Rine-
    hart & Winston, 1965. Pp. 514-543.

MEYERS, R. E. Personal Communication, 1970.

MEYERS, R. E., & SWETT, C. Social behavior deficits of free-
    ranging monkeys after anterior temporal cortex removals:
    A preliminary report. Brain Research, 1970, 19, 39.

MILLER, R. Effects of amygdalectomy on sexual behavior in
    juvenile female monkeys (M. speciosa). Masters Thesis,
    Illinois Institute of Technology, 1968.

MIRSKY, A. F., ROSVOLD, H. E., & PRIBRAM, K. Effects of cingu-
    lectomy on social behavior in monkeys. Journal of Neuro-
    physiology, 1957, 20, 588.

NARABAYASHI, H., NAGAO, T., SAITO, Y., YOSHIDA, M., & NAGAHATA,
    M. Stereotaxic amygdalotomy for behavior disorders.
    Archives of Neurology, 1963, 9, 1.

PAPEZ, J. W. A proposed mechanism of emotion. Archives of
    Neurology and Psychiatry, 1937, 38, 725.

PENFIELD, W., & JASPER, H. Epilepsy and the functional anatomy of the human brain. Boston: Little Brown & Co., 1954.

PLOTNIK, R. Changes in social behavior of squirrel monkeys after anterior temporal lobectomy. Journal of Comparative Physiological Psychiatry, 1968, 66, 369.

PRIBRAM, K. H. Emotion: Steps toward a neuropsychological theory. In D. C. Glass (Ed.), Neurophysiology and Emotion. New York: Rockefeller University Press, 1967a. Pp. 4-40.

PRIBRAM, K. H. Emotion: Steps toward a neuropsychological theory. In D. C. Glass (Ed.), Neurophysiology and Emotion. New York: Rockefeller University Press, 1967b. Pp. 3-60.

REDMOND, D. E., MAAS, J. W., KLING, A., & DEKIRMENJIAN, H. Changes in primate social behavior following treatment with alpha methyl para tyrosine. Psychosomatic Medicine, 1971, 33, pp. 97-113.

ROSENBLUM, L. A., & KAUFMAN, I. Laboratory observations of early mother-infant relations in pigtail bonnet macaques. In S. Altmann (Ed.) Social Communication Among Primates. Chicago: University of Chicago Press, 1967. Pp. 33-41.

ROSVOLD, H. E., MIRSKY, A. F., & PRIBRAM, K. H. Influence of amygdalectomy on social behavior in monkeys. Journal of Comparative and Physiological Psychology, 1954, 47, 173.

SAWA, M., VEKI, Y., ARITA, M., & HARADA, T. Preliminary report on amygdaloidectomy on psychotic patients. Folia Psychiatry and Neurology, Jap., 1954, 7, 309.

SCHWARTZBAUM, J. S. Discrimination behavior after amygdalectomy in monkeys, visual and somesthetic learning and perceptual capacity. Journal of Comparative Physiological Psychology, 1965, 3, 314.

SCOVILLE, W. B., DUNSMORE, R. H., LIBERSON, W. T., HENRY, C. E., & PEPE, A. Observations on medial temporal lobotomy uncotomy in the treatment of psychotic states. Proceedings of the Association on Research in Nervous and Mental Disease, 1953, 31, 347.

TERZIAN, H., & DALLE, O. G. Syndrome of Klüever and Bucy reproduced in man by bilateral removal of the temporal lobes. Neurology, 1955, 5, 373.

THOMPSON, C., SCHWARTZBAUM, J. S., & HARLOW, H. F.   Development of
      social fear after amygdalectomy in infant rhesus monkeys.
      Physiology and Behavior, 1969, 4, 249.

URSIN, H.   The effect of amygdaloid lesions on flight and defense
      behavior in cats.   Experimental Neurology, 1965, 11, 61.

WARD, A. A., Jr.   The cingular gyrus: Area 24.   Journal of
      Neurophysiology, 1948, 11, 13.

WEISKRANTZ, L.   Behavioral changes assisted with ablation of the
      amygdaloid complex in monkeys.   Journal of Comparative and
      Physiological Psychology, 1956, 49, 381.

WILLIAMS, D.   Man's temporal lobe.   Brain, 1968, 91, 639.

THE ROLE OF THE AMYGDALA IN ESCAPE-AVOIDANCE BEHAVIORS

Sebastian P. Grossman

Department of Psychology
University of Chicago
Chicago, Illinois

I first became interested in the influence of the amygdala on escape-avoidance behavior several years ago when I somewhat inadvertently induced temporal lobe seizures and marked "personality changes" in cats by the administration of minute quantities of acetylcholine to the basolateral portion of the amygdaloid complex (Grossman, 1963). The injections produced epileptiform spike discharges in the amygdaloid complex which spread rapidly to other portions of the temporal lobe, and involved eventually other regions of the brain. Overt psychomotor seizures typically appeared within 10-15 minutes after the injection, and persisted with minor interruptions for several hours. Afterwards, the animals appeared exhausted, but were clearly hypersensitive to any form of external stimulation, and attacked the experimenter rather than permit normal handling. Additional brief (2-5 min) epileptiform seizure attacks were observed during the next 24-48 hours. The electroencephalographic (EEG) activity of the temporal lobe remained highly abnormal for 10-15 days after the motor disturbances had subsided, and the animals continued to attack man as well as other animals (cats and rats) at the slightest provocation.

When we repeated this experiment with a cholinomimetic agent (carbachol) which is not destroyed as rapidly as acetylcholine, we obtained essentially an identical reaction except that the EEG as well as behavioral changes appeared to be permanent (Fig. 1). As late as 5 months after a single injection of carbachol into the amygdaloid complex, the cats remained extremely vicious and entirely refractory to normal handling. Most interesting, in the present context, was the fact that these animals attacked at the slightest provocation without apparent concern for their personal

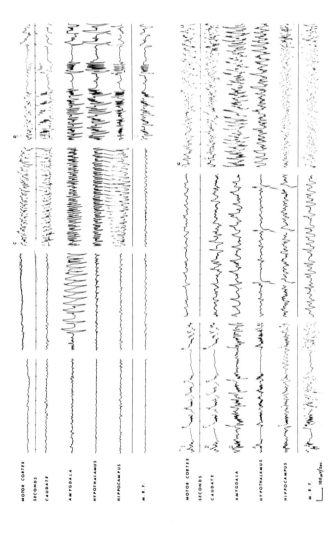

Fig. 1. Electrical activity of the brain following cholinergic (carbachol) stimulation of the basolateral nuclei of the amygdala. (A) Control period immediately preceding central stimulation; (B) Records obtained during brief "quiet" period between overt motor seizures, 35 minutes after central stimulation; (C) Electrical activity 2-1/2 hours after stimulation (no overt seizure activity); (D) Spike discharges alternating with brief periods of essentially zero electrical activity, 24 hours after stimulation (no overt seizure activity); (E) Electroencephalographic pattern 5 months after stimulation (animal vicious but otherwise normal); (F) Spike pattern, recorded within minutes after overt motor seizures, 5 months after amygdaloid stimulation. (From Grossman, 1963).

safety. They appeared incapable or unwilling to avoid or escape from even intensely painful stimulation, preferring instead to attack its source. The behavior of these animals unfortunately was so vicious that we found it impossible to test them in formal conditioned avoidance (CAR) situations.

Since lesions in the amygdala have been reported to produce opposite, taming effects, we concluded that the observed loss of fear might be due to a persisting irritation and consequent stimulation of some components of the amygdaloid complex. Cholinergic mechanisms in particular seemed to be implicated in this effect since microinjections of other neurohumors and control substances did not reproduce the effects of acetylcholine and carbachol.

Since injections of acetylcholine or carbachol into other portions of the central nervous system (including the brainstem reticular formation, hypothalamus, preoptic region, thalamus, septal area, and hippocampus) consistently produce only very short-lived, and clearly reversible effects on the EEG as well as on overt behavior, we were intrigued by the apparent permanence of the effects.

In view of the extreme viciousness of the acetylcholine- or carbachol-treated cats, we (Belluzzi and Grossman, 1969) decided to investigate the behavioral consequences of such injections further in the more easily handled and subdued rat. We observed quite comparable initial seizure reactions in this species (Fig.2), but the EEG and behavioral effects of the treatment subsided typically within a few hours, and the animals appeared to return to normal levels of reactivity.

A more careful analysis of the behavior of these animals did, however, reveal some interesting persistent effects. When we tested the rats in a one-way conditioned avoidance apparatus several weeks after all overt EEG and behavioral effects of the carbachol injections had disappeared, the experimental animals seemed incapable or unwilling to learn the simple conditioned avoidance response (opening a small door and jumping into the adjacent compartment) and, in fact, often failed to escape from the painful UCS after it had been delivered to the grid floor (Fig. 3A).

We were most surprised by this outcome in view of several reports that large lesions in the amygdaloid complex reduced the efficiency of conditioned avoidance or had no measurable effect on it (see Goddard, 1964, for a review of this literature). Since the carbachol injections produced effects on aggressive behavior which were very clearly opposite those typically seen after lesions, one might have expected opposite, i.e., facilitatory effects on avoidance behavior as well.

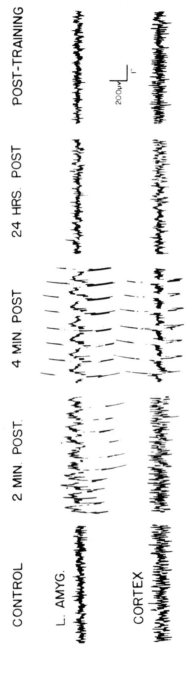

Fig. 2. Progression of EEG changes observed after a single bilateral injection of carbachol into the amygdaloid complex of rats (From Belluzzi, 1970).

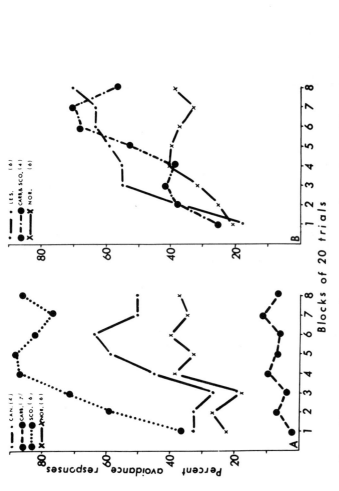

Fig. 3. Acquisition of avoidance responses in a one-way situation. (A) Comparison of the performance of rats which received a single injection of carbachol 1-3 weeks prior to training (solid circles) or daily injections of scopolamine (open circles) into the amygdaloid complex with that of normal (triangles) and cannulated (squares) controls; (B) Comparison of the performance of rats which received lesions in the vicinity of the drug implantation site (squares) or a single injection of carbachol followed by daily injections of scopolamine (solid circles) with that of normal controls (From Belluzzi and Grossman, 1969).

To investigate this matter further, we made lesions in the region of the cannula implants of several naive rats, and trained them in the same apparatus. As shown in Figure 3A, these animals made reliably more avoidance responses than normal controls at all stages of training.

To ascertain whether the observed effects could, indeed, be attributed to cholinergic components of the area, we injected small quantities of scopolamine, a substance which blocks transmission at cholinergic synapses, into the amygdala of still another group of rats a few minutes before each daily training session. As shown in Figure 3A, these animals also outperformed normal controls at all stages of training.

To put icing on the cake, we decided to examine the possibility that daily scopolamine injections might reverse the deleterious effects of prior carbachol treatments. To our delight, we found that scopolamine injections not only reversed the carbachol effect but indeed raised performance to levels comparable to those of lesioned rats (Fig. 3B).

These observations indicate that the amygdaloid complex of the rat contains a cholinergic component which exercises some influence on avoidance and escape behavior. Belluzzi (1970), working in my laboratory, has investigated further the nature of this influence. He first examined the possibility that the carbachol-induced seizures may involve temporal lobe mechanisms which are essential to learning, memory consolidation, recall, or visual functions rather than avoidance behavior per se. To do this, he trained rats which had received seizure-inducing carbachol injections several days earlier in a T-maze brightness discrimination, and compared their performance to that of normal or cannulated controls. There were no significant differences, indicating that this potential explanation can be ignored.

Next, Belluzzi (1970) asked whether the effects of carbachol on avoidance behavior might be peculiar to the intense seizure activity which is produced by the injections rather than a selective stimulation of cholinergic components in the region of direct drug action. To answer this question, he trained animals in the one-way avoidance situation after daily intra-amygdaloid injections of eserine sulfate, a substance which potentiates transmission at cholinergic synapses by blocking the destruction of acetylcholine which is released in response to normal neural activity. Figure 4 shows the results of this experiment--eserine duplicated the effects of carbachol, indicating that some cholinergic components of the amygdala are spontaneously active during the acquisition of avoidance responses.

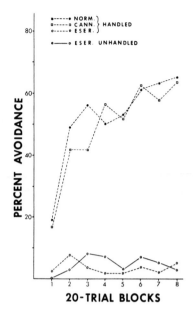

Fig. 4. Effects of daily eserine injections into the amygdaloid complex on the acquisition of conditioned avoidance responses in handled and unhandled rats (From Belluzzi, 1970).

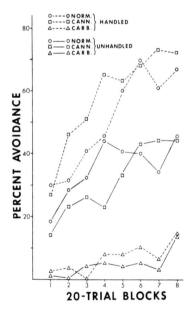

Fig. 5. Effects of a single carbachol injection, administered 1-3 weeks prior to training in a one-way avoidance apparatus in handled and unhandled rats (From Belluzzi, 1970).

What was still not clear was just how the amygdaloid influence modifies the acquisition or performance of avoidance behaviors. In thinking about this problem, we arrived at the following conclusion. If the effects of carbachol were indeed related to an increase in affective reactivity, as our earlier observations of carbachol-treated cats indicated, it might be possible to reverse or prevent its effects by treatments which decrease the animal's level of arousal or reactivity. Belluzzi (1970) tested this hypothesis by observing the effects of prolonged and repeated handling of the animals on the carbachol effect on CAR acquisition. This treatment has been shown to decrease or even eliminate such overt signs of "emotionality" as vocalizing, struggling, defecating, urinating, "freezing," crouching, etc., and to improve CAR acquisition in normal rats (Doty and O'Hare, 1966).

The results of this experiment are shown in Figure 5. Normal rats showed a marked effect of handling as expected, presumably because of a reduction in emotional reactivity which normally interferes with learning in these situations. The carbachol-treated animals, on the other hand, did not respond at all. Closer inspection of the behavior of these animals indicated that the differential effect of handling appeared to be due to the carbachol-treated animals' inability to adapt to noxious stimulation. The first time a rat is handled, the event is a traumatic one for the animal as well as the experimenter. Normal laboratory rats adapt rapidly to this procedure, and inhibit overt emotional reaction after a few days of daily contact. The carbachol-treated rats, although not as hyperreactive as carbachol-treated cats, appeared unable to show this adaptation and, consequently, failed to show the facilitatory effects seen in the controls.

These results indicate that the cholinergic components of the amygdala may mediate affective reactions. When these mechanisms are stimulated or facilitated in cats, the animals are vicious and incapable apparently of inhibiting aggressive reactions to painful stimulation. Laboratory rats which have been inbred for many generations to emphasize placidity and tameness do not show the apparently complete loss of escape and avoidance behavior, but appear to be deficient in acquiring conditioned avoidance responses. Their difficulty appears to arise, at least in part, because they fail to habituate to normal handling.

We were intrigued particularly by the observation that a blockade of this system appears to facilitate CAR acquisition and that amygdaloid lesions duplicated this effect. This, of course, is in contrast to several previous studies (see Goddard, 1964, for a review of the earlier literature) which have reported that large lesions involving the amygdaloid complex produce either inhibitory effects or no effects at all on CAR acquisition and performance.

We (Sclafani et al., 1970) have, indeed, replicated these observations (Fig. 6), and were puzzled when facilitatory effects
appeared in the present experiment. It occurred to us that the
amygdaloid complex might exert facilitatory as well as inhibitory
influences on avoidance behaviors. It is possible that damage to
the facilitatory components of this system disrupt the integration
of avoidance responses in such a fundamental fashion that con-
current damage to the inhibitory portions of the system do not produce significant facilitatory effects on behavior. In this case,
large amygdala lesions would produce only inhibitory effects. The
equivocal effects of smaller lesions might represent a more nearly
even balance of lesion effects on inhibitory and excitatory
mechanisms.

To test this hypothesis, we have investigated the effects of
very small and carefully placed lesions which destroyed selectively the cortical, central, or basolateral nuclei on CAR acquisition
in a standard shuttle box. In view of the fact that the typical
amygdala lesion often destroys a good deal of piriform cortex,
we also included a group of animals with extensive piriform
damage which did not invade any of the nuclei of the amygdala
itself. Since the results of the preceding experiments indicated
that stimulation of some of the components of the amygdala may influence CAR performance indirectly by modifying the animals' response to repeated handling, we handled all animals for 10-15
minutes each day for 15 days after surgery (prior to the onset of
behavioral testing).

The results of these experiments are summarized in Table 1
and Figure 7. It is clear that lesions restricted to the central,
basolateral, and cortical nuclei produced rather substantial
facilitatory effects on CAR acquisition. Fully 80 per cent of
the animals with central nucleus lesions reached stringent criteria of CAR performance (10 consecutive avoidance responses in
a single 15-trial test and 14 avoidance responses in a single 15-
trial test) within the 225 trial training period. A similarly
impressive 70 per cent of the animals with damage in the cortical
nucleus and 66 per cent of the animals with damage in the basolateral nucleus met the same criteria. This contrasts with the
observation that only 30 per cent of the control animals reached
one of these criteria, and only 20 per cent reached the other.
Lesions in the piriform cortex, on the other hand, seemed to
produce opposite, inhibitory effects. Whereas 60 per cent of the
control animals reached a lenient performance criterion of 5
successive avoidance responses per single 15-trial test session,
only 36 per cent of the animals with lesions in the piriform
cortex managed to do as well. Other intermediate criteria of
CAR proficiency show similar differences (Fig. 7). Nearly 70
per cent of the animals with lesions in the piriform cortex

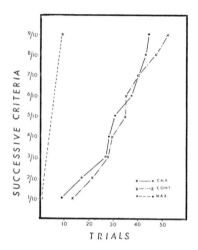

Fig. 6. Effects of large lesions in the amygdaloid complex on the acquisition of avoidance responses in a shuttle box (From Sclafani et al., 1970).

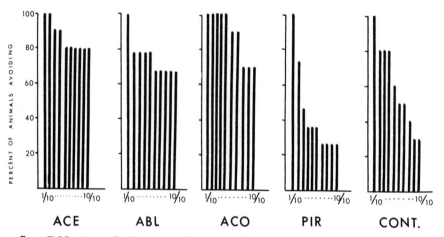

Fig. 7. Effects of discrete lesions limited to the cortical (ACO), basolateral (ABL), central (ACE) nuclei of the amygdala or of the adjacent piriform cortex (PIRI) on the acquisition of conditioned avoidance responses in a shuttle box.

Table 1. Effects of lesions in the amygdala and adjacent piriform cortex on the acquisition of shuttle-box avoidance behavior in the female rat. The first column indicates the number of animals which reached a stringent criterion (10 successive avoidance responses in a 15-trial daily test). The second column indicates the number of animals which performed 14 avoidance responses on a 15-trial test. The third column indicates the number of rats which occasionally performed avoidance responses. The last column lists the number of rats which did not learn at all in the 225 trials of the experiment.

Group:	10 successive CARs	14/15 CARs	Some CARs	No CARs
ACE lesion n=10	8	8	2	0
ABL lesion n=9	6	6	1	2
ACO lesion n=10	7	7	3	0
PIR lesion n=11	3	2	1	7
CONTROLS n=10	3	2	4	3

Table 2. Average number of spontaneous crossings between compartments in the shuttle box

Day	Condition	ACE Lesion	ABL Lesion	ACO Lesion	PIRI Lesion	Control
1	No CS or UCS	26.6	31.0	27.9	29.8	21.8
2	CS only	31.3	27.1	23.5	29.9	26.1
3	CS & UCS	22.0	16.2	21.5	17.4	17.0
4	CS & UCS	17.7	16.6	20.1	16.1	17.9
5	CS & UCS	16.5	16.4	17.4	16.8	15.9

lesions performed inconsistently (never more than 2 or 3 successive conditioned responses) to suggest that they may not, in fact, have learned the CAR at all. Only about 20 per cent of the control animals showed a similarly severe deficiency.

Since the efficiency of CAR behavior in a shuttle box can be influenced by the general level of locomotor activity, we recorded the number of spontaneous crossings between the two compartments of the apparatus (Table 2). On the first day of the experiment, when neither the CS nor the UCS were presented, experimental animals were more active than the controls. This effect was reliable statistically ($p < .05$) for all groups except in the animals with damage in the central nucleus. On day two, when the CS but not the UCS was presented periodically, this difference disappeared, largely because the control animals increased their activity, while the activity of the experimental subjects remained stable or declined somewhat. When the shock UCS was introduced on the third day, the activity of all groups showed roughly comparable decline, and no significant differences between groups remained. Similar data were collected on subsequent days, indicating that the lesion-induced changes in avoidance behavior could not be due to differences in locomotor activity.

These observations suggest the interesting possibility that the deficit in CAR acquisition which is sometimes reported after amygdaloid lesions may, in fact, be due to incidental damage to the adjacent piriform cortex. We do, however, require additional information before we can rule out the possibility that nuclear groups of the amygdala which were not damaged in this series of experiments may be responsible for the inhibitory effects often seen after larger lesions.

It also is possible that the facilitatory effects of our lesions may be related to a differential response to handling, a variable which appeared to be an important determinant of the effects of carbachol- and eserine-injections into the amygdaloid complex. In view of the fact that carbachol- and eserine-injections appeared to interfere with CAR acquisition, at least in part by preventing the gradual disappearance of overt emotional reactions to handling, it appears logically possible that destruction of the pathways which mediate this effect might have opposite, beneficial effects. We are conducting currently experiments which should shed some light on these alternatives.

It is interesting to note that the amygdaloid influences on escape-avoidance and intra-species aggressive behaviors do not appear to depend on the direct amygdalofugal pathways to the hypothalamus. I (Grossman, 1970) investigated this possibility by observing the behavioral effects of parasagittal knife cuts just

lateral to the lateral border of the hypothalamus which transected
completely this pathway. Rats with such cuts learned a shuttle-box
CAR as well as or better than normals (Fig. 8), and showed no
change in the pattern of their interaction with unoperated control
rats in a food-competition situation which elicited fighting and
permitted the establishment of stable dominance-submissiveness
relationships. Parasagittal cuts along the medial border of the
lateral hypothalamus also failed to modify aggressive reactions
in the food-competition situation but significantly inhibited CAR
acquisition. Since damage to the ventromedial nucleus itself produces opposite, facilitatory effects on CAR acquisition as well as
some aggressive reactions (Grossman, 1966; Grossman, 1971), the
amygdaloid influence on this region appears to rely on pathways
that enter the lateral hypothalamus rostrally or anteriorly, and
then turn medially near the ventromedial nucleus itself. Support
for this interpretation was obtained in additional experiments
(Grossman and Grossman, 1970) which demonstrated that knife cuts
anterior to the ventromedial nucleus itself did not affect CAR
acquisition. These cuts did inhibit aggressive behavior in the
food-competition situation, indicating that aggressive and avoidance behaviors may in part be mediated by different pathways.

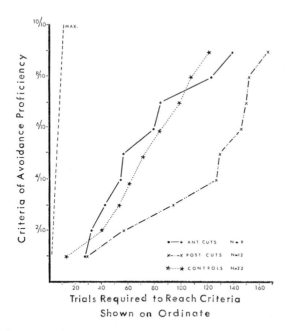

Fig. 8. Effects of parasagittal knife cuts lateral to the hypothalamus or in the medial quadrant of the lateral hypothalamus on
the acquisition of conditioned avoidance responses in a shuttle
box (From Grossman, 1970).

In summary, our experiments suggest that cholinergic components of the amygdaloid complex may be involved significantly in the mediation of escape-avoidance behavior. When these pathways are stimulated or facilitated, rats and cats find it difficult to inhibit aggressive reactions to such stimulation, and appear unable to develop normal escape or avoidance reactions. In the cat, such stimulation produces vicious attack responses to normal handling. In the normally tame laboratory rat, such overt displays of aggressive reactions rarely occur, but the animals appear unable to inhibit emotional reactions to normal handling and consequently cannot benefit from the taming effect which repeated handling exerts on normal rats. When these components of the amygdala are inhibited pharmacologically, rats appear less reactive and acquire simple conditioned avoidance responses faster than normals. Similar facilitatory effects were observed in rats with small lesions restricted to either the cortical, basolateral, or central nuclei of the amygdala. Opposite inhibitory effects were observed in animals with lesions which destroyed portions of the piriform cortex but did not invade the amygdala itself. The amygdaloid influence on escape-avoidance behavior does not appear to be related to the direct amygdalofugal pathways which interconnect it with the hypothalamus. Instead, they appear to be related to pathways which enter the lateral hypothalamus anteriorly or rostrally.

REFERENCES

BELLUZZI, J. D. Long-lasting effects of cholinergic stimulation of the amygdaloid complex in the rat. Ph.D. dissertation, University of Chicago, 1970.

BELLUZZI, J. D., & GROSSMAN, S. P. Avoidance learning motivated by high-frequency sound and electric shock. Physiology & Behavior, 1969, 4, 371-373.

GODDARD, G. V. Functions of the amygdala. Psychological Bulletin, 1964, 62, 89-109.

GROSSMAN, S. P. Chemically induced epileptiform seizures in the cat. Science, 1963, 142, 409-411.

GROSSMAN, S. P. The VMH: A center for affective reaction, satiety, or both? Physiology & Behavior, 1966, 1, 1-10.

GROSSMAN, S. P. Avoidance behavior and aggression in rats with transections of the lateral connections of the medial or lateral hypothalamus. Physiology & Behavior, 1970, 5, 1103-1108.

GROSSMAN, S. P. Effects of lesions in the ventromedial hypothala-
     mus of the female rat on intra- and inter-species aggression,
     avoidance, and behavior in novel environments. Journal of
     Comparative and Physiological Psychology, 1971, in press.

GROSSMAN, S. P., & GROSSMAN, LORE. Surgical interruption of the
     anterior or posterior connections of the hypothalamus:
     Effects on aggressive and avoidance behavior. Physiology &
     Behavior, 1970, 5, 1313-1317.

SCLAFANI, A., BELLUZZI, J., & GROSSMAN, S. P. Effects of lesions
     in the hypothalamus and amygdala on feeding behavior in the
     rat. Journal of Comparative and Physiological Psychology,
     1970, 72, 394-403.

ROLE OF THE AMYGDALA IN THE CONTROL OF

"MOUSE-KILLING" BEHAVIOR IN THE RAT

P. Karli, M. Vergnes, F. Eclancher, P. Schmitt and
J. P. Chaurand

Laboratoire de Neurophysiologie, Centre de Neurochimie
Strasbourg, France

It might be advisable to begin this paper with a few general
considerations, in order to broaden its scope which initially
seems to be a rather limited one. Such general considerations
may be considered as truisms, but we all know how easily obvious
facts get out of mind in scientific thinking no less than in
everyday life.

The central nervous system should not be conceived as a huge
"functional mosaic" composed of elements each having its own
intrinsic function. The brain is a functional entity: a given
structure or system of structures confers a further dimension on
the entity, and the functioning of the constituent part in its
turn is influenced widely by the functioning of the whole. Hence,
it follows that any role played by the amygdala hardly can be
understood fully if the relevant experimental data obtained on
this structure are not discussed against a broader background of
brain physiology. It also follows that when bringing the amygdala
into play, or when preventing it from playing its part, thereby
possibly conferring a further dimension on the functioning of the
brain or depriving the latter of that dimension, we may create
two essentially different functional entities which will express
themselves in differential behavioral characteristics. But the
experimental conditions must allow these differential characteristics to be uncovered and analyzed.

Such an holistic approach also should prevail, if we consider
the behavior of a living organism. When studying a given type of
behavior, this partial aspect obviously has to be considered
separately for experimental purposes; but one always should keep

in mind the fact that any partial aspect has to be confronted
with other aspects of the behavioral entity, if it is to be understood fully. This is true especially for the study of any kind
of social behavior which has to be evaluated in the light of
other aspects of the organism's social-emotional responsiveness.
Of course, an organism's response to a given situation is under
control of the previous history of reinforcement pertaining
specifically to that response. But when the organism is exposed,
for the first time, to a novel situation and even later on, its
behavior (fundamentally, avoidance or aggression) depends upon
its more general social-emotional responsiveness which was
shaped progressively by the sum of past experiences in more or
less similar situations.

Whether we study the characteristics of neural systems or
those of behavioral processes, we try, in either case, to un-
cover their organization, their interplays and the factors that
govern their patterning during ontogenesis. To feel comfortable,
one has to stay on one or the other side of what still appears
to be quite a gap. But it obviously is of paramount importance
for the understanding of the biological foundations of behavior
that as many bridges as possible be thrown from one conceptual
framework to the other. It would go beyond the scope of the
present paper to draw up the list of the difficulties encountered
when trying to bridge the gap. But one example of conceptual
and semantic confusion should be given, as it concerns the basic
concept of motivation.

To the psychologist, the motivation gives the answer to the
"why," to the question about the "immediate causes" of behavior.
As behavioral activities have a certain orientation and are performed with a certain intensity, the underlying motivations are
considered to have both "energizing" and "orienting" components.
In this respect, the term "motivation" is most meaningful as it
evokes both aspects of its semantic content ("motor" and
"motive"). But in most physiological studies of behavior, the
concept of motivation is used in a more limited acceptation: it
refers essentially, if not exclusively, to the "energizing"
mechanisms which realize more or less specific "drive-states";
it does not include the mechanisms through which motivating
properties are conferred upon present sensory information by
reference to the traces laid down by the previous life history.
Such a limitation of the concept of motivation to a more or less
specific behavioral "arousal" or "activation" is to some extent
responsible for the dichotomy between "biological" and "psycho-
genic" determinants of behavior which often appears in psychology.
Such a concept has been formed, as if the patterning of neural
circuits through past experience, and its consequences upon
present orientation of behavior, were less "biological" than

the mechanisms through which the sensory input and the variations of the internal milieu bring about a more or less specific behavioral arousal. We are in favour of a more total concept of motivation: the "motivational state" of an organism at any given moment of its ontogenesis should be considered as encompassing all the aspects of its physiological state that are relevant to the fact that it will respond to a given situation with a given behavioral activity. Such a total concept, which is more consonant with the one used in psychology, does not hinder the separate experimental analysis of each one of its constituent processes and underlying mechanisms. Moreover, it helps to keep in mind the fact that the separations which have to be made for obvious methodological and technical reasons merely are artificial ones and that the experimental data gathered will have to be used as building stones for a more holistic approach to the biological foundations of behavior.

As this paper deals with a certain kind of aggressive behavior, two more remarks should be made: one concerning social behavior in general, the other concerning aggressive behavior in particular.

LIFE-HISTORY AND SOCIAL BEHAVIOR

There is no known factor of the internal milieu that plays a role in the physiological control of social behavior, the essential and general role played by glycemia, osmolarity and sexual hormone levels in, respectively, the control of eating, drinking, and sexual behaviors. Hence, there lacks in the control of many kinds of social behavior, the kind of factor that is of fundamental importance for the production of a drive state arising internally which facilitates the elicitation of eating, drinking or sexual behavior. In most instances, social behavior is determined essentially by the life-history, by both the general and the more specific previous history of reinforcement, and Scott (1969) certainly is right in stressing the intimate relations between emotional and social responsiveness. If we assume that the limbic system plays an essential part in adapting present behavior to past experience, we already foreshadow the assumption that the amygdala may play an important role in the control of the rat's mouse-killing behavior.

NO UNITARY CONCEPT OF AGGRESSIVENESS

It is accepted ever more widely among biologists that any explicit or implicit reference to a unitary concept of aggressiveness and aggressive behavior is to be avoided. There are a number of different motivational states which may express themselves in different kinds of aggressive behavior. As the same organism is likely to display various kinds of aggression, it is obvious that the underlying motivational states can have many factors in common; but their relative importance may be quite different from one case to another. Moyer (1968) has proposed a valuable definition and classification of the different kinds of aggression. One may be tempted to criticize certain aspects of his classification, but one

had better refrain from doing so as long as one is unable to propose a better one.

In order to give concrete support to the above statement that
any unitary concept of aggressiveness is to be avoided, we would
like to give some of the experimental evidence which differentiates
clearly the rat's mouse-killing behavior from other kinds of
aggressive behavior:

(a) Painful stimulation elicits fighting behavior in rats
(O'Kelly and Steckle, 1939; Ulrich and Symannek, 1969). Yet, the
same stimulation does not induce mouse-killing in rats that have
never shown spontaneously this inter-specific aggressive behavior
under standard environmental conditions (Karli, 1956; Myer and
Baenninger, 1966).

(b) Ablation of the olfactory bulbs suppresses spontaneous
aggression among male mice (Ropartz, 1968), whereas the same operation is liable to induce the mouse-killing behavior in spontaneously non-killing rats (Vergnes and Karli, 1963), without modifying
intermale aggressive behavior in the same species (Bernstein and
Moyer, 1970).

(c) Social isolation from weaning does not modify significantly the incidence of the killing response in adult rats (Myer,
1969; Vergnes and Karli, 1969a). But the same condition seems to
reduce the tendency of male rats to irritable (pain-induced)
aggression (Hutchinson et al., 1965) and to strengthen spontaneous
aggressiveness among male mice (Denenberg et al., 1964; Uyeno and
White, 1967).

(d) Androgenic hormones take a prominent part in the release
of spontaneous intermale aggression in laboratory rodents (Seward,
1945; Beeman, 1947). On the other hand, castration combined with
bilateral adrenalectomy does not suppress the killing response once
it is well established; conversely, the repeated administration of
large doses of androgens does not induce mouse-killing in rats that
have never killed prior to the experiment (Karli, 1958).

(e) Septal lesions increase the occurrence of spontaneous
aggression among male rats as well as of shock-elicited fighting
(Allikmets and Ditrikh, 1965; Ahmad and Harvey, 1968), but never
induce mouse-killing in the natural non-killer (Karli, 1960a).*

(f) A bilateral transection of the amygdalo-hypothalamic
fibers passing through the ansa lenticularis abolishes the rat's
mouse-killing behavior (Vergnes and Karli, 1964), but does not seem
to have any effect on intra-species aggression (Grossman, 1970).

---

*It came out of the discussion that septal lesions may well induce
mouse-killing in a natural non-killer, but only if the rat had
little experience with mice prior to the operation; under such
circumstances, the mouse-presentation situation is still a rather
"novel" situation for the septal lesioned rat.

(g) Brain serotonin depletion provoked by administration of parachlorophenylalanine (a tryptophan hydroxylase inhibitor) increases the occurrence of inter-specific aggressive responses in the rat (Sheard, 1969; Karli et al., 1969) as well as in the cat (Ferguson et al., 1970), yet has no demonstrable effects on shock-induced fighting behavior in rats (Conner et al., 1970).

BIOLOGICAL SIGNIFICANCE OF THE RAT'S MOUSE-KILLING BEHAVIOR

It is not easy to define clearly the biological significance of the rat's mouse-killing behavior and to give a simple answer to the question of why some rats kill mice while others never do. The rat's killing response usually is considered to be a "predatory" aggression (Moyer, 1968). If the term "predatory" implies that the rat kills mice for food, then its use does not seem to be adequate. Most rats do not eat the first mouse they kill. Moreover, a large majority of both wild and domesticated spontaneously non-killing rats may starve to death without ever showing any hostility towards the mouse living in their cage: they make a clear difference between a dead mouse, which they eat almost immediately, and a live one, which they do not kill in order to obtain food (Karli, 1956). On the other hand, as we shall see later on, many experimental manipulations clearly have differential effects upon eating behavior and mouse-killing behavior. By no means do these facts exclude the possibility of a more or less important facilitation of the killing response in two ways: in a non-specific way, by food deprivation which entails a general behavioral arousal; in a more specific way, by the positive reinforcement deriving from the repeated association of mouse-killing with eating of part (most usually the brain) of the killed mice. Whatever the precise nature of the interrelation between the two kinds of behavior, it has to be stressed that in the experienced killer the availability of a mouse for killing can serve as a reinforcer in instrumental learning situations (Myer and White, 1965). And the observation that killing experience increases the resistance of the behavior to the suppressive effects of punishment (Myer, 1967) lends further support to the proposition that killing is self-strengthening.

But which are the factors that induce some rats, and only some of them, to kill the mouse when they are presented with an animal of this species for the very first time? The observation of the individual rat's behavior demonstrates that, in many instances, the rat that will kill for the first time (often with a delay of a number of hours) is quite excited and seems to be "upset" by the presence of the mouse in its cage. Conversely, the non-killing rat often seems to "accept" more easily the presence of the strange animal, and to include it rapidly into its familiar environment. A recent unpublished observation has shown that the killers eliminate significantly less urinary norepinephrine than the non-killers. This fact may be brought into a correlation with another one uncovered by Welch and Welch (1969) that the basal activity of the adrenal medulla is lowered in mice rendered more reactive and more aggressive by social isolation.

But the level of general emotional responsiveness or "irritability" hardly can be more than part of the story. For instance, there is no predictable relation between a rat's savageness towards the experimenter or its aggressiveness toward another rat and its response to a mouse. Furthermore, the level of emotional responsiveness usually is much higher in a non-killing wild rat than in a domesticated killer. On the other hand, as indicated previously, septal lesions that increase the rat's irritability do not incite a natural non-killer to start killing mice. Following olfactory bulb lesions, that also entail an increased irritability (Douglas et al., 1969; Bernstein and Moyer, 1970), some rats begin killing while others do not, and there is again no predictable relation between an animal's postoperative irritability and its behavior towards the mouse.

This leads us to the proposition that a rat's life-history is of preponderant importance in determining his behavior when presented with a mouse, as it is through the sum of his past experiences that he has learned to adapt behaviorally to the various emotion-provoking situations. The killing response is a part of the rat's behavioral repertoire; considering that the neural circuits underlying the performance of this "action pattern" are a part of the constitutional make-up of the organism, we may say that every rat is a "potential mouse-killer." But whether this part of the behavioral repertoire will be used (and then eventually be self-strengthened) or not, depends to a large extent upon the kind of behavioral-emotional adaptations shaped by previous experiences with situations somehow relevant to the mouse-presentation situation ("relevant" as regards the organization of the rat's brain and behavior, but not necessarily the organization of the experimenter's mind). Particularly relevant in this respect are, of course, early social contacts with mice, and there is clear evidence that early interaction between the two species actually reduces greatly the incidence of the killing

response in the adult rat (Denenberg et al., 1968; Myer, 1969).
On the other hand, the incidence of the killing response can be
increased in a group of rats by repeatedly exposing the animals
to both food deprivation and competition for food (Heimstra, 1965).

NEUROPHYSIOLOGY OF THE RAT'S MOUSE-KILLING BEHAVIOR

As stated previously, the amygdala is a part of a functional
entity, just as the killing response is a part of a behavioral
entity. Thus, it matters that the main neurophysiological data
concerning the rat's mouse-killing behavior be presented before
we concentrate upon the role played by the amygdala.

It is tempting to summarize the results obtained by saying
schematically that, in the adult rat, the release of the killing
response is under the control of two antagonistic systems which
also are responsible for the two fundamental and opposed tendencies shown by the organism confronted with the various stimulations and situations arising in its environment: (1) A system comprising lateral hypothalamic and ventro-medial tegmental structures; its predominant activation has arousing and rewarding or
"appetitive" effects which find their expression in a general
tendency to "approach," to "move toward" the stimuli; (2) A
system comprising peri-ventricular structures in both diencephalon and mesencephalon; its predominant activation has
"aversive" effects which find their expression in a general
tendency to "avoid," to "move away" from the stimuli arising in
the environment.

Lateral hypothalamus. Bilateral lesions placed within the
lateral hypothalamus and involving the posterior part of the
lateral hypothalamic area entail a long-lasting suppression of
the killing response. If the lesioned animal recovers oriented
behavioral activities, the recovery of the killing response invariably precedes by some days if not by a few weeks the recovery
of the feeding behavior; this means that a rat may kill mice even
though he still happens to be in a state of complete adipsia and
aphagia, never eating anything of the mice it kills (Karli and
Vergnes, 1964).

Conversely, electrical stimulation of various sites located
in the posterior two-thirds of the lateral hypothalamus (mostly
in the region of the medial forebrain bundle) elicits a clear
facilitation of the mouse-killing response in spontaneous killer-
rats (Vergnes and Karli, 1969b);

(a) The stimulation provokes an immediate killing response in
rats in which the release of the attack usually occurs with a
more or less prolonged delay;

(b) When the interspecific aggressiveness is abolished
transiently following limbic stimulation, its recovery clearly can
be speeded up by a lateral hypothalamic stimulation.

It may be added that when a mouse-killer is presented with a
mouse, its hippocampal slow wave activity becomes more regular and
increases in both amplitude and frequency, a bioelectrical change
that is also brought about by the facilitating lateral hypothalamic stimulation (Vergnes and Karli, 1968).

Elicitation of the killing response in natural non-killers
is obtained more easily with a chemical activation of cholinergic
synapses within the lateral hypothalamic area (Bandler, 1970;
Smith et al., 1970) than with an electrical stimulation affecting
the same area. The latter stimulation may have less selective
effects and may well, in the natural non-killer, involve inhibitory fibers converging upon the lateral hypothalamic structures.

Ventro-medial tegmentum. As regards the ventro-medial mesencephalic tegmentum, a few casual observations have shown that
extensive lesions destroying the ventro-medial region of the
thalamo-mesencephalic junction resulted in a complete loss of
oriented behavior, the mouse-killing response being the first
complex behavior to be recovered eventually (Karli, 1960b). In
more recent experiments, still in progress, the following preliminary results were obtained:

(a) More restricted lesions involving the ventral tip of the
central grey and extending through the ventro-medial tegmentum
close to the interpeduncular nucleus, usually suppress both
spontaneous eating and the mouse-killing response. But in some
rats the lesion abolishes the interspecific aggressive behavior
without affecting grossly the animals' eating behavior.

(b) Bilateral lesions limited to the ventral tegmental area
of Tsai suppress spontaneous eating without affecting the mouse-
killing behavior.

(c) Electrical stimulation of a number of sites within the
ventro-medial mesencephalic tegmentum facilitates the killing
response in the killer rat or even elicits such a response in a
natural non-killer. A similar facilitation of the rat's killing
behavior was observed recently by Bandler (1971a) in consequence
of chemical (carbachol as well as norepinephrine) stimulation at
a number of sites in the ventral midbrain tegmentum.

Periventricular structures. If we now turn to some periventricular structures, it appears that they effectively exert
a predominantly suppressant influence upon the release of the

killing response; the bilateral destruction of one or the other
of these structures provokes a more or less pronounced facilitation of the mouse-killing behavior: (1) Bilateral lesions destroying the ventromedial hypothalamic nuclei induce emotional
hyperreactivity in most natural non-killers, but induce mouse
killing behavior in only about 30 per cent of the lesioned
animals. The increased responsiveness is not sufficient to
provoke by itself the release of interspecific aggressive behavior. On the other hand, there is no close correlation between
two effects of the hypothalamic lesion: interspecific aggressiveness and hyperphagia (Eclancher and Karli, 1971); (2) A bilateral
destruction of the dorsomedial thalamic nuclei also can provoke
transient or long-lasting appearance of the mouse-killing behavior in natural non-killers (Eclancher and Karli, 1968, 1969).
This behavioral change does not result from the transection of
the epithalamic circuit that usually goes with the dorsomedial
thalamic lesion: bilateral lesions limited to the medial habenular nucleus or to the stria medullaris do not induce mouse-
killing behavior (Eclancher and Karli, 1968, 1969). It may be
added that local carbachol stimulation of points in medial and
midline thalamic nuclei was shown recently by Bandler (1971b) to
facilitate the rat's natural killing behavior. (3) Total destruction of the mesencephalic central grey entails a clear facilitation of the interspecific aggressive behavior as well as a
moderate but significant hyperphagia (Chaurand and Karli, 1970).
The lesion induces an immediate release of the killing response
in most rats that before the operation killed with a more or less
prolonged delay. Only in a small percentage of natural non-
killers does the lesion provoke a lasting aggressiveness toward
mice.

    Interactions between facilitating and suppressant systems.
There are most probably multiple and complex interactions at both
the diencephalic and the mesencephalic levels between the two
systems which are assumed to underlie behavioral facilitation and
behavioral suppression.

    Recently we have made the rather unexpected observation that
an electrical stimulation of the mesencephalic reticular formation invariably provokes an immediate arrest of the killing
response, whether the latter is shown spontaneously or induced
by lateral hypothalamic stimulation (Chaurand and Karli, 1971).
Since the reticular stimulation entails, in most rats, a peculiar
general tendency to avoid or to "retreat" from any kind of
somesthetic stimulation, we are inclined to think that the observed suppressant effect results from the predominant activation
of inhibitory fibers that may radiate into the ventral and lateral
tegmentum, not only from the central grey, but also from the
medial tip of the cerebral peduncle which seems to convey fibers

of an avoidance system in the rat (Stokes and Thompson, 1970).

It is to be assumed that even more complex interactions between the systems controlling purposive behavior underlie the
apparent initial, paradoxical fact (if we remind ourselves that
the interspecific aggressive behavior also can be provoked by
medial hypothalamic stimulation (Vergnes and Karli, 1970a).   The
aggressive responses induced from medial hypothalamic stimulation
sites appear to be quite different, in some respects, from those
that are released spontaneously or provoked by stimulation of
the posterior part of the lateral hypothalamic area: contrary
to the latter responses, the former ones are oriented poorly, and
they are intermingled invariably with flight reactions and an
intense emotional display.   They probably should be considered
as actual "defence" responses provoked by an aversive experience
to which the animals try to put an end.

Inhibitory role of olfactory input.   The rat being a
macrosmat, the progressive shaping of his relations with the environment is based largely upon sensory information of olfactory
nature.   Thus, it matters to recall the fact that an ablation of
the olfactory bulbs induces the mouse-killing behavior in an
important proportion of spontaneously non-killing rats (Vergnes
and Karli, 1963b; Karli et al., 1969).   The lesioned animals
must be isolated in individual cages if the behavioral change
is to appear, regardless of the age at which the ablation of the
olfactory bulbs is performed (Vergnes and Karli, 1969a).   If the
animals are kept together, the social stimuli allow a compensatory use of sensory input other than olfactory.   But the transient or lasting character of this compensatory mechanism depends upon the age at which the lesioned animals are exposed to
social stimuli:

(a) In adult animals, the inhibition of the killing response
entailed by the exposure to social stimulation is but a transient
one: the aggressive behavior appears progressively in these
animals, once they have been isolated for a few weeks after having
been kept together for two months following the operation.

(b) If, on the contrary, animals are lesioned at an early
age (4 or 7 weeks) and then kept together until they are of adult
age, the inhibition of the aggressiveness thus produced is a
very stable one: in those adult lesioned animals, isolation
never induces the killing response.

ROLE OF THE AMYGDALA

We can now examine the role of the amygdala within the outlined framework, reporting and discussing a number of results

obtained in lesion as well as in stimulation experiments.

In an early experiment, 14 rats (12 wild and 2 domesticated) out of a group of 16 animals stopped killing mice following extensive bilateral amygdaloid lesions. The wild rats that stopped killing, at the same time, lost most of their savageness and could be handled safely with bare hands; they showed a marked decrease in their responsiveness to any kind of emotion-provoking stimulus (Karli, 1956). Amygdaloid lesions abolish just as well the cat's interspecific aggressiveness (Summers and Kaelber, 1962; Cherkes, 1967).

Placing less extensive lesions in various parts of the amygdala (lateral, cortico-basal and centro-medial lesions) in a series of domesticated mouse-killers, it could be shown that involvement of the central nucleus was the crucial factor in determining the effectiveness of the lesion: 6 rats bearing a complete bilateral destruction of the central nucleus (the lesions encroaching more or less upon the medial nucleus) never exhibited again any interspecific aggressiveness during the 3 months of postoperative testing; the 18 other animals bearing centro-medial lesions recovered the killing response with delays ranging from 1 to 7 weeks, the delay of recovery being correlated grossly with the extent to which the central nucleus had been destroyed. Lesions sparing the centro-medial region and involving one or more of the cortical, basal and lateral nuclei had little or no effect upon the rat's mouse-killing behavior (Karli and Vergnes, 1965). It may be added that direct bilateral injections of imipramine or thiazesim (two antidepressant drugs) into the centromedial region of amygdala produce immediate inhibition of mouse-killing which lasts 1 to 2 hours (Horovitz and Leaf, 1967).

FACILITATION OF MOUSE-KILLING BY CENTROMEDIAL AMYGDALA

If we conclude from these experimental data that the central region of the amygdaloid nuclear complex takes part in a mechanism which exerts a facilitating influence upon the killing response, the question then arises as to which amygdalo—fugal pathway is involved predominantly in this behavioral facilitation. A bilateral transection of the ventral amygdalo-fugal fibers passing through the ansa lenticularis abolishes the mouse-killing response, whereas a bilateral interruption of the stria terminalis leaves the killer-rat's behavior unchanged (Vergnes and Karli, 1964).

Having established the fact that amygdaloid lesions reduce clearly the cat's spontaneous aggressiveness but hardly modify the "savage" behavior provoked by ventromedial hypothalamic lesions, Kling and Hutt (1958) were led to the conclusion that the amygdala acts mostly through an inhibitory influence which

it exerts upon the ventromedial hypothalamus. This observation,
that a bilateral transection of the stria terminalis, i.e. the
amygdalo-fugal pathway that distributes fibers to the ventromedial hypothalamic nucleus as dominant recipient in the rat
(Heimer and Nauta, 1969), does not modify the killer-rat's
behavior, already leads to the assumption that Kling and Hutt's
conclusion may not hold true for the control of the rat's mouse-
killing behavior. This assumption was confirmed recently: amygdaloid lesions suppress the killing behavior entailed by medial
hypothalamic lesions just as they suppress the killer-rat's
spontaneous interspecific aggressiveness (Vergnes and Karli,
1970b). Taken together, the experimental data indicate that the
centromedial amygdala* exerts its facilitating influence essentially via the diffuse ventral amygdalo-fugal fiber system. One
could imagine these fibers to act mostly in a direct way upon
ventral tegmental structures without any lateral hypothalamic
relay; but the fact that cholinergic stimulation of lateral hypothalamic sites elicits the killing response in natural non-killers
(Bandler, 1970; Smith et al., 1970) speaks strongly against such
an hypothesis.

A further question concerns the degree of specificity of the
behavioral facilitation in which the centromedial amygdala is
taking part. The following facts are relevant to this question:

(a) When placing extensive amygdaloid lesions in the
earlier experiments, it appeared that the postoperative behavior
toward mice was the same in the rats that showed changes in their
eating habits and in the rats that did not exhibit any such change.
Furthermore, the behavior of an operated animal toward mice did
not change when, after having been tube fed, it went back to the
natural way of feeding itself (Karli, 1956). On the other hand,
the more limited centromedial lesions that entailed a lasting
suppression of the killing response obviously did not affect the
animals' eating behavior (Karli and Vergnes, 1965). These observations merely confirm the fact which also comes out of the observations made following hypothalamic or tegmental lesions, namely,
the differential effects of such lesions upon eating behavior and
mouse-killing behavior.

(b) More significant for the problem under consideration
seems to be the fact that we did not succeed in eliciting a
killing response in amygdaloid-lesioned killer-rats with lateral
hypothalamic stimulations that otherwise facilitate clearly the

---

* In this study, the terms "centromedial amygdala" and "centromedial region" refer to a portion of the amygdala which includes
the central nucleus, the dorsal part of the medial nucleus, and
the medial part of the basal nucleus.

mouse-killing behavior (Vergnes and Karli, 1969b). This may mean that the contribution made by the centromedial amygdala to the facilitation of that kind of behavior is a rather specific one. It also may mean that the facilitating mechanism is not a simply descending one, but implies a more complex interplay of the centromedial amygdala with the facilitating diencephalic and mesencephalic structures.

AMYGDALA AND INHIBITORY CONTROL OF MOUSE-KILLING

Considering that, in various instances, the amygdala appears to be involved in processes that ultimately result in behavioral suppression, one is inclined to think that the olfactory input which is the main source of inhibition in the natural non-killer may well act through the amygdala. Before going into more detailed considerations about the role possibly played by the amygdala in this respect, it is to be stressed that, due to the wide direct and indirect distribution of the fibers efferent from the olfactory bulbs (Powell et al., 1965; White, 1965), the existence of such a mechanism involving the amygdala would by no means exclude the possible existence of other mechanisms in which the amygdala would not take part.

As regards the role possibly played by the amygdala in the inhibitory control of mouse-killing behavior, it may be built upon a functional as well as topographical differentiation between the facilitating centromedial region and an inhibitory region within the amygdala, i.e., the kind of differentiation shown to exist in the cat, according to behavioral (Ursin and Kaada, 1960; Egger and Flynn, 1963, 1967) as well as electrophysiological data (Egger, 1967; Dreifuss et al., 1968). If such a differentiation were shown to exist with respect to the rat's mouse-killing behavior, we would then have to make a reasoned choice between the following two hypotheses: (1) The inhibitory region of the amygdala may act by exerting a suppressant effect upon the centromedial facilitating region or (2) it may act in a more indirect way, possibly through the mediation of the stria terminalis and the inhibitory medial hypothalamic structures.

When trying to elucidate the pathways through which the inhibitory effect of the olfactory input is being mediated, we observed that in order to elicit the mouse-killing behavior in natural non-killers, both the lateral olfactory tract and the anterior commissure had to be transected on both sides. To be effective, the transection of the lateral olfactory tract had to be placed in front of and not behind the prepiriform cortex (Vergnes and Karli, 1965). Furthermore, extensive bilateral lesions of the latter did also elicit the killing response, and it was concluded that the prepiriform cortex probably acts as an

important relay station on the inhibitory pathway. From there,
fibers join the longitudinal association bundle and are distributed to the basal and lateral amygdaloid nuclei (Cowan et al.,
1965), which project in part to the ventromedial hypothalamus via
the stria terminalis. One could imagine the behavioral inhibition
of olfactory origin to be mediated predominantly by such a pathway.
But this hypothesis is to be ruled out on the basis of the following experimental data:

(a) A bilateral destruction of the lateral region of the
amygdala does not induce the killing response in rats which have
never killed mice prior to the experiment (Vergnes and Karli,
1965).

(b) A bilateral interruption of the stria terminalis does
not modify the natural non-killer's behavior (Didiergeorges et al.,
1968); nor does it modify the inhibitory effects of amygdaloid
stimulation in the killer-rat (Vergnes and Karli, 1969b).

(c) The ventromedial hypothalamic nucleus cannot be considered to act as a simple relay station within the chain of
mechanisms through which the olfactory input ultimately results
in a "non-release" of the killing response. Not only are the
ventromedial hypothalamic lesions much less often effective than
the olfactory bulb lesions, but they sometimes have a transient
effect, whereas a removal of the olfactory bulbs invariably
induces a lasting killing behavior. Furthermore, the respective
effects of the two lesions can be additive: the hypothalamic
lesions elicit the killing behavior in a high proportion of
natural non-killers whose behavior has remained unchanged following a previous removal of the olfactory bulbs (Eclancher and
Karli, 1971).

We are rather inclined to think that impulses of olfactory
origin may exert a suppressant action upon the centromedial facilitating region of the amygdala (probably without any clear topographical differentiation of an inhibitory amygdaloid region),
for the following reasons:

(a) The effects of centromedial amygdaloid lesions are the
same, whether they are carried out in "spontaneous" killers or in
rats induced to kill mice in consequence of a removal of the olfactory bulbs (Karli and Vergnes, 1965).

(b) As indicated above, imipramine was shown to produce
immediate inhibition of mouse-killing when injected directly into
the centromedial region of the amygdala (Horovitz and Leaf, 1967).
When given in systemic injections, the dose of imipramine required
for a transient suppression of the killing response appears to be

higher in animals with olfactory bulb lesions than in "spontaneous" killers (Didiergeorges et al., 1968). This is what we expect to find, if we assume that the removal of the olfactory bulbs releases largely the centromedial region of the amygdala from inhibition, whereas more inhibition still exists in the "spontaneous" killer, only being insufficient to suppress effectively the contribution made by this amygdaloid region to facilitating mechanisms.

(c) Noradrenergic synapses, which may be involved in the inhibition of the killing response by the olfactory input, appear to exist in the centromedial amygdala: norepinephrine administered locally produces some inhibition of killing in most killer-rats tested (Leaf et al., 1969).

(d) In a recent preliminary experiment, we found that the proportion of adult killers was higher in animals that had sustained centromedial amygdaloid lesions at weaning than in non-operated control animals. This rather paradoxical finding may be given the following tentative interpretation: when carried out at an early age, the centromedial amygdaloid lesions may interfere greatly with the development of behavioral inhibition based on olfactory information; and the absence of any possible inhibitory modulation of the activity of the centro-medial region may well have more pronounced effects upon the animal's adult behavior than the absence of the facilitating contribution itself which is made normally by this region.

EFFECTS OF ELECTRICAL STIMULATION OF AMYGDALA

The results obtained in the stimulation experiments will not be dealt with fully as we feel unable to draw from them any clear-cut conclusion which would throw some extra light on the problems under discussion. At first it should be indicated that we never did succeed in inducing any obvious facilitation of the mouse-killing behavior, even at sites and with stimulation parameters that otherwise reinforced self-stimulation in the absence of any (at least propagated) seizure activity. This negative finding really is not surprising, if one considers that the electrical stimulation may well activate inhibitory fibers converging upon the facilitating centromedial region, and that such a stimulation anyhow is much more likely to interfere with than to produce the kind of patterned activation which probably is involved in the natural process.

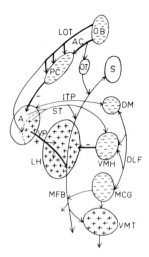

Schematic drawing showing central nervous structures predominantly
involved in the facilitation (+++) or in the suppression (---) of
the rat's mouse-killing behavior. (A: amygdala; AC: anterior
commissure; DLF: dorsal longitudinal fasciculus; DM: dorsomedial
nucleus of the thalamus; ITP: inferior thalamic peduncle;   LH:
lateral hypothalamic area; LOT: lateral olfactory tract;   MCG:
mesencephalic central grey; MFB: medial forebrain bundle; OB: olfactory bulb; OT: olfactory tubercle;   PC: prepiriform cortex;
S: septum; ST: stria terminalis; VMH: ventromedial hypothalamus;
VMT: ventromedial midbrain tegmentum; VP: ventral amygdalo-fugal
pathway).

     Every time an electrical stimulation of the amygdala proved
to be effective, it resulted in an immediate inhibition of the
mouse-killing behavior, the killer rat resuming its aggressive
behavior almost immediately when the stimulation was discontinued.
In the early experiments, which were carried out either without
recording any bioelectrical activity or with recording the
activity of the contralateral amygdala, we got the impression
that inhibition of the killing response could be obtained in the
absence of any paroxystic bioelectrical activity.   In fact, when
recording the activity of the lateral hypothalamus on the stimulated side, it appeared that only epileptogenic stimulation of
the amygdala ever interfered with the release of the killing
response (Vergnes and Karli, 1969b).   The use of various stimulation parameters as well as the use of anticonvulsant drugs (diphenylhydantoin; dipropylacetate) did not permit dissociation of
the behavioral effect from the epileptogenic effect of the amygdaloid stimulation.   It also appeared that stimulation of the
dorsal hippocampus was effective in blocking the killing behavior

only if it was epileptogenic and that the inhibition of the killing response occurred only if and when the seizure activity was
propagated to the amygdala and to the lateral hypothalamus. Under
these circumstances, it is difficult to decide whether the behavioral suppression is due to a diffuse activation of inhibitory
fibers and synapses within the amygdala or to an interference of
neuronal synchronization with the normal functioning of facilitating amygdaloid, hypothalamic and tegmental neurons.

NEUROCHEMICAL DATA

The probable existence of a suppressant action exerted by
the olfactory input (directly or through inhibitory interneurons?)
upon the facilitating centromedial amygdala, raises the question
of the chemical nature of the transmitter(s) involved. As already
indicated, norepinephrine administered locally produces some inhibition of killing in most killer-rats tested; even more effective is a local injection of D-methamphetamine which probably acts
by releasing endogenous catecholamines. Conversely, catecholamine
depletion produced by a systemic administration of α-methyl-tyrosine
induces the killing response, at least in some of the injected non-
killing rats (Leaf et al., 1969). But the partial and variable
effects obtained when activating or blocking adrenergic transmission within the amygdala suggest that mechanisms other than adrenergic may mediate inhibition of the mouse-killing behavior.
As a matter of fact, serotonin depletion produced by a systemic
administration of parachlorophenylalanine facilitates the rat's
killing behavior (Sheard, 1969; Karli et al., 1969) just as does
catecholamine depletion; conversely, a systemic administration
of 5-hydroxy-tryptophan, a precursor for endogenous biosynthesis
of serotonin, induces a transient suppression of the killing response in about 50 per cent of the treated animals (Kulkarni,
1968). The evidence concerning a possible serotoniergic mechanism
involved in a suppressant action exerted by the olfactory input
upon centromedial amygdaloid neurons is but a fragmentary and
indirect one. On the one hand, serotonin has been shown to have a
depressant action upon the activity of amygdaloid neurons
(Straughan and Legge, 1965; Eidelberg et al., 1967). On the other
hand, it was found that a removal of the olfactory bulbs reduced
greatly the serotonin content of the amygdala, the decrease being
significantly less important in animals whose behavior remained
unchanged than in those which started killing mice in consequence
of the olfactory bulb lesion (Viret, 1967 - cf. Karli et al., 1969).

POSSIBLE COMPLEXITY OF AMYGDALOID CONTROL

If we are inclined to think that the mechanisms just discussed are of major importance in the control of the mouse-killing
behavior and especially in the suppression of this behavior in
the natural non-killer, this by no means implies that other

mechanisms possibly involving the amygdala might not contribute
to such a control. For instance, even though it appears that the
contribution made by fibers of the stria terminalis to the
amygdaloid control over the killing response can only be of
minor importance, this contribution may well exist: impulses
traveling along the stria terminalis have been shown to inhibit
ventromedial neurons (Dreifuss et al., 1968) as well as to
mediate an inhibitory effect of the amygdala on lateral hypothalamic neurons (Oomura et al., 1970). Another point deserves
attention; namely, the possible interactions between the amygdala
and the dorso-medial nucleus of the thalamus. It is conceivable
that such interactions may be involved in the inhibitory effects
exerted on the killing response by sensory input other than
olfactory; i.e., the effects which are of great importance in
the compensatory reinstallment of an effective inhibition (a
lasting one in the young animal, a but transient one in the adult
animal) in rats deprived of their olfactory bulbs. Amygdaloid-
thalamic interactions also may be involved in a facilitation of
the killing response resulting from sensory (mainly somesthesic)
feedback, once the attack has been initiated.

It is always frustrating to realize that one holds in hand
no more than a scrap of truth, but we must keep in mind that even
in a species whose behavior is not as rich and shaded as it is
in higher mammals, the role played by the amygdala in the control
of social behavior hardly can be a single and simple one. As a
matter of fact, the many behavioral processes in which the amygdala has been shown to be involved probably are underlain in many
instances by one and the same common mechanism; still, when they
are all taken together, the behavioral as well as the neurophysiological and neurochemical data rather point to a variety of
underlying mechanisms. To give just one concrete example: on a
purely behavioral level, one possibly may conceive the inhibition
of the killing response and other forms of behavioral suppression
resulting from punishment or from satiation-producing reafferents,
to be underlain by an important common mechanism. But if we now
turn to some relevant data concerning the nature of this underlying mechanism, how can we easily and simply correlate the few
following facts and hypotheses?

(a) As outlined previously, norepinephrine as well as
serotonin possibly act as transmitters in the inhibition of the
mouse-killing behavior. In this instance, norepinephrine is
supposed to inhibit (centromedial) amygdaloid neurons which are
part of a behavior-facilitating system.

(b) Based on the fact that both amygdaloid lesions and local
application of norepinephrine induce passive avoidance deficits,
i.e. a marked increase in the occurrence of previously suppressed
behavior, Margules and Stein (cf. Stein, 1969) suggest that

norepinephrine released as a consequence of the activation of an
ascending component of the reward system may inhibit amygdaloid
neurons, thereby decreasing the suppressant effect of punishment.
In this instance, norepinephrine is supposed to inhibit amygdaloid
neurons which are part of a behavior-suppressant system. It must
be added that cholinergic stimulation of the amygdala also has
been shown to interfere with the suppressant effect of punishment,
as it produced severe deficits in passive avoidance and CER learning (Goddard, 1969).

(c) During eating behavior, the rewarding properties of the
various sensory reafferents have two opposed effects each having
its own time course: one which immediately supports the consumma-
tory behavior, another one which results progressively in satiety.
The data obtained by Grossman (1964) suggest that these two
effects can be mimicked with adrenergic and cholinergic stimulations of the amygdala, respectively.

CONTRIBUTION OF AMYGDALA TO EMOTIONAL AND SOCIAL RESPONSIVENESS

If one brings together the bulk of the experimental evidence
concerning the closely interacting structures of the limbic system
(cf. Karli, 1968), it appears that the amygdala contributes to,
at least, two interrelated aspects of the organism's emotional and
social responsiveness, namely, a rather general one and a more
specific one. On the one hand, the amygdala seems to be an
important link in the chain of mechanisms through which intra-
central as well as peripheral excitatory feedback amplifies and
prolongs an emotional response, and its possible consequences upon
social behavior. The fact is that extensive amygdaloid lesions
invariably induce a general leveling down of the emotional responsiveness as well as a general reduction in the amount of social
interactions exhibited. A study like the one recently carried
out by Bunnell et al. (1970) on the golden hamster demonstrates
clearly that the reduction in the amount of social behavior concerns submissive as well as aggressive behavior.

On the other hand, the amygdala seems to take an important
part in the mechanisms through which an affective significance
is conferred upon the cognitive elements of a given situation in
the laying down of more specific traces of recent experiences, and
through which these traces will later on be drawn upon so as to
adapt present behavior to the relevant aspects of the previous
life-history. A transient functional alteration of the amygdala
and related limbic structures by an experimentally produced seizure
activity interferes with some of the mechanisms through which a
previous aversive experience ultimately results in the suppression
of a given behavior (Kesner and Doty, 1968; Lidsky and Slotnick,
1970; Levine et al., 1970). Recent studies, carried out on

monkeys, have shown that if one varies the experimental situations amygdaloid lesions appear to have not only the well-known general effects, but effects that are determined more situationally (Plotnik, 1968; Thompson et al., 1969; Kling et al., 1970). The assumption that animals bearing amygdaloid lesions may be deficient in laying down the traces of recent experiences and/or in modulating the "impellance" of the present sensory information by referring the latter to the laid down traces of past experiences could well explain both the fact that the monkey does not much react to the "novelty" or to the "strangeness" of a stimulus, and the fact that it shows a tendency to be rather fearful and to withdraw from social contacts, probably being unable to adapt rapidly and continuously to the social signals arising from normal peers.

As outlined above, bilateral amygdaloid lesions abolish the rat's mouse-killing behavior. This is due to the fact that the amygdala contributes, in an essential way, to the mechanisms that facilitate this kind of behavior, its suppression being effected also essentially through the mediation of the facilitating centromedial amygdala. The situation is quite different as regards other kinds of behavior, namely eating behavior or sexual behavior. In the latter instances, a basic facilitation results from the action of humoral factors upon more or less widely distributed sensitive neurons, and this basic facilitation arises just as well in the absence of the amygdala. It then may be the main role of the limbic structures to confer individual characteristics upon a rather automatic and stereotyped behavior. If we assume that the amygdala takes an essential part in the mechanisms through which the life-history shapes progressively a selective, differentiated and adapted behavior proceeding from the basic "reflexive" kind of behavior, we better understand why amygdaloid lesions can induce hyperphagia or "hypersexuality," instead of suppressing one or the other kind of behavior. We also realize that the behavior change may not be referred to properly when using the term of "hypersexuality," as this change actually reflects a qualitative deficiency in adaptation and selectivity rather than a quantitative rise of the drive-level.

Despite the growing body of experimental data, our present knowledge of the role played by the amygdala and the interrelated limbic structures in the control of social behavior is still a most fragmentary one, and much further research is badly needed if we are to understand in more complete and precise terms the biological foundations of an organism's emotional and social responsiveness. As the limbic system is thought to give a "historical" dimension to social behavior by adapting continuously to the sum of past experiences, it is not enough to carry out studies on adult animals raised and observed under standard environmental conditions. It is just as important to study the factors that control the ontogenetic development of the emotional and social responsiveness, as well as the mechanisms through which this ontogenetic development is being modulated by the life-history. It is in this direction that we engage at present some of our research on the rat's mouse-killing behavior.

ACKNOWLEDGMENTS

This study has been supported since 1966 by grants from the Institut National de la Santé et de la Recherche Medicale (CR-66-030) and the Direction des Recherches et Moyens d'Essais (111/66, 226/67, 431/68 and 70/391).

REFERENCES

AHMAD, S. S., & HARVEY, J. A. Long-term effects of septal lesions and social experience on shock-elicited fighting in rats. Journal of Comparative and Physiological Psychology, 1968, 66, 596.

ALLIKMETS, L. K., & DITRIKH. Effects of lesions of limbic system on emotional reactions and conditioned reflexes in rats. Federation Proceedings, 1965, 24, 1003.

BANDLER, R. J. Cholinergic synapses in the lateral hypothalamus for the control of predatory aggression in the rat. Brain Research, 1970, 20, 409.

BANDLER, R. J. Chemical stimulation of the midbrain and aggressive behaviour. Nature New Biology, 1971a, 229, 222.

BANDLER, R. J. Direct chemical stimulation of the thalamus: effects on aggressive behavior in rat. Brain Research, 1971b, 26, 81.

BEEMAN, E. A. The effect of male hormone on aggressive behavior of mice. Physiological Zoology, 1947, 20, 373.

BERNSTEIN, H., & MOYER, K. E.   Aggressive behavior in the rat:
   effects of isolation and olfactory bulb lesions.
   Brain Research, 1970, 20, 75.

BUNNELL, B. N., SODETZ, F. J., & SHALLOWAY, D. I.   Amygdaloid
   lesions and social behavior in the golden hamster.
   Physiology and Behavior, 1970, 5, 153.

CHAURAND, J. P., & KARLI, P.   Effets de lesions du gris central
   du mesencephale sur le comportement d'agression interspecifique et le comportement alimentaire du Rat.
   Comptes Rendus des Seances de la Société de Biologie,
   1970, in press.

CHAURAND, J. P., & KARLI, P.   Stimulation electrique de la
   formation reticulaire du mesencephale et comportement
   d'agression interspecifique du Rat.   Comptes Rendus
   des Seances de la Société de Biologie, 1971, in press.

CHERKES, V. A.   Instinctive and conditioned responses in cats
   with removed amygdala.   Zhurnal Vysshei Nervnoi Deiatel
   Nosti Imenti I. P. Pavlova, 1967, 17, 70.

CONNER, R. L., STOLK, J. M., BARCHAS, J. D., DEMENT, W. C.,
   & LEVINE, S.   The effect of parachlorophenylalanine (PCPA)
   on shock-induced fighting behavior in rats.   Physiology
   and Behavior, 1970, 5, 1221.

COWAN, W. M., RAISMAN, G., & POWELL, T. P. S.   The connexions
   of the amygdala.   Journal of Neurology, Neurosurgery
   and Psychiatry, 1965, 28, 137.

DENENBERG, V. H., HUDGENS, G. A., & ZARROW, M. X.   Mice reared
   with rats: modification of behavior by early experience
   with another species.   Science, 1964, 143, 380.

DENENBERG, V. H., POSCHKE, R. E., & ZARROW, M. X.   Killing of
   mice by rats prevented by early interaction between the
   two species.   Psychonomic Science, 1968, 11, 39.

DIDIERGEORGES, F., VERGNES, M., & KARLI, P.   Sur le mode d'action
   d'une influence inhibitrice d'origine olfactive s'exercant
   sur l'agressivite interspecifique du Rat.   Comptes Rendus
   des Seances de la Societe de Biologie, 1968, 162,
   267.

DREIFUSS, J. J., MURPHY, J. F., & GLOOR, P. Contrasting effects of two identified amygdaloid efferent pathways on single hypothalamic neurons. Journal of Neurophysiology, 1968, 31, 237.

DOUGLAS, R. J., ISAACSON, R. L., & MOSS, R. L. Olfactory lesions, emotionality and activity. Physiology and Behavior, 1969, 4, 379.

ECLANCHER, F., & KARLI, P. Lésion du noyau dorso-médian du thalamus et comportement d'agression interspécifique Rat-Souris. Comptes Rendus des Seances de la Société de Biologie, 1968, 162, 2273.

ECLANCHER, F., & KARLI, P. Comportement d'agression inter-spécifique Rat-Souris: effets de lésions du noyau dorso-médian du thalamus et des structures épithalamiques. Journal de Physiologie (Paris), 1969, 61, 283 (F).

ECLANCHER, F., & KARLI, P. Comportement d'agression inter-spécifique et comportement alimentaire du Rat: effects de lésions des noyaux ventro-médians de l'hypothalamus. Brain Research, 1971, 26, 71.

EGGER, M. D. Responses of hypothalamic neurons to electrical stimulation in the amygdala and the hypothalamus. Electroencephalography and Clinical Neurophysiology, 1967, 23, 6.

EGGER, M. D., & FLYNN, J. P. Effects of electrical stimulation of the amygdala on hypothalamically elicited attack behavior in cats. Journal of Neurophysiology, 1963, 26, 705.

EGGER, M. D., & FLYNN, J. P. Further studies on the effects of amygdaloid stimulation and ablation on hypothalamically elicited attack behavior in cats. In W. R. Adey and T. Tokizane (Eds.) Progress in Brain Research, 27, Structure and Function of the Limbic System. Amsterdam: Elsevier, 1967. Pp. 165-182.

EIDELBERG, E., GOLDSTEIN, G. P., & DEZA, L. Evidence for serotonin as a possible inhibitory transmitter in some limbic structures. Experimental Brain Research, 1967, 4, 73.

FERGUSON, J., HENRIKSEN, S., COHEN, H., MITCHELL, G., BARCHAS, J., & DEMENT, W. Hypersexuality and behavioral changes in cats caused by administration of p.chlorophenylalanine. Science, 1970, 168, 499.

GODDARD, G. V. Analysis of avoidance conditioning following cholinergic stimulation of amygdala in rats. Journal of Comparative and Physiological Psychology, 1969, 68, 1.

GROSSMAN, S. P. Behavioral effects of chemical stimulation of the ventral amygdala. Journal of Comparative and Physiological Psychology, 1964, 57, 29.

GROSSMAN, S. P. Avoidance behavior and aggression in rats with transections of the lateral connections of the medial or lateral hypothalamus. Physiology and Behavior, 1970, 5, 1103.

HEIMER, L., & NAUTA, W. J. H. The hypothalamic distribution of the stria terminalis in the rat. Brain Research, 1969, 13, 284.

HEIMSTRA, M. W. A further investigation on the development of mouse-killing in rats. Psychonomic Science, 1965, 2, 179.

HOROVITZ, Z. P., & LEAF, R. The effects of direct injections of psychotropic drugs into the amygdala of rats, and its relationship to antidepressant site of action. In H. Brill, J. O. Cole, P. Deniker, H. Hippius, and P. B. Bradley (Eds.), Neuropharmacology, Proceedings of the Vth International Congress of the Collegium Internationale Neuropsychopharmacologicum. Amsterdam: Excerpta Medica, 1967. Pp. 1042.

HUTCHINSON, R. R., ULRICH, R. E., & AZRIN, N. H. Effects of age and related factors on the pain aggression reaction. Journal of Comparative and Physiological Psychology, 1965, 59, 365.

KARLI, P. The Norway Rat's killing response to the white mouse: an experimental analysis. Behaviour, 1956, 10, 81.

KARLI, P. Hormones stéroides et comportement d'agression interspécifique Rat-Souris. Journal de Physiologie (Paris), 1958, 50, 346.

KARLI, P. Septum, hypothalamus postérieur et agressivité interspécifique Rat-Souris. Journal de Physiologie (Paris), 1960a, 52, 135.

KARLI, P. Effets de lésions expérimentales des noyaux mamillaires sur l'agressivité interspécifique Rat-Souris. Comptes Rendus des Seances de la Société de Biologie, 1960b, 154, 1287.

KARLI, P. Système limbique et processus de motivation. Journal de Physiologie (Paris), 1968, 60 (Suppl. 1), 3 (F).

KARLI, P., & VERGNES, M. Dissociation expérimentale du comportement d'agression interspécifique Rat-Souris et du comportement alimentaire. Comptes Rendus des Seances de la Société de Biologie, 1964, 158, 650.

KARLI, P., & VERGNES, M. Rôle des différentes composantes du complexe nucléaire amygdalien dans la facilitation de l'agressivité interspécifique du Rat. Comptes Rendus des Seances de la Société de Biologie, 1965, 159, 754.

KARLI, P., VERGNES, M., & DIDIERGEORGES, F. Rat-Mouse interspecific aggressive behaviour and its manipulation by brain ablation and brain stimulation. In S. Garattini and E. B. Sigg (Eds.), Aggressive Behaviour. Amsterdam: Excerpta Medica Foundation, 1969. Pp. 47-55.

KESNER, R. P., & DOTY, R. W. Amnesia produced in cats by local seizure activity initiated from the amygdala. Experimental Neurology, 1968, 21, 58.

KLING, A., & HUTT, P. J. Effect of hypothalamic lesions on the amygdala syndrome in the cat. Archives of Neurology and Psychiatry, 1958, 79, 511.

KLING, A., LANCASTER, J., & BENITONE, J. Amygdalectomy in the freeranging vervet (Cercopithecus aetiops). Journal of Psychological Research, 1970, 7, 191.

KULKARNI, A. S. Muricidal block of 5-hydroxytryptophan and various drugs. Life Sciences, 1968, 7 (Part I), 125.

LEAF, R. C., LERNER, L., & HOROVITZ, J. P. The role of the amygdala in the pharmacological and endocrinological manipulation of aggression. In S. Garattini and E. B. Sigg (Eds.), Aggressive Behaviour. Amsterdam: Excerpta Medica Foundation, 1969. Pp. 120-131.

LEVINE, M. S., GOLDRICH, S. G., POND, F. J., LIVESEY, P., & SCHWARTZBAUM, J. S. Retrograde amnestic effects of inferotemporal and amygdaloid seizures upon conditioned suppression of lever-pressing in monkeys. Neuropsychologia, 1970, 8, 431.

LIDSKY, A., & SLOTNICK, B. M. Electrical stimulation of the hippocampus and electroconvulsive shock produce similar amnestic effects in mice. Neuropsychologia, 1970, 8, 363.

MOYER, K. E. Kinds of aggression and their physiological basis. Behavioral Biology, 1968, 2, 65.

MYER, J. S. Prior killing experience and the suppressive effects of punishment on the killing of mice by rats. Animal Behaviour, 1967, 15, 59.

MYER, J. S. Early experience and the development of mouse-killing by rats. Journal of Comparative and Physiological Psychology, 1969, 67, 46.

MYER, J. S., & WHITE, R. T. Aggressive motivation in the rat. Animal Behaviour, 1965, 13, 430.

MYER, J. S., & BAENNINGER, R. Some effects of punishment and stress on mouse-killing by rats. Journal of Comparative and Physiological Psychology, 1966, 62, 292.

O'KELLY, L. I., & STECKLE, L. C. A note on long enduring emotional responses in the rat, Journal of Psychology, 1939, 8, 125.

OOMURA, Y., ONO, T., & OOYAMA, H. Inhibitory action of the amygdala on the lateral hypothalamic area in rats. Nature, 1970, 228, 1108.

PLOTNIK, R. Changes in social behavior of squirrel monkeys after anterior temporal lobectomy. Journal of Comparative and Physiological Psychology, 1968, 66, 369.

POWELL, T. P. S., COWAN, W. M., & RAISMAN, G. The central olfactory connections. Journal of Anatomy, 1965, 99, 791.

ROPARTZ, P. The relation between olfactory stimulation and aggressive behavior in mice. Animal Behaviour, 1968, 16, 97.

SCOTT, J. P. The emotional basis of social behavior. Annals of the New York Academy of Science, 1969, 159, 777.

SEWARD, J. P. Aggressive behavior in the rat. I. General characteristics; age and sex differences. Journal of Comparative Psychology, 1945, 38, 175.

SHEARD, M. H. The effect of p-chlorophenylalanine on behavior in rats: relation to brain serotonin and 5-hydroxyindole-acetic acid. Brain Research, 1969, 15, 524.

SMITH, D. E., KING, M. B., & HOEBEL, B. G. Lateral hypothalamic control of killing: evidence for a cholinoceptive mechanism. Science, 1970, 167, 900.

STEIN, L.   Chemistry of purposive behavior.   In J. T. Tapp (Ed.),
   Reinforcement and Behavior.   New York: Academic Press,
   1969.   Pp. 328-355.

STOKES, L. D., & THOMPSON, R.   Combined damage to the medial
   cerebral peduncle and anterior hypothalamus and escape
   behavior in the rat.   Journal of Comparative and Physiological Psychology, 1970, 71, 303.

STRAUGHAN, D. W., & LEGGE, K. F.   The pharmacology of amygdaloid
   neurons.   Journal of Pharmacy and Pharmacology, 1965,
   17, 675.

SUMMERS, T. B., & KAELBER, W. W.   Amygdalectomy: effects in
   cats and a survey of its present status.   American Journal
   of Physiology, 1962, 203, 1117.

THOMPSON, C. I., SCHWARTZBAUM, J. S., & HARLOW, H. F.   Development of social fear after amygdalectomy in infant rhesus
   monkeys.   Physiology and Behavior, 1969, 4, 249.

ULRICH, R., & SYMANNECK, B.   Pain as a stimulus for aggression.
   In S. Garattini and E. B. Sigg (Ed.), Aggressive Behaviour.
   Amsterdam: Excerpta Medica Foundation, 1969.
   Pp. 59-69.

URSIN, H., & KAADA, B. R.   Functional localization within the
   amygdaloid complex in the cat.   Electroencephalography and
   Clinical Neurophysiology, 1960, 12, 1.

UYENO, E. T., & WHITE, M.   Social isolation and dominance behavior.   Journal of Comparative and Physiological Psychology,
   1967, 63, 157.

VERGNES, M., & KARLI, P.   Déclenchement du comportement d'agression interspécifique Rat-Souris par ablation bilatérale des
   bulbes olfactifs.   Action de l'hydroxyzine sur cette
   agressivité provoquée.   Comptes Rendus des Seances de la
   Société de Biologie, 1963, 157, 1061.

VERGNES, M., & KARLI, P.   Etude des voies nerveuses de l'influence facilitatrice exercée par les noyaux amygdaliens sur
   le comportement d'agression interspécifique Rat-Souris.
   Comptes Rendus des Seances de la Société de Biologie,
   1964, 158, 856.

VERGNES, M., & KARLI, P.   Etude des voies nerveuses d'une
   influence inhibitrice s'exercant sur l'agressivité interspécifique du Rat.   Comptes Rendus des Seances de la Société
   de Biologie, 1965, 159, 972.

VERGNES, M., & KARLI, P. Activité électrique de l'hippocampe et comportement d'agression interspécifique Rat-Souris. Comptes Rendus des Seances de la Société de Biologie, 1968, 162, 555.

VERGNES, M., & KARLI, P. Effets de l'ablation des bulbes olfactifs et de l'isolement sur le développement de l'agressivité interspécifique du Rat. Comptes Rendus des Seances de la Société de Biologie, 1969a, 163, 2704.

VERGNES, M., & KARLI, P. Effets de la stimulation de l'hypothalamus latéral, de l'amygdale et de l'hippocampe sur le comportement d'agression interspécifique Rat-Souris. Physiology and Behavior, 1969b, 4, 889.

VERGNES, M., & KARLI, P. Déclenchement d'un comportement d'agression par stimulation électrique de l'hypothalamus médian chez le Rat. Physiology and Behavior, 1970a, 5, 1427.

VERGNES, M., & KARLI, P. Effets des lésions amygdaliennes sur le comportement d'agression interspécifique provoqué chez le Rat par des lésions hypothalamiques médianes. Comptes Rendus des Seances de la Société de Biologie, 1970b, in press.

WELCH, B. L., & WELCH, A. S. Aggression and the biogenic amine neurohumors. In S. Garattini and E. B. Sigg (Eds.), Aggressive Behaviour. Amsterdam: Excerpta Medica Foundation, 1969. Pp. 188-202.

WHITE, L. E. Olfactory bulb projections of the rat. Anatomical Record, 1965, 152, 465.

LONG TERM ALTERATION FOLLOWING AMYGDALOID STIMULATION

Graham V. Goddard

Department of Psychology, Dalhousie University

Halifax, Canada

In the course of this symposium, both during the formal presentations and during the less formal discussions, a number of people have raised the possibility that the amygdala is somehow involved with learning processes. Thus, Gloor has suggested that the amygdala may function to modify behavior on the basis of experience, Karli has spoken of the amygdala being necessary to add an "historic dimension" to the performance of a motivated act, and Kaada has proposed that hypothalamic fear might parallel innate fear whereas amygdaloid fear is more analogous to learned fear. Murphy has laid strong emphasis on the unreliability of some of the responses observed from the amygdala and has stressed the dangers of studying a labile system with static techniques.

Several lines of evidence demonstrate that when the amygdala is stimulated, either electrically or chemically, long-lasting changes occur somewhere in the brain. These changes may be irreversible. It is not known whether normal physiological activation of the amygdala also causes permanent change. If it does, of course, it might provide part of an explanation for learning, habituation, and long term memory. However, no data directly support such a notion. The available evidence shows that artificial activation of the amygdala results in changes that can be observed in a number of ways. In all cases the effect can be related, either directly or by inference, to seizure activity present at the time of stimulation. The seizure discharge may not be a necessary condition, but change without the involvement of seizure discharge remains to be demonstrated.

As early as 1958 it was shown by Alonso-deFlorida and
Delgado that a few hours of repeated electrical stimulation of
the cat amygdala resulted in lasting electrographic and behavioral changes. In one cat, the experimental session resulted
in a limited motor seizure which continued independently for
27 days and ended in death. Subsequently, Fonberg and Delgado
(1961) observed that amygdaloid stimulation in cats had an
inhibitory effect on feeding and on learned behavior. The
inhibition outlasted the stimulation and, more importantly,
became more prolonged when the amygdaloid stimulation was
repeated on different days. In one animal the seizure "threshold
diminished, motor and electrical manifestations increased and
generalized seizure developed."

Using cholinergic stimulation of the cat amygdala, Grossman
(1963) observed very dramatic changes. A single bilateral injection of carbachol caused seizure activity that continued to
reappear two or three times daily throughout the following five-
month observation period. Pronounced changes in disposition,
including viciousness and hypersensitivity, were also observed.
Baxter (1967) was not able to replicate all of the effects
reported by Grossman, but he did observe behavioral changes
that lasted several hours after cholinergic stimulation of the
amygdala. Also one cat died overnight, unobserved, possibly as
a result of convulsions, and one cat remained resistant to
handling on the day after the injection.

Recently, in the rat, Belluzzi and Grossman (1969) have
shown that bilateral injection of carbachol into the amygdala
is followed by major alterations in avoidance learning which
persist for several weeks after the convulsions subside.
Similarly, my own studies with carbachol injected into the rat
amygdala (Goddard, 1969) have shown pronounced changes in
certain types of avoidance behavior which often last for more
than two weeks after the injection.

In man, the amygdala has been stimulated electrically by a
number of investigators, and long lasting changes have been
reported in several instances. Stevens et al. (1969) and Ervin
et al. (1969) have presented examples of long-latency long-
lasting psychological changes in both epileptic and non-epileptic
patients. In these cases, as with most studies of electrical
stimulation, only one hemisphere was stimulated at a time. The
lasting after-effects may be less noticeable if confined to one
hemisphere than if bilaterally represented.

The simplest way in which the alterations in brain can be
brought under close experimental control is to apply the same
electrical stimulus to the amygdala at intervals and record
alterations in the response to that stimulus.

Thus, in 1963, Gunne and Reis (and Reis and Gunne, 1965) found that electrical stimulation of the amygdala in cats initially caused facial twitching, turning, chewing and salivation. By the end of 3 hours of intermittent stimulation, the same stimulus resulted in complete rage reactions including clawing, snarling, hissing and attack. These responses, together with associated autonomic elements, began to continue after the termination of each train of amygdaloid stimulation.

Also, Yoshii and Yamaguchi (1963) observed the development of a rage reaction in one cat after 40 days (2,000 trials) of repeated amygdaloid stimulation. "It lasted even after the animal returned to the cage, resulting in the destruction of its own electrodes." This result was obtained despite continuous care on the part of the authors to adjust the intensity of stimulation in an effort to avoid after-discharge durations of greater than 5 sec. Even so, other cats in their experiment developed either more extensive after-discharge, progressively more widespread inter-ictal spiking, the emergence of seizure associated head turning, or frank tonic-clonic convulsions.

In man, Heath et al. (1955) reported that stimulation parameters that were initially subthreshold and caused amusement, later caused intense fear with an impulse to run. King (1961) merely stated that the results of stimulation of the amygdala in man were not consistent from trial to trial.

In the self-stimulation studies of Wurtz and Olds (1963) more than half of the rats receiving amygdaloid stimulation developed seizures. It was not reported whether the seizures developed only after repeated stimulation. Bogacz et al. (1965) have clearly shown that, with self-stimulation electrodes in anterior lateral hypothalamus and septal area, the seizure thresholds diminish over time. The authors were surprised to find that the seizure thresholds declined to a greater extent than the self-stimulation thresholds, and in some cases eventually fell below the self-stimulation thresholds.

In my own studies with low doses of carbachol, injected into the rat amygdala at two-day intervals, I observed similar changes (Goddard, 1969). Initial trials usually produced only inhibition of eating and sometimes salivation; later trials frequently induced aggression, hypersensitivity, and sometimes overt convulsions.

In all of the foregoing experiments the changes following amygdaloid stimulation and the development of various forms of seizure activity were incidental to, and sometimes an embarrassment of, the main purpose of the study. A number of recent

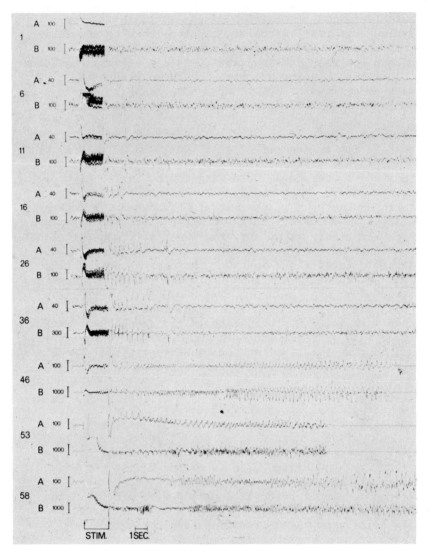

Fig. 1. Development of after-discharge from trial 1 to 58 in
a cat which received a 2 sec., 200 μa, 62.5 Hz stimulus to the
pyriform cortex once each day. A - recorded from the ipsilateral anterior hypothalamic area. B - ipsilateral ventral
hippocampus.

Note: absence of epileptiform spikes on trial 1 followed by
sequential increases in number of spikes, amplitude of spikes
and total duration of after-discharge on successive trials.
Trial 53 recorded under succinylcholine paralysis. Calibrations
in microvolts.

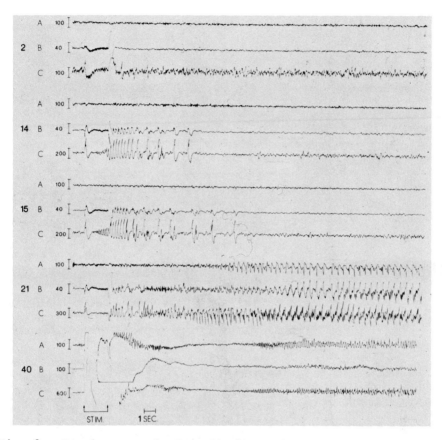

Fig. 2. Development of after-discharge in a cat which received
daily presentation of the experimental stimulus to the lateral
nucleus of the amygdala. A - recorded from contralateral
pyriform cortex. B - ipsilateral ventral hippocampus. C - ipsilateral internal capsule.

Note: after-discharge engages C on trial 2, B and C on trials 14
and 15, and does not engage A until several seconds after the
stimulus on trial 21; it is present in all channels immediately
after the stimulus on trial 40. The rhythmic spiking can be
seen to begin in C prior to the end of the 2 sec. stimulus on
trials 14 and 15. Trials 15 and 40 were recorded under succinylcholine paralysis.

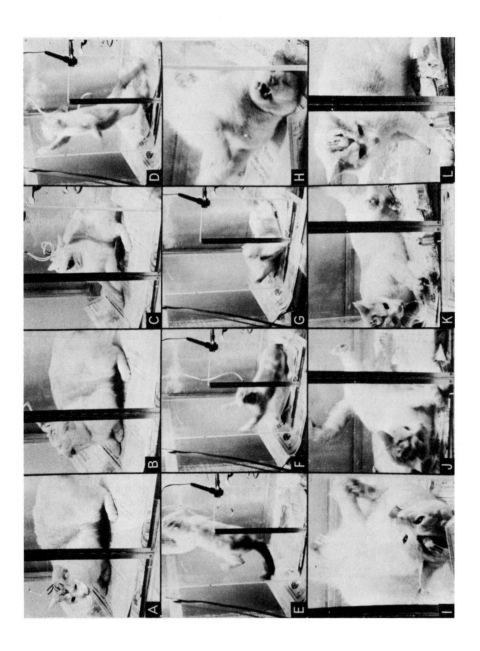

Fig. 3. An example of a complete bilateral clonic-tonic convulsion in response to the 39th stimulation trial. This was the
12th convulsion in this cat and involved leaping into the air.
A: about 5 sec. after termination of the electrical stimulation.
B-K: successive pictures taken at various intervals during the
following 40 sec. of convulsion, the photographic blur in D-G
was due to extremely rapid movement. Note tongue clonus in B
and I, and salivation in K. L: 2 min. after end of convulsion.

Note: on this particular trial the usual 25 strand EEG cable
which was used for delivering the experimental stimulus was
replaced by a lighter stimulation lead which became detached
some time between C and D.

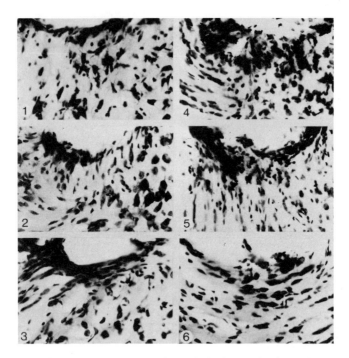

Fig. 4. Sections taken from the vicinity of the electrode tip
within the amygdala of six rats. Frozen sections, 40μ, thionine.
1-3 had been stimulated for 2 sec. each day at 75 μamp peak to
peak, 62.5 Hz, pulse pairs of 1 msec positive and 1 msec negative,
until several convulsions involving bilateral clonus of the fore-
limbs had been triggered. 4-6 were matched electrodes in other
rats which did not receive any stimulation. No gross abnormalities were observed in the kindled tissue that could not also
be found in the control tissue.

experiments have begun to explore the phenomenon directly. It
has been shown that a brief burst of subthreshold amygdaloid
stimulation will eventually lead to behavioral convulsions if
repeated once each day (Goddard, 1967; Goddard et al., 1969;
McIntyre, 1970; Racine, 1971; Walters, 1970). This progressive
development of seizure responses to a constant but repeated
stimulus has been called the kindling effect. It has been
observed to occur in mice, rats, cats and monkeys. A number of
parametric and anatomical studies in the rat have shown that the
kindling effect is a relatively permanent and tran-synaptic
change that results from electrical activation of neurons and
cannot be explained simply in terms of tissue damage, poison,
edema or gliosis.

Although the kindling effect can be obtained from stimulation
of areas outside of the amygdala, responsive areas are largely restricted to the limbic system and related structures. Within the
limbic system, the amygdala has been found to be particularly
responsive. There is also a suggestion that the responsiveness
of particular areas is related directly to the extent of their
anatomical connections with the amygdala (Goddard et al., 1969).

Electrographic evidence of the kindling process is shown for
the cat in Figures 1 and 2. The data in these figures were obtained in collaboration with Morrell and Gersch (Gersch and
Goddard, 1970; Goddard and Morrell, 1971). Stimulation was
delivered to the amygdala or subjacent pyriform cortex once each
day in six cats. The experimental current was a 2 sec. train of
biphasic 1 msec. pulses at a frequency of 62.5 Hz and an amplitude of 200 μa peak to peak.

Figure 1 shows records taken from bipolar electrodes in the
anterior hypothalamus and the ventral hippocampus. Stimulation
was in the pyriform cortex. It can be seen that on day one there
was no after-discharge. On day 6 there was a single spike in
both channels. On day 11 there were two spikes, and so on, until
each trial resulted in a prolonged epileptiform after-discharge.

Figure 2 shows another aspect of the kindled progression.
Stimulation was in lateral nucleus of amygdala. The records were
taken from contralateral pyriform cortex, ipsilateral ventral
hippocampus and ipsilateral internal capsule. It can be seen
that in the early trials the after-discharge was restricted to
channel C and on subsequent trials it began to propagate into
more remote areas of the brain.

The behavioral end point in several of these cats was the
occurrence of a bilateral clonic-tonic convulsion triggered by
the experimental stimulus. An example of such a convulsion is

shown in Figure 3. The building of the convulsion from onset
of the stimulus to the first violent contractions usually
required about 10 sec.; time to the end of clonus varied from
45 sec. to 110 sec. The violence of these convulsions was in
sharp contrast to the very slight effects of stimulation observed at the beginning of the experiment. No cat responded
with seizure activity on the first experimental trial, and the
most noticeable response in any animal was a brief arrest of
ongoing behavior.

Control stimulation of 3 Hz did not result in any of these
changes, and other controls in experiments on the rat suggest
that the kindling effect cannot result from mere presence of
the electrode, or from any aspect of net current flow that is
not dependent on pulse frequency (Goddard et al., 1969).

In none of our studies have we observed any morphological
change at the site of stimulation that cannot also be observed
in control animals beneath the tip of a nonstimulated electrode.
Figure 4 shows tissue distortions at the tip of matched electrodes located in the amygdalae of six rats. Three had been
kindled and three were nonstimulated controls. No gross
differences between the groups are apparent. Similarly, in
the above mentioned studies by other authors, in all cases
where histological data have been presented, the site of
stimulation has not appeared to be grossly abnormal.

Yet, the behavioral and electrographic effects of the
stimulation can be very long-lasting. Once bilateral behavioral
convulsions have been kindled, the animal can be left for several
weeks without stimulation. During this rest interval the animal
does not show behavioral seizures or electrographic paroxysms.
All interictal spikes that were present at the stimulation site
diminish and disappear within about one week (Walters, 1970).
When the amygdaloid stimulation is reapplied, however, the
complete behavioral convulsion reappears, usually on the first
trial and usually undiminished in any way.

It is also possible to demonstrate changes in behavior,
resulting from kindling, which persist after stimulation trials
have been discontinued. Long term changes in disposition or
behavior that result from unilateral amygdaloid stimulation are
not always easy to detect, but they can be demonstrated with the
appropriate techniques.

At a recent conference of the Canadian Psychological
Association (St. John's, 1971) both McIntyre and Adamec presented data on lasting behavioral after-effects of repeated
amygdaloid stimulation. McIntyre used rats and tested their

ability to learn a standard conditioned emotional response
(CER). The rats were chronically implanted with bilateral
amygdaloid electrodes. One group received a unilateral 60 Hz
sine wave for 5 sec. once each day until bilateral clonic convulsions had been triggered. One group received a control
stimulus of 3 Hz sine wave, and two groups were not stimulated.
Groups 1 and 2 then received an electrolytic lesion of the
amygdala on the side contralateral to stimulation. One of
the unstimulated groups received a unilateral lesion, the other
received bilateral lesions.

Bilateral lesions of the amygdala were found to disrupt
CER learning. Prior kindling of one amygdala and a lesion of
the other also disrupted CER learning. The other groups, i.e.
unoperated normal, bilateral electrode, bilateral electrode plus
unilateral lesion, and 3 Hz stimulation plus contralateral
lesion, all learned the CER without significant impairment. In
a previous study (McIntyre, 1970), it had been shown that kindling alone, with unilateral amygdaloid stimulation but no
contralateral lesion, was not sufficient to disrupt subsequent
CER learning.

In all of these studies McIntyre allowed two weeks to
elapse between the last convulsion of the kindling procedure and
the first trial of CER learning. According to Walters (1970)
this would be sufficient time for the interictal spikes to have
disappeared from the EEG.

Adamec presented data to show that predatory behavior in
the cat can be altered following daily repeated amygdaloid
stimulation. In most cases Adamec did not stimulate his cats
to the point of behavioral convulsions, but repeated the
stimulus a few times to lower the after-discharge threshold.
This was done bilaterally. The rat killing behavior of these
cats was studied before and after the threshold reduction.
Lowering the after-discharge threshold in the basal amygdala
of cats that normally had killed rats was found to inhibit
their subsequent predation.

Several other studies on the kindling effect were reported
at the 1971 conference of the Canadian Psychological Association
(Burnham, Leech, Racine, Smith). Of central importance to the
present argument are the data of Racine (1971) showing marked
alteration of evoked potentials following kindling in the
amygdala. Potentials evoked from pulse stimulation of the
amygdala were recorded in the hippocampus, preoptic area and
hypothalamus. Average evoked potentials on days following a
kindling procedure were dramatically altered (increase in late
components) from those recorded on days prior to kindling.

Control measurements showed that these changes were not due to
the passage of time alone.

Racine concluded that the kindled pathways established by
stimulation, which favour seizure propagation, also become more
efficient for conducting other patterns of neural activity
(evoked potentials). The previously cited behavioral data of
Adamec and McIntyre further suggest that normally processed
behavior patterns which utilize these pathways will also be
changed.

A difficulty is raised by these phenomena for the interpretation of some of the earlier stimulation experiments. For
example, it was reported that very low intensities of amygdaloid
stimulation can interfere selectively with the rat's ability to
learn fear-motivated tasks (Goddard, 1964). The stimulation
was delivered every second day during each test trial or for a
5 min. period which immediately followed each test trial. The
intensity of electrical stimulation was determined individually
for each rat by threshold testing on days prior to training.
In other words, each rat received repeated amygdaloid stimulation
spaced in time: obviously, a situation in which kindling will
occur.

The results were interpreted in terms of the intensity of
stimulation, without recognizing that limbic system epileptiform after-discharge may have been developing as the experiment
progressed. Lidsky et al. (1970) and McIntyre (1970) have subsequently found that the learning deficit is seen only when
the post-trial stimulation does cause seizure activity.
McDonough and Kesner (1971), on the other hand, working with
cats, are still of the opinion that post-trial amygdaloid
stimulation, without after-discharge, is sufficient to disrupt
the consolidation of learning. The situation is far from clear.
It is possible that the results were not due to propagated
seizure activity, but to prove the point will require careful
monitoring of the EEG, not only before testing, but at all times
during the experiment.

Before closing, I would like to speak briefly about the
dependence of kindling on the regime or time between each burst
of stimulation. I expect that much confusion will arise over
this issue. It has been shown (Goddard et al., 1969) that when
60 sec. bursts of low intensity amygdaloid stimulation are
separated by less than 20 minutes rats eventually adapt to the
stimulation and develop motor convulsions only rarely. Even
rats that were previously kindled, if stimulated continuously
for many hours, eventually cease having convulsions and adapt
to the stimulation.

Differences were found between groups of rats in which the 60 sec. bursts of amygdaloid stimulation were separated by 8 hr., 12 hr. and 24 hr., with the 24 hr. group requiring the fewest trials before bilateral clonic convulsions appeared. I have attempted to account for these effects by a simple two factor notion whereby the after-discharge leaves a short lasting (several hours) inhibitory effect on subsequent seizures, and also a long-lasting trace (kindling) which facilitates sub-sequent seizures. Massed trial stimulation results in a rapid building up of inhibition. Distributed trial stimulation avoids the inhibition and kindling proceeds with greater efficiency.

Racine (personal communication), working with a different strain of rat, and different stimulus conditions, has observed much less inhibition during massed trial stimulation. He believes that the inhibition affects only the local after-dis-charge threshold and can be eliminated by using higher inten-sities of electrical stimulation. Rasmusson, on the other hand, working in my laboratory, has evidence that seems to suggest a more widespread inhibition. Leech (unpublished studies) has observed that the massed trial effect differs in different strains of rat and mouse.

Much work needs to be done on this problem. However, these various studies all agree on one thing. Any stimulation, in-cluding massed trial stimulation, results in some long-lasting change in the brain. When the animals are stimulated at a later date, seizures and motor convulsions are more likely to occur than if the animals had not received prior stimulation. Distributed trial stimulation may be more efficient (depending on strain or stimulus intensity), but massed trial stimulation also leaves some permanent trace.

Delgado et al. (1971) recently have reported on massed trial stimulation of the amygdala in monkeys for periods of several days or weeks. After a few hundred repetitions the after-discharge duration became longer, spread to involve the contralateral hemisphere, and resulted in other EEG abnormalities which persisted during interstimulation periods. As the experi-ment progressed, however, the after-discharge began to decrease again and eventually stopped altogether. The experiment was discontinued at this point and the authors concluded that it is quite safe to stimulate the limbic system for indefinite periods of time. It is unfortunate that the monkeys had not been given a rest of several days or weeks and then been examined for possible long term after-effects. It is to be expected that the monkeys would show lower thresholds for after-discharge, greater probability of behavioral convulsion, subtle alterations in disposition or learning abilities, and differences in inter-limbic associations as revealed by evoked potentials.

ACKNOWLEDGMENTS

The research and preparation of this manuscript was supported by the National Research Council of Canada. The experiments on cats were conducted at Stanford University School of Medicine in collaboration with Doctors Frank Morrell and Will Gersch. A Travelling Fellowship from the Ontario Mental Health Foundation is acknowledged gratefully.

REFERENCES

ALONSO-DEFLORIDA, F., & DELGADO, J. M. R. Lasting behavioral and EEG changes in cats induced by prolonged stimulation of amygdala. American Journal of Physiology, 1958, 193, 223.

BAXTER, B. L. Comparison of the behavioral effects of electrical or chemical stimulation applied at the same brain loci. Experimental Neurology, 1967, 19, 412.

BELLUZZI, J. D., & GROSSMAN, S. P. Avoidance learning: long-lasting deficits after temporal lobe seizure. Science, 1969, 166, 1435.

BOGACZ, J., ST. LAURENT, J., & OLDS, J. Dissociation of self-stimulation and epileptiform activity. Electroencephalography and Clinical Neurophysiology, 1965, 19, 75.

DELGADO, J. M. R., RIVERA, M. L., & MIR, D. Repeated stimulation of amygdala in awake monkeys. Brain Research, 1971, 27, 111.

ERVIN, F. R., MARK, V. H., & STEVENS, J. Behavioral and affective responses to brain stimulation in man. Proceedings of the American Psychopathological Association, 1969, 58, 54.

FONBERG, E., & DELGADO, J. M. R. Avoidance and alimentary reactions during amygdala stimulation. Journal of Neurophysiology, 1961, 24, 651.

GERSCH, W., & GODDARD, G. V. Epileptic focus location: spectral analysis method. Science, 1970, 169, 701.

GODDARD, G. V. Amygdaloid stimulation and learning in the rat. Journal of Comparative and Physiological Psychology, 1964, 58, 23.

GODDARD, G. V. Development of epileptic seizures through brain stimulation at low intensity. Nature, 1967, 214, 1020.

GODDARD, G. V. Analysis of avoidance conditioning following cholinergic stimulation of amygdala in rats. Journal of Comparative and Physiological Psychology, Monograph Supplement No. 2, pt 2, 1969, 68, 1.

GODDARD, G. V., & MORRELL, F. Chronic progressive epileptogenesis induced by focal electrical stimulation of brain. Neurology, 1971, 21, 393.

GODDARD, G. V., MCINTYRE, D. C., & LEECH, C. K. A permanent change in brain function resulting from daily electrical stimulation. Experimental Neurology, 1969, 25, 295.

GROSSMAN, S. P. Chemically induced epileptiform seizures in the cat. Science, 1963, 142, 409.

GUNNE, L. M., & REIS, D. J. Changes in brain catecholamines associated with electrical stimulation of amygdaloid nucleus. Life Sciences, 1963, 11, 804.

HEATH, R. G., MONROE, R. R., & MICKLE, W. A. Stimulation of the amygdaloid nucleus in a schizophrenic patient. American Journal of Psychiatry, 1955, 111, 862.

KING, H. E. Psychological effects of excitation in the limbic system. In D. E. Shear (Ed.), Electrical Stimulation of the Brain. Austin: University of Texas Press, 1961. Pp. 477-486.

LIDSKY, T. I., LEVINE, M. S., KREINICK, C. J., & SWARTZBAUM, J. S. Retrograde effects of amygdaloid stimulation on conditioned suppression (CER) in rats. Journal of Comparative and Physiological Psychology, 1970, 73, 135.

MCINTYRE, D. C. Differential amnestic effect of cortical vs amygdaloid elicited convulsions in rats. Physiology & Behavior, 1970, 5, 747.

MCDONOUGH, J. H., JR., & KESNER, R. P. Amnesia produced by brief electrical stimulation of the amygdala or dorsal hippocampus in cats. Journal of Comparative and Physiological Psychology, 1971, in press.

RACINE, R. J. The modification of afterdischarge and convulsive behavior in the rat by electrical stimulation. Ph.D. Thesis, McGill University, 1969. Accepted for publication, Electroencephalography and Clinical Neurophysiology, 1971, in press.

REIS, D. J., & GUNNE, L. M. Brain catecholamines: relation
    to the defense reaction evoked by amygdaloid stimulation
    in cat. Science, 1965, 149, 450.

STEVENS, J. R., MARK, V. H., ERWIN, F., PACHECO, P., &
    SUEMATSU, K. Deep temporal stimulation in man. Long
    latency, long lasting psychological changes. Archives
    of Neurology, 1969, 21, 157.

WALTERS, D. J. Sporadic inter-ictal discharges in kindled
    epileptogenic foci. Unpublished M. A. Thesis, Dalhousie
    University, 1970.

WURTZ, R. H., & OLDS, J. Amygdaloid stimulation and operant
    reinforcement in the rat. Journal of Comparative and
    Physiological Psychology, 1963, 56, 941.

YOSHII, N., & YAMAGUCHI, Y. Conditioning of seizure discharges
    with electrical stimulation of the limbic structures in
    cats. Folia Psychiatrica et Neurologica Japonica
    (Niigata), 1963, 17, 276.

THE ORGANIZATION OF THE DEFENCE REACTION ELICITED

FROM AMYGDALA AND ITS CONNECTIONS

A. W. Zbrożyna

Department of Physiology, The University of Birmingham,

The Medical School, Birmingham, England

Manifestations of emotional state ranging from fear to fury evoked by electrical stimulation in the complex of amygdaloid nuclei have been demonstrated by many workers (Gastaut et al., 1951; Kaada et al., 1954; De Molina and Hunsperger, 1959; Ursin, 1960; Zbrożyna, 1963). A typical rage response in a cat to stimulation in the amygdala is shown in Figure 1. The animal shows a characteristic feline threatening posture which is always accompanied by pupillary dilatation, piloerection and threatening vocalization: growling and hissing. These postural and autonomic manifestations are usually accepted as unmistakable signs of emotional involvement and the range of these changes, particularly the extent of participation of the autonomic system, is taken as a measure of the intensity of emotion.

In the response to electrical stimulation of the amygdala the intensity of the electrical current determines the intensity of the response. In the mildest form the response consists merely of increased alertness. At the other extreme, as in intense fear, the participation of the cardiovascular, gastro-intestinal and urinary systems is dramatic. Very much the same reaction can be evoked by hypothalamic stimulation. Hess and Brügger (1943), who first described it, termed it the reaction of defence and attack (das Abwehr-Angriffreaktion) and the short version "a defence reaction" has been widely accepted. Hess considered the hypothalamic area to be an integrative centre for the threatening postural reaction. The autonomic reactions (pupillary dilation, urination, defecation or cardiovascular changes) he considered as incidental, produced by spreading of the stimulating current to adjacent areas in the hypothalamus.

Fig. 1. The effect of stimulation in the amygdalar defence centre
(basal nucleus) in cat.

Many workers (Nakao, 1958; de Molina and Hunsperger, 1959; Wassman and Flynn, 1962; Romaniuk, 1965) have confirmed Hess and Brügger's results. However, the autonomic components of the hypothalamic defence reaction, especially the cardiovascular changes, have been found to be the essential part of the hypothalamic defence reaction, and the control of the autonomic changes by the hypothalamic defence centre to be just as inevitable as its control of the postural changes (Abrahams, Hilton and Zbrożyna, 1960). The cardiovascular response is characterised by an increase in heart rate and contractility, arterial pressure and skeletal muscle blood flow, and a reduction in mesenteric and skin flow. The increase in skeletal muscle blood flow is controlled by cholinergic sympathetic vasodilator nerve fibres. The characteristics of the response to stimulation in the amygdala are strikingly similar to that evoked from hypothalamus, in both the postural and the autonomic components, including the cholinergic muscle vasodilatation (Hilton and Zbrożyna, 1962; Zbrożyna, 1963). This resemblance between the hypothalamic and amygdalar defence reactions suggests a very close relationship between the defence centres in the two regions. There are, however, interesting differences in the fashion in which the defence reaction develops following stimulation in either of these areas.

Stimulation in the hypothalamic "defence area" always produces an immediate and full defence reaction (providing that the electrode tip is positioned correctly): all postural and autonomic components appear rapidly. If the stimulating current is strong enough the response develops in its full intensity without any noticeable delay. This is never so when stimulating in the defence area of the amygdala. Even when a large stimulating current is used, the response builds up gradually: the first to appear is increased alertness and pupillary dilation, then vocalization (growling at first and later hissing), the piloerection builds up gradually and agitation is increased. This gradual building up of the response to full expression usually takes 20-40 seconds and sometimes even longer. In addition the response always outlasts the period of stimulation. By contrast the defence reaction elicited from the hypothalamus disappears promptly and completely at the moment of discontinuation of the stimulation. It takes usually from 20 sec to 2 min for the amygdalar defence reaction to disappear completely. Furthermore during stimulation in the amygdala a temporal summation occurs: for instance, giving 3-5 seconds trains of stimulation with 5 seconds intervals, produces a gradual development of the response to its full expression (Hilton and Zbrożyna, 1963). This was never found with hypothalamic stimulation. The delay of the response, the gradual building up and the ability to display summation suggest that stimulation in the amygdaloid defence area may act via a chain of internuncial neurons arranged in a system of self reverberating

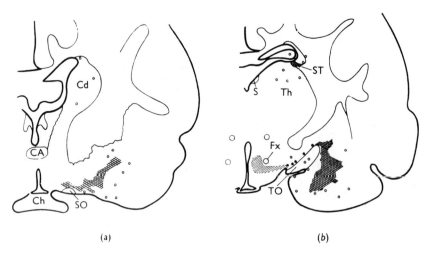

(a)   (b)

Fig. 2. Diagrammatic coronal sections of cat's brain at pre-optic
(a) and tuberal (b) levels of hypothalamus. Cross-hatching denotes
the continuous areas, and filled circles individual points, from
which full defence reaction was elicited by electrical stimulation;
open circles indicate points from which reaction was not obtained:
in all, 156 points were stimulated in 62 cats. Single hatching
shows hypothalamic area for defence reaction, as located by
Abrahams et al. (1960). CA, anterior commissure; Ch, optic chiasma;
Cd, caudate nucleus; Fx, fornix; S, stria medullaris; SO, supraoptic
nucleus; ST, stria terminalis; Th, thalamus; TO, optic tract.

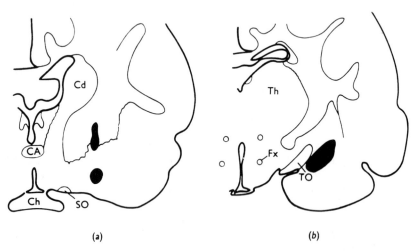

(a)   (b)

Fig. 3. Diagrammatic coronal sections of cat's brain at pre-optic
(a) and tuberal (b) levels of hypothalamus. Black regions indicate
extent of lesions which did not abolish defence reactions elicited
from amygdala. Abbreviations as in Fig. 2.

circuits which are capable of sustaining their excitatory state and having a multi-synaptic connection with the hypothalamic defence area.

The area in the amygdalar nuclear complex in cats producing a defence reaction is not located within a particular nucleus: it cuts across anatomical divisions. It includes part of the anterior amygdala, the basal nucleus (mainly its magnocellular part) and the central nucleus (Fig. 2).

Since Johnston's work (1923) the stria terminalis was thought to form the main efferent pathway from the amygdala to the hypothalamus. This was accepted on morphological evidence, without considering the functional role of the stria (Fox, 1943; Kaada et al., 1954) until it was discovered that electrical stimulation in this structure in free-moving cats produces the defence reaction or some of its components (Hunsperger, 1959; Zbrozyna, 1960). Electrical stimulation along the length of the stria produced the same effect, which shows that it is the stria itself that elicits the defence reaction. The fashion in which the manifestations of the defence reaction occur during stimulation of the stria resembles the defence reaction induced via the amygdala. As on stimulation in the amygdala, the reaction develops slowly and outlasts stimulation, and as in the case of the amygdalar response, is often accompanied by turning the head contralateral to the stimulated side. Moreover, the effect of anaesthetics is similar: they block the effects of stimulation in both stria and amygdala, while it is known that on stimulation in the hypothalamic defence area in cats anaesthetized by chloralose the autonomic components of the defence reaction (piloerection, pupillary dilation, cardiovascular changes) are readily obtained.

These features of the defence reaction evoked via the stria suggest strongly that it activates the defence area in the amygdala. There is other evidence which supports this view. Lesions in the stria terminalis have no effect on the defence reaction evoked by stimulation in the amygdala (Zbrozyna, 1960; Hilton and Zbrozyna, 1963). This undoubtedly would indicate a stria orthodromic connection with the amygdala defence centre. Some confusion arises here because de Molina and Hunsperger (1962) reported that soon after the lesion has been placed in stria terminalis the amygdala reaction may be subdued or abolished. However, the response to amygdalar stimulation recovers a few days after the lesion was placed in the stria. Therefore the amygdalar efferent fibres running in the stria terminalis cannot be concerned with the defence reaction (Zbrozyna, 1963a and b). Furthermore, when the stria was stimulated on either side of the lesion severing it in its midcourse the defence reaction could not be reproduced when the hypothalamic portion of stria was stimulated but it could still be evoked by stimulating the amygdala portion of the stria (Hilton and Zbrozyna, 1963). Dreifuss et al.

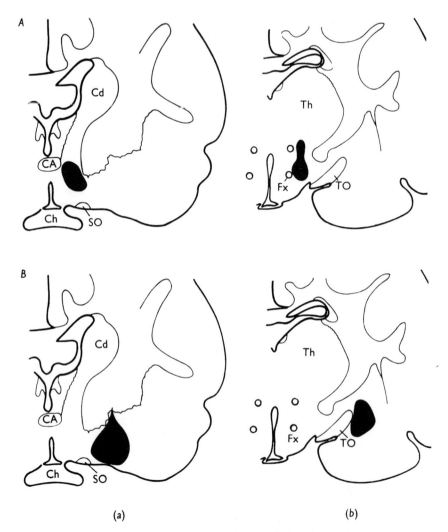

(a)   (b)

Fig. 4. Diagrammatic coronal sections of brains of cats A and B at pre-optic (a) and tuberal (b) levels of hypothalamus. Black regions indicate position of lesions which abolished defence reactions elicited from amygdala. Abbreviations as in Fig. 2.

(1968) described evoked potentials in the ventromedial hypothalamic nucleus on stimulation in amygdala. These potentials disappeared after transection of the stria terminalis. The location of these evoked potentials in the hypothalamus suggests their irrelevance to the defence reaction. It remains an open question as to what the function of the stria in the defence reaction may be. Has it any role in maintaining the prolonged response elicited via the amygdala? Another question concerns the origin of the stria fibres related to the defence reaction. Foreman and Ward (1941) described growling and piloerection evoked on stimulation in some points of the head of the caudate nucleus. It is possible that at least part of the stria fibres concerned with the defence reaction originate from this area. This is supported by the anatomical study of Johnston (1923) who maintains that the bed of the stria terminalis includes part of what is known as the head of the caudate.

The efferent connection leading from the defence area in the amygdaloid complex to the hypothalamus is composed of diffuse fibres running medially directly towards the hypothalamus. On the level of the anterior nuclei of the amygdala the fibres cross the innominate region, and from the basal and central nuclei the pathway runs dorsal to the optic tract and reaches the hypothalamic defence centre across the lateral hypothalamus. Stimulation along this pathway evokes the defence reaction (Fig. 2), the character of the response being similar to the hypothalamic defence reaction particularly when it is stimulated in the portion nearer to the hypothalamic centre (Hilton and Zbrozyna, 1963). The effect could not be due to current spread to the hypothalamic centre itself, since there was no response to stimulation outside the pathway at the same distance from the centre. Lesions interrupting part of this pathway do not affect the defence reaction elicited from the amygdala (Fig. 3). When, however, the pathway is completely interrupted by a lesion extending from the most anterior to its most posterior part (Fig. 4) the defence reaction no longer can be obtained on stimulation in the amygdala (Hilton and Zbrozyna, 1963). It may well be that the anaesthetics which abolish the effect of stimulation in the stria terminalis and in the amygdala are having their blocking effect somewhere in this pathway. This was particularly evident in the experiments on anaesthetized cats in which stimulation close to the hypothalamic defence centre was still effective, while stimulation close to the amygdala was ineffective.

Ursin and Kaada (1960) described a diffuse fibre system projecting from the amygdala to the hypothalamus in a similar fashion. They found likewise that only a complete severance of this pathway abolished the behavioural attention reaction elicited by stimulation in the amygdala.

The hypothalamic and the brain stem defence center is the final integrating centre which activates the autonomic and

postural changes in an adequate pattern. The amygdala defence
centre, however, provides a more refined control of the intensity
and timing of the display of the defence reaction. Furthermore,
it has been shown that the laterobasal nucleus of the amygdala
has an inhibitory influence on the "spontaneous" anxious behaviour
as well as on the defence reaction induced by stimulation of the
hypothalamus (Fonberg, 1963). It may well be that this inhibitory
influence plays an important role in shaping the "natural defence
reaction" which can be described as "attention," "fear-flight" or
"anger-attack" response. It remains still an open question as to
what extent learning processes are modifying the pattern of the
defence reaction in "natural" situations. There is some evidence
indicating the role the amygdala may be playing in these processes.
Discussion of this problem lies beyond the scope of this article.

ACKNOWLEDGEMENTS

My thanks are due to my friends in the Department of Physiology
who spared the time to talk to me about the problems discussed in
this article.

REFERENCES

ABRAHAMS, V. C., HILTON, S. M., & ZBROŻYNA, A. Active muscle
    vasodilatation produced by stimulation of the brain stem: its
    significance in the defence reaction. Journal of Physiology,
    1960, 154, 491-513.

DREIFUSS, J. J., MURPHY, J. T., & GLOOR, P. Contrasting effects
    of two identified amygdaloid efferent pathways. Journal of
    Neurophysiology, 1968, 31, 237-248.

FONBERG, E. The inhibitory role of Amygdala stimulation. Acta
    Biologiae Experimentalis, 1963, 23, 171-180.

FOX, C. A. The stria terminalis, longitudinal association bundle
    and precomissural fornix fibres in the cat. Journal of
    Comparative Neurology, 1943, 79, 227-295.

GASTAUT, H., NAGUET, R., VIGOUROUX, R., & CORRIOL, J. Provocation
    de comportements émotionnels divers par stimulation rhinen-
    céphalique chez le chat avec électrodes a demeure. Review of
    Neurology, 1952, 86, 319-327.

HESS, W. R., & BRÜGGER, M. 1943. Bas subkortikale Zentrum der
    affektives Abwehrreaktion. Helvetia Physiologica Acta, 1943,
    1, 33-52.

HILTON, S. M., & ZBROZYNA, A. Defence reaction from the amygdala and its afferent and efferent connections. Journal of Physiology, 1963, 165, 160-173.

HUNSPERGER, R. W. Les représentations centrales des réactions affectives dans le cerveau antérieur et dans le tronc cérébral. Neuro-Chirurgie, 1959, 5, 207-233.

JOHNSTON, J. B. Further contributions to the study of the evolution of the forebrain. Journal of Comparative Neurology, 1923, 35, 337-481.

KAADA, B. R., ANDERSEN, P., & JANSEN, J. Stimulation of the amygdaloid nuclear complex in unanaesthetized cats. Neurology (Minneapolis), 1954, 4, 48-64.

MOLINA, A. F. de, & HUNSPERGER, R. W. Central representation of affective reactions in forebrain and brain stem: electrical stimulation of amygdala, stria terminalis, and adjacent structures. Journal of Physiology, 1959, 145, 251-265.

MOLINA, A. F. de, & HUNSPERGER, R. W. Organization of the subcortical system governing defence and flight reactions in the cat. Journal of Physiology, 1962, 160, 200-213.

NAKAO, H. Emotional behaviour produced by hypothalamic stimulation. American Journal of Physiology, 1958, 194, 411-418.

ROMANIUK, A. Representation of aggression and flight reactions in the hypothalamus of the cat. Acta Biologiae Experimentalis, 1965, 15, 177-186.

URSIN, H. The temporal lobe substrate of fear and anger. Acta Psychiatrica et Neurologica Scandinavica, 1960, 35, 378-395.

URSIN, H., & KAADA, B. R. Subcortical structures mediating the attention response induced by amygdala stimulation. Experimental Neurology, 1960, 2, 109-122.

WASSMAN, M., & FLYNN, J. P. Directed attack elicited from hypothalamus. Archives of Neurology, 1962, 6, 220-227.

ZBROZYNA, A. W. Defence reactions from the amygdala and the stria terminalis. Journal of Physiology, 1960, 153, 27-28P.

ZBROZYNA, A. W. The anatomical basis of the patterns of autonomic and behavioural response effected via the amygdala. In W. Bargmann and J. P. Schade (Eds.), Progress in Brain Research Vol. 3, 1963a. Pp. 50-70.

ZBROŻYNA, A. W.   Strie terminale et réaction de défence.   Journal
    of Physiology (Paris), 1963b, 55, 703-704.

PHARMACOLOGY

ELECTRICAL ACTIVITY IN THE AMYGDALA AND ITS MODIFICATION BY
DRUGS. POSSIBLE NATURE OF SYNAPTIC TRANSMITTERS. A REVIEW.

E. Eidelberg and C. M. Woodbury

Division of Neurobiology, Barrow Neurological Institute
of St. Joseph's Hospital and Medical Center, Phoenix,
Arizona

This paper is a biased review of the literature pertaining
to the peculiarities of the electrical activity of the amygdala,
and to its modification by certain chemicals. We have concerned
ourselves less with what drugs in general do to the amygdala
than with their use as physiological tools, so that we may begin
unraveling the complexities of synaptic transmission in this locus
of the brain. The reasons for our choice of place are fairly
simple; it is there, and it must be of some evolutionary value
as it increases in relative volume, while the hippocampus slowly
shrinks and recedes into the lateral ventricle. Also, the
electrical activity of the amygdala is rather peculiar, and it
exhibits some fascinating relationships to overt behavior and
neuroendocrine and autonomic mechanisms. It receives a variety of
sensory data, and we know very little about the purpose of this
convergence. It is certainly a more exciting place for physiologists than the dull striatum so heavily favored these days.

We make no claim of comprehensiveness in our quotation of the
literature, having picked out those publications which helped
develop our case or flatly contradicted it. For the convenience
of the reader, whenever possible, we have quoted the most readable
and up-to-date review papers on matters germane to our subject,
rather than the opera prima.

This review, like Gaul, is divided in three parts sequenced
as inductively as possible from EEG to drug microelectrode, and
biochemical data. This approach makes it easier for us to establish our case, feeble though it is yet, for a role for certain
biogenic monoamines in the function of this remarkable structure.

I.   AMYGDALOID SPINDLING

McLean and Delgado (1953) were the first to record a peculiar kind of spontaneous electrical activity from the amygdaloid complex of the cat and squirrel monkey. This activity appeared in bursts (spindles) of rhythmical waves with a center frequency of 25-26 Hz and peak amplitudes of one to several hundred microvolts. The bursting pattern seemed to them closely related to respiration. Lesse (1957) reported similar spindling, but at much higher frequency (40-45 Hz). In his experiments, it was recorded from the amygdala of chronically-implanted cats and showed particularly well when they were subjected to aversive circumstances (Lesse, 1960). In a previous paper, Lesse et al. (1955), had reported spindling from the amygdala in humans with chronically implanted electrodes; this activity tended to develop maximally when the subjects were discussing emotionally charged experiences, and was minimal or absent during relaxation or while solving arithmetical problems. Schwartz and Whalen (1965) studied amygdaloid spindling in relation to the timing of copulatory behavior in the male cat. Intromission coincided with disappearance, and penile withdrawal with reappearance of the 40 Hz bursts. McLean and Delgado (1953) and, later, Freeman (1959) found similar high voltage patterns from the adjacent prepyriform cortex. Freeman (1959) proposed that the activity recorded from leads in the amygdala was generated really in the prepyriform cortex, with a rather large passive field set up in adjacent cortical and subcortical structures.

The work mentioned above raised two issues, which are the subject of some controversy: the first is whether or not this electrical activity is driven by respiration (Delgado et al., 1970; Eidelberg and Neer, 1964; McLean and Delgado, 1953). The second is whether it is generated independently by the amygdala and the prepyriform cortex or spreads from one to the other by neuronal propagation, by volume conduction, or by its properties as a large dipolar field. There were some substantial differences between the initial observations of McLean and Delgado (1953), and those of Lesse (1960), which may bear on the first question. McLean and Delgado (1953) found a center frequency of 26 Hz in the cat and monkey, i.e. rather slower than the 40-45 Hz that some of the previous investigators had observed. Second, while the bursting tended to be generally synchronous with respiration, it was not always so (see top trace of Fig. 6 of McLean and Delgado (1953). Third, the latter investigators observed that amygdaloid spindling persisted under general anesthesia, while it disappeared even under light anesthesia in Lesse's experiments. In the experience of Lesse, and our own, at least 10-15 days may elapse after implantation before clear cut high frequency spindling is observed. The experiments of McLean and Delgado (1953) started 1-2 days after implantation and ended in one week. We have seen that penetration

of an electrode into the amygdala is often signaled by a large
burst of 20-30 Hz activity; we take this to indicate an idio-
syncratic injury discharge. Gault and Leaton (1963) attempted to
solve the first problem by recording simultaneously nasal air flow
and the EEG from olfactory bulb and amygdala. Their results were
somewhat equivocal: when a burst of spindling was seen in the
amygdala it coincided with a burst of activity in the olfactory
bulb, although the reverse was not necessarily true. Also, the
association between olfactory and amygdaloid spindling was maximal
when the animals were excited.

In a recent paper, Delgado et al. (1970) recorded from
chimpanzees, using radiotelemetry, and found that only one out of
five animals showed spindling, again peaking at 25-30 Hz, and
which they claimed was synchronous with respiration. Curiously,
the spindling could be suppressed unilaterally by tegmental
stimulation and yet they insisted that spindling was driven by
respiration or nasal air flow. In another try at this problem,
we cross-correlated the "envelope" of the amygdaloid spindling with
the respiration records obtained by a chest-strap transducer. We
found that while both events recurred at similar frequencies
(0.6 ± 0.2 Hz) the correlation between them was not much better
than chance (Eidelberg and Neer, 1964).

While there is general agreement that amygdaloid activity
shows usually pronounced waxing and waning ("spindling") the
reason for this amplitude modulation, if we exclude respiratory
driving, is not entirely clear. The possibility that waxing and
waning could be produced by beat-frequency modulation was explored
in some experiments carried out several years ago with Harry Neer,
some of which were published (Eidelberg and Neer, 1964). We
analyzed the spectral content of individual spindles by playing
repeatedly the same magnetic tape loop through a filter with a
bandpass of about 1.0 Hz. We found that a relatively broad band of
frequencies was involved, rather than a single sinusoidal component. This finding suggests that beat-frequency modulation is a
possible explanation for the waxing and waning, by interaction of
closely related sinusoidal generators. This raises some very complicated questions about the nature of the generator mechanisms
involved.

We mentioned before that Freeman (1959) had raised some questions about the amygdala being the "real" source of this electrical
pattern. His reasons for assigning it to the prepyriform cortex
were based on laminar field analysis of spontaneous activity, and
of prepyriform evoked responses to lateral olfactory tract electrical stimuli. He concluded that the prepyriform spindles and
evoked responses were produced by the same elements because they
interacted with each other, reversed at the same level and had comparable time course (Freeman, 1959). We find it hard to accept

entirely Freeman's conclusions. We agree that spindling can be
recorded from the prepyriform cortex, that it looks very much
like amygdaloid spindling and that it appears in the records at
about the same time as amygdaloid spindling. We are puzzled,
however, by three problems: (1) if prepyriform evoked responses
and spontaneous spindling are generated by the same mechanisms, how
is it that the first are resistant to sleep and anesthetics, while
the second practically are suppressed by both? (2) we have found
a remarkable lack of phase correlation between simultaneous
recordings from the amygdala and prepyriform cortex (Eidelberg and
Neer, 1964), (3) the amplitude relationships between spindling in the
amygdala and prepyriform cortex ought to show substantial attenuation with distance from the cortical leads. This derives from the
isopotential maps for evoked responses derived by Freeman (1959)
and his conclusion that spindles and evoked responses are generated
by the same structures. In our experience, there is a wide
variation of amplitude of spindles, which may often be larger away
from the prepyriform cortex than near it. None of these arguments
against Freeman's hypothesis are conclusive, but they do suggest
the need for a systematic, fine grained, analysis of the topology
of spontaneous EEG activity in this part of the brain.

II.   DRUG EFFECTS ON AMYGDALOID ACTIVITY

In Lesse's laboratory, the accidental use of concentrated
(10%) solution of cocaine as a nasal mucosa anesthetic yielded some
surprising results. The first cat given intranasal cocaine underwent a period of freezing and staring, vomited, had a generalized
convulsion, and died within a few minutes. Shortly after the
application of cocaine and at the time of behavioral arrest, a
remarkable change developed in the amygdaloid EEG: the spindling
increased in amplitude, no longer waxed and waned, and became
gradually the typical record of tonic and then clonic convulsive
activity. This activity later became generalized at the time of
the grand mal convulsion (Eidelberg et al., 1963). The same
sequence of events occurred in rats, but the arrest stage was
replaced by motor hyperactivity. DeJong and Wagman (1963) reported
shortly afterwards that synthetic analogs of cocaine, such as lidocaine, produced similar effects. Thompson, Lesse and Eidelberg
(unpublished) found that previous bilateral amygdaloidectomy prevented the motor hyperactivity and often the convulsive effects of
cocaine in rats. At the same time, we found that pretreatment
with MAO inhibitors potentiated greatly the effects of low doses
of cocaine, and that adrenergic blocking agents served as
fairly effective anticonvulsants (Eidelberg et al., 1963). Later,
benzodiazepine tranquilizers were found to be potent cocaine
antagonists which depressed or eliminated altogether amygdaloid
spindling (Eidelberg et al., 1965).

The experiments with cocaine suggested what was then a novel idea: that the behavioral excitatory effects of cocaine, as well as its convulsant effects upon the limbic system might be due to changes in central monoaminergic synaptic transmission in these structures (Eidelberg et al., 1963). This was based on the known effects of cocaine upon peripheral adrenergic mechanisms, the potentiating effects of MAO inhibitors and the specific anticonvulsant effects of adrenergic blocking agents. With Michael Long and Marilyn Miller (1966), we set out to investigate systematically the EEG changes in the amygdala (and pari passu in other structures) produced by overloading with monoamine precursors, and by administering blocking agents interfering with monoamine metabolism. To quantify our data and separate out the charges in spindling activity from the rest of the EEG, we used spectrum analysis. These experiments showed that raising brain serotonin levels was associated with markedly depressed amygdaloid spindling, while raising catecholamine levels did not affect it. The results of further experiments with microelectrodes will be found later on in this review.

At the same time, and with the same collaborators, we explored the effects of psychotomimetic agents on amygdaloid electrical activity. The bases for this series of experiments were a paper by Baldwin et al., (1959) in which the behavioral effects of LSD were markedly attenuated or modified after temporal lobectomy, and the proposal by Woolley and Campbell (1962) that psychotomimetic agents act upon central serotonergic mechanisms. We found that, in support of the postulated relationship, all the agents tested (LSD, mescaline, harmine and bufotenine) affected grossly the amygdaloid EEG.

III. SYNAPTIC MECHANISMS

There generally is an agreed upon set of criteria for accepting a substance as a synaptic neurotransmitter (DeRobertis, 1969; McLennan, 1963). So far, only acetylcholine has approached closely their fulfillment. These criteria are:

(a) Positive identification of the presence of the transmitter in the presynaptic element, contained preferably in "synaptic vesicles." The last presupposes that synaptic transmission is a quantal event everywhere in the nervous system, an assumption which has good support in studies of neuromuscular transmission and of anterior horn motoneurone EPSP's, but which hasn't been demonstrated elsewhere yet, particularly in mammalian postsynaptic inhibition.

(b) Presence of enzymatic or other means of transmitter degradation in the subsynaptic membrane, so as to provide quick

termination of transmitter action.   In the case of neuromuscular
cholinergic transmission, it is clear that cholinesterase activity
at the end plate is responsible for transmitter inactivation.   It
is quite likely that enzymatic degradation may not be the only
possible means of termination, since transmitter removal by active
reuptake into storage elements may account for the same consequences
in other chemical synapses.   It has been shown that uptake serves
to reutilize a large share of norepinephrine and serotonin in the
peripheral and central nervous system.   Only a relatively small
fraction of these amines is degraded by oxidative deamination or
O-methylation following sustained stimulation (Snyder et al., 1970).

     (c) Application of the presumed transmitter to the post-
synaptic membrane must mimic the effects of the natural substance.
Again, this criterion has been met by the work of Katz and his
colleagues in the neuromuscular endplate, by iontophoretic injection of acetylcholine.   Even with the most sophisticated techniques now available, it is not yet possible to apply presumed
transmitters only and directly into subsynaptic membranes in the
mammalian central nervous system.   Non-specific effects upon non-
synaptic areas of the membrane, presynaptic structures, or glial
elements cannot be ruled out.

     (d) The presumed transmitter should be collected from
local tissue perfusates following stimulation.   This seems like
a reasonable requirement until one realizes that it conflicts
with (b) unless enough stimulation is applied to overcome the
degradative and/or removal systems.   The presumed transmitter
must also be liberated in measurable quantities into the artifac-
tual extracellular space created by the perfusion cannulae.   Apparently, this may happen under physiological conditions in the
CNS in the case of acetylcholine--it certainly did in Otto Loewi's
classical frog heart experiments--but this may not necessarily be
the case with other transmitters less resistant to degradation
or removal when outside the subsynaptic membrane.

     (e) Pharmacological agents which interfere with the opera-
tion of the neuron should similarly affect the action of the
artificially applied substance.   These actions could fall into
three groups: (1) interference with transmitter synthesis;
(2) interference with the synaptic actions of the transmitter upon
the subsynaptic membrane; (3) interference with the degradation of
the transmitter.   It is assumed that the pharmacological agents
are capable of entry into the right places in the CNS and that
their only biochemical actions are those specified.

     The above discussion is not presented to make debating points.
We are concerned over the rigid dogmatism which has dominated
synaptic physiology in the last 15 years.   Rather than follow
the inductive approach we have tended to use deductive tactics as

if we knew how synapses operate everywhere, and the problem were just to find the guilty parties. The questioning of Eccles' restricted version of Dale's principle that resulted from the work of Tauc and Gerschenfeld, on the presence of inhibitory and excitatory outputs from the same molluscan interneuron, serves as a strong reminder of the fallibility of some classical assumptions. It is just possible that some of our criteria for central transmitter identification, which are derived mostly from neuromuscular junction work, may be open to question.

Our own research has centered on the possible role of biogenic amines as transmitters in the amygdaloid complex and hippocampus of the cat. Kuntzman et al., (1961), found that these two structures are among those with the highest amounts of detectable serotonin (5-hydroxytryptamine, 5-HT) and norepinephrine (NE). While there is no discussion about the high serotonin levels, those of NE may be considerably lower than was thought originally (Bertler and Rosengren, 1959). Using the now classical formaldehyde condensation—UV microscopy technique, Andén et al. (1966), in the rat, and we in the cat (Eidelberg et al., 1967) demonstrated a rich population of monoaminergic varicosities and/or terminals in the amygdala and hippocampus. In our experimental material, the fluorescent elements were distributed rather selectively in a perisomatic arrangement. They were absent in the apical dendritic palisade of the hippocampal stratum radiatum. The fluoresence we obtained was yellow, peaking in the 570 nm band, and was interpreted as being indicative of the presence of serotonin for this reason. Some caution on this interpretation may be introduced by the later studies of Corrodi and Jonsson (1967) which indicated that microspectrofluorimetry may be needed to exclude catecholamines as possible sources of yellow fluoresence. In rats, Andén et al. (1966) found also in the amygdala and hippocampus the blue-green varicosities associated with catecholamines. We failed to demonstrate a similar picture in the cat, either because of species differences or our lesser experience with the technique. In both studies the monoamine-containing structures seemed to be presynaptic axons or boutons terminaux. We could not demonstrate the presence of serotonin inside amygdaloid or hippocampal cells, in contradistinction to the intense intracellular fluoresence of midbrain tegmental cells (Dahlström and Fuxe, 1964; Fuxe, Hökfelt and Ungerstedt, 1970. This suggests that the amines were localized at the input to amygdaloid and hippocampal cells, rather than these cells being the source of monoamine-containing axons. If these monoaminergic elements are presynaptic, the next question is that of the identity of the cells of origin, and the localization of their axons. The amygdaloid complex and the adjacent pyriform cortex receive monosynaptic or polysynaptic inputs from the olfactory bulbs, the hypothalamus, the midbrain tegmentum, and possibly the dorsomedial thalamus (Nauta, 1962; Gloor, 1960). The hippocampus receives its inputs both via the perforant pathway of

Cajal from the entorhinal gyrus and adjacent neocortex, and from elsewhere in the brain via the septum-fornix route. Both receive, directly or indirectly, afferents via the medial forebrain bundle (MFB). Harvey, Moore and Heller (cf Moore for a detailed review of their work, 1970) showed that unilateral destruction of the MFB caused a sharp decrease in the serotonin and catecholamine content of the ipsilateral hemisphere, which began one day after the placement of the lesion and continued through a month afterwards. This effect could be reproduced in cats and rats. It was greatest in the amygdala and hippocampus, where it reached a 71% decrease from control levels in the cat. It was minimal in the brain stem (6% decrease in the midbrain, and no decrease in the pons-medulla). Both their group and the Swedish group (Andén et al., 1966) also used fluoresence microscopy to locate the medial forebrain bundle endings which had lost their monoamine contents following the lesions. Their findings were in good agreement in that profound losses (from 50-75%) of monoamine elements occurred in the hippocampus and amygdala. Both groups also agreed in concluding that the cell bodies of origin of these terminals lie in the brain stem tegmentum and their axons travel primarily in the medial forebrain bundle.

Pohorecky et al., (1969) found no apparent loss in the norepinephrine levels of the hippocampus and amygdala, or the ability of these structures to take up norepinephrine after olfactory bulb removal. No fluoresence histochemical studies were carried out. These authors suggested that removal of the olfactory bulb input to the amygdala did not eliminate any great number of those elements which presumably took up the labeled NE, and that, therefore, the bulk of the monoaminergic inputs into these structures may issue from the rest of the brain rather than the olfactory system.

The second requirement for the definition of a transmitter is the presence of either enzyme systems for transmitter degradation, or mechanisms for its removal by active uptake away from the subsynaptic membrane. In the case of serotonin and catecholamines, degradation is achieved by oxidative deamination (by monoaminooxidase, MAO) and, in the case of the catecholamines, also by the activity of catechol-O-methyl-transferase (COMT) (Axelrod et al. 1959). Other alternative routes also may be present, but their localization and relative role are less well established at this time. MAO activity was demonstrated chemically in the amygdala and hippocampus by Udenfriend, Weissbach and Bogdanski (cf 33). COMT has been shown to be present also in the limbic system (Axelrod et al., 1959).

Axelrod and his colleagues, as well as von Euler's group in Stockholm, have established beyond reasonable doubt that a large share of the monoamines released at the synaptic junction is taken back into presynaptic elements rather than degraded. The work of Glowinski and Axelrod, and of Snyder and his group and others, has established the presence of reuptake mechanisms in structures other than the hippocampus and amygdala, such as the cortex, hypothalamus and striatum (Snyder et al., 1970). Although direct experimental confirmation is needed, there is no a priori reason to think that this highly efficient device for economizing transmitters would be missing in limbic structures.

Because of the relative simplicity of spinal cord reflex organization, much of the initial work on pericellular iontophoretic injection of presumed transmitters was carried out in anterior horn motoneurones. Numerous investigators have extended the use of this technique to structures above the foramen magnum but, surprisingly, only Herz and Nacimiento (1965) seem to have tried it on the hippocampus and none, to our knowledge, to the amygdaloid complex. They found that serotonin, injected next to hippocampal pyramidal cells, had powerful inhibitory effects on their ability to discharge, although it was not established whether membrane hyperpolarization or charges in membrane conductance, or both, were related to this inhibition (Herz and Nacimiento, 1965).

In our experiments, we took an easier route into the problem, that of recording the changes in single cell spontaneous firing rate induced by the intravenous administration of precursors of the suspect biogenic amines, since the amines themselves do not cross into brain readily (Eidelberg et al., 1967). This approach does not exclude indirect effects due to excitation or inhibition of neuronal systems converging into the amygdala, since the whole brain must be presumed to be affected. The final consequences--as viewed by the microelectrode--however, were the relevant issue to us, i.e. whether increasing or decreasing total brain amine levels affected the activity of these cells. We assumed that the effects of these substances on structures not related synaptically to the amygdala would not affect the cells we were recording from. What we could not exclude was the possibility of non-synaptic effects upon amygdaloid cells, i.e. directly upon their spike-generating structures, but this objection applies equally well--perhaps even more so--to localized iontophoretic injection, since the pipette tips cannot be placed exclusively upon the subsynaptic membrane.

Under these experimental conditions, using unanesthetized preparations, administration of the immediate precursor of serotonin (5-hydroxytryptophan, 5HTP) caused consistently a reduction in the probability of spontaneous firing of amygdaloid units. This effect was not present when decarboxylation of 5HTP to serotonin was prevented by blocking previously the corresponding enzyme. By contrast, administration of the catecholamine precursor amino acid, 1-3,4 dihydroxyphenylalanine (L-DOPA) caused an acceleration of firing. This effect also was prevented by blockade of the amino acid decarboxylase. The effects of the precursors were

transient in that, usually, the cells returned to, or near to their
original resting discharge within 60 to 90 minutes after a single
injection. Blocking MAO with iproniazid before injecting the
amino acid enhanced strongly and prolonged their effects (Eidelberg
et al., 1967).

We interpreted these findings as suggesting the possibility
of a dual, antagonistic, synaptic input upon amygdaloid neurons
using serotonin as the inhibitory and a catecholamine as the excitatory transmitter. This is a tentative simplification of what
must be a far more complicated story, but which has, at least, the
merit of being testable experimentally.

There is practically no evidence in the literature, except
for a paper by Stein and Wise (1969), regarding the release of the
presumed transmitters in the amygdala in vivo and in situ, following stimulation of its afferent pathways. These investigators
implanted a Gaddum "push-pull" cannula in the amygdala, kept it
under continuous perfusion and collected the washout following
electrical stimulation of the median forebrain bundle. Labelled
norepinephrine was injected intraventricularly and the radioactivity in the perfusate was measured. They found that the stimulation increased the release of NE and O-methylated metabolities.
While this paper is subject to some methodological objections
(such as the cannula being nearly as large as the rat amygdala),
their data probably are valid, and support the concept of monoaminergic transmission in the amygdala.

We have not discussed the possibility that substances other
than monoamines may be involved in amygdaloid function. This
omission is based on the unfortunate fact that not much evidence
is available in favor for or against them. There is histochemical
evidence, from Koelle's laboratory (1954), that there is abundant
cholinesterase activity in the rat amygdala, and from Lewis et al.,
(1964) that choline acetylase is present in fiber pathways into the
hippocampus. GABA (gamma-aminobutyric acid) also is present in
these structures, as determined by chromatographic assay, but
its structural distribution is yet unknown.

To conclude this brief review, it is remarkable that a
structure like the amygdala, whose relationship to hormonal and
autonomic regulation and to complex behavior is so important,
has not been the subject of much more research by cytochemists
and electrophysiologists. We do have some reasonable guesses
as to how it is connected to other central structures, but no
conclusive evidence to explain how its most peculiar activity
is generated and how it relates to synaptic and endocrine events.

ACKNOWLEDGMENTS

The work on this subject carried in our laboratory was supported by the USPHS - N.I.N.D.S. (grant NB 3496) and by the Barrow Neurological Foundation.

REFERENCES

ANDEN, N. E., DAHLSTROM, A., FUXE, K., LARSSON, K., OLSON, L., & UNGERSTEDT, U. Ascending monamine neurons to the telencephalon and discencephalon. Acta Physiologica Scandanavia, 1966, 67, 313-326.

AXELROD, J., ALBERS, W., & CLEMENTE, C. D. Distribution of catechol-O-methyl transferase in the nervous system and other tissues. Journal of Neurochemistry, 1959, 5, 68-72.

BALDWIN, M., LEWIS, S. A., & BACH, S. A. The effects of lysergic acid after cerebral ablation. Neurology (Minn.), 1959, 9, 469-474.

BERTLER, A., & ROSENGREN, E. Occurrence and distribution of catecholamines in brain. Acta Physiologica Scandanavia, 1959, 47, 350-361.

CORRODI, H., & JONSSON, G. The formaldehyde fluoresence method for the histochemical demonstration of biogenic monoamines. A review of methodology. Journal of Histochemistry and Cytochemistry, 1967, 15, 65-78.

DAHLSTROM, A., & FUXE, K. Evidence for the existence of monoamine-containing neurons in the central nervous system. Acta Physiologica Scandanavia, 1964, 62, Supplement, 232.

DE JONG, R. H., & WAGMAN, I. H. Cortical and subcortical effects of i.v. lidocaine and inhalation anesthetics. Federal Proceedings, 1963, 22, 187.

DELGADO, J. M. R., JOHNSTON, V. S., WALLACE, J. D., & BRADLEY, R. J. Operant conditioning of amygdala spindling in the free chimpanzee. Brain Research, 1970, 22, 347-362.

DE ROBERTIS, E. Structural and chemical studies on storage and receptor sites for biogenic amines in the central nervous system. Symposium of the International Society of Cell Biology, 1969, 8, 191-207.

EIDELBERG, E., LESSE, H., & GAULT, F. P.   An experimental model
      of temporal lobe epilepsy; studies of the convulsant pro-
      perties of cocaine.   In G. H. Glaser (Ed.) EEG and Behavior.
      Basic Books, 1963.   Pp. 272-283.

EIDELBERG, E., & NEER, H. M.   Electrical analysis of amygdaloid
      spindling.   Boletin Instituto Estudios Medicos y Biologicos,
      1964, 22, 71-84.

EIDELBERG, E., NEER, H. M., & MILLER, M. K.   Anticonvulsant
      properties of some benzodiazepine derivatives.   Neurology
      (Minn.), 1965, 15, 223-230.

EIDELBERG, E., LONG, M., & MILLER, M. K.   Spectrum analysis of
      EEG changes induced by psychotomimetic agents.   International
      Journal of Neuropharmacology, 1965, 4, 255-264.

EIDELBERG, E., MILLER, M. K. & LONG, M.   Spectrum analysis of
      EEG changes induced by some psychoactive agents.   Their
      possible relationship to changes in cerebral biogenic amine
      levels.   International Journal of Neuropharmacology, 1966, 5,
      59-74.

EIDELBERG, E., DEZA, L., & GOLDSTEIN, G. P.   Evidence for sero-
      tonin as a possible inhibitory transmitter in some limbic
      structures.   Experimental Brain Research, 1967, 4, 73-80.

FREEMAN, W. Distribution in time and space of prepyriform elec-
      trical activity.   Journal of Neurophysiology, 1959, 22,
      644-665.

FUXE, K., HOKFELT, T., & UNGERSTEDT, U.   Morphological and func-
      tional aspects of central monoamine neurons.   International
      Review of Neurobiology, 1970, 13, 93-126.

GAULT, F. P., & LEATON, R. N.   Electrical activity of the olfac-
      tory system.   Electroencephalography and Clinical Neuro-
      physiology, 1963, 15, 299-304.

GLOOR, P.   Amygdala.   In Handbook of Physiology, Section 1,
      Neurophysiology, Volume 2.   American Physiological Society.
      Baltimore: Williams and Wilkins, 1960.

HERZ, A., & NACIMIENTO, A. C. Uber die Wirkung von Pharmaka
      auf Neurone des Hippocampus nach mikroelektrophoretischer
      Verabfolgung.   Naunyn-Schmiedebergs Archiv für Pharmakologie
      und Experimintelle Pathologie, 1965, 251, 295-315.

KOELLE, G. B.   The histochemical localization of cholinesterases
    in the central nervous system of the rat.   Journal of Compara-
    tive Neurology, 1954, 100, 211-235.

KUNTZMAN, R., SHORE, P. A., BOGDANSKI, D., & BRODIE, B. B.
    Microanalytical procedures for fluorometric assay of brain
    DOPA-5HTP decarboxylase, norepinephrine and serotonin and a
    detailed mapping of decarboxylase activity in brain.   Journal
    of Neurochemistry, 1961, 226-232.

LESSE, H. Rhinencephalic electrophysiological activity during
    "emotional behavior" in cats.   Psychiatric Research Report,
    1960, 12, 224-237.

LESSE, H., HEATH, R. G., MICKLE, W. A., MONROE, R. R., & MILLER,
    W. H.   Rhinencephalic activity during thought.   Journal of
    Nervous and Mental Disorders, 1955, 122, 400-433.

LEWIS, P. R., SHUTE, C. C. D., & SILVER, A.   Confirmation from
    choline acetylase analyses of a massive cholinergic inner-
    vation to the hippocampus.   Journal of Physiology (London),
    1964, 172, 9-108.

MC LEAN, P. D., & DELGADO, J. M. R.   Electrical and chemical
    stimulation of frontotemporal portion of limbic system in the
    waking animal.   Electroencephalography and Clinical Neuro-
    physiology, 1953, 5, 91-100.

MC LENNAN, H.   Synaptic Transmission.   Philadelphia: W. B.
    Sanders Company, 1963.

MOORE, R. Y.   Brain lesions and amine metabolism.   International
    Review of Neurobiology, 1970, 13, 67-91.

NAUTA, W. J. H.   Neural associations of the amygdaloid complex in
    the monkey.   Brain, 1962, 85, 505-520.

PAGANO, R. R., & GAULT, F. P.   Amygdala activity.   A central
    measure of arousal.   Electroencephalography and Clinical
    Neurophysiology, 1964, 17, 255-260.

POHORECKY, L. A., ZIGMOND, M. J., HEIMER, L. & WURTMAN, R. J.
    Brain norepinephrine: effects of olfactory bulb removal.
    Federal Proceedings, 1969, 28, 795.

SCHWARTZ, A. S., & WHALEN, R. E.   Amygdala activity during
    sexual behavior in the male cat.   Life Sciences, 1965,
    4, 1359-1366.

SNYDER, S. H., KUHAR, M. J., GREEN, A. I., COYLE, J. T., &
    SHARSKAN, E. E. G.   Uptake and subcellular localization of
    neurotransmitters in the brain.   International Review of
    Neurobiology, 1970, 13, 127-159.

STEIN, L., & WISE, C. D.   Release of norepinephrine from hypoth-
    alamus and amygdala by rewarding medial forebrain bundle
    stimulation and amphetamine.   Journal of Comparative and
    Physiological Psychology, 1969, 67, 189-198.

WOOLLEY, D. W., & CAMPBELL, N. K.   Serotonin-like and anti-
    serotonin properties of psilocybin and psilocin.   Science,
    1962, 136, 777-778.

THE NEUROPHYSIOLOGICAL EFFECTS OF AMPHETAMINE
UPON THE CAT AMYGDALA

James G. Wepsic and George M. Austin

Massachusetts General Hospital

Boston, Massachusetts

INTRODUCTION

Changes in the brain catecholamines (CA), norepinephrine (NE), dopamine (DA) and serotonin, are related to changes in affective behavior. Fluctuation in the synthesis, storage and turnover of these molecules has been demonstrated in different behavioral states and, conversely, different behavioral states have been produced by administering drugs which modify the uptake, metabolism, or degradation of such amines. It is thought that behavioral changes depend upon neural activity and anatomical areas important to affective states were grouped by Paul MacLean into the Limbic System. Understanding the role of CA in the limbic system is crucial to the understanding of the neuro-chemical basis for affective behavior. To this end, we are studying the influence of amphetamine upon electrophysiological relationships between the amygdala, septal nucleus, and hypo-thalamic ventromedial nucleus (HVM). The anatomical localization of NE, its effects on behavior thought to be referable to the amygdala, and the neuro-pharmacologic relationship between amphetamine and NE will first be discussed.

PHARMACOLOGICAL BACKGROUND

The distribution of CA in the brain has been studied extensively. Glowinski (1966), utilizing radio-isotopic techniques, demonstrated that NE was highly concentrated in rat hypothalamus and medulla. This was confirmed in primates (Goldstein, 1967) utilizing the fluorescence produced when these amines are reacted with formalin vapor. Hillarp et al.

(1966) showed them to be concentrated along the tips of terminal axons in presynaptic varicosities. Large amounts of amines were shown in this histochemical fashion in the locus ceruleus and rostrally along the medial forebrain bundle (MFB) to the hypothalamus and amygdala (Fuxe, 1965). The amine in the hypothalamus and amygdala was shown to be NE (Auden et al., 1966). Lesions of the MFB depressed the concentration of NE in the diencephalon (Dahlstrom and Fuxe, 1965 and Sheard et al., 1967) showed that this change was associated with the animal's inability to avoid a shock in a learned conditional avoidance response. Studies were then undertaken to demonstrate changes in the metabolism of NE in various behavioral states or in response to drugs as amphetamine, known to influence behavior. The concentration of NE was decreased in the brains of rats forced to swim to exhaustion (Barchas and Freedman, 1963). It was also lower in animals subjected to repeated foot shocks (Maynert and Levi, 1969), and following amygdaloid stimulated "sham rage" attacks (Reis and Gunne, 1965; Gunne, 1969; Gunne and Reis, 1963; Fuxe and Gunne, 1969). Stimulation of the amygdala at sites both in the basolateral and corticomedial nuclear areas resulted in the production of excited behavior characterized by hissing, snarling, clawing, and poorly directed attacking movements. Brain NE was depressed in all cats who showed this behavioral rage. When the lateral amygdalae were stimulated, rage did not appear and NE levels did not change. When depletion of CA did occur, it did so in all forebrain regions and was accompanied by rise in brain normetanephrine indicating an increased metabolic turnover of NE. Discrete lesion in MFB have also depressed levels of NE (Heller and Moore, 1965). Reis and Fuxe (1969) have shown that rage attacks can be augmented or inhibited by pharmacologically potentiating or blocking the action of NE. Perhaps a more sensitive measure of the relationship of this amine to behavioral states is seen in study of turnover rates. Thierry (1968 showed an increase of rate of its metabolism with foot shock and Kety (1967) demonstrated that electroconvulsive shocks produced a rise in NE synthesis. These changes in concentration of NE lend credence to the theory that this endogenous amine is important in behavior.

Additional evidence favoring a NE modulated behavior system comes from psychological observation of the effects of drugs known to change the uptake, synthesis, or release of this amine (Schildkraut and Kety, 1967). Kety, in a recent monograph (1967) describes these effects in detail. Drugs which have a tendency to increase the activity of amines, as MAO inhibitors, amphetamine, and imipramine, all tend to relieve clinical depression or excite a normal animal while those drugs which produce depression have an opposite effect, decreasing amine activity. α-methyl tyrosine which blocks the synthesis of NE can block the effects of

amphetamine (Crow, 1969) while reserpine which depletes the
brain of its amines has the same ultimate effect (Stein, 1964).

BEHAVIORAL BACKGROUND

Observations of the behavioral changes produced by direct
intraventricular administration of NE are somewhat difficult to
interpret. Although early work suggested that NE tends to sedate
when administered directly into the CSF (Mandell and Spooner,
1968) a more recent study showed that activated behavior may
result when small amounts of the amine are administered chronically (Segal and Mandell, 1969). Direct injection of NE into the
brain is also of interest. On the cellular level, Salmoiraghi
and Bloom (1964) demonstrated that microinjection of this amine
into the region of single microelectrode monitored cells produced
inhibition of spontaneous electrical activity. Kety pointed out,
however, that this does not necessarily argue against a role of
NE in arousal for the important effects of the amine may be to
activate only very few selected areas. Slanger and Miller (1969)
injected 20 milli-micromoles of norepinephrine into the perifornical area of hypothalamus and produced eating in rats which
were previously satiated. This eating effect mimicked the effect
of electrical stimulation in this area. Leaf et al. (1969)
placed cannulae bilaterally in the amygdaloid nuclei of mouse-
killing rats through which they administered crystalline NE and
d-methamphetamine HCl. They found the latter agent most
effective in inhibiting the normal "killing behavior" of rats
with NE nearly as effective. Other agents used in this model
without effect were imipramine and chlorpromazine. Margules
(1968) showed that direct application of 1 or dl-NE to the medial
amygdala removed the behavioral-suppressed effect of punishment
in trained rats. Mark et al. (1971) administered small amounts
of NE into the amygdala of "enraged" cats with bilateral hypothalamic lesions and found that rather than producing greater
rage responses, as electrical stimulation of this area has been
reported to do, the amine resulted in a quieter animal.

NE is found endogenously in the parts of the brain known by
stimulation and ablation techniques to be necessary areas for
expression of emotion. We are concentrating on the amygdala and
hypothalamus. Its metabolism in these areas is influenced by
drugs that influence behavior and many have implied that these
localized changes in concentration are responsible for the
witnessed behavioral changes. Direct exogenous administration
of large amounts of NE either into the CSF or locally into the
brain has resulted in rather profound behavioral changes. A
study of these amines in a more physiological setting and at more
physiological concentration is needed before more refined roles
can be assigned to them. An understanding of the electrophysio-

logical changes produced by their action may be useful in more
clearly defining their role in behavioral states and perhaps
shedding some light upon the function of specific neural areas.

USE OF AMPHETAMINE IN STUDYING EFFECTS OF CATECHOLAMINE RELEASE

One method that we felt might be fruitful in the further
study of the physiological effects of NE in the Limbic System
is through the utilization of amphetamine, a pharmacological
agent known to specifically effect the release of NE from
vesicles in the brain. Vogt (1954) first suggested that NE was
important in the central effects of amphetamine. McLean and
McCartney (1961) showed that brain and heart NE are decreased
by amphetamine in the rat and this finding has been confirmed by
several investigators (Moore and Lariviere, 1963; Baird and
Lewis, 1964; Gunn and Lewander, 1967; Lewander, 1968). This
has been substantiated by the findings of Weissman et al. (1965)
and Randrup and Munkvad (1966) who showed that pretreatment with
α-methyl tyrosine suppresses the behavioral effects of d-amphetamine. NE release follows the intracisternal administration of
d-amphetamine (Carr, 1970). In the brain stem, the effects of
iontophoretically administered NE to single neurons was mimicked
by iontophoretically applied d-amphetamine and the effect was
blocked by pretreatment with reserpine (Figure 1) (Boakes et al.,
1971). Seventy-eight brain stem cells were studied. Short-lasting and long-lasting inhibitory effects, biphasic effects (inhibition followed by excitation) and excitatory effects were
observed when 1-norepinephrine was applied. d-Amphetamine produced the same excitatory or inhibitory effect as NE in each
case and had no effect on neurons unaffected by NE. Biochemical
and histochemical studies have shown that high doses of d-amphetamine (15mg/kgm) increases the turnover and the extraneural concentration of norepinephrine as well (Glowinski and Axelrod,
1965). In addition, it blocks the uptake of secondarily administered amine in the medulla and the hypothalamus (Glowinski et al.,
1966). It also acts as an in vivo MAO inhibitor (Blaschko,
Richter, and Schlossman, 1962). p-Hydroxynorephedrine has been
identified as a metabolite of amphetamines in the brain (Lewander,
1970; Costa and Groppetti, 1970; Groppetti and Costa, 1969;
Brodie et al., 1970) and it appears that this metabolite acts
to displace brain and heart NE (Lewander, 1971), although the
sites of this action on granular versus extra-granular NE binding
sites is not proven.

Behaviorally, amphetamine acts to facilitate the organism in
performance of operant responses (Stein, 1964), i.e., the drug
acts not to stimulate new behavior but to lower thresholds of old
behavior. This is said to be accomplished by release of NE
particularly from synapses of the MFB in the amygdala and other

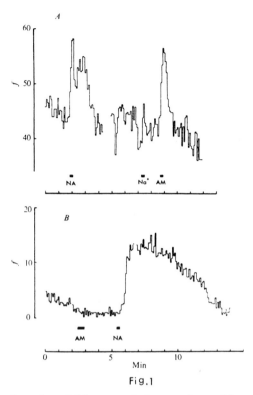

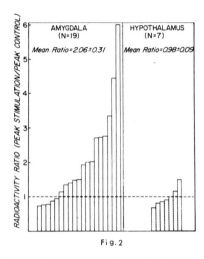

Fig. 1.   Effects of 1-noradrenalin (NA) and d-amphetamine (AM) on a
neurone in an untreated rat (A) and on a neurone in a rat pretreated
with 5 mgm/kg reserpine (B).   The firing rates of both neurones in
impulses s^{-1} are plotted against time in minutes.   Iontophoretic
applications of NA, AM, and of a current control (Na+) are indicated
by horizontal bars.   A, the excitatory response is mimicked by AM.
B, d-amphetamine is without effect on this neurone, whereas NA has a
strong excitant action.   (From Boakes et al., 1971).

Fig. 2.   Amphetamine induced release of radioactive norepinephrine
and metabolites from amygdala but not hypothalamus (one to 3 hours
after intraventricular injection of radioisotope tracer, rats received 45 minutes of rewarding electrical stimulation.   After a 1
hour rest period 3-5 mg/kg of a d-amphetamine sulfate was injected
intraperitoneally.   Radioactivity ratios were calculated from peak
values of perfusate samples collected during the 45 minute periods
before and after the amphetamine injection.   Each bar stands for
1 experiment.   (From Stein and Wise, 1969).

forebrain sites (Stein, 1967; Stein and Wise, 1969). Amphetamine
markedly facilitates the rate of self-stimulation upon electrical
MFB activation (Stein and Seifter, 1961; Olds and Milner, 1954).
This facilitation is blocked by drugs which deplete the brain of
NE. Based on the pharmacological and behavioral data, it seemed
likely to us that one of the sites of amphetamine action may be
in the amygdala (Wepsic, 1963). The experiment described below
demonstrated a neurophysiologic action there. Subsequently,
Stein and Wise made direct measurements of the effects of
amphetamine and NE in the amygdala. Push-pull Gaddam cannulae
were placed in the rostral hypothalamus and amygdala and "reward"
points stimulated in the MFB. NE levels in the perfusate of the
amygdala rose predictably in those animals who received rewarding
stimulation while little change was noted in the hypothalamus,
thalamus or cortex in control experiments. Amphetamine caused a
further rise in NE release from amygdaloid cannulae after reward
sites were stimulated but it had no effect in other areas nor did
it have as marked an effect when MFB reward areas were not activated (Figure 2) (Stein and Wise, 1969).

EXPERIMENTAL RESULTS

Three sets of experiments have been done to examine neurophysiological changes in the amygdala produced by amphetamine:

1. Effect of local instillation of d-amphetamine into
   basal amygdala (Wepsic, 1963).

Cats under local anesthesia, paralyzed and respired arti-
fically, had placement of bipolar concentric electrodes unilaterally in septum and HVM. An outer cannula was placed in the
amygdala through which a micro-injection cannula could be introduced and then exchanged for bipolar stimulating electrode.
Evoked responses were monitored oscillographically in amygdala
upon septal stimulation and in HVM upon amygdaloid stimulation.
After control injection of 0.5 μl of saline had been shown to
have no effect on the latency, threshold , or amplitude of these
responses, 2.5 μgm of freshly prepared d-amphetamine sulfate in
0.5 μl normal saline was slowly injected into the basal amygdala.
This produced little change in amplitude of the response recorded
in the amygdala with septal stimulation; however, the amplitude
of the response recorded in the HVM from amygdaloid stimulation
was reduced (Figure 3). The amplitude of this evoked response
returned to normal in about one hour following injection
(Figure 4). There was no change in latency or wave form but
a rise in threshold was seen during this period.

The absence of a change in response recorded at the site of
direct d-amphetamine or saline control argued against serious

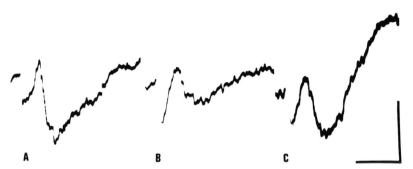

Fig. 3. Evoked response recorded in HVM on amygdaloid stimulation: a - central response; b - response 5 minutes after
injection of amphetamine; c - response 15 minutes later
(calibration - 50 msec 50 μV).

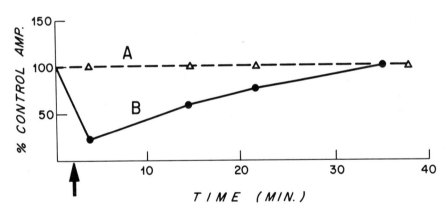

Fig. 4. Amplitude of evoked response recorded in HVM to amygdaloid stimulation (arrow indicates time of injection).

distortion of tissue and subsequent careful histological study
to demonstrate electrode location showed no greater cellular
change in the vicinity of the injection cannula than was seen
when the cannula was placed without drug injection.

  2. Effect of parenteral administration of d-amphetamine upon
      single cell firing in amygdala (Wepsic, 1963).

During a series of experiments primarily designed to map
cells in the amygdala responsive to septal or magnocellular medial
geniculate stimulation (mcMG) the effects of parenteral d-amphetamine upon spontaneously active cells and cells driven by stimulation were measured. Tungsten microelectrodes recorded extracellular potentials in the amygdala and standard bipolar stimulating electrodes were used to stimulate septum and mcMG at near
threshold levels in locally anesthetized, paralyzed, ventilated
cats. The firing rates of eleven spontaneously firing amygdaloid
units were studied before and after intravenous administration
of 30 μgm of d-amphetamine sulfate. (This dosage produced no
change in femoral arterial pressure.) All eleven cells showed
excitation with increased rate of discharge (recording rates of
at least 1.5 times control level within 1 minute of drug
administration). The amplitude and wave form of the units recorded did not change with drug administration indicating little
change in the relative geometry of recording electrode and active
cell. Eight amygdaloid cells activated by septal stimulation
that did not fire spontaneously were also studied, utilizing the
same dosage. Four of these became spontaneously active following
amphetamine administration although during control periods they
were silent. Seven of these eight cells showed an increase in
the number of spikes per stimulus and one remained unchanged. (A
ratio of mean firing rate per stimulus of 1.5 was required to
consider this an "increase" in a period of 40 single stimuli
before and after d-amphetamine administration.) Figure 5 shows
a unit before, 2 minutes after, and 12 minutes after amphetamine
administration. The number of unitary discharges per stimulus is
increased. There is no change in latency or initial wave form,
but the latter positive component of the slow wave is increased.
A plot of the firing/stimulus of this unit is shown in Figure 6.
The rate of discharge increases during the first 15 minutes after
d-amphetamine administration then levels off. (The standard
deviations for each point on Figure 6 are 0.2 to 0.7
firing/stimulus.)

  3. Effect of parenteral administration of d-amphetamine
      upon evoked responses in amygdala and HVM.

Unilateral bipolar concentric electrodes were placed in 20
locally anesthetized, paralyzed, ventilated cats in septum, amygdala and HVM (Figures 7 and 8). Evoked responses were obtained

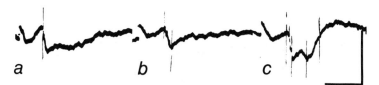

Fig. 5. Effect of amphetamine injection on cellular discharge in amygdala driven by septal stimulation: a - cortical response; b - response 2 minutes after intravenous administration of d-amphetamine sulfate (calibration: 50 msec 50 μV).

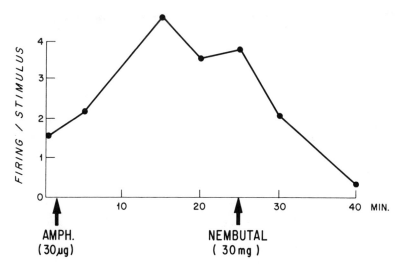

Fig. 6. Effects of d-amphetamine and Nembutol upon unitary discharge of an amygdaloid cell driven by septal stimulation.

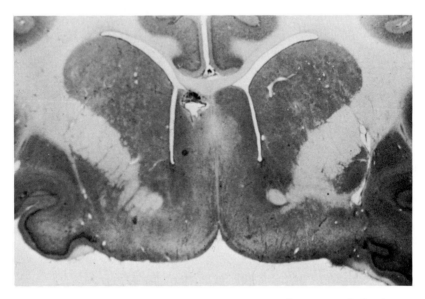

Fig. 7.   Electrode tract in septum (Cresyl violet).

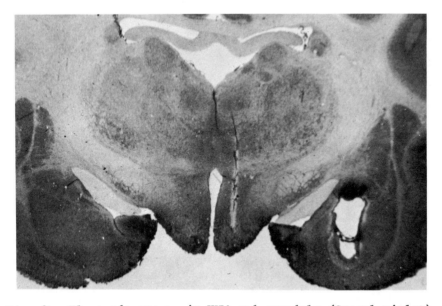

Fig. 8.   Electrode tracts in HVM and amygdala (Cresyl violet).

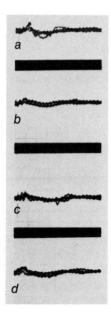

Fig. 9. Evoked response recorded in amygdala to septal stimulation: a - cortical response; b - response 5 minutes after intravenous administration of d-amphetamine; c - 15 minutes later; d - 30 minutes later.

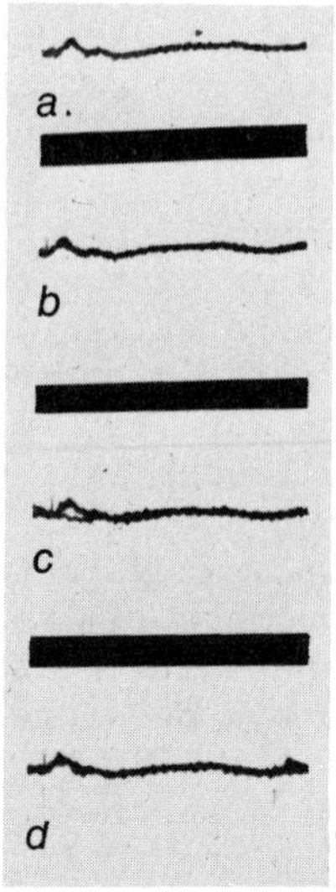

Fig. 10. Evoked potential recorded in HVM to amygdaloid stimulation at a - control; b - 5 minutes after intravenous administration of amphetamine; c - 15 minutes later; d - 30 minutes later.

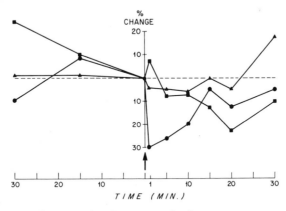

Fig. 11. Changes in amplitude of evoked responses recorded in amygdala to septal stimulation (closed circles), in HVM to septal stimulation (closed squares), and in HVM to amygdaloid stimulation (closed triangles) to intravenous administration of d-amphetamine sulfate (arrow).

in HVM with constant current single shock stimulation of septum
and amygdala not exceeding 250 μA at 100 μsec pulse duration. An
evoked response was also recorded in the amygdala to septal stimulation utilizing the same parameters. All three responses were
obtained before intravenous administration of d-amphetamine
sulfate (0.1 mgm/Kgm). Superimposed sweeps of 10 responses then
were recorded oscillographically for each evoked potential at 30
and 15 minutes prior to drug administration. If the potential
did not remain relatively stable for that period, electrode
position was altered to produce a stable response. Evoked
responses then were recorded during and at 1, 5, 10, 15, 20, and
30 minutes following intravenous d-amphetamine administration.
Samples of such changes are shown in Figures 9 and 10 for septum
to amygdala and amygdala to HVM evoked responses. The latency
and basic wave form of these responses were not changed by this
dose of d-amphetamine which did not alter the animal's temperature
or systemic blood pressure. The amplitude of the responses did
change, as is shown in Figure 11 which summarizes the changes in
averaged amplitude for each set of responses in the animals
studied, taking the amplitude at time of drug administration
as the null.

The most marked change was a drop in amplitude of responses
recorded in the amygdala upon septal stimulation (closed circles).
After one minute, an abrupt drop of about 30 per cent was seen.
The size of the potential then gradually returned to normal.
The response in HVM to septal stimulation, on the other hand,
increased at first, then showed a tendency to fall off with time
(closed squares). This effect, however, was smaller in magnitude
than that seen in amygdala. The response of hypothalamic cells
to amygdala stimulation was small with less than a 10 per cent
change over all (closed triangles) with drug administration.

DISCUSSION

Although one usually considers the effects of amphetamine
to be stimulating centrally, the neurophysiological effects of
both parenterally administered and directly administered drugs
have been shown in this study in the amygdala to decrease electrically activated evoked potentials. On the basis of previously
cited data, one would expect amphetamine to release NE from presynaptic varicosities in the amygdala. Even though the numbers
of such varicosities may be greater in the hypothalamus or MFB,
d-amphetamine in this study had little electrophysiological effect
upon ventromedial hypothalamic neurons or cells in the septum for
the amplitude of septal-HVM evoked responses was not changed by
parenteral administration. The marked reduction in activation of
neurons in the amygdala by septal stimulation after amphetamine
indicates that the drug may act in the amygdala to decrease
electrical excitability. One can not exclude the possibility

that this decreased excitability is secondary to effects in other
areas which may project to the amygdala. However, the finding
that direct application of d-amphetamine to the amygdala also
depressed the electrical activation of cells there is against
that more remote effect.

Finding a small sample of single cells in the amygdala that
increase their firing rates with parenteral administration of
d-amphetamine is more difficult to explain. Although others have
reported a wide range of single cell effects with the agent
(Boakes, 1971) we have found only excitation in the amygdala.
This could be due to an artifact of neuron sampling, i.e.,
selecting only the larger spontaneously active cells and perhaps
excluding those smaller cells which may be the origin of the
evoked responses described earlier.

We are now attempting to document these changes with amphetamine in squirrel monkeys utilizing both evoked potential and
single cell techniques, hoping to survey a larger sample of cells
in awake animals and, in addition, to examine these phenomenon in
animals pretreated with agents known to deplete the brain of NE.

Can these data be used to conclude anything specific about
the function of the amygdala? It is likely that the decrease in
evoked response in the amygdala with amphetamine is due to an
increased concentration of ME in the region of the presynaptic
vesicles. Impulses originating in the septum are not as effective
in evoking a depolarization of a population of amygdaloid cells
with localized elevation of NE concentration. However, electrical stimulation within the amygdala and propagation of an
orthodromic response presumably over this same stria terminalis
pathway to the HVM is not measurably affected.

REFERENCES

AUDEN, N. E., FUXE, K., HAMBERGER, B., & KOKFELT, T. A quantita-
    tive study on the nigro-neostriatal dopamine neuron system in
    rat. Acta Physiologica Scandinavica, 1966, 67, 306.

BAIRD, J. R. C., & LEWIS, J. J. The effects of cocaine, ampheta-
    mine, and some amphetamine-like compounds on the in vivo
    levels of noradrenalin and dopamine in the rat brain.
    Biochemical Pharmacology, 1964, 13, 1475.

BARCHAS, J. D., & FREEDMAN, D. Response to physiological stress.
    Biochemical Pharmacology, 1963, 12, 1232.

BLASCHKO, H., RICHTER, D., & SCHLOSSMAN, H. The oxydation of adrenaline and other amines. Biochemical Journal, 1962, 31, 2187.

BOAKES, R. J., BRADLEY, P. B., & CANDY, J. M. Ablation of the response of brain stem neurons to iontophoretically applied d-amphetamine by reserpine. Nature, 1971, 229, 469.

BRADLEY, P. B., HOSLI, L., & WALSTENCROFT, J. W. Synaptic transmission in the central nervous system and its relevance for drug action. International Review of Neurobiology, 1968, 11, 1.

BRODIE, B. B., CHO, A. K., & GESSA, G. G. Possible role of p-hydroxynorephedrine in the depletion of norepinephrine induced by d-amphetamine and intolerance to this drug. In E. Costa and S. Garattini (Eds.), International Symposium on Amphetamines and Related Compounds, Milano, March, 1969. New York: Raven Press, 1970.

CARR, L. A., & MOORE, K. E. Effects of amphetamine on the contents of norepinephrine and its metabolites in the effluent of perfused cerebral ventricles of the cat. Biochemical Pharmacology, 1970, 19, 2361.

COSTA, E., & GROPPETTI, A. Biosynthesis and storage of catecholamines in tissues of rats injected with various doses of d-amphetamine. In E. Costa and S. Garattini (Eds.), International Symposium on Amphetamine and Related Compounds, Milano, March, 1969. New York: Raven Press, 1970.

CROW, T. J. Mode of enhancement of self-stimulation in rats by methamphetamine. Nature, 1969, 224, 709.

DAHLSTROM, A., & FUXE, K. Evidence for the existence of monoamine neurons in the central nervous system. II. Experimentally induced changes in the intraneural amine levels of bulbospinal neuron systems. Acta Physiologica Scandinavica, 1965, 64, 1.

FUXE, K. Evidence for the existence of monoamine neurons in the central nervous system. IV. The distribution of monoamine nerve terminals in the central nervous system. Acta Physiologica Scandinavica, 1965, 64, 39.

FUXE, K., & GUNNE, L. M. Depletion of the amine stores in brain catecholamine terminals on amygdaloid stimulation. Acta Physiologica Scandinavica, 1969, 62, 493.

GLOWINSKI, J., & AXELROD, J. Effects of drugs on the uptake, release, and metabolism of H^3 norepinephrine in the rat brain. Journal of Pharmacology and Experimental Therapeutics,

1965, 149, 43.

GLOWINSKI, J., AXELROD, J., & IVERSEN, L. Regional studies of catecholamines in the rat brain. IV. Effects of drugs on the disposition and metabolism of H^3 norepinephrine and H^3 dopamine. Journal of Pharmacology, 1966, 153, 30.

GLOWINSKI, J., & IVERSEN, L. Regional studies of catecholamines in the rat brain. I. The disposition of H^3 norepinephrine, H^3 dopamine and H^3 dopa in various regions of the brain. Journal of Neurochemistry, 1966, 13, 655.

GOLDSTEIN, M. B., ANAGNOSTE, W. S., OWEN, J., & BATTISTA, A. F. Studies on the regional biosynthesis and metabolism of catecholamines in the central nervous system of the monkey. Experientia, 1967, 23, 98.

GROPPETTI, A., & COSTA, E. d-Amphetamine (A): Metabolites and depletion of brain and heart norepinephrine (NE) in guinea pig and rat. Federation Proceedings, 1969, 28, 795.

GUNNE, L. M. Brain catecholamines in the rage response evoked by intracerebral stimulation and ablation. In S. Garattini and E. B. Sigg (Eds.), Aggressive Behavior. Amsterdam, Excerpta Medica Foundation, 1969, p. 238.

GUNNE, L. M., & T. LEWANDER. Long time effects of some dependence-producing drugs on the brain monoamines. In Molecular Basis of Some Aspects of Mental Activity. London and New York: Academic Press, 2, 75.

GUNNE, L. M., & REIS, D. Changes in brain catecholamines associated with electrical stimulation of amygdaloid nucleus. Life Sciences, 1963, 1, 804.

HILLARP, N. A., FUXE, K., & DAHLSTROM, A. Demonstration and mapping of the central neurons containing dopamine, nor-adrenaline, and 5-hydroxytryptamine and their reactions to psychopharmaca. Pharmacological Review, 1966, 18, 727.

COSTA, E., & GARATTINI, S., (Eds.). International Symposium on Amphetamines and Related Compounds. New York: Raven Press, 1970.

KETY, S. S. The central physiological and pharmacological effects of the biogenic amines and their correlations with behavior. In G. C. Quarton, T. Melnechuk, and F. O. Schmitt (Eds.), The Neurosciences, A Study Program. New York: Rockefeller University Press, pp. 441-451.

KETY, S. S. The biogenic amines in the central nervous system:
    Their possible roles in arousal, emotion, and learning. In
    F. O. Schmitt (Ed.), The Neurosciences, Second Study Program.
    New York: Rockefeller University Press, 1970, p. 324.

LEAF, R. C., LERNER, L., & HOROVITZ, Z. P. The role of the amyg-
    dala in the pharmacological and endocrinological manipulation
    of aggression. In S. Garattine and E. B. Sigg (Eds.),
    Aggressive Behavior. Amsterdam, Excerpta Medical Foundation,
    1969, p. 120.

LEWANDER, T. Urinary excretion and tissue levels of catecholamines
    during chronic amphetamine intoxication. Psychopharmacologica,
    1968a, 13, 394.

LEWANDER, T. Effects of amphetamine on urinary and tissue
    catecholamines in rats after inhibition of its metabolism
    with desmethylimipramine. European Journal of Pharmacology,
    1968b, 5, 1.

LEWANDER, T. Catecholamine turn-over studies in chronic amphet-
    amine intoxication. In E. Costa and S. Garattini (Eds.),
    International Symposium on Amphetamines and Related
    Compounds, Milano, March, 1969. New York: Raven Press, 1970.

LEWANDER, T. On the presence of p-Hydroxynorephedrine in the rat
    brain and heart in relation to changes in catecholamine levels
    after administration of amphetamine. Acta Pharmacologica et
    Toxicologica, 1971, 29, 33.

MANDELL, A. J., & SPOONER, C. E. Psychochemical research studies
    in man. Science, 1968, 162, 1442.

MARGULES, P. L. Noradrenergic basis of inhibition between reward
    and punishment in amygdala. Journal of Comparative and
    Physiological Psychology, 1968, 66, 329.

MARK, V. H., TAKADA, I., TAKAMATSU, H., TOTH, E., MARK, D. B., &
    ERVIN, F. R. The effect of exogenous catecholamines in the
    amygdala of a "rage" cat. Unpublished observation, 1971.

MAYNART, E. W., & LEVI, R. Stress induced release of brain
    norepinephrine and its inhibition by drugs. Journal of
    Pharmacology and Experimental Therapeutics, 1964, 143, 90.

MOORE, K. E., & LARIVIERE, E. W. Effects of d-amphetamine and
    restraint on the content of norepinephrine and dopamine in
    rat brain. Biochemical Pharmacology, 1963, 12, 1283.

OLDS, J., & MILNER, P. Positive reinforcement produced by elec-
     trical stimulation of septal area and other regions of rat
     brain. Journal of Comparative and Physiological Psychology,
     1954, 47, 419.

RANDRUP, A., & MUNKVAD, I. Role of catecholamines in the amphet-
     amine excitatory response. Nature, 1966, 211, 540.

REIS, D. J., & GUNNE, L. M. Brain catecholamines: relation to
     the defense reaction evoked by amygdaloid stimulation in cat.
     Science, 1965, 149, 450.

REIS, D. J., & FUXE, K. Brain norepinephrine: Evidence that
     neuronal release is essential for sham rage following brain
     stem transection in cat. Proceedings of the National
     Academy of Science, 1969, 64, 108.

SALMOIRAGHI, G. C., & BLOOM, F. E. Pharmacology of individual
     neurons. Science, 1964, 144, 493.

SCHILDKRAUT, J. J., & KETY, S. Biogenic amines and emotion:
     Pharmacological studies suggest a relationship between brain
     biogenic amines and affective states. Science, 1967, 156, 21.

SEGAL, D. S., & MANDELL, A. J. Behavioral activation of rats
     during intraventricular infusion of norepinephrine.
     Proceedings of the National Academy of Science, 1970, in
     press.

SHEARD, M. H., APPEL, J. B., & FREEDMAN, D. X. The effects of
     central nervous system lesions on brain monoamines and
     behavior. Journal of Psychiatric Research, 1967, 5, 237.

SLANGER, J. L., & MILLER, N. E. Pharmacological tests for the
     function of hypothalamic norepinephrine in eating behavior.
     Physiology and Behavior, 1969, 4, 543.

STEIN, L. Amphetamine and neural reward mechanisms. In A. V. S.
     deReuch and J. Knight (Eds.), Ciba Foundation Symposium on
     Animal Behavior and Drug Action. London: Churchill, 1964a.

STEIN, L. Self-stimulation of the brain and the central stimulant
     action of amphetamine. Federation Proceedings, 1964b, 23, 836.

STEIN, L. Psychopharmacological substrates of mental depression.
     In S. Garattini and M. N. G. Dukes (Eds.), Antidepressant
     Drugs. Amsterdam: Excerpta Medica Foundation, 1967, p. 130.

STEIN, L., & SEIFTER, J.   Possible mode of antidepressive action
    of imipramine. Science, 1961, 134, 286.

STEIN, L., & WISE, C. D.   Release of norepinephrine from hypo-
    thalamus and amygdala by rewarding medial forebrain bundle
    stimulation and amphetamine. Journal of Comparative and
    Physiological Psychology, 1969, 67, 189.

VOGT, M.   Concentration of sympathin in different parts of the
    central nervous system under normal conditions and after the
    administration of drugs. Journal of Physiology, 1954,
    123, 451.

THIERRY, A. M., JAVOY, F., GLOWINSKI, J., & KETY, S. S.   Effects
    of stress on the metabolism of norepinephrine, dopamine, and
    serotonin in the central nervous system of the rat. I.
    Modifications of norepinephrine turnover. Journal of
    Pharmacology and Experimental Therapeutics, 1968, 163, 163.

WEISSMAN, A., KOE, K. B., & TEREN, S. S.   Antiamphetamine effects
    following inhibition of tyrosine hydroxylase. Journal of
    Pharmacology, 1965, 151, 339.

WEPSIC, J. G.   The electrical activity of the basal amygdaloid
    nuclei: afferent connections, sensory responses, and
    amphetamine activation. Thesis, Yale University Medical
    School.

NEUROENDOCRINOLOGY

EFFECTS OF LESIONS AND ELECTRICAL STIMULATION OF THE
AMYGDALA ON HYPOTHALAMIC-HYPOPHYSEAL-REGULATION

Andrew J. Zolovick

The Worcester Foundation for Experimental Biology

Shrewsbury, Massachusetts

INTRODUCTION

Experimental evidence supports the concept of the hypothalamus as the primary neural substrate for neuroendocrine regulation of hypophyseal hormone secretion. However, research over the last few decades has established the importance of the limbic system in the mediation of endocrine expression. Functioning as an integrative mechanism for interoceptive, exteroceptive and emotional information, the limbic system communicates with lower brain stem structures and the higher neopallial systems through extensive afferent and efferent connections. Therefore, in consideration of the enormous variety of autonomic and behavioral responses elicited by external factors or experimental manipulation of the various components of the limbic system, and of the fundamental role of the endocrines in the maintenance of homeostasis, it is not surprising that activation of the endocrine system should follow disturbances to limbic structures. Research over the last ten years has focused on elucidating the nature of the interactions of various limbic structures with the hypothalamus. Early experiments using lesioning or stimulation techniques established an intimate functional relationship between the amygdala, a component of the limbic system, and the hypothalamus in the mediation of endocrine and behavioral phenomena. A modulatory role was assigned to the amygdala on the evidence that it functioned in the regulation of these phenomena but, unlike the hypothalamus, was not necessary for their elaboration. Through interdisciplinary research, the complex nature of this modulatory mechanism is slowly being elucidated. In spite of the enormous amount of work devoted to the study of the amygdala, its functional relationship in the mediation of hypophyseal tropic hormone secretion remains unclear.

The purpose of this report is to discuss amygdaloid-hypothalamic
interactions in the mediation of hypophyseal tropic interactions and
in the mediation of hypophyseal tropic hormone secretion.

Effects of Lesions of the Amygdala on Gonadotropin Secretion

Early experiments involving lesions of the amygdala or
ablation of the temporal lobe have produced behavioral abnormalities, including hyper- or hyposexuality, with or without secondary
effects on gonadal function. The hypersexuality appears to be more
severe and more diversified in the male than in the female (Klüver
and Bucy, 1937, 1938; Schreiner and Kling, 1953, 1954, 1956;
Terzian and Ore, 1955; Green et al., 1957; Wood, 1958; Kling et al.,
1960; Anand et al., 1959), and under certain conditions is abolished
by castration or by placement of lesions in the ventral medial
hypothalamic (VMH) (Schreiner and Kling, 1954) or septal nuclei
(Kling et al., 1960). Wood (1958) and Eleftheriou and Zolovick
(1966) reported that destruction of the basolateral amygdaloid
complex was responsible for the hypersexuality, and that medial
amygdaloid lesions may, in fact, inhibit sexual behavior
(Eleftheriou and Zolovick, 1966).

Although amygdaloid lesions cause a greater amount of hypersexuality in males than in females, the reverse is true concerning
the function of the genital organs. Amygdalectomy in the adult
male rat and cat results in marked degeneration of the testes,
whereas in the female cat the ovaries remain unaffected (Greer and
Yamada, 1959; Kling et al., 1960; Yamada and Greer, 1960). In
only two studies have amygdaloid lesions resulted in increased
ovarian function. Whereas Elwers and Critchlow (1960) have shown
that medial amygdaloid lesions stimulate the release of gonadotropin in prepubertal rats, Eleftheriou and co-workers reported that
basolateral nuclear lesions are responsible for increased secretion
of gonadotropin in adult male and female deermice (Peromyscus
maniculatus bairdii) (Eleftheriou and Zolovick, 1967; Eleftheriou
et al., 1967). Only in the latter species has the endocrinology
associated with disruption of the amygdala been investigated extensively.

Shortly after lesions are placed in the basolateral amygdaloid complex of deermice, estrous cycling terminates and the
vaginal smear reflects a diestrus condition with an occasional
appearance of mucus, indicative of pseudopregnancy. The pseudo-
pregnant-like state persists from 10 to 21 days, followed by
irregular estrous cycles. Ovarian ewight gradually increases
throughout the three-week experimental period, after placement of
the lesions, while uterine weight remains essentially unchanged
from the diestrus value for the first two weeks, then gradually
increases to a third-week post-lesion value which is intermediate

Table I

Ovarian and Uterine Weight During the Estrous Cycle and
After Placement of Lesions in the Basolateral
Amygdaloid Complex

Treatment	Ovarian Weight (mg% ± SD)*	Uterine Weight (mg% ± SD)
Diestrus	91.8 ± 5.1	140.0 ± 2.5
Proestrus	112.2 ± 3.0	205.1 ± 8.1
Estrus	119.1 ± 2.8	212.3 ± 6.2
1 week post-lesion	117.4 ± 2.7	137.8 ± 4.1
2 week post-lesion	149.7 ± 4.5	131.9 ± 5.6
3 week post-lesion	160.2 ± 4.7	153.2 ± 5.8
1 week sham-control	96.8 ± 3.1	173.3 ± 4.8
2 week sham-control	83.9 ± 1.9	165.4 ± 4.7

* Tissue weight expressed as mg/100g body weight.

between that of diestrus and proestrus (Table 1). Ovaries taken from deermice two weeks after placement of the lesions are heavily luteinized without the presence of secondary follicles, indicative of increased secretion of luteinizing hormone (LH), and possibly luteotropin (LTH), and impaired secretion of follicle stimulating hormone (FSH) (Fig. 1a; Table II). The appearance of secondary follicles in the ovary between the 2 and 3 week postoperative period, accompanied by an increase in uterine weight, suggests resumption of FSH secretion (Fig. 1b). Using the pigeon-crop sac assay, Norman (1969) has demonstrated that serum and pituitary levels of LTH are depressed in the deermouse following basolateral amygdaloid lesions. At least, in the deermouse, ovarian luteinization following amygdaloid lesions results entirely from increased secretion of LH (Table II). Welsch et al. (1969) concluded that the regression in carcinogen-induced mammary tumors in rats bearing lesions in the amygdala was a result of diminished LTH and estrogen secretion. The absence of a uterine response in deermice following

Table II

Pituitary and Plasma Levels of Luteinizing Hormone (LH), Follicle Stimulating Hormone (FSH) and Hypothalamic Follicle Stimulating Hormone-Releasing-Factor (FRF) During the Estrous Cycle and After Placement of Lesions in the Basolateral Amygdaloid Complex

Treatment	Pituitary FSH (μg/mg ± SE)*	Plasma FSH (μg/ml ± SE)	FRF** Equivalent ± SD	Pituitary LH (mU/mg)***	Plasma LH (mU/ml)
Diestrus	22.3 ± 2.2	11.5 ± 1.8	0.15 ± .04	0.62	0.81
Proestrus	13.8 ± 0.6	20.4 ± 2.3	0.20 ± .03	0.54	0.71
Estrus	—	—	0.15 ± .02	0.18	1.88
1 week post-lesion	24.6 ± 1.8	11.1 ± 1.2	0.16 ± .01	0.43	0.63
2 week post-lesion	28.3 ± 1.7	9.0 ± 1.3	0.12 ± .02	0.35	1.76
3 week post-lesion	26.3 ± 1.7	10.1 ± 1.5	0.06 ± .03	0.21	2.58
1 week sham-control	14.6 ± 0.8	13.1 ± 1.1	0.18 ± .06	—	—
2 week sham-control	16.4 ± 0.4	13.8 ± 1.2	0.17 ± .04	—	—

* = Potency expressed as μg-equivalents of NIH-FSH-S3-ovine; assayed according to HCG ovarian augmentation assay of Brown, 1955.

** = Assayed according to method of Igarashi and McCann (1964); from Eleftheriou and Pattison, 1967.

*** = Potency expressed as mU-equivalent of NIH-LH-S5-ovine; adapted from Eleftheriou and Zolovick, 1967.

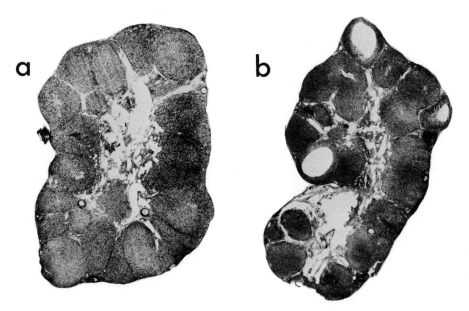

Fig. 1. Ovarian sections taken (a) two weeks and (b) three weeks
after placement of bilateral lesions in the basolateral amygdaloid
nuclei of adult deermice.

amygdaloid lesions also suggests a significant reduction in ovarian
estrogen production. While amygdaloid lesions appear to suppress
LTH secretion, local implants of estrogen into the medial amygdaloid nucleus (AME), central amygdaloid nucleus (ACE), basomedial
portion of the basolateral amygdaloid nucleus (ABL) or the stria
terminalis have been reported to induce lactogenic responses in
pseudopregnant rabbits (Tindal and Knaggs, 1966; Tindal, Knaggs
and Turvey, 1967), presumably by interfering with the secretion of
a hypothalamic prolactin-inhibiting-factor.

     Serum levels of LH also are elevated in male deermice after
placement of lesions in the basolateral amygdaloid complex
(Eleftheriou et al., 1967). Plasma and pituitary content of LH
rose significantly one week after the operation and remained
significantly higher throughout the three-week experimental period
in comparison to sham-operated or intact deermice. The increase
in plasma LH was reflected in a significant increase in sex-tissue
weight (Table III). Secretion patterns of LH and its hypothalamic
releasing-factor (LH-RF) were further studied by Eleftheriou et al.
(1970) in male deermice following placement of lesions in various
nuclei of the amygdala in an attempt to localize the regulatory
center for LH secretion. Bilateral lesions confined to the
cortical (ACO) or basolateral amygdaloid nuclei were effective in

Table III

Pituitary and Plasma Levels of Luteinizing Hormone (LH) and Weight of Sex-tissues of Male
Deermice After Placement of Bilateral Lesions in the Basolateral Amygdaloid Complex

Treatment	Time After Lesion	Pituitary LH (mU/mg)☆	Plasma LH (mU/mg)	Tissue Weight (mg%)		
				Testes	Seminal vesicles	Prostates
Intact	———	.171	.156	1198±279	632±125	83.5±26.2
Lesions	3 days	.291	.183	1194±186	689±112	106.5±42.3
	1 week	.475	.442	1335±140	835±206	119.6±24.2
	2 week	.547	.332	1369±209	916±218	133.7±23.8
	3 week	.562	.553	1363±129	778±236	124.2±28.4
Sham-control	1 week	.322	.217	1222±120	680±136	82.4±28.4

☆ = Potency expressed as mU-equivalents of NIH-LH-S5-ovine; adapted from Eleftheriou et al., 1967

Table IV

Effect of Amygdaloid Lesions on Pituitary Luteinizing Hormone (LH), Hypothalamic Luteinizing
Hormone-Releasing-Factor (LH-RF) Content, and Weight and Fructose Concentration of Seminal
Vesicles in Male Deermice

Lesion	Pituitary LH*	LH-RF Equivalent	Seminal Vesicle Weight (mg ± SE)	Seminal Vesicle Fructose Concentration (ug/100 mg ± SE)
Control	0.65	1.04	68 ± 1.4	300 ± 5.7
Sham-Control	0.61	1.05	70 ± 1.6	311 ± 5.3
Cortical	1.53**	0.72**	85 ± 2.3**	379 ± 6.2**
Basolateral	1.44**	0.46**	81 ± 2.1**	386 ± 6.6**
Medial	1.00	4.25**	71 ± 1.8	318 ± 5.8

* = Potency expressed in ug-equivalents of NIH-LH-S11; from Eleftheriou et al., 1970.

** = Significant from controls p< .01; "t" test.

Table V

Effect of Vaginal Stimulation or Electrical Stimulation of the Medial Amygdaloid Nucleus
(AME) on Hypothalamic Unit Activity and EEG After-Reaction in the Female Deermouse Under
Various Hormonal Conditions

Treatment	Stimulus	No. of Animals	Positive Response (%)	Negative Response (%)	Unresponsive (%)	After-[*] Reaction
Estrogen-Induced	Vaginal	27	32	43	25	39/43
Estrus	AME	13	22	55	23	11/12
Proestrus-	Vaginal	19	45	37	18	21/30
Estrus	AME	19	37	50	13	22/23
Diestrus	Vaginal	16	36	7	57	4/44
	AME	10	47	5	48	1/19
Pregnant	Vaginal	10	28	10	62	0/35
	AME	3	14	28	58	0/7
Ovariectomized	Vaginal	7	25	0	75	0/16
	AME	7	37	0	69	2/8
Ovariectomized + Estrogen	Vaginal	5	50	8	42	7/12
	AME	3	33	20	47	6/10

[*] = Number of after-reactions/number of units recorded; Adapted from Zolovick, 1969

producing a sustained increase in pituitary weight and LH content
and fructose content of seminal vesicles (Table IV), while pituitary content of LH was significantly reduced in animals bearing
lesions in the AME.   In addition, a significant increase in
hypothalamic content of LH-RF occurred in deermice with ACO and
ABL lesions while hypothalamic content of LH-RF was significantly
reduced in animals with lesions in the AME. Although serum
levels of LH were not determined in the present experiment,
Eleftheriou et al. (1967) had confirmed previously an increase in
serum LH in male deermice after placement of lesions in the basolateral amygdaloid complex; therefore, it must be assumed that
serum LH also is elevated following ACO lesions, especially when
seminal vesicle weight and fructose content are taken into consideration.   There exists an apparent difference in the mechanism
for LH secretion between male and female deermice after ABL lesions.
Whereas in the male, both serum and pituitary LH are elevated after
placement of the lesions, in the female, serum LH is elevated but
pituitary content is diminished (Eleftheriou and Zolovick, 1967).
Although it would appear that synthesis of LH fails to keep pace
with release in the female, it must be emphasized that serum
levels of LH are almost four times higher, and perhaps this dis-

parity only reflects a greater secretion rate of the hormone.

Extending the above studies to the adult male rat, Eleftheriou
et al. (1969) confirmed the regulatory role of the ABL in the
secretion of hypophyseal LH. Bilateral electrocoagulation of the
ABL, sub-total amygdalectomy, involving the basolateral complex,
ACE and ACO nuclei, or implants of actinomycin-D resulted in a
significant 3 to 4 fold increase in serum levels of LH three weeks
after the operation. Inhibition of protein synthesis proved more
effective in augmenting serum LH levels than either sub-total
anygdalectomy or ABL lesions, even in the presence of declining
body weight. Lesions confined to the AME nuclei or implants of
cholesterol failed to alter resting levels of LH.

The increased synthesis and release of LH following amygdaloid lesions has been confirmed recently by Lawton and Sawyer
(1970) in adult gonadectomized rats. Amygdaloid lesions produced
an additive effect on secretion of LH in the presence of already
high titers of serum LH brought about by ovariectomy. Furthermore,
implants of estrogen in the amygdala or preoptic area failed to
suppress the elevated serum levels of LH and estrogen, as evidenced
by an increase in uterine weight following gonadectomy, whereas
estrogen implanted into the arcuate-ventral medial hypothalamic
nuclei was effective in this respect. The authors concluded that
the amygdala and preoptic area represent positive loci for
estrogen in the feedback mechanism for LH. They further concluded
that the overall function of the amygdala is to exert an inhibitory influence on the hypothalamus in the secretion of LH and that
rising titers of estrogen, such as during proestrus, remove this
influence, thereby facilitating ovulation. Recently, Velasco and
taleisnik (1969) reported induction of ovulation in persistent
estrus rats, induced by continuous illumination or pharmacologic
agents, after electrolytic deposition of iron into the amygdala.
They concluded that the amygdala exerts a stimulatory influence
on the secretion of ovulatory hormone by virtue of the stimulatory
nature of the iron fragments. Lawton and Sawyer (1970) do not
dispute the possibility that metallic fragments may produce a
focal point of irritation whose potentials may spread to the hypothalamus to induce an acute discharge of ovulatory hormone, but
question the long term stimulatory effect of metal fragments.
Furthermore, in experiments with deermice, lesions were produced
by high-frequency electrocoagulation, thus precluding the deposition of iron as a major factor in long term stimulation of
LH release. In addition, studies have shown that mere insertion
of an electrode into the amygdala, without passage of current,
causes sufficient damage to induce secretion of a small but consistent amount of LH (Eleftheriou et al., 1967; Lawton and Sawyer,
1970), presumably by damaging some of the hypothalamic afferents
(Elwers and Critchlow, 1961).

In contrast to the stimulatory nature of amygdaloid lesions
on genital tissue, others have reported genital atrophy after
destruction of the amygdala or ablation of the temporal lobe,
with or without accompanying aphagia or adipsia (Klüver and Bucy,
1938; Klüver and Bartelmez, 1951; Koikegami et al., 1955; Greer
and Yamada, 1959; Kling et al., 1960; Yamada and Greer, 1960;
Kobayashi and Kobayashi, 1961; Welsch et al., 1969). Whether the
reduction in gonadal function is a secondary effect of decreased
food intake or other complicating factors due to surgery or a
specific result of the lesions remains to be determined. Of
interest is the report by Schwartz and Kling (1954) that smaller
lesions that failed to affect testicular weight in rats were,
nevertheless, sufficiently large to induce aphagia.

Destruction of other areas of the limbic system that affect
directly amygdaloid or hypothalamic function also impair gonadotropin secretion. Lesions confined to the hippocampus are known
to disrupt estrous cycling in rats (Rodriguez, 1959; Koikegami,
1964) and reduce gonadal weight and activity (Riss et al., 1958;
1963). Delay in puberty was elicited in immature rats after
placement of lesions in various parts of the olfactory-hippocampal-
hypothalamic axis, but failed to affect normal reproductive functions when the surviving animals reached adulthood (Kling and
Grove, 1963; Kling, 1964). In addition, hippocampal lesions
failed to affect testicular weight and junction. Lesions confined to the anterior neocortex of male rats resulted in degeneration in seminiferous tubules (Soulairac and Soulairac,
1958), while neocortical or amygdaloid ablation failed to affect
copulation-induced ovulation in the rabbit (Brooks, 1937; Sawyer,
1959) or estrus in cats (Bard and Rioch, 1937). Recently,
Critchlow (1958) has shown that lesions confined to the dorsal
midbrain block spontaneous ovulation in the proestrus rat, which,
however, could not be confirmed by others (Perkary et al., 1967),
while Benedetti et al. (1965) reported an increase in ovarian and
uterine weight after placement of lesions in the midbrain periaqueductal gray. Carrer and Taleisnik (1970) has shown that
electrolytic stimulation of the midbrain ventral tegmental area,
raphe nuclei or peri-aqueductal gray inhibits spontaneous ovulation and depresses serum levels of LH in proestrus rats, while
electrolytic stimulation of the dorsal tegmental area of
persistent estrus rats evokes ovulation and elevates serum
levels of LH without affecting FSH secretion. Harris (1958)
has questioned whether the reticular formation-hypothalamus-
pituitary mechanism functions in complex and discriminative endocrine responses as ovulation, suggesting instead that this
mechanism functions more for the expression of stereotyped or
uniform endocrine responses such as the adrenocorticotropic
response to stress.

Effect of Amygdaloid Stimulation on Gonadotropin Secretion

Uterine movements following electrical stimulation of the
amygdala were first observed by Yamada (1954) and Koikegami et al.
(1954) in the dog, cat and rabbit. Augmentation of frequency and
amplitude of contraction were more or less dependent upon the
stage of the sexual cycle at the time of experimentation with the
gravid uterus being particularly sensitive and the premature,
pseudopregnant or early stages of pregnancy being relatively in-
sensitive or weakly responsive. Shealy and Peele (1957) and
Spoto et al. (1961) have since confirmed Yamada's original
observations. A characteristic latent period of about 30 seconds
follows application of the stimulus, with the response terminating
about 4 to 12 minutes later. The uterine response is independent
of intact spinal cord connections and indistinguishable from the
response elicited by an injection of posterior pituitary preparation (Azuma and Kumagai, 1934). The cortical amygdaloid nucleus
appears to be the active area in these studies.

Stimulation of the cortical nucleus or the intermediate
principal nucleus has been shown to induce ovulation, hemorrhagic
follicles or newly formed corpa lutea in sexually mature rabbits
and cats (Koikegama et al., 1954; Ursi, 1955; Saul and Sawyer,
1957; Shealy and Peele, 1957). Rats in light-induced persistent
estrus also ovulate after medial amygdaloid or septal stimulation
(Bunn and Everett, 1957). Velasco and Taleisnik (1969) have
since confirmed Bunn and Everett's observation that the impulse
from the medial amygdala is carried via the stria terminalis.
Electrochemical (deposition of iron) or chemical (carbachol)
stimulation of the strial fibers or its bed nucleus induces
ovulation with a concomitant rise in serum LH and FSH, whereas
transection of the strial fibers blocks the response. Transection
of the ventral amygdalo-fugal pathway failed to block the ovulatory response to amygdaloid stimulation. The ovulatory response
was elicited from the medial and basolateral nuclei by electrolytic stimulation and from the basolateral, medial and central
amygdaloid nuclei by chemical stimulation. Since the current
used to deposit the iron from the elctrode could conceivably
produce a lesion, the specificity of the procedure is in doubt.
It would be interesting to see if there was a chronic release
of LH in these animals. The authors contend that carbachol
stimulation evoked salivation and seizures indicating amygdaloid
induced hyperarousal. Under the trauma of acute stress, animals
are known to secrete gonadotropins (Árvay, 1964; Eleftheriou
and Church, 1967) and thyrotropin (Eleftheriou et al., 1968)
along with adrenocorticotropin. Therefore, since Valasco and
Teleisnik (1969) measured an acute release of gonadotropin,
shortly after the operation, surgical trauma as well as hyperarousal and seizures must be considered in the overall ovulatory

response. Furthermore, mere insertion of an electrode into the
cerebral cortex has been shown to deplete a significant amount
of ovarian ascorbic acid in constant estrus rats (Taleisnik et
al., 1962). If iron deposition does, in fact, constitute a
genuine stimulus, then this is the first report of augmented
release of FSH following amygdaloid stimulation.

Other indices of gonadotropin activity have been used to
follow the release of an ovulatory discharge of gonadotropin
after electrical stimulation of the amygdala. Ovarian pro-
duction of 20α-OH-Δ^4-pregene-3-one (20α-OH) has been shown to
follow electrical stimulation of the medial amygdala or basal
hypothalamus (Hayward et al., 1964) of mature rabbits. The
quantity of 20α-OH released from the ovary after brain stimulation is about half the quantity released after coital or LH-
induced ovulation. Moreover, the ovarian progestin response,
as well as ovulation induced by electrical stimulation of the
brain, can be blocked by pharmacological agents that inhibit
coital-induced ovulation. Extending Hayward's experiments,
Kawakami et al. (1966, 1968) have demonstrated that electrical
stimulation applied to the intermediate principal nucleus of
the amygdala, dorsal hippocampus or hypothalamic arcuate
nucleus increases the secretion of ovarian progestin with
only a marginal increase in ovarian estrogen output. The
ovarian progestin response to hippocampal stimulation was
abolished by destruction of the dorsal fornix, septum or arcuate
nucleus but not after transection of the stria terminalis, while
transection of the latter pathway abolished the amygdaloid-in-
duced ovarian progestin response. No effect on ovarian progestin
production occurred when the basolateral nuclear complex was
stimulated. These data suggest that the hippocampus exerts its
effect on ovarian progestin secretion independently of the
amygdala. Degeneration studies have shown that information
processed in the hippocampus reaches the anterior portion of
the arcuate nucleus via the medial cortical-hypothalamic tract
(Nauta, 1956; Raisman, 1970). Whenever a subovulatory quantity
of gonadotropin followed brain stimulation, ovarian progesterone
output increased, but not the synthesis or release of 20α-OH
(Kawakami et al., 1968). Implantation of progesterone into the
hippocampus facilitated ovarian production of 20α-OH and progesterone without alteration of estradiol or estrone synthesis,
whereas progesterone implants in the amygdala or arcuate nucleus
were ineffective. These data suggest a positive feedback
mechanism for progestins in the hippocampus, and a negative
feedback mechanism in the amygdala.

The excitability of the hippocampus and amygdala to central
or peripheral stimulation is altered under the influence of
the sex-steroids. Progesterone or LH facilitates the hippo-

campally-evoked bioelectric potentials in the arcuate nucleus
in adult ovariectomized cats, while estrogen or FSH inhibits
the response. The amygdaloid evoked response recorded in the
arcuate nucleus displays an inverse relationship with the hippocampus to sex-steroids (Kawakami and Terasawa, 1967). Recording
the localized seizure threshold response in the dorsal hippocampus
and amygdala of adult cyclic rats, Terasawa and Timiras (1968)
have confirmed the reciprocal relationship in the excitability
of the above limbic structures to gonadal hormones. The seizure
threshold in the hippocampus was lowered significantly during
proestrus and increased during diestrus and estrus, while the
seizure threshold in the lateral amygdala showed an inverse
relationship to that of the dorsal hippocampus. However, the
seizure threshold pattern in the medial amygdala closely
paralleled that of the hippocampus with the important difference
that it rose abruptly during mid-proestrus. The increase in
seizure threshold at mid-proestrus coincided with the critical
period for spontaneous ovulation, that period in which ovulation
can be blocked by administration of pharmacological agents. The
difference in sensitivity between the medial and lateral
amygdala during proestrus is consistent with the interesting
observation that electrical stimulation of the medial amygdala
is more effective in eliciting ovulation in persistent estrus
and prepubertal animals than stimulation of the lateral
amygdala. In addition, only during mid-afternoon of proestrus,
the critical period, is progesterone effective in elevating
serum levels of LH in ovariectomized estrogen-treated rats
(Caligaris et al., 1968). Ovariectomy abolished the cyclic
nature of the seizure threshold, while administration of 0.1 μg
of estradiol-17β reinstated one complete cycle and 10.0 μg of
estrogen yielded multiple cycles of decreasing amplitude.

Following copulation, the cat and rabbit undergo a series
of behavioral events known as the post-coital behavioral after-
reaction (Bard, 1940; Sawyer and Kawakami, 1959). Electrophysiological correlates of this behavioral sequel have since
been observed in electrical recordings from cortical and subcortical brain structures of the cat (Porter et al., 1957),
rabbit (Sawyer and Kawakami, 1959), rat (Barraclough, 1960;
Ramirez et al., 1967) and the deermouse (Zolovick and
Eleftheriou, 1971). A uniform event in the EEG after-reaction
(EEG-AR) is the appearance of a slow wave "sleep-like" stage
followed by a stage of somatic or "paradoxical" sleep in which
there is a frequent loss of postural tone with occasional
ocular movements while the EEG displays an arousal pattern.
Unit activity studies have confirmed the hypothalamic activation
stage of the EEG-AR to vaginal stimulation (Porter et al., 1957;
Sawyer and Kawakami, 1959; Barraclough and Cross, 1963; Law and
Sackett, 1965; Ramirez et al., 1967; Chhina et al., 1968; Chhina

and Anand, 1969; Lincoln, 1969a, 1969b; Haller and Barraclough, 1970; Vincent et al., 1970; Zolovick and Eleftheriou, 1971) and Kawakami and Sawyer (1959b) have shown that sex-steroids regulate thresholds in various brain areas for the production of the EEG-AR and hypothalamic arousal, and have emphasized the importance of the reticular formation in its production. Kawakami and Sawyer (1959a) have shown that electrical stimulation of limbic structures, copulation and exogenous LH or other gonadotropins known to induce ovulation, are effective stimuli in eliciting the EEG-AR in adult animals, while genital stimulation of the immature animal fails to evoke a specific hypothalamic EEG response (Chhina et al., 1968). Estrogen appears to facilitate the EEG response in immature and adult animals (Chhina and Anand, 1969) while ovariectomy or pregnancy abolishes it (Zolovick and Eleftheriou, 1971). Closely correlated with the appearance of the EEG-AR are the post-stimulus activity patterns of hypothalamic neurons. During the proestrus-estrus phases of the estrous cycle, discharge patterns of neurons in the arcuate nucleus and lateral hypothalamus are facilitated in response to vaginal stimulation while neurons in the medial-basal hypothalamus and along the midline of the anterior hypothalamic area decrease in discharge rate (Kawakami and Saito, 1967; Ramirez et al., 1967; Chhina and Anand, 1969; Haller and Barraclough, 1970; Zolovick and Eleftheriou, 1971). In the deermouse, the decrease in discharge rate begins about 2 minutes after termination of the stimulus with a duration of about 4 to 12 minutes, which may outlast the EEG-AR. There was a significant reduction in the population of hypothalamic neurons displaying this delayed inhibitory response to vaginal stimulation in diestrus, pregnant or ovariectomized deermice (Table V). Furthermore, the post-stimulus vaginally-induced discharge patterns were similar to those elicited by injection of gonadotropin (Ramirez et al., 1967; Beyer and Sawyer, 1969). Anterior hypothalamic deafferentation, which terminates the estrous cycle, also abolishes the specific hypothalamic unit response to vaginal probing (Beyer and Sawyer, 1969).

Based on data from above experiments, Kawakami and Sawyer (1959) and Ramirez et al. (1967) proposed that ovulatory hormone released during copulation may feed back on the brain to regulate neural activity and, in the presence of ovarian hormones, inhibit secretion of gonadotropin.

Zolovick (1969) has since correlated the hypothalamic unit and EEG-AR responses evoked by amygdaloid stimulation with the hypothalamic unit and EEG responses evoked by vaginal stimulation in deermice under various hormonal conditions. Neurons were sampled randomly from various hypothalamic nuclei. Once baseline activity and the response to non-sexual stimuli were

recorded, the animal was mechanically stimulated in the vagina
for 30 seconds. The EEG and unit activity responses were
monitored for 15 to 20 minutes at which time the discharge rate
of the unit and EEG returned to their respective pre-stimulation
levels. The subject then was stimulated electrically in the
medial amygdala through bipolar electrodes (0.25 mm. tip
separation; monopolar pulses, 5 or 50/sec.; 0.5 msec. duration;
120 μA) for 30 seconds and the course of unit activity and EEG
followed once again for 15 to 20 minutes. Electrical stimulation
of the AME was more effective in eliciting the EEG-AR and altering unit activity than stimulation of the ABL or AL. Whereas
only 90 to 100 μA of current was sufficient to evoke the EEG-AR
from the AME, 300 to 480 μA were needed to elicit a similar
response from the ABL or AL. Forty-six of sixty-one hypothalamic
neurons responded in the same direction to AME stimulation as
they previously had to vaginal stimulation (Fig. 2; Table V).
More importantly, amygdaloid stimulation was more effective in
eliciting the delayed decrease in firing rate in medial-basal
hypothalamic neurons and the EEG-AR than vaginal stimulation.
Of interest is the report of Hayward et al. (1964) that medial
amygdaloid stimulation was more effective in activating more of
the "critical elements" for a sustained release of gonatropin
than stimulation of the hypothalamus. If it is assumed that
(1) amygdaloid stimulation results in LH release and (2) that
LH feeds back to the brain to alter neural activity, then, the
amygdala is capable of evoking the EEG-AR in the presence of
already high endogenous levels of LH (e.g., ovariectomized
animals). Therefore, the failure to observe the characteristic
delayed inhibitory response in the medial-basal hypothalamic
neurons in ovariectomized deermice after amygdaloid stimulation
may be attributed to (1) the high circulating levels of gonadotropin which have already depressed the spontaneous activity of
these neurons so that further release of LH is ineffective or
(2) the absence of sex-steroid hormones, which are necessary
to precondition these neurons to the effects of gonadotropin.
The latter assumption is supported in part by data from
ovariectomized estrogen-treated animals. Estrogen administration was moderately successful in reinstating the delayed
inhibitory response of hypothalamic neurons to both vaginal
and amygdaloid stimulation, while restoration of the EEG-AR was
essentially complete (Table V). Since progesterone is known to
modify the hypothalamic unit response to peripheral stimuli
(Barraclough and Cross, 1963; Lincoln, 1969c; Zolovick, 1969)
the importance of this hormone in the maintenance of neural
tissue sensitivity must be considered.

Amygdaloid neurons in the deermouse during the estrous cycle
and pregnancy were recorded after vaginal stimulation to determine if the same temporal relationship occurs in their post-

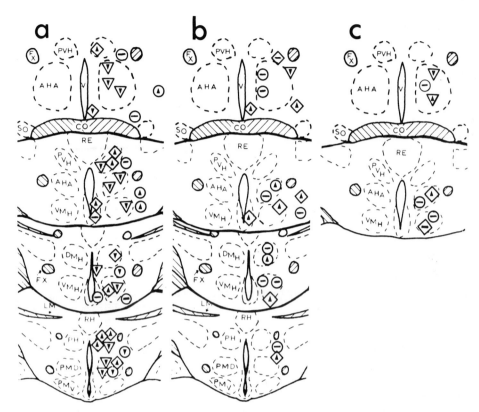

Fig. 2. Projection diagrams of the hypothalamus of adult P. m.
bairdii during (a) estrus and (b) diestrus stages of the estrous
cycle and (c) pregnancy illustrating positive, negative and
unresponsive neurons to 30 sec. of vaginal stimulation followed
by 30 sec. of electrical stimulation of the medial amygdaloid
nucleus (AME). (◇) Represent neurons facilitated; (▽) represent
neurons inhibited, and (○) represent neurons unchanged by
vaginal stimulation. (▲) Represent common neurons facilitated;
(▼) represent common neurons inhibited, and (—) represent common
neurons unchanged by electrical stimulation of the AME. See text
for description of the hypothalamic unit response. AHA = area
anterior hypothalami; ARH = nucleus arcuatus hypothalami; CO =
chiasma opticum; DMH = nucleus dorsomedialis hypothalami;
FX = fornix; LM = lemniscus medialis; PH = nucleus posterior
hypothalami; PMD = nucleus premamillaris dorsalis; PMV = nucleus
premamillaris ventralis; PVH = nucleus paraventricularis hypothalami; RE = nucleus reuniens thalami; SO = nucleus supraopticus
hypothalami; VMH = nucleus ventromedialis hypothalami; V = ventricle.

stimulus discharge patterns as the post-stimulus discharge
patterns of hypothalamic neurons. Thirty-one neurons were recorded from the medial amygdala and 25 from the basolateral amygdaloid complex in response to vaginal probing. No correlation
was found between the post-stimulus discharge patterns and the
estrous state. Six neurons, 3 located in the ABL and 2 in the
AME, decreased in activity in deermice in estrus, one in the AME
decreased in activity in the diestrus animals and none during
pregnancy. Twenty-two amygdaloid neurons were unresponsive. Of
significance was the discovery that, unlike hypothalamic neurons
which respond immediately to vaginal probing, amygdaloid neurons
characteristically fail to display the early response (from
stimulation to 2 to 3 minutes post-stimulation), and only display
the altered discharge rate once the after-reaction appears in the
EEG. Furthermore, amygdaloid neurons fail to return to pre-stimulation levels after termination of the EEG-AR and, frequently,
outlast the EEG-AR by as long as 10 to 20 minutes. The early
response to vaginal probing could not be evoked in diestrus,
estrus or ovariectomized estrogen-treated deermice, but was
elicited readily in the pregnant deermouse. The significance
of the amygdaloid response remains obscure; however, it appears
that the amygdala is not involved in the initial events leading
to LH release, but assumes an active role only after LH is
released. In the rabbit, it now appears that an ovulatory discharge of gonadotropin does not immediately follow copulation,
but that gonadotropin is released gradually from the pituitary
over a 30 to 90 minute period, with rupture of the ovarian
follicles occurring 10 to 12 hours later (Hayward et al., 1964).
Maximal release of gonadotropin appears to depend on activation
of reverberating limbic circuits, whose excitability depends on
gonadal steroids (Kawakami and Sawyer, 1959), leading to a sustained secretion of an ovulatory quantity of gonadotropin. A
comparable sequence of temporal events has been established
firmly in the intact adult cycling rat (Goldman et al., 1969)
or after stimulation of the preoptic area of proestrus rats
(Everett, 1964). Therefore, the sustained neuronal activity
noted in the amygdala of proestrus-estrus deermice after vaginal
stimulation, perhaps represents amygdaloid interaction with other
limbic structures for continual secretion of hypophyseal gonadotropins needed to complete ovulation.

Amygdaloid Regulation of Thyrotropin Secretion

The literature contains little evidence for amygdaloid
mediation of thyrotropin (TSH, thyroid stimulation hormone)
secretion. Complicating an analysis of amygdaloid-thyroid interaction is the inevitable secretion of other tropic hormones
following amygdaloid manipulation with their subsequent direct
and indirect interaction with the hypothalamic-hypophyseal-thyroid

axis. Furthermore, most effects on thyroid activity attributed to the amygdala are relatively transient and can be attributed to post-operative dietary problems or surgical procedure. Moreover, confusion arises from inconsistent and possibly inaccurate methology employed to evaluate thyroid activity.

Effects of Amygdaloid Lesions on Thyrotropin Secretion

Bilateral removal of the amygdala or temporal lobe of adult rats, cats, dogs and goats results in general atrophy of most endocrine glands, including the thyroid gland (Koikegami et al., 1955a, 1955b; Kageyama, 1960; Azzali et al., 1963), particularly if the operation is performed early in life. Histological examination of the thyroid gland six weeks after removal of the amygdala has confirmed total atrophy in the immature dog and partial atrophy in adults (Kageyama, 1960). Azzali et al. (1963) reported regression in thyroid epithelium in adult cats following ablation of the amygdala or after placement of lesions in the prepyriform cortex. Signs of regeneration were apparent in the thyroid glands of adult rats six months after amygdalectomy (Kageyama, 1960).

In contrast to the Japanese workers, Knigge (1961) and Yamada and Greer (1960) reported an insignificant reduction in thyroid weight after bilateral extirpation of the amygdala in adult rats, and Kovács et al. (1965), using I^{131} uptake as an index of thyroid activity, showed that electrocoagulation of the basolateral amygdaloid complex failed to alter thyroid function in adult rats two weeks after the operation. Eleftheriou and Zolovick (1968) examined the effect of bilateral lesions confined to the medial amygdaloid nuclei on the activity of the pituitary-thyroid axis of adult male deermice. Three days after placement of the lesions in the amygdala, thyroid and pituitary weight declined significantly from control values and remained significantly lower throughout the sixteen day experimental period (Fig. 3). Concomitant with the decline in thyroid weight was a decline in serum levels of TSH and an increase in pituitary TSH content (Fig. 4). From the above data, the authors concluded that thyroid activity must be reduced in light of the decline in thyroid weight and serum levels of TSH. The decrease in hypophyseal weight was somewhat surprising in view of TSH storage in the pituitary. However, of interest is the report of Eleftheriou et al. (1966) that lesions confined to the medial amygdaloid nuclei induce a significant mobilization of hypophyseal adreno-corticotropin and, therefore, the decline in pituitary weight probably reflects depletion of the latter hormone.

Electrical Stimulation of the Amygdala and Thyrotropin Secretion

Five minutes after application of electrical current (60 or

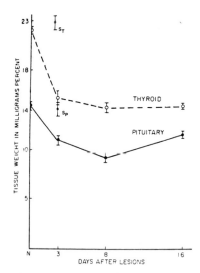

Fig. 3. Weight of pituitary and thyroid glands (mg/100g body weight ± SE) in normal, sham-operated (S_p and S_T) adult male deermice and in those with bilateral lesions in the medial amygdaloid nuclei at 3, 8, and 16 days (from Eleftheriou and Zolovick, 1968).

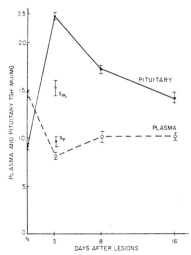

Fig. 4. Pituitary and plasma levels of thyrotropin (mU/mg or mU/ml ± SE) in normal, sham-operated (S_p and S_{PL}) adult male deermice and in those with bilateral lesions in the medial amygdaloid nuclei at 3, 8, and 16 days (from Eleftheriou and Zolovick, 1968).

100 Hz) to the medial principal nucleus of adult dogs, an increase
in colloid secretion was observed in the thyroid gland, as evidenced by histological changes in the intracellular granules and
a heightening of the follicular epithelial cells (Kageyama, 1960).
Fifteen to twenty-five minutes later, thyroid secretion began to
diminish; it returned to normal forty to fifty minutes later.
Stimulation of the lateral or intermediate principal amygdaloid
nucleus was ineffective in eliciting the thyroid response.
Amygdaloid activation of thyroid activity appears to depend on
the frequency of stimulation in the adult rat (Kovács et al.,
1965). Low frequency (15 Hz) electrical stimulation of the
medial amygdala for 30 seconds significantly increased thyroid
I^{131} uptake while high frequency (50 Hz) stimulation inhibited
thyroid activity. Bilateral adrenalectomy or administration of
cortisone abolished the inhibitory thyroid response to high
frequency stimulation. The authors attributed the inhibition
to preferential secretion of adrenocorticotropin over thyrotropin
(Mason, 1958) with a resultant suppression of thyroid activity
(Brown-Grant et al., 1954). Shizume et al. (1962) failed to
observe an increase in thyroid activity following electrical
stimulation of the amygdala in adult dogs. However, from the
description of the authors, they failed to stimulate the
important medial principal nucleus.

From previous stimulation and ablation studies, other areas
of the forebrain appear less convincing as regulatory centers
for TSH secretion. Bilateral destruction of the habenular
nucleus yielded an increase in serum levels of TSH, activated
the pituitary basophil cells and caused an increase in the size
of the thyroid epithelial cells (Szentágothai and Mess, 1958;
Szentágothai et al., 1962; Mess, 1958; Saito et al., 1960).
However, habenular lesions were ineffective in modifying release
of I^{131} from the thyroid or normal goiter production. In
addition, thyroid grafts implanted into the habenula failed to
evoke a change in thyroid activity, thus excluding the habenula
as a target tissue for thyroid hormone. Furthermore, habenular
lesions produced only a transient dysfunction in thyroid
activity, suggesting a readjustment of a more basic mechanism
for the control of TSH secretion (Mess, 1959).

The extra-pyramidal system has been implicated by
Lupulescu et al. (1962) as a chronic inhibitory center for the
control of TSH secretion. Lesions confined to the globus pallidus
and septal nuclei resulted in an increase in pituitary secretion
of TSH, as evidenced by an increase in thyroidal I^{131} uptake and
epithelial cell height in normal rats and in rats with low-
iodine-induced goiter. Inhibition of thyroid activity by direct
injection of thyroid hormone into the globus pallidus suggests
a functional feed-back mechanism in this structure for the

control of TSH secretion. However, neither removal of extensive
areas of neocortex (Greer and Shull, 1957) or entire forebrain
interfered with goiter formation or thyroid response to cold
stress; nor removal of habenula, pineal body or subcommissural
tissue (Yamada, 1961) and disease of the basal ganglia seriously
affected thyroid activity (Reichlin, 1959).

Electrical stimulation of the dorsal hippocampus or rats
was more effective in eliciting a release of TSH than stimulation
of the medial amygdala (Koikegami, 1964). Shizume and Okinaka
(1964) confirmed the facilitory effect of hippocampal stimulation
on TSH secretion by demonstrating an increase in the levels of
thyrotropin in jugular venous blood and elevated levels of
protein-bound I^{131} in thyroid venous blood of adult dogs two
hours after hippocampal stimulation, even in the presence of
exogenous adrenocorticotropin and elevated serus levels of
corticosteroids.

The reciprocal effects of electrical stimulation and lesions
of the amygdala on thyroid function suggests that a facilitory
center for regulation of hypophyseal secretion of TSH exists in
the medial portion of the basolateral amygdaloid nucleus.
Possibly, a second regulatory center exists in the dorsal hippocampus, independent of adrenocorticotropin secretion. Presumably
both limbic structures exert their effects on the thyroid through
a common pathway in the hypothalamus.

Amygdaloid Regulation of Adrenocorticotropin Secretion

The great variety of sensory, environmental, humoral and
emotional stimuli that affect adrenocortical secretion suggests
that several neural mechanisms function in the modulation of
adrenocorticotropin (ACTH) secretion. Of considerable importance
are the neural components which function in the maintenance of the
vigilant state and emotional display. Therefore, it is not surprising that manipulation of the various components of the limbic
"emotion" system affect ACTH secretion. By virtue of its close
interaction with the higher neopallial structures and the lower
brain stem structures, the limbic system is in a position to
integrate interoceptive and exteroceptive information and thus
exert considerable influence on the physiological and psychological
concomitants of emotion. In this section, emphasis will be placed
on the functional significance of the amygdala, hippocampus and
limbic midbrain area in the regulation of pituitary secretion of
ACTH in the resting state and after various stressors.

Effect of Lesions in Limbic Structures on the
       Secretion of Adrenocorticotropin

Bilateral destruction of the amygdaloid complex results in an

increase in adrenal and a decrease in thymus gland weight in
adult rats (Greer and Yamada, 1959; Yamada and Greer, 1960), cats
Kling et al., 1960), dogs (Martin et al., 1958) and deermice
(Eleftheriou et al., 1967), while destruction of the hippocampus
of the immature rat leads to impaired secretion of ACTH (Riss et
al., 1958, 1963; Koikegami, 1964). Lesions confined to the hippocampus of the adult rat results in a significant increase in
resting levels of serum corticosteroids (Knigge, 1961; Endröczi et
al., 1954). Kim and Kim (1961) demonstrated that damage to the
hippocampus attenuates the adrenal corticosteroid response to
chronic stress (repetitive skin lesions) but fails to affect the
acute compensatory adrenal response after hemiadrenalectomy. In
a study in which adult male deermice were housed and killed under
minimal stressful conditons, Eleftheriou et al. (1967) demonstrated that bilateral lesions confined to the medial amygdaloid
nuclei exert a profound influence on the basal secretion rate of
ACTH. Pituitary as well as serum levels of ACTH were found to be
significantly higher three days after placement of the lesions
and remained significantly elevated throughout the duration of
the experiment (Fig. 5). Indicative of increased ACTH secretion
was the significant elevation in adrenal and serum corticosterone
(Fig. 6). The authors suggested that the medial amygdaloid
nucleus exerts an inhibitory influence on the secretion of ACTH,
whereby removal of this inhibitory influence results in an
increase in the secretion of ACTH, independent of external stress.
The above assumption is supported in part by the experiment of
Bovard and Gloor (1961) in which they obtained a greater increase
in the secretion of adrenal corticosterone to immobilization
stress after placement of lesions in the central amygdaloid
nucleus of rats. This result is not unexpected since lesions
confined to this area of the amygdala are known to increase
aggressiveness (Wood, 1958).

More extensive damage to the amygdala or its hypothalamic
projection system (stria terminalis) attenuates the 17-hydroxy-
corticosteroid (17-OHC) response to immobilization (Knigge, 1961)
or physical stress (Knigge and Hays, 1963) while abolishing the
17-OHC response to the emotional stress experienced during
avoidance sessions (Mason, 1959). Undoubtedly, larger lesions
disrupt many other related functions. However, this result may
not be at variance with the above data in view of the placidity,
lack of effect and hypofunction reported to be associated with
extensive damage to the amygdala or removal of the temporal lobes
(for a review on the subject see Goddard, 1964). Subtotal
amygdalectomy in the adult monkey reduced but did not abolish the
17-OHC response to emotional stress (Mason et al., 1961); when
the conditioning stimulus was paired with a foot shock the hypo-
thalamic-hypophyseal-adrenal axis responded with a normal output
of 17-OHC indicating that the amygdala was not essential for the
response, but when present acts as a modulator to facilitate the

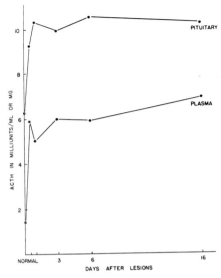

Fig. 5. Pituitary and plasma levels of adrenocorticotropin (mU/mg or mU/ml) in normal adult male deermice and in those with bilateral lesions in the medial amygdaloid nuclei (from Eleftheriou et al., 1966).

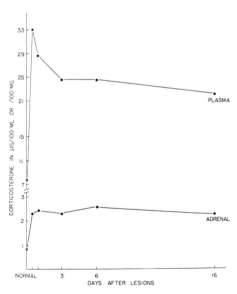

Fig. 6. Adrenal and plasma corticosterone levels (μg/100ml or μg/100mg) in normal adult deermice, and in those with bilateral lesions in the medial amygdaloid nuclei at .5, 1, 3, 6, and 16 days (from Eleftheriou et al., 1966).

acute adrenal response. Redgate (1970) has since emphasized the
importance of the amygdala-septal complex in the acute release of
ACTH to various physiological and psychological stressors.

Effect of Electrical Stimulation of Limbic Structures
on the Secretion of Adrenocorticotropin

An inverse relationship appears to exist between the amygdala
and hippocampus with respect to the secretion of ACTH in response
to electrical stimulation. Electrical stimulation of the amygdala
has been reported to augment blood levels of 17-OHC in the adult
cat (Setekleiv et al., 1961; Slusher and Hyde, 1961b; Sugano,
1963, dog (Ganong and Goldfien, 1959), monkey (Mason, 1969), and
man (Mandell et al., 1963; Rubin et al., 1966) while electrical
current applied to the hippocampus attenuates adrenal secretion of
17-OHC in cat (Porter, 1954; Endröczi and Lissák, 1962; Endröczi
et al., 1959), dog (Okinaka, 1962), rat, rabbit (Endröczi et al.,
1959), monkey (Mason, 1958), and man (Mandell et al., 1963; Rubin
et al., 1966). The adrenal corticosteroid response to amygdaloid
stimulation appears independent of psychic or emotional factors,
since it can be elicited under anesthesia (Setekleiv et al., 1960,
1961) or at a stimulus intensity which is too weak to elicit
behavioral responses (Ganong and Goldfien, 1959; Mandell et al.,
1963). Endröczi et al. (1963) have since emphasized the lack of
correlation between behavioral reactions and the activation of the
hypophyseal-adrenal axis after adrenergic or cholinergic stimulation of various limbic structures. The influence of hippocampal
stimulation on ACTH secretion is frequency-dependent in cats;
high frequency stimulation readily elicits release of ACTH from
the pituitary while low frequency stimulation inhibits the adrenal
corticosteroid response to pain (Endröczi and Lissák, 1962).

Attempts at localizing the active amygdaloid nuclei in
meidating the 17-OHC response have been disappointing. Slusher
and Hyde (1961b) report maximum secretion of 17-OHC after stimulation of the medial amygdaloid nucleus of the cat while
Setekleiv et al. (1961) found stimulation of the medial portion
of the basolateral nucleus more effective. In man, maximum
adrenal 17-OHC response occurs after stimulation of the basolateral or lateral amygdaloid nuclei (Rubin et al., 1966) while
in the monkey, stimulation of the entire amygdala appears
equally effective in this regard (Mason, 1959).

Using the EEG technique Kawakami et al. (1966) have
described an antagonistic relationship between the dorsal hippocampus and the basolateral amygdala after administration of ACTH
or cortisol in adult rabbits. Adrenocorticotropin enhanced the
2 to 13 Hz components of the EEG in the amygdala and suppressed
the bioelectrical activity in the hippocampus while cortisol

reversed this relationship. The authors proposed that the
amygdala and hippocampus function as an integrative mechanism
along with the basal hypothalamus to facilitate the secretion
of adrenocorticotropin.

Other limbic structures that facilitate the release of ACTH
after electrical stimulation are the ecto- and suprasylvian gyri,
anterior cingulate cortex, prepyriform area, posterior orbital
surface (Setekleiv et al., 1961), preoptic area (Ahren, 1962),
septum and midbrain tegmental area (Mason, 1958; Endröczi and
Lissák, 1960; Okinaka et all, 1960; Redgate, 1970). Of considerable importance in the regulation of ACTH secretion are
the reciprocal anatomical connections and functional interactions of the hypothalamus and temporal lobe structures with
the midbrain reticular formation (Nauta and Kuypers, 1958).
Recent evidence indicates that both facilitory and inhibitory
mechanisms are present in the midbrain reticular formation for
the regulation of ACTH secretion. Stimulation of the dorsal
tegmental area facilitates the release of ACTH (Endröczi and
Lissák, 1960) while stimulation of the ventral tegmental area
suppresses adrenal secretion of 17-OHC (Endröczi and Lissák,
1963; Slusher and Hyde, 1961a) and attenuates the release of
ACTH evoked by hypothalamic stimulation (Slusher and Hyde, 1966).
The latter observation is interesting, particularly in view of
the experimental results of Kawakami et al. (1966).   They discovered the interesting fact that while ACTH does not particularly
affect the electrical activity of the midbrain reticular formation,
cortisol significantly depressed the activity in this structure,
and proposed that once the organism is alerted to a stressful
condition, the primary function of the midbrain reticular
formation is to prevent hyper-excitation of the higher brain
centers and possibly conserve hypophyseal ACTH. The apparent
functional dichotomy of the midbrain limbic system in the
regulation of ACTH secretion has since been clarified by the
elegant experiments of the Italian workers and is the subject of
a recent review (Mangili et al., 1966). In summary, these authors
contend that somatic information from the peripheral nervous
system enters the midbrain limbic system, which then either
inhibits or facilitates the secretion of ACTH via hypothalamic
mechanisms. The cerebral cortex and habenula-stria medularis
exert a tonic inhibitory influence over the hypothalamus while
the limbic system, conveying emotional responses, either
facilitates, via the amygdala, or inhibits, via the hippocampus,
the hypothalamus directly or indirectly through the midbrain area.

Amygdaloid Regulation of Growth Hormone Secretion

Ever since the original report of Brown and Schäfer (1888)
that hyperphagia and obesity follow temporal lobectomy in monkeys,

many investigators have attempted to elucidate the mechanism by
which the central nervous system governs food and water intake,
their metabolic fate and subsequent effect on growth and maturation of the organism. It would be naive to assume that impaired
growth following experimental brain lesions or brain disease
results solely from dysfunction of the neural mechanism for
control of synthesis and secretion of growth hormone (GH,
somatotropin, STH). Experimental evidence has indicated that
growth and development depend on complex interactions of GH with
many other hormones, particularly insulin, thyroid hormone and
the gonadal steroids. Indeed, it would be beyond the scope of
this paper to review the extensive literature outlining the
specific hormone interactions at this time; therefore, the
reader is referred to several recent reviews on the topic
(Russell and Wilhelmi, 1958; Pecile and Müller, 1964; Reichlin,
1966).

The purpose of the present report is not to discuss
amygdaloid function in the regulation of growth per se, which
depends on numerous factors, but to focus on amygdaloid regulation of growth hormone secretion. The behavioral aspects of
amygdaloid involvement in the control of appetite and consumption
of food and water will be reviewed in other chapters. Information
regarding the interactions between feeding behavior, food intake,
thermoregulation, energy balance, hormones and growth is contained
in several recent reviews (Kennedy, 1961, 1966; Mayer and Thomas,
1967).

Experimental evidence has confirmed that growth itself is
at best an indirect measure of growth hormone secretion. Moreover,
not only is the role of GH in the regulation of growth unclear,
but knowledge is lacking concerning the role of GH in the various
stages of development of the organism from birth to senility. The
primary function of GH in the adult appears related to control of
carbohydrate and lipid metabolism, whose metabolites feed back to
neural receptors to, perhaps, regulate appetite, feeding behavior
and further secretion of hormones (Kennedy, 1966).

Role of the Amygdala in the Secretion of Growth Hormone

A review of the literature revealed only one study on extrahypothalamic regulation of HG secretion in which hypophyseal or
plasma levels of GH were actually measured. Using the tibial-
epiphyseal-width bioassay in hypophysectomized rats, Eleftheriou
et al. (1969) investigated the effects of bilateral lesions confined to various amygdaloid nuclei of adult male deermice on
hypothalamic growth-hormone-releasing-factor (GH-RF) activity and
content of hypophyseal growth hormone. A significant increase in
pituitary levels of HG occurred in deermice three weeks after

Effect of Amygdaloid Lesions on Pituitary Growth Hormone (GH)
and Hypothalamic Content of Growth Hormone-Releasing-Factor
(GH-RF) in Male Deermice

Lesion	Pituitary GH* (μg/mg $\pm$ SE)	GH-RF Equivalent $\pm$ SE
Control	16.1 $\pm$ 2.3	26.3 $\pm$ 3.4
Sham-control	17.0 $\pm$ 2.4	28.9 $\pm$ 3.8
Cortical	17.7 $\pm$ 2.3	22.5 $\pm$ 3.3
Basolateral	16.5 $\pm$ 2.2	30.4 $\pm$ 3.5
Medial	26.1 $\pm$ 1.7**	13.9 $\pm$ 3.2**

* = Potency expressed as μg-equivalent of NIH-GH-S8; adapted
from Eleftheriou et al., 1969.

** = Significant from control p <.01.

placement of lesions in the medial amygdaloid nucleus, but not in
sham-operated or unoperated deermice or in deermice bearing lesions
in the basolateral or cortical amygdaloid nuclei (Table VI). The
greatest amount of GH was mobilized from pituitaries of recipient
assay rats following injections of hypothalamic extracts from
deermice bearing lesions in the medial amygdaloid nuclei. Hypothalamic GH-RF potency values from deermice bearing lesions in
the basolateral or cortical amygdaloid nuclei did not differ from
hypothalamic GH-RF values of sham-operated or unoperated controls
(Table VI). Based on previous data (Ishida et al., 1965; Müller
et al., 1967) the authors tentatively concluded that serum levels
of GH were depressed in deermice bearing lesions in the medial
amygdaloid nuclei. However, it should be emphasized that
pituitary levels alone do not necessarily reflect secretion
rates or circulating levels of a hormone. As stated earlier,
plasma levels as well as pituitary levels of LH and its releasing-
factor were elevated in deermice bearing lesions in the basolateral
amygdaloid nuclei. Unfortunately, owing to technical problems in
obtaining and assaying for serum GH, this important index of
secretion was not evaluated. However, it must be emphasized
that the body weight of deermice with amygdaloid lesions remained
stable throughout the experimental period. These data further
indicate the lack of correlation between GH secretion and

general body growth following neural dysfunction. A complete evaluation of amygdaloid involvement in hypophyseal secretion of growth hormone must await experiments utilizing the modern sensitive and specific redioimmunoassay for the quantification of serum levels of growth hormone.

SUMMARY

Evidence is presented in this report to indicate an active participation of the amygdala in the hypophyseal secretion of luteinizing hormone, thyrotropin, adrenocorticotropin and possibly somatotropin and leutotropin. Data based on endocrinological experiments indicate that the amygdala can be divided into two functional subdivisions, medial and lateral. The medial amygdala appears to regulate hormone functions concerned with maintaining the internal milieu, adaptation, homeostatis, for the preservation of the individual, while the lateral subdivision appears to influence hormonal functions concerned with reproduction or perpetuation of the species. Evidence exists to indicate the possibility that the two subdivisions are physiologically distinct and project to separate regions of the hypothalamus (Koikegami, 1963; Egger, 1967). Evidence that hypothalamic and hypophyseal ribonucleic acid base-ratios are differentially altered after placement of lesions in the two amygdaloid subdivisions (Eleftheriou et al., 1969), indicates a division of molecular function between the two nuclear groups and adds further support to this assumption.

Evidence is presented to suggest that in the deermouse the basolateral amygdaloid complex exerts a tonic inhibitory influence on the medial amygdaloid complex, whereby the inhibitory block can be removed by either destruction of the basolateral complex or bypassed through direct electrical stimulation of the medial complex. Data also are presented to indicate that the amygdala can exert its effects on the pituitary either by direct interaction with the hypothalamus or indirectly through reverberating circuits with other limbic structures, principally the hippocampus and limbic midbrain area. A complete analysis of amygdaloid modulation of hypophyseal tropic hormone secretion must await further studies on (1) the mobilization of nucleotides for the synthesis or specific hypothalamic macromolecules governing the synthesis and releasing mechanisms of the various hormone-releasing-factors, (2) the nature and function of the neurotransmitters involved in the secretion of the hormone-releasing-factors, (3) identification of hormone feed-back receptors in the brain, and (4) the utilization of specific and sensitive radioimmunoassays for the quantification of hypophyseal tropic hormones.

REFERENCES

AHRÉN, C.   Effects of diencephalic lesions on acute and chronic
    stress responses in male rabbits.   Acta Endocrinologica,
    1962, Supplement, 69, 1.

ANAND, B. K., CHHINA, G. S., & DUA, S.   Effects of lesions in the
    limbic system on the affective behaviour and visceral
    responses in monkeys and cats.   Indian Journal of Medical
    Research, 1959, 47, 51.

ÁRVAY, A.   Cortico-hypothalamic control of gonadotropic functions.
    In E. Bajusz and G. Jasmin (Eds.), Major Problems in Neuro-
    endocrinology.   Baltimore: Williams and Wilkins Co., 1964,
    pp. 307-321.

AZUMA R., & KUMAGAI, H.   Studies on uterine activity in unanaes-
    thetized dog by means of chronic uterine fistula.   Tokyo
    Journal of Medical Science, 1934, 48, 2373.

AZZALI, G. M., CARRERAS, M., LECHI, A., & DALLA ROSA, V.   Modifi-
    cazioni morfofunzinoali del sistema ipotalamo-neuroipofisario
    e della costellazione endocrina, dopo lesioni rinencefaliche.
    Bollettino della Societa Italiana di Biologia Sperimentale,
    1963, 39, 688.

BARD, P. A.   The hypothalamus and sexual behaviour.   Research on
    Nervous and Mental Disease Processes, 1940, 20, 551.

BARD, P., & RIOCH, D. McK.   A study of four cats deprived of neo-
    cortex and additional portions of the forebrain.   Johns
    Hopkins Bulletin, 1937, 60, 73.

BARRACLOUGH, C. A.   Hypothalamic activation associated with
    stimulation of the vaginal cervix of proestrous rats.
    Anatomical Record, 1937, 136, 159.

BARRACLOUGH, C. A., & CROSS, B. A.   Unit activity in the hypo-
    thalamus of the cyclic female rat: Effect of genital
    stimuli and progesterone.   Journal of Endocrinology, 1963,
    26, 339.

BENEDETTI, W. L., APPELTAUER, L. C., REISSENWEBER, N. J.,
    DOMINGUEZ, R., GRINO, E., & SAS, J.   Ovary and uterine
    hypertrophy in the rat bearing mesencephalic lesions.
    Acta Physiologica Latino Americana, 1965, 15, 218.

BEYER, C., & SAWYER, C. H.   Hypothalamic unit activity related to
    control of the pituitary gland.   In W. F. Ganong and

L. Martini (Eds.), Frontiers in Neuroendocrinology.   London: Oxford University Press, 1969, pp. 255-287.

BOVARD, E. W., & GLOOR, P.   Effect of amygdaloid lesions on plasma corticosterone response of the albino rat to emotional stress. Experientia, 1961, 17, 521.

BROOKS, C. McC.   The role of cerebral cortex and of various sense organs in the excitation and execution of mating activity in the rabbit.   American Journal of Physiology, 1937, 120, 544.

BROWN, P. S.   The assay of gonadotrophin from urine of non-pregnant human subjects.   Journal of Endocrinology, 1955, 13, 59.

BROWN, S., & SCHÄFER, E. A.   An investigation into the functions of the occipital and temporal lobes of the monkey's brain. Transactions, Royal Philosophical Society of London, 1888, 179B, 303.

BROWN-GRANT, K., HARRIS, G. W., & REICHLIN, S.   The effect of emotional and physical stress on thyroid activity in the rabbit.   Journal of Physiology, 1954, 126, 29.

BUNN, G. P., & EVERETT, J. W.   Ovulation in persistent-estrous rats after electrical stimulation of the brain.   Proceedings of the Society for Experimental and Biological Medicine, 1957, 96, 369.

CALIGARIS, L., ASTRADA, J. J., & TALEISNIK, S.   Stimulating and inhibiting effects of progesterone on the release of luteinizing hormone.   Acta Endocrinologica Scandinavica, 1968, 59, 177.

CARRER, H. F., & TALEISNIK, S.   Effect of mesencephalic stimulation on the release of gonadotrophins.   Journal of Endocrinology, 1970, 48, 527.

CHHINA, G. S., & ANAND, B. K.   Responses of neurones in the hypothalamus and limbic system to genital stimulation in adult and immature monkeys.   Brain Research, 1969, 13, 511.

CHHINA, G. S., CHAKRABARTY, A. S., KAUR, K., & ANAND, B. K. Electroencephalographic responses produced by genital stimulation and hormone administration in sexually immature rhesus monkeys.   Physiology & Behavior, 1968, 3, 579.

CRITCHLOW, V.   Blockade of ovulation in the rat by mesencephalic lesions.   Endocrinology, 1958, 63, 596.

EGGER, M. D. Responses of hypothalamic neurons to electrical stimulation in the amygdala and the hypothalamus. Electroencephalography and Clinical Neurophysiology, 1967, 23, 6.

ELEFTHERIOU, B. E., & CHURCH, R. L. Effect of repeated exposure to aggression and defeat on plasma and pituitary levels of luteinizing hormone in C57BL/6J mice. General and Comparative Endocrinology, 1967, 9, 263.

ELEFTHERIOU, B. E., & PATTISON, M. L. Effect of amygdaloid lesions on hypothalamic follicle-stimulating hormone-releasing factor in the female deermouse. Journal of Endocrinology, 1967, 39, 613.

ELEFTHERIOU, B. E., & ZOLOVICK, A. J. Effect of amygdaloid lesions on oestrous behaviour in the deermouse. Journal of Reproduction and Fertility, 1966, 11, 451.

ELEFTHERIOU, B. E., & ZOLOVICK, A. J. Effect of amygdaloid lesions on plasma and pituitary levels of luteinizing hormone. Journal of Reproduction and Fertility, 1967, 14, 33.

ELEFTHERIOU, B. E., & ZOLOVICK, A. J. Effect of amygdaloid lesions on plasma and pituitary thyrotropin levels in deermice. Proceedings of the Society for Experimental Biology and Medicine, 1968, 127, 671.

ELEFTHERIOU, B. E., ZOLOVICK, A. J., & PEARSE, R. Effect of amygdaloid lesions on pituitary-adrenal axis in the deermouse. Proceedings of the Society for Experimental Biology and Medicine, 1966, 122, 1259.

ELEFTHERIOU, B. E., ZOLOVICK, A. J., & NORMAN, R. L. Effects of amygdaloid lesions on plasma and pituitary levels of luteinizing hormone in the male deermouse. Journal of Endocrinology, 1967, 38, 469.

ELEFTHERIOU, B. E., CHURCH, R. L., NORMAN, R. L., PATTISON, M., & ZOLOVICK, A. J. Effect of repeated exposure to aggression and defeat on plasma and pituitary levels of thyrotropin. Physiology & Behavior, 1968, 3, 467.

ELEFTHERIOU, B. E., CHURCH, R. L., ZOLOVICK, A. J., NORMAN, R. L., & PATTISON, M. L. Effects of amygdaloid lesions on regional brain RNA base ratios. Journal of Endocrinology, 1969, 45, 207.

ELEFTHERIOU, B. E., DESJARDINS, C., PATTISON, M. L., NORMAN, R. L., & ZOLOVICK, A. J. Effect of amygdaloid lesions on hypothalamic-hypophyseal growth-hormone activity. Neuroendo-crinology, 1969, 5, 132.

ELEFTHERIOU, B. E., DESJARDINS, C., & ZOLOVICK, A. J. Effects of amygdaloid lesions on hypothalamic-hypophyseal luteinizing hormone activity. Journal of Reproduction and Fertility, 1970, 21, 249.

ELWERS, M., & CRITCHLOW, V. Precocious ovarian stimulation following hypothalamic and amygdaloid lesions in rats. American Journal of Physiology, 1960, 198, 381.

ELWERS, M., & CRITCHLOW, V. Precocious ovarian stimulation follow-ing interruption of stria terminalis. American Journal of Physiology, 1961, 201, 281.

ENDRÖCZI, E., & LISSÁK, K. The role of the mesencephalon, dien-cephalon and archicortex in the activation and inhibition of the pituitary-adrenocortical system. Acta Physiologica Academiae Scientiarum Hungaricae, 1960, 17, 39.

ENDRÖCZI, E., & LISSÁK, K. Interrelations between palaeocortical activity and pituitary-adrenocortical function. Acta Physio-logica Academiae Scientiarum Hungaricae, 1962, 21, 257.

ENDRÖCZI, E., & LISSÁK, K. Effect of hypothalamic and brain stem structure stimulation on pituitary-adrenocortical function. Acta physiologica Academicae Scientiarum Hungaricae, 1963, 24, 67.

ENDRÖCZI, E., LISSÁK, K., SZEP, C., & TIGYI, A. Examination of the pituitary-adrenocortical-thyroid system after ablation of neocortical and rhinencephalic structures. Acta Phys-iologica Academiae Scientiarum Hungaricae, 1954, 6, 19.

ENDRÖCZI, E., LISSÁK, K., BOHUS, B., & KOVÁCS, S. The inhibitory influence of archicortical structures on pituitary-adrenal function. Acta Physiologica Academicae Scientiarum Hungaricae, 1959, 16, 17.

ENDRÖCZI, E., SCHREIBERG, G., & LISSÁK, K. The role of central nervous activating and inhibitory structures in the control of pituitary-adrenocortical function. Effects of intra-cerebral cholinergic and adrenergic stimulation. Acta Physiologica Academiae Scientiarum Hungaricae, 1963, 24, 211.

EVERETT, J. W. Preoptic stimulative lesions and ovulation in the rat: "Thresholds" and LH-release time in late diestrus and proestrus. In E. Bajusz and G. Jasmin (Eds.), Major Problems

in Neuroendocrinology. Basel and New York: S. Karger.
Pp. 346-366.

GANONG, W. F., & GOLDFIEN, A. Effect of diencephalic stimulation
on adrenocortical and adrenal medullary secretion in the
dog. In Atlantic City Program 41st meeting, Endocrine
Society, 1959, p. 29.

GODDARD, G. V. Functions of the amygdala. Psychological
Bulletin, 1964, 62, 89.

GOLDMAN, B. D., KAMBERI, I. A., SIITERI, P. K., & PORTER, J. C.
Temporal relationship of progestin secretion, LH release and
ovulation in rats. Endocrinology, 1964, 85, 1137.

GREEN, J. D., CLEMENTE, C. E., & DEGROOT, J.   Rhinencephalic
lesions and behavior in cats. Journal of Comparative Neurology, 1957, 108, 505.

GREER, M. A., & SHULL, H. F. Effect of ablation of neocortex on
ability of pituitary to secrete thyrotropin in the rat.
Proceedings of the Society for Experimental Biology and
Medicine, 1957, 94, 565.

GREER, M. A., & YAMADA, T. Effect of bilateral ablation of the
amygdala on endocrine function in the rat. In Atlantic City
Program 41st meeting, Endocrine Society, 1959, p. 82.

HALLER, E. W., & BARRACLOUGH, C. A. Alterations in unit activity
of hypothalamic ventromedial nuclei by stimuli which affect
gonadotropic hormone secretion. Experimental Neurology,
1970, 29, 111.

HARRIS, G. W. Reticular formation, stress and endocrine activity.
In H. H. Jasper, L. D. Proctor, R. S. Knighton, W. C. Noshay
and R. J. Costello (Eds.), Reticular Formation of the Brain.
Boston: Mittle Brown Co., 1958. Pp. 207-221.

HAYWARD, J. N., HILLIARD, J., & SAWYER, C. H. Time of release of
pituitary gonadotropin induced by electrical stimulation of
the rabbit brain. Endocrinology, 1964, 74, 108.

IGARASHI, M., & MCCANN, C. M. A new sensitive bio-assay for
follicle stimulating hormone (FSH). Endocrinology, 1964,
74, 440.

ISHIDA, Y., KUROSHIMA, A., BOWERS, C. Y., & SCHALLY, A. V. In
vivo depletion of pituitary growth hormone by hypothalamic
extracts, Endocrinology, 1965, 77, 759.

KAGEYAMA, Y. Histological alterations in the thyroid gland
    following stimulation or destruction of the amygdaloid
    nuclear complex. Niigata Medical Journal, 1960, 74, 216.

KAWAKAMI, M., & SAITO, H. Unit activity in the hypothalamus of
    the cat: Effect of genital stimuli, luteinizing hormone
    and oxytocin. Japanese Journal of Physiology, 1967, 17, 466.

KAWAKAMI, M., & SAWYER, C. H. Induction of behavioral and electro-
    encephalographic changes in the rabbit by hormone administra-
    tion or brain stimulation. Endocrinology, 1959a, 65, 631.

KAWAKAMI, M., & SAWYER, C. H. Neuroendocrine correlates to
    changes in brain activity thresholds by sex steroids and
    pituitary hormones. Endocrinology, 1959b, 65, 652.

KAWAKAMI, M., SETO, K., & YOSHIDA, K. Influence of the limbic
    system on ovulation and on progesterone and estrogen
    formation in rabbit's ovary. Japanese Journal of Physiology,
    1966, 16, 254.

KAWAKAMI, M., SETO, K., & YOSHIDA, K. Influences of the limbic
    structure on biosynthesis of ovarian steroids in rabbits.
    Japanese Journal of Physiology, 1968, 18, 356.

KAWAKAMI, M., & TERASAWA, E. Differential control of sex hormone
    and oxytocin upon evoked potentials in the hypothalamus and
    midbrain reticular formation. Japanese Journal of Phys-
    iology, 1967, 17, 65.

KAWAKAMI, M., KOSHINO, T., & HATTORI, Y. Changes in the EEG of
    the hypothalamus and limbic system after administration of
    ACTH, SU-4885 and ACH in rabbits with special reference to
    neurohumoral feedback regulation of pituitary-adrenal
    system. Japanese Journal of Physiology, 1966, 16, 551.

KENNEDY, G. C. Interactions between feeding behavior and hormones
    during growth. Annals New York Academy of Science, 1961,
    157, 1049.

KENNEDY, G. C. Food intake, energy balance and growth. British
    Medical Bulletin, 1966, 22, 216.

KIM, C., & KIM, C. U. Effect of partial hippocampal resection on
    stress mechanism in rat. American Journal of Physiology,
    1961, 201, 337.

KLING, A. Effects of rhinencephalic lesions on endocrine and
    somatic development in the rat. American Journal of
    Physiology, 1964, 206, 1395.

KLING, A., & GROVE, L. Delayed vaginal opening following lesions of the olfactory system in the neonatal rat. Federation Proceedings, 1963, 22, 573.

KLING, A., ORBACH, J., SCHWARTZ, N. B., & TOWNE, J. C. Injury to the limbic system and associated structures in cats. Archives of General Psychiatry, 1960, 3, 391.

KLÜVER, H., & BARTELMEZ, G. W. Endometriosis in a rhesus monkey. Surgery, Gynecology and Obstetrics, 1951, 92, 650.

KLÜVER, H., & BUCY, P. C. "Psychic blindness" and other symptoms following bilateral temporal lobectomy in Rhesus monkeys. American Journal of Physiology, 1937, 119, 352.

KLÜVER, H., & BUCY, P. C. An analysis of certain effects of bilateral temporal lobectomy in the rhesus monkey, with special reference to "psychic blindness." Journal of Psychology, 1938, 5, 33.

KNIGGE, K. M. Adrenocortical response to stress in rats with lesions in hippocampus and amygdala. Proceedings of the Society for Experimental Biology and Medicine, 1961, 108, 18.

KNIGGE, K. M., & HAYS, M. Evidence of inhibitive role of hippocampus in neural regulation of ACTH release. Proceedings of the Society for Experimental Biology and Medicine, 1963, 114, 67.

KOBAYASHI, T., & KOBAYASHI, T. Central nervous control over the sexual function. Folia Endocrinologica Japonica, 1961, 37, 935.

KOIKEGAMI, H. Amygdala and other related limbic structures; experimental studies on the anatomy and function. I. Anatomical researches with some neurophysiological observations. Acta Medica et Biologica (Niigata), 1963, 10, 161.

KOIKEGAMI, H. Amygdala and other related structures; experimental studies on the anatomy and function. II. Functional experiments. Acta Medica et Biologica (Niigata), 1964, 12, 73.

KOIKEGAMI, H., FUSE, S., & WATANABE, H. 1955. Studies on the amygdaloid nuclei, effects of extirpation or destruction. Acta Anatomica (Nipponese), 1955, 30, 92.

KOIKEHAMI, H., FUSE, S., YOKOYAMA, T., WATANABE, T., & WATANABE, H. Contributions to the comparative anatomy of the amygdaloid nuclei of mammals with some experiments of their destruction or stimulation. Folia Psychiatrica et Neurologica Japonica (Niigata), 1955, 8, 336.

KOIKEHAMI, H., YAMADA, T., & USUI, K. Stimulation of the amygdaloid nuclei and periamygdaloid cortex with special reference to its effects on uterine movements and ovulation. Folia Psychiatrica et Neurologica Japonica (Niigata), 1954, 8, 7.

KOVÁCS, S., SÁNDOR, A., VÉRTES, Z., & VÉRTES, M. The effect of lesions and stimulation of the amygdala on pituitary-thyroid function. Acta Physiologica Academiae Scientiarum Hungaricae, 1965, 27, 221.

LAW, O. T., & SACKETT, G. P. Hypothalamic potentials in the female rat evoked by hormones and by vaginal stimulation. Neuroendocrinology, 1965, 1, 31.

LAWTON, I. E., & SAWYER, C. H. Role of amygdala in regulating LH secretion in the adult rat. American Journal of Physiology, 1970, 218, 622.

LINCOLN, D. W. Correlation of unit activity in the hypothalamus with EEG patterns associated with the sleep cycle. Experimental Neurology, 1969a, 24, 1.

LINCOLN, D. W. Response of hypothalamic units to stimulation of the vaginal cervix: Specific versus non-specific effects. Journal of Endocrinology, 1969b, 43, 683.

LINCOLN, D. W. Effects of progesterone on the electrical activity of the forebrain. Journal of Endocrinology, 1969c, 45, 585.

LUPULESCU, A., NICOLESCU, A., GHEORGHIESCU, B., MERCULIEV, E., & LUNGU, M. Neural control of the thyroid gland: Studies on the role of extrapyramidal and rhinencephalon areas in the development of goiter. Endocrinology, 1962, 70, 517.

MANDELL, A. J., CHAPMEN, L. F., RAND, R. W., & WALKER, R. D. Plasma corticosteroids: Changes in concentration after stimulation of hippocampus and amygdala. Science, 1963, 139, 1212.

MANGILI, G., MOTTA, M., & MARTINI, L. Control of adrenocorticotropic hormone secretion. In L. Martini and W. F. Ganong (Eds.), Neuroendocrinology, Vol. 1. New York: Academic Press, 1966. Pp. 297-370.

MARTIN, J., ENDRÖCZI, E., & BATA, G. Effect of the removal of
    amygdalic nuclei on the secretion of adrenal cortical
    hormones. Acta Physiologica Academiae Scientiarum
    Hungaricae, 1958, 14, 131.

MASON, J. W. Central nervous system regulation of ACTH secretion.
    In H. H. Jasper, L. D. Proctor, A. S. Knighton, W. C. Noshay,
    and R. T. Costello (Eds.), Reticular Formation of the Brain.
    Boston: Little Brown Co., 1958. Pp. 645-670.

MASON, J. W. Plasma 17-hydroxycorticosteroid levels during
    electrical stimulation of the amygdaloid complex in
    conscious monkeys. American Journal of Physiology, 1959,
    196, 44.

MASON, J. W., NAUTA, W. J. H., BRADY, J. V., & ROBINSON, J. A.
    Limbic system influences on the pituitary-adrenal cortical
    system. In Atlantic City Program 41st meeting Endocrine
    Society, 1959, p. 29.

MASON, J. W., NAUTA, W. J. H., BRADY, J. V., ROBINSON, J. A., &
    SACHER, E. J. The role of limbic system structures in the
    regulation of ACTH secretion. Acta Neurovegetativa, 1961,
    23, 4.

MAYER, J., & THOMAS, D. W. Regulation of food intake and
    obesity. Science, 1967, 156, 328.

MESS, B. Verhinderung des Thiouracileffektes und der "Jodmangel-
    struma" durch experimentelle Zerstorung der Nuclei habenulae.
    Endokrinologie, 1958, 35, 196.

MESS, B. Die Rolle der Nuclei habenulae bei der auf erhohten
    Thyroxin-Blutspiegel eintretenden zental-nervosen Hemmung
    der thyerotrophen Aktivitat des Hypophysenvorderlappens.
    Endokrinologie, 1959, 37, 104.

MÜLLER, E. E., ARIMURA, A., SAITO, T., SCHALLY, A. V. Growth
    hormone-releasing activity in plasma of hypophysectomized
    rats. Endocrinology, 1967, 80, 77.

NAUTA, W. J. H. An experimental study of the fornix system in
    the rat. Journal of Comparative Neurology, 1956, 104, 247.

NAUTA, W. J. H., & KUYPERS, J. J. M. Some ascending pathways in
    the brain stem reticular formation. In H. H. Jasper, L. D.
    Proctor, A. S. Knighton, W. C. Noshay, & R. T. Costello
    (Eds.), Reticular Formation of the Brain. Boston: Little
    Brown Co., 1958. Pp. 3-30.

NORMAN, R. L. Effect of basolateral amygdaloid lesions on pro-
    lactin secretion in Peromyscus maniculatus bairdii. Master's
    Thesis, Kansas State University, Manhattan, 1969.

OKINAKA, S. Die Regulation der Hypophysen-Nebennierenfunktion
    durch das Limbic-system und den Mittelhirnanteil der
    Formatio n Reticularis. Acta Neurovegetativa, 1962, 23, 15.

PECILE, A., & MÜLLER, E. E. Control of growth hormone secretion.
    In L. Martini and W. F. Ganong (Eds.), Neuroendocrinology,
    Vol. 1. New York: Academic Press. Pp. 537-564.

PEKARY, A. E., DAVIDSON, J. M., & ZONDEK, B. Failure to
    demonstrate a role of midbrain-hypothalamic afferents in
    reproductive processes. Endocrinology, 1967, 80, 365.

PORTER, R. W. The central nervous system and stress-induced
    eosinopenia. Recent Progress in Hormone Research, 1954,
    10, 1.

PORTER, R. W., CAVANAUGH, E. B., CRITCHLOW, V., & SAWYER, C. H.
    Localized changes in electrical activity of the hypothalamus
    in estrous cats following vaginal stimulation. American
    Journal of Physiology, 1957, 189, 145.

RAMIREZ, V. D., KOMISARUK, B. R., WHITMOYER, D. I., & SAWYER,
    C. H. Effects of hormones and vaginal stimulation on the
    EEG and hypothalamic units in rats. American Journal of
    Physiology, 1967, 212, 1376.

RAISMAN, G. An evaluation of the basic pattern of connections
    between the limbic system and the hypothalamus. American
    Journal of Anatomy, 1970, 129, 197.

REDGATE, E. S. ACTH release evoked by electrical stimulation of
    brain stem and limbic system sites in the cat: The absence
    of ACTH release upon infundibular area stimulation. Endo-
    crinology, 1970, 86, 806.

REICHLIN, S. Peripheral thyroxine metabolism in patients with
    psychiatric and neurological diseases. A.M.A. Archives of
    General Psychiatry, 1959, 1, 434.

REICHLIN, S. Regulation of somatotrophic hormone secretion.
    In G. W. Harris and B. T. Donovan (Eds.), The Pituitary
    Gland. Los Angeles; University of California Press,
    1966. Pp. 270-298.

RISS, W. Effect of limbic damage in infancy on subsequent endo-
    crine development and running activity in rats. Anatomical
    Record, 1958, 130, 364.

RISS, W., BURSTEIN, S. D., & JOHNSON, R. W. Hippocampal or
    pyriform damage in infancy and endocrine development of
    rats. American Journal of Physiology, 1963, 204, 861.

RODRIGUES, A. Influence de l'ecorce cerebrale sur le cycle
    sexual du rat blanc. Comptes Rendus des Seances de la
    Société de Biologie, 1959, 153, 1271.

RUBIN, R. T., MANDELL, A. J., & CRANDALL, P. H. Corticosteroid
    responses to limbic stimulation in man: Localization of
    stimulus sites. Science, 1966, 153, 767.

RUSSELL, J. A., & WILHELMI, A. E. Growth (hormone regulation).
    Annual Review of Physiology, 1958, 20, 43.

SAITO, M., ISHIKAWA, A., AIBA, S., & KAWAI, T. On the central
    nervous control of the thyroid function especially on
    correlation with the habenular nucleus. 19th Nippon
    Noshinhei Gekagakki Sokai, 1960, p. 79.

SAUL, G. D., & SAWYER, C. H. EEG-monitored activation of the
    hypothalamo-hypophysial system by amygdala stimulation and
    its pharmacological blockade. Federation Proceedings, 1957,
    16, 112.

SAWYER, C. H. Effects of brain lesions on estrous behavior and
    reflexogenous ovulation in the rabbit. Journal of Experi-
    mental Zoology, 1959, 142, 227.

SAWYER, C. H., & KAWAKAMI, M. Characteristics of behavioral and
    electroencephalographic after-reactions to copulation and
    vaginal stimulation in the female rabbit. Endocrinology,
    1959, 65, 622.

SCHREINER, L., & KLING, A. Behavioral changes following rhinen-
    cephalic injury in cat. Journal of Neurophysiology, 1953,
    16, 643.

SCHREINER, L., & KLING, A. Effects of castration on hypersexual
    behavior induced by rhinencephalic injury in cat. A.M.A.
    Archives of Neurology and Psychiatry, 1954, 72, 180.

SCHREINER, L., & KLING, A. Rhinencephalon and behavior. American
    Journal of Physiology, 1956, 184, 486.

SCHWARTZ, N. B., & KLING, A. The effect of amygdaloid lesions
    on feeding, grooming and reproduction in rats. Acta
    Neurovegetative, 1964, 26, 12.

SETEKLEIV, J., SKAUG, O. E., & KAADA, B. R. Increase of plasma
    18-OH-steroids by cerebral cortical and amygdaloid stimula-
    tion in cats. Acta Physiologica Scandinavica, 1960, 50,
    supp. 175, 142.

SETEKLEIV, J., SKAUG, O. E., & KAADA, B. R. Increase of plasma
    17-hydroxycorticosteroids by cerebral cortical and amygdaloid
    stimulation in the cat. Journal of Endocrinology, 1961,
    22, 119.

SHEALY, C. N., & PEELE, T. L. Studies on amygdaloid nucleus of
    cat. Journal of Neurophysiology, 1957, 20, 125.

SHIZUME, K., MATSUZAKI, F., IINO, S., MATSUDA, K., NAGASAKI, S.,
    & OKINAKA, S. Effect of electrical stimulation of the
    limbic system on pituitary-thyroidal function. Endocrinology,
    1962, 71, 456.

SHIZUME, E., & OKINAKA, S. Control of thyroid function of the
    nervous system. In E. Bajusz and G. Jasmin (Eds.), Major
    Problems in Neuroendocrinology. Baltimore: Williams and
    Wilkins Co., 1964. Pp. 286-306.

SLUSHER, M. A., & HYDE, J. E. Inhibition of adrenal cortico-
    steroid release by brain stem stimulation in cats. Endo-
    crinology, 1961a, 68, 773.

SLUSHER, M. A., & HYDE, J. E. Effect of limbic stimulation on
    release of corticosteroids into the adrenal venous effluent
    of the cat. Endocrinology, 1961b, 69, 1080.

SLUSHER, M. A., & HYDE, J. E. Effect of diencephalic and mid-
    brain stimulation on ACTH levels in unrestrained cats.
    American Journal of Physiology, 1966, 210, 103.

SOULAIRAC, A., & SOULAIRAC, M. L. Atropie testiculaire par
    lesions du cortex cerebral chez le rat. Comptes Rendus des
    Seances de la Société de Biologie, 1958, 152, 921.

SPOTO, P., GOMIRATO, G., FERRO MILONE, F., BOCCI, F., ANGELERI, F.,
    & MANCA, R. Risposte utero-vesicali nella specie umana alla
    stimolazione dell' amigdala e dell' ippocampo. Bollettino
    della Societa Italiana di Biologia Sperimentale, 1961, 37,
    994.

SZENTÁGOTHAI, J., & MESS, B. Zur zentralen Stauerung der thyreo-
    tropen Akivitat des Hypophysenvorferlappens. Wiener Klin-
    ische Wochenschrift, 1958, 70, 285.

SZENTÁGOTHAI, J., FLERKÓ, B., MESS, B., & HALÁSZ, B. Hypothalamic
    Control of the Anterior Pituitary. Budapest: Hungarian
    Academy of Science, 1962.

TALEISNIK, S., CALIGARIS, L., & DE OLMOS, J. Luteinizing hormone
    release by cerebral cortex stimulation in rats. American
    Journal of Physiology, 1962, 203, 1109.

TERASAWA, E., & TIMIRAS, P. S. Electrical activity during the
    estrous cycle of the rat: Cyclic changes in limbic
    structures. Endocrinology, 1968, 83, 207.

TERZAIN, H., & ORE, G. D. Syndrome of Klüver and Bucy reproduced
    in man by bilateral removal of the temporal lobes. Neurology,
    1955, 5, 375.

TINDAL, J. S., & KNAGGS, G. S. Lactogenesis in the pseudopregnant
    rabbit after the local placement of oestrogen in the brain.
    Journal of Endocrinology, 1966, 34, ii.

TINDAL, J. S., KNAGGS, G. S., & TURVEY, A. Central nervous
    control of prolactin secretion in the rabbit: Effect of
    local oestrogen implants in the amygdaloid complex.
    Journal of Endocrinology, 1967, 37, 279.

URSI, K. Experimental studies on the amygdaloid nuclei and peri-
    amygdaloid cortex with special reference to ovulation.
    Niigata Medical Journal, 1955, 6, 189.

VALASCO, J. D., & TALEISNIK, S. Release of gonadotropins induced
    by amygdaloid stimulation in the rat. Endocrinology, 1969,
    84, 132.

VINCENT, J. D., DUFY, B., & FAURE, J. M. A. Effects of vaginal
    stimulation on hypothalamic single units in unrestrained
    rabbits. Experientia, 1970, 26, 1266.

WELSCH, C. W., CLEMENS, J. A., & MEITES, J. Effects of hypo-
    thalamic and amygdaloid lesions on development of growth
    of carcinogen-induced mammary tumors in the female rat.
    Cancer Research, 1969, 29, 1541.

WOOD, C. D. Behavioral changes following discrete lesions of
    temporal lobe structures. Neurology, 1958, 8, 215.

YAMADA, TAKASHI. The effect of electrical ablation of the nuclei habenulae, pineal body and subcommissural organ on endocrine function, with special reference to thyroid function. Endocrinology, 1961, 69, 706.

YAMADA, TAKASHI, & GREER, M. A. The effect of bilateral ablation of the amygdala on endocrine function in the rat. Endocrinology, 1960, 66, 565.

YAMADA, TOMOO. Experimental researches on the amygdaloid nucleus and periamygdaloid cortex, with special reference to the uterine motility. Niigata Medical Journal, 1954, 68, 682 & 788.

ZOLOVICK, A. J. Electrophysiological aspects of amygdaloid-hypothalamic interrelationships. Ph. D. Thesis, 1969. Kansas State University, Manhattan.

ZOLOVICK, A. J., & ELEFTHERIOU, B. E. Hormonal modulation of hypothalamic unit activity and EEG response to vaginal stimulation in the deermouse. Journal of Endocrinology, 1969, 49, 59.

THE AMYGDALOID NUCLEAR COMPLEX AND MECHANISMS OF RELEASE

OF VASOPRESSIN FROM THE NEUROHYPOPHYSIS

James N. Hayward
Departments of Anatomy and Neurology, School of Medicine
University of California, Los Angeles, California

INTRODUCTION

The homeostatic control of body water content depends upon the regulated release of vasopressin (antidiuretic hormone, ADH) from the neurohypophysis under 'osmometric' and 'volumetric' control and in coordination with behavior. While studies on the deafferented hypothalamus, the so-called hypothalamic island, indicate that the supraoptic nuclei and a small bit of surrounding hypothalamus connected to the neurohypophysis are the minimum amount of neural tissue necessary for 'osmometric' control of vasopressin release and water balance (Bard and Macht, 1958; Woods and Bard, 1960; Woods, Bard and Bleier, 1966; Sundsten and Sawyer, 1961), the integration of supraoptic neuronal activity with other bodily functions, such as drinking, requires intact ascending and descending pathways.

The amygdaloid nuclear complex receives neural input from diverse sites, namely the olfactory bulb, somatic and visceral afferents, visual and auditory areas (Creutzfeldt et al., 1963; Nauta, 1962; Machne and Segundo, 1955; Dell and Olson, 1951; Whitlock and Nauta, 1956), is responsive to osmotic (Sawyer and Fuller, 1960; Sawyer and Gernandt, 1956) and vagal stimuli (?volumetric) (Dell and Olson, 1951; Machne and Segundo, 1955), and has reciprocal connections to medial basal forebrain limbic structures, diencephalic and midbrain areas (Gloor, 1955 a, b; Nauta, 1958, 1961, 1962, 1963). This subcortical nuclear complex, as part of the rhinencephalon, with direct and indirect olfactory connections, has been thought, on theoretical grounds, to be involved in olfactory-gustatory reflexes as well as with the modulation of somatomotor, autonomic and endocrine activity in conjunction with 'socio-emotional' behavior (Pribram and Kruger, 1954). MacLean (MacLean, 1955; MacLean and Ploog, 1962) extended

the original Papez 'circuit' for 'emotion' (Papez, 1937) to include
medial-basal forebrain areas including the amygdala, and described
them all as belonging to the 'limbic system' or 'visceral brain.'
On the basis of their studies on penile erection in the squirrel
monkey, MacLean and Ploog (1962) distinguished between the dorsal,
anterior, and midline portions of the 'limbic system' which may be
involved in 'preservation of the species' and more lateral and
ventral portions of the anterior limbic structures related to 'self
preservation.' Nauta (1963), perhaps, extended this concept a bit
further to include a 'limbic system-midbrain circuit' which may
provide a linkage between 'limbic forebrain' and hypothalamic areas
above and the primordial lemniscal areas (brainstem reticular formation) below. On purely physiological grounds, others have proposed
a dualistic regulation of behavior by antagonistic systems involving
an excitatory-arousal component in the brainstem reticular formation
(Moruzzi and Magoun, 1949) and an inhibitory-sleep inducing region
in the medial-basal forebrain (Sterman and Clemente, 1962). Whether
or not any or all of these systematizations of behavioral-visceral
interactions represents a valid working model is difficult to say,
at present. There can be little doubt that medial forebrain and
midbrain structures, by virtue of their anatomical connections to
the amygdala on the one hand, and to the hypothalamus on the other,
are located strategically for modulation of supraoptic neuronal
activity and water balance in coordination with changes in (drinking) behavior.

In the monkey, a group of 70,000 secretory supraoptic neurons
(Magoun and Ranson, 1939) lie draped around the optic chiasm and
optic tracts (Lammers, 1969; Nauta and Haymaker, 1969) and have
the awesome task of preventing the monkey from washing its body
fluids out into the urine. These cells have the complex task of
producing the vital octapeptide, vasopressin, packaging it in
1500Å neurosecretory vesicles (Palay, 1957; Rechardt, 1969),
moving these vesicles down the unmyelinated axons (Cagal, 1911) to
the nerve endings abutting against the capillary walls in the
infundibular process (Bargmann and Scharrer, 1951; Palay, 1957),
and in dumping this antidiuretic hormone into the blood stream
for its journey to the renal distal tubule and collecting ducts
(Heller and Ginsburg, 1966). Too much hormone release can lead
to water intoxication and seizures, and too little hormone release
can lead to dehydration, cardiovascular collapse and death (Heller
and Ginsburg, 1966). The neurosecretory cells of this system,
the supraoptic and paraventricular neurons, have the morphological
(Palay, 1957; Rechardt, 1969), electrical (Cross and Green, 1959;
Kandel, 1964) and chemical (Nishioka et al., 1970; Norstrðm and
Sjðstrand, 1971; Sachs, 1967) characteristics of other nerve
cells; they are under direct neural control (Harris, 1947) and
show a close coupling between electrical and secretory activity
(Cross and Green, 1959; Dyball, 1971; Ishikawa, Koizumi and
Brooks, 1966). Impulses impinging on their dendritic, somatic

and 'axonal' membranes convey information about blood osmotic
pressure, blood volume and behavioral activities such as drinking,
muscular exercise, pain, emotional stress, and sleep-waking
cycles (Harris, 1960; Heller and Ginsburg, 1966; Lammers, 1969;
Rechardt, 1969).

The paraventricular hypothalamic nucleus generally has been
considered the site of production of oxytocin, the 'milk-letdown
factor' (Olivecrona, 1957); however, the possibility remains
that two types of neurons, one producing vasopressin and the
other only oxytocin, are distributed unequally throughout both
the supraoptic and paraventricular nuclei (Magoun and Ranson,
1939; Nishioka et al., 1970; Orkand and Palay, 1967; Rechardt,
1969; Sokol and Valtin, 1967) in mammals. Evidence exists to
suggest that the paraventricular nucleus is not a homogeneous
structure, at least as regards estrogen concentrating ability
(Stumpf, 1970) and efferent pathways (Aulsebrook and Holland,
1959 a, b; Bisset et al., 1967; Cross, 1966; Rothballer, 1966;
Woods, Holland and Powell, 1969). While many stimuli, such as
hemorrhage, injection of hypertonic solutions into carotid,
hypothalamus or III ventricle, or brain stimulation can release
both vasopressin and oxytocin from the neurohypophysis (Aulsebrook and Holland, 1969a; Andersson, 1953; Andersson and McCann,
1955; Dyball, 1971; Harris, 1947; Rothballer, 1966), natural
physiological stimuli (Heller and Ginsburg, 1966) and electrical
stimulation of particular pathways (Aulsebrook and Holland,
1969a and b; Bisset et al., 1967; Cross, 1966) can release each
hormone separately, therefore suggesting a dual and separate set
of input connections involved in vasopressin and oxytocin release.
In view of these considerations, I shall not consider either
oxytocin release or the paraventricular nucleus any further in
this paper, but limit my discussion to the supraoptic nucleus
and vasopressin release.

In the present paper, I should like to review some of the
mechanisms by which the amygdaloid nuclear complex may influence
the supraoptic neurons and the release of vasopressin from the
neurohypophysis for the integrated control of body water.

FOREBRAIN OSMORECEPTORS: SUPRAOPTIC-OSMORECEPTOR
NUCLEAR COMPLEX AND LIMBIC OSMORECEPTOR COMPLEX

Evidence for forebrain 'osmoreceptors':

Over the past twenty-five years, the search for forebrain
neural elements involved in the 'osmometric' control of water
balance and vasopressin release has focused on two areas: the
supraoptic nucleus (NSO) and its perinuclear zone (PNZ) and the
olfactory bulb-olfactory tubercle-preoptic-amygdala areas.

Evidence suggests that the basal hypothalamic 'osmoreceptors'
(NSO-PNZ) provide a direct and specific physico-chemical pathway
from the blood to neurohypophysis while the limbic 'osmoreceptors'
(amygdala and olfactory areas) provide an indirect connection
between various nonspecific noxious stimuli, including blood
hypertonicity, and behavior and the neurohypophysis (Hayward and
Vincent, 1970; Vincent and Hayward, 1970).

In 1947, Verney, on the basis of his studies of intracarotid
injections of hypertonic solutions in the unanesthetized dog,
concluded that the osmotic pressure of the blood and extracellular fluid in the brain was detected somehow by neural
elements lying in the distribution of the internal carotid
artery. In this study, Verney (1947) noted that the intracarotid
injections of hypertonic sodium chloride aroused the dog and
often resulted in behavioral responses such as licking of the
lips. Later, after selectively perfusing different parts of the
dog's brain with hypertonic solutions, Jewell and Verney (1957)
concluded that the 'osmoreceptors' for vasopressin release could
be further localized in the anterior hypothalamic-medial thalamic
area.

In 1956, Sawyer and Gernandt discovered a triphasic EEG
response in the olfactory bulb, olfactory tubercle and amygdala
in response to intracarotid (i.c.) hypertonic solutions in the
rabbit. This electrical sequence of high voltage fast waves
(40-70 c/s), depression and slow waves occurred over 20-30 sec
in the olfactory bulb with or without section of the tract or
anesthesia of the epithelium (Sawyer and Gernandt, 1956; Sawyer
and Fuller, 1960; Sundsten and Sawyer, 1959). Cutting the
olfactory bulb connections to the rest of the brain did not alter
the triphasic EEG response to i.c. hypertonic solutions in the
amygdala, olfactory tubercle or adjacent medial basal limbic
structures. There was a close association between the olfactory-
amygdala triphasic EEG response and neurohypophysial hormone
release in both cat and rabbit (Holland, Cross and Sawyer, 1959;
Sawyer and Fuller, 1960), although the triphasic electrical
response could not be recorded from the area of the supraoptic
nucleus (Sawyer and Gernandt, 1956). Unit activity in the
olfactory bulb of rabbit (Freedman, 1963) and cat (Moyano and
Brooks, 1968) in response to i.c. injections of hypertonic sodium
chloride produced marked acceleration of firing rate (70%) of
the high voltage, regular cells. Repeated injections of hypertonic saline (0.5M) or single injections of more concentrated
saline (1.0M) produced afterdischarge spiking in the olfactory
bulb (Moyano and Brooks, 1968) and in the amygdala (Sawyer and
Fuller, 1960; Sawyer and Gernandt, 1956). Sawyer and co-workers
speculated that the 'osmoreceptors' of Verney might exist in the
olfactory tubercle-amygdala region rather than in the supraoptic

nuclear area and pointed out the close temporal relationships between the triphasic EEG response in limbic structures, cortical EEG arousal and vasopressin release to i.c. hypertonic saline (Sawyer and Fuller, 1960; Sawyer and Gernandt, 1956).

Hypothalamic Island:

The critical test of their hypothesis came when Bard and his co-workers (Bard and Macht, 1958; Woods and Bard, 1960; Woods, Bard and Bleier, 1966) and Sundsten and Sawyer (1961) separated the medial basal hypothalamus and the neurohypophysis from the rest of the brain, i.e. an hypothalamic island or deafferented hypothalamus. These investigators concluded that the primary, elemental or minimal 'osmoreceptor' zone lies closely connected to the supraoptic nuclei. Presumably, the olfactory bulb-olfactory tubercle-amygdala 'osmoreceptor' elements provide an accessory and perhaps behaviorally related system. It is of interest that Sugar and Gerard (1938) described a similar triphasic EEG response in the brain secondary to brief periods of hypoxia and that Sawyer and Fuller (1960) could produce their triphasic EEG response equally well with hypoxia or with hypertonic saline. These olfactory-limbic structures thus may provide a secondary system of 'osmoreceptors' linked to noxious stimuli and behavior, and involved in the augmentation of vasopressin release during non-specific stress (Mirsky and Stein, 1963; Mirsky, Stein and Paulisch, 1954). In this regard, Moyano and Brooks (1968) found that they could drive NSO osmosensitive cells by electrical stimulation of the ipsilateral olfactory bulb.

Electrophysiological Studies:

Von Euler (1953) described slow hypothalamic 'osmo-potentials,' in response to intracarotid hypertonic sodium chloride, in the region of the supraoptic nucleus and preoptic area in the cat. He speculated that these responses might be generator potentials of central 'osmoreceptors' and that such 'osmoreceptor' cells should respond specifically to the osmotic stimulus, like peripheral receptors, and not act additionally as interneurons or secretory cells. In line with von Euler's dictum, Cross and Green (1959) and Joynt (1964) described many osmosensitive cells that seemed to qualify as 'osmoreceptor' neurons. These neurons were located in the perinuclear zone of the NSO and responded to osmotic but not to natural sensory stimuli. On the other hand, Brooks and his co-workers (Brooks, Ushiyama and Lange, 1962; Ishikawa, Koizumi and Brooks, 1966; Koizumi, Ishikawa and Brooks, 1964; Suda, Koizumi and Brooks, 1963) found that the majority of their osmosensitive cells responded to both osmotic and sensory stimuli.

These data raised the question of whether there was a
diffuse system of osmoreceptors in the anterior hypothalamus-
preoptic area, including the olfactory bulb and amygdala, involved in the regulation of vasopressin release, with afferent
input from somatic, visceral and cortical areas or whether a
small group of osmoreceptors was discretely localized in the
immediate vicinity of the supraoptic nuclei. These conflicting
characterizations of the 'osmosensitive' cell supposed to be
involved in antidiuretic hormone release remained unresolved
until we applied our stylized techniques of single unit recording in the unanesthetized monkey (Findlay and Hayward, 1969;
Hayward, 1969a, b; Hayward and Vincent, 1970; Vincent and Hayward,
1970).

Osmosensitive Cell and Behavior:

Since many of the previous workers did not monitor behavior
or EEG in their anesthetized animals (Beyer and Sawyer, 1969;
Cross and Silver, 1966), it seemed important to attempt to
correlate the changes in firing patterns of hypothalamic single
neurons in response to both osmotic and sensory stimuli in order
to document the afferent connections of the 'osmosensitive'
cells. In our earlier studies, in the hypothalamus of the behaving rabbit, we had found that 60-70 per cent of diencephalic
units showed significant changes in firing rate and patterns of
discharge in relationship to shifts in sleep-waking behavior
(Findlay and Hayward, 1969). In the present study, we used a
trained, chair-adapted monkey with a cranial platform, head-
fixing bolts and a bone-fixed adapter cylinder to support a
hydraulic microdrive which pushed the tungsten microelectrodes
into the hypothalamus (Findlay and Hayward, 1969; Hayward and
Vincent, 1970), and a chronic silicon rubber tubing in the common
carotid artery for solution injections (Baker et al., 1968). Our
system allows for simultaneous recording of single unit firing
patterns, as well as direct observations of the behavior of the
chamber isolated animal. We analyzed the cell discharges with a
digital computer, calculated the spike train statistics and
histograms, determined the cell location by a histological study
of Prussian blue spots (Findlay and Hayward, 1969; Hayward and
Vincent, 1970). Figure 1 illustrates the response of a single
septal neuron in the unanesthetized monkey to arousal from sleep
with a tenfold increase in firing rate, without a change in the
pattern of discharge, and in association with changes in EEG,
eye movements and brain and nasal temperatures.

The unanesthetized monkey responds to intracarotid injections
of sodium chloride (0.45 M, 0.25 ml/sec, 1 ml) with a complex
behavioral response that consists of EEG 'arousal,' sniffing
respiration, lip-smacking and chewing, increased facial, eye and
body movements, and head turning. It responds not at all to

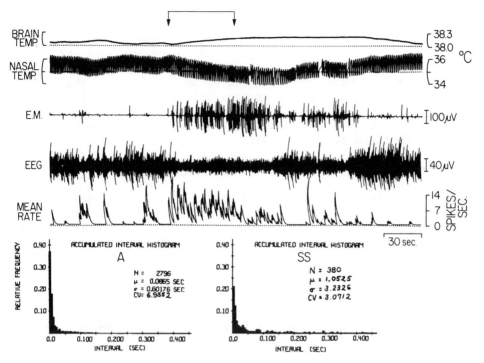

Fig. 1. Temporal relationships between the firing pattern of a
medial septal neuron and other manifestations of sleep and wakefulness in the unanesthetized monkey (<u>Macaca</u> <u>mulatta</u>). During slow
sleep the monkey sits quietly in the environmental chamber, eyelids closed, diminished eye movements, EEG synchronized high voltage and the neuron exhibits intermittent brief high frequency
bursts at an overall mean rate of 0.95/sec with an 'asymmetric'
unimodal histogram (SS). Upon waking to a tapping on the chamber
(between arrows), eyelids open, eye movements increase, EEG de-
synchronized low voltage and cell accelerates to a higher mean
rate (11.5/sec) with a sustained train of spikes and an 'asymmetric'
unimodal histogram (A). Note the slower, irregular respiration and
the nasal and cutaneous (not shown) vasoconstriction during arousal
with heat retention and elevation of brain temperature.

     Labels: Brain temp., temperature of the cerebral arterial
blood at the basilar artery; Nasal temp., temperature of the air
in the nasal cavity; EM, extraocular movements; EEG, biparietal
electrocorticogram; Mean rate, analog output proportional to the
rate of unit discharge; Accumulated interspike interval histogram,
N = number of intervals, μ = mean interspike interval, σ =
standard deviation, and CV = coefficient of variation.

intracarotid isotonic sodium chloride or distilled water (Hayward
and Vincent, 1970). These results confirm the observations of
Verney (1947), in the dog. Furthermore, the triggering of behavioral changes by intracarotid hypertonic solutions suggests the
involvement of brain structures beyond (outside) the supraoptico-
hypophysial tract, perhaps as suggested by Sawyer and co-workers
(Sawyer and Fuller, 1960; Sawyer and Gernandt, 1956; Sundsten
and Sawyer, 1959) the olfactory bulb, olfactory tubercle and
amygdala areas or possibly the medullary osmoreceptors (Clemente
et al., 1957; Holland et al., 1959).

Primate 'Osmosensitive' Cells:

We recorded two types of osmosensitive cells in the antero-
lateral hypothalamus: 1) 'specific' osmosensitive neurons responding exclusively to intracarotid osmotic stimuli and lying
in the supraoptic nucleus (NSO) and in the immediate perinuclear
zone (Fig. 2 and Fig. 3); 2) 'non-specific' osmosensitive neurons
responding to both osmotic and arousing sensory stimuli and lying
diffusely scattered in the anterolateral hypothalamus (Fig. 3).
These data support the concept of a dual system of osmoreceptor
elements in the rostral hypothalamus and medial basal forebrain.
A primary system of 'osmoreceptors' located in the supraoptic
nucleus-perinuclear zone which regulate vasopressin release just
to changes in blood-brain extracellular fluid osmolality or
sodium ion concentration. A secondary system of osmoreceptor
elements located in the preoptic area-anterior hypothalamus as
well as olfactory bulb-amygdala responds to hypoxia, osmotic
stimuli, pain-emotional stress and perhaps to drinking behavior.
I consider that our 'non-specific- osmosensitive cells are linked
directly to this 'secondary'limbic group of osmoreceptors (Hayward
and Vincent, 1970). On further study of the 'specific' osmosensitive neurons, we were able to distinguish two major sub-types
of these cells: a) single cells located in the supraoptic nucleus
and exhibiting a 'biphasic' firing pattern to the osmotic stimulus
(acceleration followed by inhibition) with no response to arousing
sensory stimuli (Fig. 2A); and b) single cells located in the
immediate perinuclear zone of the supraoptic nucleus and showing
a 'monophasic' acceleration or inhibitory response to osmotic
stimulation (Fig. 2B) with no response to arousing sensory
stimuli (Hayward and Vincent, 1970; Vincent and Hayward, 1970).
The responses of these cells to repetitive osmotic stimuli is
shown in Fig. 2A, B. The anatomical distribution of these cells
in the supraoptic nucleus, perinuclear zone and anterolateral
hypothalamus we show in Figure 3.

On the basis of the anatomical location of the cells, on
the pattern of discharge to intracarotid osmotic stimuli and on
the pattern of discharge to arousing sensory stimuli, we suggest
that the 'osmoreceptors' of Verney lie in the immediate peri-

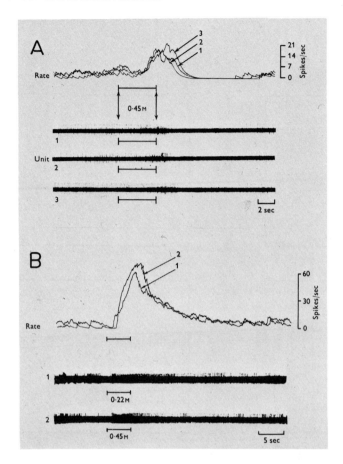

Fig. 2. Pattern of discharge of two types of 'specific' hypothalamic osmosensitive neurons during repeated intracarotid
injections of hypertonic sodium chloride in the waking rhesus
monkey.

A. 'Specific' biphasic osmosensitive <u>supraoptic neuron</u> showing an excitatory-inhibitory sequence during three consecutive
osmotic stimuli (1,2,3) of 0.45M sodium chloride at 0.25 ml/sec
injected at 2 minute intervals.

B. 'Specific' monophasic osmosensitive cell in the <u>peri-
nuclear zone</u> of the supraoptic nucleus showing only an excitatory
sequence during two consecutive osmotic stimuli (1,2) or 0.22M
(1) sodium chloride at 0.20 ml/sec and 0.45M (2) sodium chloride
at 0.10 ml/sec. In each section superimposed polygraph tracings
(above) of the mean rate of cell firing (spikes/sec) and unit
spikes photographically reproduced (below) during the repeated
osmotic stimuli. Cells did not respond to mild arousing sensory
stimuli nor to isotonic sodium chloride intracarotid. Note the
highly reproducible phase of unit acceleration (A,B) and period
of cell silence (A) following each stimulus. Reproduced from
Journal of Physiology (London), 210, 1970, by courtesy of the
Physiology Society, Cambridge University Press, Cambridge.

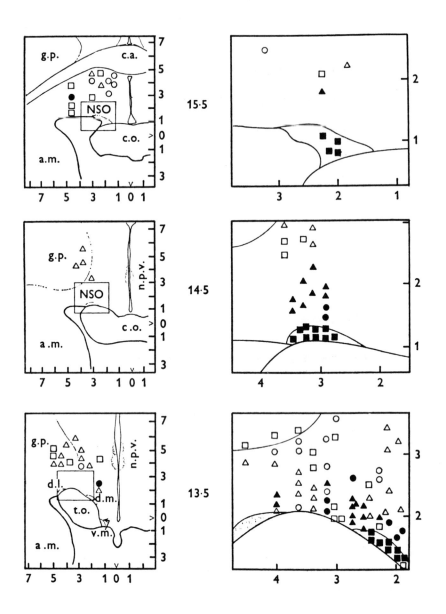

Fig. 3. Diencephalic localization of one hundred and thirty
cells recorded in eight monkeys. Frontal sections are Fr. 15.5
(above), Fr. 14.5 (middle) and Fr. 13.5 (bottom) after the atlas
of Snider and Lee (1961) with the full field on the left and an
enlarged box-insert shown on the right. Symbols:■, 'Specific'
biphasic osmosensitive cells;▲ , 'specific' monophasic acceleration osmosensitive cells; ● , 'specific' monophasic inhibitory
osmosensitive cells;△ , 'non-specific' monophasic acceleration
osmosensitive cells;○ , 'non-specific' monophasic inhibitory
osmosensitive cells;□ , 'non-osmosensitive' cells. Note the
uniform grouping of the 'specific' biphasic osmosensitive cells
in the dorsomedial and dorsolateral parts of the supraoptic
nucleus (NSO) and the localization of most of the 'specific'
monophasic neurons in the perinuclear zone of the supraoptic
nucleus. The perinuclear zone is that area immediately surrounding the NSO for 0.5-1.0 mm. The 'non-specific' osmosensitive
cells and the 'non-osmosensitive' cells lie diffusely throughout
the antero-lateral hypothalamus, a few in the perinuclear zone
but most elsewhere. We find dorsolateral, dorsomedial and ventromedial parts of the NSO in the monkey (Nauta and Haymaker, 1969).

Labels: g.p., globus pallidus; c.a., anterior commissure;
NSO, supraoptic nucleus; c.o., optic chiasm; a.m., amygdala;
n.p.v., paraventricular nucleus; d.l., pars dorsolateralis of
the supraoptic nucleus; d.m., pars dorsomedialis of the supraoptic nucleus; v.m., pars ventromedialis of the supraoptic
nucleus; t.o., optic tract. Reproduced from Journal of
Physiology (London), 210, 1970, by courtesy of the Physiology
Society, Cambridge University Press, Cambridge.

nuclear zone of the supraoptic nucleus in the primate and are
represented by our 'monophasic' specific 'osmosensitive' neurons.
We further suggest that the 'biphasic' specific 'osmosensitive'
neurons in the supraoptic nucleus represent the neuroendocrine
cells of this system (Hayward and Vincent, 1970; Vincent and
Hayward, 1970). Some of the possible synaptic inter-connections
of these neural elements which might explain their firing
patterns are shown in Figure 4. Recently, several other groups
of investigators also have suggested that a recurrent collateral
inhibitory system is present in the supraoptic (Kelly and
Dreifuss, 1970; Yamashita, Koizumi and Brooks, 1970) and preoptic nucleus (Kandel, 1964). The detailed nature of Verney's
'osmoreceptors' remain poorly understood and the subject for
future research, especially, in regard to the variability of
antidiuretic responses to different types of intracarotid hypertonic solutions (Andersson, Olsson and Warner, 1967; Eriksson,
Fernandez and Olsson, 1971; Olsson, 1969; Verney, 1947) which
may be related partly to the blood-brain barrier for these different solutes in the supraoptic capillary bed (Andersson, Olsson
and Warner, 1967; Eriksson, Fernandez and Olsson, 1971; Finley,
1939; Olsson, 1969; Yudilevich and DeRose, 1971) or, possibly,
in regard to humoral factors from the kidney (Andersson and
Eriksson, 1971). In our view, the 'non-specific' osmosensitive
neurons diffusely scattered in the antero-lateral hypothalamus
(Fig. 3) are related to the secondary system of 'osmoreceptors'
of Sawyer involving the limbic structures: olfactory bulb,
olfactory tubercle, preoptic area and amygdala. The neural
connections between the 'primary' and 'secondary' osmoreceptors
are not known. Electrical stimulation of the medial basal forebrain and midbrain may indicate some of these amygdalo-supraoptic
connections as discussed in the next section.

SUMMARY

     The amygdala forms part of the 'secondary' forebrain 'osmo-
receptors' of Sawyer which also includes the olfactory bulb,
olfactory tubercle and preoptic area. These limbic osmoreceptors
respond to osmotic, hypoxic and perhaps other noxious stimuli
with a characteristic triphasic electrical response. These
structures have direct anatomical connections to the supraoptic
nucleus and when excited electrically can alter supraoptic
neuronal activity and, in turn, the release of vasopressin from
the neurohypophysis. It seems likely that 'osmosensitive'
neurons lying in the anterior hypothalamic-preoptic region, and
which are also responsive to diverse afferent input, belong to
this 'secondary' osmoreceptor system of Sawyer. The functional
importance of these interneurons for the regulation of vasopressin release and drinking is not known at the present time.
However, in view of the well known release of vasopressin from

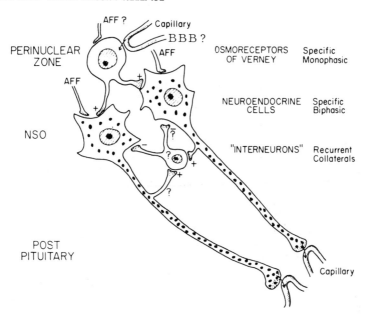

Fig. 4. A schematic interpretation of the possible cellular
elements and their synaptic interconnections in the osmoreceptor-
supraoptic nuclear complex of the monkey. We propose that the
'osmoreceptors' of Verney are distinct neurons lying the perinuclear zone of the supraoptic nucleus (PNZ) and respond
'specifically' to an osmotic stimulus with a 'monophasic' discharge which is transferred directly (excitatory axosomatic
synapses) to the neuroendocrine cells of NSO. The functional
nature of Verney's 'osmoreceptors' in relation to the local PNZ
blood-brain barrier for ions and molecules remains obscure at
present (Eriksson, Fernandex and Olsson, 1971; Yudilevich and
deRose, 1971). We further suggest that the supraoptic neuroendocrine cells respond to an osmotic stimulus with an initial
acceleration discharge, due to the synaptic driving by the
'osmoreceptors,' followed quickly by an inhibitory phase possibly
due to recurrent collateral activation of 'interneurons' or via
direct action on these neuroendocrine cells with slow inhibitory
postsynaptic potentials. These supraoptic cells show a specific
'biphasic' discharge to an osmotic stimulus. Limbic 'osmoreceptors' of Sawyer, perhaps our 'non-specific' cells (not
shown) may provide input to NSO from noxious and behaviorally
related (drinking) stimuli. Further physiological and morphological studies are needed to support our hypothesis. Labels:
BBB, blood-brain barrier; NSO, supraoptic nucleus; AFF, afferent
fiber connections; +, excitatory synaptic action; -, inhibitory
synaptic action; ?, an unknown entity. Modified from Brain
Research 23, 1970, by courtesy of the Elsevier Publishing Co.,
Amsterdam.

emotional, painful or conditional stimuli, it seems likely that
such excitatory pathways may engage these 'osmosensitive' limbic
interneurons.

The 'primary' forebrain 'osmoreceptors' of Verney probably
are single cells lying in the perinuclear zone of the supraoptic
nucleus, responding to an intracarotid osmotic stimulus with a
monophasic discharge and not responding to mild arousing sensory
stimuli. These cells may have direct monosynaptic connections
to the supraoptic neurons and are probably influenced principally
by the solute concentration in their immediate extracellular
space. Supraoptic neurons respond to intracarotid hypertonic
saline with an initial acceleration followed quickly by a period
of silence. Such a 'biphasic' pattern of firing suggests that
the supraoptic neurons have a recurrent collateral system with
or without an interneuron (a neuroendocrine 'Renshaw cell') to
account for this excitatory-inhibitory sequence. Other afferent
input from the 'secondary' osmoreceptor system as well as from
olfactory, vagal, glossopharyngeal, cholinergic and adrenergic
pathways, provide additional input to this final common endocrine
motor pathway for regulation of body water, the supraoptic neuron.

AMYGDALO-SUPRAOPTIC CONNECTIONS: SOME MECHANISMS OF VASOPRESSIN RELEASE BASED ON ELECTRICAL STIMULATION OF THE BRAIN

In order to obtain direct evidence that the amygdaloid
nuclear complex is involved in modulating body water balance
and vasopressin release, we stimulated electrically the amygdala
of the behaving monkey with chronically implanted bipolar concentric stainless steel electrodes fixed to a cranial platform
(Hayward and Smith, 1963, 1964). In trained, adult, rhesus
monkeys sitting in a chair-type restraining apparatus with
stomach tube and urethral catheter and undergoing a sustained
water diuresis, unilateral biphasic square wave electrical
stimuli (30 Hz, 0.2-0.8 mA, 1-5 msec pulse duration, 30 sec
trains with 2 min intervals) delivered to histological identified points in the amygdaloid nuclear complex resulted in
behavioral and antidiuretic responses usually without local
or generalized afterdischarge.

Amygdaloid Nuclear Complex:

Figure 5 shows vasopressin release from the neurohypophysis
upon stimulation of the accessory basal nucleus of the amygdala.
Our criteria for a neurohypophysial antidiuresis, as shown in
Figure 5, in contrast to a renal-vasoconstrictor antidiuresis,
is a greater than 50 per cent drop in urine flow (V) and free
water clearance (C_{H_2O}), a rise in urine osmolality to over

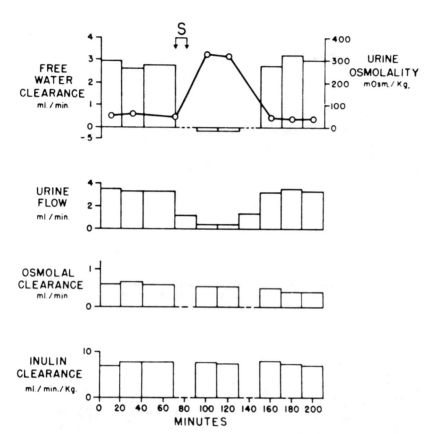

Fig. 5. Antidiuretic hormone release from the neurohypophysis
during electrical stimulation of the amygdala and measurement
of renal clearances in the behaving rhesus monkey. Stimulation
of the accessory basal nucleus of the amygdala for 10 min.
(0.3 ma, 30 Hz, 30 sec on - 2 min off) in the hydrated, waking
monkey resulted in an antidiuresis lasting for 80 minutes with
a negative free-water clearance, a urine osmolality rise to
325 mOm/kg without significant change in osmolal and inulin
clearances. Interruption of the baseline indicates that no
values were obtained because of the dead space effect.
Labels: S, period of electrical stimulation of the amygdala.
Reproduced from Archives of Neurology (Chicago), 9, 1963, by
courtesy of the American Medical Association, Chicago.

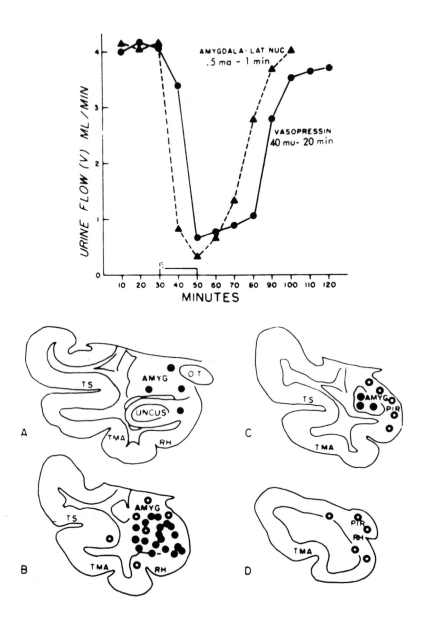

Fig. 6. Vasopressin release from the neurohypophysis during electrical stimulation of the amygdaloid nuclear complex and hippocampus in the unanesthetized monkey (Macaca mulatta).
UPPER: Antidiuretic responses (acute drop in urine flow, V, ml/min) in a waking, hydrated monkey sitting in a sound-shielded chamber during electrical stimulation of the lateral nucleus of the amygdala (0.5 ma, 30 Hz, 1 min) shown by dashed line, and during intravenous infusion of vasopressin (40 mu, 20 min, chronic right atrial cannula) shown by continuous line. The associated rise in urine osmolality, fall in free water clearance and lack of significant change in osmolal and inulin clearances are not shown. Note the similarities between the timing and extent of the antidiuretic responses to stimulation of the amygdaloid complex and to vasopressin infusion.
LOWER: Histological locations of 42 electrode sites in six monkeys in the medial temporal lobes in transverse sections at A. Fr.15, B. Fr.17, C. Fr.19, D. Fr.22 as adapted from Olszewski (1952). Electrode tip locations shown in closed circles in the amygdaloid nuclear complex (25 points in medial, lateral, basal and accessory basal nuclei) and hippocampus (1-uncus) where unilateral electrical stimulation repeatedly released antidiuretic hormone from the neurohypophysis. Electrode tip locations shown in open circles in the temporal neocortex (2 points in middle and inferior temporal gyri); periamygdaloid cortex (10 points in temporal pole, temporal prepyriform area and entorhinal area; and amygdaloid complex (4 points in cortical nucleus and anterior area) repeated stimulation produced behavioral responses but no changes in water excretion. Labels: OT, optic tract; AMYGD, amygdala; RH, rhinal sulcus; PIR, piriform cortex; TMA, medial anterior temporal sulcus; TS, superior temporal sulcus. Reproduced from Arch. Neurol. (Chicago) 9, 1963 by courtesy of the American Medical Assoc., Chicago.

300 mOsm/kg, without significant change in osmolal clearance
(C_{OSM}) or in glomerular filtration rate (C_{IN}). A comparison of
the antidiuretic responses from endogenous vasopressin release
due to stimulation of the lateral nucleus of the amygdala and
exogenous vasopressin infusion is shown in Figure 6 (upper).
Vasopressin release from the neurohypophysis occurred with
excitation of twenty-five points in the medial, lateral, basal
and accessory basal nuclei of the amygdala and one point in the
uncus of the hippocampus (Fig. 6, lower). Behavioral responses
elicited from these sites ranged from no change in behavior to
arousal from sleep, attentiveness, and orienting reaction, licking, chewing, salivation, coughing and rarely gagging or vomiting.
At times, the animal remained immobile, but, at other times,
there occurred rotation of the body, smacking of the lips,
shaking of the head and generalized restlessness. There was no
good correlation between the behavioral response and the antidiuretic response. A number of adjacent points in the cortical
nucleus and anterior area of the amygdala, middle and inferior
temporal gyri, in the temporal pole and in the piriform cortex,
including the temporal prepiriform area and the entorhinal area,
produced no change in water excretion on repeated stimulation
(Fig. 6, lower). The behavioral effects produced from stimulation
of these areas were identical to those produced from sites yielding antidiuretic responses (Hayward and Smith, 1963).

Rothballer and co-workers (Rothballer, 1966; Slotnick and
Rothballer, 1964) found in the acutely prepared cat, using the
less sensitive pressor response, that electrical stimulation in
the medial and central nuclei of the amygdala produced vasopressin release. Dingman et al. (1959) obtained antidiuretic
responses in man by electrical stimulation of the amygdaloid
nuclear complex. In contrast to our negative results from stimulation of the prepiriform area in the monkey, Yoshida et al.
(1965) obtained antidiuretic hormone release from this area in
the dog. Species differences, anesthesia, high resting levels
of ADH in acutely prepared animals and different 'bioassay'
techniques for detecting vasopressin release probably account
for these differences between monkey, cat and dog. Of greater
interest are the results of Matheson and Sundsten (1969) in the
unanesthetized Macaca mulatta where ACTH-cortisol release
occurred upon electrical stimulation of the basal, accessory
basal and lateral nuclei of the amygdala. Inhibition of ACTH
release occurred upon stimulation of cortical, medial amygdaloid
nuclei, the anterior area and stria terminalis. Our experimental
design, maximal inhibition of supraoptic neurons with water
loading, did not allow us to study inhibition of ADH release
but the similarity between excitatory (release facilitated)
points for ACTH and ADH is striking and suggests a common
mechanism, possibly a linkage with the secondary 'osmoreceptor'

system via a stress-noxious stimuli responsive pathway. If one
considers our 'negative' points in the cortical and anterior
areas of amygdala and piriform cortex as possibly being inhibitory,
then two general endocrine areas could be delimited in the amygdala, supporting the concept of specific localization (Ursin and
Kaada, 1960; Kaada, 1951).

Mechanisms of ADH Release: Behavior

What physiological responses could be triggered by the
synchronous electrical firing of amygdaloid neural elements
which would then lead to vasopressin release in the monkey? In
view of the well recognized relationship between vasopressin
release and painful-stressful and 'emotional' stimuli (Corson,
1966; Rydin and Verney, 1938), and recognized involvement of the
amygdala in 'emotional' behavior (Gloor, 1960) it seems reasonable to ask whether the behavioral effects of stimulation can
explain the ADH release? We found no clear correlation between
the minor behavioral effects of our electrical stimulation and
ADH release. We saw no sham rage, no extreme 'emotional' display
in our monkeys and found similar behavioral effects from points
both releasing and not releasing vasopressin from the neurohypophysis.

Mechanisms of ADH Release: Direct Neural Pathways

Activation of direct, excitatory, polysynaptic neural pathways to the supraoptic nuclei could account for amygdaloid
initiation of vasopressin release. Amygdala stimulation produces
evoked potentials (Gloor, 1955a, b) and acceleration of unit
activity (Egger, 1967; Stuart et al., 1964) in the vicinity of
the supraoptic nucleus, but no monosynaptic connections have been
described in rat or monkey (deOlmos, 1971; Nauta, 1961). When we
stimulated points along the known direct efferent pathways (Gloor,
1955a, b; Nauta, 1961, 1962) to the hypothalamus in the olfactory
tubercle, diagonal band of Broca, and ventral amygdalo-hypothalamic tract (Hayward and Smith, 1963, 1964), we obtained
release of vasopressin from the neurohypophysis (Fig. 7). Others
also have released vasopressin by electrical stimulation along
these direct amygdalo-hypothalamic pathways (Andersson and McCann,
1955; Aulsebrook and Holland, 1969a; Dingman, 1966; Rothballer,
1966).

Mechanisms of ADH Release: Indirect Neural Pathways

The amygdaloid nuclear complex is associated directly and
indirectly via multisynaptic pathways with other areas in the
'limbic system' and in the 'limbic midbrain area' in the monkey
(Nauta, 1961, 1962, 1963). When we stimulated the ventral

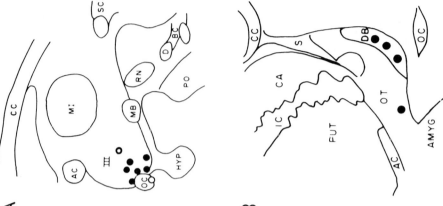

Fig. 7. Vasopressin release from the neurohypophysis during electrical stimulation of the hypothalamus and medial basal forebrain in the unanesthetized monkey (Macaca mulatta). Histological localization of 12 electrode points in 6 monkeys. UPPER: Midsagittal view of the hypothalamus. Antidiuretic hormone released upon unilateral stimulation of points in the supraoptic nucleus, supraopticoneurohypophysial tract, the ventromedial hypothalamic nucleus, the lateral hypothalamic nucleus, the medial foregrain bundle and the ventral amygdalohypothalamic tract, closed circles. No change in urine flow, urine osmolality or free water clearance with repeated stimulation of the optic chiasma or dorsolateral hypothalamus, open circles. LOWER: Transverse section of the septal region at Fr.19 adapted from Olszewski (1952). Antidiuretic responses to repeated unilateral stimulation of 3 points in the diagonal band of Broca and 1 point in the olfactory tubercle, closed circles. No change in glomerular filtration rate (inulin clearance) nor renal solute excretion (osmolal clearance) during these antidiuretic responses. Labels: OC, optic chiasm; AC, anterior commissure; Hyp, hypophysis; MB, mamillary body; MI, massa intermedia; CC, corpus callosum; RN, red nucleus; PO, pons; BC, brachium conjunctivum; D, decussation of brachium conjunctivum; SC, superior colliculus; III, third ventricle; CA, caudate nucleus; IC, internal capsule; S, septum; DB, diagonal band of Broca and its nucleus; OT, olfactory tubercle; PUT, putamen; AMYGD, amygdala. Reproduced from Am. J. Physiol. 206, 1964, (upper) courtesy of Amer. Physiol. Soc., Bethesda and from Arch. Neurol. (Chicago) 9, 1963, (lower) courtesy of American Medical Assoc., Chicago.

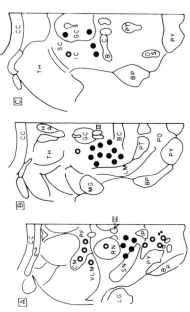

Fig. 8. Vasopressin release from the neurohypophysis during electrical stimulation of the brain stem in the unanesthetized monkey (Macaca mulatta). UPPER: Antidiuretic responses in a waking monkey during 50 min. excitation of the central tegmental tract and mesencephalic reticular formation (0.3 ma, 30 Hz, 30 sec on–2 min off) shown by the dashed line, and during intravenous infusion of vasopressin (40 mu, 10 min) shown by continuous line. No change in GRF or COSM. Note the greater antidiuretic response to brain stem stimulation. LOWER: Histological locations of electrode tips in the pons, mesencephalon and diencephalon in transverse sections at A. Fr. + 6, B. + 3, C. + 0 in six monkeys as adapted from Olszewski (1952). Antidiuretic responses to repeated unilateral electrical stimulation of 12 points in the mesencephalic reticular formation and the central tegmental tract, 5 points in the ventral tegmental area of Tsai and 2 points in the periaqueductal central gray shown by closed circles. Stimulation of the thalamus (2 points in centre median, 1–parafascicularis nucleus, 1–medial part of nuc. ventralis lateralis), red nucleus (1 point), tectum (1–superior colliculus, 1–inferior colliculus) and pons (4 points in tegmentum & pyramidal tract) produced no change in water excretion upon repeated stimulation as shown in open circles. Labels: LG, lat. geniculate; MG, medial geniculate; TH, thalamus; CM, centre median; Pf, nuc. parafasciculus; VLm, medial part of nuc. ventralis lateralis; Ha, habenular complex; CC, corpus callosum; SC, superior colliculus; IC, inferior colliculus; GC, central grey; RN, red nucleus; III, oculomotor nucleus; SN, substantia nigra; BC, brachium conjunctivum; LM, medial lemniscus; IP, nuc. interpedunculus; PY, pyramid; PT, pterygoid nuc.; BP, brachium pontis; PO, pons; OS, superior olive. Reproduced from Am. J. Physiol. 206, 1964, by courtesy of the American Physiological Society, Bethesda.

tegmental area of Tsai, periaqueductal grey matter and the
central tegmental tract and midbrain reticular formation (Fig. 8,
lower) we released vasopressin from the neurohypophysis, producing an antidiuretic response equivalent to 4.0 mμ/min of vasopressin release (Fig. 8, upper). Other sites in the thalamus,
centre median, parafascicular nucleus and medial part of ventralis
lateralis, tectum and pyramidal tract, failed to release any vasopressin (Fig. 8, lower). Those sites releasing vasopressin in
the midbrain may be related to 'limbic circuits,' but also to
ascending pathways from the vagus-nucleus tractus solitarius
complex to the supraoptic nucleus (Chang et al., 1937; Barker,
Crayton and Nicoll, 1971a), cholinergic and monoaminergic
systems (Anden et al., 1966; Fuxe, 1965; Shute and Lewis, 1966),
and spinothalamic pathways (Rothballer, 1966). Others have
released vasopressin from the neurohypophysis from stimulation
of the septal-preoptic-diagonal band-olfactory tubercle areas
(Andersson and McCann, 1955; Aulsebrook and Holland, 1969a;
Dingman, 1959, 1966; Rothballer, 1966); central tegmental tract
and midbrain reticular formation (Aulsebrook and Holland, 1969a;
Mills and Wang, 1964; Rothballer, 1966; Sharpless and Rothballer,
1961), periaqueductal grey (Aulsebrook and Holland, 1969a; Mills
and Wang, 1964; Rothballer, 1966) and ventral tegmental area
Rothballer, 1966). Other limbic-midbrain sites found to release
ADH in species other than the monkey include the cingulate gyrus
(Aulsebrook and Holland, 1969a; Rothballer, 1966; Yoshida et al.,
1966), hippocampus (Dingman et al., 1966), anterior nucleus of
the thalamus (Rothballer, 1966) and medullary reticular formation
(Rothballer, 1966). Acceleration of supraoptic unit firing has
been described from stimulation of the cingulate gyrus (Koizumi
et al., 1964), midbrain reticular formation (Ishikama et al.,
1966) and vagus and carotid sinus nerve (Barker, Crayton and
Nicoll, 1971a). Many of these limbic-midbrain sites can release
or inhibit the release of oxytocin from the neurohypophysis
(Beyer et al., 1961; Aulsebrook and Holland, 1969a, b;
Rothballer, 1966; Tindal et al., 1967, 1969) but the nature of
the two overlapping systems for vasopressin and oxytocin regulation is not well understood. Many of these same limbic-hypo-
thalamo-midbrain areas also yield drinking responses upon osmotic,
electrical and cholinergic stimulation (Andersson, 1953; Andersson
and Eriksson, 1971; Andersson and McCann, 1955; Andersson, Olsson
and Warner, 1967; Grossman, 1969; Grossman and Grossman, 1963).

Mechanisms of ADH Release: Indirect Humoral Pathways

A fourth mechanism for the release of vasopressin from the
neurohypophysis not dependent upon direct or indirect neural
pathways to the supraoptic neurons is humoral: blood levels of
oxygen, carbon dioxide, hydrocortisone, epinephrine, norepinephrine and angiotensin II. How can amygdala stimulation alter the

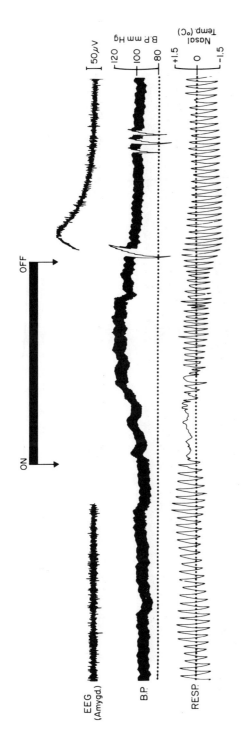

Fig. 9. Respiratory and cardiovascular effects of amygdaloid stimulation in the unanesthetized monkey. Electrical excitation of the basal nucleus of the amygdala with monophasic square wave pulses at 300 μamp, 1 msec pulse duration, 60 Hz for 68 sec (on-off) between arrows. Respiratory rates: control, 25/min; 25 seconds of apnea during initial period of stimulation with break-through; 40/min during the later period of stimulation; post-stimulation 25/min. Note the onset of apnea begins with the electrical stimulus, 20 mmHg rise in arterial blood pressure lags behind 15-20 sec. No afterdischarge seen. Labels: EEG (amygd.), electrical activity in basal nucleus of amygdala (stimulation site); BP, arterial blood pressure measured with chronically implanted silicon cannula in aortic arch; Resp., respirations measured with a thermocouple in the external nasal cavity.

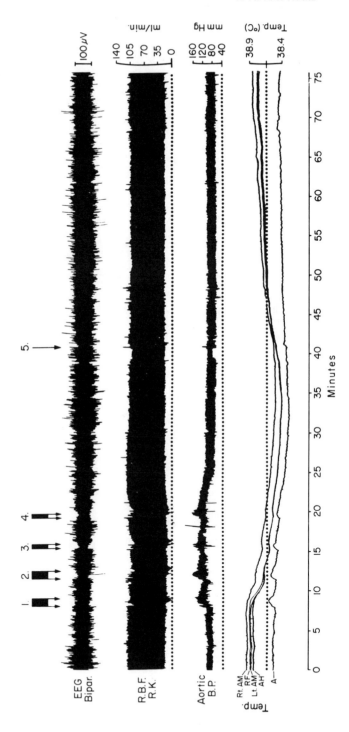

Fig. 10. Cardiovascular, renal and brain temperature changes following amygdala stimulation in the unanesthetized monkey (Macaca mulatta). Air. temp. 35C. Electrical excitation of the basal nucleus of the right amygdala with monophasic square wave pulses at 500 μamp, 60 Hz, 1 msec pulse duration for 60 sec (1), 60 sec (2), 30 sec (3) and 30 sec (4). During stimulation there was tonic turning of the head and eyes to the left with looking upward and smacking of the lips 1-2 min. after cessation of stimulation. Tap on the environmental chamber at (5). Monkey dozing during most of the study with arousal during amygdala stimulation and tap on chamber. During amygdala stimulation note EEG low voltage, brief drop in renal blood flow, abrupt elevation in arterial blood pressure, abrupt rise in arterial blood temperature. Sustained post-stimulation effects consist of a 40 mmHg elevation of arterial blood pressure with return to normal in 5 min. and a -0.3C narrowing of the brain-blood gradients, suggesting increased cerebral blood flow, with a return to normal in 30 min.

Labels: EEG, bipar., biparietal electrocorticogram; RBF, RK, pulsatile blood flow in the right renal artery measured by a chronically implanted non-cannulating flow probe (3mm diam.) of a pulsed-field electromagnetic flowmeter (Statham Instruments, Co., Model 0-5000) with electronic zero flow and lead wires in a small subcutaneous tunnel from subcostal region to platform on skull; Aortic BP, silicon cannula in aortic arch; Temp., thermocouples implanted in: Rt. AM, right amygdala; Lt. AM, left amygdala; RF, midbrain reticular formation; AH, anterior hypothalamus; A, aortic arch arterial blood.

blood levels of these gases and hormones and what effect, if
any, do they have on supraoptico-neurohypophysial activity?   It
is well known that electrical stimulation of the amygdaloid
nuclear complex and other limbic structures can cause apnea
(Hayward and Baker, 1968b; Kaada, 1951; Reis and McHugh, 1968;
Smith, 1938) with ensuing hypoxia and hypercapnia; also amygdala
stimulation produces elevation of arterial blood pressure with
tachycardia or bradycardia and increased sympathetic discharge
as a primary response or secondary to hypoxia (Angell, James
and Daly, 1969; Hayward and Baker, 1968b; Reis and McHugh, 1968;
Reis and Oliphant, 1964) and increased cerebral blood flow
(Hayward, 1967; Hayward and Baker, 1968b, 1969; Reivich, 1964).
Amygdala stimulation can release ACTH-corticol (Mason, 1959;
Matheson and Sundsten, 1969), norepinephrine-epinephrine (Gunne
and Reis, 1963; Reis and Gunne, 1965) from adrenal medulla and
possibly via direct sympathetic activation or secondary to
hypoxia, produce vasoconstriction in the kidney and activation
of the renin-angiotensin system (Buney et al, 1966; Peart, 1965;
Vander, 1967).   What effects will these humoral agents have on
the supraoptic neuronal activity and ADH release? Hypoxia, via
the olfactory-amygdala 'secondary' system of osmoreceptors, can
enhance vasopressin release (Mirsky and Stein, 1953; Mirsky et
al, 1954; Sawyer and Fuller, 1960).   Hypoxia, via the chemo-
receptor induced sympathetic discharge, can release angiotensin II with stimulation of vasopressin release (Bonjour and
Malvin, 1970).   Hypercapnia and hydrocortisone inhibit vasopressin release, perhaps partly by changes in right atrial
volume (Aubry et al., 1965; Heller and Ginsburg, 1966; Share,
1969).   Catecholamines, epinephrine and norepinephrine, inhibit
vasopressin release (Abrahams and Pickford, 1956; O'Connor and
Verney, 1945) and inhibit supraoptic neuronal activity (Barker,
Crayton and Nicoll, 1971b; Bloom and Salmoiraghi, 1963).   Elevation of arterial blood pressure with stretch of carotid and
aortic baroreceptors inhibits vasopressin release (Share, 1969;
Share and Levy, 1962).   Increased sympathetic discharge, directly
or secondary to hypoxia, can shift intra-renal blood flow
(Pomeranz et al., 1968) away from the renal cortex with activation of the renin-angiotensin system (Bunay et al., 1966; Peart,
1965) with elevation of angiotensin II and stimulation of antidiuretic hormone release (Bonjour and Malvin, 1970) and drinking
(Epstein et al., 1970; Fitzsimmons and Simons, 1969).

As shown in Figures 9 and 10 (see Hayward and Baker, 1968b)
electrical stimulation of the amygdala can indeed produce apnea,
elevation of the arterial blood pressure, decreased renal blood
flow and reduction in the temperature gradient between the
arterial blood and the amygdala and other brain sites.   These
changes indicate probable hypoxia and hypercapnia along with
sympathetic discharge and increased cerebral blood flow.   It is

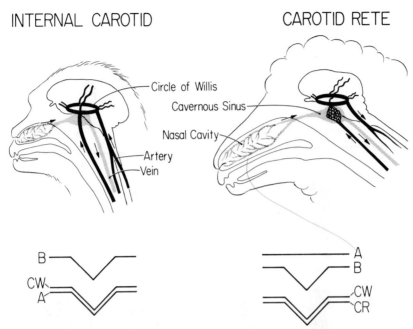

Fig. 11.   Regulation of brain temperature by cerebral arterial
blood in prototype 'internal carotid' and 'carotid rete' species.
Internal carotid arterial blood in the monkey (left) circulates
through a large single vessel which courses through the cooler
blood of the cavernous sinus without heat loss (upper).   Cavernous
sinus venous blood is cooler than central arterial blood because
of the drainage of blood from the nasal cavity and skin of the
head, sites of heat loss.   Accelerated peripheral heat loss from
the skin and decreased heat production cools, sequentially,
systemic venous blood, arterial blood in the aorta and carotid
arteries (A), cerebral arterial blood at the circle of Willis (CW)
and the amygdala (B) (lower).   Carotid rete arterial blood in the
(right) cooled by countercurrent heat exchange between the cool
venous blood in the cavernous sinus and the numerous small vessels
of the rete with 100 fold increase in surface area (upper); accelerated heat loss from the nasal mucosa and the skin of the head
causes increased cooling of venous blood draining into the
cavernous sinus, accelerated heat exchange with the retial
arterial blood (CR), a drop in the cerebral arterial blood
temperature at the circle of Willis (CW) and in the amygdala
(B) without a change in the warmer aortic or carotid arterial
blood temperature (A) (lower).   Reproduced from Brain Research
16, 1969, by courtesy of the Elsevier Publishing Company,
Amsterdam.

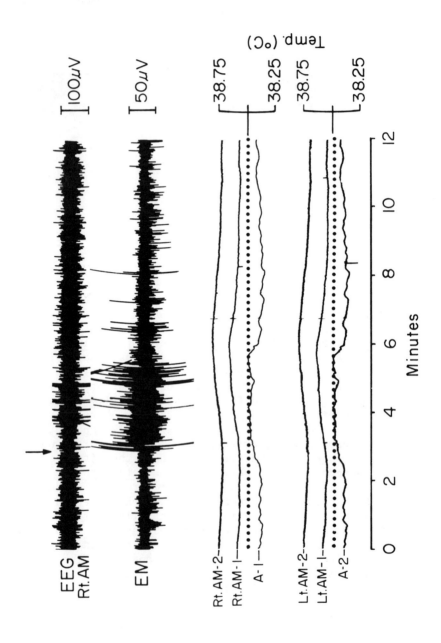

Fig. 12. Behavioral arousal and elevation of arterial blood and brain temperature in the unanesthetized monkey (Macaca mulatta). Monkey sitting quietly awake in a lighted environmental chamber at 35°C air temperature. At the arrow a loud tap on the chamber (novel stimulus) arouses the animal which looks around with lower voltage EEG, increased extraocular movements, and rise in aortic, amygdala and globus pallidus temperatures. Vasoconstriction of the cutaneous and nasal mucosal surfaces not shown. The two sites in the aortic arterial blood (A-1, A-2) show immediate and parallel temperature rises of 0.2C lasting for the 3 minutes of arousal and then drop abruptly to control levels. The slower, parallel rise in the bilateral amygdala and globus pallidus brain sites by 0.2C took 2-3 minutes due to thermal inertia and showed a comparable lag during return to control levels. During a long period of steady behavior and without change in cerebral blood flow ($\triangle P_A O_2$ or $P_A CO_2$) the temperature gradients between perfusing arterial blood and amygdala remain constant in the monkey. Note that the smaller temperature oscillations seen in each aortic site are not present in the brain sites due to thermal inertia. See text for further description.

Labels: EEG, Rt. AM, electrical activity from right amygdala; EM, extraocular movements; A-1-2, temperature of moving arterial blood in arch of the aorta at sites 10 mm apart; AM-1, temperatures at histologically similar sites in right and left basolateral amygdala; AM-2, temperatures at histologically similar sites in right and left globus pallidus.

evident that stimulation of the amygdala in the monkey can activate a series of parallel physiological responses which may lead
to vasopressin release. Future studies on amygdaloid modulation
of supraoptic neuronal activity and vasopressin release will need
to examine each of these possible mechanisms in order to establish
the role of the amygdala in body water balance.

SUMMARY

    Amygdalo-supraoptic neuronal pathways and mechanisms of
vasopressin release can be determined by electrical stimulation
of the amygdala and other brain sites. Electrical stimulation of
a number of points in the medial and basolateral nuclei of the
amygdala of the awake rhesus monkey can release vasopressin from
the neurohypophysis by pathways that probably reach the supraoptic
nuclei directly over the ventral amygdalo-hypothalamic tract and/or
indirectly with synaptic delays via the diagonal band of Broca,
olfactory tubercle, lateral hypothalamus-medial forebrain bundle,
ventral tegmental area of Tsai, periaqueductal grey matter and the
midbrain reticular formation. Studies in other species generally
support these results. Of the possible physiological systems
activated by amygdala stimulation which might augment direct
amygdalo-supraoptic excitatory pathways, the most likely are:
via respiratory inhibition, hypoxia, activation of 'secondary'
osmoreceptors with ADH release; via sympathetic discharge (induced
directly by amygdala stimulation or secondary to hypoxia), renal
vasoconstriction and production of angiotensin II with ADH release.

FOREBRAIN THERMORECEPTORS: CENTRAL THERMAL INHIBITION OF
VASOPRESSIN RELEASE AND 'VOLUMETRIC' CONTROL MECHANISMS

Amygdala and Thermoregulation:

    Amygdaloid modulation of a number of hypothalamic behavioral,
autonomic and endocrine functions does not seem to be shared by
the neural structures in the preoptic area involved in temperature
regulation (Gloor, 1960). The studies of Pinkston, Bard and Rioch
(1934) indicated that in the decorticate cat, including removal of
the amygdala, the gross thermoregulatory reflexes were intact.
Subsequent studies by a number of workers using bilateral amygdala
lesions showed only slight hypothermia with relative poikilo-
thermia (Gloor, 1960). Some of the minor changes in body temperature following stimulation or lesions in the amygdala could be due
to a conflict in homeostasis where primary respiratory effects
(apnea or hyperventilation) caused a secondary thermal change. In
view of the newer concepts of thermoregulation based on intraventricular alterations of norepinephrine or serotonin (von Euler,
1961; Feldberg and Myers, 1963), Eleftheriou (1970) placed bilateral amygdaloid lesions in the deermouse and examined hypo-

thalamic norepinephrine under heat stress. He found that heat
stress in normal and sham operated mice caused a rise in hypothalamic norepinephrine levels, and that amygdaloid lesions
caused a similar norepinephrine rise. When he then exposed the
lesioned mice to heat stress, there was no further change in the
already elevated norepinephrine in the hypothalamus. Using turnover techniques, Simmonds (1969) found an increased norepinephrine turnover in the hypothalamus of the rat both during heat
and cold stress. Taken together, these results suggest that
perhaps a common non-specific stress, namely thermal or amygdalectomy, may contribute to changes in hypothalamic catecholamines.
Another possibility is that Eleftheriou's amygdala lesions somehow activated the parent noradrenergic cell body in the pons
(Anden et al., 1966; Fuxe, 1965) with resultant increased
norepinephrine delivery by the non-amygdala axonal branches of
these cells to the hypothalamus. In any event, at the present
time, there is no good experimental evidence of amygdaloid
involvement in thermoregulation.

Amygdala and Cerebral Arterial Blood Temperature:

       Local changes in temperature in the amygdala and other deep
brain sites during various behavioral states have been attributed
by some authors (Delgade and Hanai, 1966; Tachibana, 1969) to
changes in cerebral blood flow and/or cerebral metabolic heat
production in the cat, dog and sheep (for review see Hayward and
Baker, 1969). We find in five mammalian species that the temperature of the cerebral arterial blood, perfusing the amygdala and
other deep sites, is the major factor determining shifts in brain
temperature during changes in behavior (Baker and Hayward, 1967a,
b; Baker and Hayward, 1968; Hayward, 1968; Hayward and Baker,
1968b; Hayward et al., 1966). Whether the amygdala plays some
role in these changes is not known, at this time. There are
species differences, however, in the thermal relationships between
temperature of the arterial blood in the large vessels in the neck
and those at the circle of Willis due to a countercurrent heat
exchange in the carotid rete. In the monkey, a species having an
internal carotid artery, the direct vascular connection between
heart and brain is demonstrated by the fact that temperature
changes in the central arterial blood are followed quickly by
parallel shifts in blood temperature at the circle of Willis,
in the amygdala and at other brain sites (Fig. 11, left; Hayward
and Baker, 1968b, 1969). In contrast to the monkey, the carotid-
rete-bearing species, such as cat, dog and sheep (Fig. 11, right),
show heat exchange between warm arterial blood in the small
vessels of the rete and the surrounding cool blood in the cavernous sinus or pterygoid plexus (Baker and Hayward, 1967b; Baker
and Hayward, 1968; Hayward, 1968; Hayward and Baker, 1969). In
these species with carotid rete, there may be little change in

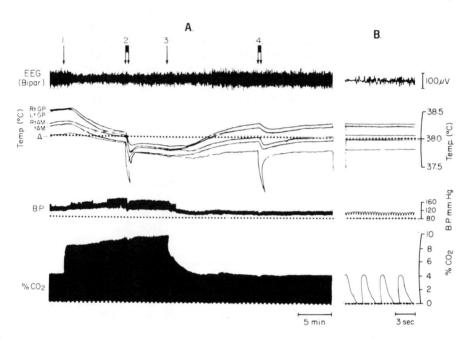

Fig. 13. Hypercapnia, cerebral cooling and accelerated cerebral
blood flow and heat transfer in the lightly anesthetized, paralyzed
chronic monkey at 35°C air temperature. A. Animal resting quietly
under controlled ventilation in an environmental chamber: (1) begin
inhalation of 8-10% CO_2 in air; (2) intra-atrial injection of
10 ml cold (5°C) isotonic saline in 20 sec.; (3) resume inhalation
of air; (4) intra-atrial injection of 10 ml cold (5°C) isotonic
saline in 20 sec. B. Fast paper speed. During the period of
hypercapnia note the accelerated heat removal from the brain by
the cerebral arterial blood with reduction of brain-blood temperature gradients (T_b-T_a) at amygdala (AM) and globus pallidus (GP)
sites and enhanced thermal dilution curve in brain without change
in arterial blood thermal dilution curves (compare 2A with 4A).
Note the EEG arousal pattern, elevation of arterial blood pressure
and cardiac irregularity seen during hypercapnia. Labels: EEG,
Bipar, biparietal electrocorticogram; GP, right and left globus
pallidus; AM, right and left amygdala; A, arterial blood temperature
in the aortic arch; BP, arterial blood pressure; % CO_2, percent end-
expired carbon dioxide. Reproduced from American Journal of
Physiology, 215, 1968, by courtesy of the American Physiological
Society, Bethesda.

central arterial temperature during behavioral events despite a
marked and independent thermal shift at the circle of Willis, in
the amygdala and in other brain sites. These findings further
suggest that cerebral arterial blood temperature is the major
determinant of changes in brain temperature in all mammals and
that such changes reflect the general thermoregulatory
activities related to behavior. While our results rule out
any local basis (heat production or blood flow) for these oscillations of amygdala or preoptic temperatures, it is apparent that,
in the species with carotid rete, preoptic temperatures often
times are cooler and show changes independent of central arterial
blood (deep body core, spinal cord) temperatures. In terms of
classical thermoregulation theory, this implies that the preoptic
thermodetectors do not compute deep body temperature in the
carotid-rete-species, which raises some problems for the
supposed body thermostat and its setting and which has not been
considered in discussions of these problems (Bligh, 1966; von
Euler, 1961; Hammel, 1968; Hardy, 1961). In the dog, a species
with carotid-rete-like characteristics (Hayward, 1968; Hayward
and Baker, 1969), Simon, Rautenberg Thauer and Iriki (1963)
found a deep body core temperature detector system in the spinal
cord. Recently, Jessen and co-workers (Jessen and Mayer, 1971;
Jessen and Ludwig, 1971; Jessen and Simon, 1971) have described
convincingly the equivalence of responses, the addition of signals,
and the identity of functions of these spinal cord and preoptic
core sensors of temperature in the conscious dog. These results
further strengthen the concept of a functional role of the countercurrent heat exchange and thermal dissociation of brain, and deep
core, in these "carotid rete" bearing mammals (Hayward and Baker,
1969).

Amygdala Temperature and Blood Flow:

In the monkey, amygdaloid temperature increases with arousal
and decreases with sleep (Fig. 12; Hayward and Baker, 1968b), due
to an earlier change in the temperature of the arterial blood
circulating to the brain. Temperature of the amygdala is higher
than the entering arterial blood: medial amygdala, +0.27°C;
lateral amygdala, +0.40°C (Hayward and Baker, 1968b). The
amygdala is warmer than the blood because of an incomplete removal of the metabolically produced heat by the flow of cooler
blood into the amygdala and, consequently, a thermal gradient
develops, perhaps partly due to countercurrent heat exchange
between arterioles and venules in the amygdala. An increased
flow of this cooler arterial blood through the amygdala might be
expected, therefore, to remove more efficiently the heat produced
in the amygdala and narrow the thermal gradient between the
arterial blood and the amygdala. Such thermal changes are shown
in Figure 13 (Hayward, 1967; Hayward and Baker, 1968b; Hayward

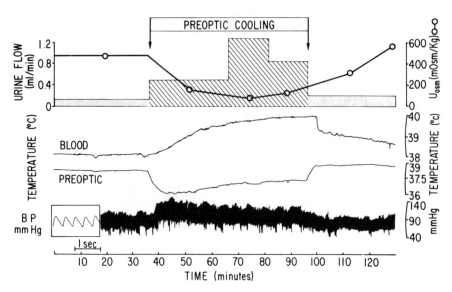

Fig. 14.   Inhibition of vasopressin release from the neurohypophysis
during preoptic cooling in the unanesthetized monkey (Macaca
mulatta).   Mid-preoptic cooling for 60 minutes (between arrows) produces behavioral arousal, increased eye & general body activity and
shivering (not shown), elevation of arterial blood pressure and a
rise in arterial blood temperature in the waking monkey.   Inhibition of vasopressin release is indicated by increased urine flow
and free water clearance (not shown), decreased urine osmolality
(U_{OSM}) with no change in osmolal or inulin clearances (not shown).
Bilateral mid-preoptic thermodes were cooled by 8C producing a preoptic field cooling of 2C below control preoptic temperatures.
Note that the abrupt rise in blood temperature occurs during the
first thirty minutes of preoptic cooling while the peak level of
water excretion occurs during the second thirty minutes of preoptic
cooling at a time of relatively little change in blood temperature.
Further discussion in the text.   Labels: Urine flow, V, in ml/min,
control period dotted on graph, preoptic cooling cross hatched;
U_{OSM}, urine osmolality in mOsm/kg H_2O by freezing point depression
method; Blood, temperature of the arterial blood at the aortic
arch; Preoptic, temperature of the mid-preoptic area 3 mm lateral
to the thermode; BP, aortic arterial blood pressure in mmHg.
Reproduced from American Journal of Physiology, 214, 1968, by
courtesy of the American Physiological Society, Bethesda.

and Baker, 1969). To confirm this hypothesis, we injected a
bolus of cold isotonic saline into the right atrium both during
eucapnia with 'normal' resting amygdala blood flow (Fig. 13, A-4),
and again during hypercapnia with 'high' amygdala blood flow
(Fig. 13, A-2; Reivich, 1964). Such levels of hypercapnia may
produce roughly a one-hundred per cent increase in cerebral blood
flow (Reivich, 1964) in the monkey. As shown in Figure 13, amygdaloid temperature can be manipulated predictably in the monkey
by altering the temperature of the blood and the rate of flow of
blood through the amygdala in this 'internal carotid' species.
If we now look back at Figure 10 and the changes in amygdala and
other brain temperatures following electrical stimulation of the
amygdala, it is apparent that the narrowing of the amygdala-blood
thermal gradient during the post-stimulus period (30 min) probably was initiated by hypoxia and/or hypercapnia from the apnea
induced in the amygdala. Whether the prolonged nature of this
increased amygdala blood flow is due to local tissue or systemic
humoral factors is not known at present.

Amygdala and Thermal Inhibition of Supraoptic Neurons:

        In view of the well known behavioral (drinking), endocrine
and autonomic effects produced by preoptic cooling in the monkey
and other species (Andersson et al., 1962, 1964a, 1964b; Hayward
and Baker, 1968a; Sundsten, 1969), and the known modulation of
these same 'hypothalamic' functions by the amygdala(Gloor, 1960),
it is necessary to examine such studies in more detail to see if
the amygdala may be involved. In our own work, we have studied
central cold diuresis in the monkey, using preoptic thermodes
(Hayward et al., 1965), renal clearance techniques and measurements of behavior (Hayward and Baker, 1968a). In the unanesthetized rhesus monkey, abrupt cooling of a mid-preoptic field by
2°C produces EEG low voltage fast activity, behavioral arousal,
increased bodily movements, shivering (Hayward and Baker, 1968a),
and in the baboon decreased drinking (Sundsten, 1969). In the
autonomic sphere, preoptic cooling produces cutaneous vasoconstriction, increased arterial blood pressure, and a rapidly
rising arterial blood and amygdaloid temperature (Fig. 14). In
our monkeys, an increased urine flow and free water clearance,
decreased urine osmolality with no change in glomerular filtration
rate or solute excretion indicated to us that one of the endocrine
responses to preoptic cooling in the primate was an inhibition of
vasopressin release from the neurohypophysis (Hayward and Baker,
1968a).

        What physiological mechanisms could be thrown into action by
preoptic cooling in the primate which would lead to inhibition of
vasopressin release and is the amygdala involved in any way in
this diuretic response? Figure 15 summarizes some of the known

THE AMYGDALA AND CENTRAL THERMAL INHIBITION OF VASOPRESSIN
RELEASE FROM THE NEUROHYPOPHYSIS

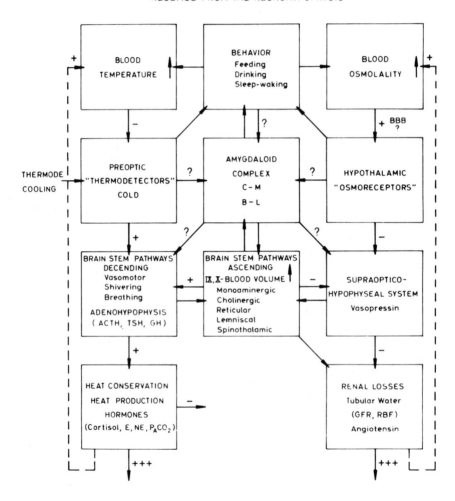

Fig. 15. Summary diagram of some of the major amygdalo-hypothalamic mechanisms possibly involved in central cold diuresis in the monkey as described in the text. The question marks (?) placed on the arrows leading to and from the amygdala merely indicate the lack of experimental data to support or refute amygdalofugal or amygdal-optic osmoregulatory neurons. Three negative feedback loops are diagrammed: 1) Preoptic thermodectector-thermoregulatory-blood temperature system; 2) Osmoreceptor-supraoptic complex-renal blood osmolality system; 3) Volume receptor (high and low pressure)-vagal-glosso-pharyngeal-brainstem pathway-blood volume system. Thermode cooling interrupts the preoptic thermoregulatory loop and triggers descending somatomotor, autonomic and endocrine responses with concomitant behavioral effects. Inhibition of supraoptic neuronal activity may result from multiple factors such as increased central blood volume, increased baroreceptor discharge, increased levels of cortisol, epinephrine, norepinephrine and possibly hypercapnia.

Labels: BBB, blood-brain barrier; ACTH, adrenocorticotrophin; $P_A CO_2$, partial pressure of oxygen in arterial blood; GFR, glomerular filtration rate; RBF, renal blood flow; C-M, cortico-medial amygdaloid nuclear complex; B-L, basolateral amygdaloid nuclear complex; IX, glossopharyngeal cranial nerve; X, vagus cranial nerve.

effects of preoptic cooling in mammals as related to vasopressin
release and the amygdala. A major effect of preoptic cooling is
a shift of water out of the blood, probably by first shifting
blood to central venous sites with activation of central 'volume'
receptors. This increased central blood volume triggers, via a
vagal pathway, inhibition of supraoptic neuronal activity, and
vasopressin release with a consequent water diuresis (Henry,
Gauer and Reeves, 1956; Share, 1969). In regard to the elevation
of arterial blood pressure in our monkeys, the high pressure
volume receptors, baroreceptor afferents, provide an additional
inhibitory input to the supraoptic nucleus (Share, 1969; Share
and Levy, 1962). The recent studies by Johnson, Zehr and Moore
(1970), in the unanesthetized sheep, have shown clearly the
equipotent and interacting effects of changes in blood volume
and blood osmolality on vasopressin release. While the inhibitory
neural pathway from the nucleus tractus solitarius to the supraoptic nucleus is not known, an excitatory vago-neurohypophysial
pathway has been known for many years (Chang et al., 1937). The
more recent studies using brain stem electrical stimulation
(Aulsebrook and Holland, 1969a; Hayward and Smith, 1963; Mills
and Wang, 1964; Rothballer, 1966) may have involved some of these
ascending fibers. The recent study of Barker, Crayton and Nicoll
(1971a) has confirmed, at the cellular level, such an excitatory
vago-solitario-supraoptic pathway. Since vagal stimulation can
evoke electrical potential (Dell and Olson, 1953) and unitary
changes (Machne and Segundo, 1955) in the amygdaloid nuclear
complex, and since stimulation of the amygdala can alter vagal
efferent activity (Gloor, 1960), modulation of this vago-neurohypophysial reflex by the amygdala could occur. No evidence
exists to support or refute such an hypothesis.

    The humoral effects of preoptic cooling include increased
secretion of: ACTH-hydrocortisone (Andersson et al., 1964a;
Chower's et al, 1964), TSH-thyroid hormone (Andersson et al.,
1962), epinephrine-norepinephrine (Andersson et al., 1964a) and
growth hormone, in some species (Glick, 1969). The diuresis of
chill and similar acute hormone responses to acute cold exposure
have been described in man (see Hayward and Baker, 1968a; Suzuki
et al., 1967). All of these humoral factors may contribute
directly or indirectly to inhibition of vasopressin release, and
many of these hormones also can be released by stimulation of the
amygdaloid complex. For instance, hydrocortisone can elevate
the threshold for osmotic release of vasopressin from the neurohypophysis in man (Aubry et al., 1965; Share, 1969) and ACTH-hydrocortisone release occurs after stimulation of the amygdala in the
unanesthetized monkey (Mason, 1959; Matheson and Sundsten, 1969).
Similarly, epinephrine-norepinephrine inhibits vasopressin release
from the neurohypophysis (Abrahams and Pickford, 1956; Harris,
1960; Heller and Ginsburg, 1966; O'Connor and Verney, 1945),
inhibits firing of supraoptic neurons (Bloom and Salmoirhagi, 1963;

Barker, Crayton and Nicoll, 1971b), and can be released from the
stimulation of the amygdala (Gunne and Reis, 1963; Reis and Gunne,
1965). Whether such hormones as TSH-thyroid hormone, growth
hormone and angiotensin II are released following amygdaloid
stimulation, and play a role in inhibition of vasopressin during
central cold diuresis in the unanesthetized monkey is not known.
Preoptic cooling could alter respiration, but such effects are
not known at present. Certainly, amygdala induced hypopnea
(Gloor, 1960; Hayward and Baker, 1968b; Kaada, 1951; Reis and
McHugh) with hypercapnia can inhibit vasopressin release (Heller
and Ginsberg, 1966; Share, 1969). It is of some interest that
in the 'carotid rete' species, the goat, preoptic cooling causes
release of vasopressin from the neurohypophysis, not inhibition as
in the primate (Olsson, 1969), while in other species 'cold
diuresis' has a renal rather than a neurohypophysial basis (see
Hayward and Baker, 1968a).

SUMMARY

Amygdalo-preoptic thermoregulatory interactions could play a
role in the central thermal inhibition of vasopressin release in
the monkey. There is little evidence that the amygdala can
modulate thermoregulatory reflexes in the unanesthetized mammal.
Oscillations in amygdaloid temperature during sleep-waking behavior in both 'internal carotid' and 'carotid rete' bearing species
are due to changes in the temperature of the cerebral arterial
blood. The counter-current heat exchanger, the carotid rete-
cavernous sinus complex, causes a dissociation between preoptic
and spinal cord temperatures in cat, dog, sheep and makes a dual
set of thermoreceptors (preoptic and spinal) a necessity for
thermoregulation in these species. Direct alteration in cerebral
blood flow can change amygdala temperature independently of
cerebral arterial blood temperature. Despite a number of parallel
hormonal effects produced by preoptic cooling, on the one hand,
and by stimulation of the amygdala, on the other, many of which
can inhibit vasopressin release from the neurohypophysis, the
link between the amygdala and preoptic-thermoregulatory systems
in the regulation of body water content is unproven at present.

EPILOGUE

This paper reviews the role of the amygdaloid nuclear complex
in the modulation of supraoptic neuronal activity and the release
of vasopressin from the neurohypophysis under 'osmometric,'
'volumetric' and 'behavioral' control. The amygdala, along with
the olfactory bulb, olfactory tubercle and the preoptic area, is
part of the 'secondary' forebrain 'osmoreceptor' system of
Sawyer. Diverse noxious stimuli, such as pain, 'emotional' stress,

conditional states, hypoxia and hypertonicity of the carotid blood may activate these limbic interneurons with a resultant vasopressin release. In contrast, the 'osmoreceptors' of Verney and the supraoptic neurons are much less influenced by non-osmotic afferent input. Amygdalo-supraoptic neuronal pathways, as determined by electrical stimulation of the brain, probably include the following: direct excitatory neural pathway via the ventral amygdalo-hypothalamic tract; indirect excitatory neural routes via medial forebrain and limbic-midbrain areas; and indirect excitatory humoral pathways. These latter may include hypoxia and release of ADH by angiotensin II. While preoptic cooling and amygdala stimulation can each trigger autonomic, endocrine and behavioral responses via a basic hypothalamic substrate which takes part in inhibition of vasopressin release to cold stress, at the present time, a direct link between the amygdala and preoptic-thermoregulatory effector mechanisms for regulation of water balance is unproven. Future research, using the techniques of antidromic identification of supraoptic neurons (Barker et al., 1971a, b; Dyball, 1971; Kelly and Dreifuss, 1970; Novin et al., 1970; Sundsten, 1971; Sundsten et al., 1970; Yamashita et al., 1970) and the radioimmunoassay for vasopressin (Oyama et al., 1970), undoubtedly will extend our present limited understanding of the amygdalo-supraoptic neuronal connections and their importance for the control of body water content.

ACKNOWLEDGMENTS

Supported in part by the Ford Foundation, The Los Angeles County Heart Association and USPHS Grants NB-05638 and Special Fellowship NS-02277.

I thank Mrs. R. Lawrence for valuable technical assistance and the illustration division of the Nobel Institute for Neurophysiology, Karolinska Institutet, Stockholm, Sweden, for aid with the Figures.

REFERENCES

ABRAHAMS, V. C., & PICKFORD, M. Observations on a central antagonism between adrenaline and acetylcholine. Journal of Physiology (London), 1956, 131, 712-718.

ANDEN, N. E., DAHLSTROM, A., FUXE, K., LARSSON, K., OLSON, L., & UNGERSTEDT, U. Ascending monoamine neurons to the telencephalon and diencephalon. Acta Physiologica Scandinavica, 1966, 67, 313-326.

ANDERSSON, B. The effect of injections of hypertonic NaCl-solutions into different parts of the hypothalamus of goats. Acta Physiologica Scandinavica, 1953, 28, 188-201.

ANDERSSON, B., & ERIKSSON, L. Conjoint action of sodium and angiotensin on brain mechanisms controlling water and salt balances. Acta Physiologica Scandinavica, 1971, 81, 18-29.

ANDERSSON, B., & McCANN, S. M. Drinking, antidiuresis and milk ejection from electrical stimulation within the hypothalamus of the goat. Acta Physiologica Scandinavica, 1955, 35, 191-201.

ANDERSSON, B., EKMAN, L., GALE, C. C., & SUNDSTEN, J. W. Activation of the thyroid gland by cooling the preoptic area in the goat. Acta Physiological Scandinavica, 1962, 54, 191-192.

ANDERSSON, B., GALE, C. C., HOKFELT, B., & OGHA, A. Relation of preoptic temperatures to the function of the sympathico adreno-medullary system and the adrenal cortex. Acta Physiological Scandinavica, 1964a, 61, 182-191.

ANDERSSON, B., GALE, C. C., & SUNDSTEN, J. W. Preoptic influences on water intake. In M. J. Wayner (Ed.), Thirst: Proceedings of the First International Symposium on Thirst in the Regulation of Body Water. Oxford: Pergamon Press, 1964b. Pp 361-377.

ANDERSSON, B., OLSSON, K., & WARNER, R. G. Dissimilarities between the central control of thirst and the release of antidiuretic hormone (ADH). Acta Physiologica Scandinavica, 1967, 71, 57-63.

ANGELL, J. E., & DALY, M. deB. Cardiovascular responses in apnoeic asphyxia: role of arterial chemoreceptors and the modification of their effects by a pulmonary vagal inflation reflex. Journal of Physiology (London), 1969, 210, 87-104.

AUBRY, R. H., NANKIN, H. R., MOSES, A. M., & STREETEN, D. H. P. Measurement of the osmotic threshold for vasopressin release in human subjects, and its modification by cortisol Journal of Clinical Endocrinology, 1965, 25, 1481-1492.

AULSEBROOK, L. H., & HOLLAND, R. C. Central regulation of oxytocin release with and without vasopressin release. American Journal of Physiology, 1969a, 216, 818-829.

AULSEBROOK, L. H., & HOLLAND, R. C.   Central inhibition of oxy-
    tocin release. American Journal of Physiology, 1969b, 216,
    830-842.

BAKER, M. A., & HAYWARD, J. N.   Autonomic basis for the rise in
    brain temperature during paradoxical sleep. Science, 1967a,
    157, 1586-1588.

BAKER, M. A., & HAYWARD, J. N.   Carotid rete and brain temperature
    of cat. Nature (London), 1967b, 216, 139-141.

BAKER, M. A., & HAYWARD, J. N.   The influence of the nasal mucosa
    and the carotid rete upon hypothalamic temperature in sheep.
    Journal of Physiology (London), 1968, 198, 561-579.

BAKER, M. A., BURRELL, E., PENKHUS, J., & HAYWARD, J. N.   Capping
    and stabilizing chronic intravascular cannulae. Journal
    of Applied Physiology, 1968, 24, 577-579.

BARD, P., & MACHT, M. B.   The behavior of chronically decerebrate
    cats. In Ciba Foundation Symposium on the Neurological
    Basis of Behavior. London: Churchill, 1958. Pp. 155-175.

BARGMANN, W., & SCHARRER, E.   The site of origin of the hormones
    of the posterior pituitary. American Scientist, 1951, 39,
    255-259.

BARKER, J. L., CRAYTON, J. W., & NICOLL, R. A.   Supraoptic neuro-
    secretory cells: autonomic modulation. Science, 1971a,
    171, 206-207.

BARKER, J. L., CRAYTON, J. W., & NICOLL, R. A.   Supraoptic neuro-
    secretory cells: adrenergic and cholinergic sensitivity.
    Science, 1971b, 171, 208-209.

BEYER, C., & SAWYER, C. H.   Hypothalamic unit activity related to
    control of the pituitary. In W. F. Ganong and L. Martini
    (Eds.) Frontiers in Neuroendocrinology. New York: Oxford
    Univ. Press, 1969. Chap. 7, pp. 255-287.

BEYER, C., ANGUIANO, L. G., & MENA, J. F.   Oxytocin release in
    response to stimulation of the cingulate gyrus. American
    Journal of Physiology, 1961, 200, 625-627.

BISSET, G. W., HILTON, S. M., & POISNER, A. M.   Hypothalamic
    pathways for independent release of vasopressin and oxytocin.
    Proceedings of the Royal Society, Series B, 1967, 166, 422-
    442.

BLIGH, J.   Thermosensitivity of the hypothalamus and thermoregu-
    lation in mammals. Biological Review, 1966, 41, 317-367.

BLOOM, F. E., OLIVER, A. P., & SALMOIRAGHI, G. C. The responsive-
    ness of individual hypothalamic neurons to microelectro-
    phoretically administered endogenous amines. International
    Journal of Neuropharmacology, 1963, 2, 181-193.

BONJOUR, J. P., & MALVIN, R. L. Stimulation of ADH release by
    the reninangiotensin system. American Journal of Physiology,
    1970, 218, 1555-1559.

BROOKS, C. McC., USHIYAMA, J., & LANGE, G. Reactions of neurons
    in or near the supraoptic nuclei. American Journal of
    Physiology, 1962, 202, 487-490.

BUNAY, R. D., PAGE, I. H., & McCUBBIN, J. W. Neural stimulation
    of release of renin. Circulation Research, 1966, 19, 851-
    858.

CAJAL, S. R. Y. Histologie du Systeme Nerveus de l'Homme et des
    Vertebres. Paris: A. Maloine.

CHANG, H. C., CHIA, K-F, HSU, C. H., & LIM, R. K. S. A vagus
    post pituitary reflex. I. Pressor component. Chinese
    Journal of Physiology, 1937, 12, 309-326.

CHOWERS, I., HAMMEL, H. T., STROME, S. B., & McCANN, S. M. Com-
    parison of effect of environmental and preoptic cooling on
    plasma cortisol levels. American Journal of Physiology,
    1964, 207, 577-582.

CLEMENTE, C. D., SUTIN, J., & SILVERSTONE, J. T. Changes in
    electrical activity of the medulla on the intravenous in-
    jection of hypertonic solutions. American Journal of Physi-
    ology, 1957, 188, 193-198.

CREUTZFELDT, O. D., BELL, F. R., & ADEY, W. R. The activity of
    neurons in the amygdala of the cat following afferent stimu-
    lation. In W. Bargmann and J. P. Schade (Eds.) The Rhinen-
    cephalon and Related Structures, Progress in Brain Research.
    Amsterdam: Elsevier. Vol. 3, pp. 31-49.

CROSS, B. A. Neural control of oxytocin secretion. In L. Martini
    and W. F. Ganong (Eds.) Neuroendocrinology. New York:
    Academic Press. Vol. I, Chap. 7, pp. 217-259.

CROSS, B. A., & GREEN, J. D. Activity of single neurones in the
    hypothalamus: effect of osmotic and other stimuli. Jour-
    nal of Physiology (London) 1959, 148, 554-569.

CROSS, B. A., & SILVER, I. A. Electrophysiological studies on the
    hypothalamus. British Medical Bulletin, 1966, 22, 254-260.

CORSON, S. A. Conditioning of water and electrolyte excretion.
    Research Publications of the Association for Research in
    Nervous and Mental Disease, 1966, 43, 140-198.

DELGADO, J. M. R., & HANAI, T. Intracerebral temperatures in free
    moving cats. American Journal of Physiology, 1966, 211, 755-
    769.

DELL, P., & OLSON, R. Projections "secondaires" mesencepháliques,
    diencéphaliques et amygdaliénnes des afférences visceráles
    vagáles. Comptes Rendus des Seances de la Societe de
    Biologie, 1951, 145, 1088-1091.

DE OLMOS, J. Personal communication, 1971.

DINGMAN, J. F., & GAITAN, E. Subcortical stimulation of the
    brain and release of antidiuretic hormone in man. Journal
    of Clinical Endocrinology, 1959, 19, 1346-1349.

DINGMAN, J. F., GAITAN, E., ARIMURA, A., & HEATH, R. G. Cerebral
    regulation of vasopressin secretion in the monkey. Personal
    communication. In A. B. Rothballer, Pathways of secretion
    and regulation of posterior pituitary factors. Research
    Publications of the Association for Research in Nervous and
    Mental Disease, 1966, 43, 86-131.

DYBALL, R. E. J. Oxytocin and ADH secretion in relation to
    electrical activity in antidromically identified supraoptic
    and paraventricular units. Journal of Physiology (London),
    1971, 214, 245-256.

EGGER, M. D. Responses of hypothalamic neurons to electrical
    stimulation in the amygdala and the hypothalamus. Electro-
    encephalography & Clinical Neurophysiology, 1967, 23, 6-15.

ELEFTHERIOU, B. E. Effects of amygdaloid lesions on hypothalamic
    norepinephrine response to increased ambient temperature.
    Neuroendocrinology, 1970, 6, 175-179.

EPSTEIN, A. N., FITZSIMONS, J. T., & ROLLS, B. J. Drinking in-
    duced by injection of angiotensin into the brain of the rat.
    Journal of Physiology (London), 1970, 210, 457-474.

ERIKSSON, L., FERNANDEZ, O., & OLSSON, K. Central regulation of ADH-release in the conscious goat. Acta Physiologica Scandinavica, Abstract. Meeting of the Scandinavia Physiological Society, Bergen, Norway, 7-9 May, 1971.

EULER, C. von. A repliminary note on slow hypothalamic "Osmopotentials." Acta Physiologica Scandinavica, 1953, 29, 133-136.

EULER, C. von. Physiology and pharmacology of temperature regulation. Pharmacological Review, 1961, 13, 361-398.

FELDBERG, W., & MYERS, R. D. A new concept of temperature regulation by amines in the hypothalamus. Nature (London), 1963, 200, 1325.

FINDLAY, A. L. R., & HAYWARD, J. N. Spontaneous activity of single neurones in the hypothalamus of rabbits during sleep and waking. Journal of Physiology (London), 1969, 201, 237-258.

FINLEY, K. H. Angio-architecture of the hypothalamus and its peculiarities. Research Publications of the Association for Research in Nervous and Mental Disease, 1939, 20, 286-309.

FITZSIMONS, J. T., & SIMONS, B. J. The effect on drinking in the rat of intravenous infusion of angiotensin given alone or in combination with other stimuli of thirst. Journal of Physiology (London), 1969, 203, 45-57.

FREEDMAN, S. Effects of osmotic stimuli on unit activity in the rabbit olfactory bulb. Anatomical Record, 1963, 145, 229-230.

FUXE, K. Evidence for the existence of monoamine neurons in the central nervous system IV. The distribution of monoamine nerve terminals in the central nervous system. Acta Physiologica Scandinavica, 1965, 64: Suppl. 247, 39-85.

GLICK, S. M. The regulation of growth hormone secretion. In W. F. Ganong and L. Martini (Eds.), Frontiers in Neuroendocrinology. New York: Oxford Univ. Press, 1969. Chap. 4, pp. 141-182.

GLOOR, P. Electrophysiological studies on the connections of the amygdaloid nucleus of the cat. I. The neuronal organization of the amygdaloid projection system. Electroencephalography & Clinical Neurophysiology, 1955a, 7, 223-242.

GLOOR, P. Electrophysiological studies on the connections of the
    amygdaloid nucleus of the cat. II. The electrophysiological
    properties of the amygdaloid projection system. Electro-
    encephalography & Clinical Neurophysiology, 1955b, 7, 243-262.

GLOOR, P. Amygdala. Handbook of Physiology, Sect. I, Neuro-
    physiology, 1960, 2, 1395-1420.

GROSSMAN, S. P. A neuropharmacological analysis of hypothalamic
    and extrahypothalamic mechanisms concerned with the regula-
    tion of food and water intake. Annals of the New York Acad-
    emy of Science, 1969, 157, 902-912.

GROSSMAN, S. P., & GROSSMAN, L. Food and water intake following
    lesions or electrical stimulation of the amygdala. American
    Journal of Physiology, 1963, 205, 761-765.

GUNNE, L. N., & REIS, D. J. Changes in brain catecholamines
    associated with electrical stimulation of amygdaloid nucleus.
    Life Sciences, 1963, 11, 804-809.

HAMMEL, H. T. Regulation of internal body temperature. Annual
    Review of Physiology, 1968, 30, 641-710.

HARDY, J. D. Physiology of temperature regulation. Physiological
    Review, 1961, 41, 521-606.

HARRIS, G. W. The innervation and actions of the neurohypophysis:
    an investigation using the method of remote-control stimula-
    tion. Philosophical Transactions of the Royal Society,
    Series B, 1947, 232, 385-391.

HARRIS, G. W. Central control of pituitary secretion. Handbook
    of Physiology, Sect. I, Neurophysiology, 1960, 2, 1007-1038.

HAYWARD, J. N. Cerebral cooling during increased cerebral blood
    flow in the monkey. Proceedings of the Society for Experi-
    mental Biology and Medicine, 1967, 124, 555-557.

HAYWARD, J. N. Brain temperature regulation during sleep and
    arousal in the dog. Experimental Neurology, 1968, 21, 201-
    212.

HAYWARD, J. N. Hypothalamic single cell activity during the
    thermoregulatory adjustments of sleep and waking in the
    monkey. Anatomical Record, 1969a, 163, 197.

HAYWARD, J. N. Brain temperature and thermosensitive nerve cells
    in the monkey. Transactions of the American Neurological
    Association, 1969b, 94, 157-159.

HAYWARD, J. N., & BAKER, M. A. Diuretic and thermoregulatory responses during preoptic cooling in the monkey. American Journal of Physiology, 1968a, 214, 843-850.

HAYWARD, J. N., & BAKER, M. A. The role of the cerebral arterial blood in the regulation of brain temperature in the monkey. American Journal of Physiology, 1968b, 215, 389-403.

HAYWARD, J. N., & BAKER, M. A. A comparative study of the role of the cerebral arterial blood in the regulation of brain temperature in five mammals. Brain Research, 1969, 16, 417-440.

HAYWARD, J. N., & SMITH, W. K. Influence of limbic system on neurohypophysis. Archives of Neurology, 1963, 9, 171-177.

HAYWARD, J. N., & SMITH, W. K. Antidiuretic response to electrical stimulation in brain stem of the monkey. American Journal of Physiology, 1964, 206, 15-20.

HAYWARD, J. N., & VINCENT, J. D. Osmosensitive single neurones in the hypothalamus of unanesthetized monkeys. Journal of Physiology (London), 1970, 210, 947-972.

HAYWARD, J. N., OTT, L. H., STUART, D. G., & CHESHIRE, F. C. Peltier biothermodes. American Journal of Medical Electronics, 1964, 206, 15-20.

HAYWARD, J. N., SMITH, E., & STUART, D. G. Temperature gradients between arterial blood and brain in the monkey. Proceedings of the Society for Experimental Biology and Medicine, 1966, 121, 547-551.

HELLER, H., & GINSBURG, M. Secretion, metabolism and fate of the posterior pituitary hormones. In G. W. Harris and B. T. Donovan (Eds.) The Pituitary Gland. London: Butterworths, 1966. Vol. 3, pp. 330-373.

HENRY, J. P., GAUER, O. H., & REEVES, J. L. Evidence of the atrial location of receptors influencing urine flow. Circulation Research, 1956, 4, 85-90.

HOLLAND, R. C., CROSS, B. A., & SAWYER, C. H. EEG correlates of osmotic activation of the neurohypophyseal milk-ejection mechanism. American Journal of Physiology, 1959a, 196, 796-802.

HOLLAND, R. C., SUNDSTEN, J. W., & SAWYER, C. H. Effects of
    intracarotid injections of hypertonic solutions on arterial
    pressure in the rabbit. Circulation Research, 1959b, 7, 712-
    720.

ISHIKAWA, T., KOIZUMI, K., & BROOKS, C. McC. Electrical activity
    recorded from the pituitary stalk of the cat. American Jour-
    nal of Physiology, 1966, 210, 427-431.

JESSEN, C., & LUDWIG, O. Spinal cord and hypothalamus as core
    sensors of temperature in the conscious dog. II. Addition
    of signals. Pflugers Archives fur die Gesamte Physiologie,
    1971, 324, 205-216.

JESSEN, C., MAYER, E. TH. Spinal cord and hypothalamus as core
    sensors of temperature in the conscious dog. I. Equilavence
    of responses. Pflugers Archives fur die Gesamte Physiologie,
    I971, 324, 189-204.

JESSEN, C., & SIMON, E. Spinal cord and hypothalamus as core
    sensors of temperature in the conscious dog. III. Identity
    of function. Pflugers Archives fur die Gesamte Physiologie,
    1971, 324, 205-216.

JEWELL, P. A., & VERNEY, E. B. An experimental attempt to deter-
    mine the site of neurohypophysial osmoreceptors in the dog.
    Philosophical Transactions of the Royal Society, Series B,
    1957, 240, 197-324.

JOHNSON, J. A. ZEHR, J. E., & MOORE, W. W. Effects of separate
    and concurrent osmotic and volume stimuli on plasma ADH in
    sheep. American Journal of Physiology, 1970, 218, 1273-1280.

JOYNT, R. J. Functional significance of osmosensitive units in the
    anterior hypothalamus. Neurology, 1964, 14, 584-590.

KAADA, B. R. Somato-motor, autonomic and electrocorticographic
    responses to electrical stimulation of "rhinencephalic" and
    other structures in primates, cat and dog. Acta Physiologica
    Scandinavica, 1951, 23, Suppl. 83, 1-100.

KANDEL, E. R. Electrical properties of hypothalamic neuroendocrine
    cells. Journal of General Physiology, 1964, 47, 691-717.

KELLEY, J. S., & DREIFUSS, J. J. Antidromic inhibition of iden-
    tified rat supraoptic neurones. Brain Research, 1970, 22,
    406-409.

KOIZUMI, K., ISHIKAWA, T., & BROOKS, C. Mc. Control of activity or neurones in the supraoptic nucleus. Journal of Neurophysiology, 1964, 27, 878-892.

LAMMERS, H. J. The neuronal connexions of the hypothalamic neurosecretory nuclei in mammals. Journal of Neuro-Visceral Relations, 1969, Suppl. 9, 311-328.

MACHNE, X., & SEGUNDO, J. P. Unitary responses to afferent volleys in amygdaloid complex. Journal of Neurophysiology, 1955, 19, 232-240.

MacLEAN, P. D. The limbic system ("visceral brain") in relation to central gray and reticulum of the brain stem. Psychosomatic Medicine, 1955, 17, 355-366.

MacLEAN, P. D., & PLOOG, D. W. Cerebral representation of penile erection. Journal of Neurophysiology, 1962, 25, 29-55.

MAGOUN, H. W., & RANSON, S. W. Retrograde degeneration of the supraoptic nuclei after section of the infundibular stalk in the monkey. Anatomical Record, 1939, 75, 107-122.

MASON, J. W. Plasma 17-hydroxycorticosteroid levels during electrical stimulation of the amygdaloid complex in conscious monkeys. American Journal of Physiology, 1959, 196, 44-48.

MATHESON, G. K., & SUNDSTEN, J. W. Changes in plasma cortisol levels in conscious primates after forebrain stimulation. Anatomical Record, 1969, 163, 227.

MILLS, E., & WANG, S. C. Liberation of antidiuretic hormone: location of ascending pathways. American Journal of Physiology, 1964, 207, 1399-1404.

MIRSKY, I. A., & STEIN, M. The effect of a noxious stimulus in man on the antidiuretic activity of the blood. Science, 1953, 118, 602-603.

MIRSKY, I. A., STEIN, M., & PAULISCH, G. The secretion of an antidiuretic substance into the circulation of adrenalectomized and hypophysectomized rats exposed to noxious stimuli. Endocrinology, 1954, 55, 28-39.

MORUZZI, G., & MAGOUN, H. W. Brainstem reticular formation and activation of the EEG. Electroencephalography and Clinical Neurophysiology, 1949, 1, 455-473.

MOYANO, H. F., & BROOKS, C. Mc. Unit and EEG osmosensitive
    responses in cat olfactory bulb. Federation Proceedings,
    1968, 27, 1320.

NAUTA, W. J. H. Hippocampal projections and related neural path-
    ways to the midbrain in the cat. Brain, 1958, 81, 319-340.

NAUTA, W. J. H. Fibre degeneration following lesions of the
    amygdaloid complex in the monkey. Journal of Anatomy, 1961,
    95, 515-531.

NAUTA, W. J. H. Neural associations of the amygdaloid complex
    in the monkey. Brain, 1962, 85, 505-520.

NAUTA, W. J. H. Central nervous organization and the endocrine
    motor system. In A.V. Nalbandov (Ed.), Advances in Neuro-
    endocrinology. Urbana: Univ. Illinois Press, 1963. Chap. 2,
    pp. 5-21.

NAUTA, W. J. H., & HAYMAKER, W. Hypothalamic nuclei and fiber
    connections. In W. Haymaker, E. Anderson and W. J. H.
    Nauta (Eds.), The Hypothalamus. Springfield, Ill.: C. C.
    Thomas, 1969. Pp. 136-209.

NISHIOKA, R. S., ZAMBRANO, D., & BERN, H. A.   Electron micro-
    scope autoradiography of amino acid incorporation by supra-
    optic neurons of the rat. General and Comparative Endo-
    crinology, 1970, 15, 477-495.

NORSTRÖM, A., and SJÖSTRAND, J. Axonal transport of proteins in
    the hypothalamo-neurohypophysial system of the rat. Journal
    of Neurochemistry, 1971, 18, 29-39.

NOVIN, D., SUNDSTEN, J. W., & CROSS, B. A. Some properties of
    antidromically activated units in the paraventricular nucleus
    of the hypothalamus. Experimental Neurology, 1970, 26, 330-
    341.

O'CONNOR, W. J., & VERNEY, E. B. The effect of increased activity
    of the sympathetic system in the inhibition of water diuresis
    by emotional stress. Quarterly Journal of Experimental
    Physiology, 1945, 33, 77-90.

OLIVECRONA, H. Paraventricular nucleus and pituitary gland.
    Acta Physiological Scandinavica, 1957, Suppl. 136, 1-178.

OLSSON, K. Studies on central regulation of secretion of anti-
    diuretic hormone (ADH) in the goat. Acta Physiological
    Scandinavica, 1969, 77, 465-474.

OLSZEWSKI, J.   The Thalamus of the Macaca Mulatta.   New York:
    Karger, 1952.

ORKAND, P. M., & PALAY, S. L.   Effects of treatment with exogenous
    vasopressin on the structural alterations in the hypothalamo-
    neurohypophysial system of rats with hereditary diabetes
    insipidus.   Anatomical Record, 1967, 157, 295.

OYAMA, S. N., KAGAN, A., & GLICK, S. M.   Radioimmunoassay study
    of urinary vasopressin during hydration and dehydration.
    Program 52nd Meeting of the Endocrine Society, St. Louis,
    Mo., 1970.   Pp. 691.

PALAY, S. L.   The fine structure of the neurohypophysis.   In
    H. Waelsch (Ed.), Ultrastructure and Cellular Chemistry of
    Neural Tissue.   New York: Hoeber, 1957.   Pp. 31-49.

PAPEZ, J. W.   A proposed mechanism of emotion.   Archives of
    Neurology and Psychiatry, 1937, 38, 725-749.

PEART, W. S.   The renin-angiotensin system.   Pharmacological
    Review, 1965, 17, 143-182.

PINKSTON, J. O., BARD, P., & RIOCH, D. McK.   The responses to
    changes in environmental temperature after removal of por-
    tions of the forebrain.   American Journal of Physiology,
    1934, 109, 515-531.

POMERANZ, B. H., BIRTCH, A. G., & BARGER, A. C.   Neural control
    of intra-renal blood flow.   American Journal of Physiology,
    1968, 215, 1067-1081.

PRIBRAM, K. H., & KRUGER, L.   Functions of the "olfactory brain,"
    Annals of the New York Academy of Science, 1954, 58, 109-138.

RECHARDT, L.   Electron microscopic and histochemical observations
    on the supraoptic nucleus of normal and dehydrated rats.
    Acta Physiologica Scandinavica, 1969, Suppl. 329, 1-25.

REIS, D. J., & GUNNE, L. M.   Brain catecholamines:relation to
    the defense reaction evoked by amygdaloid stimulation in
    the cat.   Science, 1965, 149, 450-451.

REIS, D. J., & McHUGH, P. R.   Hypoxia as a cause of bradycardia
    curing amygdala stimulation in monkey.   American Journal of
    Physiology, 1968, 214, 601-610.

REIS, D. J., & OLIPHANT, M. C. Bradycardia and tachycardia
    following electrical stimulation of the amygdaloid region in
    the monkey. Journal of Neurophysiology, 1964, 27, 893-912.

REIVICH, M. Arterial P_{CO_2} and cerebral hemodynamics. American
    Journal of Physiology, 1964, 206, 25-35.

ROTHBALLER, A. B. Pathways of secretion and regulation of
    posterior pituitary factors. Research Publications of the
    Association for Research in Nervous and Mental Disease,
    1966, 43, 86-131.

RYDIN, H., & VERNEY, E. B. The inhibition of water diuresis by
    emotional stress and muscular exercise. Quarterly Journal
    of Experimental Physiology, 1938, 27, 343-374.

SACHS, H. Biosynthesis and release of vasopressin. American
    Journal of Medicine, 1967, 42, 687-700.

SAWYER, C. H., & FULLER, G. R. Electroencephalographic correlates
    of reflex activation of the neurohypophysial antidiuretic
    mechanism. Electroencephalography and Clinical Neuro-
    physiology, 1960, 12, 83-93.

SAWYER, C. H., & GERNANDT, B. E. Effects of intracarotid and
    intraventricular injections of hypertonic solutions on
    electrical activity of the rabbit brain. American Journal
    of Physiology, 1956, 185, 209-216.

SHARE, L. Extracellular fluid volume and vasopressin secretion.
    In W. F. Ganong and L. Martini (Eds.), Frontiers in Neuro-
    endocrinology. New York: Oxford Univ. Press, 1969.
    Pp. 183-210.

SHARE, L., & LEVY, M. N. Cardiovascular receptors and blood
    titer of antidiuretic hormone. American Journal of Physiology,
    1962, 203, 425-428.

SHARPLESS, S. K., & ROTHBALLER, A. B. Humoral factors released
    from intracranial sources during stimulation of the reticu-
    lar formation. American Journal of Physiology, 1961, 200,
    909-915.

SHUTE, C. C. D., & LEWIS, P. R. Cholinergic and monoaminergic
    pathways in the hypothalamus. British Medical Bulletin,
    1966, 22, 221-226.

SIMMONDS, M. A.   Effect of environmental temperature on the turn-over of noradrenaline in hypothalamus and other areas of rat brain.   Journal of Physiology (London), 1969, 203, 199-210.

SIMON, E., RAUTENBERG, W., THAUER, R., & IRIKI, M.   Auslösung thermo-regulatorischer Reaktionen durch lokale Kühlung im Vertebralkanal.   Naturwissenschaften, 1963, 50, 337.

SLOTNICK, B. M., & ROTHBALLER, A. B.   Vasopressin release following stimulation of limbic forebrain structures in the cat.   Federation Proceedings, 1964, 23, 150.

SMITH, W. K.   The representation of respiratory movements in the cerebral cortex.   Journal of Neurophysiology, 1938, 1, 55-68.

SNIDER, R. S., & LEE, J. C.   A Stereotaxic Atlas of the Monkey Brain.   Chicago: Univ. Chicago Press, 1961.

SOKOL, H. W., & VALTIN, H.   Evidence for the synthesis of oxytocin and vasopressin in separate neurons.   Nature (London), 1967, 214, 314-316.

STERMAN, M. B., & CLEMENTE, C. D.   Forebrain inhibitory mechanisms: sleep patterns induced by basal forebrain stimulation in the behaving cat.   Experimental Neurology, 1952, 6, 103-117.

STUART, D. G., PORTER, R. W., ADEY, W. R.   Hypothalamic unit activity.   II. Central and peripheral influences.   Electroencephalography and Clinical Neurophysiology, 1964, 16, 248-258.

STUMPF, W. E.   Estrogen neurons and estrogen-neuron systems in the periventricular brain.   American Journal of Anatomy, 1970, 129, 207-218.

SUDA, I., KOIZUMI, K., & BROOKS, C. Mc.   Study of unitary activity in the supraoptic nucleus of the hypothalamus.   Japanese Journal of Physiology, 1963, 13, 374-385.

SUGAR, O., & GERARD, R. W.   Anoxia and brain potentials.   Journal of Neurophysiology, 1938, 1, 558-572.

SUNDSTEN, J. W.   Alterations in water intake and core temperature in baboons during hypothalamic thermal stimulation.   Annals of the New York Academy of Science, 1969, 157, 1018-1029.

SUNDSTEN, J. W.   Septal inhibition of antidromically activated hypothalamic paraventricular neurons in the monkey. Anatomical Record, 1971, 169, 439.

SUNDSTEN, J. W., & SAWYER, C. H.   Electroencephalographic evidence of osmosensitive elements in olfactory bulb of dog brain. Proceedings of the Society for Experimental Biology and Medicine, 1959, 101, 524-527.

SUNDSTEN, J. W., & SAWYER, C. H.   Osmotic activation of neurohypophysial hormone release in rabbits with hypothalamic islands. Experimental Neurology, 1961, 4, 548-561.

SUNDSTEN, J. W., NOVIN, D., & CROSS, B. A.   Identification and distribution of paraventricular units excited by stimulation of the neural lobe of the hypophysis. Experimental Neurology, 1970, 26, 316-329.

SUZUKI, M., TONOUE, T., MATSUZAKI, S., & YAMAMOTO, K.   Initial response of human thyroid, adrenal cortex and adrenal medulla to acute cold exposure. Canadian Journal of Physiology and Pharmacology, 1967, 45, 423-432.

TACHIBANA, S.   Relation between hypothalamic heat production and intra- and extracranial circulatory factors. Brain Research, 1969, 16, 405-416.

TINDAL, J. S., KNAGGS, G. S., & TURVEY, A.   The afferent path of the milk-ejection reflex in the brain of the rabbit.   Journal of Endocrinology, 1969, 43, 663-671.

URSIN, H., & KAADA, B. R.   Functional localization within the amygdaloid complex in the cat. Electroencephalography and Clinical Neurophysiology, 1960, 12, 1-20.

VANDER, A. J.   Control of renin release.   Physiological Reviews, 1967, 47, 359-382.

VERNEY, E. B.   The antidiuretic hormone and the factors which determine its release. Philosophical Transactions of the Royal Society, Series B, 1947, 135, 25-106.

VINCENT, J. D., & HAYWARD, J. N.   Activity of single cells in the osmoreceptor-supraoptic nuclear complex in the hypothalamus of the waking rhesus monkey. Brain Research, 1970, 23, 105-108.

WHITLOCK, D. G., & NAUTA, W. J. H. Subcortical projections from
   the temporal neocortex in Macaca mulatta. Journal of Comparative Neurology, 1956, 106, 183-191.

WOODS, J. W., & BARD, P. Antidiuretic hormone secretion in the
   cat with a chronically denervated hypothalamus. Proceedings of the International Congress of Endocrinology, Ist.
   Copenhagen, 1960, p. 113.

WOODS, J. W., BARD, P., & BLEIER, R. Functional capacity of
   the deafferented hypothalamus: water balance and responses
   to osmotic stimuli in the decerebrate cat and rat. Journal
   of Neurophysiology, 1966, 29, 751-767.

WOODS, W. H., HOLLAND, R. C., & POWELL, E. W. Connections of
   cerebral structures functioning in neurohypophysial hormone
   release. Brain Research, 1969, 12, 26-46.

YAMASHITA, H., KOIZUMI, K., & BROOKS, C. Mc. Electrophysiological studies of neurosecretory cells in the cat hypothalamus. Brain Research, 1970, 20, 462-466.

YOSHIDA, S., IBAYASHI, H., MURAKAWA, S., & NAKAO, K. Cerebral
   control of secretion of antidiuretic hormone: effect of
   electrical stimulation of the prepyriform area on the
   neurohypophysis in the dog. Endocrinology, 1965, 77, 597-
   601.

YOSHIDA, S., IBAYASHI, H., MURAKAWA, S., & NAKAO, K. Cerebral
   control of antidiuretic hormone release: effect of electrical stimulation of the medial aspect of the dog brain.
   Endocrinology, 1966, 79, 871-874.

YUDILEVICH, D. L., & DeROSE, N. Blood-brain transfer of glucose
   and other molecules measured by rapid indicator dilution.
   American Journal of Physiology, 1971, 220, 841-846.

DO HYPOTHALAMIC NEUROENDOCRINE CELLS HAVE A SYNAPTIC OUTPUT?*

J. J. Dreifuss and J. S. Kelly**

Department of Physiology, University of Geneva Medical

School, Geneva, Switzerland

The main problems concerning the synthesis of vasopressin
and oxytocin in the cell bodies of the supraoptic and paraventricular nuclei of the hypothalamus, the axonal transport
of the neurosecretory material to the neurohypophysis, and its
release into the general circulation have been clarified to
some extent. Hypothalamic neuroendocrine cells have a dual
function: as secreting cells, they produce, store and release
octapeptide hormones, while as neurons they are capable of
receiving, integrating and conveying neural information.
Below, we present some evidence which suggests that they share
still another property with conventional neurones, namely that
they establish synaptic contacts with nerve cells.

A transpharyngeal approach to the rat hypothalamus permitted the positioning, under microscopic control, of bipolar
stimulating electrodes across the exposed pituitary stalk.
Micropipettes were used for extracellular recording and were
inserted into the supraoptic nucleus (Fig. 1A). All-or-none
action potentials were recorded following electrical stimulation of the pituitary stalk. A spike potential was positively
identified as arising from a supraoptic neurone when antidromic
action potentials, which occurred at a fixed latency following
the stimulus artifact, were cancelled by a properly timed,
spontaneously occurring action potential (Fig. 1B-D).

In spontaneously active supraoptic neurones, action
potentials travelling antidromically along the fibres of the
supraoptico-neurohypophysial tract not only invaded the cell
bodies, but also produced a period of reduction of the spon-

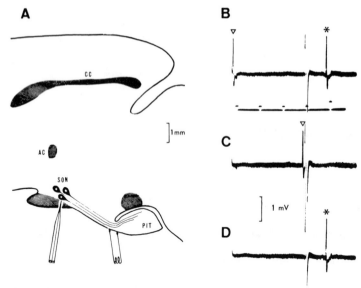

Fig. 1.   A, schematic, parasagittal section of the rat brain, to
illustrate the experimental arrangement: a recording micropipette
is located in the supraoptic nucleus (SON); bipolar stimulating
electrodes lie across the pituitary stalk, where the axons of
supraoptic neurones pass.   CC, corpus callosum; AC, anterior
commissure; PIT, pituitary gland.
B-D, records from a spontaneously active supraoptic neurone.   ▽,
spontaneous action potentials, *, antidromic action potentials,
following stimulation of the pituitary stalk at a constant latency
of 9 msec.   Note that the antidromic action potential has been
cancelled in C, but not in B, by a spontaneously occurring action
potential.   Time base: 10 msec.

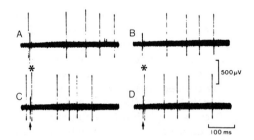

Fig. 2.   Effects of thraddling threshold stimulation of the pituitary stalk on the firing of another identified supraoptic neurone.
In C & D, antidromic action potentials (*) follow the stimulation
artifact (arrows); in A & B, the same stimulus as in C & D was
infra-liminar for the axon of this neurone.   Note that a reduction
in cell firing lasting approximately 100 msec follows the stimulation artifacts in all 4 single oscilloscope traces.

taneous discharge, which lasted approximately 100 msec. (Fig 2,
C & D). Part, at least, of this period of reduced probability of
discharge following stimulation of the pituitary stalk was probably
mediated synaptically, since it was also observed when the
stimuli applied to the stalk were too weak to excite the axon of
the cell under study (Fig. 2., A, B). Moreover, above threshold
for antidromic invasion both the intensity and duration of the
period of reduced probability of discharge increased with increases of stimulation intensity (Kelly & Dreifuss, 1970).

The existence of this antidromic inhibition constitutes
electrophysiological evidence for the presence of recurrent
collaterals of the supraoptico-neurohypophysial tract (cf. Christ,
1966), and suggests that axons of this tract end not only on blood
vessels, but also establish synaptic junctions with neurones.

In the goldfish, Kandel (1964) succeeded in recording intracellularly from neuroendocrine cells in the hypothalamic preoptic
nucleus, which is the functional equivalent of the two nuclei
found in mammals. Goldfish preoptic cells develop full sized
action potentials and generate an antidromic post-synaptic inhibitory potential in response to electrical stimulation of the
pituitary (Fig. 3). We interpret our results, obtained by extracellular recording in the rat as indicating the existence in
rodents of a recurrent collateral pathway similar to the one
described in the goldfish. If true, this introduces the alternative either that vasopressin (and/or oxytocin) may serve as
inhibitory transmitter substances at the level of recurrent

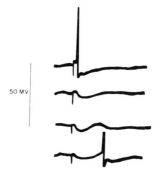

50 MV

150 MSEC.

Fig. 3. Intracellular records from a goldfish preoptic neurone.
Four responses obtained after stimulation of the pituitary at
decreasing intensity (from above downwards) are shown. Note that
when the stimuli are infra-liminar for antidromic invasion, a
hyperpolarizing, inhibitory post-synaptic potential follows the
stimulation artifacts (From Kandel, 1964).

collaterals, or that supraoptic neurones may be capable of reseasing both neurohormones and a synaptic transmitter substance.

ACKNOWLEDGMENTS

* Supported by grants from the Swiss National Science Foundation and the F. Hoffmann-La Roche Foundation.

** Canadian MRC Scholar and IBRO/UNESCO Fellow. Present address: Department of Research in Anaesthesia, McGill University, Montreal, Canada.

REFERENCES

CHRIST, J. F. Nerve supply, blood supply and cytology of the neurohypophysis. In G. W. Harris & B. T. Donovan (Eds.), The Pituitary Gland, Vol. 3, Pars intermedia and Neuro-hypophysis. London: Butterworth, 1966. Pp. 62-130.

KANDEL, E. R. Electrical properties of hypothalamic neuroendocrine cells. Journal of General Physiology, 1964, 47, 691-717.

KELLY, J. S. & J. J. DREIFUSS. Antidromic inhibition of identified rat supraoptic neurones. Brain Research, 1970, 22, 406-409.

FUNCTIONS OF THE AMYGDALA RELATED TO THE FEEDBACK ACTIONS

OF GONADAL STEROID HORMONES

Charles H. Sawyer

Department of Anatomy and Brain Research
Institute UCLA 90024 and Long Beach VA Hospital
Long Beach, California

The hormones of the pituitary target organs, which include
the gonads, adrenals and thyroid glands, exert profound feedback
influences on brain-pituitary function to control pituitary
trophic secretions. The target organ hormones also influence behavior. As a model of this feedback circuit, we shall emphasize
brain-pituitary-ovarian interactions in which the ovarian steroids,
estrogen and progesterone, influence sex behavior and the secretion of pituitary luteinizing hormone, follicle stimulating hormone and prolactin. The latter hormones, in turn, control ovulation and the secretion of ovarian steroids. This circuit will be
emphasized, because there is evidence that the action of estrogen
may involve the amygdala, the subject of this conference, whereas
the adrenal steroids seem to exert an important extrahypothalamic
influence on the hippocampus.

With such experts as Stumpf and Pfaff present to discuss
autoradiographic localization of steroids and Zolovick scheduled
to cover effects of amygdaloid stimulation and lesions on hypothalamo-hypophysial function, our discussion will be somewhat
limited in scope. We shall define loosely a hormone receptor as
a system of neurons which bind the hormone, and react to its
local application with a specific response. A relatively local
influence of the hormone also is revealed, following systemic administration, by an alteration in threshold of a local neural
response to a locally applied stimulus, e.g., a change in local
seizure threshold.

Before discussing steroid hormone receptors in the amygdala,
we should point out that even in the hypothalamus receptor neurons

745

related to pituitary-gonadal function anatomically are discrete
from those concerned with sex behavior. Lisk (1967) has reviewed
his pioneer work in rats in which he found that estradiol implants
in the basal hypothalamus exerted a negative feedback influence
on pituitary gonadotrophic secretion whereas preoptic implants
induced estrous behavior in the ovariectomized animal. We
(Sawyer, 1967) have reviewed similarly our work in the rabbit
with Davidson, Kanematsu and Palka, showing that basal midline
hypothalamic implants of estrogen depressed gonadotrophic function
as in rats, but that the ovariectomized doe became behaviorally
estrous only when estrogen was deposited discretely in her ventromedial nucleus rather than in her preoptic area. In the cat, both
lesion (Sawyer, 1960) and estrogen implant data (Sawyer, 1963;
Michael, 1965) place the sex behavioral estrogen receptors in the
anterior hypothalamus-preoptic region as in the rat.

Electrophysiological studies have revealed that gonadal
steroids injected systemically can alter neural thresholds involving the amygdala and rhinencephalic pathways, but they usually
have left the location(s) of the steroid receptor(s) in doubt.
For example, Kawakami and Sawyer (1959) showed that estrogen,
progesterone and testosterone could alter the threshold of a
response in rabbits which they called an "EEG afterreaction"--a
sleep sequence evoked by low frequency stimulation of the amygdala,
hippocampus, septum or olfactory projections. Similar observations were made at that time, in France, by Faure, who has reviewed recently the field (Faure et al., 1968). Kawakami et al.
(1967) have shown that progesterone influences evoked potentials
in the rabbit stimulated in amygdala or hippocampus and recorded
in the arcuate nucleus of the hypothalamus. Endröczi (1971)
found that conditioning stimuli, applied to the medial amygdala,
facilitated evoked responses in the basal hypothalamus from
stimulation of the reticular formation, and that the facilitation
was enhanced by estrogen and testosterone and suppressed by progesterone. Terasawa and Timiras (1968a) described cyclic changes
in thresholds of local amygdaloid and hippocampal seizures related to the estrous cycle in female rats (Fig. 1), which changes
were lost after ovariectomy and restored by exogenous estrogen
treatment. Since the afterdischarges did not spread far from the
stimulating electrode the estrogen must have been influencing
neurons in the proximity of the stimulating-recording electrode.

The olfactory system, with its projections to the amygdala
and other rhinencephalic areas, has been the subject of many
studies in reproductive endocrinology and physiology. The
"Bruce effect" in which the urine odor of a strange male mouse
will interrupt early pregnancy in a female by blocking prolactin
secretion, and the "Whitten effect" in which similar stimuli will
coordinate estrous cycles in non-pregnant female mice are examples
of pheromonal influences (Bronson, 1968). Estrogen alters

olfactory thresholds in vertebrate forms as lowly as the goldfish
(Hara, 1967). Electrophysiological unit recording studies in male
rats show that units in the preoptic area respond more differentially to the odor of estrous female rat urine than do units in
the olfactory bulb itself (Pfaff and Gregory, 1971). Testosterone
administered systemically or directly into the preoptic area in
male rats influenced the electrical activity of preoptic units
and their response to olfactory bulb stimulation (Pfaff and Pfaffman, 1969).

The ovarian steroids exert both stimulatory and inhibitory
feedback influences on pituitary-gonadal function in rats
(Everett, 1961) and rabbits (Sawyer et al., 1966). Ovulation
can be advanced, delayed or inhibited altogether depending on the
hormone administered and the timing of its injection. In accomplishing these feats, the steroids appear to act at multiple sites
within the brain and in the pituitary gland (Sawyer and Gorski,
1971). The inhibitory effect of estrogen is not prevented by
hypothalamic deafferentation (Taleisnik et al., 1970) and, presumably, it involves primarily the hypothalamo-pituitary complex.
Implant data in rats suggest that the stimulatory influence of
estrogen is exerted in the median eminence (Palka et al., 1966)
or directly on the pituitary (Weick and Davidson, 1970). However,
deafferentation experiments show that the rostral afferents to
the hypothalamus are essential for the stimulatory feedback influence of both progesterone (Taleisnik et al., 1970) and estrogen
(Caligaris et al., 1971). Consistent with these findings are the
observations of Terasawa and Sawyer (1970) that multiple-unit
electrical changes in the rat median eminence which appear to be
correlated with pituitary activation are lost following anterior
deafferentation. The results indicate that extrahypothalamic
influences are involved importantly in the stimulatory feedback
mechanisms.

A widely discussed example of stimulatory feedback action of
estrogen is the induction of precocious puberty in the female rat.
Hohlweg (1934) stimulated ovulation in immature rats with single
massive doses of estrogen. Ramirez and Sawyer (1965a) advanced
puberty (vaginal opening, rise in plasma LH, drop in pituitary
LH, and initiation of estrous cycles) a full week by injecting
low (0.05 μ estradiol benzoate) doses on days 26-30. The
presence of the ovary (to supply progestin ?) was necessary to
activate release of pituitary LH, and testosterone was ineffective
as a substitute for estrogen even though it did induce vaginal
opening. Treatment with norethindrone suppressed not only the
estrogen advancement of puberty, but the natural onset of puberty
as well (Ramirez and Sawyer, 1965b).

   Searching for the site(s) of stimulatory action of estrogen
on puberty, Smith and Davidson (1968) made temporary removable
implants of estrogen on day 26 into various regions of the brain.
Unilateral implants into the medial amygdala gave negative results
(Davidson, 1969) as did basal hypothalamic placements, but estrogen in the anterior hypothalamus-preoptic region, beneath the anterior commissure and columns of the fornix, advanced puberty as
readily as Ramirez' systemic injections. The results were interpreted as either a stimulatory action of estrogen on anterior hypothalamic neurons, or a blockade of "an inhibitory influence
originating in this area or relayed through it from another area
such as the amygdala."   It should be remembered that the stria
terminalis projects through the anterior hypothalamus and a localized estrogen focus on their axons might influence more amygdaloid
neurons than a similar implant in the amygdala itself. Electrical
stimulation and lesion data supporting the inhibition hypothesis
are summarized in the review of Critchlow and Bar-Sela (1967).
In contrast to the findings of Smith and Davidson, Motta et al.
(1968) reported that basal hypothalamic estrogen implants, on day
26, induced precocious puberty. However, implants in the habenular region delayed it and inhibited ovulation as had been observed
in the rabbit by Faure et al. (1968).

   Ramaley and Gorski (1967) induced precocious vaginal opening
by anterior deafferentation of the hypothalamus, an effect they
attributed to transection of inhibitory pathways. However, ovulation did not occur, and this was interpreted as a simultaneous
loss of ovulatory pathways, which were severed completely by deafferentation but not necessarily by electrolytic lesions in the
amygdala (Critchlow and Bar-Sela, 1967). Even treatment with
PMS-gonadotrophin was not followed by ovulation in anterior deafferented rats, whose ovaries obviously were producing estrogen
which apparently could not exert its full stimulatory feedback
action in the absence of rostral ovulatory pathways to the hypothalamus.

   Further evidence, suggesting the possible involvement of the
amygdala in hypothalamo-pituitary changes at puberty, comes from
more recent experiments of Terasawa and Timiras (1968b). They
described a lowering of local seizure thresholds in the medial
amygdala, but not the hippocampus, associated with both natural
puberty and precocious ovulation, and uterine development following treatment with PMS. They proposed that ovarian estrogen secretion is augmented to the point that it disinhibits the amygdala
and breaks down the earlier negative feedback influence of the
steroid. Recently Zarrow et al. (1969) have reported that brief
treatment (e.g., days 21 and 23) with testosterone causes true
precocious puberty in female rats, perhaps by hastening the decrease in sensitivity to the inhibitory feedback influence of

estrogen. Precocious puberty here was blocked by ventral hippocampal lesions of a type which also delayed natural puberty
(Riss et al., 1963).

In adult female rats, stereotaxic implants of estrogen into
amygdaloid nuclei also have produced a limited number of positive
effects. Preliminary results reported by Littlejohn and de Groot
(1963) suggested a stimulatory influence on gonadotrophic secretion: 6 out of 8 cases bearing unilateral estradiol implants in
anterior or anteromedial nuclei, for a week, exhibited greater
than normal compensatory ovarian hypertrophy three weeks after
hemi-ovariectomy. This is consistent with the findings of Halasz
and Gorski (1967) that extrahypothalamic pathways are involved in
the process of compensatory hypertrophy (OCH). Smith et al.
(1971) have reported recently that bilateral lesions of the stria
terminalis or the cortical amygdaloid nuclei (ACO) inhibited
completely the OCH response, and they suggested that this might
be due to an inhibitory influence on FSH secretion. The lesions
did not block ovulation in the remaining ovary nor interfere with
the estrous cycle, but the corpora lutea were unusual in that
they each retained a central lumen (Smith and Lawton, personal
communication).

Neither estrogen implants nor lesions in the cortical amygdala (ACO) inhibit LH secretion (Lawton and Sawyer, 1970); in
fact, they both appear to facilitate it. Three weeks after
ovariectomy, combined with ACO lesions or implants of estradiol,
plasma LH levels relatively were much higher in lesioned or
estrogen implanted animals than in ovariectomized controls (Fig.
2 and 3). Sham-lesioned rats, which also showed a significant
rise in LH, were found to have striae terminales damaged by mechanical insertion of the sham electrodes into the amygdala
(Fig. 2). Estrogen implants in the preoptic area (POA) also
exerted a stimulatory action on LH secretion. Implants of undiluted estradiol at the tips of 27 gauge (27G) needles exerted
some systemic effects such as promoting uterine growth, but not
enough negative feedback potency to counteract the stimulatory
influence on LH secretion (Fig. 3).

In my laboratory, Drs. Ellendorff and Blake have started
to repeat these estrogen implantation experiments with periodic
measurements of plasma LH and prolactin by radioimmunoassay.
Initial results have shown a consistent drop in plasma LH during
the first few days after implantation of dilute estrogen
(E:C-1:5) into ACO followed by recovery and some elevated values
at the end of two weeks. A few cases have shown markedly elevated plasma prolactin values by the end of the first week after
implantation of estrogen. Two rats have shown brief periods of
vaginal cornification during the first week and the initial

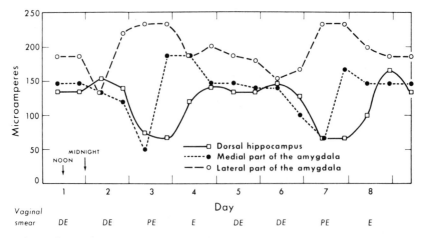

Fig. 1. Changes in local seizure thresholds in medial and
lateral amygdala and dorsal hippocampus through two estrous cycles
in female rats. DE, diestrus; PE, proestrus; E, estrus. From
Terasawa and Timiras (1968a). Courtesy of Endocrinology and
J. B. Lippincott Co.

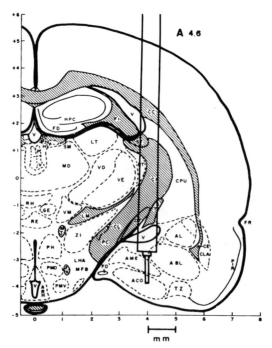

Fig.2. Diagrammatic drawing of stereotaxic electrode tract to
cortical amygdaloid nucleus (ACO) on plane A 4.6 of the de Groot
Atlas, showing how lowering the electrode (sham lesion) might
damage the stria terminalis. From Lawton and Sawyer (1970).
Courtesy of American Journal of Physiology.

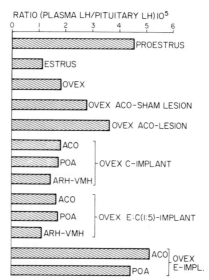

Fig. 3. Graphic representation of Lawton and Sawyer's (1970) tabulated data on the effects of cortical amygdaloid (ACO) lesions and estrogen implants on LH secretion in ovariectomized (OVEX) rats. POA, preoptic area; ARH-VMH, arcuate nucleus-ventromedial region of hypothalamus; E:C (1:5), one part estradiol diluted with 5 parts cholesterol; E-IMPL, implant of undiluted estradiol.

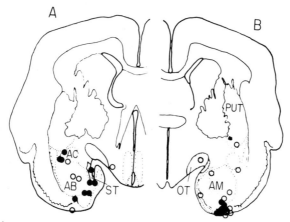

Fig. 4.   Sites of estrogen implants (closed circles) which induced lactogenesis in pseudopregnant rabbits. Open circles represent impalnts which failed to cause lactogenesis. Implants were positioned bilaterally. Amygdaloid nuclei: AB, basal; AC, central; AM, medial nuclei. OT, optic tract; PUT, putamen; ST, stria terminalis. Section A lies 1 mm caudal to section B. Adapted from Tindal et al. (1967). Courtesy of the Journal of Endocrinology.

effects on pituitary secretion may therefore represent non-specific
influences as far as the amygdala is concerned.

Lactogenesis in the pseudopregnant rabbit has been induced
with bilateral implants of estradiol benzoate (EB) in the amygdaloid complex (Tindal et al., 1967). Solid EB in 27G steel tubes
was implanted on the 7th day of pseudopregnancy and the animals
were autopsied 11 days later. Of 76 rabbits implanted, the 20
which exhibited positive lactogenic responses had their implants
in the medial nucleus, the central nucleus and the basomedial part
of the basal nucleus of the amygdala as well as in the stria terminalis (Fig. 4). The authors suggested that the lactogenic responses might have been the result of estrogen-sensitive amygdaloid
neurons acting via the stria terminalis on the preoptic area
and/or hypothalamus to cause the release of prolactin. It is of
interest here to note that EB implants in the basal hypothalamus
did not cause lactogenesis (Kanematsu and Sawyer, 1963) but implants directly into the pituitary gland resulted in activated
mammary glands and a pituitary depleted of prolactin.

Kawakami et al. (1969) have studied the effects of unilateral
implants of estrogen or progesterone into various regions of the
rabbit brain on the biosynthetic capacity of the ovaries at autopsy three weeks later. They determined the in vitro capacity of
ovarian homogenates to incorporate $1(^{14}C)$ acetate into progesterone, 20α-hydroxy-pregn-4-en-3-one, 17-hydroxyprogesterone, an-
drostendione and estradiol plus estrone. Progesterone implants
in the amygdala exerted little influence on ovarian biosynthetic
capacity, whereas implants in the arcuate nucleus elevated synthesis of both estrogen and progestins, and hippocampal implants
stimulated synthesis of progestins, but not of estrogen. In contrast, estrogen implants in the hippocampus essentially failed to
influence ovarian function while implants throughout the amygdala
as in the arcuate nucleus depressed synthesis of all of the
steroids tested. Ventromedial and anterior hypothalamic estrogen
implants stimulated ovarian biosynthesis of progestins while de-
pressing estrogen uptake of $1(^{14}C)$ acetate. The authors are aware
of limitations of the work--that they are testing at a late stage
following steroid implantation, and that the systemic feedback
actions of endogenous ovarian steroids, stimulated by the intracerebral implant, may have influenced the synthesizing ability of
the gonads by the time of autopsy at 21 days.

The hypersexuality in male animals resulting from lesions of
the amygdala and its underlying cortex requires gonadal hormones
to sustain it (Schreiner and Kling, 1954; Green et al., 1957),
but this does not imply that the steroids are exerting their
effects at the site of the lesion. There have been few reports of
lesion-induced hypersexuality in female animals, but de Groot and

Critchlow (1960) observed that their female rats with bilateral
lesions in the amygdala or stria terminalis would accept the male
during diestrus or the "anestrus" of gestation (14 cases). Unilateral lesions were ineffective, suggesting that the "release
from inhibition" must be complete to foster sex behavior under
less than optimal hormonal conditions. The effects on sex behavior of implanting estrogen into the amygdala was not reported
in the studies summarized above. Lisk and his associates recently have attacked the problem (personal communication): in adult
rats ovariectomized for 11 days they have placed unilateral estrogen implants in 27G tubing into the cortical or medial amygdala.
The estrogen implant was ineffective by itself in evoking estrous
behavior, but by 11 days after implantation the addition of a
large subcutaneous implant of progesterone (Lisk, 1969) brought
8/20 of the rats into heat within a few hours. With smaller estrogen implants (30G tubing) none of 18 rats showed the lordosis
response even with the synergistic action of progesterone. It
would, therefore, appear to be a problem of quantity of estrogen
present as well as its localization, and in view of the slight
systemic effects of 27 gauge implants mentioned above one must
consider the possibility that these results were not specific for
the amygdala.

An important approach to the problem of steroid localization
and function in brain has been the biochemical demonstration of
steroid-binding macromolecules in brain cell nuclei with the use
of radioactive hormones. Earlier biochemical studies of steroid
binding by the hypothalamus have been summarized in the review of
McEwen et al. (1970d). The Rockefeller University scientists
have themselves expanded the scope of these investigations to include such limbic areas as the amygdala, hippocampus, septum and
olfactory bulb. McEwen and Pfaff (1970) listed the areas retain-
ing estrogen in descending order as pituitary, hypothalamus, preoptic area, septum and amygdala. These were all more consistent
than hippocampus, brain stem, cerebellum and olfactory bulbs.
The brain of the ovariectomized adult rat retained estrogen more
effectively than that of the castrated male or early androgenized
female. This last finding confirmed an earlier report of Flerko
et al. (1969). (^{3}H)-testosterone was taken up by the same areas
of the brain (McEwen et al., 1970a), but was not bound as tightly
as estrogen, and uptake was reduced by competition with unlabeled
estrogen, testosterone or the antiandrogen cyproterone (McEwen
et al., 1970b).

With elaborate methods of homogenization, cell fractionation,
and extraction of brain samples developed for earlier studies on
adrenal steroids (McEwen et al., 1970c) Zigmond and McEwen (1970)
have described a selective retention of estradiol by cell nuclei
from the preoptic-hypothalamic area and secondarily, the amygdaloid region of the ovariectomized rat's brain (Fig. 5).

Pre-treatment with unlabeled estradiol saturates the nuclei and
blocks the uptake of the tritiated steroid, but treatment with
testosterone does not interfere with the process. In somewhat
related studies in the hamster Lisk and his associates (personal
communication) have found that during the estrous cycle less
^{3}H-estradiol is taken up by the amygdala during late diestrus and
early proestrus than during other parts of the cycle, presumably
due to the competition afforded by endogenous estrogen. Treatment with progesterone does not reduce the uptake of labeled estrogen but pretreatment with exogenous estradiol drastically
lowers retention of the labeled steroid.

Although it would be inappropriate to discuss, at length,
the results of electrical stimulation experiments, we should
mention in the context of the present topic that electrical stimulation of the release of pituitary ovulating hormone(s) fails
unless the titer of natural or exogenous estrogen is adequate.
When they induced ovulation by electrical stimulation of the
amygdala, Koikegami et al. (1954) employed naturally estrous
rabbits, and Shealy and Peele (1957) primed their cats with estrogen and PMS according to the regimen of Sawyer and Everett
(1953). In our laboratory, we have used rabbits primed with
estrogen (Saul and Sawyer, 1961; Hayward et al., 1964), and
stimulation of the amygdala was performed in the unanesthetized
state with chronically implanted electrodes. In unpublished
experiments with Saul, we found that stimulation of the amygdala
did not induce ovulation in acute experiments in which the rabbit
was under stress of restraint in a stereotaxic instrument--condi-
tions under which hypothalamic stimulation had been effective
(Saul and Sawyer, 1957). Similar limitations appear to hold in
the rat: Bunn and Everett (1957) and Velasco and Taleisnik (1969)
employed persistent estrous rats in their ovulation experiments,
and Kawakami and Terasawa (personal communication) use proestrous
pentobarbital-blocked rats (Fig. 6). Velasco and Taleisnik (1969)
report that, in male rats, electrochemical stimulation of the
amygdala at parameters effective in the preoptic area fails to
cause release of pituitary LH even if the rat is castrated and
steroid-primed. Kawakami and Terasawa (personal communication)
have failed to induce ovulation by electrochemical stimulation
of the amygdala in the androgen sterilized rat: again stimulation
of the preoptic area or median eminence was effective (Terasawa
et al., 1969). One is reminded that Clayton et al. (1970)
found that the medial amygdala and the medial preoptic regions
of the newborn female rat's brain responded to androgen differently from the rest of the brain relative to RNA metabolism as
reflected in the uptake of ^{3}H-uridine.

From the viewpoint of the reproductive neuroendocrinologist,
the corticomedial amygdala of the female rat appears to contain

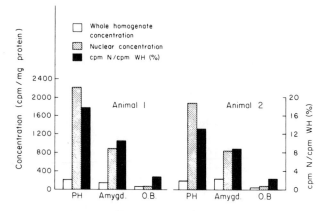

Fig. 5. Concentration of (^{3}H) estradiol in the whole homogenate (WH) and nuclear pellets (N) from different regions of the brains of two ovariectomized rats 2 h after an intraperitoneal injection of 0.1 mCi of (^{3}H) estradiol. The amounts of radioactivity recovered in the nuclear pellets is also expressed as a percentage of the total counts in the whole homogenate (right ordinate. PH, preoptic-hypothalamic area; Amygd., amygdala; O.B., olfactory bulb. From Zigmond and McEwen (1970). Courtesy of the Journal of Neurochemistry.

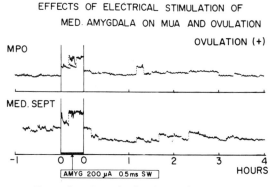

Fig. 6. Effects of ovulation-inducing electrical stimulation of the medial amygdala (200 μ Amps, 0.5 msec square waves, 100 Hz, 30 sec on and off, 30 min during the 2-4 PM critical period of proestrus) on multiple unit activity (MUA) of the medial preoptic area (MPO) and medial septum. "Spontaneous" activation of the pituitary was blocked with pentobarbital. Elevated MUA was observed during the stimulation period in rats destined to ovulate the following morning, suggesting that amygdaloid neurons may activate preoptic and septal elements en route to the hypophysiotropic area (unpublished work of Kawakami and Terasawa, with permission of the authors).

two functional groups of neurons: (1) cells <u>inhibitory</u> to gonadotrophic function in general and (2) cells <u>facilitatory</u> to the ovulatory surge of pituitary LH release.   Both appear to project to
the hypothalamus via the stria terminalis (Fig. 7).   Anterior
hypothalamic deafferentation would interrupt both projections and
induce precocious puberty without ovulation (Ramaley and Gorski,
1967).   Estrogen may simultaneously suppress group 1 and facilitate group 2 to cause precocious puberty or permit electrical
or electrochemical stimulation to induce ovulation.   The preoptic
area appears to contain additional stimulatory neurons which
respond positively to estrogen, and permit ovulation in precocious
puberty fostered by amygdaloid or stria terminalis lesions.   In
the immature female rat, electrical stimulation of the amygdala
may activate only inhibitory neurons since the hormonal conditions
are inappropriate for activation of group 2 neurons.   Such an
explanation would account for the delayed puberty described by
Critchlow and Bar-Sela (1967) following electrical stimulation
of the amygdala.   In the adult female rat, lesions in the cortical

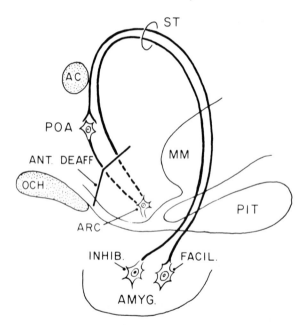

Fig. 7.   Diagram of hypothetical stria terminalis (ST) projections
of inhibitory (INHIB) and facilitatory (FACIL) neurons from the
cortico-medial amygdala (AMYG).   Synapses are indicated with
neurons in the preoptic area (POA) which lies beneath the anterior
commissure (AC) and above the optic chiasma (CH).   Anterior deafferentation (ANT-DEAFF) would prevent both inhibitory and
facilitatory elements from reaching the basal hypothalamic region
of the arcuate nucleus (ARC).   MM, mammillary body; PIT, pituitary.

nucleus or stria terminalis which eliminate compensatory ovarian
hypertrophy, perhaps by interfering with FSH secretion (Smith
et al., 1971), foster levels of pituitary and plasma LH higher
than the elevated castration value, while estrogen implants
stimulate release of LH into the plasma (Lawton and Sawyer, 1970).
It is hoped that newly available methods of measuring pituitary
and ovarian hormones in blood by radioimmunoassay, with careful
application of the experimental approaches outlined above, soon
will clarify the role of the amygdala in the control of pituitary
gonadotrophic functions.

ACKNOWLEDGMENTS

Research from the author's laboratory presented in this re-
view was supported by grants from the National Institutes of
Health (NB 01162) and the Ford Foundation. The assistance of the
National Library of Medicine for a MEDLARS search and the UCLA
Brain Information Service for current references (NINDB contract
43-66-59) is gratefully acknowledged. Thanks are also due Mrs.
Frances Smith for bibliographic and secretarial help.

REFERENCES

BRONSON, F. H. Pheromonal influences on mammalian reproduction.
    In M. Diamond (Ed.), Perspectives in Reproduction and Sexual
    Behavior. Bloomington: Indiana University Press, 1968.
    Pp. 341-361.

BUNN, J. P., & EVERETT, J. W. Ovulation in persistent estrous
    rats after electrical stimulation of the brain. Proceedings
    of the Society for Experimental Biology and Medicine, 1957,
    96, 369.

CALIGARIS, L., ASTRADA, J. J., & TALEISNIK, S. Release of lutein-
    izing hormone induced by estrogen injection into ovariecto-
    mized rats. Endocrinology, 1971, 88, 810.

CLAYTON, R. B., KOGURA, J., & KRAEMER, H. C. Sexual differentia-
    tion of the brain: effects of testosterone on brain RNA
    metabolism in newborn male rats. Nature, 1970, 226, 810.

CRITCHLOW, V., & BAR-SELA, M. E. Control of the onset of puberty.
    In L. Martine and W. F. Ganong (Eds.) Neuroendocrinology.
    New York: Academic Press, 1967. Vol. 2, pp. 101-162.

DAVIDSON, J. M. Feedback control of gonadotropin secretion. In
    W. F. Ganong and L. Martini (Eds.) Frontiers in Neuroendo-
    crinology, 1969. Oxford University Press, 1969. Pp. 343-388.

de GROOT, J., & CRITCHLOW, V.   Effects of "limbic system" lesions on reproductive functions of female rats.   The Physiologist, 1960, 3, 49.

ENDROCZI, E.   The role of brainstem and limbic structures in regulation of sexual behavioural patterns.   Journal of Neuro-Visceral Relations, 1971, Supplement 10, 263.

EVERETT, J. W.   The mammalian female reproductive cycle and its controlling mechanisms.   In W. C. Young (Ed.) Sex and Internal Secretions.   Baltimore: The Williams and Wilkins Co., 1961.   Vol.1, pp. 497-555.

FAURE, J. M. A., VINCENT, J. D., & BENSCH, C.   Interdependances entre le niveau de vigilance et les fonctions gonadotropes chez le lapin femelle.   Revue Europeenne d'Endocrinologie, 1968, 5, 25.

FLERKO, B., MESS, B., & ILLEI-DONHOFFER, A.   On the mechanism of androgen sterilization.   Neuroendocrinology, 1969, 4, 164.

GREEN, J. D., CLEMENTE, C. D., & de GROOT, J.   Rhinencephalic lesions and behavior in cats.   Journal of Comparative Neurology, 1957, 108, 505.

HALASZ, B., & GORSKI, R. A.   Gonadotrophic hormone secretion in female rats after partial or total interruption of neural afferents to the medial basal hypothalamus.   Endocrinology, 1967, 80, 608.

HARA, T. J.   Electrophysiological studies of the olfactory system of the goldfish, Carassius auratus L.- III Effects of sex hormones on olfactory activity.   Comparative Biochemistry and Physiology, 1967, 22, 209.

HAYWARD, J. N., HILLIARD, J., & SAWYER, C. H.   Time of release of pituitary gonadotropin induced by electrical stimulation of the rabbit brain.   Endocrinology, 1964, 74, 108.

HOHLWEG, W.   Veränderungen des Hypophysenvorderlappens und des Ovariums nach Behandlung mit grossen Dosen von Follikel-hormon.   Klinische Wochenschrift, 1934, 13, 92.

KANEMATSU, S., & SAWYER, C. H.   Effects of intrahypothalamic and intrahypophysial estrogen implants on pituitary prolactin and lactation in the rabbit.   Endocrinology, 1963, 72, 243.

KAWAKAMI, M., & SAWYER, C. H.   Neuroendocrine correlates of
    changes in brain activity thresholds by sex steroids and
    pituitary hormones.   Endocrinology, 1959, 65, 652.

KAWAKAMI, M., SETO, K., YOSHIDA, K., & MIYAMOTO, T.   Biosynthesis
    of ovarian steroids in the rabbit: influence of progesterone
    or estradiol implantation into the hypothalamus and limbic
    system.   Neuroendocrinology, 1969, 5, 303.

KAWAKAMI, M., SETO, K., TERASAWA, E., & YOSHIDA, K.   Mechanisms
    in the limbic system controlling reproductive functions of
    the ovary with special reference to the positive feedback of
    progestin to the hippocampus.   In W. R. Adey and T. Tokizane
    (Eds.) Structure and Function of the Limbic System.
    Progress in Brain Research, 1967, 27, 70.

KOIKEGAMI, H., YAMADA, T., & USUI, K.   Stimulation of amygdaloid
    nuclei and periamygdaloid cortex with special reference to
    its effects on uterine movements and ovulation.   Folia
    Psychiatrica et Neurologica Japonica (Niigata), 1954, 8, 7.

LAWTON, I. E., & SAWYER, C. H.   Role of amygdala in regulating
    LH secretion in the adult female rat.   American Journal of
    Physiology, 1970, 218, 622.

LISK, R. D.   Sexual behavior: hormonal control.   In L. Martini
    and W. F. Ganong (Eds.) Neuroendocrinology.   New York:
    Academic Press, 1967.   Vol. 2, pp. 197-239.

LISK, R. D.   Progesterone: biphasic effects on the lordosis
    response in adult or neonatally gonadectomized rats.
    Neuroendocrinology, 1969, 5, 149.

LITTLEJOHN, B. M., & de GROOT, J.   Estrogen-sensitive areas in
    the rat brain.   Federation Proceedings, 1963, 22, 571.

McEWEN, B. S., & PFAFF, D. W.   Factors influencing sex hormone
    uptake by rat brain regions.   I. Effects of neonatal treat-
    ment, hypophysectomy and competing steroid on estradiol up-
    take.   Brain Research, 1970, 21, 1.

McEWEN, B. S., PFAFF, D. W., & ZIGMOND, R. E.   Factors influencing
    sex hormone uptake by rat brain regions. II. Effects of neo-
    natal treatment and hypophysectomy on testosterone uptake.
    Brain Research, 1970a, 21, 17.

McEWEN, B. S., PFAFF, D. W., & ZIGMOND, R. E. Factors influ-
    encing sex hormone uptake by rat brain regions. III. Effects
    of competing steroids on testosterone uptake. Brain Research,
    1970b, 21, 29.

McEWEN, B. S., WEISS, J. M., & SCHWARTZ, L. S. Retention of
    corticosterone by cell nuclei from brain regions of adrenal-
    ectomized rats. Brain Research, 1970c, 17, 471.

McEWEN, B. S., ZIGMOND, R. E., AZMITIA, E. C., & WEISS, J. M.
    Steroid hormone interaction with specific brain regions.
    In R. E. Bowman and S. P. Datta (Eds.) Biochemistry of Brain
    and Behavior. New York: Plenum Press, 1970d. Pp. 123-167.

MICHAEL, R. P. Oestrogens in the central nervous system.
    British Medical Bulletin, 1965, 21, 87.

MOTTA, M., FRASCHINI, F., GIULIANI, G., & MARTINI, L. The central
    nervous system, estrogen and puberty. Endocrinology, 1968,
    83, 1101.

PALKA, Y. S., RAMIREZ, V. D., & SAWYER, C. H. Distribution and
    biological effects of tritiated estradiol implanted in the
    hypothalamo-hypophysial region of female rats. Endocrin-
    ology, 1966, 78, 487.

PFAFF, D. W., & GREGORY, E. Olfactory coding in olfactory bulb
    and medial forebrain bundle of normal and castrated male
    rats. Journal of Neurophysiology, 1971, 34, 208.

PFAFF, D. W., & PFAFFMAN, C. Olfactory and hormonal influences
    on the basal forebrain of the male rat. Brain Research,
    1969, 15, 137.

RAMALEY, J. A., & GORSKI, R. A. The effect of hypothalamic de-
    afferentation upon puberty in the female rat. Acta Endo-
    crinology, 1967, 56, 661.

RAMIREZ, V. D., & SAWYER, C. H. Advancement of puberty in the
    female rat by estrogen. Endocrinology, 1965a, 76, 1158.

RAMIREZ, V. D., & SAWYER, C. H. Suppression of the initiation of
    natural puberty and of pubertas praecox induced by estrogen
    in rats treated with norethindrone. Excerpta Medica Inter-
    national Congress Series No. 99, 1965b. P. E175.

RISS, W., BURSTEIN, S. D., & JOHNSON, R. W. Hippocampal or pyri-
    form lobe damage in infancy and endocrine development of rats.
    American Journal of Physiology, 1963, 204, 861.

SAUL, G. D., & SAWYER, C. H. Atropine blockade of electrically induced hypothalamic activation of the rabbit adenohypo-physes. Federation Proceedings, 1957, 16, 112.

SAUL, G. D., & SAWYER, C. H. EEG-monitored activation of the hypothalamo-hypophysial system by amygdala stimulation and its pharmacological blockade. Electroencephalography and Clinical Neurophysiology, 1961, 13, 307.

SAWYER, C. H. Reproductive behavior. In J. Field, H. W. Magoun and V. E. Hall (Eds.) Handbook of Physiology, Neurophysiology. Washington, D.C.: American Physiological Society, 1960. Vol. 2, Chapter 49, pp. 1225-1240.

SAWYER, C. H. Induction of estrus in the ovariectomized cat by local hypothalamic treatment with estrogen. Anatomical Record, 1963, 145, 280.

SAWYER, C. H. Effects of hormonal steroids on certain mechanisms in the adult brain. In M. Martini, F. Frachini and M. Motta (Eds.) Hormonal Steroids. Excerpta Medica International Congress Series, 1967, No. 132, p. 123.

SAWYER, C. H., & EVERETT, J. W. Priming the anestrous cat for reflex discharge of pituitary ovulating hormone. Proceedings of the Society for Experimental Biology and Medicine, 1953, 83, 820.

SAWYER, C. H., & GORSKI, R. A. (Eds.) Steroid Hormones and Brain Function. Los Angeles: University of California Press, 1971 (in press).

SAWYER, C. H., KAWAKAMI, M., & KANEMATSU, S. Neuroendocrine aspects of reproduction. In R. Levine (Ed.) Endocrines and the Central Nervous System. Research Publications of the Association for Research in Nervous and Mental Disease, 1966, 43, 59.

SCHREINER, L., & KLING, A. Effects of castration on hypersexual behavior induced by rhinencephalic injury in the cat. A.M.A. Archives of Neurology and Psychiatry, 1954, 72, 180.

SHEALY, C. N., & PEELE, T. L. Studies on the amygdaloid nucleus of the cat. Journal of Neurophysiology, 1957, 20, 125.

SMITH, E. R., & DAVIDSON, J. M. Role of estrogen in the cerebral control of puberty in female rats. Endocrinology, 1968, 82, 100.

SMITH, S. W., ADANIYA, J., GORSKI, M., & LAWTON, I. E. Absence
    of the ovarian compensatory response in rats with lesions in
    the cortical amygdaloid nucleus and stria terminalis.
    Federation Proceedings, 1971, 30, 253. (Abstr.)

TALEISNIK, S., VELASCO, M. E., & ASTRADA, J. J. Effect of hypo-
    thalamic deafferentation on the control of luteinizing hor-
    mone secretion. Journal of Endocrinology, 1970, 46, 1.

TERASAWA, E., KAWAKAMI, M., & SAWYER, C. H. Induction of ovula-
    tion by electrochemical stimulation in androgenized and spon-
    taneously constant-estrous rats. Proceedings of the Society
    for Experimental Biology and Medicine, 1969, 132, 497.

TERASAWA, E., & SAWYER, C. H. Diurnal variation in the effects
    of progesterone on multiple unit activity in the rat hypo-
    thalamus. Experimental Neurology, 1970, 27, 359.

TERASAWA, E., & TIMIRAS, P. S. Electrical activity during the
    estrous cycle of the rat; cyclic changes in limbic structures.
    Endocrinology, 1968a, 83, 207.

TERASAWA, E., & TIMIRAS, P. S. Electrophysiological study of
    the limbic system in the rat at onset of puberty. American
    Journal of Physiology, 1968b, 215, 1462.

TINDAL, J. S., KNAGGS, G. S., & TURVEY, A. Central nervous
    control of prolactin secretion in the rabbit: effect of
    local oestrogen implants in the amygdaloid complex. Journal
    of Endocrinology, 1967, 37, 279.

VELASCO, M. E., & TALEISNIK, S. Release of gonadotropins induced
    by amygdaloid stimulation in the rat. Endocrinology, 1969,
    84, 132.

WEICK, R. F., & DAVIDSON, J. M. Localization of the stimulatory
    feedback effect of estrogen on ovulation in the rat.
    Endocrinology, 1970, 87, 693.

ZARROW, M. X., NAQVI, R. H., & DENENBERG, V. H. Androgen-induced
    precocious puberty in the female rat and its inhibition by
    hippocampal lesions. Endocrinology, 1969, 84, 14.

ZIGMOND, R. E., & McEWEN, B. S. Selective retention of oestra-
    diol by cell nuclei in specific brain regions of the ovariec-
    tomized rat. Journal of Neurochemistry, 1970, 17, 889.

ESTROGEN, ANDROGEN, AND GLUCOCORTICOSTEROID CONCENTRATING NEURONS
IN THE AMYGDALA, STUDIED BY DRY AUTORADIOGRAPHY

Walter E. Stumpf

Laboratories for Reproductive Biology, Departments of
Anatomy and Pharmacology, University of North Carolina

Chapel Hill, North Carolina

INTRODUCTION

The topographic distribution of estradiol concentrating neurons
in the hypothalamus and preoptic region--which corresponds to a
large degree to known terminations of the stria terminalis--suggest-
ed an amygdaloid-hypothalamic-pituitary endocrine interrelationship
(Stumpf, 1968a). This concept is supported by autoradiographic
findings of estradiol target cells in the anterior lobe of the
pituitary (Stumpf, 1968b) as well as reports in the literature regarding the effect of lesions and hormone implantations in the
amygdala, the preoptic regions, and the hypothalamus on gonadotropin secretion (e.g., Koikegami et al., 1954; Kawakami and
Sawyer, 1959; Elwers and Critchlow, 1960; Yamada and Greer, 1960;
Zouhar and de Groot, 1963; Eleftheriou and Zolovick, 1966; Tindal
et al., 1967). As expected, subsequent dry-autoradiographic
studies of the amygdala revealed a distinct pattern of estradiol
concentrating hormone-neurons in this area (Stumpf and Sar, 1969;
Stumpf and Sar, 1971). However, not only in the amygdala, but also
in other selective parts of the phylogenetically older periventricular brain (Stumpf, 1970b), estradiol concentrating neurons
(estrogen-neurons) were observed. Since these hormone neurons
appear to be associated with certain nerve fiber tracts, the concept of estrogen-neuron systems (Stumpf, 1970b, 1971a) was advanced
and the validity of the generally held concept of a single or dual
sex center questioned. Autoradiographic studies with tritium
labeled testosterone (Stumpf and Sar, 1971b; Stumpf, 1971d),
cortisol (Stumpf and Sar, 1971b; Sar and Stumpf, 1971) and
corticosterone (Stumpf, 1971b) showed, similar to estradiol, a

763

retention and concentration of radioactivity in groups of neurons as well as scattered neurons in distinct areas of the brain, including the amygdala. Although it appears not justified, from functional viewpoints, to consider the amygdala alone, tribute is made to the topic of the conference in reviewing the present state of our efforts to map this area of the brain regarding the distribution of steroid hormone concentrating neurons.

Due to progress in autoradiographic techniques for the localization of diffusible compounds (Stumpf, 1971b; Stumpf and Roth, 1964), useful data on the cellular and subcellular distribution of hormones could be obtained. New detailed anatomical information regarding certain populations of neurons is now provided with directional influence on research in neuroendocrinology, neurophysiology, biopsychology, and other areas. It may be stated that results available to date on hormone-neuron distribution must still be considered incomplete. The study and interpretation of the autoradiographic cellular and subcellular localization of hormones in the brain are fraught with numerous difficulties. While there is increasing confirmation of our data (Anderson and Greenwald, 1969; Warembourg, 1970; Tuohimaa, 1970; Attramadal, 1970), considerable discrepancies still exist between the results published by different investigators. The importance of technique in this endeavor has been investigated at the beginning of our studies (Stumpf and Roth, 1966) and discussed detail (Stumpf, 1969, 1970a, 1971b).

METHODS

Experiments were conducted with intact immature female and male and gonadectomized or adrenalectomized mature Sprague Dawley rats; at least two rats for each hormonal state. $6,7-^3H$-estradiol-17β, $1,2-^3H$ testosterone, $1,2-^3H$ cortisol, or $1,2-^3H$ corticosterone--dissolved in 10% alcohol in isotonic saline--was injected subcutaneously with a single pulse at a dose of 0.1 to 1.0 μg (in the case of ^3H corticosterone up to 4.6 μg) of the steroid per 100 gram body weight. The dose range for the specific compound was selected as "physiologic" or "near physiologic" in order to exclude or minimize occupation of possibly existing unspecific binding sites, or of binding to proteins with high affinity to and physiologic specificity for other steroids. The specific activity of the tritium labeled steroids was between 10 and 40 Ci/mM.

At different time intervals after the injection, preferentially at 1 or 2 hours, the animals were killed by decapitation. The brain was removed and areas of 2 to 3 mm^3 size were excised, mounted on a tissue holder and frozen in liquified propane, cooled by liquid nitrogen. The samples were cut in a Wide-Range

Cryostat (Harris Mfg. Co., Cambridge, Mass.), at 2 μ or 3 μ
thickness, freeze-dried with a Cryo-Pump (Thermovac Industries,
Inc., Copiague, L. I., New York) within the cryostat and then
dry-mounted on a desiccated emulsion precoated slide. After
photographic exposure for a period between 6 to 17 months,
slides were processed photographically, stained with methylgreen
pyronin, air dried, and mounted with Permount and a coverglass.

Detailed descriptions of the technique have been published
(Stumpf, 1970c, 1971c).

RESULTS

I. RADIOACTIVITY CONCENTRATION AFTER ^{3}H ESTRADIOL INJECTION

Of the hormones mentioned in this paper, estradiol has been
studied most extensively. Radioactivity was found to be concentrated in nuclei of certain neurons, while not in others, and not
in glia cells and not in ependyma cells. A detailed description
of estrogen-neuron topography in the amygdala has been published
(Stumpf and Sar, 1969; Stumpf, 1970b) and a map provided (Fig. 1).
No qualitative differences were found to exist between ovariectomized female and intact male, and intact immature female and male
rats. From collaborative studies with Drs. M. Sar and A. Eisenfeld
(unpublished) it can be concluded that the radioactivity retained
in the amygdaloid region is likely to represent ^{3}H-estradiol.

The topography of estrogen-neurons in the amygdala is depicted
in the schematic drawings of fronto-caudal coronal sections (Fig. 1).
In the area amygdala anterior, a few single labeled neurons are
found, while the nucleus (n.) of the lateral olfactory tract is
unlabeled. In fronto-caudal direction, the first accumulation of
estrogen-neurons appears in a ventral portion of the n. corticalis,
in the pars anterior of the n. lateralis, and, less concentrated,
in the n. centralis.

The most intensive neuronal labeling with the highest labeling
index is observed in portions of the n. medialis, the n. corticalis,
and the n. basalis, pars anterior. The n. corticalis and n. medialis,
however, are not labeled throughout their entire course, as can be
seen in Figure 1.

Unlabeled in our experiments are the n. lateralis, pars
posterior; the n. basalis, pars lateralis; the n. corticalis parvocellularis; and the cells of the massa intercalata. A few labeled
neurons are found in the bordering subiculum and occasionally in
sections of the piriform cortex, the endorhinal cortex and along
the path of the alveus. In the neighboring regions of the hippocampus, a few scattered intensely labeled neurons are found com-

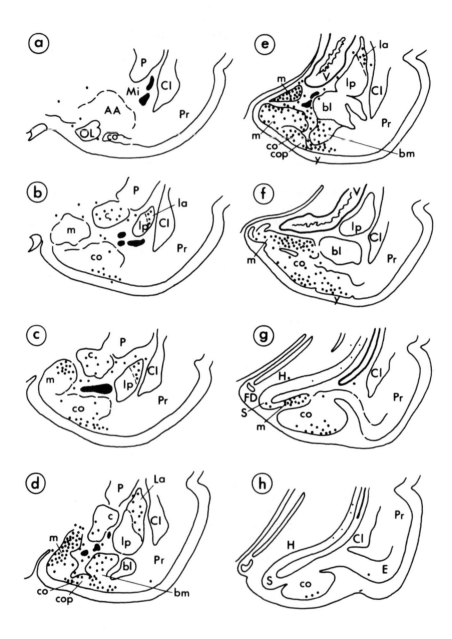

Fig. 1a-h. Schematic demonstration of the localization of estrogen-neurons in the amygdala of immature intact female, male, and mature ovariectomized rats prepared after the distribution of radioactivity in frontocaudal serial section dry-mount autoradiograms at 1 hr. after the injection of 0.1 μg of 6,7-^{3}H-estradiol-17β/100g of body weight. No qualitative differences in the topographic pattern of estrogen-neurons were found to exist between the different hormonal conditions. The intensity of the stippling represents the respective accumu_ation of estradiol concentrating neurons. Cells of the massa intercalata were un-labeled and are indicated in black. The schematic drawings were prepared according to the description of the amygdala by Brodal (1947). AA, area amygdala anterior; bl, n. amygdaloideus basalis, pars lateralis; bm, n. amygdaloideus basalis, pars medialis; c, n. amygdaloideus centralis; CL, claustrum; co, n. amygdaloideus corticalis; cop, n. corticalis parvocellularis; E, entorhinal area, area 28; fascia dentata; H, hippocampus; la, n. amygdaloideus lateralis, pars anterior; lp, n. amygdaloideus lateralis, pars posterior; m, n. amygdaloideus medialis; MI, massa intercalata; Pr, piriform cortex; S, subiculum; V, ventriculus lateralis; Y, zona transitionalis. [Reproduced from Stumpf, W. E. and Sar, M. Proc. Soc. Exptl. Biol. Med. 136, 102-106 (1971)].

---

parable to those in the n medialis of the amygdala--while a larger number of neurons of the fronto-ventral hippocampus show only a comparatively weak labeling, with about 10x lower silver grain density in their nuclei, if compared with nuclei of labeled neurons of the amygdala. The amount of radioactivity in this area is so low that it could not be detected in some animals treated with lower doses, or shorter exposure times, although neurons of the amygdala, preoptic region and hypothalamus were clearly labeled. The significance of these differences is unclear and further studies are required for its clarification.

In some areas of the amygdala, the definition of anatomical nuclei appeared difficult, or impossible, in our own studies. These difficulties in subdividing the amygdala are reflected in the literature of the descriptive neuroanatomy of this area. The estrogen-neuron distribution as obtained from our experiments in the rat follows "classical" boundaries, e.g., those offered by Brodal (1947) or Koenig and Klippel (1965), only to a limited degree.

The topography of estrogen-neurons in other parts of the rat brain has been reported (Stumpf, 1968a, 1970b).

II. RADIOACTIVITY CONCENTRATION AFTER ^{3}H CORTISOL INJECTION

The selectivity of uptake and retention of radioactivity was less pronounced after the injection of ^{3}H cortisol when compared to the relatively high gradient between estrogen target cells and surrounding structures. This impression prevailed in all animals after injection of 0.5 to 1.0 μg per 100 g body weight, previously adrenalectomized, which were killed at 30 minutes, 1 hour, or 2 hours after the subcutaneous administration of the hormone. Despite these differences, a clear nuclear concentration was observed in nuclei of neurons of the dentate gyrus, the hippocampus, the indusium griseum, the dorsal septum, and the nucleus medialis of the amygdala.

The studies with ^{3}H cortisol are preliminary and incomplete. Other areas, not mentioned here, may contain "target" neurons and also concentrate the hormone or a metabolite of it. The conditions of the experiment do not preclude such possibilities. The Topographic distribution of radioactivity after ^{3}H corticosterone injection agrees with the areas listed for ^{3}H cortisol, but also shows additional loci. It can be assumed, from the relationships of functions, that the binding sites for cortisol and corticosterone are the same.

III. RADIOACTIVITY CONCENTRATION AFTER ^{3}H CORTICOSTERONE INJECTION

At 30 minutes, 1 and 2 hours after the subcutaneous injection of 1.4 to 4.6 μg per 100 g body weight of the hormone, a distinct subcellular retention and concentration of radioactivity was seen in the nuclei and, to a lesser degree, in the cytoplasm of certain neurons--while not in others--similar to the studies with tritium labeled estrogen, androgen, and cortisol. Although the studies are still incomplete, a distinct topographic pattern of "glucocorticoid-neurons" begins to arise. Extensive labeling of neurons exists in outer layers of the piriform cortex from its medial contact with the zona transitionalis to and beyond the rhinal fissure. Scattered labeled neurons are found also in deeper layers of this part of the cortex. In the area of the fissura rhinalis, an increased number of neurons, in deeper cortical layers toward the corpus callosum, is labeled. The area of neuronal labeling does not terminate at the rhinal fissure but extends beyond it dorsally, characterizing a lower segment of this part of the hemisphere. In this area scattered labeled neurons also exist in deeper layers. Other parts of the cortex are also labeled, including especially neurons of the cingulum and the indusium griseum. In addition, scattered labeled neurons with weak radioactivity are found in superficial and deeper parts of the frontal pole, in portions of the cortex of the dorsal hemisphere and the endorhinal cortex.

In the amygdala, a heavy concentration of labeled neurons exists in the nucleus centralis. Neurons are labeled also in the nucleus corticalis, including probably neurons of the n. basalis, pars medialis. The topographic distribution in this area needs to be detailed and the studies extended, since, thus far, only 1 rat for each time interval has been evaluated.

In the hippocampus, the gyrus dentatus, the hippocampus anterior, and the indusium griseum, neurons concentrate radio activity most intensely. In addition, the radioactivity content in the cytoplasm of these cells and the nerve fibers of this area is higher than in the surrounding parts of the brain. It can be assumed that the radioactivity in these brain regions is corticosterone (McEwen et al., 1970). The studies with ^{3}H corticosterone of this area as well as other parts of the brain are still incomplete. A larger number of animals, different dose levels, different time intervals, and different hormonal conditions will have to be considered.

IV. RADIOACTIVITY CONCENTRATION AFTER ^{3}H TESTOSTERONE INJECTION

The distribution of radioactivity in the brain was studied at 1 and 2 hours after the injection of the hormone with doses ranging between 0.2 and 1.0 μg per 100 g body weight.

In the amygdala, accumulations of labeled neurons are observed in the nucleus medialis and the nucleus corticalis, probably including the nucleus basalis, pars medialis. Labeled neurons also exist in the hippocampus and in other parts of the brain, which has been reported elsewhere (Stumpf and Sar, 1971b; Stumpf, 1971d). The autoradiographic studies with androgens are still underway and no final judgment on the distribution of androgens can be made at present.

The chemical nature of the radioactivity in the amygdala has not been identified yet. It may be assumed that it is dihydrotestosterone (Kniewald et al., 1970).

CONCLUSIONS

From the results of the localization of radioactively labeled estrogen, androgen, and glucocorticoid, it becomes apparent that the brain contains a population of neurons which is characterized by a selective binding affinity for specific steroids. Certain characteristics arise:

1. These steroid hormones are taken up by the brain apparently without impairment by a so-called blood-brain-barrier.

2. They are concentrated and retained in nuclei of certain
    neurons, which may be called target-neurons, while not in
    others. Ependyma cells and glia cells do not appear to
    concentrate steroids.

3. There are differences in the affinity of "target"-neurons
    or groups of "target"-neurons in the capacity to bind
    steroids, although, in general, the subcellular binding
    of radioactivity in neurons is similar to the one observed
    in classicial peripheral "target"-tissues. The different
    affinities of "target"-neurons to a given steroid may
    reflect functional properties which are important in the
    feedback control of endocrine gland functions and the
    regulation of certain vegetative functions and behavior.

4. The anatomical target areas in the brain for estrogen,
    androgen, and glucocorticoids are in part different from
    each other, but in part overlap or are identical.

5. Whether or not some neurons can bind not only one but
    different hormones can not yet be answered with certainty.
    In the case of estradiol and androgen a high labeling index
    exists in identical areas, e.g., the nucleus medialis of
    the amygdala, the bed nucleus of the stria terminalis, and
    the nucleus preopticus medialis. This suggests that
    neurons in certain areas can be addressed by different
    steroids.

6. "Sex steroid" attracting neurons are found, probably exclu-
    sively, in the phylogenetically older part of the mammalian
    brain, i.e., the periventricular brain (Stumpf, 1970b).
    Glucocorticoid concentrating neurons are similarly found
    preferentially in parts of the periventricular brain,
    while corticosterone attracting neurons with weaker
    affinity appear to exist also outside of it.

7. The distribution pattern of "hormone-neurons" seems to
    follow the distribution of certain nerve fiber tracts
    (Stumpf, 1970b). This suggests the existence of hormone-
    neuron circuits. The validity of the concept of a single
    or dual sex center can be questioned in view of the
    results from the autoradiographic studies. The generally
    held view of a dominating neuroendocrine role of hypo-
    thalamic neurons may be reconsidered.

8. The nuclear concentration suggests nuclear effects of the
    steroid hormones, similar to the ones reported for periph-
    eral target tissues. It is therefore possible that hormone
    neurons are hypophyseotropic neurons and that they produce

polypeptide messengers that can be extracted from the tuber cinereum.

9. There is fair agreement between the topography of hormone neurons and the sites in the brain, which have been reported, however controversial and less well defined, by other investigators, using lesions, implants, and electrical stimulation, to be involved in the physiological mechanisms of hormone action such as the regulation of endocrine gland function and behavior.

ACKNOWLEDGMENT

Supported by PHS Grant 14929 and a grant of the Rockefeller Foundation to the Laboratories for Reproductive Biology, Chapel Hill.

REFERENCES

ANDERSON, C. H., & GREENWALD, G. S. Autoradiographic analysis of estradiol uptake in the brain and pituitary of the female rat. Endocrinology, 1969, 85, 1160-1165.

ATTRAMADAL, A. The uptake of ^{3}H-oestradiol by the anterior hypophysis and hypothalamus of male and female rats. Zeitschrift fur Zellforschung und Mikroskopische Anatomie, 1970, 104, 582-596.

BRODAL, A. The amygdaloid nucleus in the rat. Journal of Comparative Neurology, 1947, 87, 1-16.

ELEFTHERIOU, B. E., & ZOLOVICK, A. T. Effect of amygdaloid lesions on oestrous behaviour in the deermouse. Journal of Reproduction and Fertility, 1966, 11, 451-453.

ELWERS, M., & CRITCHLOW, V. Precocious ovarian stimulation following hypothalamic and amygdaloid lesions in rats. American Journal of Physiology, 1960, 198, 380-385.

KAWAKAMI, M., & SAWYER, C. H. Neuroendocrine correlates of changes in brain activity thresholds by sex steroids and pituitary hormones. Endocrinology, 1959, 65, 652-668.

KNIEWALD, Z., MASSA, R., & MARTINI, L. The transformation of testosterone into dehydrotestosterone by the anterior pituitary and the hypothalamus. Excerpta Medica, 1970, 210, 59 (Abstract).

KOENIG, J. F. R., & KLIPPEL, R. A. The Rat Brain. Baltimore, Williams and Wilkins Co., 1965.

KOIKEGAMI, H., YAMADA, T., & USUI, K. Stimulation of amygdaloid nuclei and periamygdaloid cortex with special reference to its effect on uterine movements and ovulation. Folia Psychiatrica et Neurologica Japonica (Niigata), 1954, 8, 7-31.

MACLEAN, P. D. Some psychiatric implications of physiological studies on frontotemporal portion of limbic system (visceral brain). Electroencephalography and Clinical Neurophysiology, 1952, 4, 407-418.

MCEWEN, B. S., WEISS, T. M., & SCHWARTS, L. S. Retention of corticosterone by cell nuclei from brain regions of adrenal-ectomized rats. Brain Research, 1970, 17, 471-482.

SAR, M., & STUMPF, W. E. Androgen localization in the brain and pituitary. Federation Proceedings, 1971, 30, 363 (Abstract).

STUMPF, W. E. Estradiol concentrating neurons: topography in the hypothalamus by dry-mount autoradiography. Science, 1968a, 162, 1001-1003.

STUMPF, W. E. Cellular and subcellular ^{3}H estradiol localization in the pituitary by autoradiography. Zeitschrift fur Zell-forschung und Mikroskopische Anatomie, 1968b, 92, 23-33.

STUMPF, W. E. Too much noise in the autoradiogram? Science, 1969, 958-959.

STUMPF, W. E. Localization of hormones by autoradiography and other histochemical techniques, a critical review. Journal of Histochemistry and Cytochemistry, 1970a, 18, 21-29.

STUMPF, W. E. Estrogen-neurons and estrogen-neuron systems in the periventricular brain. American Journal of Anatomy, 1970b, 129, 207-218.

STUMPF, W. E. Tissue preparation for the autoradiographic locali-zation of hormones. In G. L. Wied and G. F. Bahr (Eds.), Introduction of Quantitative Cytochemistry-II. New York: Academic Press, 1970c. Pp. 507-526.

STUMPF, W. E. Probable sites for estrogen receptors in brain and pituitary. Journal of Neuro-Visceral Relations, 1971a, Supplement 10, 51-64.

STUMPF, W. E. Autoradiographic techniques and the localization
      of estrogen, androgen, and glucocorticoid in pituitary and
      brain. American Zoologist, 1971b, in press.

STUMPF, W. E. Autoradiographic techniques for the localization
      of hormones and drugs at the cellular and subcellular level.
      Acta Endocrinologica, 1971c, Supplement 153, 205-222.

STUMPF, W. E. Estrogen, androgen, and adrenal hormone attracting
      neurons in the periventricular brain. Federation Proceedings,
      1971d, 30, 309 (Abstract).

STUMPF, W. E., & ROTH, L. J. Vacuum freeze-drying of frozen
      sections for dry-mounting, high-resolution autoradiography.
      Stain Technology, 1964, 39, 219-223.

STUMPF, W. E., & ROTH, L. J. High resolution autoradiography
      with dry-mounted, freeze-dried, frozen sections. Comparative
      study of six methods using two diffusible compounds, ^{3}H estra-
      diol and ^{3}H mesobilirubinogen. Journal of Histochemistry and
      Cytochemistry, 1966, 14, 274-287.

STUMPF, W. E., & SAR, M. Distribution of radioactivity in hippo-
      campus and amygdala after injection of ^{3}H estradiol by dry-
      mount autoradiography. Physiologist, 1969, 12, 368 (Abstract).

STUMPF, W. E., & SAR, M. Estradiol concentrating neurons in the
      amygdala. Proceedings of the Society for Experimental
      Biology and Medicine, 1971a, 136, 102-106.

STUMPF, W. E., & SAR, M. Localization of steroid hormones in the
      brain. Proceedings of the Third International Congress on
      Hormonal Steroids. Amsterdam: Excerpta Medica, 1971b.
      Pp. 503-507.

TINDAL, T. S., KNAGGS, G. S., & TURVEY, A. Central nervous
      control of prolactin secretion in the rabbit: Effect of
      local oestrogen implants in the amygdaloid complex. Journal
      of Endocrinology, 1967, 37, 279-287.

TUOHIMAA, P. Radioautography of tritiated sex steroids in the
      rat. Histochemie, 1970, 23, 349-357.

YAMADA, T., & GREER, M. A. The effect of bilateral ablation of
      the amygdala on endocrine function in the rat. Endocrinology,
      1960, 66, 565-574.

ZOUHAR, R. L., & DEGROOT, T. Effects of limbic brain lesions on
    aspects of reproduction in female rats. Anatomical Record,
    1963, 145, 358 (Abstract).

ESTRADIOL-CONCENTRATING CELLS IN THE RAT AMYGDALA AS
PART OF A LIMBIC-HYPOTHALAMIC HORMONE-SENSITIVE SYSTEM

Donald W. Pfaff and Melvyn Keiner

Rockefeller University

New York, New York

INTRODUCTION

Earlier autoradiographic work resulted in the description of
a limbic-hypothalamic system of cells in the female rat brain,
which concentrated estradiol-H^3 more highly than cells elsewhere
in the brain (Pfaff, 1968a). This description was supported by
scintillation counting of finely dissected brain regions after
systemic estradiol-H^3 injection (McEwen and Pfaff, 1970). The
initial autoradiographic approach employed a combination of
fixatives - formalin and osmium tetroxide - which had been found
successful in preventing tritiated sex steroid translocation in
thin brain sections. In the present experiment a new autoradiographic procedure, involving the direct mounting of unfixed, unembedded frozen sections to emulsion-coated slides, was used to
test the earlier conclusion that cells in certain limbic and medial
hypothalamic structures show the highest estradiol retention in rat
brain.

METHODS

Nine young adult female rats were injected intraperitoneally
with physiologic doses of estradiol-17β-H^3 (95 curies/millimole,
New England Nuclear). The rats had been ovariectomized 2 days
before use, in order to reduce endogenous estrogen levels. Two
hours after injection, the rats were sacrificed by decapitation,
and their brains were quickly removed, blocked and quickly
frozen onto specimen holders for the cryostat. Fresh frozen
sections - unfixed and unembedded - were cut on the cryostat at
-16 to -19°C, in the darkroom, at a thickness of 6 μ or 8 μ.

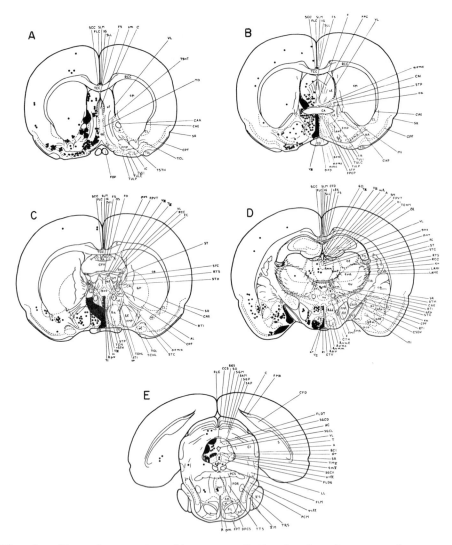

Fig. 1. Maps from autoradiograms of the brain of an ovariectomized female rat, injected intraperitoneally with estra-
diol-17β-H^3 two hours before sacrifice. Autoradiograms were
prepared by mounting unfixed, unembedded frozen sections
directly onto emulsion-coated slides in the darkroom. The
five sections shown include regions of highest estradiol
uptake, drawn on the rat brain atlas of König and Klippel
(1963). Section A is from König and Klippel's Fig. 16b;
B, from 19b; C, from 25b; D, from 33b; E from 54b. Each
dot shows the position of an estradiol-concentrating neuron.
Where many dots would overlap, that area is filled in with
solid black (Pfaff, Keiner and Warren, unpublished observations).

The sections were mounted directly onto slides precoated with
Kodak nuclear emulsion NTB-3. Darkroom histology procedures were
slightly modified from those reported by Anderson and Greenwald
(1969). The slides were then stored in light-tight black plastic
slide boxes equipped with dessicant and sealed with plastic tape.
These slide boxes were, in turn, stored in a lead box equipped
with dessicant and sealed against humidity, in a cold room at
4°C. Exposure times before development were from 5 to 7 months.
Autoradiograms were developed in D19 for 2 minutes at 16°C,
rinsed for 45 sec. in water, and fixed for 18 minutes in 2 changes
of Kodak Fixer. Then the sections were stained with cresyl violet
acetate, dehydrated and cover-slipped.

RESULTS

Cell bodies covered with high concentrations of reduced
grains - indicating high estradiol-H^3 concentration - were
found in a limbic-hypothalamic system confirming the main features
of the distribution reported previously with another autoradiographic method (Pfaff, 1968a).

High numbers of estradiol-concentrating cells were found in
the medial and cortical nuclei of the amygdala, and also in the
lateral septum, the medial preoptic area, the nucleus of the
stria terminalis, the olfactory tubercle, the medial anterior
hypothalamus, the arcuate nucleus of the hypothalamus, the ventromedial nucleus of the hypothalamus, and in more posterior periventricular structures going as far back as structures just
lateral and ventrolateral to the cerebral aqueduct in the mesencephalon. The main features of the distribution were similar
in all 9 animals. The maps in Figure 1 show quantitative and
detailed anatomic features of estradiol concentration in one
representative brain.

Figure 2 shows a more detailed map of estradiol uptake in
the amygdala, based on all the brains. Highest numbers of
heavily labeled cells were found in the medial and cortical
amygdaloid nuclei. Also labeled were cells in parts of the
central and anterior lateral amygdaloid nuclei.

DISCUSSION

Three different experimental approaches have yielded consistent results in showing peak estradiol-H^3 retention in specific
limbic, preoptic and hypothalamic structures. Autoradiography using a successful fixation technique (Pfaff, (1968a), scintillation
counting of dissected brain regions (McEwen and Pfaff, 1970); and
autoradiography on unfixed frozen sections, described above, all
support the same conclusion, represented pictorially in Figure 1.

Other autoradiographic approaches have given less reliable, less consistent reports of the distribution of estradiol-concentrating cells. A technique in which frozen sections are pressed onto the emulsion using finger pressure (Stumpf, 1968) is susceptible inherently to pressure artifacts and to variable sensitivity depending on the degree of adhesion. Moreover, the histologic limitations of that technique - requiring very small tissue blocks - leads to inadequate sampling of different brain regions and subsequent misimpressions of the full distribution of the radioactive hormone. Thus, the temporary claim that estradiol uptake was limited exclusively to medial preoptic and hypothalamic structures in the distribution of the stria terminalis (Stumpf, 1968) was later revised (Stumpf, 1970) to a description of a limbic-hypothalamic system which confirmed earlier (Pfaff, 1968a) results. Even in this later report (Stumpf, 1970), the misnomer "periventricular brain" is given to the estrogen-system; this name does not apply because (a) some of the best estradiol-concentrating regions are not periventricular (e.g. the cortical amygdaloid nucleus and the olfactory tubercle) and (b) several periventricular structures (e.g. the caudate nucleus) do not concentrate estradiol highly. The actual distribution reported in Stumpf's (1970) revised description confirm the conclusion that peak concentrations of estradiol-retaining cells reside in the limbic and hypothalamic structures described in this report and previous reports (Pfaff, 1968a; McEwen and Pfaff, 1970) from our laboratory.

COMPARISON OF ESTRADIOL AND TESTOSTERONE RETENTION

Testosterone-H^3 and estradiol-H^3 retention share several features, as demonstrated by autoradiographic (Pfaff, 1968b) and scintillation counting (McEwen et al., 1970a, b) techniques. First, both hormones are taken up in both male and female rat brains, with similar patterns of distribution. Second, the brain regions which concentrate testosterone tend to be among those which have the highest number of estradiol-concentrating cells: the overlap between the estradiol and testosterone systems is most striking in the septum, preoptic area and the hypothalamus. Third, significant binding site competition effects for the two hormones tend to occur in the same brain regions, all within the limbic-hypothalamic system described above. Finally, autoradiographic evidence (Pfaff, 1968a, b) shows that both hormones are concentrated by neuron cell bodies, within regions of highest uptake. Further autoradiographic work with the new method employed above is being carried out in our laboratory to confirm the description of testosterone-H^3 distribution in rat brain and to extend this description to the bird brain.

That the septal-preoptic-hypothalamic continuum is not a

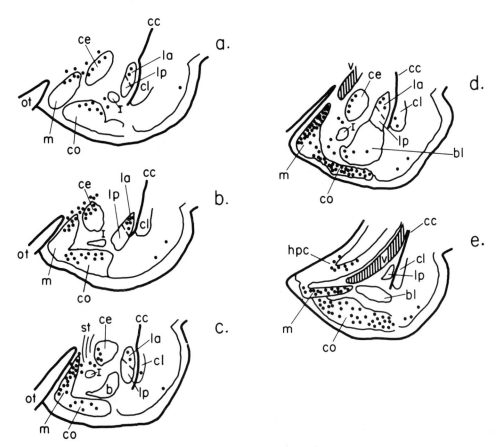

Fig. 2. Maps representing estradiol-17β-H^3 uptake in the amygdala of ovariectomized female rats sacrificed 2 hours after intraperitoneal injection with the radioactive hormone. Autoradiograms were prepared as described in text. The conc trations of dots in the maps indicate the approximate relative density of labelled cells in each subdivision of the amygdala. Nomenclature follows that of Brodal (1947): section a represents Brodal's level c; b, Brodal's level d; c, Brodal's level e; d, Brodal's level g; e, Brodal's level j.

Abbreviations: b, basal complex; bl, basolateral complex; cc, corpus callosum; ce, central nucleus; cl, claustrum; co, cortical nucleus; hpc, hippocampus; I, massa intercalata; la, lateral nucleus pars anterior; lp, lateral nucleus pars posterior; m, medial nucleus; ot, optic tract; st, stria terminalis; v, lateral ventricle.

non-specific steroid hormone depot - but may rather be specific
for sex steroids - is indicated by the high concentration of
corticosterone-H^3 in the hippocampus, and relatively lower corticosterone uptake in preoptic area and hypothalamus (McEwen, Weiss
and Schwartz, 1970).

Some features of testosterone retention indicate that, even
in areas of highest uptake, there may be fewer binding sites or
weaker binding for testosterone than for estradiol. Testosterone
is concentrated from the blood less highly than estradiol and
less specifically into basal forebrain structures (McEwen et al.,
1970a). Moreover, the competition effects of nonradioactive
estradiol are quantitatively more striking than those of testosterone (McEwen et al., 1970b). Finally, with cell fractionation
methods (Zigmond and McEwen, 1970), it is apparently easier to
demonstrate concentration of estradiol in the cell nucleus than
of testosterone.

If, indeed, estradiol binding in the limbic-hypothalamic
system is stronger than testosterone binding, this could be the
possible explanation of two well-established comparisons at the
behavioral level: (a) It is well known that much more testosterone
than estradiol must be injected, and for a longer time, to initiate
sex behavior effects in gonadectomized rats. These differences
may also reflect differing normal blood levels of the two hormones.
(b) In parametric behavioral observations of testosterone and
estradiol effects on male and female behavior in male and female
rats (Pfaff, 1970), the most specific and strongest causal
connection seems to be the estradiol effect on female behavior
in female rats. In the face of these comparisons in which
estradiol binding and estradiol's behavioral effects appear
more specific than those of testosterone, the similarity in the
pattern of retention between the two hormones still exists, and
may be the basis of the limited ability of the two to be substi-
tuted for each other during comparisons in parametric behavioral
experiments (Pfaff, 1970).

FUNCTIONAL EFFECTS OF ESTRADIOL ON ESTRADIOL-CONCENTRATING
NEURONS AND THEIR OUTPUTS

Although it is presumed that neurons which highly concentrate
estradiol - as described above - are functionally altered in a
meaningful way by the hormone, other techniques than those of
hormone-binding analysis are required to establish this fact. In
this context, estradiol retention in the corticomedial amygdala
and in other parts of the limbic-hypothalamic system pictured in
Figure 1 fulfills only one step in the research strategy summarized in Table 1. That estradiol-17β is effective in altering
pituitary output and female mating behavior is well established
(step A), and its concentration by cells in a limbic-hypothalamic

system is described above (step B).

TABLE I

Steps in the Physiological Analysis of Hormone Effects on Brain

A.   Determine chemical identity of a hormone involved in
     effects on pituitary or behavior.

B.   Determine if and where hormone is concentrated by cells
     in brain tissue.

C.   Search for and characterize physiological effects of
     the hormone in sites where it is concentrated.

D.   Investigate how the physiological effects of the hormone
     on individual neurons are related to the mechanism by
     which the hormone alters pituitary or behavioral function.

A functional role for estradiol in the cells which concentrate it is strongly suggested by the anatomical correspondence between the sites of estradiol-concentrating cells and the sites of estradiol implant effects. Implants of estradiol in the arcuate-ventromedial region of the hypothalamus cause ovarian atrophy in the rat (Lisk, 1960), while preoptic implants can trigger female sex behavior in ovariectomized female rats (Lisk, 1962; Pfaff, unpublished observations). Studies using other techniques such as lesioning and electrical stimulation (Sawyer, 1960; Lisk, 1967; Everett and Radford, 1961) also have implicated medial preoptic and hypothalamic regions which have peak numbers of estradiol-concentrating cells in the hormonal control of pituitary and mating behavior. Finally, effects of estradiol (Lincoln and Cross, 1967; Beyer, 1971), testosterone (Pfaff and Pfaffmann, 1969; Pfaff and Gregory, 1971) and progesterone (Barraclough and Cross, 1963; Komisaruk et al., 1967) on the electrophysiological activity of preoptic and hypothalamic neurons have been described and partially characterized. All of these studies demonstrate physiological effects of steroid sex hormones (step C in Table 1). However, none of them relate directly the hormone effects on individual neurons to the functional mechanisms by which the steroids influence pituitary function and mating behavior (step D in Table 1).

In order to attack this last problem, more information must be gathered about the place and participation of steroid-concentrating neurons in the neural circuits which control pituitary

function and mating behavior. Only with this information can the
effects of estradiol and other hormones on the electrical (or
chemical) activity cf single neurons be interpreted in terms of
those neurons' participation in the control circuits of interest.
With such information it may be possible to understand how estradiol effects on individual estradiol-concentrating neurons comprise part of the mechanism of estradiol's effects on pituitary
or behavioral function. This reasoning directs attention not
only towards hormone effects on brain function, but also towards
the neurophysiological properties of the neural circuits themselves which mediate the control functions to be studied. For
the investigation of pituitary control, attention is focussed on
the relatively short pathways between estradiol-concentrating
neurons in the preoptic-hypothalamic region and the pituitary,
while for the study of mating behavior longer pathways mediating
sensory input and motor output must be considered.

In this context, analysis of mating behavior mechanisms
includes the description of sensory and motor control pathways.
A convenient behavior to study is the lordosis reflex of female
rats, which has been chosen not only for the strength of its
control by estradiol but also for its relative reflexological
simplicity and stereotypy. Circuits mediating this reflex must
be present in the anestrous or ovariectomized female, since
lordosis can occasionally be observed in such animals (although
much less frequently than during estrus). Classically, therefore,
estradiol is conceived as modulating the activity of pre-existing
circuits to facilitate the occurrence of the lordosis response
following adequate stimulation from the male (Bard, 1940; Beach,
1948, 1967). The reflex itself - and the stimulation from the
male which triggers it - has been described by analysis of movie
films (Pfaff, unpublished observations). The elevation of the
female's rump and tail base has been identified as an early
component of the reflex and, from behavioral studies, appears
biologically important by way of facilitating intromission by
the male (Pfaff and Diakow, unpublished observations). Therefore,
rump elevation and tail movements proved to be a heuristic subject
for further physiological analysis. Points effective in stimulating rump and tail movements have been found in the lower brainstem, running in a ventrolateral position, lateral to the inferior
olive and ventral to the spinal trigeminal complex. In the midbrain, we have found some effective points just lateral to the
central grey, very near positions in which estradiol-concentrating
cells - the most posterior aspect of the limbic-hypothalamic
estrogen-concentrating system - have been demonstrated (Figure 1).
Studying the effects of estradiol administration and of preoptic-
hypothalamic manipulations on the operation of this brainstem
pathway will help to specify further the place of estradiol-concentrating cells in lordosis-control circuits.

REFERENCES

ANDERSON, C. H., & GREENWALD, G. S. Autoradiographic analysis
    of estradiol uptake in the brain and pituitary of the female
    rat. Endocrinology, 1969, 85, 1160-1165.

BARD, P. The hypothalamus and sexual behavior. In The Hypothala-
    mus. Research Publications of the Association for Research
    in Nervous and Mental Disease, 1940, 20, 551-579.

BARRACLOUGH, C., & CROSS, B. Unit activity in the hypothalamus
    of the cyclic female rat: Effect of genital stimuli and
    progesterone. Journal of Endocrinology, 1963, 26, 339-359.

BEACH, F. A. Hormones and Behavior. New York: Harper, 1948.

BEACH, F. A. Cerebral and hormonal control of reflexive mechan-
    isms involved in copulatory behavior. Physiological Reviews,
    1967, 47, 289-316.

BEYER, C. Changes in neuronal activity by estrogen in the female
    cat. In C. H. Sawyer and R. A. Gorsky (Eds.), Steroid
    Hormones and Brain Function. Berkeley: University of
    California Press, 1971, in press.

BRODAL, A. The amygdaloid nucleus in the rat. Journal of
    Comparative Neurology, 1947, 87, 1-16.

EVERETT, J. W., & RADFORD, H. M. Irritative deposits from stain-
    less steel electrodes in the preoptic rat brain causing
    release of pituitary gonadotropin. Proceedings of the
    Society for Experimental Biology and Medicine, 1961, 108,
    604-609.

KOMISARUK, B. R., McDONALD, P. G., WHITMOYER, D. I., & SAWYER,
    C. H. Effects of progesterone and sensory stimulation on
    EEG and neuronal activity in the rat. Experimental Neurology,
    1967, 19, 494-507.

KÖNIG, J. F. R., & KLIPPEL, R. A. The Rat Brain. Baltimore:
    Williams & Wilkins, 1963.

LINCOLN, D., & CROSS, B. Effect of oestrogen on the responsive-
    ness of neurones in the hypothalamus, septum and preoptic
    area of rats with light-induced persistent oestrus.
    Journal of Endocrinology, 1967, 37, 191-203.

LISK, R. D. Estrogen-sensitive centers in the hypothalamus of
    the rat. Journal of Experimental Zoology, 1960, 145,
    197-205.

LISK, R. D. Diencephalic placement of estradiol and sexual
    receptivity in the female rat. American Journal of
    Physiology, 1962, 203, 493-496.

LISK, R. D. Sexual behavior: hormonal control. In L. Martini
    and W. F. Ganong (Eds.), Neuroendocrinology, Vol. 2. New
    York: Academic Press, 1967. Pp. 197-239.

McEWEN, B. S., & PFAFF, D. W. Factors influencing sex hormone
    uptake by rat brain regions. I. Effects of neonatal treat-
    ment, hypophysectomy and competing steroid on estradiol
    uptake. Brain Research, 1970, 21, 1-16.

McEWEN, B. S., PFAFF, D. W., & ZIGMOND, R. E. Factors influencing
    sex hormone uptake by rat brain regions. II. Effects of
    neonatal treatment and hypophysectomy on testosterone up-
    take. Brain Research, 1970a, 21, 17-28.

McEWEN, B. S., PFAFF, D. W., & ZIGMOND, R. E. Factors influencing
    sex hormone uptake by rat brain regions. III. Effects of
    competing steroids on testosterone uptake. Brain Research,
    1970b, 21, 29-38.

McEWEN, B. S., WEISS, J. M., & SCHWARTZ, L. S. Retention of
    corticosterone by cell nuclei from brain regions of adrenal-
    ectomized rats. Brain Research, 1970, 17, 471-482.

PFAFF, D. W. Uptake of ^{3}H-estradiol by the female rat brain. An
    autoradiographic study. Endocrinology, 1968a, 82, 1149-1155.

PFAFF, D. W. Autoradiographic localization of radioactivity in
    rat brain after injection of tritiated sex hormones. Science,
    1968b, 161, 1355-1356.

PFAFF, D. W. Nature of sex hormone effects on rat sex behavior:
    specificity of effects and individual patterns of response.
    Journal of Comparative and Physiological Psychology, 1970,
    73, 349-358.

PFAFF, D. W., & GREGORY, E. Correlation between preoptic area
    unit activity and the EEG: difference between normal and
    castrated male rats. Electroencephalography and Clinical
    Neurophysiology, 1971, Vol. 31, pp. 223-230.

PFAFF, D. W., & PFAFFMANN, C. Olfactory and hormonal influences
    on the basal forebrain of the male rat. Brain Research,
    1969, 15, 137-156.

SAWYER, C. H. Reproductive behavior. In J. Field (Ed.), Hand-
    book of Physiology: Neurophysiology, Vol. II. Washington:
    American Physiological Society, 1960. Pp. 1225-1240.

STUMPF, W. E. Estradiol-concentrating neurons: topography in
    the hypothalamus by dry-mount autoradiography. Science, 1968,
    162, 1001-1003.

STUMPF, W. E. Estrogen-neurons and estrogen-neuron systems in the
    periventricular brain. American Journal of Anatomy, 1970,
    129, 207-218.

ZIGMOND, R. E., & McEWEN, B. S. Selective retention of estradiol
    by cell nuclei in specific brain regions of the ovariecto-
    mized rat. Journal of Neurochemistry, 1970, 17, 889-899.

REMARKS TO DR. PFAFF

W. E. Stumpf

In the above discussion paper of Dr. Pfaff (together with
M. Keiner), my publications are incorrectly quoted. Responding
to only a few misquotes: I have defined the term <u>periventricular</u>
<u>brain</u> (Stumpf, 1970a); it includes the phylogenetically older
parts of the mammalina brain such as the amygdala, the hippocampus,
the olfactory tubercle. The term periventricular brain is preferred over the variably defined "limbic system" or "rhinencephalon"; it reflects a more unifying concept and hints at the neuroendocrine importance of the ventricular system.

I never stated that estradiol uptake was limited to preoptic
and hypothalamic structures. In 1968, I published about estradiol
localization in the diencephalon. Only this part of the brain was
studied carefully at that time. In 1969, I published, together
with M. Sar, detailed localization in the amygdala and, in 1970,
I published a map of the rat brain and reported that estradiol is
concentrated in neurons "<u>in defined parts</u>" of the periventricular
brain, and not in the caudate nucleus as Dr. Pfaff erroneously
mentioned above.

If Dr. Pfaff would provide a map of estradiol concentrating
cells in the brain according to his published work, which he
still accepts as valid, everything would be dotted in his map.
He stated (1968):"...grains were found regularly over diverse
types of neurons and glial cells throughout the brain. No
evidence was seen for exclusive uptake by or absence of uptake
from any particular type of nerve cell or glial cell." We have
provided evidence that this reflects redistribution artifacts.
The work of Dr. Pfaff, in which he ignored earlier published
progress and warnings in the field of the autoradiography of
diffusible substances, was criticized by myself in an article
in <u>Science</u> (1969). This article was specifically aimed at such
published work as Dr. Pfaff's in order to alert editors and
reviewers of journals and to reduce the acceptance of papers--and
the confusion caused by them--with results based on non-controlled
techniques. Dr. Pfaff's work also was criticized and could not be
reproduced by P. Tuohimaa, Finland (1970); A. Attramadal, Norway
(1970); and Anderson and Greenwald, U.S.A. (1969).

It is to be noted that Dr. Pfaff now has changed and improved his autoradiographic technique. The technique he uses now has been reported and described as "technique no. 2"--and later as "thaw-mount" procedure--by W. E. Stumpf and L. J. Roth (1966) and W. E. Stumpf (1970b). The difference is, however, that Dr. Pfaff cuts his tissue at -16° to -19°C, instead of -40°C or below, and that his sections are 8μ or 12μ thick, instead of 1μ to 3μ. These changes reduce considerably the resolution of the autoradiograms and limit the utility of the thaw-mount procedure, due to destructive ice crystal growth, increased thaw-diffusion, and superimposition of structures (Stumpf, 1968b).

The schematic drawings provided now by Dr. Pfaff appear to approach the ones provided by myself two and four years ago.

I am restating: contrary to Dr. Pfaff's previous reports, there is strong evidence for a high selectivity in the localization and retention of estradiol in certain nerve cells in certain areas of the rat brain, but not in the whole brain and also not in the whole "limbic system," not in glia cells and not in ependyma cells.

REFERENCES

ANDERSON, C. H., & GREENWALD, G. S. Autoradiographic analysis of estradiol uptake in the brain and pituitary of the female rat. Endocrinology, 1969, 85, 1160-1165.

ATTRAMADAL, A. The uptake of ^{3}H-oestradiol by the anterior hypophysis and hypothalamus of male and female rats. Zeitschrift fur Zellforschung und Mikroskopische Anatomie, 1970, 104, 582-596.

PFAFF, D. W. Autoradiographic localization of radioactivity in rat brain after injection of tritiated sex hormones. Science, 1968a, 161, 1355-1356.

STUMPF, W. E. Estradiol concentrating neurons: topography in the hypothalamus by dry-mount autoradiography. Science, 1968a, 162, 1001-1003.

STUMPF, W. E. High-resolution autoradiography and its application to In Vitro experiments: subcellular localization of ^{3}H-estradiol in rat uterus. In R. L. Hayes, F. A. Goswitz, and B. E. P. Murphy (Eds.), Radioisotopes in Medicine: In Vitro Studies. Oak Ridge, Tennessee: AEC Symposium Series No. 13 (CONF-671111), 1968b. Pp. 633-660.

STUMPF, W. E. Too much noise in the autoradiogram? Science, 1969, 163, 948-959.

STUMPF, W. E. Estrogen-neurons and estrogen-neuron systems in
the periventricular brain. American Journal of Anatomy,
1970a, 129, 207-218.

STUMPF, W. E. Localization of hormones by autoradiography and
other histochemical techniques, a critical review. Journal
of Histochemistry and Cytochemistry, 1970b, 18, 21-29.

STUMPF, W. E., & ROTH, L. J. High resolution autoradiography with
dry-mounted, freeze-dried, frozen sections. Comparative study
of six methods using two diffusible compounds, ^{3}H estradiol
and ^{3}H mesobilirubinogen. Journal of Histochemistry and
Cytochemistry, 1966, 14, 274-287.

STUMPF, W. E., & SAR, M. Distribution of radioactivity in hippocampus and amygdala after injection of ^{3}H estradiol by dry-
mount autoradiography. Physiologist, 1969, 12, 368 (Abstract).

TUOHIMAA, P. Radioautography of tritiated sex steroids in the rat.
Histochemie, 1970, 23, 349-357.

MOLECULAR BIOLOGY

EFFECTS OF AMYGDALOID LESIONS ON HYPOTHALAMIC MACROMOLECULES

Basil E. Eleftheriou

The Jackson Laboratory

Bar Harbor, Maine

INTRODUCTION

The role of the amygdaloid nuclear complex in endocrinology did not become apparent until the late 1950s when a number of detailed studies investigated the relationship of electrical stimulation and electrolytic lesions of the amygdala on sexual behavior, fear, and arousal. As a result of these studies, it was found that electrical stimulation of the amygdala in female rats, rabbits, cats, and dogs produces ovulation and increased uterine contractility (Bunn and Everett, 1957; Koikegami, Yamada, and Usui, 1953; Shealy and Peele, 1957). The major psychoendocrine effect of lesions in the amygdala is the disruption of maternal behavior (Masserman et al., 1958; Walker, Thompson and McQueen, 1953). One of the unexplained phenomena has been the differential effects of amygdalectomy in the two sexes. Paradoxically, although amygdaloid lesions cause more hypersexuality in males than in females, the effects on the genital organs show the opposite picture. Amygdalectomy in adult male rats and cats causes a significant degeneration of the testes, whereas in adult female cats the ovaries remain unaffected (Greer and Yamada, 1959; Kling et al., 1960; Yamada and Greer, 1960), and in the young female cat, amygdalectomy causes a precocious development of the ovaries, uterus, and vagina (Elwers and Critchlow, 1961; Lundberg, 1962).

The effect of amygdalectomy on the endocrine system appear to be generally depressed in rats, dogs, and goats, but primarily if the removal is made early in life (Koikegami et al., 1953; Koikegami et al., 1955). Amygdalectomized animals have impaired

793

growth and general atrophy of the parathyroid, thyroid, pituitary, adrenal, and pancreatic glands. If amygdaloid lesions are placed in adult animals (rat, rabbit, cat, and monkey), the endocrine effects are less severe. Apart from testicular degeneration, the only gland which is much affected is the adrenal gland. The actual weight of the adrenal gland does not change appreciably except for a small increase during the first week or two post-operatively (Greer and Yamada, 1959; Kling et al., 1960; Yamada and Greer, 1960). However, Martin et al. (1958) found that amygdalectomy in cats and dogs dramatically increased the level of adrenal corticoids in the blood, and sometimes a new type of corticoid was produced.

Other authors have concerned themselves with the adrenal response to stress in amygdalectomized animals. Knigge (1961) found that amygdalectomy in rats caused a delay in the cortico-sterone response to immobilization. Mason, Nauta, Brady, and Robinson (1959) found that the 17-hydroxycorticosteroid response to avoidance training was abolished by total amygdaloid removal. This type of result would be expected from the ensuing placidity. However, Anand et al. (1957b) report the eosinophil response to subcutaneous injection of saline is not affected by lesions including the amygdala, and Bovard and Gloor (1961) found that small lesions in the central nucleus of the amygdala increased the corticosterone response to immobilization in rats. This is the location where Wood (1958) found that small lesions increased aggressiveness so perhaps this latter finding is not really a contradiction of the other studies.

Electrical stimulation of the amygdala has been shown to cause an increase in the production of 17-hydroxycorticosteroids. Mason (1958, 1959) has shown this in monkeys and Mandell et al. (1963) obtained similar results in humans using an intensity which was too low to produce any detectable behavioral or sub-jective changes.

The other major endocrine effect is for the stimulation to cause an increase in gastric secretions (Anand and Dua, 1956c; Sen and Anand, 1957; Shealy and Peele, 1957). Gastric and intestinal movements may be either increased or decreased (Eliasson, 1952; Gastaut, 1952; Koikegami, Kimoto and Kido, 1953; Koikegami, Kushiro and Kimoto, 1953; Shealy and Peele, 1957). Repeated amygdaloid stimulation has been shown to cause gastric bleeding and ulceration of the gastroduodenal junction (Sen and Anand, 1957). Bleeding also has been observed in the lungs (Koikegami and Fuse, 1952a, 1952b). Such changes are suggestively similar to the psychosomatic diseases that are supposed to result from psychological stress.

Generally, the amygdala must now be accepted as a neuro-endocrine modulator, especially of hypothalamic neurohumoral and tropic releasing factor activity (see chapters by Sawyer, Zolovick, Stumpf, and Hayward in this volume). Possibly, one of the set-backs in this connection has been the conflicting results obtained in the various stimulation studies. However, it must be kept in mind that all experimenters have not used identical power instru-mentation or identical and minimal current discharge. Thus, although we should be very cautious about accepting any functional localization within the amygdala before extensive replication of the research work, we cannot deny that any such localization exists within this nuclear region. The latter alternative is re-inforced by the singularly outstanding lack of contradiction among experimenters using lesion techniques. Whether this is due to the limited number of studies, to the species used, or to a reliable topographic organization of function within the amyg-dala is not clear at this time. However, it must be kept in mind that it is usually easier to place discrete lesions than to propagate artificial current through nervous tissue.

My own work was begun with the thought of clarifying some of the topographic localization of neuroendocrine functions within the amygdala, its relationship to hypothalamic control of the hypophysis, and of introducing a new species, the deermouse (Peromyscus maniculatus bairdii), in hopes of broadening the functional generalization. After considerable effort, on my part, to demonstrate the modulating effects of the amygdala on hypo-physeal tropic hormone secretion by the use of lesions, Velasco and Taleisnik (1969) summarized existing experimental data to support the view that the amygdala exerts a stimulatory effect on gonadotropin secretion. My own work demonstrated clearly an inhibitory control over gonadotropin secretion. These authors (Velasco and Taleisnik, 1969) contended that our findings and those of Elwers and Critchlow (1960, 1961), who maintain that the amygdala is inhibitory, "can be interpreted as the result of stimulation by iron deposition rather than suppression of the amygdala by a destructive lesion." However, it must be kept in mind that, in acute lesion studies, utilizing stainless steel electrodes, the deposition of iron probably induces a rapid ovulation-inducing discharge of gonadotropin (if one assumes a priori that the amygdala has a stimulatory role). However, our work and that of Elwers and Critchlow (1960, 1961) and Lawton and Sawyer (1970) represent long-term, chronic studies in which the prolonged effect of iron deposition is questionable. I am unaware of any published reports which demonstrate chronic stimulatory effects of iron, copper or other metallic deposits. The observation, in all studies (Elwers and Critchlow, 1960, 1961; Lawton and Sawyer, 1970; Eleftheriou and Zolovick, 1967; Eleftheriou, Desjardins and Zolovick, 1970), that sham lesions

significantly stimulated LH synthesis and release rules out
electrochemical stimulation of the amygdala and favors strongly
the alternative hypothesis--destruction of an inhibitory central
neural site. Finally, it should be emphasized that we have used
not only stainless steel electrodes but also platinum electrodes
with similar results.

The topographic localization of hypophyseal tropic hormone
secretion within the amygdala, the neuroendocrine interrelationships to the hypothalamus, demonstrated by myself, and its
general significant role in neuroendocrine feedback mechanisms
are discussed extensively by Zolovick, Stumpf and Hayward elsewhere in this conference, and I shall not go into these. However, I should like to emphasize the fact that my work has
demonstrated both inhibitory and stimulatory influences on
hypophyseal secretion depending on particular tropic hormones
and particular location of lesions within the amygdaloid complex.
The latter finding demonstrates clearly the extent, versatility,
and diversity of the amygdala impressed upon hormonal secretion
of the hypophysis, through the hypothalamus, and the significant
modulating role of this brain region. We now are at a juncture
in neuroendocrinology where we must accept the amygdala as an
important neuroendocrine regulator possessing behavioral and
neuroendocrine functions similar to those of the hypothalamus.
The works of Stumpf and of Sawyer, presented previously in this
conference, emphasize this view and support my previous work,
which indicates specific topographic localization of classic
endocrine feedback mechanisms for adrenal and gonadal steroids
and regulation of their respective hypothalamic neurohumoral
releasing factors and hypophyseal tropic hormones.

Recently, I decided to abandon the classic approach of
placing lesions in various amygdaloid nuclear groups followed by
measurement of either hypothalamic releasing factor activity or
hypophyseal as well as blood tropic hormone activity. I felt
that this latter approach would not yield an answer to the mode
of mediation of amygdaloid neuroendocrine influence on hypothalamus and hypophysis. Rather, such an answer would be obtained
by combining the approaches of Stumpf, Raisman, and Hayward. However, a specific part of the total answer may be realized by a
molecular approach combining lesion and stimulation techniques
with analyses of specific RNA molecules within distinct areas of
the hypothalamus. Ideally, it would be of great significance to
demonstrate that a discrete lesion or electrical stimulation of
a distinct nuclear group within the amygdala gives rise to increased production of a particular species of RNA in a distinct
area of the hypothalamus. Following such demonstration, one may
ultimately synthesize in vitro a specific protein by combining
specific amino acids and a specific RNA under the appropriate

conditions. Such protein may be shown ultimately to be an intermediate or final product of an hormonal-releasing factor, thus
demonstrating the topographic interrelationship between amygdala
and hypothalamus at the neuromolecular level. Although such an
approach is somewhat ambitious and accompanied by great problems
in the chemical techniques, nevertheless, it is worthy of persistent pursuit.

    Essentially, this is my present approach and the work I am
about to present represents the initial and somewhat embryonic
phases of this work.

MATERIALS AND METHODS

    Initially, my work involved the isolation of RNA from various
brain regions and the testing of this substance for chemical purity, optical density at different wave lengths, as well as concentration and RNA base composition. These data have appeared
previously and will not be discussed here (Eleftheriou, 1971;
Eleftheriou et al., 1970).

    In all instances, the biochemical techniques employed in
various phases of my work are those that have been published
previously by a number of investigators and now have become
acceptable and rather routine for neurochemical analyses. For
this reason, I shall not go into them.

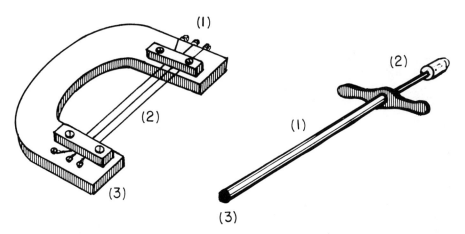

Fig. 1. (A) Brain slicer made of plexiglass yoke (3) with 3 stainless steel wires (2) whose tension is controlled by screws (1);
(B) Trochar needle (1) with grip and plunger (2). Diameter and
shape of bore may vary (3) according to particular needs.

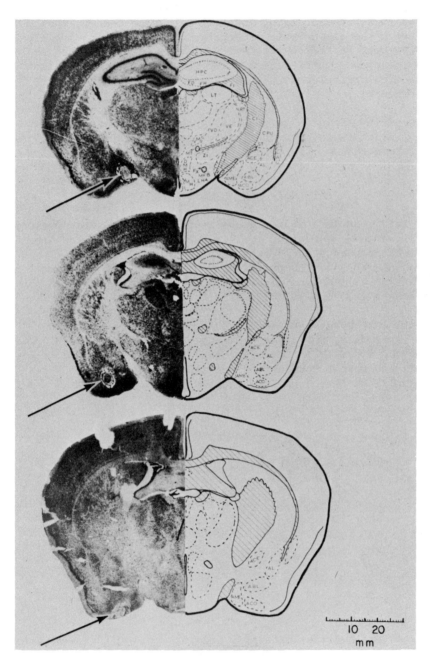

Fig. 2. Photomicrographic and schematic representation of the diencephalon of P. m. bairdii indicating location of various lesions.

However, one detail of our procedure need slight elaboration. This detail involves the dissection technique for taking out major regions such as the hypothalamus as well as the taking out of regions within the hypothalamus. Figure 1A represents the "brain slicer." This is an instrument composed of a "yoke" made of plexiglass which has several stainless steel wires strung across it. The size of the yoke and the number of wires may vary according to the size of the brain of a particular species or whether one is dealing with a young or adult brain. At the time of sacrifice, animals are killed by cervical dislocation, the brain is recovered rapidly and placed on a glass slide which rests on an aluminum box containing either liquid nitrogen or dry ice and acetone, depending on the demands for freezing the tissue for a particular analysis. The slicer is used so that the most anterior wire is placed at the anterior tip of the frontal neocortex (posterior to the olfactory lobes) and pressed so that one obtains slabs of brain in an anterior posterior order sequence. Figure 1B represents modified trochar needles with plungers which then are used to remove particular brain regions. An appropriate trochar is used depending on the size of the tissue to be retrieved. For example, the trochar for the removal of the hippocampus is somewhat semilunar. For removal of particular regions within the hypothalamus, the trochar is small in comparison to the trochar for removal of the entire hypothalamus. This method gives us an excellent procedure for consistency in the removal of brain regions from one animal to the next. In addition, it is very rapid in a situation where rapid removal of tissue is critical. It is true that the method does not ensure 100 per cent removal of a particular region, but there is about 100 per cent reliability in the removal of identical tissue samples of a particular brain region within a certain brain size. This technique of removal of brain regions has been used in our laboratory for a number of years and there is significant reliability in the tissue removed.

RESULTS

The initial molecular study dealing with effects of amygdaloid lesions was carried out using the rat as the experimental animal (Eleftheriou et al., 1969). Subsequent work was carried out on the deermouse (Peromyscus maniculatus bairdii), and the location of lesions is represented in Figure 2. The basolateral (ABL), medial (AME), and cortical (ACO) amygdaloid nuclear complexes were chosen because these three were the nuclear groups whose damage gave us the most consistent and divergent results on the endocrine system (Eleftheriou and Zolovick, 1966a; Eleftheriou et al., 1966b; Eleftheriou and Zolovick, 1967; Eleftheriou et al., 1967; Eleftheriou and Pattison, 1967; Eleftheriou, 1967, 1970; Eleftheriou and Zolovick, 1968; Eleftheriou et al., 1969, 1970).

Table 1. *Effects of lesions in the amygdaloid nuclei, implants of actinomycin D or cholesterol on pituitary RNA base % composition* ✱

Site of lesion or treatment	Nucleotides (Means ± s.d.)				G+C:A+U ratio	A+G:C+U ratio
	AMP	CMP	GMP	UMP		
Controls	18·10 ± 0·63	26·00 ± 0·88	38·52 ± 0·62	17·36 ± 0·22	1·81	1·23
Basolateral	17·23 ± 0·18	22·13 ± 0·71	39·40 ± 0·50	21·22 ± 0·21	1·60	1·30
Subtotal amygdalectomy	6·49 ± 0·32	18·32 ± 0·20	35·72 ± 0·66	38·92 ± 1·20	1·19	0·73
Medial	11·19 ± 0·23	21·11 ± 0·82	24·62 ± 1·09	43·08 ± 1·22	0·84	0·55
[³H]Actinomycin D	20·00 ± 0·32	22·17 ± 0·41	32·00 ± 0·66	25·82 ± 0·18	1·18	1·09
Cholesterol	19·82 ± 1·32	23·25 ± 0·62	30·72 ± 1·20	26·18 ± 0·16	1·17	1·02

Abbreviations: AMP = adenosine monophosphate; CMP = cytidine monophosphate; GMP = guanosine monophosphate; UMP = uridine monophosphate; G+C:A+U = guanosine+cytidine:adenosine+uridine ratio; A+G:C+U = adenosine+guanosine:cytidine+uridine ratio.

Table 2. *Effects of lesions in the amygdaloid nuclei, implants of actinomycin D or cholesterol on hypothalamic RNA base % composition*

Site of lesion or treatment	Nucleotides (Means ± s.d.)				G+C:A+U ratio	A+G:C+U ratio
	AMP	CMP	GMP	UMP		
Controls	15·06 ± 0·88	23·18 ± 2·15	38·19 ± 1·29	23·56 ± 1·09	1·58	1·14
Basolateral	7·77 ± 0·42	17·92 ± 0·76	40·52 ± 2·30	33·76 ± 0·62	1·40	0·93
Subtotal amygdalectomy	3·80 ± 0·92	14·18 ± 0·82	43·92 ± 1·96	38·08 ± 0·62	1·38	0·91
Medial	27·71 ± 1·23	14·94 ± 0·55	14·42 ± 0·76	51·61 ± 0·44	0·37	0·63
[³H]Actinomycin D	19·00 ± 0·76	20·27 ± 0·86	35·23 ± 0·84	25·48 ± 0·86	1·24	1·19
Cholesterol	15·37 ± 0·92	19·43 ± 1·42	40·43 ± 1·09	24·76 ± 1·07	1·49	1·26

Table 3. *Effects of lesions in the amygdaloid nuclei, implants of actinomycin D or cholesterol on frontal cortical RNA base % composition*

Site of lesion or treatment	Nucleotides (Means ± s.d.)				G+C:A+U ratio	A+G:C+U ratio
	AMP	CMP	GMP	UMP		
Controls	18·91±0·26	23·06±1·32	39·56±0·88	18·44±0·82	1·67	1·40
Basolateral	16·73±0·42	21·49±0·23	39·08±0·20	22·69±0·76	1·53	1·27
Subtotal amygdalectomy	12·81±0·76	17·87±0·86	36·80±0·32	34·07±0·20	1·16	0·95
Medial	6·82±0·68	7·24±0·72	23·34±0·46	64·69±0·93	0·42	0·41
[³H]Actinomycin D	19·48±0·32	21·45±0·44	36·25±1·20	22·81±1·06	1·36	1·27
Cholesterol	16·68±0·44	27·41±2·42	35·08±1·33	27·57±0·23	1·41	0·93

Table 4. *Effects of lesions in the amygdaloid nuclei, implants of actinomycin D or cholesterol on cerebellar RNA base % composition*

Site of lesion or treatment	Nucleotides (Means ± s.d.)				G+C:A+U ratio	A+G:C+U ratio
	AMP	CMP	GMP	UMP		
Controls	20·41±0·76	24·98±0·62	37·62±0·20	16·96±0·20	1·07	1·38
Basolateral	19·06±0·32	17·58±0·32	39·29±0·77	24·05±0·62	1·31	1·40
Subtotal amygdalectomy	12·81±0·63	16·90±1·09	35·52±0·23	34·77±0·89	1·10	0·93
Medial	—	—	—	—	—	—
[³H]Actinomycin D	23·78±0·71	16·94±0·16	34·19±0·33	25·07±0·62	1·04	1·37
Cholesterol	19·52±0·92	25·79±0·86	33·32±2·60	21·35±0·88	1·44	1·12

*Tables 1 - 4 reprinted from Eleftheriou et al., Journal of Endocrinology, 45, 207-214, 1969.

Tables 1 through 4 summarize the results obtained in the rat. The ribonucleic acid (RNA) obtained by extraction with cold phenol and ethanol precipitation from rat brain gave an absorption curve typical of nucleic acids. The maximal absorption was at 230 nm. The 280:260 nm. ratio was 0.48, thus demonstrating the purity of the product. These results agree well with those of Popa et al. (1967) who found the maximal absorption of mouse brain RNA at 258 nm., minimal absorption at 234 nm., and the 280:260 nm. ratio equal to 0.47. The most significant effect of these treatments occurred in the pituitary RNA (Table 1) base percentages and ratios after subtotal amygdalectomy and medial nuclear lesions. In all cases, AMP and UMP showed the greatest change. The guanosine + cytidine:adenosine + uridine (G + C:A + U) ratio showed the most significant decline (P <0.001) after lesions had been placed in the medial amygdaloid complex, although a decrease in this ratio occurred with all treatments, including cholesterol implants.

The hypothalamic G + C:A + U ratio (Table 2) showed the most significant (P <0.001) decrease to 0.37 from a control level of 1.58 after medial lesions. The change in the ratio of the baso-laterally lesioned animals and rats implanted with cholesterol was not significant. Generally, the overall changes in the AMP and UMP fractions were the most significant (P <0.01).

In the frontal cortex, no significant changes occurred in the G + C:A + U ratios after lesions were placed in the baso-lateral amygdaloid complex or after cholesterol implants (Table 3). Significant changes (P <0.01) occurred after partial amygdalectomy, actinomycin D implants, and after lesions had been placed in the medial amygdaloid complex with the greatest reduction to 0.42 from a control value of 1.67 after the latter treatment.

The effects of all treatments on the cerebellum were negative (Table 4. Unfortunately, the cerebellar regions of all animals with medial lesions had thawed and could not be used for analysis. Generally, some significant changes did occur in isolated instances such as in AMP levels after subtotal amyg-dalectomy, and levels of CMP after actinomycin D implants, and after lesions had been placed in the basolateral area or after subtotal amygdalectomy. Some changes also occurred in UMP. However, they were insignificant in relation to the overall effects.

Following this experiment, in the rat, we applied the micro-dissection technique to the hypothalamus of P. m. bairdii. In order to test the accuracy of dissecting areas within the hypo-thalamus, we conducted a pilot experiment on the incorporation of uridine-H^3 into RNA in males that were either intact or castrated or castrated and subsequently treated with testosterone

Table 5.

Uridine-H incorporation into RNA in intact males and in males after 7 or 14 days of castration or castration with replacement therapy of testosterone propionate. Region involved is indicated. Radioactivity is expressed as cpm/mg of tissue for unincorporated uridine-H^3 divided by cpm/mg of tissue of uridine-H^3 incorporated into RNA ± standard error.

Treatment	N	Pituitary	HYPOTHALAMUS					
			n. mamillary	n. posterior	n. preoptic	n. ventromedial	n. dorsomedial	n. paraventricularis
Intact ♂	10	2.33±0.18	2.73±0.21	2.34±0.08	2.25±0.15	2.61±0.13	2.74±0.19	2.63±0.26
Castrated ♂ (7 days)	10	4.43±0.08*	2.85±0.07	2.76±0.14	3.74±0.19*	3.89±0.11*	3.17±0.09*	3.03±0.14
Castrated ♂ (14 days)	10	4.35±0.26*	3.30±0.15	3.21±0.10*	3.65±0.24*	3.89±0.20*	3.26±0.16*	3.46±0.29
Castrated + TP	10	2.58±0.15	2.87±0.13	2.71±0.10	2.54±0.12	2.67±0.10	2.69±0.10	3.00±0.17

*Comparison to intact. Level of significance = <0.05.

Table 6.

DL-leucine-C^{14} incorporation into acid precipitable fractions in the pituitary and various hypothalamic regions of male P. m. bairdii after amygdaloid lesions (± SE) at 19 days after the operation.

Treatment	N	Pituitary	HYPOTHALAMUS					
			n. mamillary	n. posterior	n. preoptic	n. ventromedial	n. dorsomedial	n. paraventricularis
Sham-operated	5	41.53±2.29	32.55±2.11	37.55±3.58	37.21±2.45	36.29±2.78	40.43±2.74	37.73±4.24
AME lesions	5	62.05±3.42*	50.61±2.16*	43.75±2.54	68.69±3.87*	39.83±1.17	38.18±2.35	63.99±2.26*
ACO "	5	61.11±4.49*	36.31±1.67	40.55±1.80	59.77±3.71*	52.97±2.15*	36.83±2.40	41.67±2.39
ABL "	5	66.45±2.35*	36.77±1.55	38.68±2.15	64.79±2.29*	68.57±1.99*	38.07±3.11	39.97±1.58

*Level of significance = < 0.05. Comparison to sham-operated.

propionate (TP). From the data obtained (Table 5), it appeared
that there was considerable reliability in the technique applied
for the removal of the various hypothalamic regions. This was
based on replicability of the biochemical analyses and statistical
analyses of the data. It should be pointed out that least reliable removal was noted with the removal of the nucleus paraventricularis. The very best reliability of removal was obtained with
the preoptic and the ventromedial nuclei.

Once the reliability of this technique was established, we
conducted an experiment on the incorporation of uridine-H^3 into
RNA and the incorporation of DL-leucine-C^{14} into acid precipitable
fractions in the pituitary and various hypothalamic regions following the placement of lesions in the medial (AME), basolateral (ABL)
and cortical (ACO) amygdaloid nuclear groups. These data are
summarized in Tables 6 and 7. In all instances, the injection of
the isotopically labeled material, appropriately buffered, was
made intracisternally in the total amount of 0.01 ml, and the
animals were killed 60 minutes after injections.

The incorporation of DL-leucine-C^{14} into acid precipitable
fractions increased significantly ($P < 0.05$) in the pituitary gland
and the preoptic hypothalamic region in all treatments. While
animals bearing AME lesions exhibited increases in the mammillary
and paraventricular nuclear regions, animals with ACO and ABL
lesions exhibited significant increases in the ventromedial nuclear
region (Table 6). With the exception of the increases in the preoptic region, there was no overlap of effects in the other hypothalamic nuclear groups. The posterior hypothalamus and the
dorsomedial nucleus did not exhibit any significant changes.

Generally, the incorporation of uridine-H^3 into RNA followed
a similar pattern to that of DL-leucine-C^{14} incorporation into
acid precipitable fraction (Table 7). However, in only one
instance, after AME lesions, was there a significant incorporation
of radioactively labeled uridine into RNA of the pituitary. This
indicates that either incorporation did not take place with the
other treatments or, more reasonably, the time sequence is different in animals bearing lesions in different amygdaloid nuclear
groups. At any rate, there is a great need for a time study to
arrive at any meaningful assumptions.

                                DISCUSSION

Since this was the first attempt at correlating effects of
brain lesions with RNA base ratios in other regions of the brain,
some of the significant changes may be difficult to interpret due
to the need for additional information. Although the data are

Table 7.

Uridine-H^3 incorporation into RNA in various hypothalamic regions of P. m. bairdii after amygdaloid lesions. Radioactivity expressed in cpm/mg of tissue for unincorporated uridine-H^3 divided by cpm/mg of tissue of uridine-H^3 incorporated into RNA ± standard error. Animals killed on day 19 after the operation

Treatment	N.	Pituitary	n. mamillary	n. posterior	n. preoptic	n. ventromedial	n. dorsomedial	n. paraventricularis
						HYPOTHALAMUS		
Sham-operated	8	2.42±0.16	2.72±0.13	2.68±0.13	2.52±0.06	2.85±0.08	2.64±0.16	2.46±0.11
AME lesions	8	1.36±0.14*	1.51±0.11*	2.58±0.09	1.29±0.07*	2.58±0.08	2.20±0.12	1.18±0.06*
ABL "	8	2.20±0.07	2.90±0.09	2.67±0.14	1.14±0.08*	1.10±0.05*	2.60±0.09	2.67±0.10
ACO "	8	2.17±0.05	2.27±0.08*	2.60±0.09	1.73±0.09*	2.03±0.08	2.68±0.09	2.52±0.09

*Level of significance < 0.05. Comparison to sham-operated group.

being collected, the microchemical techniques for brain analyses
of RNA are not sufficiently refined to permit rapid accumulation
of the necessary additional information. Nevertheless, the
present results allow some generalizations. The persistent
decline in the $G + C:A + U$ ratio found in all areas examined,
except the cerebellum, in animals that had medial lesions is of
some significance inasmuch as the new RNA synthesized undoubtedly
was of the messenger type (Hydén and Egyházi, 1964). All treatments uniformly produced a decline in the ratio in the pituitary
indicating that the adenohypophysis responded by an increased
synthesis of m-RNA, showing that any disturbance in the amygdala
is reflected in this gland.

Since it is known, from our previous results, that the treatments used in the present study result in the release of a number
of tropic hormones, it may be assumed that the pituitary m-RNA
synthesized under the influence of each treatment reflects m-RNA
for a particular tropic hormone released as a consequence of
amygdaloid lesions. The obvious next step is to characterize
the m-RNA involved and, if possible, to associate it with the
synthesis and/or release of a particular tropic hormone.

Generally, the $A + G:C + U$ ratios decreased significantly
after total amygdalectomy and after lesions in the medial amygdaloid area. Thus, the decline in the ratio of $G + C:A + U$ after
the various lesions would support the hypothesis that some of
these operations resulted in stimulation of a genomic nature in
the various neural networks. The pituitary effects may reflect
the profound effects in the hypothalamus which probably are
accompanied by a release of various releasing factors. In
instances when the pituitary was affected by releasers, the RNA
produced in nervous tissue was uniformly deficient in AMP and
CMP but exceptionally high in UMP. Whether this reflects
synthesis of a particular type of RNA responsible for the synthesis of releasing factors or their release is difficult to establish since the m-RNA or r-RNA have not been isolated for these
compounds.

Lesions in the medial amygdaloid complex had a more profound
effect than similar lesions in the basolateral region. We now
have obtained electrophysiological data (Zolovick, 1968) which
indicate that hypothalamic unit activity is changed significantly
on medial amygdaloid stimulation, and that whatever changes occur
as a result of stimulation of all other amygdaloid areas, usually
they are mediated by way of the medial amygdaloid complex.
Furthermore, it was found that the lateral and basolateral
amygdaloid areas exert an inhibitory effect on the medial area
and this inhibition can be removed by electrical stimulation of
the medial nuclear complex. Thus, we must assume that, because

of its relationship to the hypothalamus, the medial amygdaloid complex mediates or modulates all regulatory influences arising in, or passing through, the amygdala to the ultimate goal, the hypothalamus.

The finding that lesions placed in the basolateral amygdaloid complex only, and not in the medial, result in LH release does not conflict with the results of previous studies. Lesions placed in the medial area already have been shown to result in continuous release of ACTH (Eleftheriou et al., 1966). Thus, the assumption can be made that, depending on the site of the lesion, the result probably is a stimulation of a genomic character resulting in the production of RNA with highly specific base ratios in the neurons immediately involved. In this manner, lesions in the basolateral-lateral complex result in alterations of synthesis of LH-RF in the hypothalamus, while lesions in the medial amygdaloid complex result in alterations of CRF synthesis. One may hazard an hypothesis and propose that the changes in areas other than the hypothalamus may reflect neurophysiological disturbances in the balance of some neurohumors, perhaps serotonin and noradrenaline, which undoubtedly occur as a consequence of the lesions placed in the amygdala.

Indeed, it is somewhat unfortunate that, at this stage of the research, the incompleteness of the data allows for such great assumptions. It is hoped that we will obtain shortly additional information to further clarify our initial data. The present data, however, appear encouraging. Taken together with data from electrophysiological, cytochemical and neuroendocrine experiments, the molecular data tend to support the early thesis that certain amygdaloid nuclear groups exert, directly or indirectly, an influence on specific hypothalamic nuclear groups. The nature of the molecular changes within the hypothalamus is not clear at this time. Presently, these changes may reflect any number of events dealing with simple electrophysiological conduction to more complex events such as enzyme induction, protein synthesis or modification of neural encoding patterns within certain brain regions. Information now being gathered should begin to enlighten this situation of amygdaloid-hypothalamic molecular interrelationship.

Figure 3 represents an initial attempt to summarize and coordinate the various interrelationships of the amygdala with other brain regions and, particularly, the hypothalamus. This working model attempts to bring together information from the various fields that have contributed to the understanding of the amygdaloid nuclear complex, but emphasizes mainly the neuroendocrine information and interrelationships.

WORKING MODEL DEMONSTRATING THE EFFECTS OF EXPERIMENTAL MANIPULATION AND ENVIRONMENTAL INFLUENCES MEDIATED

THROUGH THE AMYGDALA AND REFLECTED ON OTHER BRAIN AREAS AND THE HYPOPHYSEAL-ADRENAL-GONADAL AXES

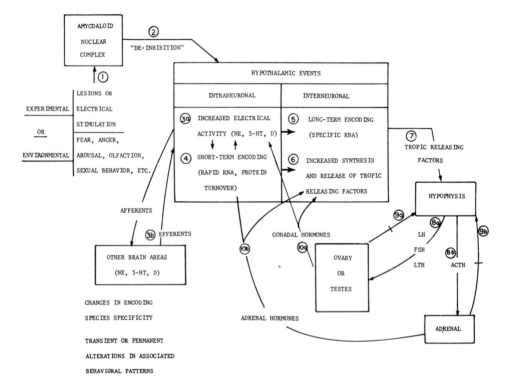

For example, environmental or experimental manipulations
(1) which affect amygdaloid function result, directly or indirectly, upon electrical changes in the hypothalamus (2) which are
accompanied by changes in concentration, binding, turnover, and
inactivation of the biogenic amines, norepinephrine (NE), serotonin (5-HT) or dopamine (D), and, certainly, acetylcholine (3a).
Based on the majority of the available data, the assumption is
made that the amygdala exerts an inhibitory influence on the
hypothalamus so that, upon experimental or environmental manipulation, this inhibition is eliminated or "de-inhibited."
Almost simultaneous to the hypothalamic effect, there may occur
through various pathways a reciprocal influence on other brain
regions (3b). The increased electrophysiological and neurochemical activities may result in short-term neuromolecular events
(4) which may contribute to transient (tonic) discharges of
hypothalamic releasing factors (6) or permanent (5) and highly
specific hypothalamic activity contributing to long lasting
release of the tropic releasing factors (7). For the sake of
brevity, I have omitted other effects. Based on available data,
we now know that the various releasing factors affect hypophyseal
secretions especially of gonadotropins (8a) and adrenocorticotropin (8b) which through their respective target organ hormones
feed back to the hypophysis (9a and 9b), but simultaneously
affect the electrophysiology (10a and 10b) as well as the neurochemistry of the hypothalamus to either perpetuate or terminate
hypothalamic events depending on the strength of the original (1)
driving stimulus. For example, it is a well-established fact
that steroids alter significantly the electrophysiology of the
brain and, particularly, the hypothalamus.

I realize that there are a number of missing "links" in
this scheme. But, until we obtain additional data to clarify
some of these events, this working model is plausible. In short,
we now have some understanding of some of the molecular interrelationships between the hypothalamus and amygdala.

ACKNOWLEDGMENTS

The work reported in this chapter was supported by an
allocation from General Research Support Grant FR-05545 from
the Division of Research Resources of The Jackson Laboratory.

Special thanks are due my assistant, Mrs. Jacqueline McLeod.

REFERENCES

ANAND, B. K., DUA, S., & CHHINA, G. S.   Changes in the affective
    behaviour produced by lesions in the frontal and temporal
    lobes.   Indian Journal of Medical Research, 1957, 45,
    353-358. (a)

ANAND, B. K., DUA, S., & CHHINA, G. S.   Changes in visceral and
    metabolic activities after frontal and temporal lobe lesions.
    Indian Journal of Medical Research, 1957, 45, 345-362. (b)

ANAND, B. K., & DUA, S.   Effect of electrical stimulation of the
    limbic system ("visceral brain") on gastric secretion and
    motility.   Indian Journal of Medical Research, 1956, 44,
    125-130. (c)

ANAND, B. K., & DUA, S.   Electrical stimulation of the limbic
    system of brain ("visceral brain") in the waking animals.
    Indian Journal of Medical Research, 1956, 44, 107-119. (d)

BOVARD, E. W., & GLOOR, P.   Effect of amygdaloid lesions on plasma
    corticosterone response of the albino rat to emotional stress.
    Experientia, 1961, 17, 521-526.

BUNN, J. P., & EVERETT, J. W.   Ovulation in persistent-estrous
    rats after electrical stimulation of the brain.   Proceedings
    of the Society for Experimental Biology and Medicine, 1957,
    96, 369-372.

ELEFTHERIOU, B. E.   Effect of amygdaloid nuclear lesions on hypo-
    thalamic LH-RF in the male deermouse.   Journal of Endocrinology,
    38, 479-480.

ELEFTHERIOU, B. E., & PATTISON, M. L.   Effect of amygdaloid lesions
    on hypothalamic FSH-RF in the female deermouse.   Journal of
    Endocrinology, 1967, 39, 613-614.

ELEFTHERIOU, B. E., & ZOLOVICK, A. J.   Effect of amygdaloid lesions
    on oestrous behaviour in the deermouse.   Journal of Reproduc-
    tion and Fertility, 1966, 11, 451-453.

ELEFTHERIOU, B. E., & ZOLOVICK, A. J.   Effect of amygdaloid lesions
    on plasma and pituitary levels of luteinizing hormone.
    Journal of Reproduction and Fertility, 1967, 14, 33-37.

ELEFTHERIOU, B. E., & ZOLOVICK, A. J.   Effect of amygdaloid lesions
    on plasma and pituitary thyrotropin levels in the deermouse.
    Proceedings of the Society for Experimental Biology and
    Medicine, 1968, 127, 671-674.

ELEFTHERIOU, B. E., CHURCH, R. L., ZOLOVICK, A. J., NORMAN, R. L., & PATTISON, M. L.   Effects of amygdaloid lesions on regional brain RNA ratios.   Journal of Endocrinology, 1969, 45, 207-214.

ELEFTHERIOU, B. E., ZOLOVICK, A. J., & NORMAN, R. L.   Effects of amygdaloid lesions on plasma and pituitary levels of luteinizing hormone in the male deermouse.   Journal of Endocrinology, 1967, 38, 469-474.

ELEFTHERIOU, B. E., ZOLOVICK, A. J., & PEARSE, R.   Effect of amygdaloid lesions on the pituitary-adrenal axis in the deermouse.   Proceedings of the Society for Experimental Biology and Medicine, 1966, 122, 1259-1262.

ELIASSON, S.   Cerebral influence on gastric motility in the cat.   Acta Physiologica Scandinavica, 1952, 26 (Supplement No. 95), 1-70.

ELWERS, M., & CRITCHLOW, V.   Precocious ovarian stimulation following hypothalamic and amygdaloid lesions in rats.   American Journal of Physiology, 1960, 198, 381-385.

ELWERS, M., & CRITCHLOW, V.   Precocious ovarian stimulation following interruption of stria terminalis.   American Journal of Physiology, 1961, 201, 281-284.

GASTAUT, H.   Corrélations entre le système nerveux végétatif et la vie de relation dans le rhinencéphale.   Journal de Physiologie et Pathologie Général, 1952, 44, 431-470.

GREER, M. A., & YAMADA, T.   Effect of bilateral ablation of the amygdala on endocrine function in the rat.   Atlantic City Program, 41st Meeting, Endocrine Society, 1959, 82 (Abstract).

KLING, A., ORBACH, J., SCHWARTZ, N. B., & TOWNE, J. C.   Injury to the limbic system and associated structures in cats.   Archives of General Psychiatry, 1960, 3, 391-420.

KNIGGE, K. M.   Adrenocortical response to immobilization in rats with lesions in hippocampus and amygdala.   Federation Proceedings, 1961, 20, 185 (Abstract).

KOIKEGAMI, H., & FUSE, S.   Studies on the functions and fiber connections of the amygdaloid nuclei and periamygdaloid cortex: Part 2.   Experiment on the respiratory movements.   Folia Psychiatrica et Neurologica Japonica (Niigata), 1952a, 5, 188-197.

KOIKEGAMI, H., & FUSE, S.   Studies on the functions and fiber
    connections of the amygdaloid nuclei and periamygdaloid cor-
    tex: Part 2. Experiment on the respiratory movements. Folia
    Psychiatrica et Neurologica Japonica, 1952b, 6, 94-103.

KOIKEGAMI, H., FUSE, S., YOKOYAMA, T., WATANABE, T., & WATANABE, H.
    Contributions to the comparative anatomy of the amygdaloid
    nuclei of mammals with some experiments of their destruction
    or stimulation. Folia Psychiatrica et Neurologica Japonica
    (Niigata), 1955, 8, 336-370.

KOIKEGAMI, H., KIMOTO, A., & KIDO, C.   Studies on the amygdaloid
    nuclei and periamygdaloid cortex: Experiments on the influ-
    ence of their stimulation upon motility of small intestine
    and blood pressure. Folia Psychiatrica et Neurologica
    Japonica (Niigata), 1953, 7, 86-108.

KOIKEGAMI, H., JUSHIRO, H., & KIMOTO, A.   Studies on the functions
    and fiber connections of the amygdaloid nuclei and periamyg-
    daloid cortex: Experiments on gastrointestinal motility and
    body temperature in cats. Folia Psychiatrica et Neurologica
    Japonica (Niigata), 1953, 6, 76-93.

KOIKEGAMI, H., YAMADA, T., & USUI, K.   Stimulation of amygdaloid
    nuclei and periamygdaloid cortex with special reference to its
    effects on uterine movements and ovulation. Folia Psychiat-
    rica et Neurologica Japonica (Niigata), 1953, 8, 7-31.

LAWTON, I. E., & SAWYER, C. H.   Role of amygdala in regulating LH
    secretion in the adult female rat. American Journal of
    Physiology, 1970, 218, 622-626.

LESSE, H. Amygdaloid electrical activity during a conditioned
    response. Proceedings of the International Congress of Neuro-
    logical Science, Brussels, 1957, 1.

LUNDBERG, P. O. Extrahypothalamic regions of the central nervous
    system and gonadotrophin secretion. Proceedings of the Inter-
    national Union of Physiological Science, XXII Congress, 1962,
    1, 615-619.

MANDELL, A. J., CHAPMAN, L. F., RAND, R. W., & WALTER, R. D.   Plasma
    corticosteroids: Changes in concentration after stimulation of
    hippocampus and amygdala. Science, 1963, 139, 1212.

MARTIN, J., ENDRÖCZI, E., & BATA, G.   Effect of the removal of amyg-
    dalic nuclei on the secretion of adrenal cortical hormones.
    Acta Physiologica Academiae Scientiarum Hungaricae, 1958, 14,
    131-134.

MASON, J. W. The central nervous system regulation of ACTH
    secretion. In Jasper et al. (Eds.), Reticular Formation of
    the Brain. Boston: Little, Brown, 1958, pp. 645-670.

MASON, J. W. Plasma 17-hydroxycorticosteroid levels during
    electrical stimulation in the amygdaloid complex in conscious
    monkeys. American Journal of Physiology, 1959, 196, 44-48.

MASON, J. W., NAUTA, W. J. H., BRADY, J. V., & ROBINSON, J. A.
    Limbic system influences on the pituitary-adrenal cortical
    system. Atlantic City Program, 41st Meeting, Endocrine
    Society, 1959, 29 (Abstract).

MASSERMAN, J. H., LEVITT, M., MCAVOY, T., KLING, A., & PECHTEL, C.
    The amygdalae and behaviour. American Journal of Psychiatry,
    1958, 115, 14-17.

POPA, L., CRUCEANU, A., & LACTATUS, V. Some physiochemical prop-
    erties of mouse brain RNA. Revue Roumaine de Biochimie, 1967,
    4, 137-142.

SEN, R. N., & ANAND, B. K. Effect of electrical stimulation of
    the limbic system of brain ("visceral brain") on gastric
    secretory activity and ulceration. Indian Journal of
    Medical Research, 1957, 45, 515-521.

SHEALY, N. C., & PEELE, T. L. Studies on amygdaloid nucleus of
    the cat. Journal of Neurophysiology, 1957, 20, 125-139.

VELASCO, M. E., & TALEISNIK, S. Release of gonadotropins induced
    by amygdaloid stimulation in the rat. Endocrinology, 1969,
    84, 132-139.

WALKER, A. E., THOMPSON, A. F., & MCQUEEN, J. D. Behavior and the
    temporal rhinencephalon in the monkey. Johns Hopkins Bulletin,
    1953, 93, 65-93.

WOOD, C. D. Behavioral changes following discrete lesions of
    temporal lobe structures.

YAMADA, T., & GREER, M. A. The effect of bilateral ablation of
    the amygdala on endocrine function in the rat. Endocrinology,
    1960, 66, 565-574.

ZOLOVICK, A. J. Electrophysiological aspects of amygdaloid-hypo-
    thalamic interrelations. Ph.D. Thesis, Kansas State Univer-
    sity, Manhattan.

SUBJECT INDEX

Acetylcholine, 228, 537, 539,
    613
Acetylcholinesterase, 97,
    105-107, 210
Adenosine monophosphate, 800
Adipsia, 247, 651
Adrenocorticotropin, 658,
    662-666, 722
Aggressive behavior, 433, 445,
    518, 539, 554
  acetylcholinesterase, 288-289
  amygdala, 283-292
  concepts, 555-557
Agonistic behavior, 212, 220
Alarm, 337
Amnesia, 425
Amphetamine, 431, 624, 626,
    630, 634
Amygdala
  active avoidance, 234-237
  adrenocortical responses,
    237
  aggression, 220-233, 288
  cardiovascular system,
    238-240
  chemoarchitecture
    acetylcholinesterase,
      105-107, 283-292
    monoamine oxidase, 107
    dithizone stain, 107-111
    cupric-silver stain,
      145-196
  cytoarchitecture, 284
    Golgi, 103-105
    Nissl, 96-103, 210
  defense, 223, 597-603

drugs, 609-618
embryonic development
  amygdalocortical transition
    area, 51
  amygdalohippocampal transi-
    tion area, 47, 57, 62
  basal, 25, 31, 47, 62
  bat, 26, 66
  caudate putamen, 52, 58, 63
  central, 25, 36
  cortical, 25, 36, 39, 45, 50
  human, 21-72
  lateral, 25, 31, 47, 55, 62
  medial, 37, 39, 45, 50
  olfactory tract, 25, 45, 50,
    66, 173
  shrew, 27, 66, 71, 72
epilepsy, 440-445, 463-480
escape-avoidance, 537-550
flight, 223
food intake, 243-249
gastric activity, 250-252
gonadotropins, 644-658
hypothalamic relationships,
    295-312, 319-338, 343-367,
    373-391
maternal behavior, 512-514
micturition, 241
mouse-killing, 562-573
neocortical afferents, 110-112,
    128-130
olfaction, 431-436
olfactory projections, 124-127
ovulation, 255
paleocortical connections,
    128-129